Numerical Techniques in Electromagnetics

Numerical Techniques in Electromagnetics

MATTHEW N. O. SADIKU

CRC Press
Boca Raton   Ann Arbor   London   Tokyo

Library of Congress Cataloging-in-Publication Data

Catalog record is available from the Library of Congress.

This book represents information obtained from authentic and highly regarded sources. Reprinted material is quoted with permission, and sources are indicated. A wide variety of references are listed. Every reasonable effort has been made to give reliable data and information, but the authors and the publisher cannot assume responsibility for the validity of all materials or for the consequences of their use.

All rights reserved. This book, or any parts thereof, may not be reproduced in any form without written consent from the publisher.

Direct all inquiries to CRC Press, Inc., 2000 Corporate Blvd., N.W., Boca Raton, Florida, 33431.

© 1992 by CRC Press, Inc.

International Standard Book Number 0-8493-4232-5

Printed in the United States of America   2 3 4 5 6 7 8 9 0

Printed on acid-free paper

To my teacher
 Carl A. Ventrice
and my parents
 Ayisat and Solomon Sadiku

Contents

Preface		xv
Acknowledgments		xvii
A Note to Students		xviii

1 Fundamental Concepts — 1
- 1.1 Introduction — 1
- 1.2 Review of Electromagnetic Theory — 2
    - 1.2.1 Electrostatic Fields — 3
    - 1.2.2 Magnetostatic Fields — 4
    - 1.2.3 Time-varying Fields — 5
    - 1.2.4 Boundary Conditions — 7
    - 1.2.5 Wave Equations — 8
    - 1.2.6 Time-varying Potentials — 9
    - 1.2.7 Time-harmonic Fields — 11
- 1.3 Classification of EM Problems — 15
    - 1.3.1 Classification of Solution Regions — 15
    - 1.3.2 Classification of Differential Equations — 16
    - 1.3.3 Classification of Boundary Conditions — 19
- 1.4 Some Important Theorems — 22
    - 1.4.1 Superposition Principle — 22
    - 1.4.2 Uniqueness Theorem — 23
- References — 24
- Problems — 25

2 Analytical Methods — 29
- 2.1 Introduction 29
- 2.2 Separation of Variables 30
- 2.3 Separation of Variables in Rectangular Coordinates 32
  - 2.3.1 Laplace's Equations 32
  - 2.3.2 Wave Equation 37
- 2.4 Separation of Variables in Cylindrical Coordinates 43
  - 2.4.1 Laplace's Equation 44
  - 2.4.2 Wave Equation 47
- 2.5 Separation of Variables in Spherical Coordinates . 59
  - 2.5.1 Laplace's Equation 59
  - 2.5.2 Wave Equation 64
- 2.6 Some Useful Orthogonal Functions 76
- 2.7 Series Expansion 86
  - 2.7.1 Poisson's Equation in a Cube 87
  - 2.7.2 Poisson's Equation in a Sphere 88
  - 2.7.3 Strip Transmission Line 90
- 2.8 Practical Applications 96
  - 2.8.1 Scattering by Dielectric Sphere 96
  - 2.8.2 Scattering Cross Sections 101
  - 2.8.3 Attenuation Due to Raindrops 104
- 2.9 Concluding Remarks 115
- References 117
- Problems 118

3 Finite Difference Methods — 135
- 3.1 Introduction 135
- 3.2 Finite Difference Schemes 136
- 3.3 Finite Differencing of Parabolic PDEs 139
- 3.4 Finite Differencing of Hyperbolic PDEs 146
- 3.5 Finite Differencing of Elliptic PDEs 152
  - 3.5.1 Band Matrix Method 154
  - 3.5.2 Iterative Methods 154
- 3.6 Accuracy and Stability of FD Solutions 161
- 3.7 Practical Applications I—Guided Structures . . . 165
  - 3.7.1 Transmission Lines 166
  - 3.7.2 Waveguides 173
- 3.8 Practical Applications II—Wave Scattering (FD-TD) 179
  - 3.8.1 Yee's Finite Difference Algorithm 179
  - 3.8.2 Accuracy and Stability 184
  - 3.8.3 Lattice Truncation Conditions 184
  - 3.8.4 Initial Fields 188

		3.8.5 Programming Aspects	188
	3.9	Finite Differencing for Nonrectangular Systems	204
		3.9.1 Cylindrical Coordinates	205
		3.9.2 Spherical Coordinates	209
	3.10	Numerical Integration	214
		3.10.1 Euler's Rule	214
		3.10.2 Trapezoidal Rule	216
		3.10.3 Simpson's Rule	217
		3.10.4 Newton-Cotes Rules	218
		3.10.5 Gaussian Rules	221
		3.10.6 Multiple Integration	224
	3.11	Concluding Remarks	230
		References	231
		Problems	238
4	**Variational Methods**		**255**
	4.1	Introduction	255
	4.2	Operators in Linear Spaces	256
	4.3	Calculus of Variations	259
	4.4	Construction of Functionals from PDEs	263
	4.5	Rayleigh-Ritz Method	266
	4.6	Weighted Residual Method	273
		4.6.1 Collocation Method	275
		4.6.2 Subdomain Method	275
		4.6.3 Galerkin Method	276
		4.6.4 Least Squares Method	277
	4.7	Eigenvalue Problems	283
	4.8	Practical Applications	290
	4.9	Concluding Remarks	297
		References	298
		Problems	302
5	**Moment Methods**		**309**
	5.1	Introduction	309
	5.2	Integral Equations	310
		5.2.1 Classification of Integral Equations	310
		5.2.2 Connection Between Differential and Integral Equations	311
	5.3	Green's Functions	315
		5.3.1 For Free Space	317
		5.3.2 For Domain with Conducting Boundaries	321
	5.4	Applications I—Quasi-Static Problems	334

x   Numerical Techniques in Electromagnetics

    5.5 Applications II—Scattering Problems 340
          5.5.1 Scattering by Conducting Cylinder 340
          5.5.2 Scattering by an Arbitrary
                 Array of Parallel Wires 344
    5.6 Applications III—Radiation Problems 350
          5.6.1 Hallen's Integral Equation 353
          5.6.2 Pocklington's Integral Equation 354
          5.6.3 Expansion and Weighting Functions . . . 354
    5.7 Applications IV—EM Absorption in the Human Body 366
          5.7.1 Derivation of Integral Equations 367
          5.7.2 Transformation to Matrix
                 Equation (discretization) 370
          5.7.3 Evaluation of Matrix Elements 372
          5.7.4 Solution of the Matrix Equation 373
    5.8 Concluding Remarks 381
        References   387
        Problems 394

6 Finite Element Method   407
    6.1 Introduction 407
    6.2 Solution of Laplace's Equation 408
          6.2.1 Finite Element Discretization 409
          6.2.2 Element Governing Equations 410
          6.2.3 Assembling of All Elements 414
          6.2.4 Solving the Resulting Equations 417
    6.3 Solution of Poisson's Equation 429
          6.3.1 Deriving Element-governing Equations . . 429
          6.3.2 Solving the Resulting Equations 431
    6.4 Solution of the Wave Equation 432
    6.5 Automatic Mesh Generation I—Rectangular Domains 439
    6.6 Automatic Mesh Generation II—Arbitrary Domains 443
          6.6.1 Definition of the Blocks 443
          6.6.2 Subdivision of Each Block 445
          6.6.3 Connection of Individual Blocks 445
    6.7 Bandwidth Reduction 454
    6.8 Higher Order Elements 457
          6.8.1 Pascal Triangle 458
          6.8.2 Local Coordinates 459
          6.8.3 Shape Functions 461
          6.8.4 Fundamental Matrices 463
    6.9 Three-dimensional Elements 474
    6.10 Hybrid Finite Element Methods 479

			Contents	xi

		6.10.1 Infinite Element Method	479
		6.10.2 Boundary Element Method	481
	6.11	Concluding Remarks	482
		References	483
		Problems	492

7 Transmission-line-matrix Method — 501

	7.1	Introduction	501
	7.2	Transmission-line Equations	503
	7.3	Solution of Diffusion Equation	507
	7.4	Solution of Wave Equations	512
		7.4.1 Equivalence Between Network and Field Parameters	512
		7.4.2 Dispersion Relation of Propagation Velocity	517
		7.4.3 Scattering Matrix	519
		7.4.4 Boundary Representation	521
		7.4.5 Computation of Fields and Frequency Response	522
		7.4.6 Output Response and Accuracy of Results	523
	7.5	Inhomogeneous and Lossy Media in TLM	530
		7.5.1 General Two-dimensional Shunt Node	530
		7.5.2 Scattering Matrix	533
		7.5.3 Representation of Lossy Boundaries	534
	7.6	Three-dimensional TLM Mesh	539
		7.6.1 Series Nodes	539
		7.6.2 Three-dimensional Node	541
		7.6.3 Boundary Conditions	545
	7.7	Error Sources and Correction	555
		7.7.1 Truncation Error	555
		7.7.2 Coarseness Error	556
		7.7.3 Velocity Error	556
		7.7.4 Misalignment Error	557
	7.8	Concluding Remarks	557
		References	558
		Problems	562

8 Monte Carlo Methods — 571

	8.1	Introduction	571
	8.2	Generation of Random Numbers and Variables	572
	8.3	Evaluation of Error	576
	8.4	Numerical Integration	581

xii   Numerical Techniques in Electromagnetics

		8.4.1 Crude Monte Carlo Integration	582
		8.4.2 Monte Carlo Integration with Antithetic Variates	583
		8.4.3 Improper Integrals	584
	8.5	Solution of Potential Problems	587
		8.5.1 Fixed Random Walk	587
		8.5.2 Floating Random Walk	591
		8.5.3 Exodus Method	593
		8.5.4 Regional Monte Carlo Method	595
	8.6	Inverse Monte Carlo Analysis	608
	8.7	Concluding Remarks	610
		References	611
		Problems	616

Appendices

A Vector Relations   625
A.1 Vector Identities 625
A.2 Vector Theorems 626
A.3 Orthogonal Coordinates 626

B Review of FORTRAN   629
B.1 Characters 629
B.2 Variables 629
B.3 Constants 630
B.4 Statement Declaration 630
B.5 Arithmetic Statements and Mathematical Functions   631
B.6 Transfer of Control 632
B.7 DO Loops 634
B.8 Input/output Statement 634
B.9 Subprograms 635
B.10 Other Statements 636
B.11 Program Organization 637
B.12 FORTRAN Library Functions 638
     B.12.1 Integer Functions 638
     B.12.2 Real Functions 638
     B.12.3 Double Precision Functions 640
     B.12.4 Complex Functions 641
     References 641

C	**How to Write Good Programs**	**643**
	C.1 Documentation	643
	C.2 Size Reduction	644
	C.3 Accuracy Enhancement	644
	C.4 Speed Enhancement	644
	References	646
D	**Solution of Simultaneous Equations**	**647**
	D.1 Elimination Methods	647
	D.1.1 Gauss's Method	648
	D.1.2 Cholesky's Method	650
	D.2 Iterative Methods	653
	D.2.1 Jacobi's Method	653
	D.2.2 Gauss-Seidel Method	655
	D.2.3 Relaxation Method	656
	D.2.4 Gradient Methods	656
	D.3 Matrix Inversion	660
	D.4 Eigenvalue Problems	662
	D.4.1 Iteration (or Power) Method	663
	D.4.2 Jacobi's Method	666
	References	670
E	**Answers to Odd-numbered Problems**	**673**
	Index	**685**

Preface

For too long, the electromagnetic (EM) community has suffered without a suitable text on computational techniques commonly used in solving EM-related problems. Although there have been monographs on one particular technique or the other, the monographs are written for the experts rather than being student-oriented. There has not been a single text that covers all the major techniques and does so in a manner suitable for classroom use. It seems experts in this area are familiar with one or few techniques, and no one seems to be familiar with all the common techniques. This text attempts to fill the gap.

The text is intended for seniors or graduate students and may be used for a one-semester or two-semester course. The main requirements for students taking a course based on this text are an introductory EM course and a knowledge of a high-level computer language, preferably FORTRAN. Although familiarity with linear algebra and numerical analysis is useful, it is not required.

In writing this book, three major objectives were kept in mind. First, the book is intended to teach the students how to pose, numerically analyze, and solve EM problems. Second, it is designed to give them the ability to expand their problem-solving skills using a variety of available numerical methods. Third, it is meant to prepare graduate students for research in EM. The aim throughout has been simplicity of presentation so that the text can be useful for both teaching and self-study. In striving for simplicity, however, the reader is referred to the references for more information. Toward the end of each chapter, the techniques covered in the chapter are applied to real life problems. Since the application of the technique is as vast as EM and the author's experience is limited, the choice of application is selective.

The first chapter covers some fundamental concepts in EM. For the sake of clarity, EM problems are classified in terms of the equations describing them (differential or integral or both), the boundary conditions (Dirichlet, Neumann, or mixed), and the form of the solution regions (interior/closed or exterior/open).

Chapter 2 is intended to put numerical methods in a proper perspective. Analytical methods such as separation of variables and series expansion are covered. A practical problem of attenuation and phase shift of EM wave due to raindrops is presented to illustrate the concepts developed in the chapter. Chapter 3 on the finite difference methods begins with the derivation of difference equation from a partial differential equation (PDE) using forward, backward, and central differences. Various methods of solving elliptic partial differential equations (iteration, relaxation, and band matrix) are considered. Stability of finite difference schemes and discretization error are discussed. The finite-difference time-domain technique involving Yee's algorithm is presented and applied to scattering problems. Numerical integration is covered using trapezoidal, Simpson's, Newton-Cotes rules, and Gaussian quadratures.

The fourth chapter on variational methods serves as a preparation for the next two major topics: moment methods and finite element methods. Basic concepts such as inner product, self-adjoint operator, functionals, and the Euler equation are covered. Derivation of a functional from a given partial differential equation and vice versa are presented. The classical Rayleigh-Ritz method with its limitations is considered as a direct variational method. Along with the concepts of expansion and weighting functions, the weighted residual methods (collocation/point matching, least squares, and Galerkin) are covered. A simple application of determining the characteristic impedance of a strip transmission line is used in illustrating the concepts developed in the chapter.

Chapter 5 on moment methods focuses on the solution of integral equations. The interrelationship between differential and integral equation, Green's functions for two- and three-dimensional problems, and some typical applications in antenna, scattering, and EM absorption by human body are presented.

Chapter 6 covers the basic steps involved in using the finite element method. Solutions of Laplace's, Poisson's, and wave equations using the finite element method are covered. The problems of proper node numbering, bandwidth reduction, and mesh generation are treated. Higher-order and three-dimensional elements are briefly discussed.

Chapter 7 is devoted to transmission-line matrix or modeling (TLM). The method is applied to diffusion problem and scattering problems.

The last chapter is on Monte Carlo methods. The technique of generating random numbers is presented. Typical applications of the method to electrostatic problems are presented.

The appendices contain useful vector relations, a brief review of FORTRAN, some hints on how to develop good code, various techniques for solving simultaneous equations, and answers to some problems. A solutions manual is available from the publisher.

Acknowledgments

I began this book while at Florida Atlantic University and completed it at Temple University. I would like to express my appreciation to my students, Lawrence Agba, Vincent Bemmel, Susan Dervain, and Raymond Jongakiem, for graciously allowing me to incorporate in this text some computer programs from their masters theses. Professor Roger Messenger, the former chairman of Electrical and Computer Engineering Department at Florida Atlantic University, deserves special thanks. I am grateful to Professor Charles Alexander, dean of College of Engineering, Computer Science, and Architecture at Temple University. I am greatly indebted to Dr. Brian Butz, my chairman, for his support and patience while I was busy working on the book. I wish to thank Vernell Ross, Michelle Ayers, Carol Dahlberg, and Pamela Sullivan for their secretarial support. Special thanks are due to Dr. Parveen Wahid for her comments on a portion of the manuscript, to Dr. Saroj Biswas for his excellent instruction on how to use TeX, and to Robert Herman and Alan Sanders for their assistance with the computer programs. To my wife, Chris, and our daughter, Ann, I express my deepest gratitude. Without their understanding, sacrifices, and prayers the dream would never have come to reality.

A Note to Students

Before you embark on writing your own computer program or using the ones in this text, you should try to understand all relevant theoretical background. A computer is no more than a tool used in the analysis of a program. For this reason, you should be as clear as possible what the machine is really being asked to do before setting it off on several hours of expensive computations.

It has been well said by A. C. Doyle that "It is a capital mistake to theorize before you have all the evidence. It biases the judgment." Therefore, you should never trust the results of a numerical computation unless they are validated, at least in part. You validate the results by comparing them with those obtained by previous investigators or with similar results obtained using a different approach which may be analytical or numerical. For this reason, it is advisable that you become familiar with as many numerical techniques as possible.

The references provided at the end of each chapter are by no means exhaustive but are meant to serve as the starting point for further reading.

Chapter 1

Fundamental Concepts

"Thy friend has a friend, and thy friend's friend has a friend; be discreet."   The Talmud

1.1 Introduction

Scientists and engineers use several techniques in solving continuum or field problems. Loosely speaking, these techniques can be classified as experimental, analytical, or numerical. Experiments are expensive, time consuming, sometimes hazardous, and usually do not allow much flexibility in parameter variation. However, every numerical method, as we shall see, involves an analytic simplification to the point where it is easy to apply the numerical method. Notwithstanding this fact, the following methods are among the most commonly used in electromagnetics (EM).

   A. Analytical Methods (exact solutions)
      (1) Separation of variables
      (2) Series expansion
      (3) Conformal mapping
      (4) Integral solutions, e.g., Laplace and Fourier transforms
      (5) Pertubation methods
   B. Numerical methods (approximate solutions)
      (1) Finite difference method
      (2) Method of weighted residuals
      (3) Moment method
      (4) Finite element method
      (5) Transmission-line modeling
      (6) Monte Carlo method

Application of these methods is not limited to EM-related problems; they find applications in other continuum problems such as in fluid, heat transfer, and acoustics [1].

As we shall see, some of the numerical methods are related and they all generally give approximate solutions of sufficient accuracy for engineering purposes. Since our objective is to study these methods in detail in the subsequent chapters, it may be premature to say more than this at this point.

The need for numerical solution of electromagnetic problems is best expressed in the words of Paris and Hurd: "Most problems that can be solved formally (analytically) have been solved."[1] Until the 1940s, most EM problems were solved using the classical methods of separation of variables and integral equation solutions. Besides the fact that a high degree of ingenuity, experience, and effort were required to apply those methods, only a narrow range of practical problems could be investigated due to the complex geometries defining the problems.

Numerical solution of EM problems started in the mid-1960s with the availability of modern high-speed digital computers. Since then considerable effort has been expended on solving practical, complex EM-related problems for which closed form analytical solutions are either intractable or do not exist. The numerical approach has the advantage of allowing the actual work to be carried out by operators without a knowledge of higher mathematics or physics, with a resulting economy of labor on the part of the highly trained personnel.

Before we set out to study the various techniques used in analyzing EM problems, it is expedient to remind ourselves of the physical laws governing EM phenomena in general. This will be done in Section 1.2. In Section 1.3, we shall be acquainted with different ways EM problems are categorized. The principle of superposition and uniqueness theorem will be covered in Section 1.4.

1.2 Review of Electromagnetic Theory

The whole subject of EM unfolds as a logical deduction from seven postulated equations, namely, Maxwell's four field equations and three medium-dependent equations [2–5]. Before we briefly review these equations, it may be helpful to state two important theorems commonly used in EM. These are the divergence (or Gauss's) theorem,

---

[1] *Basic Electromagnetic Theory*, D. T. Paris and F. K. Hurd, McGraw-Hill, New York, 1969, p. 166.

$$\oint_S \mathbf{F} \cdot d\mathbf{S} = \int_v \nabla \cdot \mathbf{F} \, dv \tag{1.1}$$

and Stokes's theorem

$$\oint_L \mathbf{F} \cdot d\mathbf{l} = \int_S \nabla \times \mathbf{F} \cdot d\mathbf{S}. \tag{1.2}$$

Perhaps the best way to review EM theory is by using the fundamental concept of electric charge. EM theory can be regarded as the study of fields produced by electric charges at rest and in motion. Electrostatic fields are usually produced by static electric charges, whereas magnetostatic fields are due to motion of electric charges with uniform velocity (direct current). Dynamic or time-varying fields are usually due to accelerated charges or time-varying currents.

1.2.1 Electrostatic Fields

The two fundamental laws governing this type of field are Gauss's law,

$$\oint \mathbf{D} \cdot d\mathbf{S} = \int \rho_v \, dv \tag{1.3}$$

which is a direct consequence of Coulomb's force law, and the law describing electrostatic fields as conservative,

$$\oint \mathbf{E} \cdot d\mathbf{l} = 0. \tag{1.4}$$

In Eqs. (1.3) and (1.4), $\mathbf{D}$ is the electric flux density (in coulombs/meter^2), ρ_v is the volume charge density (in coulombs/meter^3), and $\mathbf{E}$ is the electric field intensity (in volts/meter). The integral form of the laws in Eqs. (1.3) and (1.4) can be expressed in the differential form by applying Eq. (1.1) to Eq. (1.3) and Eq. (1.2) to Eq. (1.4). We obtain

$$\nabla \cdot \mathbf{D} = \rho_v \tag{1.5}$$

and

$$\nabla \times \mathbf{E} = 0. \tag{1.6}$$

The vector fields $\mathbf{D}$ and $\mathbf{E}$ are related as

$$\mathbf{D} = \epsilon \mathbf{E} \tag{1.7}$$

4 Numerical Techniques in Electromagnetics

where ϵ is the dielectric permittivity (in farads/meter) of the medium. In terms of the electric potential V (in volts), $\mathbf{E}$ is expressed as

$$\mathbf{E} = -\nabla V \tag{1.8}$$

or

$$V = -\int \mathbf{E} \cdot d\mathbf{l}. \tag{1.9}$$

Combining Eqs. (1.5), (1.7), and (1.8) gives Poisson's equation:

$$\nabla \cdot \epsilon \nabla V = -\rho_v \tag{1.10a}$$

or, if ϵ is constant,

$$\boxed{\nabla^2 V = -\frac{\rho_v}{\epsilon}} \tag{1.10b}$$

When $\rho_v = 0$, Eq. (1.10) becomes Laplace's equation:

$$\nabla \cdot \epsilon \nabla V = 0 \tag{1.11a}$$

or for constant ϵ

$$\boxed{\nabla^2 V = 0} \tag{1.11b}$$

1.2.2 Magnetostatic Fields

The basic laws of magnetostatic fields are Ampere's law

$$\oint_L \mathbf{H} \cdot d\mathbf{l} = \int_S \mathbf{J} \cdot d\mathbf{S} \tag{1.12}$$

which is related to Biot-Savart law, and the law of conservation of magnetic flux (also called Gauss's law for magnetostatics)

$$\oint \mathbf{B} \cdot d\mathbf{S} = 0 \tag{1.13}$$

where $\mathbf{H}$ is the magnetic field intensity (in amperes/meter), $\mathbf{J}$ is the electric current density (in amperes/meter^2), and $\mathbf{B}$ is the magnetic flux density (in tesla or webers/meter^2). Applying Eq. (1.2) to Eq. (1.12) and Eq. (1.1) to Eq. (1.13) yields their differential form as

$$\nabla \times \mathbf{H} = \mathbf{J} \tag{1.14}$$

Fundamental Concepts 5

and
$$\nabla \cdot \mathbf{B} = 0. \tag{1.15}$$

The vector fields $\mathbf{B}$ and $\mathbf{H}$ are related through the permeability μ (in henries/meter) of the medium as

$$\mathbf{B} = \mu \mathbf{H}. \tag{1.16}$$

Also, $\mathbf{J}$ is related to $\mathbf{E}$ through the conductivity σ (in mhos/meter) of the medium as

$$\mathbf{J} = \sigma \mathbf{E}. \tag{1.17}$$

This is usually referred to as Ohm's law. In terms of the magnetic vector potential $\mathbf{A}$ (in Wb/meter)

$$\mathbf{B} = \nabla \times \mathbf{A}. \tag{1.18}$$

Applying the vector identity

$$\nabla \times (\nabla \times \mathbf{F}) = \nabla(\nabla \cdot \mathbf{F}) - \nabla^2 \mathbf{F} \tag{1.19}$$

to Eqs. (1.14) and (1.18) leads to Poisson's equation for magnetostatic fields:

$$\boxed{\nabla^2 \mathbf{A} = -\mu \mathbf{J}} \tag{1.20}$$

When $\mathbf{J} = 0$, Eq. (1.20) becomes Laplace's equation

$$\boxed{\nabla^2 \mathbf{A} = 0} \tag{1.21}$$

1.2.3 Time-varying Fields

In this case, electric and magnetic fields exist simultaneously. Equations (1.5) and (1.15) remain the same whereas Eqs. (1.6) and (1.14) require some modification for dynamic fields. Modification of Eq. (1.6) is necessary to incorporate Faraday's law of induction, and that of Eq. (1.14) is warranted to allow for displacement current. The time-varying EM fields are governed by physical laws expressed mathematically as

$$\nabla \cdot \mathbf{D} = \rho_v \tag{1.22a}$$

$$\nabla \cdot \mathbf{B} = 0 \tag{1.22b}$$

$$\nabla \times \mathbf{E} = -\frac{\partial \mathbf{B}}{\partial t} \tag{1.22c}$$

$$\nabla \times \mathbf{H} = \mathbf{J} + \frac{\partial \mathbf{D}}{\partial t}. \tag{1.22d}$$

6   Numerical Techniques in Electromagnetics

These equations are referred to as Maxwell's equations in the generalized form. They are first-order linear coupled differential equations relating the vector field quantities to each other. The equivalent integral form of Eq.(1.22) is

$$\oint_S \mathbf{D} \cdot d\mathbf{S} = \int_v \rho_v \, dv \tag{1.23a}$$

$$\oint_S \mathbf{B} \cdot d\mathbf{S} = 0 \tag{1.23b}$$

$$\oint_L \mathbf{E} \cdot d\mathbf{l} = -\int_S \frac{\partial \mathbf{B}}{\partial t} \cdot d\mathbf{S} \tag{1.23c}$$

$$\oint_L \mathbf{H} \cdot d\mathbf{l} = \int_S (\mathbf{J} + \frac{\partial \mathbf{D}}{\partial t}) \cdot d\mathbf{S}. \tag{1.23d}$$

In addition to these four Maxwell's equations, there are three medium-dependent equations:

$$\mathbf{D} = \epsilon \mathbf{E} \tag{1.24a}$$

$$\mathbf{E} = \mu H \tag{1.24b}$$

$$\mathbf{J} = \sigma E. \tag{1.24c}$$

These are called *constitutive relations* for the medium in which the fields exist. Equations (1.22) and (1.24) form the seven postulated equations on which EM theory unfolds itself. We must note that in the region where Maxwellian fields exist, the fields are assumed to be:

(1) single valued,
(2) bounded, and
(3) continuous functions of space and time with continuous derivatives.

It is worthwhile to mention two other fundamental equations that go hand in hand with Maxwell's equations. One is the Lorentz force equation

$$\mathbf{F} = Q(\mathbf{E} + \mathbf{u} \times \mathbf{B}) \tag{1.25}$$

where **F** is the force experienced by a particle with charge Q moving at velocity **u** in an EM field; the Lorentz force equation constitutes a link between EM and mechanics. The other is the continuity equation

$$\boxed{\nabla \cdot \mathbf{J} = -\frac{\partial \rho_v}{\partial t}} \tag{1.26}$$

which expresses the conservation (or indestructibility) of electric charge. The continuity equation is implicit in Maxwell's equations (see Example 1.2). Equation (1.26) is not peculiar to EM. In fluid mechanics, where **J** corresponds with velocity and ρ_v with mass, Eq. (1.26) expresses the law of conservation of mass.

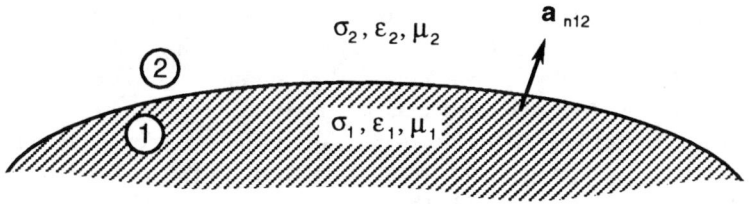

Figure 1.1 Interface between two media.

1.2.4 Boundary Conditions

The material medium in which an EM field exists is usually characterized by its constitutive parameters $\sigma, \epsilon,$ and μ. The medium is said to be *linear* if $\sigma, \epsilon,$ and μ are independent of **E** and **H** or nonlinear otherwise. It is *homogeneous* if $\sigma, \epsilon,$ and μ are not functions of space variables or inhomogeneous otherwise. It is *isotropic* if $\sigma, \epsilon,$ and μ are independent of direction (scalars) or anisotropic otherwise.

The boundary conditions at the interface separating two different media 1 and 2, with parameters $(\sigma_1, \epsilon_1, \mu_1)$ and $(\sigma_2, \epsilon_2, \mu_2)$ as shown in Fig. 1.1, are easily derived from the integral form of Maxwell's equations. They are

$$E_{1t} = E_{2t} \text{ or } (\mathbf{E}_1 - \mathbf{E}_2) \times \mathbf{a}_{n12} = 0 \tag{1.27a}$$

$$H_{1t} - H_{21} = K \text{ or } (\mathbf{H}_1 - \mathbf{H}_2) \times \mathbf{a}_{n12} = \mathbf{K} \tag{1.27b}$$

$$D_{1n} - D_{2n} = \rho_S \text{ or } (\mathbf{D}_1 - \mathbf{D}_2) \cdot \mathbf{a}_{n12} = \rho_S \tag{1.27c}$$

$$B_{1n} - B_{2n} = 0 \text{ or } (\mathbf{B}_2 - \mathbf{B}_1) \cdot \mathbf{a}_{n12} = 0 \tag{1.27d}$$

where $\mathbf{a}_{n12}$ is a unit normal vector directed from medium 1 to medium 2, subscripts 1 and 2 denote fields in regions 1 and 2, and subscripts t and n respectively denote tangent and normal components of the fields. Equations (1.27a) and (1.27d) state that the tangential components of **E** and the normal components of **B** are continuous across the boundary. Equation (1.27b) states that the tangential component of **H** is discontinuous by the surface current density **K** on the boundary. Equation (1.27c) states that the discontinuity in the normal component of **D** is the same as the surface charge density ρ_s on the boundary.

In practice, only two of Maxwell's equations are used (Eqs. (1.22c,d)) when a medium is source-free ($\mathbf{J} = 0, \rho_v = 0$), since the other two are implied (see Prob. 1.3). Also, in practice, it is sufficient to make the tangential components of the fields satisfy the necessary boundary conditions since the normal components implicitly satisfy their corresponding boundary conditions.

8 Numerical Techniques in Electromagnetics

1.2.5 Wave Equations

As mentioned earlier, Maxwell's equations are coupled first-order differential equations which are difficult to apply when solving boundary-value problems. The difficulty is overcome by decoupling the first-order equations, thereby obtaining the wave equation, a second-order differential equation which is useful for solving problems. To obtain the wave equation for a linear, isotropic, homogeneous, source-free medium ($\rho_v = 0, \mathbf{J} = 0$) from Eq. (1.22), we take the curl of both sides of Eq.(1.22c). This gives

$$\nabla \times \nabla \times \mathbf{E} = -\mu \frac{\partial}{\partial t}(\nabla \times \mathbf{H}). \tag{1.28}$$

From (1.22d),

$$\nabla \times \mathbf{H} = \epsilon \frac{\partial \mathbf{E}}{\partial t}$$

since $\mathbf{J} = 0$, so that Eq. (1.28) becomes

$$\nabla \times \nabla \times \mathbf{E} = -\mu\epsilon \frac{\partial^2 \mathbf{E}}{\partial t^2}. \tag{1.29}$$

Applying the vector identity

$$\nabla \times \nabla \times \mathbf{F} = \nabla(\nabla \cdot \mathbf{F}) - \nabla^2 \mathbf{F} \tag{1.30}$$

in Eq.(1.29),

$$\nabla(\nabla \cdot \mathbf{E}) - \nabla^2 \mathbf{E} = -\mu\epsilon \frac{\partial^2 \mathbf{E}}{\partial t^2}.$$

Since $\rho_v = 0, \nabla \cdot \mathbf{E} = 0$ from Eq. (1.22a), and hence we obtain

$$\boxed{\nabla^2 \mathbf{E} - \mu\epsilon \frac{\partial^2 \mathbf{E}}{\partial t^2} = 0} \tag{1.31}$$

which is the time-dependent vector Helmholtz equation or simply wave equation. If we had started the derivation with Eq.(1.22d), we would obtain the wave equation for $\mathbf{H}$ as

$$\boxed{\nabla^2 \mathbf{H} - \mu\epsilon \frac{\partial^2 \mathbf{H}}{\partial t^2} = 0} \tag{1.32}$$

Equations (1.31) and (1.32) are the equations of motion of EM waves in the medium under consideration. These waves are represented by the coupled $\mathbf{E}$ and $\mathbf{H}$ fields. The velocity (in m/s) of wave propagation is

$$u = \frac{1}{\sqrt{\mu\epsilon}} \tag{1.33}$$

where $u = c \approx 3 \times 10^8$ m/s in free space. It should be noted that each of the vector equations in (1.31) and (1.32) has three scalar components, so that altogether we have six scalar equations for $E_x, E_y, E_z, H_x, H_y,$ and H_z. Thus each component of the wave equations has the form

$$\nabla^2 \Psi - \frac{1}{u^2}\frac{\partial^2 \Psi}{\partial t^2} = 0 \qquad (1.34)$$

which is the scalar wave equation.

1.2.6 Time-varying Potentials

Although we are often interested in electric and magnetic field intensities (**E** and **H**), which are physically measurable quantities, it is often convenient to use auxiliary functions in analyzing an EM field. These auxiliary functions are the scalar electric potential V and vector magnetic potential **A**. Although these potential functions are arbitrary, they are required to satisfy Maxwell's equations. Their derivation is based on two fundamental vector identities (see Prob. 1.1),

$$\nabla \times \nabla \Phi = 0 \qquad (1.35)$$

and

$$\nabla \cdot \nabla \times \mathbf{F} = 0 \qquad (1.36)$$

which an arbitrary scalar field Φ and vector field **F** must satisfy. Maxwell's equation (1.22b) along with Eq. (1.36) is satisfied if we define **A** such that

$$\boxed{\mathbf{B} = \nabla \times \mathbf{A}} \qquad (1.37)$$

Substituting this into Eq. (1.22c) gives

$$-\nabla \times (\mathbf{E} + \frac{\partial \mathbf{A}}{\partial t}) = 0.$$

Since this equation has to be compatible with Eq. (1.35), we can choose the scalar field V such that

$$\mathbf{E} + \frac{\partial \mathbf{A}}{\partial t} = -\nabla V$$

or

$$\boxed{\mathbf{E} = -\nabla V - \frac{\partial \mathbf{A}}{\partial t}} \qquad (1.38)$$

10   Numerical Techniques in Electromagnetics

Thus, if we knew the potential functions V and $\mathbf{A}$, the fields $\mathbf{E}$ and $\mathbf{B}$ could be obtained from Eqs. (1.37) and (1.38). However, we still need to find the solution for the potential functions. Substituting Eqs. (1.37) and (1.38) into Eq. (1.22d) and assuming linear, homogeneous medium,

$$\nabla \times \nabla \times \mathbf{A} = \mu \mathbf{J} + \epsilon\mu \frac{\partial}{\partial t}\left(-\nabla V - \frac{\partial \mathbf{A}}{\partial t}\right).$$

Applying the vector identity in Eq. (1.30) leads to

$$\nabla^2 \mathbf{A} - \nabla(\nabla \cdot \mathbf{A}) = -\mu \mathbf{J} + \mu\epsilon \nabla \frac{\partial^2 \mathbf{A}}{\partial t^2} + \mu\epsilon \nabla \frac{\partial V}{\partial t}. \tag{1.39}$$

Substituting Eq. (1.38) into Eq. (1.22a) gives

$$\nabla \cdot \mathbf{E} = \frac{\rho}{\epsilon} = -\nabla^2 V - \frac{\partial (\nabla \cdot \mathbf{A})}{\partial t}$$

or

$$\nabla^2 V + \frac{\partial}{\partial t} \nabla \cdot \mathbf{A} = -\frac{\rho_v}{\epsilon}. \tag{1.40}$$

According to the Helmholtz theorem of vector analysis, a vector is uniquely defined if and only if both its curl and divergence are specified. We have only specified the curl of $\mathbf{A}$ in Eq. (1.37); we may choose the divergence of $\mathbf{A}$ so that the differential equations (1.39) and (1.40) have the simplest forms possible. We achieve this in the so-called *Lorentz condition:*

$$\nabla \cdot \mathbf{A} = -\mu\epsilon \frac{\partial V}{\partial t}. \tag{1.41}$$

Incorporating this condition into Eqs. (1.39) and (1.40) results in

$$\boxed{\nabla^2 \mathbf{A} - \mu\epsilon \frac{\partial^2 \mathbf{A}}{\partial t^2} = -\mu \mathbf{J}} \tag{1.42}$$

and

$$\boxed{\nabla^2 V - \mu\epsilon \frac{\partial^2 V}{\partial t^2} = -\frac{\rho_v}{\epsilon}} \tag{1.43}$$

which are inhomogeneous wave equations. Thus Maxwell's equations in terms of the potentials V and $\mathbf{A}$ reduce to the three equations (1.41) to (1.43). In other words, the three equations are equivalent to the ordinary form of Maxwell's equations in that potentials satisfying these equations always lead to a solution of Maxwell's equations for $\mathbf{E}$ and $\mathbf{B}$ when used

Fundamental Concepts   11

with Eqs. (1.37) and (1.38). Integral solutions to Eqs. (1.42) and (1.43) are the so-called *retarded* potentials

$$\mathbf{A} = \int \frac{\mu[\mathbf{J}]dv}{4\pi R} \tag{1.44}$$

and

$$V = \int \frac{[\rho_v]dv}{4\pi\epsilon R} \tag{1.45}$$

where R is the distance from the source point to the field point, and the square brackets denote ρ_v and $\mathbf{J}$ are specified at a time $R(\mu\epsilon)^{1/2}$ earlier than for which $\mathbf{A}$ or V is being determined.

1.2.7 Time-harmonic Fields

Up to this point, we have considered the general case of arbitrary time variation of EM fields. In many practical situations, especially at low frequencies, it is sufficient to deal with only the steady-state (or equilibrium) solution of EM fields when produced by sinusoidal currents. Such fields are said to be sinusoidal time-varying or time-harmonic, that is, they vary at a sinusoidal frequency ω. An arbitrary time-dependent field $\mathbf{F}(x,y,z,t)$ or $\mathbf{F}(\mathbf{r},t)$ can be expressed as

$$\boxed{\mathbf{F}(\mathbf{r},t) = \text{Re}[\mathbf{F}_s(\mathbf{r})e^{j\omega t}]} \tag{1.46}$$

where $\mathbf{F}_s(\mathbf{r}) = \mathbf{F}_s(x,y,z)$ is the phasor form of $\mathbf{F}(\mathbf{r},t)$ and is in general complex, Re[ ] indicates "taking the real part of" quantity in brackets, and ω is the angular frequency (in rad/s) of the sinusoidal excitation. The EM field quantities can be represented in phasor notation as

$$\begin{bmatrix} \mathbf{E}(\mathbf{r},t) \\ \mathbf{D}(\mathbf{r},t) \\ \mathbf{H}(\mathbf{r},t) \\ \mathbf{B}(\mathbf{r},t) \end{bmatrix} = \begin{bmatrix} \mathbf{E}_s(\mathbf{r}) \\ \mathbf{D}_s(\mathbf{r}) \\ \mathbf{H}_s(\mathbf{r}) \\ \mathbf{B}_s(\mathbf{r}) \end{bmatrix} e^{j\omega t}. \tag{1.47}$$

Using the phasor representation allows us to replace the time derivations $\partial/\partial t$ by $j\omega$ since

$$\frac{\partial e^{j\omega t}}{\partial t} = j\omega e^{j\omega t}.$$

Thus Maxwell's equations, in sinusoidal steady state, become

$$\nabla \cdot \mathbf{D}_s = \rho_{vs} \tag{1.48a}$$
$$\nabla \cdot \mathbf{B}_s = 0 \tag{1.48b}$$
$$\nabla \times \mathbf{E}_s = -j\omega \mathbf{B}_s \tag{1.48c}$$
$$\nabla \times \mathbf{H}_s = \mathbf{J}_s + j\omega \mathbf{D}_s. \tag{1.48d}$$

12   Numerical Techniques in Electromagnetics

We should observe that the effect of the time-harmonic assumption is to eliminate the time dependence from Maxwell's equations, thereby reducing the time-space dependence to space dependence only. This simplification does not exclude more general time-varying fields if we consider ω to be one element of an entire frequency spectrum, with all the Fourier components superposed. In other words, a nonsinusoidal field can be represented as

$$\mathbf{F}(\mathbf{r},t) = Re\left[\int_{-\infty}^{\infty}\mathbf{F}_s(\mathbf{r},\omega)e^{j\omega t}d\omega\right]. \tag{1.49}$$

Thus the solutions to Maxwell's equations for a nonsinusoidal field can be obtained by summing all the Fourier components $\mathbf{F}_s(\mathbf{r},\omega)$ over ω. Henceforth we drop the subscript s denoting phasor quantity when no confusion results.

Replacing the time derivative in Eq. (1.34) by $(j\omega)^2$ yields the scalar wave equation in phasor representation as

$$\nabla^2\Psi + k^2\Psi = 0 \tag{1.50}$$

where k is the propagation constant (in rad/m), given by

$$k = \frac{\omega}{u} = \frac{2\pi f}{u} = \frac{2\pi}{\lambda}. \tag{1.51}$$

We recall that Eqs.(1.31) to (1.34) were obtained assuming that $\rho_v = 0 = \mathbf{J}$. If $\rho_v \neq 0 \neq \mathbf{J}$, Eq. (1.50) will have the general form (see Prob. 1.4)

$$\boxed{\nabla^2\Psi + k^2\Psi = g} \tag{1.52}$$

We notice that this Helmholtz equation reduces to:
(1) Poisson's equation

$$\nabla^2\Psi = g \tag{1.53}$$

when $k = 0$ (i.e., $\omega = 0$ for static case).
(2) Laplace's equation

$$\nabla^2\Psi = 0 \tag{1.54}$$

when $k = 0 = g$.

Thus Poisson's and Laplace's equations are special cases of the Helmholtz equation. Note that function Ψ is said to be *harmonic* if it satisfies Laplace's equation.

Example 1.1

From the divergence theorem, derive Green's theorem

$$\int_v (U\nabla^2 V - V\nabla^2 U)\, dv = \oint_S (U\frac{\partial V}{\partial n} - V\frac{\partial U}{\partial n})\cdot d\mathbf{S}$$

where $\frac{\partial \Phi}{\partial n} = \nabla\Phi \cdot \mathbf{a}_n$ is the directional derivation of Φ along the outward normal to S.

Solution

In Eq.(1.1), let $\mathbf{F} = U\nabla V$, then

$$\int_v \nabla\cdot(U\nabla V)\, dv = \oint_S U\nabla V \cdot d\mathbf{S}. \tag{1.1.1}$$

But

$$\nabla\cdot(U\nabla V) = U\nabla\cdot\nabla V + \nabla V \cdot \nabla U$$
$$= U\nabla^2 V + \nabla U \cdot \nabla V.$$

Substituting this into Eq. (1.1.1) gives *Green's first identity:*

$$\int_v (U\nabla^2 V + \nabla U \cdot \nabla V)\, dv = \oint_S U\nabla V \cdot d\mathbf{S}. \tag{1.1.2}$$

By interchanging U and V in Eq. (1.1.2), we obtain

$$\int_v (V\nabla^2 U + \nabla V \cdot \nabla U)\, dv = \oint_S V\nabla U \cdot d\mathbf{S} \tag{1.1.3}$$

Subtracting Eq. (1.1.3) from Eq. (1.1.2) leads to *Green's second identity* or Green's theorem:

$$\int_v (U\nabla^2 V - V\nabla^2 U)\, dv = \oint_S (U\nabla V - V\nabla U)\cdot d\mathbf{S}.$$

Example 1.2

Show that the continuity equation is implicit (or incorporated) in Maxwell's equations.

14   Numerical Techniques in Electromagnetics

Solution

According to Eq. (1.36), the divergence of the curl of any vector field is zero. Hence, taking the divergence of Eq. (1.22d) gives

$$0 = \nabla \cdot \nabla \times \mathbf{H} = \nabla \cdot \mathbf{J} + \frac{\partial}{\partial t}\nabla \cdot \mathbf{D}.$$

But $\nabla \cdot \mathbf{D} = \rho_v$ from Eq. (1.22a). Thus,

$$0 = \nabla \cdot \mathbf{J} + \frac{\partial \rho_v}{\partial t}$$

which is the continuity equation.

Example 1.3

Express:
 (a) $\mathbf{E} = 10\sin(\omega t - kz)\mathbf{a}_x + 20\cos(\omega t - kz)\mathbf{a}_y$ in phasor form.
 (b) $\mathbf{H}_s = (4 - j3)\sin x \mathbf{a}_x + \frac{e^{j10°}}{x}\mathbf{a}_z$ in instantaneous form.

Solution

(a) We can express $\sin\theta$ as $\cos(\theta - \pi/2)$. Hence,

$$\begin{aligned}\mathbf{E} &= 10\cos(\omega t - kz - \pi/2)\mathbf{a}_x + 20\cos(\omega t - kz)\mathbf{a}_y \\ &= \text{Re}\left[(10e^{-jkz}e^{-j\pi/2}\mathbf{a}_x + 20e^{-jkz}\mathbf{a}_y)e^{j\omega t}\right] \\ &= Re[\mathbf{E}_s e^{j\omega t}].\end{aligned}$$

Thus,

$$\begin{aligned}\mathbf{E}_s &= 10e^{-jkz}e^{-j\pi/2}\mathbf{a}_x + 20e^{-jkz}\mathbf{a}_y \\ &= (-j10\mathbf{a}_x + 20\mathbf{a}_y)e^{-jkz}.\end{aligned}$$

(b) Since

$$\begin{aligned}\mathbf{H} &= \text{Re}[\mathbf{H}_s e^{j\omega t}] \\ &= \text{Re}[5\sin x e^{j(\omega t - 36.87°)}\mathbf{a}_x + \frac{1}{x}e^{j(\omega t + 10°)}\mathbf{a}_z] \\ &= 5\sin x \cos(\omega t - 36.78°)\mathbf{a}_x + \frac{1}{x}\cos(\omega t + 10°)\mathbf{a}_z.\end{aligned}$$

Fundamental Concepts   15

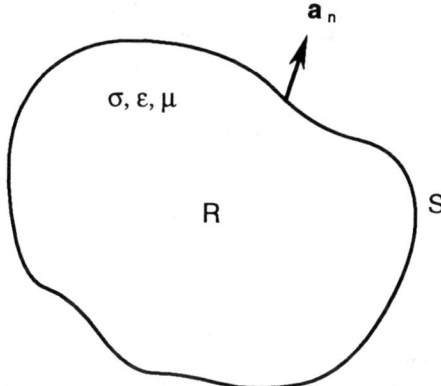

Figure 1.2 Solution region R with boundary S.

1.3 Classification of EM Problems

Problem classification is an important concept because the general theory and methods of solution usually apply only to a given class of problems. Classifying EM problems will help us later to answer the question of what method is best for solving a given problem. Continuum problems are categorized differently depending on the particular item of interest, which could be one of these:

(1) the solution region of the problem,
(2) the nature of the equation describing the problem,
(3) the associated boundary conditions.

It will soon become evident that these classifications are sometimes not independent of each other.

1.3.1 Classification of Solution Regions

In terms of the solution region or problem domain, the problem could be an interior problem, also variably called an inner, closed, or bounded problem, or an exterior problem, also variably called an outer, open, or unbounded problem.

Consider the solution region R with boundary S, as shown in Fig. 1.2. If part or all of S is at infinity, R is exterior/open, otherwise R is interior/closed. For example, wave propagation in a waveguide is an interior problem, whereas while wave propagation in free space, scattering of EM waves by raindrops, and radiation from a dipole antenna are exterior.

16 Numerical Techniques in Electromagnetics

A problem can also be classified in terms of the electrical, constitutive properties (σ, ϵ, μ) of the solution region. As mentioned in Section 1.2.4, the solution region could be linear (or nonlinear), homogeneous (or inhomogeneous), and isotropic (or anisotropic). We shall be concerned, for the most part, with linear, homogeneous, isotropic media in this text.

1.3.2 Classification of Differential Equations

EM problems are classified in terms of the equations describing them. The equations could be differential or integral or both. Most EM problems can be stated in terms of an operator equation

$$\boxed{L\Phi = g} \qquad (1.55)$$

where L is an operator (differential, integral, or integro-differential), g is the known excitation or source, and Φ is the unknown function to be determined. A typical example is the electrostatic problem involving Poisson's equation. In differential form, Eq. (1.55) becomes

$$-\nabla^2 V = \frac{\rho_v}{\epsilon} \qquad (1.56)$$

so that $L = -\nabla^2$ is the Laplacian operator, $g = \rho_v/\epsilon$ is the source term, and $\Phi = V$ is the electric potential. In integral form, Poisson's equation is of the form

$$V = \int \frac{\rho_v dv}{4\pi \epsilon r^2} \qquad (1.57)$$

so that

$$L = \int \frac{dv}{4\pi r^2}, \; g = V, \text{and } \Phi = \rho_v/\epsilon.$$

In this section, we shall limit our discussion to differential equations; integral equations will be considered in detail in Chapter 5.

As observed in Eqs. (1.52) to (1.54), EM problems involve linear, second-order differential equations. In general, a second-order partial differential equation (PDE) is given by

$$a\frac{\partial^2 \Phi}{\partial x^2} + b\frac{\partial^2 \Phi}{\partial x \partial y} + c\frac{\partial^2 \Phi}{\partial y^2} + d\frac{\partial \Phi}{\partial x} + e\frac{\partial \Phi}{\partial y} + f\Phi = g$$

or simply

$$\boxed{a\Phi_{xx} + b\Phi_{xy} + c\Phi_{yy} + d\Phi_x + e\Phi_y + f\Phi = g} \qquad (1.58)$$

The coefficients a, b, and c in general are functions of x and y; they may also depend on Φ itself, in which case the PDE is said to be *nonlinear*. Since most EM problems involve linear PDE, a, b, and c will be regarded as constants. A PDE in which $g(x,y)$ in Eq. (1.58) equals zero is termed *homogeneous*; it is *inhomogeneous* if $g(x,y) \neq 0$. Notice that Eq. (1.58) has the same form as Eq. (1.55), where L is now a differential operator given by

$$L = a\frac{\partial^2}{\partial x^2} + b\frac{\partial^2}{\partial x \partial y} + c\frac{\partial^2}{\partial y^2} + d\frac{\partial}{\partial x} + e\frac{\partial}{\partial y} + f. \tag{1.59}$$

A PDE in general can have both boundary values and initial values. PDEs whose boundary conditions are specified are called *steady-state equations*. If only initial values are specified, they are called *transient equations*.

Any linear second-order PDE can be classified as elliptic, hyperbolic, or parabolic depending on the coefficients a, b, and c. Equation (1.58) is said to be:

$$\boxed{\begin{array}{ll} \text{elliptic if} & b^2 - 4ac < 0 \\ \text{hyperbolic if} & b^2 - 4ac > 0 \\ \text{parabolic if} & b^2 - 4ac = 0 \end{array}} \tag{1.60}$$

The terms *hyperbolic*, *parabolic*, and *elliptic* are derived from the fact that the quadratic equation

$$ax^2 + bxy + cy^2 + dx + ey + f = 0$$

represents a hyperbola, parabola, or ellipse if $b^2 - 4ac$ is positive, zero, or negative, respectively. In each of these categories, there are PDEs that model certain physical phenomena. Such phenomena are not limited to EM but extend to almost all areas of science and engineering. Thus the mathematical model specified in Eq. (1.58) arises in problems involving heat transfer, boundary-layer flow, vibrations, elasticity, electrostatic, wave propagation, and so on.

Elliptic PDEs are associated with steady-state phenomena, i.e., boundary-value problems. Typical examples of this type of PDE include Laplace's equation

$$\frac{\partial^2 \Phi}{\partial x^2} + \frac{\partial^2 \Phi}{\partial y^2} = 0 \tag{1.61}$$

and Poisson's equation

$$\frac{\partial^2 \Phi}{\partial x^2} + \frac{\partial^2 \Phi}{\partial y^2} = g(x,y) \tag{1.62}$$

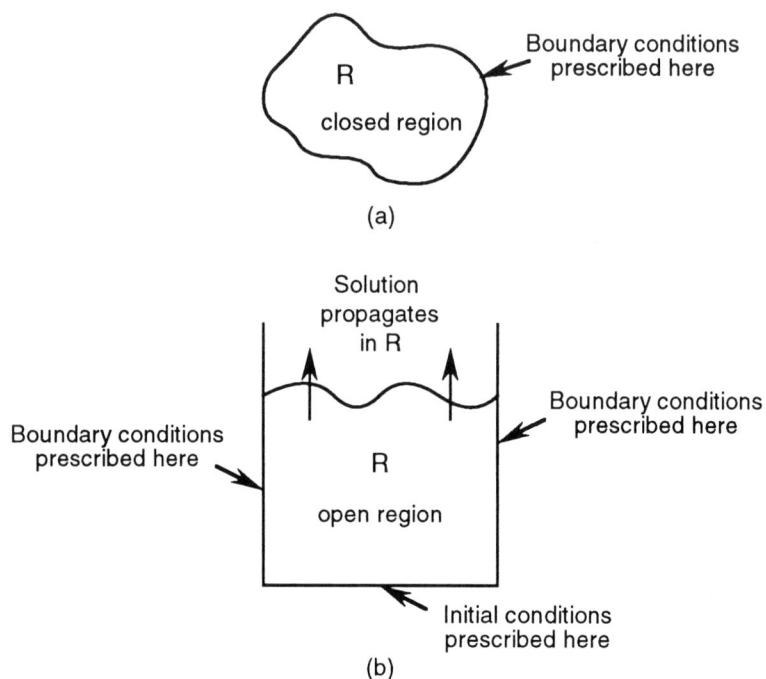

Figure 1.3 (a) Elliptic, (b) parabolic, or hyperbolic problem.

where in both cases $a = c = 1, b = 0$. An elliptic PDE usually models an interior problem, and hence the solution region is usually closed or bounded as in Fig. 1.3(a).

Hyperbolic PDEs arise in propagation problems. The solution region is usually open so that a solution advances outward indefinitely from initial conditions while always satisfying specified boundary conditions. A typical example of hyperbolic PDE is the wave equation in one dimension

$$\frac{\partial^2 \Phi}{\partial x^2} = \frac{1}{u^2} \frac{\partial^2 \Phi}{\partial t^2} \qquad (1.63)$$

where $a = u^2, b = 0, c = -1$. Notice that the wave equation in (1.50) is not hyperbolic but elliptic, since the time-dependence has been suppressed and the equation is merely the steady-state solution of Eq. (1.34).

Parabolic PDEs are generally associated with problems in which the quantity of interest varies slowly in comparison with the random motions

Fundamental Concepts   19

which produce the variations. The most common parabolic PDE is the diffusion (or heat) equation in one dimension

$$\frac{\partial^2 \Phi}{\partial x^2} = k\frac{\partial \Phi}{\partial t} \tag{1.64}$$

where $a = 1, b = 0 = c$. Like hyperbolic PDE, the solution region for parabolic PDE is usually open, as in Fig. 1.3(b). The initial and boundary conditions typically associated with parabolic equations resemble those for hyperbolic problems except that only one initial condition at $t = 0$ is necessary since Eq. (1.64) is only first order in time. Also, parabolic and hyperbolic equations are solved using similar techniques, whereas elliptic equations are usually more difficult and require different techniques.

Note that: (1) since the coefficients $a, b,$ and c are in general functions of x and y, the classification of Eq. (1.58) may change from point to point in the solution region, and (2) PDEs with more than two independent variables $(x, y, z, t, \ldots)$ may not fit as neatly into the classification above. A summary of our discussion so far in this section is shown in Table 1.1.

The type of problem represented by Eq. (1.55) is said to be *deterministic*, since the quantity of interest can be determined directly. Another type of problem where the quantity is found indirectly is called *nondeterministic* or *eigenvalue*. The *standard eigenproblem* is of the form

$$L\Phi = \lambda \Phi \tag{1.65}$$

where the source term in Eq. (1.55) has been replaced by $\lambda \Phi$. A more general version is the *generalized eigenproblem* having the form

$$\boxed{L\Phi = \lambda M \Phi} \tag{1.66}$$

where M, like L, is a linear operator for EM problems. In Eqs. (1.65) and (1.66), only some particular values of λ called *eigenvalues* are permissible; associated with these values are the corresponding solutions Φ called *eigenfunctions*. Eigenproblems are usually encountered in vibration and waveguide problems where the eigenvalues λ correspond to physical quantities such as resonance and cutoff frequencies, respectively.

1.3.3 Classification of Boundary Conditions

Our problem consists of finding the unknown function Φ of a partial differential equation. In addition to the fact that Φ satisfies Eq. (1.55) within a prescribed solution region R, Φ must satisfy certain conditions on S, the boundary of R. Usually these boundary conditions are of the Dirichlet and

Numerical Techniques in Electromagnetics

Table 1.1 Classification of Partial Differential Equations

Type	Sign of $b^2 - 4ac$	Example	Solution region
Elliptic	$-$	Laplace's equation: $\Phi_{xx} + \Phi_{yy} = 0$	Closed
Hyperbolic	$+$	Wave equation: $u^2 \Phi_{xx} = \Phi_{tt}$	Open
Parabolic	0	Diffusion equation: $\Phi_{xx} = k\Phi_t$	Open

Neumann types. Where a boundary has both, a mixed boundary condition is said to exist.

(1) Dirichlet boundary condition:

$$\Phi(\mathbf{r}) = 0, \quad \mathbf{r} \text{ on } S. \tag{1.67}$$

(2) Neumann boundary condition:

$$\frac{\partial \Phi(\mathbf{r})}{\partial n} = 0, \quad \mathbf{r} \text{ on } S, \tag{1.68}$$

i.e., the normal derivative of Φ vanishes on S.

(3) Mixed boundary condition:

$$\frac{\partial \Phi(\mathbf{r})}{\partial n} + h(\mathbf{r})\Phi(\mathbf{r}) = 0, \quad \mathbf{r} \text{ on } S \tag{1.69}$$

where $h(\mathbf{r})$ is a known function and $\frac{\partial \Phi}{\partial n}$ is the directional derivative of Φ along the outward normal to the boundary S, i.e.,

$$\frac{\partial \Phi}{\partial n} = \nabla \Phi \cdot \mathbf{a}_n \tag{1.70}$$

where $\mathbf{a}_n$ is a unit normal directed out of R, as shown in Fig. 1.2. Note that the Neumann boundary condition is a special case of the mixed condition with $h(\mathbf{r}) = 0$.

The conditions in Eq. (1.67) to (1.69) are called *homogeneous boundary conditions*. The more general ones are the inhomogeneous:

Dirichlet:

$$\Phi(\mathbf{r}) = p(\mathbf{r}), \quad \mathbf{r} \text{ on } S \tag{1.71}$$

Neumann:
$$\frac{\partial \Phi(\mathbf{r})}{\partial n} = q(\mathbf{r}), \quad \mathbf{r} \text{ on S} \tag{1.72}$$

Mixed:
$$\frac{\partial \Phi(\mathbf{r})}{\partial n} + h(\mathbf{r})\Phi(\mathbf{r}) = w(\mathbf{r}), \quad \mathbf{r} \text{ on S} \tag{1.73}$$

where $p(\mathbf{r})$, $q(\mathbf{r})$, and $w(\mathbf{r})$ are explicitly known functions on the boundary S. For example, $\Phi(0) = 1$ is an inhomogeneous Dirichlet boundary condition, and the associated homogeneous counterpart is $\Phi(0) = 0$. Also $\Phi'(1) = 2$ and $\Phi'(1) = 0$ are, respectively, inhomogeneous and homogeneous Neumann boundary conditions. In electrostatics, for example, if the value of electric potential is specified on S, we have Dirichlet boundary condition, whereas if the surface charge ($\rho_s = D_n = \epsilon \frac{\partial V}{\partial n}$) is specified, the boundary condition is Neumann. The problem of finding a function Φ that is harmonic in a region is called *Dirichlet problem* (or *Neumann problem*) if Φ (or $\frac{\partial \Phi}{\partial n}$) is prescribed on the boundary of the region.

It is worth observing that the term "homogeneous" has been used to mean different things. The solution region could be homogeneous meaning that $\sigma, \epsilon,$ and μ are constant within R; the PDE could be homogeneous if $g = 0$ so that $L\Phi = 0$; and the boundary conditions are homogeneous when $p(\mathbf{r}) = q(\mathbf{r}) = w(\mathbf{r}) = 0$.

Example 1.4

Classify these equations as elliptic, hyperbolic, or parabolic:

(a) $4\Phi_{xx} + 2\Phi_x + \Phi_y + x + y = 0$

(b) $e^x \dfrac{\partial^2 V}{\partial x^2} + \cos y \dfrac{\partial^2 V}{\partial x \partial y} - \dfrac{\partial^2 V}{\partial y^2} = 0$

State whether the equations are homogeneous or inhomogeneous.

Solution

(a) In this PDE, $a = 4, b = 0 = c$. Hence
$$b^2 - 4ac = 0,$$
i.e., the PDE is parabolic. Since $g = -x - y$, the PDE is inhomogeneous.

(b) For this PDE, $a = e^x, b = \cos y, c = -1$. Hence
$$b^2 - 4ac = \cos^2 y + 4e^x > 0$$
and the PDE is hyperbolic. Since $g = 0$, the PDE is homogeneous.

1.4 Some Important Theorems

Two theorems are of fundamental importance in solving EM problems. These are the principle of superposition and the uniqueness theorem.

1.4.1 Superposition Principle

The principle of superposition is applied in several ways. We shall consider two of these. If each member of a set of functions $\Phi_n, n = 1, 2, \ldots, N$, is a solution to the PDE $L\Phi = 0$ with some prescribed boundary conditions, then a linear combination

$$\Phi_N = \Phi_0 + \sum_{n=1}^{N} a_n \Phi_n \qquad (1.74)$$

also satisfies $L\Phi = g$.

Given a problem described by the PDE

$$L\Phi = g \qquad (1.75)$$

subject to the boundary conditions

$$M_1(s) = h_1 \qquad (1.76)$$
$$M_2(s) = h_2$$
$$\vdots$$
$$M_N(s) = h_N,$$

as long as L is linear, we may divide the problem into a series of problems as follows:

$$\begin{array}{llll} L\Phi_0 = g & L\Phi_1 = 0 & \cdots & L\Phi_N = 0 \\ M_1(s) = 0 & M_1(s) = h_1 & \cdots & M_1(s) = 0 \\ M_2(s) = 0 & M_2(s) = 0 & \cdots & M_2(s) = 0 \\ \vdots & \vdots & & \vdots \\ M_N(s) = 0 & M_N(s) = 0 & \cdots & M_N(s) = h_N \end{array} \qquad (1.77)$$

where $\Phi_0, \Phi_1, \ldots, \Phi_N$ are the solutions to the reduced problems, which are easier to solve than the original problem. The solution to the original problem is given by

$$\Phi = \sum_{n=0}^{N} \Phi_n. \qquad (1.78)$$

1.4.2 Uniqueness Theorem

This theorem guarantees that the solution obtained for a PDE with some prescribed boundary conditions is the only one possible. For EM problems, the theorem may be stated as follows: If in any way a set of fields $(\mathbf{E}, \mathbf{H})$ is found which satisfies simultaneously Maxwell's equations and the prescribed boundary conditions, this set is unique. Therefore, a field is uniquely specified by the sources $(\rho_v, \mathbf{J})$ within the medium plus the tangential components of $\mathbf{E}$ or $\mathbf{H}$ over the boundary.

To prove the uniqueness theorem, suppose there exist two solutions (with subscripts 1 and 2) that satisfy Maxwell's equations

$$\nabla \cdot \epsilon \mathbf{E}_{1,2} = \rho_v \tag{1.79a}$$

$$\nabla \cdot \mathbf{H}_{1,2} = 0 \tag{1.79b}$$

$$\nabla \times \mathbf{E}_{1,2} = -\mu \frac{\partial \mathbf{H}_{1,2}}{\partial t} \tag{1.79c}$$

$$\nabla \times \mathbf{H}_{1,2} = \mathbf{J} + \sigma \mathbf{E}_{1,2} + \epsilon \frac{\partial \mathbf{E}_{1,2}}{\partial t}. \tag{1.79d}$$

If we denote the difference of the two fields as $\Delta \mathbf{E} = \mathbf{E}_2 - \mathbf{E}_1$ and $\Delta \mathbf{H} = \mathbf{H}_2 - \mathbf{H}_1$, $\Delta \mathbf{E}$ and $\Delta \mathbf{H}$ must satisfy the source-free Maxwell's equations, i.e.,

$$\nabla \cdot \epsilon \Delta \mathbf{E} = 0 \tag{1.80a}$$

$$\nabla \cdot \Delta \mathbf{H} = 0 \tag{1.80b}$$

$$\nabla \times \Delta \mathbf{E} = -\mu \frac{\partial \Delta \mathbf{H}}{\partial t} \tag{1.80c}$$

$$\nabla \times \Delta \mathbf{H} = \sigma \Delta \mathbf{E} + \epsilon \frac{\partial \Delta \mathbf{E}}{\partial t}. \tag{1.80d}$$

Dotting both sides of Eq. (1.80d) with $\Delta \mathbf{E}$ gives

$$\Delta \mathbf{E} \cdot \nabla \times \Delta \mathbf{H} = \sigma |\Delta \mathbf{E}|^2 + \epsilon \nabla \mathbf{E} \cdot \frac{\partial \Delta \mathbf{E}}{\partial t}. \tag{1.81}$$

Using the vector identity

$$\mathbf{A} \cdot (\nabla \times \mathbf{B}) = \mathbf{B} \cdot (\nabla \times \mathbf{A}) - \nabla \cdot (\mathbf{A} \times \mathbf{B})$$

and Eq. (1.80c), Eq. (1.81) becomes

$$\nabla \cdot (\Delta \mathbf{E} \times \Delta \mathbf{H}) = -\frac{1}{2} \frac{\partial}{\partial t} (\mu |\Delta \mathbf{H}|^2 + \epsilon |\Delta \mathbf{E}|^2) - \sigma |\Delta \mathbf{E}|^2.$$

Integrating over volume v bounded by surface S and applying divergence theorem to the left-hand side, we obtain

$$\oint_S (\Delta \mathbf{E} \times \Delta \mathbf{H}) \cdot d\mathbf{S} = -\frac{\partial}{\partial t} \int_v \left[\frac{1}{2}\epsilon|\Delta \mathbf{E}|^2 + \frac{1}{2}\mu|\Delta \mathbf{H}|^2\right] dv - \int_v \sigma|\Delta \mathbf{E}| \, dv. \quad (1.82)$$

showing that $\Delta \mathbf{E}$ and $\Delta \mathbf{H}$ satisfy the Poynting theorem just as $\mathbf{E}_{1,2}$ and $\mathbf{H}_{1,2}$. Only the tangential components of $\Delta \mathbf{E}$ and $\Delta \mathbf{H}$ contribute to the surface integral on the left side of Eq. (1.82). Therefore, if the tangential components of $\mathbf{E}_1$ and $\mathbf{E}_2$ or $\mathbf{H}_1$ and $\mathbf{H}_2$ are equal over S (thereby satisfying Eq. (1.27)), the tangential components of $\Delta \mathbf{E}$ and $\Delta \mathbf{H}$ vanish on S. Consequently, the surface integral in Eq. (1.82) is identically zero, and hence the right side of the equation must vanish also. It follows that $\Delta \mathbf{E} = 0$ due to the second integral on the right side and hence also $\Delta \mathbf{H} = 0$ throughout the volume. Thus $\mathbf{E}_1 = \mathbf{E}_2$ and $\mathbf{H}_1 = \mathbf{H}_2$, confirming that the solution is unique.

The theorem just proved for time-varying fields also holds for static fields as a special case. In terms of electrostatic potential V, the uniqueness theorem may be stated as follows: A solution to $\nabla^2 V = 0$ is uniquely determined by specifying either the value of V or the normal component of ∇V at each point on the boundary surface. For a magnetostatic field, the theorem becomes: A solution of $\nabla^2 \mathbf{A} = 0$ (and $\nabla \cdot \mathbf{A} = 0$) is uniquely determined by specifying the value of $\mathbf{A}$ or the tangential component of $\mathbf{B}$ $(= \nabla \times \mathbf{A})$ at each point on the boundary surface.

References

[1] K. H. Huebner and E. A. Thornton, *The Finite Element Method for Engineers*. New York: John Wiley and Sons, 1982, Chap. 3, pp. 62-107.

[2] P. P. Silvester and R. L. Ferrari, *Finite Elements for Electrical Engineers*. Cambridge, UK: Cambridge University Press, 1983, Chap. 2, pp. 33-72.

[3] J. A. Kong, *Electromagnetic Wave Theory*. New York: John Wiley and Sons, 1986, Chap. 1, pp. 1-41.

[4] R. E. Collins, *Foundations of Microwave Engineering*. New York: McGraw-Hill, 1966, Chap. 2, pp. 11-63.

[5] M. N. O. Sadiku, *Elements of Electromagnetics*. New York: Holt, Rinehart and Winston, 1989, Chap. 9, pp. 389-429.

Problems

1.1 In a coordinate system of your choice, prove that:
   (a) $\nabla \times \nabla \Phi = 0$,
   (b) $\nabla \cdot \nabla \times \mathbf{F} = 0$,
   (c) $\nabla \times \nabla \times \mathbf{F} = \nabla(\nabla \cdot \mathbf{F}) - \nabla^2 \mathbf{F}$,

   where Φ and $\mathbf{F}$ are scalar and vector fields, respectively.

1.2 Derive each of Maxwell's equations from the corresponding physical laws.

1.3 Show that in a source-free region $(\mathbf{J} = 0, \rho_v = 0)$, Maxwell's equations can be reduced to the two curl equations.

1.4 In deriving the wave equations (1.31) and (1.32), we assumed a source-free medium $(\mathbf{J} = 0, \rho_v = 0)$. Show that if $\rho_v \neq 0, \mathbf{J} \neq 0$, the equations become

$$\nabla^2 \mathbf{E} - \frac{1}{c^2} \frac{\partial^2 \mathbf{E}}{\partial t^2} = \nabla(\rho_v/\epsilon) + \mu \frac{\partial \mathbf{J}}{\partial t},$$

$$\nabla^2 \mathbf{H} - \frac{1}{c^2} \frac{\partial^2 \mathbf{H}}{\partial t^2} = -\nabla \times \mathbf{J}.$$

What assumptions have you made to arrive at these expressions?

1.5 Determine whether the fields

$$\mathbf{E} = 20\sin(\omega t - kz)\mathbf{a}_x - 10\cos(\omega t + kz)\mathbf{a}_y$$

$$\mathbf{H} = \frac{k}{\omega \mu_o}\left[-10\cos(\omega t + kz)\mathbf{a}_x + 20\sin(\omega t - kz)\mathbf{a}_y\right],$$

where $k = \omega\sqrt{\mu_o \epsilon_o}$, satisfy Maxwell's equations.

1.6 Given that in free space

$$\mathbf{E} = \sin(\omega t - \beta z)(\mathbf{a}_x + \mathbf{a}_y) \ V/m,$$

find the relationship between ω and β such that $\mathbf{E}$ satisfies Helmholtz homogeneous equation. Find the corresponding $\mathbf{H}$ field.

26 Numerical Techniques in Electromagnetics

1.7 Show that the electric field

$$\mathbf{E}_s = 20\sin(k_x x)\cos(k_y y)\mathbf{a}_z,$$

where $k_x^2 + k_y^2 = \omega^2\mu_0\epsilon_0$, can be represented as the superposition of four propagating plane waves. Find the corresponding $\mathbf{H}_s$ field.

1.8 Find the phasor form of the following time-harmonic fields:
  (a) $\mathbf{E}(y,t) = e^{-\alpha y}\cos(\omega t - \beta y)\mathbf{a}_x,$
  (b) $\mathbf{H}(z,t) = 10\sin(\omega t - \beta z)\mathbf{a}_x - 5\cos(\omega t - \beta z)\mathbf{a}_y.$

1.9 Determine the instantaneous quantities corresponding to the following phasors:
  (a) $\mathbf{A}_s = x^2 e^{-jkz}\mathbf{a}_y,$
  (b) $\mathbf{B}_s = e^{j(x-y)}\sin x\, \mathbf{a}_z,$
  (c) $\mathbf{C}_s = 10 e^{-jkz}\mathbf{a}_x + j5 e^{jkz + \frac{\pi}{4}}\mathbf{a}_y.$

1.10 Show that a time-harmonic EM field in a conducting medium ($\sigma \gg \omega\epsilon$) satisfies the diffusion equation

$$\nabla^2 \mathbf{E}_s - j\omega\mu\sigma \mathbf{E}_s = 0.$$

1.11 Show that in an inhomogeneous medium, the wave equations become

$$\nabla \times \left(\frac{1}{j\omega\mu}\nabla \times \mathbf{E}_s\right) + j\omega\epsilon \mathbf{E}_s = 0,$$

$$\nabla \times \left(\frac{1}{j\omega\epsilon}\nabla \times \mathbf{H}_s\right) + j\omega\mu \mathbf{H}_s = 0.$$

1.12 Classify the following PDEs as elliptic, parabolic, or hyperbolic:
  (a) $\Phi_{xx} + 2\Phi_{xy} + 3\Phi_{yy} + 4\Phi_x + 5\Phi = xy,$
  (b) $\Phi_{xx} = 2\Phi_{yy} + \Phi - y$
  (c) $-2y\Phi_{xx} + x\Phi_{yy} - \Phi_{yy} = 0,$
  (d) $(1 - y^2)\Phi_{xx} - 2xy\Phi_{xy} + (1 - x^2)\Phi_{yy} + \Phi = \sin x,$
  (e) $\Phi_{xx} + y\Phi_{xy} - x^2\Phi_x + \cos y\Phi_y + e^x\Phi = 0.$

1.13 Repeat Prob. 1.12 for the following PDEs:
  (a) $\alpha\dfrac{\partial^2 \Phi}{\partial x^2} = \beta\dfrac{\partial \Phi}{\partial x} + \dfrac{\partial \Phi}{\partial t}$ \quad ($\alpha, \beta =$ constant)

which is called convective heat equation.

(b) $\nabla^2 \phi + \lambda \Phi = 0$,

which is the Helmholtz equation.

(c) $\nabla^2 \Phi + [\lambda - \rho(x)]\Phi = 0$

which is the time-independent Schrodinger equation.

(d) $\dfrac{\partial^4 \Phi}{\partial x^4} + 2\dfrac{\partial^4 \Phi}{\partial x^2 \partial y^2} + \dfrac{\partial^4 \Phi}{\partial y^4} = 0$

which is the biharmonic equation.

Chapter 2

Analytical Methods

"The important thing about a problem is not its solution, but the strength we gain in finding the solution."   Anonymous

2.1 Introduction

The most satisfactory solution of a field problem is an exact mathematical one. Although in many practical cases such an analytical solution cannot be obtained and we must resort to numerical approximate solution, analytical solution is useful in checking solutions obtained from numerical methods. Also, one would hardly appreciate the need for numerical methods without first seeing the limitations of the classical analytical methods. Hence our objective in this chapter is to briefly examine the common analytical methods and thereby put numerical methods in proper perspective.

The most commonly used analytical methods in solving EM-related problems include:

(1) separation of variables,
(2) series expansion,
(3) conformal mapping,
(4) integral methods.

Perhaps the most powerful analytical method is the separation of variables; it is the one that will be emphasized in this chapter. Since the application of conformal mapping is restricted to certain EM problems, it will not be discussed here. The interested reader is referred to Gibbs [1]. The integral methods will be covered in Chapter 5.

2.2 Separation of Variables

The method of separation of variables (sometimes called the method of Fourier) is a convenient method for solving a partial differential equation (PDE). Basically, it entails seeking a solution which breaks up into a product of functions, each of which involves only one of the variables. For example, if we are seeking a solution $\Phi(x,y,z,t)$ to some PDE, we require that it has the product form

$$\Phi(x,y,z,t) = X(x)Y(y)Z(z)T(t). \tag{2.1}$$

A solution of the form in Eq. (2.1) is said to be separable in x, y, z, and t. For example, consider the functions

(1) $x^2 yz \sin 10t$,
(2) $xy^2 + \frac{2}{t}$,
(3) $(2x + y^2)z \cos 10t$.

(1) is completely separable, (2) is not separable, while (3) is separable only in z and t.

To determine whether the method of independent separation of variables can be applied to a given physical problem, we must consider the PDE describing the problem, the shape of the solution region, and the boundary conditions. For example, to apply the method to a problem involving two variables x and y (or ρ and ϕ, etc.), three things must be considered [2]:

(i) The differential operator L must be separable, i.e., it must be a function of $\Phi(x,y)$ such that

$$\frac{L\{X(x)Y(y)\}}{\Phi(x,y)X(x)Y(y)}$$

is a sum of a function of x only and a function of y only.

(ii) All initial and boundary conditions must be on constant-coordinate surfaces, i.e., $x = $ constant, $y = $ constant.

(iii) The linear operators defining the boundary conditions at $x = $ constant (or $y = $ constant) must involve no partial derivatives of Φ with respect to y (or x), and their coefficient must be independent of y (or x).

For example, the operator equation

$$L\Phi = \frac{\partial^2 \Phi}{\partial x^2} + \frac{\partial^2 \Phi}{\partial x \partial y} + \frac{\partial^2 \Phi}{\partial y^2}$$

violates (i). If the solution region R is not a rectangle with sides parallel to the x and y axes, (ii) is violated. With a boundary condition $\Phi = 0$ on a part of $x = 0$ and $\partial \Phi / \partial x = 0$ on another part, (iii) is violated.

With this preliminary discussion, we will now apply the method of separation of variables to PDEs in rectangular, circular cylindrical, and spherical coordinate systems. In each of these applications, we shall always take these three major steps:

(1) separate the (independent) variables,
(2) find particular solutions of the separated equations, which satisfy some of the boundary conditions,
(3) combine these solutions to satisfy the remaining boundary conditions.

We begin the application of separation of variables by finding the product solution of the homogeneous scalar wave equation

$$\nabla^2 \Phi - \frac{1}{c^2} \frac{\partial^2 \Phi}{\partial t^2} = 0 \tag{2.2}$$

Solution to Laplace's equation can be derived as a special case of the wave equation. Diffusion and heat equation can be handled in the same manner as we will treat wave equation. To solve Eq. (2.2), it is expedient that we first separate the time dependence. We let

$$\Phi(\mathbf{r}, t) = U(\mathbf{r}) T(t) \tag{2.3}$$

Substituting this in Eq. (2.2),

$$T \nabla^2 U - \frac{1}{c^2} U T'' = 0.$$

Dividing by UT gives

$$\frac{\nabla^2 U}{U} = \frac{T''}{c^2 T}. \tag{2.4}$$

The left side is independent of t, while the right side is independent of $\mathbf{r}$; the equality can be true only if each side is independent of both variables. If we let an arbitrary constant $-k^2$ be the common value of the two sides, Eq. (2.4) reduces to

$$T'' + c^2 k^2 T = 0, \tag{2.5a}$$

$$\nabla^2 U + k^2 U = 0. \tag{2.5b}$$

Thus we have been able to separate the space variable $\mathbf{r}$ from the time variable t. The arbitrary constant $-k^2$ introduced in the course of the separation of variables is called the *separation constant*. We shall see that in general the total number of independent separation constants in a given problem is one less than the number of independent variable involved.

Equation (2.5a) is an ordinary differential equation with solution

$$T(t) = a_1 e^{jckt} + a_2 e^{-jckt} \tag{2.6a}$$

or

$$T(t) = b_1 \cos(ckt) + b_2 \sin(ckt). \tag{2.6b}$$

Since the time dependence does not change with coordinate system, the time dependence expressed in Eq. (2.6) is the same for all coordinate systems. Therefore, we shall henceforth restrict our effort to seeking solution to Eq. (2.5b). Notice that if $k = 0$, the time dependence disappears and Eq. (2.5b) becomes Laplace's equation.

2.3 Separation of Variables in Rectangular Coordinates

In order not to complicate things, we shall first consider Laplace's equation in two dimensions and later extend the idea to wave equation in three dimensions.

2.3.1 Laplace's Equations

Consider the Dirichlet problem of an infinitely long conducting rectangular trough whose cross section is shown in Fig. 2.1. For simplicity, let three of its sides be maintained at zero potential while the fourth side is at a fixed potential V_o. This is a boundary value problem. The PDE to be solved is

$$\frac{\partial^2 V}{\partial x^2} + \frac{\partial^2 V}{\partial y^2} = 0 \tag{2.7}$$

subject to (Dirichlet) boundary conditions

$$V(0, y) = 0 \tag{2.8a}$$
$$V(a, y) = 0 \tag{2.8b}$$
$$V(x, 0) = 0 \tag{2.8c}$$
$$V(x, b) = V_o. \tag{2.8d}$$

We let

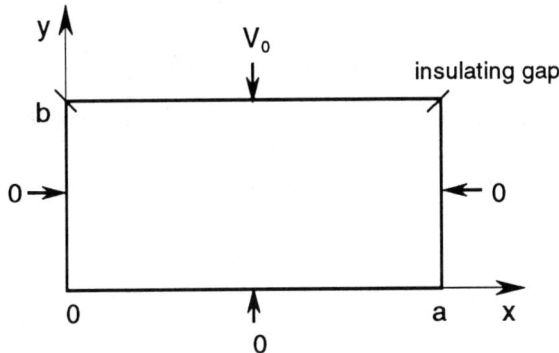

Figure 2.1 Cross section of the rectangular conducting trough.

$$V(x,y) = X(x)Y(y). \qquad (2.9)$$

Substitute this into Eq. (2.7) and divide by XY. This leads to

$$\frac{X''}{X} + \frac{Y''}{Y} = 0$$

or

$$\frac{X''}{X} = -\frac{Y''}{Y} = \lambda \qquad (2.10)$$

where λ is the separation constant. Thus the separated equations are

$$X'' - \lambda X = 0 \qquad (2.11)$$

$$Y'' + \lambda Y = 0. \qquad (2.12)$$

To solve the ordinary differential equations (2.11) and (2.12), we must impose the boundary conditions in Eq. (2.8). However, these boundary conditions must be transformed so that they can be applied directly to the separated equations. Since $V = XY$,

34   Numerical Techniques in Electromagnetics

$$V(0,y) = 0 \quad \rightarrow \quad X(0) = 0 \quad (2.13a)$$
$$V(a,y) = 0 \quad \rightarrow \quad X(a) = 0 \quad (2.13b)$$
$$V(x,0) = 0 \quad \rightarrow \quad Y(0) = 0 \quad (2.13c)$$
$$V(x,b) = V_o \quad \rightarrow \quad X(x)Y(b) = V_o. \quad (2.13d)$$

Notice that only the homogeneous conditions are separable. To solve Eq. (2.11), we distinguish the three possible cases: $\lambda = 0$, $\lambda > 0$, and $\lambda < 0$.
Case 1: If $\lambda = 0$, Eq. (2.11) reduces to

$$X'' = 0 \quad \text{or} \quad \frac{d^2 X}{dx^2} = 0 \quad (2.14)$$

which has the solution

$$X(x) = a_1 x + a_2 \quad (2.15)$$

where a_1 and a_2 are constants. Imposing the conditions in Eq. (2.13a) and Eq. (2.13b),

$$X(0) = 0 \quad \rightarrow \quad a_2 = 0$$
$$X(a) = 0 \quad \rightarrow \quad a_1 = 0.$$

Hence $X(x) = 0$, a trivial solution. This renders case $\lambda = 0$ as unacceptable.
Case 2:   If $\lambda > 0$, say $\lambda = \alpha^2$, Eq. (2.11) becomes

$$X'' - \alpha^2 X = 0 \quad (2.16)$$

with the corresponding auxiliary equations $m^2 - \alpha^2 = 0$ or $m = \pm \alpha$. Hence the general solution is

$$X = b_1 e^{-\alpha x} + b_2 e^{\alpha x} \quad (2.17)$$

or

$$X = b_3 \sinh \alpha x + b_4 \cosh \alpha x. \quad (2.18)$$

The boundary conditions are applied to determine b_3 and b_4.

$$X(0) = 0 \quad \rightarrow \quad b_4 = 0$$
$$X(a) = 0 \quad \rightarrow \quad b_3 = 0$$

since $\sinh \alpha x$ is never zero for $\alpha > 0$. Hence $X(x) = 0$, a trivial solution, and we conclude that case $\lambda > 0$ is not valid.

Analytical Methods   35

Case 3:   If $\lambda < 0$, say $\lambda = -\beta^2$,

$$X'' + \beta^2 X = 0 \tag{2.19}$$

with the auxiliary equation $m^2 + \beta^2 = 0$ or $m = \pm j\beta$. The solution to Eq. (2.19) is

$$X = A_1 e^{j\beta x} + A_2 e^{j\beta x} \tag{2.20a}$$

or

$$X = B_1 \sin \beta x + B_2 \cos \beta x. \tag{2.20b}$$

Again,

$$X(0) = 0 \;\; \rightarrow \;\; B_2 = 0$$
$$X(a) = 0 \;\; \rightarrow \;\; \sin \beta a = 0 = \sin n\pi$$

or

$$\beta = \frac{n\pi}{a}, \qquad n = 1, 2, 3, \cdots \tag{2.21}$$

since B_1 cannot vanish for nontrivial solution, whereas $\sin \beta a$ can vanish without its argument being zero. Thus we have found an infinite set of discrete values of λ for which Eq. (2.11) has nontrivial solutions, i.e.,

$$\lambda = -\beta^2 = \frac{-n^2 \pi^2}{a^2}, \qquad n = 1, 2, 3, \cdots \tag{2.22}$$

These are the eigenvalues of the problem and the corresponding eigenfunctions are

$$X_n(x) = \sin \beta x = \sin \frac{n\pi x}{a}. \tag{2.23}$$

From Eq. (2.22) note that it is not necessary to include negative values of n since they lead to the same set of eigenvalues. Also we exclude $n = 0$ since it yields the trivial solution $X = 0$ as shown under case 1 when $\lambda = 0$. Having determined λ, we can solve Eq. (2.12) to find $Y_n(y)$ corresponding to $X_n(x)$. That is, we solve

$$Y'' - \beta^2 Y = 0, \tag{2.24}$$

which is similar to Eq. (2.16), whose solution is in Eq. (2.18). Hence the solution to Eq. (2.24) has the form

36   Numerical Techniques in Electromagnetics

$$Y_n(y) = a_n \sinh \frac{n\pi y}{a} + b_n \cosh \frac{n\pi y}{a}. \tag{2.25}$$

Imposing the boundary condition in Eq. (2.13c),

$$Y(0) = 0 \quad \rightarrow \quad b_n = 0$$

so that

$$Y_n(y) = a_n \sinh \frac{n\pi y}{a}. \tag{2.26}$$

Substituting Eqs. (2.23) and (2.26) into Eq. (2.9), we obtain

$$V_n(x,y) = X_n(x)Y_n(y) = a_n \sin \frac{n\pi x}{a} \sinh \frac{n\pi y}{a}, \tag{2.27}$$

which satisfies Eq. (2.7) and the three homogeneous boundary conditions in Eqs. (2.8a), (2.8b), and (2.8c). By the superposition principle, a linear combination of the solutions V_n, each with different values of n and arbitrary coefficient a_n, is also a solution of Eq. (2.7). Thus we may represent the solution V of Eq. (2.7) as an infinite series in the function V_n, i.e.,

$$V(x,y) = \sum_{n=1}^{\infty} a_n \sin \frac{n\pi x}{a} \sinh \frac{n\pi y}{a}. \tag{2.28}$$

We now determine the coefficient a_n by imposing the inhomogeneous boundary condition in Eq. (2.8d) on Eq. (2.28). We get

$$V(x,b) = V_o = \sum_{n=1}^{\infty} a_n \sin \frac{n\pi x}{a} \sinh \frac{n\pi b}{a}, \tag{2.29}$$

which is Fourier sine expansion of V_o. Hence,

$$a_n \sinh \frac{n\pi b}{a} = \frac{2}{b} \int_0^b V_o \sin \frac{n\pi x}{a} dx = \frac{2V_o}{n\pi}(1 - \cos n\pi)$$

or

$$a_n = \begin{cases} \dfrac{4V_o}{n\pi} \dfrac{1}{\sinh \dfrac{n\pi b}{a}}, & n = \text{odd}, \\ 0, & n = \text{even}. \end{cases} \tag{2.30}$$

Substitution of Eq. (2.30) into Eq. (2.28) gives the complete solution as

$$V(x,y) = \frac{4V_o}{\pi} \sum_{n=\text{odd}}^{\infty} \frac{\sin \dfrac{n\pi x}{a} \sinh \dfrac{n\pi y}{a}}{\sinh \dfrac{n\pi b}{a}}. \tag{2.31a}$$

Analytical Methods 37

By replacing n by $2k - 1$, Eq. (2.31a) may be written as

$$V(x,y) = \frac{4V_o}{\pi} \sum_{k=1}^{\infty} \frac{\sin\frac{n\pi x}{a} \sinh\frac{n\pi y}{a}}{\sinh\frac{n\pi b}{a}}, \qquad n = 2k - 1 \qquad (2.31b)$$

2.3.2 Wave Equation

The time dependence has been taken care of in Section 2.2. We are left with solving the Helmholtz equation

$$\nabla^2 U + k^2 U = 0. \qquad (2.5b)$$

In rectangular coordinates, Eq. (2.5b) becomes

$$\frac{\partial^2 U}{\partial x^2} + \frac{\partial^2 U}{\partial y^2} + \frac{\partial^2 U}{\partial z^2} + k^2 U = 0. \qquad (2.32)$$

We let

$$U(x,y,z) = X(x)Y(y)Z(z). \qquad (2.33)$$

Substituting Eq. (2.33) into Eq. (2.32) and dividing by XYZ, we obtain

$$\frac{X''}{X} + \frac{Y''}{Y} + \frac{Z''}{Z} + k^2 = 0. \qquad (2.34)$$

Each term must be equal to a constant since each term depends only on the corresponding variable; X on x, etc. We conclude that

$$\frac{X''}{X} = -k_x^2, \quad \frac{Y''}{Y} = -k_y^2, \quad \frac{Z''}{Z} = -k_z^2 \qquad (2.35)$$

so that Eq. (2.34) reduces to

$$k_x^2 + k_y^2 + k_z^2 = k^2. \qquad (2.36)$$

Notice that there are four separation constants k, k_x, k_y, and k_z since we have four variables t, x, y, and z. But from Eq. (2.36), one is related to the other three so that only three separation constants are independent. As mentioned earlier, the number of independent separation constants is generally one less than the number of independent variables involved. The ordinary differential equations in Eq. (2.35) have solutions

$$X = A_1 e^{jk_x x} + A_2 e^{-jk_x x} \qquad (2.37a)$$

or

$$X = B_1 \sin k_x x + B_2 \cos k_x x, \qquad (2.37b)$$

$$Y = A_3 e^{jk_y y} + A_4 e^{-jk_y y} \qquad (2.37c)$$

or

$$Y = B_3 \sin k_y y + B_4 \cos k_y y, \qquad (2.37d)$$

$$Z = A_5 e^{jk_z z} + A_6 e^{-jk_z z} \qquad (2.37e)$$

or

$$Z = B_5 \sin k_z z + B_6 \cos k_z z, \qquad (2.37f)$$

Various combinations of X, Y, and Z will satisfy Eq. (2.5b). Suppose we choose

$$X = A_1 e^{jk_x x}, \qquad Y = A_3 e^{jk_y y}, \qquad Z = A_5 e^{jk_z z}, \qquad (2.38)$$

then

$$U(x,y,z) = A e^{j(k_x x + k_y y + k_z z)} \qquad (2.39)$$

or

$$U(\mathbf{r}) = A e^{j\mathbf{k}\cdot\mathbf{r}}. \qquad (2.40)$$

Introducing the time dependence of Eq. (2.6a) gives

$$\boxed{\Phi(x,y,z,t) = A e^{j(\mathbf{k}\cdot\mathbf{r}+\omega t)}} \qquad (2.41)$$

where $\omega = kc$ is the angular frequency of the wave and k is given by Eq. (2.36). The solution in Eq. (2.41) represents a plane wave of amplitude A propagating in the direction of the wave vector $\mathbf{k} = k_x \mathbf{a}_x + k_y \mathbf{a}_y + k_z \mathbf{a}_z$ with velocity c.

Analytical Methods   39

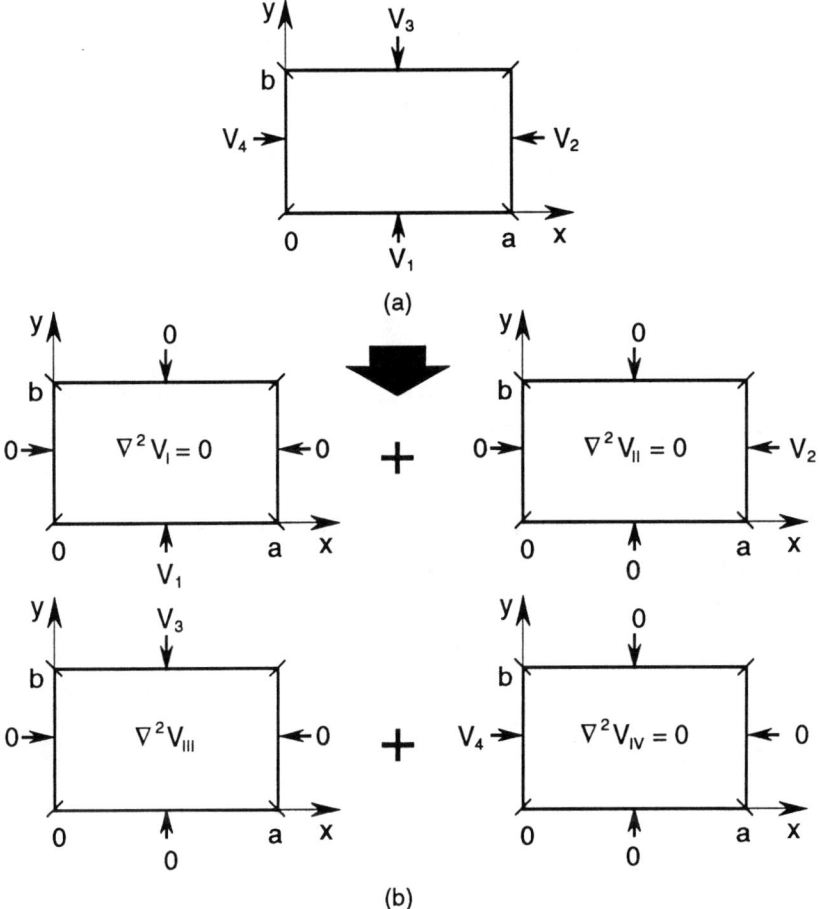

Figure 2.2   Applying the principle of superposition reduces the problem in (a) to those in (b).

Example 2.1

In this example, we would like to show that the method of separation of variables is not limited to problem with only one inhomogeneous boundary conditions as presented in Section 2.3(a). We reconsider the problem of Fig. 2.1, but with four inhomogeneous boundary conditions as in Fig. 2.2(a).

Solution

The problem can be stated as solving Laplace's equation

$$\frac{\partial^2 V}{\partial x^2} + \frac{\partial^2 V}{\partial y^2} = 0 \qquad (2.1.1)$$

subject to the following inhomogeneous Dirichlet conditions:

$$\begin{aligned} V(x,0) &= V_1 \\ V(x,b) &= V_3 \\ V(0,y) &= V_4 \\ V(a,y) &= V_2. \end{aligned} \qquad (2.1.2)$$

Since Laplace's equation is a linear homogeneous equation, the problem can be simplified by applying the superposition principle. If we let

$$V = V_I + V_{II} + V_{III} + V_{IV}, \qquad (2.1.3)$$

we may reduce the problem to four simpler problems, each of which is associated with one of the inhomogeneous conditions. The reduced, simpler problems are illustrated in Fig. 2.2(b) and stated as follows:

$$\frac{\partial^2 V_I}{\partial x^2} + \frac{\partial^2 V_I}{\partial y^2} = 0 \qquad (2.1.4)$$

subject to

$$\begin{aligned} V_I(x,0) &= V_1 \\ V_I(x,b) &= 0 \\ V_I(0,y) &= 0 \\ V_I(a,y) &= 0; \end{aligned} \qquad (2.1.5)$$

$$\frac{\partial^2 V_{II}}{\partial x^2} + \frac{\partial^2 V_{II}}{\partial y^2} = 0 \qquad (2.1.6)$$

subject to

$$\begin{aligned} V_{II}(x,0) &= 0 \\ V_{II}(x,b) &= 0 \\ V_{II}(0,y) &= 0 \\ V_{II}(a,y) &= V_2; \end{aligned} \qquad (2.1.7)$$

Analytical Methods   41

$$\frac{\partial^2 V_{III}}{\partial x^2} + \frac{\partial^2 V_{III}}{\partial y^2} = 0 \qquad (2.1.8)$$

subject to

$$\begin{aligned} V_{III}(x,0) &= 0 \\ V_{III}(x,b) &= V_3 \\ V_{III}(0,y) &= 0 \\ V_{III}(a,y) &= 0; \end{aligned} \qquad (2.1.9)$$

and

$$\frac{\partial^2 V_{IV}}{\partial x^2} + \frac{\partial^2 V_{IV}}{\partial y^2} = 0 \qquad (2.1.10)$$

subject to

$$\begin{aligned} V_{IV}(x,0) &= 0 \\ V_{IV}(x,b) &= 0 \\ V_{IV}(0,y) &= V_4 \\ V_{IV}(a,y) &= 0. \end{aligned} \qquad (2.1.11)$$

It is obvious that the reduced problem in Eqs. (2.1.8) and (2.1.9) with solution V_{III} is the same as that in Fig. 2.1. The other three reduced problems are quite similar. Hence the solutions V_I, V_{II}, and V_{IV} can be obtained by taking the same steps as in Section 2.3 or by a proper exchange of variables in Eq. (2.31). Thus

$$V_I = \frac{4V_1}{\pi} \sum_{n=\text{odd}}^{\infty} \frac{\sin \frac{n\pi x}{a} \sinh \frac{n\pi(b-y)}{a}}{\sinh \frac{n\pi b}{a}}, \qquad (2.1.12)$$

$$V_{II} = \frac{4V_2}{\pi} \sum_{n=\text{odd}}^{\infty} \frac{\sin \frac{n\pi x}{b} \sinh \frac{n\pi y}{b}}{\sinh \frac{n\pi a}{b}}, \qquad (2.1.13)$$

$$V_{III} = \frac{4V_3}{\pi} \sum_{n=\text{odd}}^{\infty} \frac{\sin \frac{n\pi x}{a} \sinh \frac{n\pi y}{a}}{\sinh \frac{n\pi b}{a}}, \qquad (2.1.14)$$

$$V_{IV} = \frac{4V_4}{\pi} \sum_{n=\text{odd}}^{\infty} \frac{\sin \frac{n\pi(a-x)}{b} \sinh \frac{n\pi y}{b}}{\sinh \frac{n\pi a}{b}}. \qquad (2.1.15)$$

Example 2.2

Find the product solution of the diffusion equation

$$\Phi_t = k\Phi_{xx}, \qquad 0 < x < 1, \ t > 0 \qquad (2.2.1)$$

subject to the boundary conditions

$$\Phi(0,t) = 0 = \Phi(1,t), \qquad t > 0 \qquad (2.2.2)$$

and initial condition

$$\Phi(x,0) = 5\sin 2\pi x, \qquad 0 < x < 1. \qquad (2.2.3)$$

Solution

Let
$$\Phi(x,t) = X(x)T(t), \qquad (2.2.4)$$

substitute this into Eq. (2.2.1) and divide by kXT to obtain

$$\frac{T'}{kT} = \frac{X''}{X} = \lambda$$

where λ is the separation constant. Thus

$$X'' - \lambda X = 0 \qquad (2.2.5)$$

$$T' - \lambda k T = 0. \qquad (2.2.6)$$

As usual, in order for the solution of Eq. (2.2.5) to satisfy Eq. (2.2.2), we must choose $\lambda = -\beta^2 = -n^2\pi^2$ so that $n = 1, 2, 3, \cdots$ and

$$X_n(x) = \sin n\pi x. \qquad (2.2.7)$$

Equation (2.2.6) becomes

$$T' + kn^2\pi^2 T = 0,$$

which has solution

$$T_n(t) = e^{-kn^2\pi^2 t}. \qquad (2.2.8)$$

Substituting Eqs. (2.2.7) and (2.2.8) into Eq. (2.2.4),

$$\Phi_n(x,t) = a_n \sin n\pi x \, \exp(-kn^2\pi^2 t)$$

where the coefficients a_n are to be determined from the initial conditions of Eq. (2.2.3). The complete solution is a linear combination of Φ_n, i.e.,

$$\Phi(x,t) = \sum_{n=1}^{\infty} a_n \sin n\pi x \exp(-kn^2\pi^2 t).$$

This satisfies Eq. (2.2.3) if

$$\Phi(x,0) = \sum_{n=1}^{\infty} a_n \sin n\pi x = 5 \sin 2\pi x. \qquad (2.2.9)$$

The coefficients a_n are determined as $(T = 1)$

$$a_n = \frac{2}{T} \int_0^1 5 \sin 2\pi x \, \sin n\pi x \, dx = \begin{cases} 5, & n = 2 \\ 0, & n \neq 0. \end{cases}$$

Alternatively, by comparing the middle term in Eq. (2.2.9) with the last term, the two are equal only when $n = 2$, $a_n = 5$, otherwise $a_n = 0$. Hence the solution of the diffusion problem becomes

$$\Phi(x,t) = 5 \sin 2\pi t \, exp(-4k\pi^2 t).$$

2.4 Separation of Variables in Cylindrical Coordinates

Coordinate geometries other than rectangular Cartesian are used to describe many EM problems whenever it is necessary and convenient. For example, a problem having cylindrical symmetry is best solved in cylindrical system where the coordinate variables (ρ, ϕ, z) are related as shown in Fig. 2.3 and $0 \leq \rho \leq \infty$, $0 \leq \phi < 2\pi$, $-\infty \leq z \leq \infty$. In this system, the wave equation (2.5b) becomes

$$\nabla^2 U + k^2 U = \frac{1}{\rho} \frac{\partial}{\partial \rho}\left(\rho \frac{\partial U}{\partial \rho}\right) + \frac{1}{\rho^2} \frac{\partial^2 U}{\partial \phi^2} + \frac{\partial^2 U}{\partial z^2} + k^2 U = 0. \qquad (2.42)$$

As we did in the previous section, we shall first solve Laplace's equation $(k = 0)$ in two dimensions before we solve the wave equation.

44   Numerical Techniques in Electromagnetics

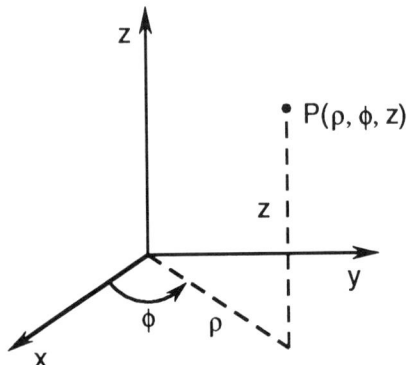

Figure 2.3   Coordinate relations in a cylindrical system.

2.4.1 Laplace's Equation

Consider an infinitely long conducting cylinder of radius a with the cross section shown in Fig. 2.4. Assume that the upper half of the cylinder is maintained at potential V_o while the lower half is maintained at potential $-V_o$. This is a Laplacian problem in two dimensions. Hence we need to solve for $V(\rho,\phi)$ in Laplace's equation

$$\nabla^2 V = \frac{1}{\rho}\frac{\partial}{\partial \rho}\left(\rho\frac{\partial V}{\partial \rho}\right) + \frac{1}{\rho^2}\frac{\partial^2 V}{\partial \phi^2} = 0 \tag{2.43}$$

subject to the inhomogeneous Dirichlet boundary condition

$$V(a,\phi) = \begin{cases} V_o, & 0 < \phi < \pi \\ -V_o, & \pi < \phi < 2\pi. \end{cases} \tag{2.44}$$

We let
$$V(\rho,\phi) = R(\rho)F(\phi). \tag{2.45}$$

Substituting Eq. (2.45) into Eq. (2.43) and dividing through by RF/ρ^2 result in

$$\frac{\rho}{R}\frac{d}{d\rho}\left(\rho\frac{dR}{d\rho}\right) + \frac{1}{F}\frac{d^2F}{d\phi^2}$$

or

$$\frac{\rho^2}{R}\frac{d^2R}{d\rho^2} + \frac{\rho}{R}\frac{dR}{d\rho} = -\frac{1}{F}\frac{d^2F}{d\phi^2} = \lambda^2 \tag{2.46}$$

Analytical Methods 45

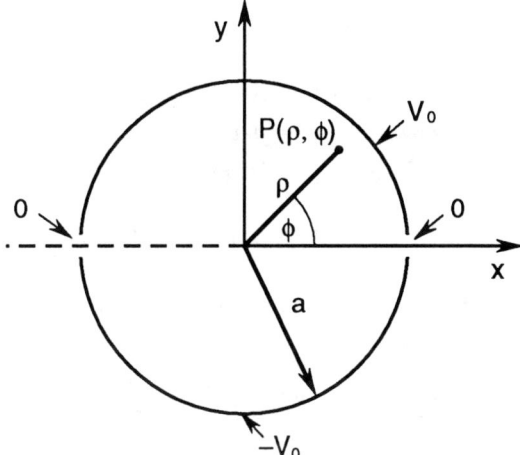

Figure 2.4 A two-dimensional Laplacian problem in cylindrical coordinates.

where λ is the separation constant. Thus the separated equations are:

$$F'' + \lambda^2 F = 0 \tag{2.47a}$$

$$\rho^2 R'' + \rho R' - \lambda^2 R = 0. \tag{2.47b}$$

It is evident that Eq. (2.47a) has the general solution of the form

$$F(\phi) = c_1 \cos(\lambda\phi) + c_2 \sin(\lambda\phi). \tag{2.48}$$

From the boundary conditions of Eq. (2.44), we observe that $F(\phi)$ must be a periodic, odd function. Thus $c_1 = 0$, $\lambda = n$, a real integer, and hence Eq. (2.48) becomes

$$F_n(\phi) = c_2 \sin n\phi. \tag{2.49}$$

Equation (2.47b), known as the *Cauchy-Euler equation*, can be solved by making a substitution $\rho = e^u$ and reducing it to an equation with constant coefficients. This leads to

$$R_n(\rho) = c_3 \rho^n + c_4 \rho^{-n}, \quad n = 1, 2, \cdots \tag{2.50}$$

Note that case $n = 0$ is excluded; if $n = 0$, we obtain $R(\rho) = \ln \rho +$ constant, which is not finite at $\rho = 0$. For the problem of a coaxial cable, $a < \rho < b, \rho \neq 0$ so that case $n = 0$ is the only solution. However, for the problem at hand, $n = 0$ is not acceptable.

Substitution of Eqs. (2.49) and (2.50) into Eq. (2.45) yields

$$V_n(\rho, \phi) = \sin n\phi \, (A_n \rho^n + B_n \rho^{-n}) \tag{2.51}$$

where A_n and B_n are constants to be determined. As usual, it is possible by the superposition principle to form a complete series solution

$$V(\rho, \phi) = \sum_{n=1}^{\infty} (A_n \rho^n + B_n \rho^{-n}) \sin n\phi. \tag{2.52}$$

For $\rho < a$, inside the cylinder, V must be finite as $\rho \to 0$ so that $B_n = 0$. At $\rho = a$,

$$V(a, \phi) = \sum_{n=1}^{\infty} A_n a^n \sin n\phi = \begin{cases} V_o, & 0 < \phi < \pi \\ -V_o, & \pi < \phi < 2\pi. \end{cases} \tag{2.53}$$

Multiplying both sides by $\sin m\phi$ and integrating over $0 < \phi < 2\pi$, we get

$$\int_0^{\pi} V_o \sin m\phi \, d\phi - \int_{\pi}^{2\pi} V_o \sin m\phi \, d\phi = \sum_{n=1}^{\infty} A_n a^n \int_0^{2\pi} \sin n\phi \sin m\phi \, d\phi.$$

All terms in the right-hand side vanish except when $m = n$. Hence

$$\frac{2V_o}{\pi}(1 - \cos n\pi) = A_n a^n \int_0^{2\pi} \sin^2 \phi \, d\phi = \pi A_n a^n$$

or

$$A_n = \begin{cases} \dfrac{4V_o}{n\pi a^n}, & n = \text{odd} \\ 0, & n = \text{even}. \end{cases} \tag{2.54}$$

Thus,

$$V(\rho, \phi) = \frac{4V_o}{\pi} \sum_{n=\text{odd}}^{\infty} \frac{\rho^n \sin n\phi}{n a^n}, \quad \rho < a. \tag{2.55}$$

For $\rho > a$, outside the cylinder, V must be finite as $\rho \to \infty$ so that $A_n = 0$ in Eq. (2.52) for this case. By imposing the boundary condition in Eq. (2.44) and following the same steps as for case $\rho < a$, we obtain

$$B_n = \begin{cases} \dfrac{4V_o a^n}{n\pi}, & n = \text{odd} \\ 0, & n = \text{even}. \end{cases} \tag{2.56}$$

Hence,

$$V(\rho, \phi) = \frac{4V_o}{\pi} \sum_{n=\text{odd}}^{\infty} \frac{a^n \sin n\phi}{n\rho^n}, \quad \rho > a. \tag{2.57}$$

2.4.2 Wave Equation

Having taken care of the time-dependence in Section 2.2, we now solve Helmholtz's equation (2.42), i.e.,

$$\frac{1}{\rho}\frac{\partial}{\partial \rho}\left(\rho\frac{\partial U}{\partial \rho}\right) + \frac{1}{\rho^2}\frac{\partial^2 U}{\partial \phi^2} + \frac{\partial^2 U}{\partial z^2} + k^2 U = 0. \tag{2.42}$$

Let

$$U(\rho, \phi, z) = R(\rho)F(\phi)Z(z). \tag{2.58}$$

Substituting Eq. (2.58) into Eq. (2.42) and dividing by RFZ/ρ^2 yields

$$\frac{\rho}{R}\frac{d}{d\rho}\left(\rho\frac{dR}{d\rho}\right) + \frac{\rho^2}{Z}\frac{d^2 Z}{dz^2} + k^2 \rho^2 = -\frac{1}{F}\frac{d^2 F}{d\phi^2} = n^2$$

where $n = 0, 1, 2, \cdots$ and n^2 is the separation constant. Thus

$$F'' + n^2 F = 0 \tag{2.59}$$

and

$$\frac{\rho}{R}\frac{d}{d\rho}\left(\rho\frac{dR}{d\rho}\right) + \frac{\rho^2}{Z}\frac{d^2 Z}{dz^2} + k^2 \rho^2 = n^2. \tag{2.60}$$

Dividing both sides of Eq. (2.60) by ρ^2 leads to

$$\frac{1}{\rho R}\frac{d}{d\rho}\left(\rho\frac{dR}{d\rho}\right) + \left(k^2 - \frac{n^2}{\rho^2}\right) = -\frac{1}{Z}\frac{d^2 Z}{dz^2} = \mu^2$$

where μ^2 is another separation constant. Hence

$$-\frac{1}{Z}\frac{d^2 Z}{dz^2} = \mu^2 \tag{2.61}$$

and

$$\frac{1}{\rho R}\frac{d}{d\rho}\left(\rho\frac{dR}{d\rho}\right) + \left(k^2 - \mu^2 - \frac{n^2}{\rho^2}\right) = 0. \tag{2.62}$$

If we let

$$\lambda^2 = k^2 - \mu^2, \tag{2.63}$$

the three separated equations (2.59), (2.61), and (2.62) become

$$F'' + n^2 F = 0, \tag{2.64}$$

$$Z'' + \mu^2 Z = 0, \tag{2.65}$$

$$\rho^2 R'' + \rho R + (\lambda^2 \rho^2 - n^2)R = 0. \tag{2.66}$$

The solution to Eq. (2.64) is given by

$$F(\phi) = c_1 e^{jn\phi} + c_2 e^{-jn\phi} \tag{2.67a}$$

or

$$F(\phi) = c_3 \sin n\phi + c_4 \cos n\phi. \tag{2.67b}$$

Similarly, Eq. (2.65) has the solution

$$Z(z) = c_5 e^{jn\mu} + c_6 e^{-jn\mu} \tag{2.68a}$$

or

$$Z(z) = c_7 \sin n\mu + c_8 \cos n\mu. \tag{2.68b}$$

To solve Eq. (2.66), we let $x = \lambda \rho$ and replace R by y; the equation becomes

$$x^2 y'' + x y' + (x^2 - n^2) y = 0. \tag{2.69}$$

This is called *Bessel's equation*. It has a general solution of the form

$$y(x) = b_1 J_n(x) + b_2 Y_n(x) \tag{2.70}$$

where $J_n(x)$ and $Y_n(x)$ are, respectively, *Bessel functions* of the first and second kinds of order n and real argument x. Y_n is also called the *Neumann function*. If x in Eq. (2.69) is imaginary so that we may replace x by jx, the equation becomes

$$x^2 y'' + x y' - (x^2 + n^2) y = 0 \tag{2.71}$$

which is called *modified Bessel's equation*. This equation has a solution of the form

$$y(x) = b_3 I_n(x) + b_4 K_n(x) \tag{2.72}$$

where $I_n(x)$ and $K_n(x)$ are respectively *modified Bessel functions* of the first and second kind of order n. For small values of x, Fig. 2.5 shows the sketch of some typical Bessel functions (or cylindrical functions) $J_n(x)$, $Y_n(x)$, $I_n(x)$, and $K_n(x)$.

To obtain the Bessel functions from Eqs. (2.69) and (2.71), the method of Frobenius is applied. A detailed discussion is found in Kersten [3] and Myint-U [4]. For the Bessel function of the first kind,

$$\boxed{y = J_n(x) = \sum_{m=0}^{\infty} \frac{(-1)^m (x/2)^{n+2m}}{m!\, \Gamma(n+m+1)}} \tag{2.73}$$

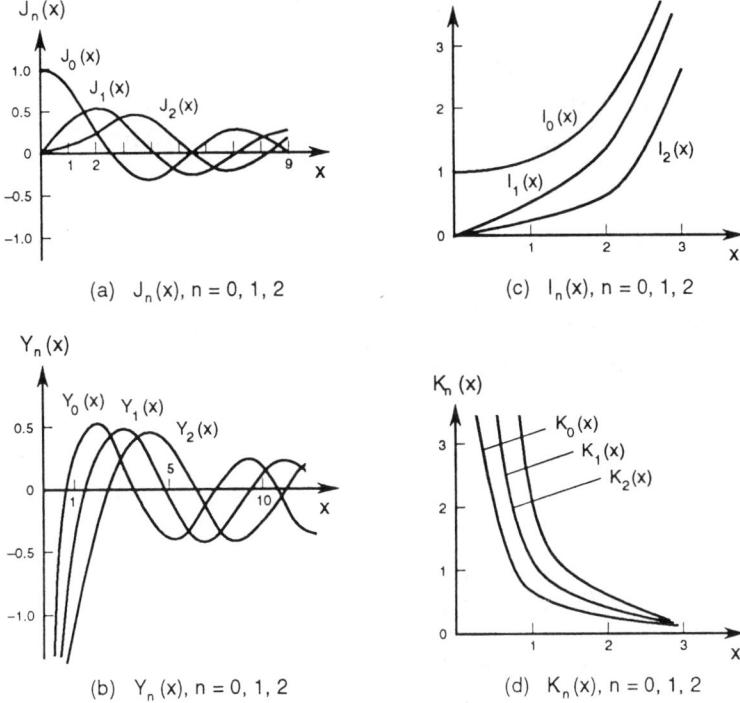

Figure 2.5 Bessel functions.

where $\Gamma(k+1) = k!$ is the Gamma function. This is the most useful of all Bessel functions. Some of its important properties and identities [5] are listed in Table 2.1. For the modified Bessel function of the second kind

$$I_n(x) = j^{-n} J_n(jx) = \sum_{m=0}^{\infty} \frac{(x/2)^{n+2m}}{m!\, \Gamma(n+m+1)}. \tag{2.74}$$

For the Neumann functions, when $n > 0$

$$Y_n(x) = \frac{2}{\pi} J_n(x) \ln \frac{\gamma x}{2} - \frac{1}{\pi} \sum_{m=0}^{n-1} \frac{(n-m-1)!\,(x/2)^{2m-n}}{m!}$$

$$- \frac{1}{\pi} \sum_{m=0}^{\infty} \frac{(-1)^m\,(x/2)^{n+2m}}{m!\,\Gamma(n+m+1)} \Big[p(m) + p(n+m)\Big] \tag{2.75}$$

Table 2.1 Properties and Identities of Bessel Functions[1] $J_n(x)$

(a) $J_{-n}(x) = (-1)^n J_n(x)$

(b) $J_n(-x) = (-1)^n J_n(x)$

(c) $J_{n+1}(x) = \frac{2n}{x} J_n(x) - J_{n-1}(x)$ (recurrence formula)

(d) $\frac{d}{dx} J_n(x) = \frac{1}{2}[J_{n-1}(x) - J_{n+1}(x)]$

(e) $\frac{d}{dx}[x^n J_n(x)] = x^n J_{n-1}(x)$

(f) $\frac{d}{dx}[x^{-n} J_n(x)] = -x^{-n} J_{n+1}(x)$

(g) $J_n(x) = \frac{1}{\pi} \int_0^\pi \cos(n\theta - x\sin\theta)\, d\theta$, $n \geq 0$

(h) Fourier-Bessel expansion of $f(x)$:

$$f(x) = \sum_{k=1}^{\infty} A_k J_n(\lambda_k x), \quad n \geq 0$$

$$A_k = \frac{2}{[a J_{n+1}(\lambda_i a)]^2} \int_0^a x f(x) J_n(\lambda_k x)\, dx, \quad 0 < x < a$$

where λ_k are the positive roots in ascending order of magnitude of $J_n(\lambda_i a) = 0$.

(i) $\int_0^a \rho J_n(\lambda_i \rho) J_n(\lambda_j \rho)\, d\rho = \frac{a^2}{2}[J_{n+1}(\lambda_i a)]^2 \delta_{ij}$

where λ_i and λ_j are the positive roots of $J_n(\lambda a) = 0$.

---

1. Properties (a) to (f) also hold for $Y_n(x)$.

where $\gamma = 1.781$ is Euler's constant and

$$p(m) = \sum_{k=1}^{m} \frac{1}{k}, \quad p(0) = 0. \tag{2.76}$$

If $n = 0$,

$$Y_0(x) = \frac{2}{\pi} J_0(x) \ln \frac{\gamma x}{2} + \frac{2}{\pi} \sum_{m=0}^{\infty} \frac{(-1)^{m+1} (x/2)^{2m}}{(m!)^2} p(m). \tag{2.77}$$

For the modified Bessel function of the second kind,

$$K_n(x) = \frac{\pi}{2} j^{n+1} [J_n(jx) + j Y_n(jx)]. \tag{2.78}$$

If $n > 0$,

$$K_n(x) = \frac{1}{2}\sum_{m=0}^{n-1}\frac{(-1)^m\,(n-m-1)!\,(x/2)^{2m-n}}{m!}$$
$$+ (-1)^n\frac{1}{2}\sum_{m=0}^{\infty}\frac{(x/2)^{n+2m}}{m!\,(n+m)!}\left[p(m)+p(n+m)-2\ln\frac{\gamma x}{2}\right] \quad (2.79)$$

and if $n = 0$,

$$K_0(x) = -I_0(x)\ln\frac{\gamma x}{2} + \sum_{m=0}^{\infty}\frac{(x/2)^{2m}}{(m!)^2}p(m). \quad (2.80)$$

Other functions closely related to Bessel functions are *Hankel functions* of the first and second kinds, defined respectively by

$$H_n^{(1)}(x) = J_n(x) + jY_n(x) \quad (2.81a)$$

$$H_n^{(2)}(x) = J_n(x) - jY_n(x). \quad (2.81b)$$

Hankel functions are analogous to functions $\exp(\pm jx)$ just as J_n and Y_n are analogous to cosine and sine functions. This is evident from asymptotic expressions

$$J_n(x) \quad x\to\infty \quad \sqrt{\frac{2}{\pi x}}\cos(x-n\pi/2-\pi/4), \quad (2.82a)$$

$$Y_n(x) \quad x\to\infty \quad \sqrt{\frac{2}{\pi x}}\sin(x-n\pi/2-\pi/4), \quad (2.82b)$$

$$H_n^{(1)}(x) \quad x\to\infty \quad \sqrt{\frac{2}{\pi x}}\exp[j(x-n\pi/2-\pi/4)], \quad (2.82c)$$

$$H_n^{(2)}(x) \quad x\to\infty \quad \sqrt{\frac{2}{\pi x}}\exp[-j(x-n\pi/2-\pi/4)], \quad (2.82d)$$

$$I_n(x) \quad x\to\infty \quad \frac{1}{\sqrt{2\pi x}}e^x, \quad (2.82e)$$

$$K_n(x) \quad x\to\infty \quad \frac{1}{\sqrt{2\pi x}}e^{-x}. \quad (2.82f)$$

With the time factor $e^{j\omega t}$, $H_n^{(1)}(x)$ and $H_n^{(2)}(x)$ represent inward and outward traveling waves, respectively, while $J_n(x)$ or $Y_n(x)$ represents a standing wave. With the time factor $e^{-j\omega t}$, the roles of $H_n^{(1)}(x)$ and $H_n^{(2)}(x)$ are

52   Numerical Techniques in Electromagnetics

reversed. For further treatment of Bessel and related functions, refer to the works of Watson [6] and Bell [7].

Any of the Bessel functions or related functions can be a solution to Eq. (2.66) depending on the problem. If we choose $R(\rho) = J_n(x) = J_n(\lambda\rho)$ with Eqs. (2.67) and (2.68) and apply the superposition theorem, the solution to Eq. (2.42) is

$$U(\rho,\phi,z) = \sum_n \sum_\mu A_{n\mu} J_n(\lambda\rho) \exp(\pm jn\phi \pm j\mu z). \qquad (2.83)$$

Introducing the time dependence of Eq. (2.6a), we finally get

$$\Phi(\rho,\phi,z,t) = \sum_m \sum_n \sum_\mu A_{mn\mu} J_n(\lambda\rho) \exp(\pm jn\phi \pm j\mu z \pm \omega t), \qquad (2.84)$$

where $\omega = kc$.

Example 2.3

Consider the skin effect on a solid cylindrical conductor. The current density distribution within a good conducting wire ($\sigma/\omega\epsilon \gg 1$) obeys the diffusion equation

$$\nabla^2 J = \mu\sigma \frac{\partial J}{\partial t}.$$

We want to solve this equation for a long conducting wire of radius a.

Solution

We may derive the diffusion equation directly from Maxwell's equation. Since

$$\nabla \times \mathbf{H} = \mathbf{J} + \mathbf{J}_d$$

where $\mathbf{J} = \sigma\mathbf{E}$ is the conduction current density and $\mathbf{J}_d = \frac{\partial \mathbf{D}}{\partial t}$ is the displacement current density. For $\sigma/\omega\epsilon \gg 1$, $\mathbf{J}_d$ is negligibly small compared with $\mathbf{J}$. Hence

$$\nabla \times \mathbf{H} \simeq \mathbf{J}. \qquad (2.3.1)$$

Also,

$$\nabla \times \mathbf{E} = -\mu \frac{\partial \mathbf{H}}{\partial t}$$

$$\nabla \times \nabla \times \mathbf{E} = \nabla\nabla \cdot \mathbf{E} - \nabla^2 \mathbf{E} = -\mu \frac{\partial}{\partial t} \nabla \times \mathbf{H}.$$

Since $\nabla \cdot \mathbf{E} = 0$, introducing Eq. (2.3.1), we obtain

$$\nabla^2 \mathbf{E} = \mu \frac{\partial \mathbf{J}}{\partial t}. \qquad (2.3.2)$$

Replacing **E** with **J**/σ, Eq. (2.3.2) becomes

$$\nabla^2 \mathbf{J} = \mu\sigma \frac{\partial \mathbf{J}}{\partial t}, \tag{2.3.3}$$

which is the diffusion equation.

Assuming harmonic field with time factor $e^{j\omega t}$,

$$\nabla^2 \mathbf{J} = j\omega\mu\sigma \mathbf{J}. \tag{2.3.4}$$

For infinitely long wire, Eq. (2.3.4) reduces to a one-dimensional problem in cylindrical coordinates:

$$\frac{1}{\rho}\frac{\partial}{\partial \rho}\left(\rho \frac{\partial J_z}{\partial \rho}\right) = j\omega\mu\sigma J_z$$

or

$$\rho^2 J_z'' + \rho J_z' - j\omega\mu\sigma\rho^2 J_z = 0. \tag{2.3.5}$$

Comparing this with Eq. (2.71) shows that Eq. (2.3.5) is the modified Bessel equation of zero order. Hence the solution to Eq. (2.3.5) is

$$J_z(\rho) = c_1 I_0(\lambda\rho) + c_2 K_0(\lambda\rho) \tag{2.3.6}$$

where c_1 and c_2 are constants and

$$\lambda = \sqrt{j\omega\mu\sigma} = j^{1/2}\frac{\sqrt{2}}{\delta} \tag{2.3.7}$$

and $\delta = \sqrt{\frac{2}{\sigma\mu\omega}}$ is the skin depth. Constant c_2 must vanish if J_z is to be finite at $\rho = 0$. At $\rho = a$,

$$J_z(a) = c_1 I_0(\lambda a) \to c_1 = J_z(a)/I_0(\lambda a).$$

Thus

$$J_z(\rho) = J_z(a)\frac{I_0(\lambda\rho)}{I_0(\lambda a)}. \tag{2.3.8}$$

If we let $\lambda\rho = j^{1/2}\frac{\sqrt{2}}{\delta}\rho = j^{1/2}x$, it is convenient to replace

$$I_0(\lambda\rho) = I_0(j^{1/2}x) = J_0(xe^{j3\pi/4})$$
$$= ber_0(x) + jbei_0(x) \tag{2.3.9}$$

54   Numerical Techniques in Electromagnetics

where ber_0 and bei_0 are ber and bei functions of zero order. Ber and ber functions are also known as *Kelvin functions*. For zero order, they are given by

$$ber_0(x) = \sum_{m=0}^{\infty} \frac{\cos(m\pi/2)\,(x/2)^{2m}}{(m!)^2}, \qquad (2.3.10)$$

$$bei_0(x) = \sum_{m=0}^{\infty} \frac{\sin(m\pi/2)\,(x/2)^{2m}}{(m!)^2}. \qquad (2.3.11)$$

Using ber and bei functions, Eq. (2.3.8) may be written as

$$J_z(\rho) = J_z(a)\frac{ber_0(x) + jbei_0(x)}{ber_0(y) + jbei_0(y)} \qquad (2.3.12)$$

where $x = \sqrt{2}\rho/\delta$, $y = \sqrt{2}a/\delta$.

Example 2.4

A semi-infinitely long cylinder ($z \geq 0$) of radius a has its end at $z = 0$ maintained at $V_0(a^2 - \rho^2)$, $0 \leq \rho \leq a$. Find the potential distribution within the cylinder.

Solution

The problem is that of finding a function $V(\rho, z)$ satisfying the PDE

$$\nabla^2 V = \frac{\partial^2 V}{\partial \rho^2} + \frac{1}{\rho}\frac{\partial V}{\partial \rho} + \frac{\partial^2 V}{\partial z^2} = 0 \qquad (2.4.1)$$

subject to the boundary conditions:

(i) $V = V_0(a^2 - \rho^2)$, $z = 0$, $0 \leq \rho \leq a$,
(ii) $V \to 0$ as $z \to \infty$, i.e., V is bounded,
(iii) $V = 0$ on $\rho = a$,
(iv) V is finite on $\rho = 0$.

Let $V = R(\rho)Z(z)$ and obtain the separated equations

$$Z'' - \lambda Z = 0 \qquad (2.4.2a)$$

and

$$\rho^2 R'' + \rho R' + \lambda^2 \rho^2 R = 0 \qquad (2.4.2b)$$

where λ is the separated constant. The solution to Eq. (2.4.2a) is

$$Z_1 = c_1 e^{-\lambda z} + c_2 e^{\lambda z}. \qquad (2.4.3)$$

Analytical Methods

Comparing Eq. (2.4.2b) with Eq. (2.69) shows that $n = 0$ so that Eq. (2.4.2b) is Bessel's equation with solution

$$R = c_3 J_0(\lambda \rho) + c_4 Y_0(\lambda \rho). \tag{2.4.4}$$

Condition (ii) forces $c_2 = 0$, while condition (iv) implies $c_4 = 0$ since $Y_0(\lambda \rho)$ blows up when $\rho = 0$. Hence the solution to Eq. (2.4.1) is

$$V(\rho, z) = \sum_{n=0}^{\infty} A_n e^{-\lambda_n z} J_0(\lambda_n \rho) \tag{2.4.5}$$

where A_n and λ_n are constants to be determined using conditions (i) and (iii). Imposing condition (iii) on Eq. (2.4.5) yields the transcendent equation

$$J_0(\lambda_n a) = 0. \tag{2.4.6}$$

Thus λ_n are the positive roots of $J_0(\lambda_n a)$. If we take λ_1 as the first root, λ_2 as the second root, etc., n must start from 1 in Eq. (2.4.5). Imposing condition (i) on Eq. (2.4.5), we obtain

$$V(\rho, 0) = V_o(a^2 - \rho^2) = \sum_{n=1}^{\infty} A_n J_0(\lambda_n \rho)$$

which is simply the Fourier-Bessel expansion of $V_o(a^2 - \rho^2)$. From Table 2.1, property (h),

$$A_n = \frac{2}{a^2 [J_1(\lambda_n a)]^2} \int_0^a \rho V_o(a^2 - \rho^2) J_0(\lambda_n \rho) \, d\rho. \tag{2.4.7}$$

To evaluate the integral, we utilize property (e) in Table 2.1:

$$\int_0^a x^n J_{n-1}(x) \, dx = x^n J_n(x) \Big|_0^a = a^n J_n(a), \quad n > 0.$$

By changing variables, $x = \lambda \rho$,

$$\int_0^a \rho^n J_{n-1}(\lambda \rho) \, d\rho = \frac{a^n}{\lambda} J_n(\lambda a). \tag{2.4.8}$$

If $n = 1$,

$$\int_0^a \rho J_0(\lambda \rho) \, d\rho = \frac{a}{\lambda} J_1(\lambda a). \tag{2.4.9}$$

Similarly, using property (e) in Table 2.1, we may write

$$\int_0^a \rho^3 J_0(\lambda\rho)\, d\rho = \int_0^a \frac{\rho^2}{\lambda} \frac{\partial}{\partial \rho}[\rho J_1(\lambda\rho)]\, d\rho.$$

Integrating the right-hand side by parts and applying Eq. (2.4.8),

$$\int_0^a \rho^3 J_0(\lambda\rho)\, d\rho = \frac{a^3}{\lambda} J_1(\lambda a) - \frac{2}{\lambda}\int_0^a \rho^2 J_1(\lambda\rho)\, d\rho$$

$$= \frac{a^3}{\lambda} J_1(\lambda a) - \frac{2a^2}{\lambda^2} J_2(\lambda a).$$

$J_2(x)$ can be expressed in terms of $J_0(x)$ and $J_1(x)$ using the recurrence relations, i.e., property (c) in Table 2.1:

$$J_2(x) = \frac{2}{x} J_1(x) - J_0(x).$$

Hence

$$\int_0^a \rho^3 J_0(\lambda_n \rho)\, d\rho = \frac{2a^2}{\lambda_n^2}\left[J_0(\lambda_n a) + \left(\frac{a\lambda_n}{2} - \frac{2}{a\lambda_n}\right) J_1(\lambda_n a)\right]. \quad (2.4.10)$$

Substitution of Eqs. (2.4.9) and (2.4.10) into Eq. (2.4.7) gives

$$A_n = \frac{2V_o}{a^2[J_1(\lambda_n a)]^2}\left[\frac{4a}{\lambda_n^3} J_1(\lambda_n a) - \frac{2a^2}{\lambda_n^2} J_0(\lambda_n a)\right]$$

$$= \frac{8V_o}{a\lambda_n^3 J_1(\lambda_n a)}$$

since $J_0(\lambda_n a) = 0$ from Eq. (2.4.6). Thus the potential distribution is given by

$$V(\rho, z) = \frac{8V_o}{a}\sum_{n=1}^{\infty} \frac{e^{-\lambda_n z} J_0(\lambda_n \rho)}{\lambda_n^3 J_1(\lambda_n a)}.$$

Example 2.5

A plane wave $\mathbf{E} = E_o e^{j(\omega t - kx)} \mathbf{a}_z$ is incident on an infinitely long conducting cylinder of radius a. Determine the scattered field.

Analytical Methods

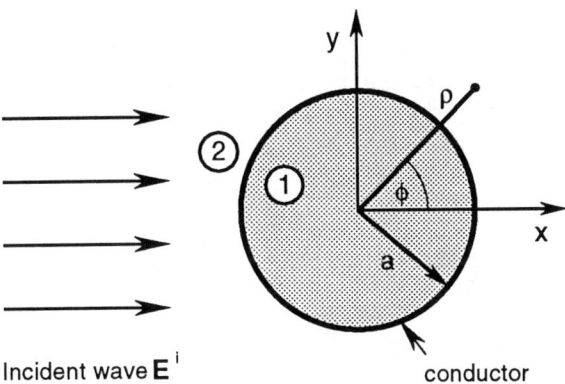

Figure 2.6 Scattering by a conducting cylinder.

Solution

Since the cylinder is infinitely long, the problem is two-dimensional as shown in Fig. 2.6. We shall suppress the time factor $e^{j\omega t}$ throughout the analysis. For the sake of convenience, we need to express the plane wave in terms of cylindrical waves. We let

$$e^{-jx} = e^{-j\rho\cos\phi} = \sum_{n=-\infty}^{\infty} a_n J_n(\rho) e^{jn\phi} \qquad (2.5.1)$$

where a_n are expansion coefficients to be determined. Since $e^{jn\phi}$ are orthogonal functions, multiplying both sides of Eq. (2.5.1) by $e^{jm\phi}$ and integrating over $0 \le \phi \le 2\pi$ gives

$$\int_0^{2\pi} e^{-j\rho\cos\phi} e^{jm\phi} = 2\pi a_m J_m(\rho).$$

Taking the mth derivative of both sides with respect to ρ and evaluating at $\rho = 0$ leads to

$$2\pi \frac{j^{-m}}{2^m} = 2\pi a_m \frac{1}{2^m} \rightarrow a_m = j^{-m}.$$

Substituting this into Eq. (2.5.1), we obtain

$$e^{-jx} = \sum_{n=-\infty}^{\infty} j^{-n} J_n(\rho) e^{jn\phi}.$$

(An alternative, easier way of obtaining this is using the generating function for $J_n(x)$ in Table 2.7 .) Thus the incident wave may be written as

$$E_z^i = E_o e^{-jkx} = E_o \sum_{n=-\infty}^{\infty} (-j)^n J_n(k\rho) e^{jn\phi}. \quad (2.5.2)$$

Since the scattered field E_z^s must consist of outgoing waves that vanish at infinity, it contains

$$J_n(k\rho) - jY_n(k\rho) = H_n^{(2)}(k\rho).$$

Hence

$$E_z^s = \sum_{n=-\infty}^{\infty} A_n H_n^{(2)}(k\rho) e^{jn\phi}. \quad (2.5.3)$$

The total field in medium 2 is

$$E_2 = E_z^i + E_z^s$$

while the total field in medium 1 is $E_1 = 0$ since medium 1 is conducting. At $\rho = a$, the boundary condition requires that the tangential components of E_1 and E_2 be equal. Hence

$$E_z^i(\rho = a) + E_z^s(\rho = a) = 0. \quad (2.5.4)$$

Substituting Eqs. (2.5.2) and (2.5.3) into Eq. (2.5.4),

$$\sum_{n=-\infty}^{\infty} \left[E_o(-j)^n J_n(ka) + A_n H_n^{(2)}(ka) \right] e^{jn\phi} = 0.$$

From this, we obtain

$$A_n = -\frac{E_o(-j)^n J_n(ka)}{H_n^{(2)}(ka)}.$$

Finally, substituting A_n into Eq. (2.5.3) and introducing the time factor leads to the scattered wave as

$$\mathbf{E}_z^s = -E_o e^{j\omega t} \mathbf{a}_z \sum_{n=-\infty}^{\infty} (-j)^n \frac{J_n(ka) H_n^{(2)}(k\rho) e^{jn\phi}}{H_n^{(2)}(ka)}.$$

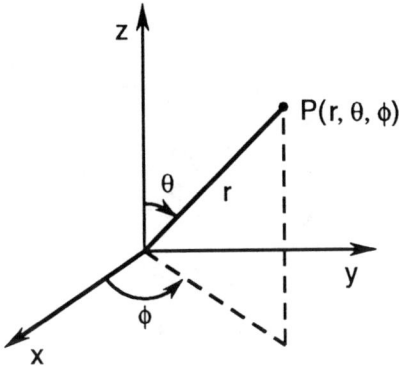

Figure 2.7 Coordinate relation in a spherical system.

2.5 Separation of Variables in Spherical Coordinates

Spherical coordinates (r, θ, ϕ) may be defined as in Fig. 2.7 where $0 \leq r \leq \infty$, $0 \leq \theta \leq \pi$, $0 \leq \phi < 2\pi$. In this system, the wave equation (2.5b) becomes

$$\nabla^2 U + k^2 U = \frac{1}{r^2}\frac{\partial}{\partial r}\left(r^2 \frac{\partial U}{\partial r}\right) + \frac{1}{r^2 \sin\theta}\frac{\partial}{\partial \theta}\left(\sin\theta \frac{\partial U}{\partial \theta}\right) + \frac{1}{r^2 \sin^2\theta}\frac{\partial^2 U}{\partial \phi^2} + k^2 U = 0. \tag{2.85}$$

As usual, we shall first solve Laplace's equation in two dimensions and later solve the wave equation in three dimensions.

2.5.1 Laplace's Equation

Consider the problem of finding the potential distribution due to an uncharged conducting sphere of radius a located in an external uniform electric field as in Fig. 2.8. The external electric field can be described as

$$\mathbf{E} = E_o \mathbf{a}_z \tag{2.86}$$

while the corresponding electric potential can be described as

$$V = -\int \mathbf{E} \cdot d\mathbf{l} = -E_o z$$

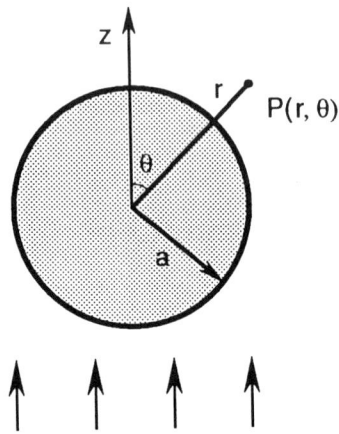

Figure 2.8 An uncharged conducting sphere in a uniform external electric field.

or
$$V = -E_o r \cos\theta \tag{2.87}$$

where $V(\theta = \pi/2) = 0$ has been assumed. From Eq. (2.87), it is evident that V is independent of ϕ, and hence our problem is solving Laplace's equation in two dimensions, namely,

$$\nabla^2 V = \frac{1}{r^2}\frac{\partial}{\partial r}\left(r^2 \frac{\partial V}{\partial r}\right) + \frac{1}{r^2 \sin\theta}\frac{\partial}{\partial \theta}\left(\sin\theta \frac{\partial V}{\partial \theta}\right) = 0 \tag{2.88}$$

subject to the conditions

$$V(r,\theta) = -E_o r \cos\theta \quad \text{as} \quad r \to \infty, \tag{2.89a}$$
$$V(a,\theta) = 0. \tag{2.89b}$$

We let
$$V(r,\theta) = R(r)H(\theta) \tag{2.90}$$

so that Eq. (2.88) becomes

$$\frac{1}{R}\frac{d}{dr}(r^2 R') = -\frac{1}{H \sin\theta}\frac{d}{d\theta}(\sin\theta H') = \lambda \tag{2.91}$$

where λ is the separation constant. Thus the separated equations are

$$r^2 R'' + 2rR' - \lambda R = 0 \tag{2.92}$$

and

$$\frac{d}{d\theta}(\sin\theta H') + \lambda \sin\theta H = 0. \tag{2.93}$$

Equation (2.92) is the *Cauchy-Euler equation*. It can be solved by making the substitution $R = r^k$. This leads to the solution

$$R_n(r) = A_n r^n + B_n r^{-(n+1)}, \quad n = 0, 1, 2, \cdots \tag{2.94}$$

with $\lambda = n(n+1)$. To solve Eq. (2.93), we may replace H by y and $\cos\theta$ by x so that

$$\frac{d}{d\theta} = \frac{dx}{d\theta}\frac{d}{dx} = -\sin\theta \frac{d}{dx}$$

$$\frac{d}{d\theta}\left(\sin\theta \frac{dH}{d\theta}\right) = -\sin\theta \frac{d}{dx}\left(\sin\theta \frac{dx}{d\theta}\frac{dH}{dx}\right)$$
$$= \sin\theta \frac{d}{dx}\left(\sin^2\theta \frac{dy}{dx}\right)$$
$$= \sqrt{1-x^2}\frac{d}{dx}\left[(1-x^2)\frac{dy}{dx}\right].$$

Making these substitutions in Eq. (2.93) yields

$$\frac{d}{dx}\left[(1-x^2)\frac{dy}{dx}\right] + n(n+1)y = 0$$

or

$$(1-x^2)y'' - 2xy' + n(n+1)y = 0 \tag{2.95}$$

which is the *Legendre differential equation*. Its solution is obtained by the method of Frobenius [4] as

$$y = c_n P_n(x) + d_n Q_n(x) \tag{2.96}$$

where $P_n(x)$ and $Q_n(x)$ are Legendre functions of the first and second kind, respectively.

$$\boxed{P_n(x) = \sum_{k=0}^{N} \frac{(-1)^k (2n-2k)! x^{n-2k}}{2^n k!(n-k)!(n-2k)!}} \tag{2.97}$$

62   Numerical Techniques in Electromagnetics

where $N = n/2$ if n is even and $N = (n-1)/2$ if n is odd. For example,

$P_0(x) = 1$

$P_1(x) = x = \cos\theta$

$P_2(x) = \frac{1}{2}(3x^2 - 1) = \frac{1}{4}(3\cos 2\theta + 1)$

$P_3(x) = \frac{1}{2}(5x^3 - 3x) = \frac{1}{8}(5\cos 3\theta + 3\cos\theta)$

$P_4(x) = \frac{1}{8}(35x^4 - 30x^2 + 3) = \frac{1}{64}(35\cos 4\theta + 20\cos 2\theta + 9)$

$P_5(x) = \frac{1}{8}(63x^5 - 70x^3 + 15x) = \frac{1}{128}(30\cos\theta + 35\cos 3\theta + 63\cos 5\theta).$

Some useful identities and properties [5] of Legendre functions are listed in Table 2.2. The Legendre functions of the second kind are given by

$$Q_n(x) = P_n(x)\left[\frac{1}{2}\ln\frac{1+x}{1-x} - p(n)\right]$$
$$+ \sum_{k=1}^{n}\frac{(-1)^k(n+k)!}{(k!)^2(n-k)!}p(k)\left[\frac{1-x}{2}\right]^k \qquad (2.98)$$

where $p(k)$ is as defined in Eq. (2.76). Typical graphs of $P_n(x)$ and $Q_n(x)$ are shown in Fig. 2.9. Q_n are not as useful as P_n since they are singular at $x = \pm 1$ (or $\theta = 0, \pi$) due to the logarithmic term in Eq. (2.98). We use Q_n only when $x \neq \pm 1$ (or $\theta \neq 0, \pi$), e.g., in problems having conical boundaries that exclude the axis from the solution region. For the problem at hand, $\theta = 0, \pi$ is included so that the solution to Eq. (2.93) is

$$H_n(\theta) = P_n(\cos\theta). \qquad (2.99)$$

Substituting Eqs. (2.94) and (2.99) into Eq. (2.90) gives

$$V_n(r,\theta) = \left[A_n r^n + B_n r^{-(n+1)}\right]P_n(\cos\theta). \qquad (2.100)$$

To determine A_n and B_n we apply the boundary conditions in Eq. (2.89). Since as $r \to \infty$, $V = -E_o r\cos\theta$, it follows that $n = 1$ and $A_1 = -E_o$, i.e.,

$$V(r,\theta) = (-E_o r + \frac{B_1}{r^2})\cos\theta.$$

Also since $V = 0$ when $r = a$, $B_1 = E_o a^3$. Hence the complete solution is

$$V(r,\theta) = -E_o(r - \frac{a^3}{r^2})\cos\theta. \qquad (2.101)$$

Table 2.2 Properties and Identities of Legendre Functions[1]

(a) For $n \geq 1$, $P_n(1) = 1$, $P_n(-1) = (-1)^n$, $P_{2n+1} = 0$, $P_{2n}(0) = (-1)^n \dfrac{(2n)!}{2^{2n}(n!)^2}$

(b) $P_n(-x) = (-1)^n P_n(x)$

(c) $P_n(x) = \dfrac{1}{2^n n!} \dfrac{d^n}{dx^n}(x^2 - 1)^n$, $n \geq 0$ (Rodriguez formula)

(d) $(n+1)P_{n+1}(x) = (2n+1)xP_n(x) - nP_{n-1}(x)$, $n \geq 1$ (recurrence relation)

(e) $P_n'(x) = xP_{n-1}'(x) + nP_{n-1}(x)$, $n \geq 1$

(f) $P_n(x) = xP_{n-1}(x) + \dfrac{x^2-1}{n} P_{n-1}'(x)$, $n \geq 1$

(g) $P_{n+1}'(x) - P_{n-1}'(x) = (2n+1)P_n(x)$, $n \geq 1$

or $\int P_n(x)\, dx = \dfrac{P_{n+1} - P_{n-1}}{2n+1}$.

(h) Legendre series expansion of $f(x)$:

$$f(x) = \sum_{n=0}^{\infty} A_n P_n(x), \quad -1 \leq x \leq 1$$

where

$$A_n = \dfrac{2n+1}{2} \int_{-1}^{1} f(x) P_n(x)\, dx, \quad n \geq 0.$$

If $f(x)$ is odd,

$$A_n = (2n+1) \int_0^1 f(x) P_n(x)\, dx, \; n = 0, 2, 4, \cdots$$

and if $f(x)$ is even,

$$A_n = (2n+1) \int_0^1 f(x) P_n(x)\, dx, \; n = 1, 3, 5, \cdots$$

(i) Orthogonality property

$$\int_{-1}^{1} P_n(x) P_m(x)\, dx = \begin{cases} 0, & n \neq m \\ \dfrac{2}{2n+1}, & n = m \end{cases}$$

---

[1] Properties (d) to (g) are also valid for $Q_n(x)$.

64 Numerical Techniques in Electromagnetics

The electric field intensity is given by

$$\mathbf{E} = -\nabla V = -\frac{\partial V}{\partial r}\mathbf{a}_r - \frac{1}{r}\frac{\partial V}{\partial \theta}\mathbf{a}_\theta$$
$$= E_o\left[1 + \frac{2a^3}{r^3}\right]\cos\theta\,\mathbf{a}_r + E_o\left[1 - \frac{a^3}{r^3}\right]\sin\theta\,\mathbf{a}_\theta. \qquad (2.102)$$

2.5.2 Wave Equation

To solve the wave equation (2.85), we substitute

$$U(r,\theta,\phi) = R(r)H(\theta)F(\phi) \qquad (2.103)$$

into the equation. Multiplying the result by $r^2 \sin^2\theta/RHF$ gives

$$\frac{\sin^2\theta}{R}\frac{d}{dr}\left(r^2\frac{dR}{dr}\right) + \frac{\sin\theta}{H}\frac{d}{d\theta}\left(\sin\theta\frac{dH}{d\theta}\right) + k^2 r^2 \sin^2\theta = -\frac{1}{F}\frac{d^2 F}{d\phi^2}. \qquad (2.104)$$

Since the left-hand side of this equation is independent of ϕ, we let

$$-\frac{1}{F}\frac{d^2 F}{d\phi^2} = m^2, \qquad m = 0, 1, 2, \cdots$$

where m, the first separation constant, is chosen to be nonnegative integer such that U is periodic in ϕ. This requirement is necessary for physical reasons that will be evident later. Thus Eq. (2.104) reduces to

$$\frac{1}{R}\frac{d}{dr}\left(r^2\frac{dR}{dr}\right) + k^2 r^2 = -\frac{1}{H\sin\theta}\frac{d}{d\theta}\left(\sin\theta\frac{dH}{d\theta}\right) + \frac{m^2}{\sin^2\theta} = \lambda$$

where λ is the second separation constant. As in Eqs. (2.91) to (2.94), $\lambda = n(n+1)$ so that the separated equations are now

$$F'' + m^2 F = 0, \qquad (2.105)$$

$$R'' + \frac{2}{r}R' + \left[k^2 - \frac{n(n+1)}{r^2}\right]R = 0, \qquad (2.106)$$

and

$$\frac{1}{\sin\theta}\frac{d}{d\theta}(\sin\theta\, H') + \left[n(n+1) - \frac{m^2}{\sin^2\theta}\right]H = 0. \qquad (2.107)$$

As usual, the solution to Eq. (2.105) is

Analytical Methods   65

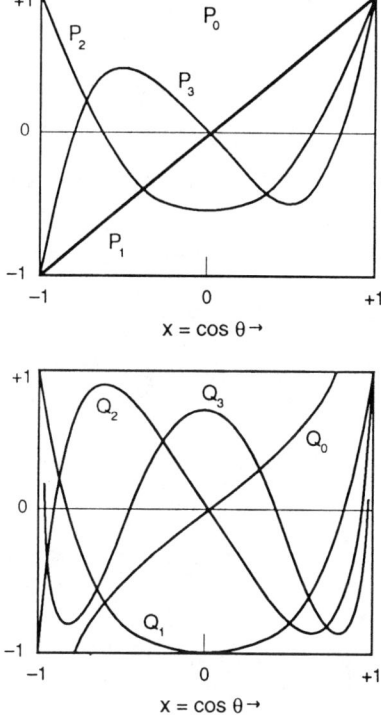

Figure 2.9 Typical Legendre functions of the first and second kinds.

$$F(\phi) = c_1 e^{jm\phi} + c_2 e^{-jm\phi} \qquad (2.108a)$$

or

$$F(\phi) = c_3 \sin m\phi + c_4 \cos m\phi. \qquad (2.108b))$$

If we let $R(r) = r^{1/2}\widetilde{R}(r)$, Eq. (2.106) becomes

$$\widetilde{R}'' + \frac{1}{r}\widetilde{R}' + \left[k^2 - \frac{(n+1/2)^2}{r^2}\right]\widetilde{R} = 0,$$

which has the solution

$$\widetilde{R} = A\ r^{1/2} z_n(kr) = B\ Z_{n+1/2}(kr). \qquad (2.109)$$

Functions $z_n(x)$ are *spherical Bessel functions* and are related to ordinary Bessel functions $Z_{n+1/2}$ according to

$$z_n(x) = \sqrt{\frac{\pi}{2x}} Z_{n+1/2}(x). \tag{2.110}$$

In Eq. (2.110), $Z_{n+1/2}(x)$ may be any of the ordinary Bessel functions of half-integer order, $J_{n+1/2}(x)$, $Y_{n+1/2}(x)$, $I_{n+1/2}(x)$, $K_{n+1/2}(x)$, $H^{(1)}_{n+1/2}(x)$, and $H^{(2)}_{n+1/2}(x)$, while $z_n(x)$ may be any of the corresponding spherical Bessel functions $j_n(x)$, $y_n(x)$, $i_n(x)$, $k_n(x)$, $h^{(1)}_n(x)$, and $h^{(2)}_n(x)$. Bessel functions of fractional order are, in general, given by

$$J_\nu(x) = \sum_{k=0}^{\infty} \frac{(-1)^k x^{2k+\nu}}{2^{2k+\nu} k! \Gamma(\nu+k+1)} \tag{2.111}$$

$$Y_\nu(x) = \frac{J_\nu(x)\cos(\nu\pi) - J_{-\nu}}{\sin(\nu\pi)} \tag{2.112}$$

$$I_\nu(x) = (-j)^\nu J_\nu(jx) \tag{2.113}$$

$$K_\nu(x) = \frac{\pi}{2} \left[\frac{I_{-\nu} - I_\nu}{\sin(\nu\pi)} \right] \tag{2.114}$$

where $J_{-\nu}$ and $I_{-\nu}$ are, respectively, obtained from Eqs. (2.111) and (2.113) by replacing ν with $-\nu$. Although ν in Eqs. (2.111) to (2.114) can assume any fractional value, in our specific problem, $\nu = n + 1/2$. Since Gamma function of half-integer order is needed in Eq. (2.111), it is necessary to add that

$$\Gamma(n+1/2) = \begin{cases} \dfrac{(2n)!}{2^{2n} n!} \sqrt{\pi}, & n \geq 0 \\[2mm] \dfrac{(-1)^n 2^{2n} n!}{(2n)!} \sqrt{\pi}, & n < 0. \end{cases} \tag{2.115}$$

Thus the lower order spherical Bessel functions are as follows:

$$j_0(x) = \frac{\sin x}{x}, \quad y_0(x) = -\frac{\cos x}{x},$$

$$h^{(1)}_0(x) = \frac{e^{jx}}{jx}, \quad h^{(2)}_0(x) = \frac{e^{-jx}}{-jx},$$

$$i_0(x) = \frac{\sinh x}{x}, \quad k_0(x) = \frac{e^{-x}}{x},$$

Table 2.3 Properties and Identities of Spherical Bessel Functions.

(a) $z_{n+1} = \dfrac{(2n+1)}{x} z_n(x) - z_{n-1}(x)$ (recurrence relation)

(b) $\dfrac{d}{dx} z_n(x) = \dfrac{1}{2n+1}[n z_{n-1} - (n+1) z_{n+1}(x)]$

(c) $\dfrac{d}{dx}[x z_n(x)] = -n z_n(x) + x z_{n-1}(x)$

(d) $\dfrac{d}{dx}[x^{n+1} z_n(x)] = -x^{n+1} z_{n-1}(x)$

(e) $\dfrac{d}{dx}[x^{-n} z_n(x)] = -x^{-n} z_{n+1}(x)$

(f) $\int x^{n+2} z_n(x) dx = x^{n+2} z_{n+1}(x)$

(g) $\int x^{1-n} z_n(x) dx = -x^{1-n} z_{n-1}(x)$

(h) $\int x^2 [z_n(x)]^2\, dx = \tfrac{1}{2} x^3 \left[z_n(x) - z_{n-1}(x) z_{n+1}(x) \right]$

$$j_1(x) = \frac{\sin x}{x^2} - \frac{\cos x}{x}, \quad y_1(x) = -\frac{\cos x}{x^2} - \frac{\sin x}{x},$$

$$h_1^{(1)} = -\frac{(x+j)}{x^2} e^{jx}, \quad h_1^{(2)}(x) = -\frac{(x-j)}{x^2} e^{-jx}.$$

Other $z_n(x)$ can be obtained from the series expansion in Eqs. (2.111) and (2.112) or the recurrence relations and properties of $z_n(x)$ presented in Table 2.3.

By replacing H in Eq. (2.107) with y, $\cos\theta$ by x, and making other substitutions as we did for Eq. (2.93), we obtain

$$(1-x^2)y'' - 2xy' + \left[n(n+1) - \frac{m^2}{1-x^2}\right] y = 0, \quad (2.116)$$

which is Legendre's associated differential equation. Its general solution is of the form

$$y(x) = a_{mn} P_n^m(x) + d_{mn} Q_n^m(x) \quad (2.117)$$

where $P_n^m(x)$ and $Q_n^m(x)$ are called associated Legendre functions of the first and second kind, respectively. Equation (2.96) is a special case of Eq. (2.117) when $m=0$. $P_n^m(x)$ and $Q_n^m(x)$ can be obtained from ordinary Legendre functions $P_n(x)$ and $Q_n(x)$ using

$$P_n^m(x) = [1-x^2]^{m/2} \frac{d^m}{dx^m} P_n(x) \quad (2.118)$$

and

$$Q_n^m(x) = [1-x^2]^{m/2} \frac{d^m}{dx^m} Q_n(x) \tag{2.119}$$

where $-1 < x < 1$. We note that

$$\begin{aligned} P_n^0(x) &= P_n(x), \\ Q_n^0(x) &= Q_n(x), \\ P_n^m(x) &= 0 \quad \text{for} \quad m > n. \end{aligned} \tag{2.120}$$

Typical associated Legendre functions are:

$$P_1^1(x) = (1-x^2)^{1/2} = \sin\theta,$$
$$P_2^1(x) = 3x(1-x^2)^{1/2} = 3\cos\theta\sin\theta,$$
$$P_2^2(x) = 3(1-x^2) = 3\sin^2\theta,$$
$$P_3^1(x) = \frac{3}{2}(1-x^2)^{1/2}(5x-1) = \frac{3}{2}\sin\theta(5\cos\theta - 1),$$
$$Q_1^1(x) = (1-x^2)^{1/2}\left[\frac{1}{2}\ln\frac{1+x}{1-x} + \frac{x}{1-x^2}\right],$$

$$Q_2^1 = (1-x^2)^{1/2}\left[\frac{3x}{2}\ln\frac{1+x}{1-x} + \frac{3x^2-2}{1-x^2}\right],$$

$$Q_2^2 = (1-x^2)^{1/2}\left[\frac{3}{2}\ln\frac{1+x}{1-x} + \frac{5x^2-3x^2}{[1-x^2]^2}\right].$$

Higher-order associated Legendre functions can be obtained using Eqs. (2.118) and (2.119) along with the properties in Table 2.4. As mentioned earlier, $Q_n^m(x)$ is unbounded at $x = \pm 1$, and hence it is only used when $x = \pm 1$ is excluded. Substituting Eqs. (2.108), (2.109), and (2.117) into Eq. (2.103) and applying superposition theorem, we obtain

$$U(r,\theta,\phi,t) = \sum_{n=0}^{\infty}\sum_{m=0}^{n}\sum_{\ell=0}^{\infty} A_{mn\ell}\, z_n(k_{m\ell}r)\, P_n^m(\cos\theta) \exp(\pm jm\phi \pm j\omega t). \tag{2.121}$$

Note that the products $H(\theta)F(\phi)$ are known as spherical harmonics.

Table 2.4 Properties and Identities of Associated Legendre Functions[1].

(a) $P_n^m(x) = 0, \quad m > n$

(b) $P_n^m(x) = \dfrac{(2n-1)xP_{n-1}^m(x) - (n+m-1)P_{n-2}^m(x)}{n-m}$

(recurrence relations for fixed m)

(c) $P_n^m(x) = \dfrac{2(m-1)x}{(1-x^2)^{1/2}} P_n^{m-1}(x) - (n-m+2)(n+m-1)P_n^{m-2}$

(recurrence relations for fixed n)

(d) $P_n^m(x) = \dfrac{[1-x^2]^{m/2}}{2^n} \displaystyle\sum_{k=0}^{\left[\frac{m-n}{2}\right]} \dfrac{(-1)^k (2n-2k)! x^{n-2k-m}}{k!(n-k)!(n-2k-m)!}$

where $[t]$ is the bracket or greatest integer function, e.g., $[3.54] = 3$.

(e) $\dfrac{d}{dx}P_n^m(x) = \dfrac{(n+m)P_{n-1}^m(x) - nxP_n^m(x)}{1-x^2}$

(f) $\dfrac{d}{d\theta}P_n^m(x) = \dfrac{1}{2}\left[(n-m+1)(n+m)P_n^{m-1}(x) - P_n^{m+1}(x)\right]$

(g) $\dfrac{d}{dx}P_n^m(x) = -\dfrac{mxP_n^m(x)}{1-x^2}$

$+ \dfrac{(1-x^2)^{m/2}}{2^n} \displaystyle\sum_{k=0}^{\left[\frac{m-n}{2}\right]} \dfrac{(-1)^k (2n-2k)! \, x^{n-2k-m-1}}{k! \, (n-k)! \, (n-2k-m)!}$

(h) $\dfrac{d}{d\theta}P_n^m(x) = -(1-x^2)^{1/2}\dfrac{d}{dx}P_n^m(x)$

(i) The series expansion of $f(x)$:

$$f(x) = \sum_{n=0}^{\infty} A_n P_n^m(x),$$

where $A_n = \dfrac{(2n+1)(n-m)!}{2(n+m)!} \displaystyle\int_{-1}^{1} f(x)P_n^m(x)dx$

(j) $\left.\dfrac{d^m}{dx^m}P_n(x)\right|_{x=1} = \dfrac{(n+m)!}{2^m \, m! \, (n-m)!}$

$\left.\dfrac{d^m}{dx^m}P_n(x)\right|_{x=-1} = \dfrac{(-1)^{n+m}(n+m)!}{2^m \, m! \, (n-m)!}$

(k) $P_n^{-m}(x) = (-1)^m \dfrac{(n-m)!}{(n+m)!} P_n^m(x)$, $m = 0, 1, \cdots, n$

(l) $\displaystyle\int_{-1}^{1} P_n^m(x) P_n^m(x)\, dx = \dfrac{2}{2n+1} \dfrac{(n-m)!}{(n+m)!} \delta_{nk}$,

where δ_{nk} is the kronecker delta defined by $\delta_{nk} = \begin{cases} 0, & n \neq k \\ 1, & n = k \end{cases}$

---

[1] Properties (b) and (c) are also valid for $Q_n^m(x)$.

Example 2.6

A thin ring of radius a carries charge of density ρ. Find the potential at: (a) point $P(0,0,z)$ on the axis of the ring, (b) point $P(r,\theta,\phi)$ in space.

Solution

Consider the thin ring as in Fig. 2.10.
(a) From elementary electrostatics, at $P(0,0,z)$

$$V = \int \dfrac{\rho\, dl}{4\pi\epsilon R}$$

where $dl = a\,d\phi$, $R = \sqrt{a^2 + z^2}$. Hence

$$V = \int_0^{2\pi} \dfrac{\rho a\, d\phi}{4\pi\epsilon[a^2 + z^2]^{1/2}} = \dfrac{a\rho}{2\epsilon[a^2 + z^2]^{1/2}}. \tag{2.6.1}$$

(b) To find the potential at $P(r,\theta,\phi)$, we may evaluate the integral for the potential as we did in part (a). However, it turns out that the boundary-value solution is simpler. So we solve Laplace's equation $\nabla^2 V = 0$ where $V(0,0,z)$ must conform with the result in part (a). From Fig. 2.10, it is evident that V is invariant with ϕ. Hence the solution to Laplace's equation is

$$V = \sum_{n=0}^{\infty} \left[A_n r^n + \dfrac{B_n}{r^{n+1}} \right] [A_n' P_n(u) + B_n' Q_n(u)]$$

where $u = \cos\theta$. Since Q_n is singular at $\theta = 0, \pi$, $B_n' = 0$. Thus

$$V = \sum_{n=0}^{\infty} \left[C_n' r^n + \dfrac{D_n'}{r^{n+1}} \right] P_n(u). \tag{2.6.2}$$

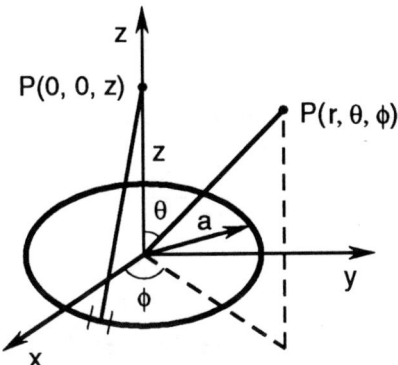

Figure 2.10 Charged ring of Example 2.6.

For $0 \leq r \leq a$, $D'_n = 0$ since V must be finite at $r = 0$.

$$V = \sum_{n=0}^{\infty} C'_n r^n P_n(u). \qquad (2.6.3)$$

To determine the coefficients C'_n, we set $\theta = 0$ and equate V to the result in part (a). But when $\theta = 0$, $u = 1, P_n(1) = 1$, and $r = z$. Hence

$$V(0,0,z) = \frac{a\rho}{2\epsilon[a^2 + z^2]^{1/2}} = \frac{a\rho}{2\epsilon} \sum_{n=0}^{\infty} C_n z^n. \qquad (2.6.4)$$

Using the binomial expansion, the term $[a^2 + z^2]^{-1/2}$ can be written as

$$\frac{1}{a}[1 + \frac{z^2}{a^2}]^{-1/2} = \frac{1}{a}\left[1 - \frac{1}{2}(z/a)^2 + \frac{1\cdot 3}{2\cdot 4}(z/a)^4 - \frac{1\cdot 3\cdot 5}{2\cdot 4\cdot 6}(z/a)^6 + \cdots \right].$$

Comparing this with the last term in Eq. (2.6.4), we obtain

$$C_0 = 1, \quad C_1 = 0, \quad C_2 = -\frac{1}{2a^2}, \quad C_3 = 0,$$

$$C_4 = \frac{1\cdot 3}{2\cdot 4}\frac{1}{a^4}, \quad C_5 = 0, \quad C_6 = -\frac{1\cdot 3\cdot 5}{2\cdot 4\cdot 6}\frac{1}{a^6}, \quad \cdots$$

or in general,

$$C_{2n} = (-1)^n \frac{(2n)!}{[n!2^n]^2 a^{2n}}.$$

Substituting these into Eq. (2.6.3) gives

$$V = \frac{a\rho}{2\epsilon} \sum_{n=0}^{\infty} \frac{(-1)^n (2n)!}{[n! 2^n]^2} (r/a)^{2n} P_{2n}(\cos\theta), \quad 0 \le r \le a. \tag{2.6.5}$$

For $r \ge a$, $C'_n = 0$ since V must be finite as $r \to \infty$, and

$$V = \sum_{n=0}^{\infty} \frac{D'_n}{r^{n+1}} P_n(u). \tag{2.6.6}$$

Again, when $\theta = 0$, $u = 1$, $P_n(1) = 1$, $r = z$,

$$V(0,0,z) = \frac{a\rho}{2\epsilon [a^2 + z^2]^{1/2}} = \frac{a\rho}{2\epsilon} \sum_{n=0}^{\infty} D_n z^{-(n+1)}. \tag{2.6.7}$$

Using the binomial expansion, the middle term $[a^2 + z^2]^{-1/2}$ can be written as

$$\frac{1}{z}[1 + \frac{a^2}{z^2}]^{-1/2} = \frac{1}{z}\left[1 - \frac{1}{2}(a/z)^2 + \frac{1\cdot 3}{2\cdot 4}(a/z)^4 - \frac{1\cdot 3\cdot 5}{2\cdot 4\cdot 6}(a/z)^6 + \cdots\right].$$

Comparing this with the last term in Eq. (2.6.7), we obtain

$$D_0 = 1, \quad D_1 = 0, \quad D_2 = -\frac{a^2}{2}, \quad D_3 = 0,$$

$$D_4 = \frac{1\cdot 3}{2\cdot 4} a^4, \quad D_5 = 0, \quad D_6 = -\frac{1\cdot 3\cdot 5}{2\cdot 4\cdot 6} a^6, \quad \cdots$$

or in general,

$$D_{2n} = (-1)^n \frac{(2n)!}{[n! 2^n]^2} a^{2n}.$$

Substituting these into Eq. (2.6.6) gives

$$V = \frac{a\rho}{2\epsilon r} \sum_{n=0}^{\infty} \frac{(-1)^n (2n)!}{[n! 2^n]^2} (a/r)^{2n} P_{2n}(\cos\theta), \quad r \ge a. \tag{2.6.8}$$

We may combine Eqs. (2.6.5) and (2.6.8) to get

$$V = \begin{cases} a \sum_{n=0}^{\infty} g_n \, (r/a)^{2n} \, P_{2n}(\cos\theta), & 0 \le r \le a \\ \sum_{n=0}^{\infty} g_n \, (a/r)^{2n+1} \, P_{2n}(\cos\theta), & r \ge a \end{cases}$$

where

$$g_n = (-1)^n \frac{p}{2\epsilon} \frac{2n!}{[n!2^n]^2}.$$

Example 2.7

A conducting spherical shell of radius a is maintained at potential $V_o \cos 2\phi$; determine the potential at any point inside the sphere.

Solution

The solution to this problem is somewhat similar to that of the previous problem except that V is a function of ϕ. Hence the solution to Laplace's equation for $0 \leq r \leq a$ is of the form

$$V = \sum_{n=0}^{\infty} \sum_{m=0}^{\infty} \left(a_{mn} \cos m\phi + b_{mn} \sin m\phi\right)(r/a)^n P_n^m(\cos \theta).$$

Since $\cos m\phi$ and $\sin m\phi$ are orthogonal functions, $a_{mn} = 0 = b_{mn}$ except that $a_{n2} \neq 0$. Hence at $r = a$

$$V_o \cos 2\phi = \cos 2\phi \sum_{n=2}^{\infty} a_{n2} P_n^2(\cos \theta)$$

or

$$V_o = \sum_{n=2}^{\infty} a_{n2} P_n^2(x), \quad x = \cos \theta$$

which is the Legendre expansion of V_o. Multiplying both sides by $P_m^2(x)$ gives

$$\frac{2}{2n+1} \frac{(n+2)!}{(n-2)!} a_{n2} = V_o \int_{-1}^{1} P_n^2(x) \, dx = V_o \int_{-1}^{1} (1-x^2) \frac{d^2}{dx^2} P_n(x) \, dx.$$

Integrating by parts twice yields

$$a_{n2} = V_o \frac{2n+1}{2} \frac{(n-2)!}{(n+2)!} \left(2P_n(1) - 2P_n(-1) - 2 \int_{-1}^{1} P_n(x) \, dx \right).$$

Using the generating functions for $P_n(x)$ (see Table 2.7 and Example 2.10) it is readily shown that

$$P_n(1) = 1, \quad P_n(-1) = (-1)^n.$$

Also
$$\int_{-1}^{1} P_n(x)\, dx = \int_{-1}^{1} P_0(x) P_n(x)\, dx = 0$$

by the orthogonality property of $P_n(x)$. Hence

$$a_{n2} = V_o\, (2n+1)\frac{(n-2)!}{(n+2)!}[1-(-1)^n]$$

and

$$V = V_o \cos 2\phi \sum_{n=2}^{\infty}(2n+1)\frac{(n-2)!}{(n+2)!}[1-(-1)^n](r/a)^n P_n^2(\cos\theta).$$

Example 2.8

Express: (a) the plane wave e^{jz} and (b) the cylindrical wave $J_0(\rho)$ in terms of spherical wave functions.

Solution

(a) Since $e^{jz} = e^{jr\cos\theta}$ is independent of ϕ and finite at the origin, we let

$$e^{jz} = e^{jr\cos\theta} = \sum_{n=0}^{\infty} a_n j_n(r) P_n(\cos\theta) \qquad (2.8.1)$$

where a_n are the expansion coefficients. To determine a_n, we multiply both sides of Eq. (2.8.1) by $P_m(\cos\theta)\sin\theta$ and integrate over $0 < \theta < \pi$:

$$\int_0^{\pi} e^{jr\cos\theta} P_m(\cos\theta) \sin\theta\, d\theta = \sum_{n=0}^{\infty} a_n j_n(r) \int_{-1}^{1} P_n(x) P_m(x)\, dx$$

$$= \begin{cases} 0, & n \neq m \\ \dfrac{2}{2n+1} a_n j_n(r), & n = m \end{cases}$$

where the orthogonality property (i) of Table 2.2 has been utilized. Taking the nth derivative of both sides and evaluating at $r = 0$ gives

$$j^n \int_0^{\pi} \cos^n\theta P_n(\cos\theta) \sin\theta\, d\theta = \frac{2}{2n+1} a_n \frac{d^n}{dr^n} j_n(r)\bigg|_{r=0}. \qquad (2.8.2)$$

The left-hand side of Eq. (2.8.2) yields

$$j^n \int_{-1}^{1} x^n P_n(x)\, dx = \frac{2^{n+1}(n!)^2}{(2n+1)!} j^n. \tag{2.8.3}$$

To evaluate the right-hand side of Eq. (2.8.2), we recall that

$$j_n(r) = \sqrt{\frac{\pi}{2r}} J_{n+1/2}(r) = \sqrt{\frac{\pi}{2}} \sum_{m=0}^{\infty} \frac{(-1)^m r^{2m+n}}{m!\Gamma(m+n+3/2)2^{2m+n+1/2}}.$$

Hence

$$\left.\frac{d^n}{dr^n} j_n(r)\right|_{r=0} = \sqrt{\frac{\pi}{2}} \frac{n!}{\Gamma(n+3/2)2^{n+1/2}} = \frac{2^n(n!)^2}{(2n+1)!}. \tag{2.8.4}$$

Substituting Eqs. (2.8.3) and (2.8.4) into Eq. (2.8.2) gives

$$a_n = j^n(2n+1).$$

Thus

$$e^{jz} = e^{jr\cos\theta} = \sum_{n=0}^{\infty} j^n(2n+1) j_n(r) P_n(\cos\theta). \tag{2.8.5}$$

(b) Since $J_0(\rho) = J_0(r\sin\theta)$ is even, independent of ϕ, and finite at the origin,

$$J_0(\rho) = J_0(r\sin\theta) = \sum_{n=0}^{\infty} b_n j_{2n}(r) P_{2n}(\cos\theta). \tag{2.8.6}$$

To determine the coefficients of expansion b_n, we multiply both sides by $P_m(\cos\theta)\sin\theta$ and integrate over $0 < \theta < \pi$. We obtain

$$\int_0^{\pi} J_0(r\sin\theta) P_m(\cos\theta) \sin\theta\, d\theta = \begin{cases} 0, & m \neq 2n \\ \dfrac{2b_n}{4n+1} j_{2n}(r), & m = 2n. \end{cases}$$

Differentiating both sides $2n$ times with respect to r and setting $r = 0$ gives

$$b_n = \frac{(-1)^n (4n+1)(2n-1)!}{2^{2n-1} n!(n-1)!}.$$

Hence

$$J_0(\rho) = \sum_{n=0}^{\infty} \frac{(-1)^n (4n+1)(2n-1)!}{2^{2n-1} n!\, (n-1)!} j_{2n}(r) P_{2n}(\cos\theta).$$

2.6 Some Useful Orthogonal Functions

Orthogonal functions are of great importance in mathematical physics and engineering. A system of real functions Φ_n $(n = 0, 1, 2, \cdots)$ is said to be *orthogonal with weight $w(x)$* on the interval (a, b) if

$$\int_a^b w(x)\Phi_m(x)\Phi_n(x)\, dx = 0 \tag{2.122}$$

for every $m \neq n$. For example, the system of functions $\cos(nx)$ is orthogonal with weight 1 on the interval $(0, \pi)$ since

$$\int_0^\pi \cos mx \cos nx\, dx = 0, \quad m \neq n.$$

Orthogonal functions usually arise in the solution of partial differential equations governing the behavior of certain physical phenomena. These include Bessel, Legendre, Hermite, Laguerre, and Chebyshev functions. In addition to the orthogonality properties in Eq. (2.122), these functions have many other general properties, which will be discussed briefly in this section. They are very useful in series expansion of functions belonging to very general classes, e.g., Fourier-Bessel series, Legendre series, etc. Although Hermite, Laguerre, and Chebyshev functions are of less importance in EM problems than Bessel and Legendre functions, they are sometimes useful and therefore deserve some attention.

An arbitrary function $f(x)$, defined over interval (a, b), can be expressed in terms of any complete, orthogonal set of functions:

$$f(x) = \sum_{n=0}^\infty A_n \Phi_n(x) \tag{2.123}$$

where the expansion coefficients are given by

$$A_n = \frac{1}{N_n} \int_a^b w(x) f(x) \Phi_n(x)\, dx \tag{2.124}$$

and the (weighted) norm N_n is defined as

$$N_n = \int_a^b w(x) \Phi_n^2(x)\, dx. \tag{2.125}$$

Simple orthogonality results when $w(x) = 1$ in Eqs. (2.122) to (2.125).

Table 2.5 Differential Equations with Solutions

Equations		Solutions
$x^2 y'' + xy' + (x^2 - n^2)y = 0$	$J_n(x)$	Bessel functions of the first kind
	$Y_n(x)$	Bessel functions of the second kind
	$H_n^{(1)}(x)$	Hankel functions of the first kind
	$H_n^{(2)}(x)$	Hankel functions of the second kind
$x^2 y'' + xy' - (x^2 + n^2)y = 0$	$I_n(x)$	Modified Bessel functions of the first kind
	$K_n(x)$	Modified Bessel functions of the second kind
$x^2 y'' + 2xy' + [x^2 - n(n+1)]y = 0$	$j_n(x)$	Spherical Bessel functions of the first kind
	$y_n(x)$	Spherical Bessel functions of the second kind
$(1 - x^2)y'' - 2xy + n(n+1)y = 0$	$P_n(x)$	Legendre polynomials
	$Q_n(x)$	Legendre functions of the second kind
$(1 - x^2)y'' - 2xy' + \left[n(n+1) - \dfrac{m^2}{1-x^2}\right]y = 0$	$P_n^m(x)$	Associated Legendre polynomials
	$Q_n^m(x)$	Associated Legendre functions of the second kind
$y'' - 2xy' + 2ny = 0$	$H_n(x)$	Hermite polynomials
$xy'' + (1 - x)y' + ny = 0$	$L_n(x)$	Laguerre polynomials
$xy'' + (m + 1 - x)y' + ny = 0$	$L_n^m(x)$	Associated Laguerre polynomials
$(1 - x^2)y'' - xy' + n^2 y = 0$	$T_n(x)$	Chebyshev polynomials of the first kind
	$U_n(x)$	Chebyshev polynomials of the second kind

Perhaps the best way to briefly describe the orthogonal functions is in table form. This is done in Tables 2.5 to 2.7. The differential equations giving rise to each function are provided in Table 2.5. The orthogonality relations in Table 2.6 are necessary for expanding a given arbitrary function $f(x)$ in terms of the orthogonal functions as in Eqs. (2.123) to (2.125).

78 Numerical Techniques in Electromagnetics

Table 2.6 Orthogonality Relations

Functions	Relations
Bessel functions	$\int_0^a x J_n(\lambda_i x) J_n(\lambda_j x)\, dx = \dfrac{a^2}{2}[J_{n+1}(\lambda_i a)]^2 \delta_{ij}$ where λ_i and λ_j are the roots of $J_n(\lambda a) = 0$
Spherical Bessel functions	$\int_{-\infty}^{\infty} j_n(x) j_m(x)\, dx = \dfrac{\pi}{2n+1}\delta_{mn}$
Legendre polynomials	$\int_{-1}^{1} P_n(x) P_m(x)\, dx = \dfrac{2}{2n+1}\delta_{mn}$
Associated Legendre polynomials	$\int_{-1}^{1} P_n^k(x) P_m^k(x)\, dx = \dfrac{2(n+k)!}{(2n+1)(n-k)}\delta_{mn}$ $\int_{-1}^{1} \dfrac{P_n^m(x) P_n^k(x)}{1-x^2}\, dx = \dfrac{(n+m)!}{m(n-m)!}\delta_{mk}$
Hermite polynomials	$\int_{-\infty}^{\infty} e^{-x^2} H_n(x) H_m(x)\, dx = 2^n n!(\sqrt{\pi})\delta_{mn}$
Laguerre polynomials	$\int_0^{\infty} e^{-x} L_n(x) L_m(x)\, dx = \delta_{mn}$
Associated Laguerre polynomials	$\int_0^{\infty} e^{-x} x^k L_n^k(x) L_m^k(x)\, dx = \dfrac{(n+k)!}{n!}\delta_{mn}$
Chebyshev polynomials	$\int_{-1}^{1} \dfrac{T_n(x) T_m(x)}{(1-x^2)^{1/2}}\, dx = \begin{cases} 0, & m \ne n \\ \pi/2, & m = n \ne 0 \\ \pi, & m = n = 0 \end{cases}$ $\int_{-1}^{1} \dfrac{U_n(x) U_m(x)}{(1-x^2)^{1/2}}\, dx = \begin{cases} 0, & m \ne n \\ \pi/2, & m = n \ne 0 \\ \pi, & m = n = 0 \end{cases}$

Most of the properties of the orthogonal functions can be proved using the generating functions of Table 2.7. To the properties in Tables 2.5 to 2.7 we may add the recurrence relations and series expansion formulas for calculating the functions for specific argument x and order n. These have been provided for $J_n(x)$ and $Y_n(x)$ in Table 2.1 and Eqs. (2.73) and (2.75), for $P_n(x)$ and $Q_n(x)$ in Table 2.2 and Eqs. (2.97) and (2.98), for $j_n(x)$ and $y_n(x)$ in Table 2.3 and Eq. (2.110), and for $P_n^m(x)$ and $Q_n^m(x)$ in Table 2.4 and Eqs. (2.118) and (2.119). For Hermite polynomials, the series expansion formula is

$$H_n(x) = \sum_{k=0}^{[n/2]} \frac{(-1)^k\, n!\, (2x)^{n-2k}}{k!\,(n-2k)!} \qquad (2.126)$$

Table 2.7 Generating Functions

Functions	Generating function $R = [1 - 2xt + t^2]^{1/2}$
Bessel function	$\exp\left[\dfrac{x}{2}\left(t - \dfrac{1}{t}\right)\right] = \sum_{n=-\infty}^{\infty} t^n J_n(x)$
Legendre polynomial	$\dfrac{1}{R} = \sum_{n=0}^{\infty} t^n P_n(x)$
Associated Legendre polynomial	$\dfrac{(2m)!\,(1-x^2)^{m/2}}{2^m\, m!\, R^{m+1}} = \sum_{n=0}^{\infty} t^n P_{n+m}^m(x)$
Hermite polynomial	$\exp(2tx - t^2) = \sum_{n=0}^{\infty} \dfrac{t^n}{n!} H_n(x)$
Laguerre polynomial	$\dfrac{\exp[-xt/(1-t)]}{1-t} = \sum_{n=0}^{\infty} t^n L_n(x)$
Associated Laguerre polynomial	$\dfrac{\exp[-xt/(1-t)]}{(1-t)^{m+1}} = \sum_{n=0}^{\infty} t^n L_n^m(x)$
Chebyshev polynomial	$\dfrac{1-t^2}{R^2} = T_0(x) + 2\sum_{n=1}^{\infty} t^n T_n(x)$ $\dfrac{\sqrt{1-x^2}}{R^2} = \sum_{n=0}^{\infty} t^n U_{n+1}(x)$

where $[n/2] = N$ is the largest even integer $\leq n$ or simply the greatest integer function. Thus,

$$H_0(x) = 1, \quad H_1(x) = 2x, \quad H_2(x) = 4x^2 - 2, \quad \text{etc.}$$

The recurrence relations are

$$H_{n+1}(x) = 2xH_n(x) - 2nH_{n-1}(x) \qquad (2.127a)$$

and

$$H_n'(x) = 2nH_{n-1}(x). \qquad (2.127b)$$

For Laguerre polynomials,

$$L_n(x) = \sum_{k=0}^{n} \frac{n! \, (-x)^k}{(k!)^2 \, (n-k)!} \qquad (2.128)$$

so that

$$L_0(x) = 1, \quad L_1(x) = -x + 1, \quad L_2(x) = \frac{1}{2!}(x^2 - 4x + 2), \quad \text{etc.}$$

The recurrence relations are

$$L_{n+1}(x) = (2n + 1 - x)L_n(x) - n^2 L_{n-1}(x) \qquad (2.129a)$$

and

$$\frac{d}{dx} L_n(x) = \frac{1}{x}\left[n L_n(x) - n^2 L_{n+1}(x) \right]. \qquad (2.129b)$$

For the associated Laguerre polynomials,

$$L_n^m(x) = (-1)^m \frac{d^m}{dx^m} L_{n+m}(x) = \sum_{k=0}^{n} \frac{(m+n)! \, (-x)^k}{k! \, (n-k)! \, (m+k)!} \qquad (2.130)$$

so that

$$L_1^1(x) = -x + 2, \quad L_2^1(x) = \frac{x^2}{2} - 3x + 3, \quad L_2^2(x) = \frac{x^2}{2} - 4x + 6, \quad \text{etc.}$$

Note that $L_n^m(x) = 0, \quad m > n$. The recurrence relations are

$$L_{n+1}^m(x) = \frac{1}{n+1}\left[(2n + m + 1 - x)L_n^m(x) - (n + m)L_{n-1}^m(x) \right]. \qquad (2.131)$$

For Chebyshev polynomials of the first kind,

$$T_n(x) = \sum_{k=0}^{[n/2]} \frac{(-1)^k \, n! \, x^{n-2k} \, (1 - x^2)^k}{(2k)! \, (n - 2k)!}, \quad -1 \leq x \leq 1 \qquad (2.132)$$

so that

$$T_0(x) = 1, \quad T_1(x) = x, \quad T_2(x) = 2x^2 - 1, \quad \text{etc.}$$

The recurrence relation is

$$T_{n+1}(x) = 2x T_n(x) - T_{n-1}(x). \qquad (2.133)$$

Analytical Methods 81

For Chebyshev polynomials of the second kind,

$$U_n(x) = \sum_{k=0}^{N} \frac{(-1)^{k-1} (n+1)! \, x^{n-2k+2} (1-x^2)^{k-1}}{(2k+1)! \, (n-2k+2)!}, \quad -1 \leq x \leq 1 \quad (2.134)$$

where $N = \left[\frac{n+1}{2}\right]$ so that

$$U_0(x) = 1, \quad U_1(x) = 2x, \quad U_2(x) = 4x^2 - 1, \quad \text{etc.}$$

The recurrence relation is the same as that in Eq. (2.133).

For example, if a function $f(x)$ is to be expanded on the interval $(0, \infty)$, Laguerre functions can be used as the orthogonal functions with an exponential weighting function, i.e., $w(x) = e^{-x}$. If $f(x)$ is to be expanded on the interval $(-\infty, \infty)$, we may use Hermite functions with $w(x) = e^{-x^2}$. As we have noticed earlier, if $f(x)$ is defined on the interval $(-1, 1)$, we may choose Legendre functions with $w(x) = 1$. For more detailed treatment of these functions, see Bell [7] or Johnson and Johnson [9].

Example 2.9

Expand the function

$$f(x) = |x|, \quad -1 \leq x \leq 1$$

in a series of Chebyshev polynomials.

Solution

The given function can be written as

$$f(x) = \begin{cases} -x, & -1 \leq x < 0 \\ x, & 0 < x \leq 1. \end{cases}$$

Let

$$f(x) = \sum_{n=0}^{\infty} A_n T_n(x)$$

where A_n are expansion coefficients to be determined. Since $f(x)$ is an even function, the odd terms in the expansion vanish. Hence

$$f(x) = A_0 + \sum_{n=1}^{\infty} A_{2n} T_{2n}(x).$$

82 Numerical Techniques in Electromagnetics

If we multiply both sides by $w(x) = \dfrac{T_{2m}}{\sqrt{1-x^2}}$ and integrate over $-1 \le x \le 1$, all terms in the summation vanish except when $m = n$. That is, from Table 2.6, the orthogonality property of $T_n(x)$ requires that

$$\int_{-1}^{1} \frac{T_m(x)T_n(x)}{(1-x^2)^{1/2}} dx = \begin{cases} 0, & m \ne n \\ \pi/2, & m = n \ne 0 \\ \pi, & m = n = 0. \end{cases}$$

Hence

$$A_0 = \frac{1}{\pi} \int_{-1}^{1} \frac{f(x)\, T_0(x)}{(1-x^2)^{1/2}} dx = \frac{2}{\pi} \int_{0}^{1} \frac{x}{(1-x^2)^{1/2}} dx = \frac{2}{\pi},$$

$$A_{2n} = \frac{2}{\pi} \int_{-1}^{1} \frac{f(x)\, T_{2n}(x)}{(1-x^2)^{1/2}} dx = \frac{4}{\pi} \int_{0}^{1} \frac{x T_{2n}}{(1-x^2)^{1/2}} dx.$$

Since $T_n(x) = \cos(n \cos^{-1} x)$, it is convenient to let $x = \cos\theta$ so that

$$A_{2n} = \frac{4}{\pi} \int_{\pi/2}^{0} \frac{\cos\theta \cos 2n\theta}{\sin\theta} (-\sin\theta\, d\theta) = \frac{4}{\pi} \int_{0}^{\pi/2} \cos\theta \cos 2n\theta\, d\theta$$

$$= \frac{4}{\pi} \int_{0}^{\pi/2} \frac{1}{2}\left[\cos(2n+1)\theta + \cos(2n-1)\theta\right] d\theta = \frac{4}{\pi} \frac{(-1)^{n+1}}{4n^2 - 1}.$$

Hence

$$f(x) = \frac{2}{\pi} + \frac{4}{\pi} \sum_{n=1}^{\infty} \frac{(-1)^{n+1}}{4n^2 - 1} T_{2n}(x).$$

Example 2.10

Evaluate $\dfrac{P_n^1(x)}{\sin\theta}$ at $x = 1$ and $x = -1$.

Solution

This example serves to illustrate how the generating functions are useful in deriving some properties of the corresponding orthogonal functions. Since

$$\frac{P_n^1(x)}{\sin\theta} = \frac{P_n^1(x)}{\sqrt{1-x^2}},$$

direct substitution of $x = 1$ or $x = -1$ gives $0/0$, which is indeterminate. But $P_n^1(x) = (1-x^2)^{1/2}\frac{d}{dx}P_n$ by definition. Hence

$$\frac{P_n^1(x)}{\sin\theta} = \frac{d}{dx}P_n,$$

i.e., the problem is reduced to evaluating dP_n/dx at $x = \pm 1$. We use the generating function for P_n, namely,

$$(1 - 2xt + t^2)^{-1/2} = \sum_{n=0}^{\infty} t^n P_n(x).$$

Differentiating both sides with respect to x,

$$\frac{t}{(1-2xt+t^2)^{3/2}} = \sum_{n=0}^{\infty} t^n \frac{d}{dx}P_n. \qquad (2.10.1)$$

When $x = 1$,

$$\frac{1}{(1-t)^3} = \sum_{n=0}^{\infty} t^{n-1} \frac{d}{dx}P_n \bigg|_{x=1} \qquad (2.10.2)$$

But

$$(1-t)^{-3} = 1 + 3t + 6t^2 + 10t^3 + 15t^4 + \cdots = \sum_{n=1}^{\infty} \frac{n}{2}(n+1)t^{n-1}. \qquad (2.10.3)$$

Comparing this with Eq. (2.10.2) clearly shows that

$$\frac{d}{dx}P_n\bigg|_{x=1} = n(n+1)/2.$$

Similarly, when $x = -1$, Eq. (2.10.1) becomes

$$\frac{1}{(1+t)^3} = \sum_{n=0}^{\infty} t^{n-1} \frac{d}{dx}P_n \bigg|_{x=-1} \qquad (2.10.4)$$

But

$$(1+t)^{-3} = 1 - 3t + 6t^2 - 10t^3 + 15t^4 - \cdots = \sum_{n=1}^{\infty}(-1)^{n+1}\frac{n}{2}(n+1)t^{n-1}.$$

84   Numerical Techniques in Electromagnetics

Hence
$$\left.\frac{d}{dx}P_n\right|_{x=-1} = (-1)^{n+1}n(n+1)/2.$$

Example 2.11

Write a program to generate Hermite functions $H_n(x)$ for any argument x and order n. Use the series expansion and recurrence formulas and compare your results. Take $x = 0.5$, $0 \leq n \leq 15$.

Solution

The program is shown in Fig. 2.11. Equation (2.126) is used for the series expansion method, while Eq. (2.127a) with $H_0(x) = 1$ and $H_1(x) = 2x$ is used for the recurrence formula. Note that in the program, we have replaced n by $n-1$ in Eq. (2.127) so that

$$H_n(x) = 2xH_{n-1}(x) - 2(n-1)H_{n-2}(x).$$

The result of the computation is in Table 2.8. In this case, the two methods give identical results. In general, the series expansion method gives results of greater accuracy since error in one computation is not propagated to the next as is the case when using recurrence relations.

```
0001    C     ***********************************************************
0002    C     THIS PROGRAM GENERATES HERMITE'S FUNCTIONS HN(X) IN
0003    C     TWO WAYS USING:   1) SERIES EXPANSION
0004    C                       2) RECURRENCE RELATION
0005    C     THE TWO METHODS ARE COMPARED
0006    C     X = ARGUMENT (FIXED IN THIS PROGRAM)
0007    C     N = ORDER OF THE FUNCTION  ( 0 < N < 15 IN THIS PROGRAM)
0008    C     ***********************************************************
0009
0010          DIMENSION    HS(0:50), HR(0:50)
0011
0012          X = 0.5
0013          NMAX = 15
0014          WRITE(6,1)
0015    1     FORMAT(2X,68('-'),/)
0016          WRITE(6,2)
0017    2     FORMAT(3X,'N',14X,'SERIES HS(N)',7X,'RECURRENCE HR(N)',
0018         2       7X,'DIFFERENCE',/)
0019          WRITE(6,1)
0020          DO 60 N=0,NMAX
0021
0022    C     METHOD 1:   SERIES EXPANSION FORMULA
0023
0024          SUM = 0.0
0025          CALL FACTORIAL(N,FN)
```

Table 2.8 Results of the Program in Fig. 2.11.

Values of $H_n(x)$ for $x = 0.5$, $0 \leq n \leq 15$			
N	Series Expansion	Recurrence	Difference
0	1.00	1.00	0.00
1	1.00	1.00	0.00
2	-1.00	-1.00	0.00
3	-5.00	-5.00	0.00
4	1.00	1.00	0.00
5	11.00	1.00	0.00
6	31.00	31.00	0.00
7	-461.00	-461.00	0.00
8	-895.00	-895.00	0.00
9	6181.00	6181.00	0.00
10	22591.00	22591.00	0.00
11	-107029.00	-107029.00	0.00
12	-604031.00	-604031.00	0.00
13	1964665.00	1964665.00	0.00
14	17669472.00	17669472.00	0.00
15	-37341152.00	-37341148.00	-4.00

```
0026              CALL GREATEST(N,I)
0027              DO 10 K=0,I
0028              M = N - 2*K
0029              CALL FACTORIAL(M,FM)
0030              CALL FACTORIAL(K,FK)
0031              A = ( ((-1.)**K)*FN*((2.*X)**M) )/( FK*FM )
0032       C      FK*FM MAY BE TOO LARGE IF N IS LARGE
0033              SUM = SUM + A
0034       10     CONTINUE
0035              HS(N) = SUM
0036
0037       C      METHOD 2:  RECURRENCE FORMULA
0038
0039              HR(0) = 1.0
0040              HR(1) = 2.*X
0041              IF(N-1) 40,40,20
0042       20     DO 30 I=2,N
0043              HR(I) = 2.0*X*HR(I-1) - 2.*FLOAT(I-1)*HR(I-2)
0044       30     CONTINUE
0045       40     CONTINUE
0046              DIFFERENCE = HS(N) - HR(N)
0047              WRITE(6,50) N,HS(N),HR(N),DIFFERENCE
0048              PRINT *,N,HS(N),HR(N),DIFFERENCE
0049       50     FORMAT(2X,I2,9X,F12.2,9X,F12.2,9X,F10.2,/)
```

```
0050      60       CONTINUE
0051               WRITE(6,1)
0052               STOP
0053               END


0001      C****************************************************************
0002      C
0003      C        SUBROUTINE FOR CALCULATING  N!
0004               SUBROUTINE FACTORIAL(N,FN)
0005
0006               FN = 1.0
0007               IF(N.EQ.0) GO TO 20
0008               DO 10 I=1,N
0009               FN = FN*FLOAT(I)
0010      10       CONTINUE
0011      20       RETURN
0012               END


0001      C****************************************************************
0002      C        SUBROUTINE FOR CALCULATING THE GREATEST INTEGER FUNCTION
0003      C        M = [X]  WHERE X = N/2 IN THIS PARTICULAR CASE
0004               SUBROUTINE GREATEST(MAX,M)
0005
0006               A = MAX/2
0007               M = IFIX(A)
0008               IF(M) 10,20,20
0009      10       M = M - 1
0010      20       RETURN
0011               END
```

Figure 2.11 Program for Hermite function $H_n(x)$.

Generating functions such as this is sometimes needed in numerical computations. This example has served to illustrate how this can be done in two ways. Special techniques may be required for very large or very small values of x or n.

2.7 Series Expansion

As we have noticed in earlier sections, partial differential equations can be solved with the aid of infinite series and, more generally, with the aid of series of orthogonal functions. In this section we apply the idea of infinite series expansion to those PDEs in which the independent variables are not separable or, if they are separable, the boundary conditions are not satisfied by the particular solutions. We will illustrate the technique in the following three examples.

2.7.1 Poisson's Equation in a Cube

Consider the problem

$$\nabla^2 V = \frac{\partial^2 V}{\partial x^2} + \frac{\partial^2 V}{\partial y^2} + \frac{\partial^2 V}{\partial z^2} = -f(x,y,z) \tag{2.135}$$

subject to the boundary conditions

$$V(0,y,z) = V(a,y,z) = V(x,0,z) = 0$$
$$V(x,b,z) = V(x,y,0) = V(x,y,c) = 0 \tag{2.136}$$

where $f(x,y,z)$, the source term, is given. We should note that the independent variables in Eq. (2.135) are not separable. However, in Laplace's equation, $f(x,y,z) = 0$, and the variables are separable. Although the problem defined by Eqs. (2.135) and (2.136) can be solved in several ways, we stress the use of series expansion in this section.

Let the solution be of the form

$$V(x,y,z) = \sum_{m=1}^{\infty}\sum_{n=1}^{\infty}\sum_{p=1}^{\infty} A_{mnp} \sin\frac{m\pi x}{a} \sin\frac{n\pi y}{b} \sin\frac{p\pi z}{c} \tag{2.137}$$

where the triple sine series is chosen so that the individual terms and the entire series would satisfy the boundary conditions of Eq. (2.136). However, the individual terms do not satisfy either Poisson's or Laplace's equation. Since the expansion coefficients A_{mnp} are arbitrary, they can be chosen such that Eq. (2.137) satisfies Eq. (2.135). We achieve this by substituting Eq. (2.137) into Eq. (2.135). We obtain

$$-\sum\sum\sum A_{mnp}(m\pi/a)^2 \sin\frac{m\pi x}{a} \sin\frac{n\pi y}{b} \sin\frac{p\pi z}{c}$$

$$-\sum\sum\sum A_{mnp}(n\pi/b)^2 \sin\frac{m\pi x}{a} \sin\frac{n\pi y}{b} \sin\frac{p\pi z}{c}$$

$$-\sum\sum\sum A_{mnp}(p\pi/c)^2 \sin\frac{m\pi x}{a} \sin\frac{n\pi y}{b} \sin\frac{p\pi z}{c} = -f(x,y,z).$$

Multiplying both sides by $\sin(i\pi x/a)\sin(j\pi y/b)\sin(k\pi z/c)$ and integrating over $0 < x < a$, $0 < y < b$, $0 < z < c$ gives

$$\sum\sum\sum A_{mnp}\left[(m\pi/a)^2 + (n\pi/b)^2 + (p\pi/c)^2\right].$$

$$\int_0^a \sin\frac{m\pi x}{a} \sin\frac{i\pi x}{a} \, dx \int_0^b \sin\frac{n\pi y}{b} \sin\frac{j\pi y}{b} \, dy \int_0^c \sin\frac{p\pi z}{c} \sin\frac{k\pi z}{c} \, dz$$

$$= \int_0^a \int_0^b \int_0^c f(x,y,z) \sin\frac{i\pi x}{a} \sin\frac{j\pi y}{b} \sin\frac{k\pi z}{c} \, dx dy dz.$$

88   Numerical Techniques in Electromagnetics

Each of the integrals on the left-hand side vanishes except when $m = i$, $n = j$, and $p = k$. Hence

$$A_{mnp}\left[(m\pi/a)^2 + (n\pi/b)^2 + (p\pi/c)^2\right]\frac{a}{2}\cdot\frac{b}{2}\cdot\frac{c}{2} = $$
$$\int_0^a \int_0^b \int_0^c f(x,y,z)\sin\frac{i\pi x}{a}\sin\frac{j\pi y}{b}\sin\frac{k\pi z}{c}\,dx\,dy\,dz$$

or

$$A_{mnp} = \frac{8}{abc}\left[(m\pi/a)^2 + (n\pi/b)^2 + (p\pi/c)^2\right]^{-1}$$
$$\cdot \int_0^a \int_0^b \int_0^c f(x,y,z)\sin\frac{i\pi x}{a}\sin\frac{j\pi y}{b}\sin\frac{k\pi z}{c}\,dx\,dy\,dz. \tag{2.138}$$

Thus the series expansion solution to the problem is in Eq. (2.137) with A_{mnp} given by Eq. (2.138).

2.7.2 Poisson's Equation in a Sphere

The problem to be solved is described as follows:

$$\nabla^2 V = \frac{\partial^2 V}{\partial r^2} + \frac{2}{r}\frac{\partial V}{\partial r} + \frac{1}{r^2 \sin\theta}\frac{\partial}{\partial \theta}\left(\sin\theta \frac{\partial V}{\partial \theta}\right) + \frac{1}{r^2 \sin^2\theta}\frac{\partial^2 V}{\partial \phi^2}$$

$$= -f(r,\theta,\phi), \qquad 0 < r < a \tag{2.139a}$$

$$V(a,\theta,\phi) = 0. \tag{2.139b}$$

We may follow the same procedure taken to solve Eq. (2.135). However, the approach we will take here will be slightly different from that. (On a close examination, the two approaches are essentially the same.) We expand both V and f in Fourier series:

$$V(r,\theta,\phi) = \sum_{n=0}^{\infty}\sum_{m=0}^{\infty}\left[A_{nm}(r)\cos m\phi + B_{nm}(r)\sin m\phi\right]P_n^m(\cos\theta), \tag{2.140}$$

$$f(r,\theta,\phi) = \sum_{n=0}^{\infty}\sum_{m=0}^{\infty}\left[C_{nm}(r)\cos m\phi + D_{nm}(r)\sin m\phi\right]P_n^m(\cos\theta). \tag{2.141}$$

From the orthogonality properties of $P_n^m(\cos\theta)$, $\cos m\phi$, and $\sin m\phi$, the expansion coefficients are obtained as

$$A_{nm}(r) = K_{nm} \int_0^{2\pi}\int_0^{\pi} V(r,\theta,\phi)P_n^m(\cos\theta)\cos m\phi \sin\theta\, d\theta d\phi, \quad (2.142a)$$

$$B_{nm}(r) = K_{nm} \int_0^{2\pi}\int_0^{\pi} V(r,\theta,\phi)P_n^m(\cos\theta)\sin m\phi \sin\theta\, d\theta d\phi, \quad (2.142b)$$

$$C_{nm}(r) = K_{nm} \int_0^{2\pi}\int_0^{\pi} f(r,\theta,\phi)P_n^m(\cos\theta)\cos m\phi \sin\theta\, d\theta d\phi, \quad (2.142c)$$

$$D_{nm}(r) = K_{nm} \int_0^{2\pi}\int_0^{\pi} f(r,\theta,\phi)P_n^m(\cos\theta)\sin m\phi \sin\theta\, d\theta d\phi, \quad (2.142d)$$

where

$$K_{nm} = \frac{(2n+1)(n-m)!}{2\pi(n+m)!}. \quad (2.143)$$

Substituting Eqs. (2.140) and (2.141) into Eq. (2.139) and integrating by parts leads to

$$A''_{nm} + \frac{2}{r}A'_{nm} - \frac{n(n+1)}{r^2}A_{nm} = -C_{nm}, \quad (2.144a)$$

$$B''_{nm} + \frac{2}{r}B'_{nm} - \frac{n(n+1)}{r^2}B_{nm} = -D_{nm}, \quad (2.144b)$$

$$A_{nm}(a) = 0 = B_{nm}(a), \quad (2.144c)$$

$$A_{nm}(r), B_{nm}(r) \quad \text{are bounded}. \quad (2.144d)$$

The solutions to Eq. (2.144) are [2]

$$A_{nm}(r) = \int_0^a G_n(r,\xi)C_{nm}(\xi)\xi^2\, d\xi \quad (2.145a)$$

$$B_{nm}(r) = \int_0^a G_n(r,\xi)D_{nm}(\xi)\xi^2\, d\xi \quad (2.145b)$$

where

$$G_n(r,\xi) = \begin{cases} \dfrac{1}{(2n+1)a}(\xi/a)^n\left[(r/a)^{-n-1}-(r/a)^n\right], & \xi < r \\ \dfrac{1}{(2n+1)a}(r/a)^n\left[(\xi/a)^{-n-1}-(\xi/a)^n\right], & \xi \geq r \end{cases} \quad (2.145c)$$

Thus the solution of Eq. (2.139) is of the form

$$V(r,\theta,\phi) = \sum_{n=0}^{\infty}\left[\frac{1}{2}A_{n0}(r)P_n(\cos\theta)\right.$$
$$+ \sum_{m=1}^{\infty}\{A_{nm}(r)\cos m\phi$$
$$\left. + B_{nm}(r)\sin m\phi\}P_n^m(\cos\theta)\right] \quad (2.146)$$

where A_{nm} and B_{nm} are given by Eq. (2.145).

2.7.3 Strip Transmission Line

Consider a strip conductor enclosed in a shielded box containing homogeneous medium as shown in Fig. 2.12(a). If TEM mode of propagation is assumed, our problem is reduced to finding V satisfying Laplace's equation $\nabla^2 V = 0$. Due to symmetry, we need only consider one quarter-section of the line as in Fig. 2.12(b). This quadrant can be subdivided into regions 1 and 2, where region 1 is under the center conductor and region 2 is not. We now seek solutions V_1 and V_2 for regions 1 and 2, respectively.

If $w \gg b$, region 1 is similar to parallel-plate problem. Thus, we have a one-dimensional problem similar to Eq. (2.14) with solution

$$V_1 = a_1 y + a_2.$$

Since $V_1(y=0) = 0$ and $V_2(y=-b/2) = V_o$, $a_2 = 0$, $a_1 = -2V_o/b$. Hence

$$V_1(x,y) = \frac{-2V_o}{b}y. \quad (2.147)$$

For region 2, the series expansion solution is of the form

$$V_2(x,y) = \sum_{n=1,3,5}^{\infty} A_n \sin\frac{n\pi y}{b}\sinh\frac{n\pi}{b}(a/2-x), \quad (2.148)$$

Analytical Methods 91

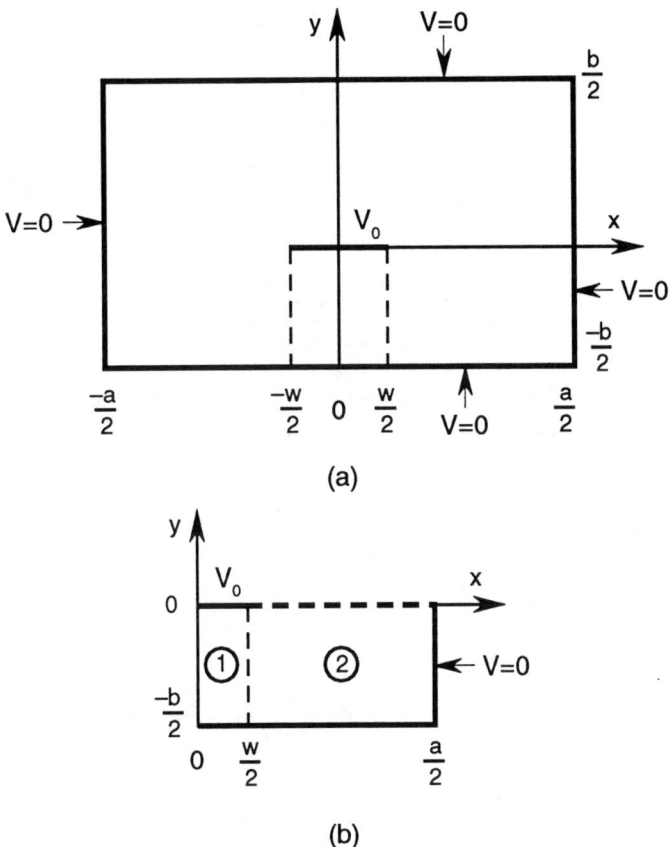

Figure 2.12 Strip line example.

which satisfies Laplace's equation and the boundary condition along the box. Notice that the even-numbered terms could not be included because they do not satisfy the boundary condition requirements about line $y = 0$, i.e., $E_y(y = 0) = -\partial V_2/\partial y\big|_{y=0} \neq 0$. To determine the expansion coefficients A_n in Eq. (2.148), we utilize the fact that V must be continuous at the interface $x = w/2$ between regions 1 and 2, i.e.,

$$V_1(x = w/2, y) = V_2(x = w/2, y)$$

or

92   Numerical Techniques in Electromagnetics

$$-\frac{2V_o y}{b} = \sum_{n=\text{odd}}^{\infty} A_n \sin\frac{n\pi y}{b} \sinh\frac{n\pi}{2b}(a-w),$$

which is Fourier series. Thus,

$$A_n \sinh\frac{n\pi}{2b}(a-w) = -\frac{2}{b}\int_{-b/2}^{b/2} \frac{2V_o y}{b}\sin\frac{n\pi y}{b}\,dy = -\frac{8V_o \sin\frac{n\pi}{2}}{n^2\pi^2}.$$

Hence

$$A_n = -\frac{8V_o \sin\frac{n\pi}{2}}{n^2\pi^2 \sinh\frac{n\pi}{2b}(a-w)}. \tag{2.149}$$

It is instructive to find the capacitance C of the strip line using the fact that the energy stored per length is related to C according to

$$W = \frac{1}{2}CV_o^2 \tag{2.150}$$

where

$$W = \frac{1}{2}\int \mathbf{D}\cdot\mathbf{E}\,dS = \frac{1}{2}\epsilon\int |\mathbf{E}|^2\,dv. \tag{2.151}$$

For region 1,

$$\mathbf{E} = -\nabla V = -\frac{\partial V}{\partial x}\mathbf{a}_x - \frac{\partial V}{\partial y}\mathbf{a}_y = \frac{2V_o}{b}\mathbf{a}_y.$$

Hence

$$W_1 = \frac{1}{2}\epsilon\int_{x=0}^{w/2}\int_{y=-b/2}^{0} \frac{4V_o^2}{b^2}\,dy dx = \frac{\epsilon V_o^2 w}{2b}. \tag{2.152}$$

For region 2,

$$E_x = -\frac{\partial V}{\partial x} = \sum \frac{n\pi}{b} A_n \cosh\frac{n\pi}{b}(a/2-x)\sin\frac{n\pi y}{b}$$

$$E_y = -\frac{\partial V}{\partial y} = -\sum \frac{n\pi}{b} A_n \sinh\frac{n\pi}{b}(a/2-x)\cos\frac{n\pi y}{b}$$

and

$$W_2 = \frac{1}{2}\epsilon \iint (E_x^2 + E_y^2)\, dx\, dy$$

$$= \frac{1}{2}\epsilon \int_{y=-b/2}^{0} \int_{x=w/2}^{a/2} \sum_n \sum_m \frac{mn\pi^2}{b^2} A_n A_m \cdot$$

$$\left[\sinh^2 \frac{m\pi}{b}(a/2 - x) \sinh^2 \frac{n\pi}{b}(a/2 - x) \cos \frac{m\pi y}{b} \cos \frac{n\pi y}{b}\right.$$

$$\left. + \cosh^2 \frac{m\pi}{b}(a/2 - x) \cosh^2 \frac{n\pi}{b}(a/2 - x) \sin \frac{m\pi y}{b} \sin \frac{n\pi y}{b}\right] dx\, dy$$

where the double summation is used to show that we are multiplying two series which may have different indices m and n. Due to the orthogonality properties of sine and cosine functions, all terms vanish except when $m = n$. Thus

$$W_2 = \frac{1}{2}\epsilon \sum_{n=\text{odd}}^{\infty} \frac{n^2\pi^2 A_n^2}{b^2} \cdot \frac{b/2}{2} \int_{w/2}^{a/2} \left[\sinh^2 \frac{n\pi}{b}(a/2-x)\right.$$

$$\left. + \cosh^2 \frac{n\pi}{b}(a/2-x)\right] dx$$

$$= \frac{1}{2}\epsilon \sum_{n=\text{odd}}^{\infty} \frac{n^2\pi^2 A_n^2}{4b} \frac{b}{n\pi} \cosh \frac{n\pi}{2b}(a-w) \sinh \frac{n\pi}{2b}(a-w).$$

Substituting for A_n gives

$$W_2 = \sum_{n=1,3,5}^{\infty} \frac{8\epsilon V_o^2}{n^3 \pi^3} \coth \frac{n\pi}{2b}(a-w). \qquad (2.153)$$

The total energy in the four quadrants is

$$W = 4(W_1 + W_2).$$

Thus

$$C = \frac{2W}{V_o^2} = \frac{8}{V_o^2}(W_1 + W_2)$$

$$= \epsilon \left[\frac{4w}{b} + \frac{64}{\pi^3} \sum_{n=1,3,5}^{\infty} \frac{1}{n^3} \coth \frac{n\pi}{2b}(a-w)\right]. \qquad (2.154)$$

The characteristic impedance of the lossless line is given by

$$Z_o = \frac{\sqrt{\mu\epsilon}}{C} = \frac{\sqrt{\mu_r \epsilon_r}}{cC} = \sqrt{\frac{\mu}{\epsilon}}\frac{1}{C/\epsilon}$$

or

$$Z_o = \frac{120\pi}{\sqrt{\epsilon_r}\left[\frac{4w}{b} + \frac{64}{\pi^3}\sum_{n=1,3,5}^{\infty}\frac{1}{n^3}\coth\frac{n\pi}{2b}(a-w)\right]} \qquad (2.155)$$

where $c = 3 \times 10^8$ m/s, the speed of light in vacuum, and $\mu_r = 1$ is assumed.

Example 2.12

Solve the two-dimensional problem

$$\nabla^2 V = -\frac{\rho_s}{\epsilon_o}$$

where

$$\rho_s = x(y-1) \text{ nC/m}^2$$

subject to

$$V(x,0) = 0, \quad V(x,b) = V_o, \quad V(0,y) = 0 = V(a,y).$$

Solution

If we let

$$\nabla^2 V_1 = 0, \qquad (2.12.1a)$$

subject to

$$V_1(x,0) = 0, \quad V_1(x,b) = V_o, \quad V_1(0,y) = 0 = V(a,y) \qquad (2.12.1b)$$

and

$$\nabla^2 V_2 = -\frac{\rho_s}{\epsilon_o}, \qquad (2.12.2a)$$

subject to

$$V_2(x,0) = 0, \quad V_2(x,b) = 0, \quad V_2(0,y) = 0 = V(a,y). \qquad (2.12.2b)$$

By the superposition principle, the solution to the given problem is

$$V = V_1 + V_2. \qquad (2.12.3)$$

The solution to Eq. (2.12.1) is already found in Section 2.3.1, i.e.,

$$V_1(x,y) = \frac{4V_o}{\pi} \sum_{n=1,3,5}^{\infty} \frac{\sin\frac{n\pi x}{a} \sinh\frac{n\pi y}{a}}{n \sinh\frac{n\pi b}{a}}. \qquad (2.12.4)$$

The solution to Eq. (2.12.2) is a special case of that of Eq. (2.137). The only difference between this problem and that of Eqs. (2.135) and (2.136) is that this problem is two-dimensional while that of Eqs. (2.135) and (2.136) is three-dimensional. Hence

$$V_2(x,y) = \sum_{m=1}^{\infty}\sum_{n=1}^{\infty} A_{mn} \sin\frac{n\pi x}{a} \sin\frac{n\pi y}{b} \qquad (2.12.5)$$

where, according to Eq. (2.138), A_{mn} is given by

$$A_{mn} = \frac{4}{ab}\left[(m\pi/a)^2 + (n\pi/b)^2\right]^{-1} \cdot \int_0^b \int_0^a f(x,y) \sin\frac{n\pi x}{a} \sin\frac{n\pi y}{b}\, dx\, dy. \qquad (2.12.6)$$

But $f(x,y) = x(y-1)/\epsilon_0$ nC/m^2,

$$\int_0^b \int_0^a f(x,y) \sin\frac{n\pi x}{a} \sin\frac{n\pi y}{a}\, dx\, dy$$

$$= \frac{10^{-9}}{\epsilon_0} \int_0^a x \sin\frac{n\pi x}{a}\, dx \int_0^b (y-1) \sin\frac{n\pi y}{b}\, dy$$

$$= \frac{10^{-9}}{10^{-9}/36\pi}\left(-\frac{a^2 \cos m\pi}{m\pi}\right)\left(-\frac{b^2 \cos n\pi}{n\pi} + \frac{b}{n\pi}[\cos n\pi - 1]\right)$$

$$= \frac{36\pi(-1)^{m+n}a^2b^2}{mn\pi^2}\left(1 - \frac{1}{b}[1-(-1)^n]\right) \qquad (2.12.7)$$

since $\cos n\pi = (-1)^n$. Substitution of Eq. (2.12.7) into Eq. (2.12.6) leads to

$$A_{mn} = \left[(m\pi/a)^2 + (n\pi/b)^2\right]^{-1} \frac{(-1)^{m+n}\, 144ab}{mn\pi}\left(1 - \frac{1}{b}[1-(-1)^n]\right). \qquad (2.12.8)$$

Substituting Eqs. (2.12.4) and (2.12.5) into Eq. (2.12.3) gives the complete solution as

$$V(x,y) = \frac{4V_o}{\pi} \sum_{n=1,3,5}^{\infty} \frac{\sin\frac{n\pi x}{a} \sinh\frac{n\pi y}{a}}{n \sinh\frac{n\pi b}{a}} + \sum_{m=1}^{\infty}\sum_{n=1}^{\infty} A_{mn} \sin\frac{n\pi x}{a} \sin\frac{n\pi y}{b}$$
(2.12.9)

where A_{mn} is in Eq. (2.12.8).

2.8 Practical Applications

The scattering of EM waves by a dielectric sphere, known as the Mie scattering problem due to its first investigator in 1908, is an important problem whose analytic solution is usually referred to in assessing some numerical computations. Though the analysis of the problem is more rigorous, the procedure is similar to that of Example 2.5, where scattering due to a conducting cylinder was treated. Our treatment here will be brief; for an in-depth treatment, consult Stratton [12].

2.8.1 Scattering by Dielectric Sphere

Consider a dielectric sphere illuminated by a plane wave propagating in the z direction and **E** polarized in the x direction as shown in Fig. 2.13. The incident wave is described by

$$\mathbf{E}^i = E_o e^{j(\omega t - kz)} \mathbf{a}_x \quad (2.156a)$$

$$\mathbf{H}^i = \frac{E_o}{\eta} e^{j(\omega t - kz)} \mathbf{a}_y. \quad (2.156b)$$

The first step is to express this incident wave in terms of spherical wave functions as in Example 2.8. Since

$$\mathbf{a}_x = \sin\theta\cos\phi\,\mathbf{a}_r + \cos\theta\cos\phi\,\mathbf{a}_\theta - \sin\phi\,\mathbf{a}_\phi,$$

the r-component of $\mathbf{E}^i$, for example, is

$$E_r^i = \cos\phi\sin\theta E_x^i = E_o e^{j\omega t} \frac{\cos\phi}{jkr} \frac{\partial}{\partial\theta}\left(e^{-jkr\cos\theta}\right).$$

Introducing Eq. (2.8.5),

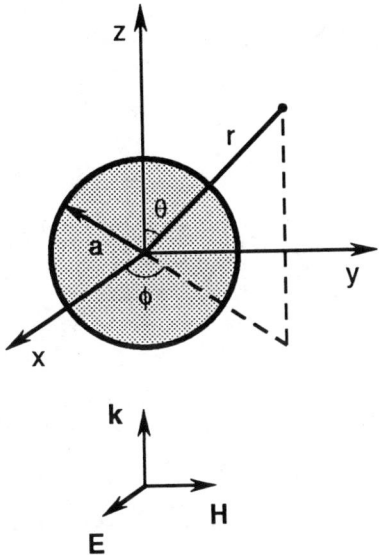

Figure 2.13 Incident EM plane wave on a dielectric sphere.

$$E_r^i = E_o e^{j\omega t} \frac{\cos\phi}{jkr} \sum_{n=0}^{\infty} (-j)^n (2n+1) j_n(kr) \frac{\partial}{\partial \theta} P_n(\cos\theta).$$

But

$$\frac{\partial P_n}{\partial \theta} = P_n^1.$$

hence

$$E_r^i = E_o e^{j\omega t} \frac{\cos\phi}{jkr} \sum_{n=1}^{\infty} (-j)^n (2n+1) j_n(kr) P_n^1(\cos\theta) \qquad (2.157)$$

where the $n=0$ term has been dropped since $P_0^1 = 0$. The same steps can be taken to express E_θ^i and E_ϕ^i in terms of the spherical wave functions. The result is

98 Numerical Techniques in Electromagnetics

$$\mathbf{E}^i = \mathbf{a}_x E_o e^{j(\omega t - kz)}$$
$$= E_o e^{j\omega t} \sum_{n=1}^{\infty} (-j)^n \frac{2n+1}{n(n+1)} \left[\mathbf{M}_n^{(1)}(k) + j\mathbf{N}_n^{(1)}(k) \right] \tag{2.158a}$$

$$\mathbf{H}^i = \mathbf{a}_y H_o e^{j(\omega t - kz)}$$
$$= -\frac{kE_o}{\mu\omega} e^{j\omega t} \sum_{n=1}^{\infty} (-j)^n \frac{2n+1}{n(n+1)} \left[\mathbf{M}_n^{(1)}(k) - j\mathbf{N}_n^{(1)}(k) \right] \tag{2.158b}$$

where

$$\mathbf{M}_n(k) = \frac{1}{\sin\theta} z_n(kr) P_n^1(\cos\theta) \cos\phi \, \mathbf{a}_\theta$$
$$- z_n(kr) \frac{\partial P_n^1(\cos\theta)}{\partial\theta} \sin\phi \, \mathbf{a}_\phi \tag{2.159}$$

$$\mathbf{N}_n(k) = \frac{n(n+1)}{kr} z_n(kr) P_n^1(\cos\theta) \cos\phi \, \mathbf{a}_r$$
$$+ \frac{1}{kr} \frac{\partial}{\partial r} [z_n(kr)] \frac{\partial P_n^1(\cos\theta)}{\partial\theta} \cos\phi \, \mathbf{a}_\theta$$
$$+ \frac{1}{kr\sin\theta} \frac{\partial}{\partial r} [z_n(kr)] P_n^1(\cos\theta) \sin\phi \, \mathbf{a}_\phi. \tag{2.160}$$

The superscript (1) on the spherical vector functions **M** and **N** in Eq. (2.158) denotes that these functions are constructed with spherical Bessel function of the first kind; i.e., $z_n(kr)$ in Eqs. (2.159) and (2.160) is replaced by $j_n(kr)$ when **M** and **N** are substituted in Eq. (2.158).

The induced secondary field consists of two parts. One part applies to the interior of the sphere and is referred to as the transmitted field, while the other applies to the exterior of the sphere and is called the scattered field. Thus the total field outside the sphere is the sum of the incident and scattered fields. We now construct these fields in a fashion similar to that of the incident field. For the scattered field, we let

$$\mathbf{E}^s = E_o e^{j\omega t} \sum_{n=1}^{\infty} (-j)^n \frac{2n+1}{n(n+1)} \left[a_n \mathbf{M}_n^{(4)}(k) + j b_n \mathbf{N}_n^{(4)}(k) \right] \qquad (2.161a)$$

$$\mathbf{H}^s = -\frac{kE_o}{\mu\omega} e^{j\omega t} \sum_{n=1}^{\infty} (-j)^n \frac{2n+1}{n(n+1)} \left[a_n \mathbf{M}_n^{(4)}(k) - j b_n \mathbf{N}_n^{(4)}(k) \right] \qquad (2.161b)$$

where a_n and b_n are expansion coefficients and the superscript (4) on $\mathbf{M}$ and $\mathbf{N}$ shows that these functions are constructed with spherical Bessel function of the fourth kind (or Hankel function of the second kind); i.e., $z_n(kr)$ in Eqs. (2.159) and (2.160) is replaced by $h_n^{(2)}(kr)$ when $\mathbf{M}$ and $\mathbf{N}$ are substituted in Eq. (2.161). The spherical Hankel function has been chosen to satisfy the radiation condition. In other words, the asymptotic behavior of $h_n^{(2)}(kr)$, namely,

$$h_n^{(2)}(kr) \sim j^{n+1} \frac{e^{-kr}}{kr}, \qquad (2.162)$$

when combined with the time factor $e^{j\omega t}$, represents an outgoing spherical wave (see Eq. (2.82d)). Similarly, the transmitted field inside the sphere can be constructed as

$$\mathbf{E}^t = E_o e^{j\omega t} \sum_{n=1}^{\infty} (-j)^n \frac{2n+1}{n(n+1)} \left[c_n \mathbf{M}_n^{(1)}(k_1) + j d_n \mathbf{N}_n^{(1)}(k_1) \right] \qquad (2.163a)$$

$$\mathbf{H}^t = -\frac{kE_o}{\mu\omega} e^{j\omega t} \sum_{n=1}^{\infty} (-j)^n \frac{2n+1}{n(n+1)} \left[c_n \mathbf{M}_n^{(1)}(k_1) - j d_n \mathbf{N}_n^{(1)}(k_1) \right] \qquad (2.163b)$$

where c_n and d_n are expansion coefficients, k_1 is the propagation constant in the sphere. The functions $M_n^{(1)}$ and $N_n^{(1)}$ in Eq. (2.163) are obtained by replacing $z_n(kr)$ in Eq. (2.160) by $j_n(k_1 r)$; j_n is the only solution in this case since the field must be finite at the origin, the center of the sphere.

The unknown expansion coefficients a_n, b_n, c_n, and d_n are determined by letting the fields satisfy the boundary conditions, namely, the continuity of the tangential components of the total electric and magnetic fields at surface of the sphere. Thus at $r = a$,

$$\mathbf{a}_r \times (\mathbf{E}^i + \mathbf{E}^s - \mathbf{E}^t) = 0 \qquad (2.164a)$$
$$\mathbf{a}_r \times (\mathbf{H}^i + \mathbf{H}^s - \mathbf{H}^t) = 0. \qquad (2.164b)$$

This is equivalent to

$$E_\theta^i + E_\theta^s = E_\theta^t, \quad r = a \qquad (2.165a)$$
$$E_\phi^i + E_\phi^s = E_\phi^t, \quad r = a \qquad (2.165b)$$
$$H_\theta^i + H_\theta^s = H_\theta^t, \quad r = a \qquad (2.165c)$$
$$H_\phi^i + H_\phi^s = H_\phi^t, \quad r = a. \qquad (2.165d)$$

Substituting Eqs. (2.158), (2.161), and (2.163) into Eq. (2.165), multiplying the resulting equations by $\cos\phi$ or $\sin\phi$ and integrating over $0 \leq \phi < 2\pi$, and then multiplying by $\dfrac{dP_m^1}{d\theta}$ or $\dfrac{dP_m^1}{\sin\theta}$ and integrating over $0 \leq \theta \leq \pi$, we obtain

$$j_n(ka) + a_n h_n^{(2)}(ka) = c_n j_n(k_1 a) \qquad (2.166a)$$

$$\mu_1[kaj_n(ka)]' + a_n\mu_1[kah_n^{(2)}(ka)]' = c_n\mu[k_1 a j_n(k_1 a)]' \qquad (2.166b)$$

$$\mu_1 j_n(ka) + b_n \mu_1 h_n^{(2)}(ka) = d_n \mu j_n(k_1 a) \qquad (2.166c)$$

$$k[kaj_n(ka)]' + b_n k[kah_n^{(2)}(ka)]' = d_n k_1 [k_1 a j_n(k_1 a)]'. \qquad (2.166d)$$

Solving Eqs. (2.166a) and (2.166b) gives a_n and c_n, while solving Eqs. (2.166c) and (2.166d) gives b_n and d_n. Thus for $\mu = \mu_o = \mu_1$,

$$a_n = \frac{j_n(m\alpha)[\alpha j_n(\alpha)]' - j_n(\alpha)[m\alpha j_n(m\alpha)]'}{j_n(m\alpha)[\alpha h_n^{(2)}(\alpha)]' - h_n^{(2)}(\alpha)[m\alpha j_n(m\alpha)]'} \qquad (2.167a)$$

$$b_n = \frac{j_n(\alpha)[m\alpha j_n(m\alpha)]' - m^2 j_n(m\alpha)[\alpha j_n(\alpha)]'}{h_n^{(2)}(\alpha)[m\alpha j_n(m\alpha)]' - m^2 j_n(m\alpha)[\alpha h_n^{(2)}(\alpha)]'} \qquad (2.167b)$$

$$c_n = -\frac{j/\alpha}{h_n^{(2)}(\alpha)[m\alpha j_n(m\alpha)]' - j_n(m\alpha)[\alpha h_n^{(2)}(\alpha)]'} \qquad (2.167c)$$

$$d_n = -\frac{j/\alpha}{h_n^{(2)}(\alpha)[m\alpha j_n(m\alpha)]' - m^2 j_n(m\alpha)[\alpha h_n^{(2)}(\alpha)]'} \qquad (2.167d)$$

Analytical Methods 101

where $\alpha = ka = 2\pi a/\lambda$ and $m = k_1/k$ is the refractive index of the dielectric, which may be real or complex depending on whether the dielectric is lossless or lossy. The primes at the square brackets indicate differentiation with respect to the argument of the Bessel function inside the brackets, i.e., $[xz_n(x)]' = \frac{\partial}{\partial x}[xz_n(x)]$. To obtain Eqs. (2.167c) and (2.167d), we have made use of the Wronskian relationship

$$j_n(x)[xh_n^{(2)}(x)]' - h_n^{(2)}(x)[xj_n(x)]' = -j/x. \tag{2.168}$$

If the dielectric is lossy and its surrounding medium is free space,

$$k_1^2 = \omega\mu_o(\omega\epsilon_1 - j\sigma), \quad k^2 = \omega^2\mu_o\epsilon_o \tag{2.169}$$

so that the (complex) refractive index m becomes

$$m = \frac{k_1}{k} = \sqrt{\epsilon_c} = \sqrt{\epsilon_{r1} - j\frac{\sigma_1}{\omega\epsilon_o}} = m' - jm''. \tag{2.170}$$

The problem of scattering by a conducting sphere can be obtained as a special case of the problem considered above. Since the EM fields must vanish inside the conducting sphere, the right-hand sides of Eqs. (2.165a), (2.165b), (2.166a), and (2.166d) must be equal to zero so that ($c_n = 0 = d_n$)

$$a_n = -\frac{j_n(\alpha)}{h_n^{(2)}(\alpha)} \tag{2.171a}$$

$$b_n = -\frac{[\alpha j_n(\alpha)]'}{[\alpha h_n^{(2)}(\alpha)]'}. \tag{2.171b}$$

Thus we have completed the Mie solution; the field at any point inside or outside the sphere can now be determined. We will now apply the solution to problems of practical interest.

2.8.2 Scattering Cross Sections

Often scattered radiation is most conveniently measured by the scattering cross section Q_{sca} (in meter^2) which may be defined as the ratio of the total energy scattered per second W_s to the energy density P of the incident wave, i.e.,

$$Q_{\text{sca}} = \frac{W_s}{P}. \tag{2.172}$$

The energy density of the incident wave is given by

$$P = \frac{E_o^2}{2\eta} = \frac{1}{2}E_o^2\sqrt{\frac{\epsilon}{\mu}}. \tag{2.173}$$

102 Numerical Techniques in Electromagnetics

The scattered energy from the sphere is

$$W_s = \frac{1}{2}\text{Re} \int_0^{2\pi} \int_0^{\pi} [E_\theta H_\phi^* - E_\phi H_\theta^*] r^2 \sin\theta \, d\theta d\phi$$

where the star sign denotes complex conjugation and field components are evaluated at far field ($r \gg a$). By using the asymptotic expressions for spherical Bessel functions, we can write the resulting field components as

$$E_\theta^s = \eta H_\phi^s = -\frac{j}{kr} E_o e^{j(\omega t - kr)} \cos\phi \, S_2(\theta) \quad (2.174a)$$

$$-E_\phi^s = \eta H_\theta^s = -\frac{j}{kr} E_o e^{j(\omega t - kr)} \sin\phi \, S_1(\theta) \quad (2.174b)$$

where the amplitude functions $S_1(\theta)$ and $S_2(\theta)$ are given by [13]

$$S_1(\theta) = \sum_{n=1}^{\infty} \frac{2n+1}{n(n+1)} \left(\frac{a_n}{\sin\theta} P_n^1(\cos\theta) + b_n \frac{dP_n^1(\cos\theta)}{d\theta} \right) \quad (2.175a)$$

$$S_2(\theta) = \sum_{n=1}^{\infty} \frac{2n+1}{n(n+1)} \left(\frac{b_n}{\sin\theta} P_n^1(\cos\theta) + a_n \frac{dP_n^1(\cos\theta)}{d\theta} \right). \quad (2.175b)$$

Thus,

$$W_s = \frac{\pi E_o^2}{2k^2 \eta} \text{Re} \int_0^{\pi} \left(|S_1(\theta)|^2 + |S_2(\theta)|^2 \right) \sin\theta \, d\theta.$$

This is evaluated with the help of the identities [12]

$$\int_0^{\pi} \left(\frac{dP_n^1}{d\theta} \frac{dP_m^1}{d\theta} + \frac{1}{\sin^2\theta} P_n^1 P_m^1 \right) \sin\theta \, d\theta =$$

$$\begin{cases} 0, & n \neq m \\ \dfrac{2}{2n+1} \dfrac{(n+1)!}{(n-1)!} n(n+1), & n = m \end{cases}$$

and

$$\int_0^{\pi} \left(\frac{dP_m^1}{\sin\theta} \frac{dP_n^1}{d\theta} + \frac{P_n^1}{\sin\theta} \frac{P_m^1}{d\theta} \right) \sin\theta \, d\theta = 0.$$

We obtain

$$W_s = \frac{\pi E_o^2}{k^2 \eta} \sum_{n=1}^{\infty} (2n+1)(|a_n|^2 + |b_n|^2). \qquad (2.176)$$

Substituting Eqs. (2.173) and (2.176) into Eq. (2.172), the scattering cross section is found to be

$$Q_{\text{sca}} = \frac{2\pi}{k^2} \sum_{n=1}^{\infty} (2n+1)(|a_n|^2 + |b_n|^2). \qquad (2.177)$$

Similarly, the *cross section for extinction* Q_{ext} (in meter^2) is obtained [13] from the amplitude functions for $\theta = 0$, i.e.,

$$Q_{\text{ext}} = \frac{4\pi}{k^2} \text{ Re } S(0)$$

or

$$Q_{\text{ext}} = \frac{2\pi}{k^2} \text{ Re } \sum_{n=1}^{\infty} (2n+1)(a_n + b_n) \qquad (2.178)$$

where

$$S(0) = S_1(0^\circ) = S_2(0^\circ) = \frac{1}{2} \sum_{n=1}^{\infty} (2n+1)(a_n + b_n). \qquad (2.179)$$

In obtaining Eq. (2.179), we have made use of

$$\left.\frac{P_n^1}{\sin\theta}\right|_{\theta=0} = \left.\frac{dP_n^1}{d\theta}\right|_{\theta=0} = n(n+1)/2.$$

If the sphere is absorbing, the *absorption cross section* Q_{abs} (in meter^2) is obtained from

$$Q_{\text{abs}} = Q_{\text{ext}} - Q_{\text{sca}} \qquad (2.180)$$

since the energy removed is partly scattered and partly absorbed.

A useful, measurable quantity in radar communications is the *radar cross section* or *back-scattering cross section* σ_b of a scattering obstacle. It is a lump measure of the efficiency of the obstacle in scattering radiation back to the source ($\theta = 180^\circ$). It is defined in terms of the far zone scattered field as

$$\sigma_b = 4\pi r^2 \frac{|\mathbf{E}^s|^2}{E_o^2}, \qquad \theta = \pi. \qquad (2.181)$$

From Eq. (2.174),

$$\sigma_b = \frac{2\pi}{k^2}\left[|S_1(\pi)|^2 + |S_2(\pi)|^2\right].$$

But

$$-S_1(\pi) = S_2(\pi) = \frac{1}{2}\sum_{n=1}^{\infty}(-1)^n(2n+1)(a_n - b_n)$$

where we have used

$$-\left.\frac{P_n^1}{\sin\theta}\right|_{\theta=\pi} = \left.\frac{dP_n^1}{d\theta}\right|_{\theta=\pi} = (-1)^n n(n+1)/2.$$

Thus

$$\sigma_b = \frac{\pi}{k^2}\left|\sum_{n=1}^{\infty}(-1)^n(2n+1)(a_n - b_n)\right|^2. \qquad (2.182)$$

Similarly, we may determine the *forward-scattering cross section* ($\theta = 0^o$) as

$$\sigma_f = \frac{2\pi}{k^2}\left[|S_1(0)|^2 + |S_2(0)|^2\right].$$

Substituting Eq. (2.180) into this yields

$$\sigma_f = \frac{\pi}{k^2}\left|\sum_{n=1}^{\infty}(2n+1)(a_n + b_n)\right|^2. \qquad (2.183)$$

2.8.3 Attenuation Due to Raindrops

The rapid growth in demand for additional communication capacity has put pressure on engineers to develop microwave systems operating at higher frequencies. It turns out, however, that at frequencies above 10 GHz attenuation caused by atmospheric particles can reduce the reliability and performance of radar and space communication links. Such particles include oxygen, ice crystals, rain, fog, and snow. Prediction of the effect of these precipitates on the performance of a system becomes important. In this final subsection, we will examine attenuation and phase shift of an EM wave propagating through rain drops. We will assume that raindrops are spherical so that Mie rigorous solution can be applied. This assumption is valid if the rate intensity is low. For high rain intensity, an oblate spheroidal model would be more realistic [14].

The magnitude of an EM wave traveling through a homogeneous medium (with N identical spherical particles per unit volume) in a distance ℓ is given by $e^{-\gamma \ell}$, where γ is the attenuation coefficient given by [13]

$$\gamma = N Q_{\text{ext}}$$

or

$$\gamma = \frac{N\lambda^2}{\pi} \text{ Re } S(0). \tag{2.184}$$

Thus the wave is attenuated by

$$A = 10 \, \log_{10} \frac{1}{e^{-\gamma \ell}} = \gamma \ell \, 10 \log_{10} e$$

or

$$A = 4.343 \gamma \ell \quad \text{(in dB)}.$$

The attenuation per length (in dB/m) is

$$A = 4.343 \gamma$$

or

$$A = 4.343 \frac{\lambda^2 N}{\pi} \text{ Re } S(0). \tag{2.185}$$

Similarly, it can be shown [13] that the phase shift of the EM wave caused by the medium is

$$\Phi = -\frac{\lambda^2 N}{2\pi} \text{ Im } S(0) \quad \text{(in radians/unit length)}$$

or

$$\Phi = -\frac{\lambda^2 N}{2\pi} \text{ Im } S(0) \frac{180}{\pi} \quad \text{(in deg/m)}. \tag{2.186}$$

To relate attenuation and phase shift to a realistic rainfall rather than identical drops assumed so far, it is necessary to know the drop-size distribution for a given rate intensity. Representative distributions were obtained by Laws and Parsons [15] as shown in Table 2.9. To evaluate the effect of the drop-size distribution, suppose for a particular rain rate R, p is the percent of the total volume of water reaching the ground (as in Table 2.9), which consists of drops whose diameters fall in the interval centered in D cm ($D = 2a$), the number of drops in that interval is given by

$$N_c = p N(D). \tag{2.187}$$

The total attenuation and phase shift over the entire volume become

Table 2.9 Laws and Parsons Drop-size Distributions for Various Rain Rates

Drop diameter (cm)	Rain Rate (mm/hour)								
	0.25	1.25	2.5	5	12.5	25	50	100	150
	Percent of total volume								
0.05	28.0	10.9	7.3	4.7	2.6	1.7	1.2	1.0	1.0
0.1	50.1	37.1	27.8	20.3	11.5	7.6	5.4	4.6	4.1
0.15	18.2	31.3	32.8	31.0	24.5	18.4	12.5	8.8	7.6
0.2	3.0	13.5	19.0	22.2	25.4	23.9	19.9	13.9	11.7
0.25	0.7	4.9	7.9	11.8	17.3	19.9	20.9	17.1	13.9
0.3		1.5	3.3	5.7	10.1	12.8	15.6	18.4	17.7
0.35		0.6	1.1	2.5	4.3	8.2	10.9	15.0	16.1
0.4		0.2	0.6	1.0	2.3	3.5	6.7	9.0	11.9
0.45			0.2	0.5	1.2	2.1	3.3	5.8	7.7
0.5				0.3	0.6	1.1	1.8	3.0	3.6
0.55					0.2	0.5	1.1	1.7	2.2
0.6						0.3	0.5	1.0	1.2
0.65							0.2	0.7	1.0
0.7									0.3

$$A = 0.4343 \frac{\lambda^2}{\pi} \cdot 10^6 \sum pN(D) \text{ Re } S(0) \quad \text{(dB/km)} \tag{2.188}$$

$$\Phi = -\frac{9\lambda^2}{\pi^2} \cdot 10^6 \sum pN(D) \text{ Im } S(0) \quad \text{(deg/km)} \tag{2.189}$$

where λ is the wavelength in cm and $N(D)$ is the number of raindrops with equivolumic diameter D per cm^3. The summations are taken over all drop sizes. In order to relate the attenuation and phase shift to the rain intensity measured in rain rate R (in mm/hour), it is necessary to have a relationship between N and R. The relationship obtained by Best [15], shown in Table 2.10, involves the terminal velocity u (in m/s) of the rain drops, i.e.,

$$R = u \cdot N \cdot \text{ (volume of a drop)}$$
$$= uN \frac{4\pi a^3}{3} \quad \text{(in m/s)}$$

or

$$R = 6\pi N u D^3 \cdot 10^5 \quad \text{(mm/hr)}.$$

Thus

$$N(D) = \frac{R}{6\pi u D^3} 10^{-5}. \tag{2.190}$$

Table 2.10 Raindrop Terminal Velocity

Radius (cm)	Velocity (m/s)
0.025	2.1
0.05	3.9
0.075	5.3
0.10	6.4
0.125	7.3
0.15	7.9
0.175	8.35
0.20	8.7
0.225	9.0
0.25	9.2
0.275	9.35
0.30	9.5
0.325	9.6

Substituting this into Eqs. (2.188) and (2.189) leads to

$$A = 4.343 \frac{\lambda^2}{\pi^2} R \sum \frac{p}{6uD^3} \text{ Re } S(0) \quad \text{(dB/km)} \tag{2.191}$$

$$\Phi = -90 \frac{\lambda^2}{\pi^3} R \sum \frac{p}{6uD^3} \text{ Im } S(0) \quad \text{(deg/km)}, \tag{2.192}$$

where $N(D)$ is in per cm^3, D and λ are in cm, u is in m/s, p is in percent, and $S(0)$ is the complex forward-scattering amplitude defined in Eq. (2.179). The complex refractive index of raindrops [16] at 20°C required in calculating attenuation and phase shift is shown in Table 2.11.

Example 2.13

For ice spheres, plot the normalized back-scattering cross section, $\sigma_b/\pi a^2$, as a function of the normalized circumference, $\alpha = 2\pi a/\lambda$. Assume that the refractive index of ice is independent of wavelength, making the normalized cross section for ice applicable over the entire microwave region. Take $m = 1.78 - j2.4 \times 10^{-3}$ at 0°C.

Table 2.11 Refractive Index of Water at 20°C

Frequency (GHz)	Refractive index ($m = m' - jm''$)
0.6	8.960 - j0.1713
0.8	8.956 - j0.2172
1.0	8.952 - j0.2648
1.6	8.933 - j0.4105
2.0	8.915 - j0.5078
3.0	8.858 - j0.7471
4.0	8.780 - j0.9771
6.0	8.574 - j1.399
11	7.884 - j2.184
16	7.148 - j2.614
20	6.614 - j2.780
30	5.581 - j2.848
40	4.886 - j2.725
60	4.052 - j2.393
80	3.581 - j2.100
100	3.282 - j1.864
160	2.820 - j1.382
200	2.668 - j1.174
300	2.481 - j0.8466

Solution

From Eq. (2.182),

$$\sigma_b = \frac{\pi}{k^2} \left| \sum_{n=1}^{\infty} (-1)^n (2n+1)(a_n - b_n) \right|^2.$$

Since $\alpha = ka$, the normalized back-scattering cross section is

$$\frac{\sigma_b}{\pi a^2} = \frac{1}{\alpha^2} \left| \sum_{n=1}^{\infty} (-1)^n (2n+1)(a_n - b_n) \right|^2. \quad (2.13.1)$$

Using this expression in conjunction with Eq. (2.167), the subroutine SCATTERING in the FORTRAN code of Fig. 2.15 was used as the main

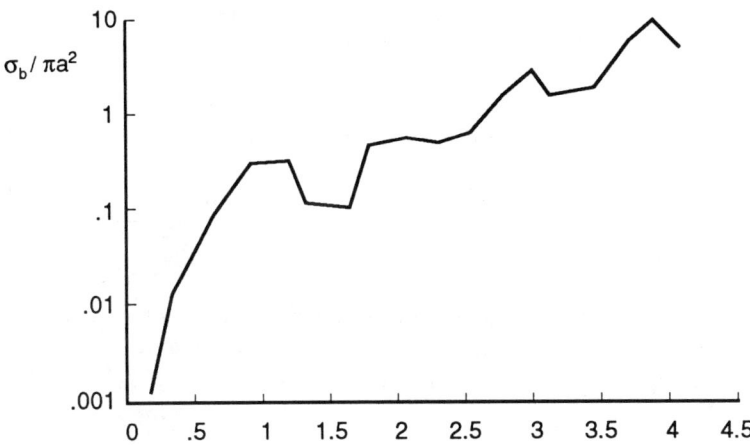

Figure 2.14 Normalized back-scattering (radar) cross sections $\alpha = 2\pi a/\lambda$ for ice at $0°$ C.

program to determine $\sigma_b/\pi a^2$ for $0.2 < \alpha < 4$. Details on the program will be explained in the next example. It suffices to mention that the maximum number of terms of the infinite series in Eq. (2.13.1) was ten. It has been found that truncating the series at $n = 2\alpha$ provides sufficient accuracy [18]. The plot of the normalized radar cross section versus α is shown in Fig. 2.14. From the plot, we note that back-scattering oscillates between very large and small values. If α is increased further, the normalized radar cross section increases rapidly. The unexpectedly large cross sections have been attributed to a lens effect; the ice sphere acts like a lens which focuses the incoming wave on the back side from which it is reflected backwards in a concentrated beam. This is recognized as a crude description, but it at least permits visualization of a physical process which may have some reality.

110   Numerical Techniques in Electromagnetics

Table 2.12   Attenuation and Phase Shift at 11 GHz

Rain rate (mm/hr)	Attenuation (dB/km)	Phase shift (deg/km)
0.25	2.56×10^{-3}	0.4119
1.25	1.702×10^{-3}	1.655
2.5	4.072×10^{-3}	3.040
5.0	9.878×10^{-3}	5.601
12.5	0.3155	12.58
25	0.7513	23.19
50	1.740	42.74
100	3.947	78.59
150	6.189	112.16

Example 2.14

Assuming the Laws and Parsons' rain drop-size distribution, calculate the attenuation in dB/km for rain rates of $0.25, 1.25, 2.5, 5.0, 12.5, 50.0, 100.0,$ and 150.0 mm/hr. Consider the incident microwave frequencies of 6, 11, and 30 GHz.

Solution

The FORTRAN code developed for calculating attenuation and phase shift of microwaves due to rain is shown in Fig. 2.15. The main program calculates attenuation and phase shift for given values of frequency and rain rate by employing Eqs. (2.191) and (2.192). For each frequency, the corresponding value of the refractive index of water at 20°C is taken from Table 2.11. The data in Tables 2.9 and 2.10 on the drop-size distributions and terminal velocity are incorporated in the main program.

Seven subroutines are involved. The subroutine SCATTERING calculates the expansion coefficients a_n, b_n, c_n and d_n using Eq. (2.167) and also the forward-scattering amplitude $S(0)$ using Eq. (2.179). The same subroutine was used as the main program in the previous example to calculate the radar cross section of ice spheres. Enough comments are inserted to make the program self-explanatory. Subroutine BESSEL and BESSELCMPLX are exactly the same except that the former is for real argument, while the latter is for complex argument. They both employ Eq. (2.110) to find $j_n(x)$. Subroutine HANKEL employs Eq. (2.112) to find $y_n(x)$, which involves calling subroutine BESSELN to calculate $j_{-n}(x)$. The derivative of Bessel-Riccati function $[xz_n(x)]$ is obtained from (see Prob. 2.20)

$$[xz_n(x)]' = -nz_n(x) + xz_{n-1}(x)$$

where z_n is j_n, j_{-n}, y_n or $h_n(x)$. Subroutine GAMMA calculates $\Gamma(n + 1/2)$ using Eq. (2.115), while subroutine FACTORIAL determines $n!$. All computations were done in double precision arithmetic, although it was observed that using single precision would only alter results slightly.

Typical results for 11 GHz are tabulated in Table 2.12. A graph of attenuation versus rain rate is portrayed in Fig. 2.16. The plot shows that attenuation increases with rain rate and conforms with the common rule of thumb. We must note that the underlying assumption of spherical raindrops renders the result as only a first order approximation of the practical rainfall situation.

```
0001      C  ************************************************************
0002      C  MAIN PROGRAM
0003      C
0004      C  FOR SPHERICAL RAIN DROPS,
0005      C  THIS PROGRAM CALCULATES ATTENUATION IN dB/KM AND
0006      C  PHASE SHIFT IN DEG/KM FOR A GIVEN RAIN RATE
0007      C
0008      C  R = RAIN RATE IN MM/HR
0009      C  D = DROP DIAMETER IN CM
0010      C  F = FREQUENCY IN GHZ
0011      C  AT = ATTENUATION IN dB/KM
0012      C  PH = PHASE SHIFT IN DEG/KM
0013      C  V = TERMINAL VELOCITY OF RAIN DROPS
0014      C  P = PERCENT OF TOTAL VOLUME AS MEASURED
0015      C  BY LAWS AND PARSONS
0016      C  M = COMPLEX REFRACTIVE INDEX OF WATER AT T = 20 C
0017      C  X = ALPHA = K*A
0018
0019            DIMENSION R(9),P(14,9),V(14)
0020            REAL*8 LAM,MR,MI
0021            COMPLEX*8 M,SO,SOO(15)
0022            DATA V/2.1,3.9,5.3,6.4,7.3,7.9,8.35,8.7,9.0,9.2,
0023         1   9.35,9.5,9.6,9.6/
0024            DATA R/0.25,1.25,2.5,5.0,12.5,25.0,50.0,100.0,150.0/
0025            DATA (( P(I,J),I=1,14),J=1,9) /28.0,50.1,18.2,3.0,0.7,
0026         1     9*0.0,  10.9,37.1,31.3,13.5,4.9,1.5,0.6,0.2,6*0.0,
0027         1    7.3,27.8,32.8,19.0,7.9,3.3,1.1,0.6,0.2,5*0.0,
0028         2    4.7,20.3,31.0,22.2,11.8,5.7,2.5,1.0,0.5,0.3,4*0.0,
0029         3    2.6,11.5,24.5,25.4,17.3,10.1,4.3,2.3,1.2,0.6,
0030         4    0.2,3*0.0,  1.7,7.6,18.4,23.9,19.9,12.8,8.2,3.5,2.1,
0031         5    1.1,0.5,0.3,2*0.0,  1.2,5.4,12.5,19.9,20.9,15.6,
0032         6    10.9,6.7,3.3,1.8,1.1,0.5,0.2,0.0,  1.0,4.6,8.8,
0033         7    13.9,17.1,18.4,15.0,9.0,5.8,3.0,1.7,1.0,0.7,0.0,
0034         8    1.0,4.1,7.6,11.7,13.9,17.7,16.1,11.9,7.7,3.6,2.2,
0035         9    1.2,1.0,0.3 /
0036
0037            F = 30.0
0038            LAM = 30.0/F          ! WAVELENGTH IN CM
0039            MR = 5.581
0040            MI = 2.848
0041            M = CMPLX(MR,-MI)
0042            PIE = 3.141592654
0043            DO 10 I = 1,14
0044            D = 0.05*DFLOAT(I)
0045            X = PIE*D/LAM
0046            CALL SCATTERING(X,M,SO,SB)
0047            SOO(I) = SO
0048      10    CONTINUE
```

```
0049            DO 50 N = 1,9
0050            SUMAT = 0.0
0051            SUMPH = 0.0
0052            WRITE(6,20) R(N),F
0053   20       FORMAT(5X,'RAIN RATE =',F8.4,1X,'MM/HR',4X
0054          1              'FREQUENCY =',F10.5,1X,'GHZ',/)
0055            DO 30 K=1,14
0056            D = 0.05*DFLOAT(K)
0057            X = PIE*D/LAM
0058            SO = SOO(K)
0059   C
0060   C        CALCULATE ATTENUATION AND PHASE SHIFT
0061   C
0062            ACO = REAL(SO)/(6.0*X*X)
0063            PCO = AIMAG(SO)/(12.0*X*X)
0064            ATO = ACO*( P(K,N)/100.0 )/ (V(K)*D )
0065            SUMAT = SUMAT + ATO
0066            PHO = PCO*( P(K,N)/100.0 )/ (V(K)*D )
0067            SUMPH = SUMPH + PHO
0068   30       CONTINUE
0069            AT = 4.343*R(N)*SUMAT
0070            PH = 180.0*R(N)*SUMPH/PIE
0071            PRINT *,R(N),AT,PH
0072            WRITE(6,40) AT,PH
0073   40       FORMAT(3X,'ATTENUATION =',F12.6,1X,'dB/KM',3X,
0074          1              'PHASE SHIFT =',F12.6,1X,'DEG/KM',/)
0075   50     CONTINUE
0076   60       FORMAT(I5,F15.8,F15.8)
0077            STOP
0078            END


0001
0002   C ************************************************************
0003   C        USING MIE'S SOLUTION,
0004   C        THIS SUBROUTINE CALCULATES THE SCATTERING COEFFIENTS
0005   C        AND THE FORWARD SCATTERING FUNCTION
0006   C
0007            SUBROUTINE SCATTERING(X,M,SO,SB)
0008
0009            IMPLICIT COMPLEX*8 (A-D)
0010            COMPLEX*8 M,SO,Y,JC,SUM,SUMB
0011            COMPLEX*8 JM(0:20),JMD(0:20),H(0:20),HD(0:20)
0012            REAL*8 J(0:20),JD(0:20)
0013            DIMENSION A(20),B(20),C(20),D(20)
0014
0015            Y = M*X
0016            JC=(0.0,1.0)
0017            SUM = (0.0,0.0)
0018            SUMB = (0.0,0.0)
0019            NMAX = 10
0020   C
0021   C        FIRST, CALCULATE THE SCATTERING COEFFICIENTS an and bn
0022   C
0023            DO 10 N=1,NMAX
0024            CALL BESSEL(X,N,J,JD)
0025            CALL BESSELCMPLX(Y,N,JM,JMD)
0026            CALL HANKEL(X,N,H,HD)
0027            A1 = JM(N)*JD(N) - J(N)*JMD(N)
0028            A2 = JM(N)*HD(N) - H(N)*JMD(N)
0029            A(N) = A1/A2
0030            B1 = J(N)*JMD(N) - (M**2)*JM(N)*JD(N)
```

```
0031            B2 = H(N)*JMD(N) - (M**2)*JM(N)*HD(N)
0032            B(N) =  B1/B2
0033            C(N) = JC/(X*A2)
0034            D(N) = - JC*M/(X*B2)
0035      C
0036      C     CALCULATE THE FORWARD SCATTERING AMPLITUDE FUNCTION S(0)
0037      C
0038            F = 2.0*FLOAT(N) + 1.0
0039            SUM = SUM + F*( A(N) + B(N) )
0040            SUMB = SUMB + F*((-1.)**N)*( A(N) - B(N) )
0041      10    CONTINUE
0042            S0 = SUM/2.0            ! FORWARD SCATTERING AMPLITUDE FUNCTION
0043            SB=(CABS( SUMB )/X)**2  ! NORMALIZED RADAR CROSS-SECTION
0044            RETURN
0045            END


0001      C ***********************************************************
0002      C     SUBROUTINE FOR SPHERICAL HANKEL FUNCTIONS H AND HD
0003      C     OF REAL ARGUMENT (SERIES EXPANSION METHOD)
0004
0005            SUBROUTINE HANKEL(X,N,H,HD)
0006            IMPLICIT REAL*8(A-G,J,O-Z),COMPLEX*8(H)
0007            DIMENSION J(0:20),JD(0:20),Y(0:20),YD(0:20)
0008            DIMENSION JN(0:20),JND(0:20),H(0:20),HD(0:20)
0009
0010            PIE = 3.141592654
0011            CALL BESSEL(X,N,J,JD)
0012            CALL BESSELN(X,N,JN,JND)
0013            V = DFLOAT(N) + 0.5
0014            Y(N) = ( DCOS(V*PIE)*J(N) - JN(N) )/DSIN(V*PIE)
0015            H(N) = CMPLX( J(N), - Y(N) )       ! HANKEL OF 2ND KIND
0016            YD(N) = ( DCOS(V*PIE)*JD(N) - JND(N) )/DSIN(V*PIE)
0017            HD(N) = CMPLX( JD(N), - YD(N) )
0018            RETURN
0019            END


0001      C ***********************************************************
0002      C     SUBROUTINE FOR SPHERICAL BESSEL FUNCTION J AND JD
0003      C     OF REAL ARGUMENT (SERIES EXPANSION METHOD)
0004      C
0005            SUBROUTINE BESSEL(X,N,J,JD)
0006            IMPLICIT REAL*8(A-H,J,O-Z)
0007            DIMENSION J(0:20),JD(0:20)
0008
0009            TOL = 0.00000001          ! TOLERANCE
0010            PIE = 3.141592654
0011            V = DFLOAT(N) + 0.5
0012            K = 0
0013            SUM1 = 0.0
0014            SUM2 = 0.0
0015      10    PN = V + DFLOAT(K) + 1.0
0016            CALL GAMMA(PN,GN)
0017            CALL FACTORIAL(K,FK)
0018            A = ((-1.)**K)*((.5)**(V+2*K))*(X**(V+2*K-.5))/(GN*FK)
0019            SUM1 = SUM1 + A
0020            B = A*(V + FLOAT(2*K) + 0.5)
0021            SUM2 = SUM2 + B
0022            K = K + 1
0023            IF(ABS(A).GE.TOL) GO TO 10
0024            CONTINUE
```

```
0025            Q = DSQRT(PIE/2.0)
0026            J(N) = Q*SUM1
0027            JD(N) = Q*SUM2
0028            RETURN
0029            END


0001    C ************************************************************
0002    C       SUBROUTINE FOR SPHERICAL BESSEL FUNCTIONS JN AND JND
0003    C       OF REAL ARGUMENT BUT NEGATIVE ORDER (SERIES EXPANSION)
0004    C
0005            SUBROUTINE BESSELN(X,N,JN,JND)
0006            IMPLICIT REAL*8(A-H,J,O-Z)
0007            DIMENSION JN(0:20),JND(0:20)
0008
0009            TOL = 0.00000001          ! TOLERANCE
0010            PIE = 3.141592654
0011            V = DFLOAT(N) + 0.5
0012            K = 0
0013            SUM1 = 0.0
0014            SUM2 = 0.0
0015    10      PN = -V + DFLOAT(K) + 1.0
0016            CALL GAMMA(PN,GN)
0017            CALL FACTORIAL(K,FK)
0018            A = ((-1.)**K)*((.5)**(-V+2*K))*(X**(-V+2*K-.5))/(GN*FK)
0019            SUM1 = SUM1 + A
0020            B = A*(-V + FLOAT(2*K) + 0.5)
0021            SUM2 = SUM2 + B
0022            K = K + 1
0023            IF(ABS(A).GE.TOL) GO TO 10
0024            CONTINUE
0025            Q = DSQRT(PIE/2.0)
0026            JN(N) = Q*SUM1
0027            JND(N) = Q*SUM2
0028            RETURN
0029            END


0001    C ************************************************************
0002    C       SUBROUTINE FOR SPHERICAL BESSEL FUNCTIONS JM AND JMD
0003    C       OF COMPLEX ARGUMENT   (SERIES EXPANSION)
0004    C
0005            SUBROUTINE BESSELCMPLX(Z,N,JM,JMD)
0006            IMPLICIT COMPLEX*8(A-D,J,S,Z),REAL*8(G,P-R,V)
0007            DIMENSION JM(0:20),JMD(0:20)
0008
0009            TOL = 0.001              ! TOLERANCE
0010            PIE = 3.141592654
0011            V = DFLOAT(N) + 0.5
0012            K = 0
0013            SUM1 = (0.0,0.0)
0014            SUM2 = (0.0,0.0)
0015    10      PN = V + DFLOAT(K) + 1.0
0016            CALL GAMMA(PN,GN)
0017            CALL FACTORIAL(K,FK)
0018            A = ((-1.)**K)*((.5)**(V+2*K))*(Z**(V+2*K-.5))/(GN*FK)
0019            SUM1 = SUM1 + A
0020            B = A*(V + FLOAT(2*K) + 0.5)
0021            SUM2 = SUM2 + B
0022            K = K + 1
0023            IF( CABS(A).GE.TOL) GO TO 10
0024            CONTINUE
```

```
0025            Q = DSQRT(PIE/2.0)
0026            JN(N) = Q*SUM1
0027            JND(N) = Q*SUM2
0028            RETURN
0029            END


0001    C ************************************************************
0002    C       SUBROUTINE FOR GAMMA FUNCTION
0003    C
0004            SUBROUTINE GAMMA(V,G)
0005            IMPLICIT REAL*8(A-H,O-Z)
0006
0007            PIE = 3.1415927
0008            IF(V-0.5) 10,20,20
0009    10      N = -V + 0.5
0010            N2 = 2*N
0011            CALL FACTORIAL(N,FN)
0012            CALL FACTORIAL(N2,FN2)
0013            G = ((-4.)**N)*DSQRT(PIE)*FN/FN2
0014            RETURN
0015    20      N = V - 0.5
0016            N2 = 2*N
0017            CALL FACTORIAL(N,FN)
0018            CALL FACTORIAL(N2,FN2)
0019            G = FN2*DSQRT(PIE)/(FN*(2.**N2))
0020            RETURN
0021            END


0001    C ************************************************************
0002    C       SUBROUTINE FOR FACTORIAL OF N, i.e. N!
0003    C
0004            SUBROUTINE FACTORIAL(N,F)
0005            REAL*8 F
0006
0007            F = 1.0
0008            IF(N.EQ.0) GO TO 20
0009            DO 10 I=1,N
0010    10      F = F*DFLOAT(I)
0011    20      RETURN
0012            END
```

Figure 2.15   FORTRAN program for Examples 2.13 and 2.14.

2.9 Concluding Remarks

We have reviewed analytic methods of solving partial differential equations. Analytic solutions are of major interest as test models for comparison with numerical techniques. The emphasis has been on the method of separation of variables that is the most powerful. For an excellent, more in-depth exposition of this method, consult Myint-U [4]. In the course of applying the method of separation of variables, we have encountered some mathematical functions such as Bessel functions and Legendre polynomials. For

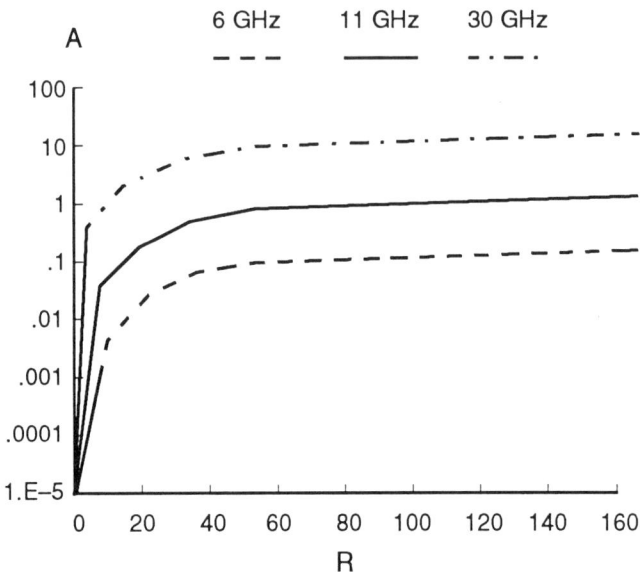

Figure 2.16 Attenuation versus rain rate.

a thorough treatment of these functions and their properties, Bell [7] and Johnson and Johnson [9] are recommended. The mathematical handbook by Abramowitz and Stegun [17], as well as References [25] and [26], provides tabulated values of these functions for specific orders and arguments. A few useful texts on the topics covered in this chapter are also listed in the references.

As an example of real life problems, we have applied the analytical techniques developed in this chapter to the problem of attenuation of microwaves due to spherical raindrops. Spherical models have also been used to assess the absorption characteristics of the human skull exposed to EM plane waves [19-24] (see Probs. 2.40 to 2.43).

We conclude this chapter by remarking that the most satisfactory solution of a field problem is an exact analytical one. In many practical situations, no solution can be obtained by the analytical methods now available, and one must therefore resort to numerical approximation, graphical or experimental solutions. (Experimental solutions are usually very expensive, while graphical solutions are not so accurate.) The remainder of this book will be devoted to a study of the numerical methods commonly used in EM.

References

[1] W. J. Gibbs, *Conformal Transformation in Electrical Engineering*. London: Chapman & Hall, 1958.

[2] H. F. Weinberger, *A First Course in Partial Differential Equations*. New York: John Wiley, 1965, Chap. IV, pp. 63-116.

[3] R. D. Kersten, *Engineering Differential Systems*. New York: McGraw-Hill, 1969, Chap. 5, pp. 66-106.

[4] T. Myint-U, *Partial Differential Equations of Mathematical Physics*. New York: North-Holland, 1980.

[5] M. R. Spiegel, *Mathematical Handbook of Formulas and Tables* (Schaum's Outline Series). New York: McGraw-Hill, 1968.

[6] G. N. Watson, *Theory of Bessel Functions*. London: Cambridge University Press, 1966.

[7] W. W. Bell, *Special Functions for Scientists and Engineers*. London: D. Van Nostrand, 1968.

[8] S. A. Schelkunoff, *Applied Mathematics for Engineers and Scientists*, 2nd ed. New York: D. Van Nostrand, 1965.

[9] D. E. Johnson and J. R. Johnson, *Mathematical Methods in Engineering and Physics*. Englewood Cliffs, NJ: Prentice-Hall, 1982.

[10] L. D. Kovach, *Boundary-value Problems*. Reading, MA: Addison-Wesley, 1984.

[11] P. M. Morse and H. Feshback, *Methods of Theoretical Physics*. New York: McGraw-Hill, Parts I and II, 1953.

[12] J. A. Stratton, *Electromagnetic Theory*. New York: McGraw-Hill, 1941, pp. 394-421, 563-573.

[13] H. C. van de Hulst, *Light Scattering of Small Particles*. New York: John Wiley, 1957, pp. 28-37, 114-136, 284.

[14] J. Morrison and M. J. Cross, "Scattering of a plane electromagnetic wave by axisymmetric raindrops," *Bell Syst. Tech. J.*, vol. 53, no. 6, July-Aug. 1974, pp. 955-1019.

[15] D. E. Setzer, "Computed transmission through rain at microwave and visible frequencies," *Bell Syst. Tech. J.*, vol. 49, no. 8, Oct. 1970, pp. 1873-1892.

[16] M. N. 0. Sadiku, "Refractive index of snow at microwave frequencies," *Appl. Optics*, vol. 24, no. 4, Feb. 1985, pp. 572-575.

[17] M. Abramowitz and I. A. Stegun, *Handbook of Mathematical Functions*. New York: Dover, 1965.

[18] J. J. Stephens, "Radar cross sections for water and ice spheres," *J. Meteor.*, vol. 18, 1961, pp. 348-359.

[19] C. H. Durney, "Electromagnetic dosimetry for models of humans and animals: a review of theoretical and numerical techniques," *Proc. IEEE.*, vol. 68, no. 1, Jan. 1980, pp. 33-40.

[20] M. A. Morgan, "Finite element calculation of microwave absorption by the cranial structure," *IEEE Trans. Bio. Engr.*, vol. BME-28, no. 10, Oct. 1981, pp. 687-695.

[21] J. W. Hand, "Microwave heating patterns in simple tissue models," *Phys. Med. Biol.*, vol. 22, no. 5, 1977, pp. 981-987.

[22] G. H. Wong et al., "Probing electromagnetic fields in lossy spheres and cylinders," *IEEE Trans. Micro. Theo. Tech.*, vol. MTT-32, no. 8, Aug. 1984, pp. 824-828.

[23] W. T. Joines and R. J. Spiegel, "Resonance absorption of microwaves by the human skull," *IEEE Trans. Bio. Engr.*, vol. BME-21, Jan. 1974, pp. 46-48.

[24] C. M. Weil, "Absorption characteristics of multilayered sphere models exposed to UHF/microwave radiation," *IEEE Trans. Bio. Engr.*, vol. BME-22, no. 6, Nov. 1975, pp. 468-476.

[25] National Bureau of Standards, *Tables of Associated Legendre Functions*. New York: Columbia Press, 1945.

[26] National Bureau of Standards, *Tables of Spherical Bessel Functions*, vol. 1. New York: Columbia Press, 1947.

Problems

2.1 Consider the PDE

$$a\Phi_{xx} + b\Phi_{xy} + c\Phi_{yy} + d\Phi_x + e\Phi_y + f\Phi = 0$$

where the coefficients a, b, c, d, e, and f are in general functions of x and y. Under what conditions is the PDE separable ?

2.2 Solve
$$\Phi_{xx} - 4\Phi_x + \Phi_y = 0$$

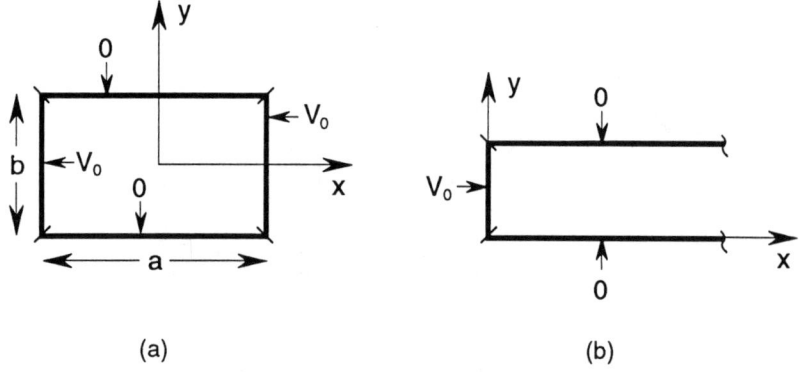

Figure 2.17 For problem 2.3.

by the method of separation of variables.

2.3 Determine the distribution of electrostatic potential inside the conducting rectangular boxes with cross sections shown in Fig. 2.17.

2.4 Find the two-dimensional electrostatic potential field due to each of the electrodes in Fig. 2.18.

2.5 Find the solution U of:

(a) Laplace equation

$$\nabla^2 U = 0, \quad 0 < x, y < \pi$$
$$U_x(0,y) = 0 = U_x(\pi,y), \quad U(x,0) = 0,$$
$$U(x,\pi) = x, \quad 0 < x < \pi.$$

(b) Heat equation

$$kU_{xx} = U_t, \quad 0 \le x \le 1, \ t > 0.$$
$$U(0,t) = 0, \ t > 0, \quad U(1,t) = 1, \ t > 0$$
$$U(x,0) = 0, 0 \le x \le 1.$$

(c) Wave equation

$$a^2 U_{xx} = U_{tt}, \quad 0 \le x \le 1, \ t > 0$$
$$U(0,t) = 0 = U(1,t), \ t > 0$$
$$U(x,0) = 0, \quad U_t(x,0) = x.$$

120   Numerical Techniques in Electromagnetics

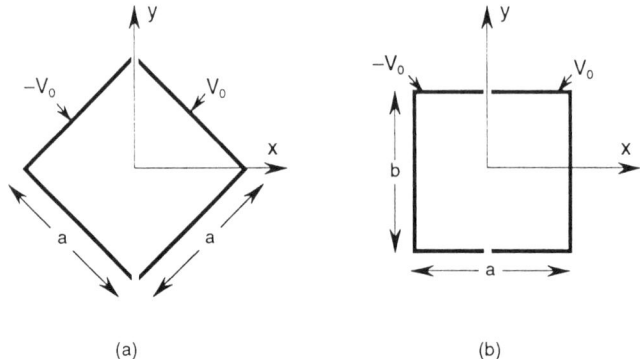

Figure 2.18 For problem 2.4.

2.6 Solve the PDE

$$4\frac{\partial^4 \Phi}{\partial x^4} + \frac{\partial^2 \Phi}{\partial t^2} = 0, \qquad 0 < x < 1,\ t > 0$$

subject to the boundary conditions

$$\Phi(0,t) = 0 = \Phi(1,t) = \Phi_{xx}(0,t) = \Phi_{xx}(1,t)$$

and initial conditions

$$\Phi_t(x,0) = 0, \quad \Phi(x,0) = x.$$

2.7 Find the solution Φ of:

(a) Laplace equation

$$\nabla^2 \Phi = 0, \qquad \rho \geq 1,\ 0 < \phi < \pi$$
$$\Phi(1,\phi) = \sin\phi, \qquad \Phi(\rho,0) = \Phi(\rho,\pi) = 0.$$

(b) Laplace equation

$$\nabla^2 \Phi = 0, \qquad 0 < \rho < 1,\ 0 < z < L$$
$$\Phi(\rho,\phi,0) = 0 = \Phi(\rho,\phi,L),\ \Phi(a,\phi,z) = 1.$$

(c) Heat equation

$$\Phi_t = k\nabla^2 \Phi, \qquad 0 \leq \rho \leq 1,\ -\infty < z < \infty,\ t > 0$$
$$\Phi(a,\phi,t) = 0,\ t > 0,\quad \Phi(\rho,\phi,0) = \rho^2 \cos 2\phi,\ 0 \leq \phi < 2\pi.$$

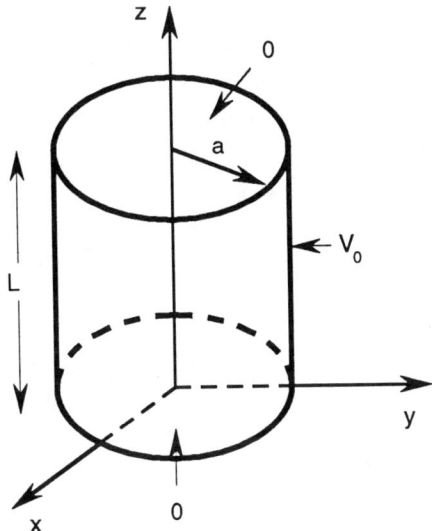

Figure 2.19 For problem 2.8.

2.8 Determine the potential distribution in a cylinder of radius a and length L with ends held at zero potential while the lateral surface is held at potential V_o as in Fig. 2.19. Calculate the potential along the axis of the cylinder when $L = 2a$.

2.9 Solve the PDE

$$\frac{\partial^2 \Phi}{\partial \rho^2} + \frac{1}{\rho}\frac{\partial \Phi}{\partial \rho} = \frac{\partial^2 \Phi}{\partial t^2}, \quad 0 \le \rho \le a,\ t \ge 0$$

under the conditions

$$\Phi(0,t) \text{ is bounded}, \quad \Phi(a,t) = 0,\ t > 0,$$

$$\Phi(\rho,0) = (1 - \rho^2/a^2), \quad \left.\frac{\partial \Phi}{\partial t}\right|_{t=0} = 0,\ 0 \le \rho \le a.$$

2.10 Show that the solution to

$$\frac{\partial^2 V}{\partial x^2} + \frac{\partial^2 V}{\partial y^2} + V = 0$$

is
$$V = (A_n \cos n\phi + B_n \sin n\phi)(-2\rho)^n \frac{d^n J_0(\rho)}{d(\rho^2)^n}$$
where $x = \rho \cos\phi$ and $y = \rho \sin\phi$.

2.11 (a) Prove that
$$e^{\pm j\rho \sin\phi} = \sum_{n=-\infty}^{\infty} (\pm 1)^n J_n(\rho) e^{jn\phi}.$$

(b) Show that
$$\cos(\rho \sin\phi) = J_0(\rho) + 2\sum_{n=1}^{\infty} J_{2n}(\rho) \cos 2n\phi$$

and
$$\sin(\rho \sin\phi) = 2\sum_{n=1}^{\infty} J_{2n-1}(\rho) \sin(2n-1)\phi$$

which demonstrates the close tie between Bessel function and trigonometric functions.

(c) Derive the *Bessel's integral formula*
$$J_n(\rho) = \frac{1}{\pi} \int_0^{\pi} \cos(n\theta - \rho \sin\theta) d\theta.$$

2.12 In Example 2.3, show that if a/δ is large,
$$J_z(\rho) = J_z(a) \sqrt{\frac{a}{\rho}} e^{-(1+j)(a-\rho)/\delta}.$$

2.13 Evaluate:

(a)
$$\int_{-1}^{1} P_1(x) P_2(x) \, dx,$$

(b)
$$\int_{-1}^{1} [P_4(x)]^2 \, dx,$$

(c)
$$\int_{0}^{1} x^2 P_3(x) \, dx.$$

2.14 In Legendre series of the form $\sum_{n=0}^{\infty} A_n P_n(x)$, expand:

(a) $f(x) = \begin{cases} 0, & -1 < x < 0, \\ 1, & 0 < x < 1 \end{cases}$

(b) $f(x) = x^3, \quad -1 < x < 1,$

(c) $f(x) = \begin{cases} 0, & -1 < x < 0, \\ x, & 0 < x < 1 \end{cases}$

(d) $f(x) = \begin{cases} 1 + x, & -1 < x < 0, \\ 1 - x, & 0 < x < 1 \end{cases}$

2.15 Solve Laplace's equation:

(a) $\nabla^2 U = 0, \quad 0 \leq r \leq a, \quad U(a, \theta) = \begin{cases} 1, & 0 < \theta < \pi/2, \\ 0, & \text{otherwise.} \end{cases}$

(b) $\nabla^2 U = 0, \quad r > a, \quad \left.\dfrac{\partial U}{\partial r}\right|_{r=a} = \cos\theta + 3\cos^3\theta, \quad 0 < \theta < \pi,$

(c) $\nabla^2 U = 0, \quad r < a, \quad 0 < \theta < \pi, \quad 0 < \phi < 2\pi,$
$U(a, \theta, \phi) = \sin^2\theta.$

2.16 A hollow conducting sphere of radius a has its upper half charged to potential V_o while its lower half is grounded. Find the potential distribution inside and outside the sphere.

2.17 A circular disk of radius a carries charge of surface charge density ρ_o. Show that the potential at point $(0, 0, z)$ on its axis $\theta = 0$ is
$$V = \frac{\rho_o}{2\epsilon}[(z^2 + a^2)^{1/2} - z].$$

124   Numerical Techniques in Electromagnetics

From this deduce the potential at any point (r,θ,ϕ).

2.18 Find: (a) $P_7(x)$,   (b) $P_3^2(x)$,   (c) $P_4^1(x)$.

2.19 Verify the following identities:

(a) $\int_{-1}^{1} P_n(x) P_m(x)\, dx = \frac{2}{2n+1}\delta_{nm}$,

(b) $\int_{-1}^{1} P_n^m(x) P_k^m(x)\, dx = \frac{2}{2n+1}\frac{(n+m)!}{(n-m)!}\delta_{nk}$.

2.20 Prove that:

(a) $J_{1/2}(x) = \sqrt{\frac{2}{\pi x}}\sin x$,

(b) $J_{-1/2}(x) = \sqrt{\frac{2}{\pi x}}\cos x$,

(c) $\frac{d}{dx}[x^{-n} J_n(x)] = -x^{-n} J_{n+1}(x)$,

(d) $\left.\frac{d^n}{dx^n} J_n(x)\right|_{x=0} = \frac{1}{2^n}$,

(e) $\frac{d}{dx}[x z_n(x)] = -n z_n(x) + x z_{n-1}(x) = (n+1) z_n(x) + x z_{n+1}(x)$.

2.21 Rework the problem in Fig. 2.8 if the boundary conditions are now

$$V(r=a) = V_o, \quad V(r\to\infty) = E_o r \cos\theta + V_o.$$

Find V and $\mathbf{E}$ everywhere. Determine the maximum value of the field strength.

2.22 Find the potential distribution inside and outside a dielectric sphere of radius a placed in a uniform electric field E_o.

Hint: The problem to be solved is $\nabla^2 V = 0$ subject to

$$\epsilon_r \frac{\partial V_1}{\partial r} = \frac{\partial V_2}{\partial r} \quad \text{on } r=a, \quad V_1 = V_2 \quad \text{on } r=a,$$
$$V_2 = -E_o r \cos\theta \quad \text{as } r\to\infty.$$

2.23 Verify the following identities:

(a) $$\frac{dP_n^m(x)}{dx} = \frac{(n+m)P_{n-1}^m(x) - nxP_n^m(x)}{(1-x^2)}.$$

(b)
$$\frac{dP_n^m(x)}{d\theta} = \frac{1}{2}\left[(n-m+1)(n+m)P_n^{m-1}(x) - P_n^{m+1}(x)\right], \quad x = \cos\theta.$$

2.24 Show that
$$\left.\frac{d^m P_n}{dx^m}\right|_{x=1} = \frac{(n+m)!}{2^m\, m!\,(n-m)!}$$
$$\left.\frac{d^m P_n}{dx^m}\right|_{x=-1} = (-1)^m \frac{(n+m)!}{2^m\, m!\,(n-m)!}.$$

2.25 Expand $V = \cos 2\phi \sin^2 \phi$ in terms of the spherical harmonics $P_n^m(\cos\theta)\sin m\phi$ and $P_n^m(\cos\theta)\cos m\phi$.

2.26 In the prolate spheroidal coordinates (ξ, η, ϕ), the equation
$$\nabla^2 \Phi + k^2 \Phi = 0$$
assumes the form
$$\frac{\partial}{\partial \xi}\left[(\xi^2-1)\frac{\partial \Phi}{\partial \xi}\right] + \frac{\partial}{\partial \eta}\left[(1-\eta^2)\frac{\partial \Phi}{\partial \eta}\right] + \left[\frac{1}{\xi^2-1} + \frac{1}{1-\eta^2}\right]\frac{\partial^2 \Phi}{\partial \phi^2} + k^2 d^2(\xi^2-\eta^2)\Phi = 0.$$

Show that the separated equations are
$$\frac{d}{d\xi}\left[(\xi^2+1)\frac{d\Psi_1}{d\xi}\right] + \left[k^2 d^2 \xi^2 - \frac{m^2}{\xi^2-1} - c\right]\Psi_1 = 0$$
$$\frac{d}{d\eta}\left[(1-\eta^2)\frac{d\Psi_2}{d\eta}\right] - \left[k^2 d^2 \eta^2 + \frac{m^2}{1-\eta^2} - c\right]\Psi_2 = 0$$

$$\frac{d^2\Psi_3}{d\phi^2} + m^2\Psi_3 = 0$$

where m and c are separation constants.

2.27 Solve Eq. (2.135) if $a = b = c = \pi$ and :

(a) $f(x,y,z) = e^{-x}$, (b) $f(x,y,z) = \sin^2 x$.

2.28 Solve Eq. (2.139) when:

(a) $f(r,\theta,\phi) = \rho_o$, (b) $f(r,\theta,\phi) = r^2 \sin^2 \theta$.

2.29 Solve the inhomogeneous PDE

$$\frac{\partial^2 \Phi}{\partial \rho^2} + \frac{1}{\rho}\frac{\partial \Phi}{\partial \rho} - \frac{\partial^2 \Phi}{\partial t^2} = -\Phi_o \sin \omega t, \quad 0 \leq \rho \leq a,\ t \geq 0$$

subject to the conditions $\Phi(a,t) = 0$, $\Phi(\rho,0) = 0$, $\Phi_t(\rho,0) = 0$, Φ is finite for all $0 \leq \rho \leq a$. Take Φ_o as a constant and $a\omega$ not being a zero of $J_0(x)$.

2.30 Figure 2.20 illustrates a cylindrical metal tank partially filled by charged liquid. To find the potential distribution in the tank, we need to solve a two-dimensional problem;

$$\frac{1}{\rho}\frac{\partial}{\partial \rho}\left(\rho \frac{\partial V_g}{\partial \rho}\right) + \frac{\partial^2 V_g}{\partial z^2} = 0 \quad \text{for gas space}$$

$$\frac{1}{\rho}\frac{\partial}{\partial \rho}\left(\rho \frac{\partial V_\ell}{\partial \rho}\right) + \frac{\partial^2 V_\ell}{\partial z^2} = \frac{-\rho_v}{\epsilon} \quad \text{for liquid space}$$

subject to

$$V = 0 \quad \text{at the wall,}$$
$$V_g = V_\ell \quad \text{at the gas-liquid interface,}$$
$$\frac{\partial V_g}{\partial z} = \epsilon_r \frac{\partial V_\ell}{\partial z} \quad \text{the gas-liquid interface.}$$

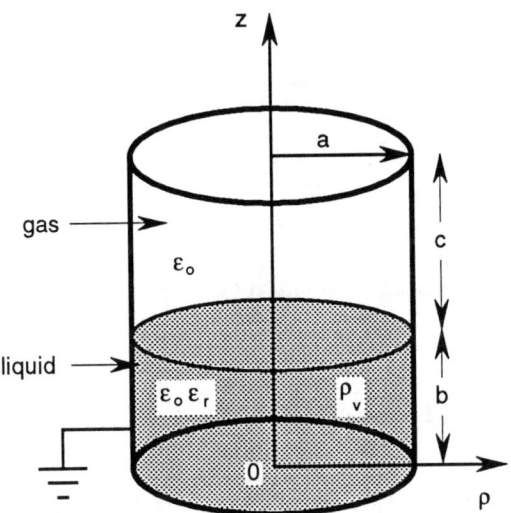

Figure 2.20 For problem 2.30—A cylindrical metal tank partially filled with charged liquid.

where V_g and V_ℓ are the potentials in gas and liquid space, respectively. Using series expansion techniques, let

$$V_g = \sum_{n=1}^{\infty} J_0(\lambda_n \rho) \Big[A_n \sinh[\lambda_n(b+c-z)] + B_n \cosh[\lambda_n(b+c-z)] \Big]$$

$$V_\ell = \sum_{n=1}^{\infty} J_0(\lambda_n \rho) F_n(z)$$

where $F_n(z)$ is a function of z only. Find A_n, B_n and $F_n(z)$ and show that

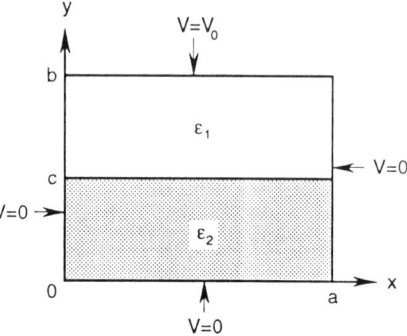

Figure 2.21 Potential system for problem 2.31.

$$V_g = \sum_{n=1}^{\infty} \frac{2\rho_v}{C_n K_n}[\cosh(\lambda_n b) - 1] J_0(\lambda_n \rho) \sinh[\lambda_n(b + c - z)]$$

$$V_\ell = \sum_{n=1}^{\infty} \frac{2\rho_v}{C_n \epsilon_r} J_0(\lambda_n \rho) \left[\frac{\sinh(\lambda_n z)}{K_n} (\cosh(\lambda_n b) \cosh(\lambda_n c) \right.$$
$$\left. + \epsilon_r \sinh(\lambda_n b) \sinh(\lambda_n c) - \cosh(\lambda_n c)) - \cosh(\lambda_n z) + 1 \right]$$

where

$$C_n = \epsilon_o a \lambda_n^3 J_1(\lambda_n a), \quad K_n = \sinh(\lambda_n b)\cosh(\lambda_n c) + \epsilon_r \cosh(\lambda_n b)\sinh(\lambda_n c),$$

and λ_n are the roots of $J_0(\lambda a) = 0$. Find $\mathbf{E}_g$ and $\mathbf{E}_\ell$.

2.31 Consider the potential problem shown in Fig. 2.21. The potentials at $x = 0$, $x = a$, and $y = 0$ sides are zero while the potential at $y = b$ side is V_o. Using the series expansion technique similar to that used in Prob. 2.30, find the potential distribution in the solution region.

2.32 Consider a grounded rectangular pipe with the cross section shown in Fig. 2.22. Assuming that the pipe is partially filled with hydrocarbons with charge density ρ_o, apply the same series expansion technique used in Prob. 2.30 to find the potential distribution in the pipe.

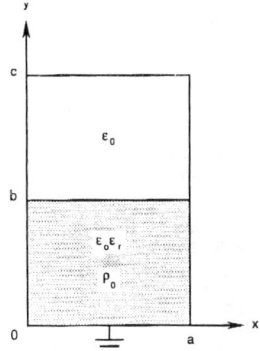

Figure 2.22 Earthed rectangular pipe partially filled with charged liquid—for problem 2.32.

2.33 Write a program to generate associated Legendre polynomial, with $x = \cos\theta = 0.5$. You may use either series expansion or recurrence relations. Take $0 \le n \le 15$, $0 \le m \le n$. Compare your results with those tabulated in [25].

2.34 The FORTRAN program of Fig. 2.15 uses the series expansion method to generate $j_n(x)$. Write a subroutine for generating $j_n(x)$ using recurrence relations. For $x = 2.0$ and $0 \le n \le 10$, compare your result with that obtained using the subroutine BESSEL of Fig. 2.15 and the tabulated values in [26]. Which result do you consider to be more accurate? Explain.

2.35 Use the generating functions to prove that:

(a) $J_n(x+y) = \sum\limits_{k=-\infty}^{\infty} J_k(x) J_{n-k}(y)$,

(b) $\dfrac{1}{R} = \dfrac{1}{r_o} \sum\limits_{n=0}^{\infty} (r/r_o)^n P_n(\cos\theta), \quad r < r_o,$

$\dfrac{1}{R} = \dfrac{1}{r} \sum\limits_{n=0}^{\infty} (r_o/r)^n P_n(\cos\theta), \quad r > r_o$, where $R = |\mathbf{r} - \mathbf{r}_o| = [r^2 - r_o^2 - 2rr_o\cos\alpha]^{1/2}$ and α is the angle between $\mathbf{r}$ and $\mathbf{r}_o$.

130   Numerical Techniques in Electromagnetics

2.36 Show that

$$\int T_0(x)\,dx = T_1(x)$$
$$\int T_1(x)\,dx = \frac{1}{4}T_2(x) + \frac{1}{4}$$
$$\int T_n(x)\,dx = \frac{1}{2}\left(\frac{T_{n+1}(x)}{n+1} - \frac{T_{n-1}(x)}{n-1}\right),\ n > 1$$

so that integration can be done directly in Chebyshev polynomials.

2.37 In $-1 \leq x \leq 1$, establish the formulas

$$|x| = \frac{1}{2} + \sum_{n=1}^{\infty}(-1)^{n+1}\frac{(2n-2)!}{(n-1)!(n+1)!}\frac{4n+1}{2^{2n}}P_{2n}$$
$$= \frac{2}{\pi} + \frac{4}{\pi}\sum_{n=1}^{\infty}(-1)^{n+1}\frac{T_{2n}(x)}{4n^2-1}.$$

Show that

$$\frac{\pi}{4} = 1 - 2\left[\frac{1}{1^2\cdot 3^2} - \frac{1}{3^2\cdot 5^2} + \frac{1}{5^2\cdot 7^2} - \cdots\right].$$

2.38 A function is defined by

$$f(x) = \begin{cases} 1, & -1 \leq x \leq 1 \\ 0, & \text{otherwise}. \end{cases}$$

(a) Expand $f(x)$ in a series of Hermite functions,

(b) expand $f(x)$ in a series of Laguerre functions.

2.39 By expressing E_θ^i and E_ϕ^i in terms of the spherical wave functions, show that Eq. (2.158) is valid.

2.40 By defining

$$\rho_n(x) = \frac{d}{dx}\ln[xh_n^{(2)}(x)], \quad \sigma_n(x) = \frac{d}{dx}\ln[xj_n(x)],$$

show that the scattering amplitude coefficients can be written as

$$a_n = \frac{j_n(\alpha)}{h_n^{(2)}(\alpha)} \left[\frac{\sigma_n(\alpha) - m\sigma_n(m\alpha)}{\rho_n(\alpha) - m\sigma_n(m\alpha)} \right]$$

$$b_n = \frac{j_n(\alpha)}{h_n^{(2)}(\alpha)} \left[\frac{\sigma_n(m\alpha) - m\sigma_n(\alpha)}{\sigma_n(m\alpha) - m\rho_n(\alpha)} \right].$$

2.41 For the problem in Fig. 2.13, plot $|E_z^t|/|E_x^i|$ for $-a < z < a$ along the axis of the dielectric sphere of radius $a = 9$ cm in the $x-z$ plane. Take $E_o = 1$, $\omega = 2\pi \times 5 \times 10^9$ rad/s, $\epsilon_1 = 4\epsilon_o$, $\mu_1 = \mu_o$, $\sigma_1 = 0$. You may modify the program in Fig. 2.15 or write your own.

2.42 In analytical treatment of the radio-frequency radiation effect on the human body, the human skull is frequently modeled as a lossy sphere. Of major concern is the calculation of the normalized heating potential

$$\Phi(r) = \frac{1}{2}\sigma \frac{|E^t(r)|^2}{|E_o|^2} \quad (\Omega \cdot m)^{-1},$$

where E^t is the internal electric field strength and E_o is the peak incident field strength. If the human skull can be represented by a homogeneous sphere of radius $a = 10$ cm, plot $\Phi(r)$ against the radial distance $-10 \leq r = z \leq 10$ cm. Assume an incident field as in Fig. 2.13 with $f = 1$ GHz, $\mu_r = 1$, $\epsilon_r = 60$, $\sigma = 0.9$ mhos/m, $E_o = 1$.

2.43 Instead of the homogeneous spherical model assumed in the previous problem, consider the multilayered spherical model shown in Fig. 2.23 with each region labeled by an integer p, such that $p = 1$ represents the central core region and $p = 4$ represents air. At $f = 2.45$ GHz, plot the heating potential along x axis, y axis, and z axis. Assume the data given below.

Region p	Tissue	Radius (mm)	ϵ_r	σ(mho/m)
1	muscle	18.5	46	2.5
2	fat	19	6.95	0.29
3	skin	20	43	2.5
4	air		1	0

$$\mu_r = 1$$

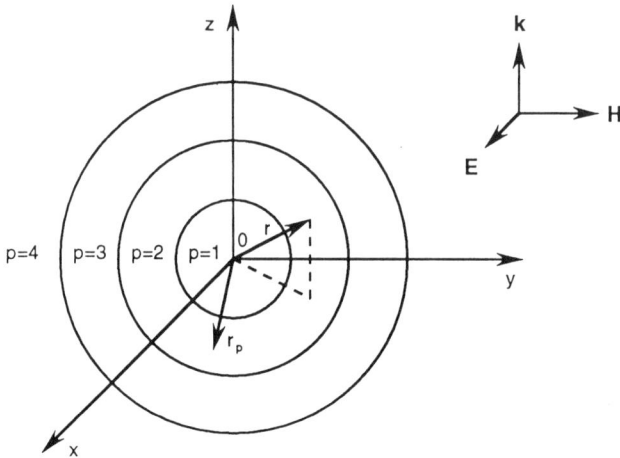

Figure 2.23 For problem 2.43, a multilayered spherical model of the human skull.

Note that for each region p, the resultant field consists of the transmitted and scattered fields and is in general given by

$$E_p(r,\theta,\phi) = E_o e^{j\omega t} \sum_{n=1}^{\infty} (-j)^n \frac{2n+1}{n(n+1)} \Big[a_{np} \mathbf{M}_{np}^{(4)}(k) + j b_{np} \mathbf{N}_{np}^{(4)}(k) + c_{np} \mathbf{M}_{np}^{(1)}(k_1) + j d_{np} \mathbf{N}_{np}^{(1)}(k_1) \Big].$$

2.44 The absorption characteristic of biological bodies is determined in terms of the specific absorption rate (SAR) defined as the total power absorbed divided as the power incident on the geometrical cross section. For an incident power density of 1 mW/cm² in a spherical model of the human head,

$$\text{SAR} = 2 \frac{Q_{abs}}{\pi a} \quad \text{mW/cm}^3$$

where a is in centimeters. Using the above relation, plot SAR against frequency for $0.1 < f < 3$ GHz, $a = 10$ cm assuming frequency-dependent and dielectric properties of head as

$$\epsilon_r = 5 \left(\frac{12 + (f/f_o)^2}{1 + (f/f_o)^2} \right)$$

$$\sigma = 6 \left(\frac{1 + 62(f/f_o)^2}{1 + (f/f_o)^2} \right)$$

where f is in GHz and $f_o = 20$ GHz.

2.45 For the previous problem, repeat the calculations of SAR assuming a six-layered spherical model of the human skull (similar to that of Fig. 2.23) of outer radius $a = 10$ cm. Plot P_a/P_i versus frequency for $0.1 < f < 3$ GHz where

$$\frac{P_a}{P_i} = \frac{2}{\alpha^2} \sum (2n+1)\left[\text{Re}(a_n + b_n) - (|a_n|^2 + |b_n|^2)\right],$$

P_a = absorbed power, P_i = incident power, $\alpha = 2\pi a/\lambda$, λ is the wavelength in the external medium. Use the dimensions and electrical properties shown below.

Layer p	Tissue	Radius (mm)	ϵ_r	σ(mho/m)
1	brain	9	$5\nabla(f)$	$6\Delta(f)$
2	CSF	12	$7\nabla(f)$	$8\Delta(f)$
3	dura	13	$4\nabla(f)$	$8\Delta(f)$
4	bone	17.3	5	62
5	fat	18.5	6.95	0.29
6	skin	20	43	2.5

where $\mu_r = 1$,

$$\nabla(f) = \frac{1 + 12(f/f_o)^2}{1 + (f/f_o)^2},$$
$$\Delta(f) = \frac{1 + 62(f/f_o)^2}{1 + (f/f_o)^2},$$

f is in GHz, and $f_o = 20$ GHz. Compare your result with that from the previous problem.

Chapter 3

Finite Difference Methods

"Research is to see what everybody else has seen, and think what nobody has thought." Albert Szent-Gyorgyi

3.1 Introduction

It is rare for EM problems to fall neatly into a class that can be solved by the analytical methods presented in the preceding chapter. Classical approaches may fail if [1]:

- The PDE is not linear and cannot be linearized without seriously affecting the result.
- The solution region is complex.
- The boundary conditions are of mixed types.
- The boundary conditions are time-dependent.
- The medium is inhomogeneous or anisotropic.

Whenever a problem with such complexity arises, numerical solutions must be employed. Of the numerical methods available for solving PDEs, those employing finite differences are more easily understood, more frequently used, and more universally applicable than any other.

The finite difference method (FDM) was first developed by A. Thom [2] in the 1920s under the title "the method of squares" to solve nonlinear hydrodynamic equations. Since then, the method has found applications in solving different field problems. The finite difference techniques are based upon approximations which permit replacing differential equations by finite difference equations. These finite difference approximations are algebraic in form; they relate the value of the dependent variable at a point in the

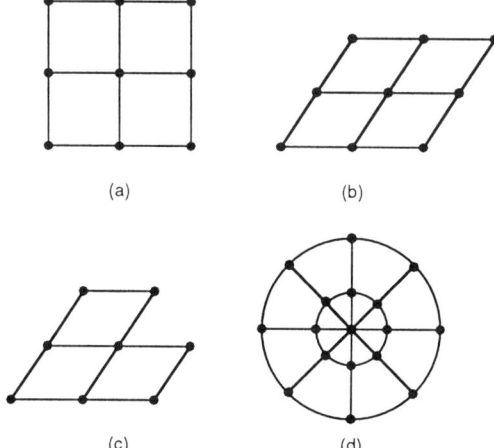

Figure 3.1 Common grid patterns: (a) rectangular grid, (b) skew grid, (c) triangular grid, (d) circular grid.

solution region to the values at some neighboring points. Thus a finite difference solution basically involves three steps:

(1) Dividing the solution region into a grid of nodes.
(2) Approximating the given differential equation by finite difference equivalent that relates the dependent variable at a point in the solution region to its values at the neighboring points.
(3) Solving the difference equations subject to the prescribed boundary conditions and/or initial conditions.

The course of action taken in three steps is dictated by the nature of the problem being solved, the solution region, and the boundary conditions. The most commonly used grid patterns for two-dimensional problems are shown in Fig. 3.1. A three-dimensional grid pattern will be considered later in the chapter.

3.2 Finite Difference Schemes

Before finding the finite difference solutions to specific PDEs, we will look at how one constructs finite difference approximations from a given differential equation. This essentially involves estimating derivatives numerically.

Given a function $f(x)$ shown in Fig. 3.2, we can approximate its derivative, slope or the tangent at P by the slope of the arc PB, giving the *forward-difference* formula,

Finite Difference Methods

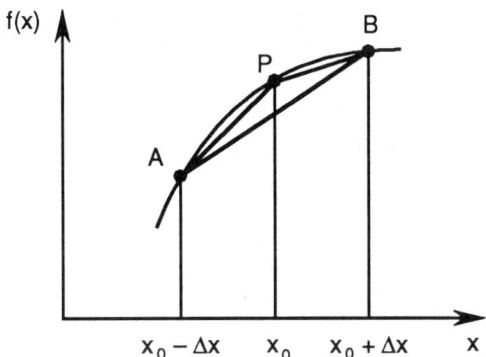

Figure 3.2 Estimates for the derivative of $f(x)$ at P using forward, backward, and central differences.

$$f'(x_o) \simeq \frac{f(x_o + \Delta x) - f(x_o)}{\Delta x} \quad (3.1)$$

or the slope of the arc AP, yielding the *backward-difference* formula,

$$f'(x_o) \simeq \frac{f(x_o) - f(x_o - \Delta x)}{\Delta x} \quad (3.2)$$

or the slope of the arc AB, resulting in the *central-difference* formula,

$$f'(x_o) \simeq \frac{f(x_o + \Delta x) - f(x_o - \Delta x)}{2\Delta x} \quad (3.3)$$

We can also estimate the second derivative of $f(x)$ at P as

$$f''(x_o) \simeq \frac{f'(x_o + \Delta x/2) - f'(x_o - \Delta x/2)}{\Delta x}$$

$$= \frac{1}{\Delta x}\left[\frac{f(x_o + \Delta x) - f(x_o)}{\Delta x} - \frac{f(x_o) - f(x_o - \Delta x)}{\Delta x}\right]$$

or

$$f''(x_o) \simeq \frac{f(x_o + \Delta x) - 2f(x_o) + f(x_o - \Delta x)}{(\Delta x)^2}. \quad (3.4)$$

Any approximation of a derivative in terms of values at a discrete set of points is called *finite difference* approximation.

The approach used above in obtaining finite difference approximations is rather intuitive. A more general approach is using Taylor's series. According to the well-known expansion,

$$f(x_o + \Delta x) = f(x_o) + \Delta x f'(x_o) + \frac{1}{2!}(\Delta x)^2 f''(x_o) + \frac{1}{3!}(\Delta x)^3 f'''(x_o) + \cdots \quad (3.5)$$

and

$$f(x_o - \Delta x) = f(x_o) - \Delta x f'(x_o) + \frac{1}{2!}(\Delta x)^2 f''(x_o) - \frac{1}{3!}(\Delta x)^3 f'''(x_o) + \cdots. \quad (3.6)$$

Upon adding these expansions,

$$f(x_o + \Delta x) + f(x_o - \Delta x) = 2f(x_o) + (\Delta x)^2 f''(x_o) + O(\Delta x)^4 \quad (3.7)$$

where $O(\Delta x)^4$ is the error introduced by truncating the series. We say that this error is of the order $(\Delta x)^4$ or simply $O(\Delta x)^4$. Therefore, $O(\Delta x)^4$ represents terms that are not greater than $(\Delta x)^4$. Assuming that these terms are negligible,

$$f''(x_o) \simeq \frac{f(x_o + \Delta x) - 2f(x_o) + f(x_o - \Delta x)}{(\Delta x)^2}$$

which is Eq. (3.4). Subtracting Eq. (3.6) from Eq. (3.5) and neglecting terms of the order $(\Delta x)^3$ yields

$$f'(x_o) \simeq \frac{f(x_o + \Delta x) - f(x_o - \Delta x)}{2\Delta x}$$

which is Eq. (3.3). This shows that the leading errors in Eqs. (3.3) and (3.4) are of the order $(\Delta x)^2$. Similarly, the difference formula in Eqs. (3.1) and (3.2) have truncation errors of $O(\Delta x)$. Higher order finite difference approximations can be obtained by taking more terms in Taylor series expansion. If the infinite Taylor series were retained, an exact solution would be realized for the problem. However, for practical reasons, the infinite series is usually truncated after the second-order term. This imposes an error which exists in all finite difference solutions.

To apply the difference method to find the solution of a function $\Phi(x,t)$, we divide the solution region in the $x - t$ plane into equal rectangles or meshes of sides Δx and Δt as in Fig. 3.3. We let the coordinates (x,t) of a typical grid point or node be

$$x = i\Delta x, \quad i = 0,1,2,...$$
$$t = j\Delta t, \quad j = 0,1,2,... \quad (3.8a)$$

and the value if Φ at P be

$$\Phi_P = \Phi(i\Delta x, j\Delta t) = \Phi(i,j). \quad (3.8b)$$

With this notation, the central difference approximations of the derivatives of Φ at the (i,j)th node are

$$\Phi_x|_{i,j} \simeq \frac{\Phi(i+1,j) - \Phi(i-1,j)}{2\Delta x}, \quad (3.9a)$$

$$\Phi_t|_{i,j} \simeq \frac{\Phi(i,j+1) - \Phi(i,j-1)}{2\Delta t}, \quad (3.9b)$$

$$\Phi_{xx}|_{i,j} \simeq \frac{\Phi(i+1,j) - 2\Phi(i,j) + \Phi(i-1,j)}{(\Delta x)^2}, \quad (3.9c)$$

$$\Phi_{tt}|_{i,j} \simeq \frac{\Phi(i,j+1) - 2\Phi(i,j) + \Phi(i,j-1)}{(\Delta t)^2} \quad (3.9d)$$

Table 3.1 gives some useful finite difference approximations for Φ_x and Φ_{xx}.

3.3 Finite Differencing of Parabolic PDEs

Consider a simple example of a parabolic (or diffusion) partial differential equation with one spatial independent variable

$$k\frac{\partial \Phi}{\partial t} = \frac{\partial^2 \Phi}{\partial x^2} \quad (3.10)$$

where k is a constant. The equivalent finite difference approximation is

$$k\frac{\Phi(i,j+1) - \Phi(i,j)}{\Delta t} = \frac{\Phi(i+1,j) - 2\Phi(i,j) + \Phi(i-1,j)}{(\Delta x)^2} \quad (3.11)$$

Table 3.1 Finite Difference Approximations for Φ_x and Φ_{xx}, where FD = Forward Difference, BD = Backward Difference, and CD = Central Difference

Derivative	Finite Difference Approximation	Type	Error
Φ_x	$\dfrac{\Phi_{i+1} - \Phi_i}{\Delta x}$	FD	$O(\Delta x)$
	$\dfrac{\Phi_i - \Phi_{i+1}}{\Delta x}$	BD	$O(\Delta x)$
	$\dfrac{\Phi_{i+1} - \Phi_{i-1}}{\Delta x}$	CD	$O(\Delta x)^2$
	$\dfrac{-\Phi_{i+2} + 4\Phi_{i+1} - 3\Phi_i}{2\Delta x}$	FD	$O(\Delta x)^2$
	$\dfrac{3\Phi_i - 4\Phi_{i-1} + \Phi_{i-2}}{2\Delta x}$	BD	$O(\Delta x)^2$
	$\dfrac{-\Phi_{i+2} + 8\Phi_{i+1} - 8\Phi_{i-1} + \Phi_{i-2}}{12\Delta x}$	CD	$O(\Delta x)^4$
Φ_{xx}	$\dfrac{\Phi_{i+2} - 2\Phi_{i+1} + \Phi_i}{(\Delta x)^2}$	FD	$O(\Delta x)^2$
	$\dfrac{\Phi_i - 2\Phi_{i-1} + \Phi_{i-2}}{(\Delta x)^2}$	BD	$O(\Delta x)^2$
	$\dfrac{\Phi_{i+1} - 2\Phi_i + \Phi_{i-1}}{(\Delta x)^2}$	CD	$O(\Delta x)^2$
	$\dfrac{-\Phi_{i+2} + 16\Phi_{i+1} - 30\Phi_i + 16\Phi_{i-1} - \Phi_{i-2}}{(\Delta x)^2}$	CD	$O(\Delta x)^4$

where $x = i\Delta x$, $i = 0, 1, 2, \cdots, n$, $t = j\Delta t$, $j = 0, 1, 2, \cdots$. In Eq. (3.11), we have used the forward difference formula for the derivative with respect to t and central difference formula for that with respect to x. If we let

$$r = \frac{\Delta t}{k(\Delta x)^2}, \tag{3.12}$$

Eq. (3.11) can be written as

Finite Difference Methods 141

$$\boxed{\Phi(i,j+1) = r\,\Phi(i+1,j) + (1-2r)\Phi(i,j) + r\,\Phi(i-1,j)} \qquad (3.13)$$

This *explicit formula* can be used to compute $\Phi(x, t+\Delta t)$ explicitly in terms of $\Phi(x,t)$. Thus the values of Φ along the first time row (see Fig. 3.3), $t = \Delta t$, can be calculated in terms of the boundary and initial conditions, then the values of Φ along the second time row, $t = 2\Delta t$, are calculated in terms of the first time row, and so on.

A graphic way of describing the difference formula of Eq. (3.13) is through the *computational molecule* of Fig. 3.4(a), where the square is used to represent the grid point where Φ is presumed known and a circle where Φ is unknown.

In order to ensure a stable solution or reduce errors, care must be exercised in selecting the value of r in Eqs. (3.12) and (3.13). It will be shown in Section 3.6 that Eq. (3.13) is valid only if the coefficient $(1-2r)$ in Eq. (3.13) is nonnegative or $0 < r \leq 1/2$. If we choose $r = 1/2$, Eq. (3.13) becomes

$$\Phi(i,j+1) = \frac{1}{2}[\Phi(i+1,j) + \Phi(i-1,j)] \qquad (3.14)$$

so that the computational molecule becomes that shown in Fig. 3.4(b).

The fact that obtaining stable solutions depends on r or the size of the time step Δt renders the explicit formula of Eq. (3.13) inefficient. Although the formula is simple to implement, its computation is slow. An *implicit formula*, proposed by Crank and Nicholson in 1974, is valid for all finite values of r. We replace $\partial^2\Phi/\partial x^2$ in Eq. (3.10) by the average of the central difference formulas on the jth and $(j+1)$th time rows so that

$$k\frac{\Phi(i,j+1) - \Phi(i,j)}{\Delta t} = \frac{1}{2}\Big[\frac{\Phi(i+1,j) - 2\Phi(i,j) + \Phi(i-1,j)}{(\Delta x)^2}$$
$$+ \frac{\Phi(i+1,j+1) - 2\Phi(i,j+1) + \Phi(i-1,j+1)}{(\Delta x)^2}\Big].$$

This can be rewritten as

$$-r\Phi(i-1,j+1) + 2(1+r)\Phi(i,j+1) - r\Phi(i+1,j+1)$$
$$= r\Phi(i-1,j) + 2(1-r)\Phi(i,j) + r\Phi(i+1,j) \qquad (3.15)$$

where r is given by Eq. (3.12). The left side of Eq. (3.15) consists of three known values, while the right side has the three unknown values of Φ. This is illustrated in the computational molecule of Fig. 3.5(a). Thus if there

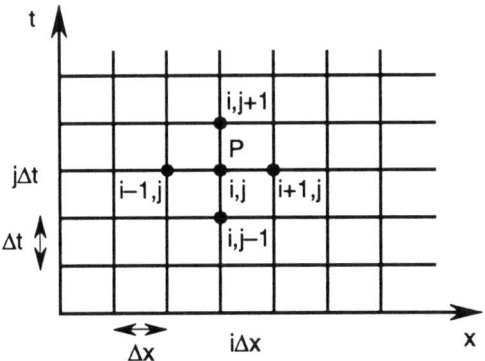

Figure 3.3 Finite difference mesh for two independent variables x and t.

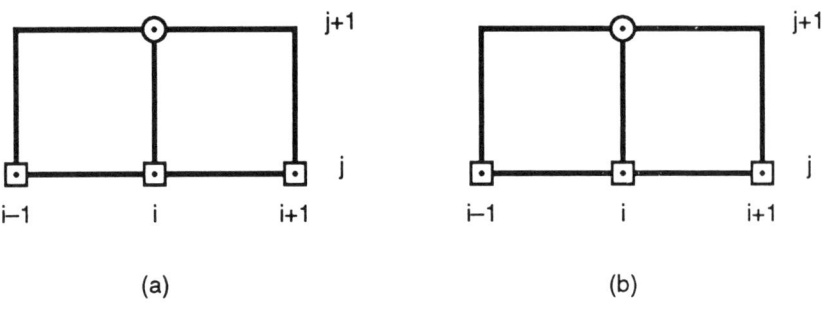

Figure 3.4 Computational molecule for parabolic PDE: (a) for $0 < r \leq 1/2$, (b) for $r = 1/2$.

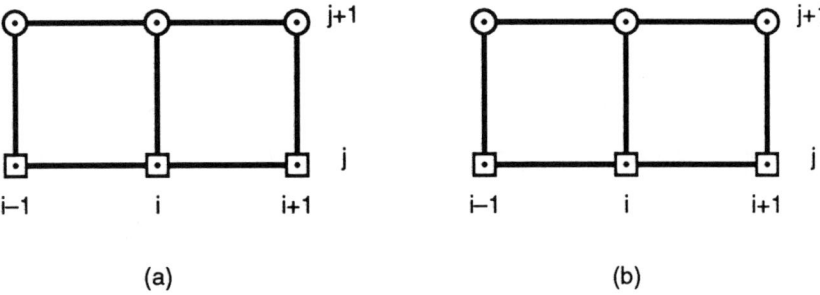

Figure 3.5 Computational molecule for Crank-Nicholson method: (a) for finite values of r, (b) for $r = 1$.

are n free nodes along each time row, then for $j = 0$, applying Eq. (3.15) to nodes $i = 1, 2, \cdots, n$ results in n simultaneous equations with n unknown values of Φ and known initial and boundary values of Φ. Similarly, for $j = 1$, we obtain n simultaneous equations for n unknown values of Φ in terms of the known values $j = 0$, and so on. The combination of accuracy and unconditional stability allows the use of a much larger time step with Crank-Nicholson method than is possible with the explicit formula. Although the method is valid for all finite values of r, a convenient choice of $r = 1$ reduces Eq. (3.15) to

$$-\Phi(i-1, j+1) + 4\Phi(i, j+1) - \Phi(i+1, j+1) = \Phi(i-1, j) + \Phi(i+1, j) \quad (3.16)$$

with the computational molecule of Fig. 3.5(b).

More complex finite difference schemes can be developed by applying the same principles discussed above. Two of such schemes are the Leapfrog method and the Dufort-Frankel method [3,4]. These and those discussed earlier are summarized in Table 3.2. Notice that the last two methods are two-step finite difference schemes in that finding Φ at time $j + 1$ requires knowing Φ at two previous time steps j and $j - 1$, whereas the first two methods are one-step schemes. For further treatment on the finite difference solution of parabolic PDEs, see Smith [5] and Ferziger [6].

Example 3.1

Solve the diffusion equation

$$\frac{\partial^2 \Phi}{\partial x^2} = \frac{\partial \Phi}{\partial t}, \qquad 0 \leq x \leq 1 \qquad (3.1.1)$$

144 Numerical Techniques in Electromagnetics

Table 3.2 Finite Difference Approximation to the Parabolic Equation: $\frac{\partial \Phi}{\partial t} = \frac{1}{k}\frac{\partial^2 \Phi}{\partial x^2}$, $k > 0$

Method	Algorithm	Molecule
1. First order (Euler)	$\frac{\Phi_i^{j+1} - \Phi_i^j}{\Delta t} = \frac{\Phi_{i+1}^j - 2\Phi_i^j + \Phi_{i-1}^j}{k(\Delta x)^2}$ explicit, stable for $r = \Delta t / k(\Delta x)^2 \le 0.5$	
2. Crank-Nicholson	$\frac{\Phi_i^{j+1} - \Phi_i^{j+1}}{\Delta t} = \frac{\Phi_{i+1}^{j+1} - 2\Phi_i^{j+1} + \Phi_{i-1}^{j+1}}{k(\Delta x)^2}$ $+ \frac{\Phi_{i+1}^j - 2\Phi_i^j + \Phi_{i-1}^j}{k(\Delta x)^2}$ implicit, always stable	
3. Leapfrog	$\frac{\Phi_i^{j+1} - \Phi_i^{j-1}}{2\Delta t} = \frac{\Phi_{i+1}^j - 2\Phi_i^j + \Phi_{i-1}^j}{k(\Delta x)^2}$ explicit, always unstable	
4. Dufort-Frankel	$\frac{\Phi_i^{j+1} - \Phi_i^{j-1}}{2\Delta t} = \frac{\Phi_{i+1}^j - \Phi_i^{j+1} - \Phi_i^{j-1} + \Phi_{i-1}^j}{k(\Delta x)^2}$ explicit, unconditionally stable	

subject to the boundary conditions

$$\Phi(0,t) = 0 = \Phi(1,t) = 0, \quad t > 0 \qquad (3.1.2a)$$

and initial condition

$$\Phi(x,0) = 100. \qquad (3.1.2b)$$

Solution

This problem may be regarded as a mathematical model of the temperature distribution in a rod of length L = 1 m with its end in contacts with ice blocks (or held at 0°C) and the rod initially at 100°C. With that physical interpretation, our problem is finding the internal temperature Φ as a function of position and time. We will solve this problem using both explicit and implicit methods.

(a) Explicit Method

For easy hand calculations, let us choose $\Delta x = 0.1, r = 1/2$ so that

$$\Delta t = \frac{r(\Delta x)^2}{k} = 0.05$$

since $k = 1$. We need the solution for only $0 \leq x \leq 0.5$ due to the fact that the problem is symmetric with respect to $x = 0.5$. First we calculate the initial and boundary values using Eq. (3.1.2). These values of Φ at the fixed nodes are shown in Table 3.3 for $x = 0, x = 1$, and $t = 0$. Notice that the values of $\Phi(0,0)$ and $\Phi(1,0)$ are taken as the average of 0 and 100. We now calculate Φ at the free nodes using Eq. (3.14) or the molecule of Fig. 3.4(b). The result is shown in Table 3.3. The analytic solution to Eq. (3.1.1) subject to Eq. (3.1.2) is

$$\Phi(x,t) = \frac{400}{\pi} \sum_{k=0}^{\infty} \frac{1}{n} \sin n\pi x \, \exp(-n^2\pi^2 t), \; n = 2k+1.$$

Comparison of the explicit finite difference solution with the analytic solution at $x = 0.4$ is shown in Table 3.4. The table shows that the finite difference solution is reasonably accurate. Greater accuracy can be achieved by choosing smaller values of Δx and Δt.

(b) Implicit Method

Let us choose $\Delta x = 0.2, r = 1$ so that $\Delta t = 0.04$. The values of Φ at the fixed nodes are calculated as in part (a) (see Table 3.3). For the free nodes, we apply Eq. (3.16) or the molecule of Fig. 3.5(b). If we denote $\Phi(i, j+1)$ by Φ_i ($i = 1, 2, 3, 4$), the values of Φ for the first time step (Fig. 3.6) can be obtained by solving the following simultaneous equations

$$-0 + 4\Phi_1 - \Phi_2 = 50 + 100$$
$$-\Phi_1 + 4\Phi_2 + \Phi_3 = 100 + 100$$
$$-\Phi_2 + 4\Phi_3 - \Phi_4 = 100 + 100$$
$$-\Phi_3 + 4\Phi_4 - 0 = 100 + 50.$$

We obtain

$$\Phi_1 = 58.13, \quad \Phi_2 = 82.54, \quad \Phi_3 = 72, \quad \Phi_4 = 55.5$$

at $t = 0.04$. Using these values of Φ, we apply Eq. (3.16) to obtain another set of simultaneous equations for $t = 0.08$ as

Table 3.3 Result for Example 3.1

x	0	0.1	0.2	0.3	0.4	0.5	0.6	...	1.0
t									
0	50	100	100	100	100	100	100		50
0.005	0	75.0	100	100	100	100	100		0
0.01	0	50.0	87.5	100	100	100	100		0
0.015	0	43.75	75	93.75	100	100	100		0
0.02	0	37.5	68.75	87.5	96.87	100	96.87		0
0.025	0	34.37	62.5	82.81	93.75	96.87	93.75		0
0.03	0	31.25	58.59	78.21	89.84	93.75	89.84		0
⋮									
0.1	0	14.66	27.92	38.39	45.18	47.44	45.18		0

$$-0 + 4\Phi_1 - \Phi_2 = 0 + 82.54$$
$$-\Phi_1 + 4\Phi_2 - \Phi_3 = 58.13 + 72$$
$$-\Phi_2 + 4\Phi_3 - \Phi_4 = 82.54 + 55.5$$
$$-\Phi_3 + 4\Phi_4 - 0 = 72 + 0$$

which results in

$$\Phi_1 = 34.44, \quad \Phi_2 = 55.23, \quad \Phi_3 = 56.33, \quad \Phi_4 = 32.08.$$

This procedure can be programmed and accuracy can be increased by choosing more points for each time step.

3.4 Finite Differencing of Hyperbolic PDEs

The simplest hyperbolic partial differential equation is the wave equation of the form

$$u^2 \frac{\partial^2 \Phi}{\partial x^2} = \frac{\partial^2 \Phi}{\partial t^2} \tag{3.17}$$

where u is the speed of the wave. An equivalent finite difference formula is

Table 3.4 Comparison of Explicit Finite Difference Solution with Analytic Solution; for Example 3.1

t	Finite difference solution at $x = 0.4$	Analytic solution at $x = 0.4$	Percentage error
0.005	100	99.99	0.01
0.01	100	99.53	0.47
0.015	100	97.85	2.2
0.02	96.87	95.18	1.8
0.025	93.75	91.91	2.0
0.03	89.84	88.32	1.7
0.035	85.94	84.61	1.6
0.04	82.03	80.88	1.4
⋮			
0.10	45.18	45.13	0.11

$$u^2 \frac{\Phi(i+1,j) - \Phi(i,j) + \Phi(i-1,j)}{(\Delta x)^2} = \frac{\Phi(i,j+1) - 2\Phi(i,j) + \Phi(i,j-1)}{(\Delta t)^2}$$

where $x = i\Delta x$, $t = j\Delta t$, $i,j = 0, 1, 2, \ldots$. This equation can be written as

$$\Phi(i,j+1) = 2(1-r)\Phi(i,j) + r[\Phi(i+1,j) + \Phi(i-1,j)] - \Phi(i,j-1)$$
(3.18)

where $\Phi(i, j+1)$ is an approximation to $\Phi(x,t)$ and r is the "aspect ratio" given by

$$r = \left(\frac{u\Delta t}{\Delta x}\right)^2. \qquad (3.19)$$

Equation (3.18) is an explicit formula for the wave equation. The corresponding computational molecule is shown in Fig. 3.7(a). For the solution algorithm in Eq. (3.18) to be stable, the aspect ratio $r \leq 1$, as will be shown in Example 3.5. If we choose $r = 1$, Eq. (3.18) becomes

$$\Phi(i,j+1) = \Phi(i+1,j) + \Phi(i-1,j) - \Phi(i,j-1) \qquad (3.20)$$

148 Numerical Techniques in Electromagnetics

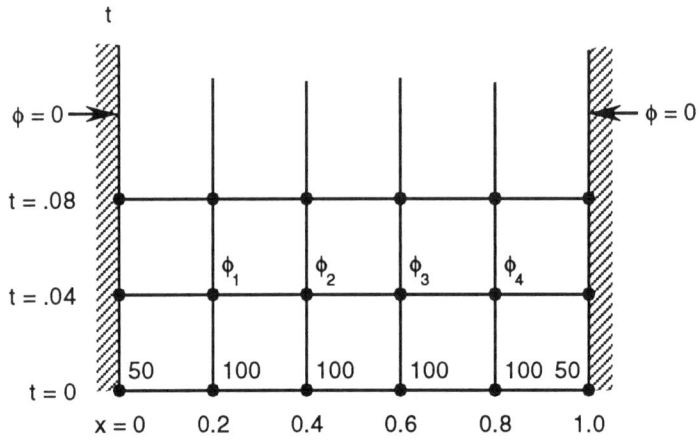

Figure 3.6 For example 3.1, part (b).

with the computational molecule in Fig. 3.7(b). Unlike the single-step schemes of Eqs. (3.13) and (3.15), the two-step schemes of Eqs. (3.18) and (3.20) require that the values of Φ at times j and $j-1$ be known to get Φ at time $j+1$. Thus, we must derive a separate algorithm to "start" the solution of Eq. (3.18) or (3.20); that is, we must compute $\Phi(i, 1)$ and $\Phi(i, 2)$. To do this, we utilize the prescribed initial condition. For example, suppose the initial condition on the PDE in Eq. (3.17) is

$$\left.\frac{\partial \Phi}{\partial t}\right|_{t=0} = 0.$$

We use the backward-difference formula

$$\frac{\partial \Phi(x,0)}{\partial t} \simeq \frac{\Phi(i,1) - \Phi(i,-1)}{2\Delta t} = 0$$

or

$$\Phi(i, 1) = \Phi(i, -1). \tag{3.21}$$

Substituting Eq. (3.21) into Eq. (3.18) and taking $j = 0$ (i.e., at $t = 0$), we get

$$\Phi(i, 1) = 2(1 - r)\Phi(i, 0) + r\big[\Phi(i-1, 0) + \Phi(i+1, 0)\big] - \Phi(i, 1)$$

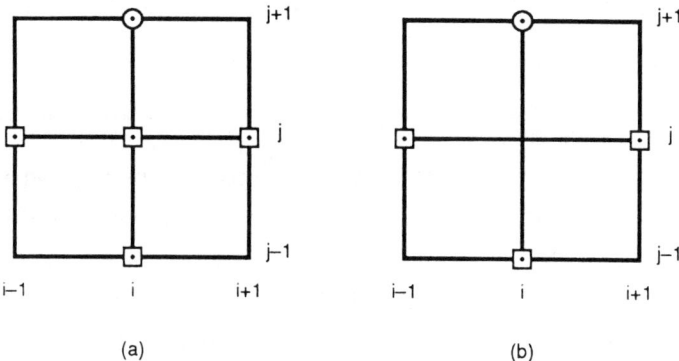

Figure 3.7 Computational molecule for wave equation: (a) for arbitrary $r \leq 1$, (b) for $r = 1$.

or

$$\Phi(i,1) = (1-r)\Phi(i,0) + \frac{r}{2}[\Phi(i-1,0) + \Phi(i+1,0)]. \tag{3.22}$$

Using the starting formula in Eq. (3.22) together with the prescribed boundary and initial conditions, the value of Φ at any grid point (i,j) can be obtained directly from Eq. (3.18).

There are implicit methods for solving hyperbolic PDEs just as we have implicit methods for parabolic PDEs. However, for hyperbolic PDEs, implicit methods result in an infinite number of simultaneous equations to be solved and therefore cannot be used without making some simplifying assumptions. Interested readers are referred to Smith [5] or Ferziger [6].

Example 3.2

Solve the wave equation

$$\Phi_{tt} = \Phi_{xx}, \qquad 0 < x < 1, \; t \geq 0$$

subject to the boundary conditions

$$\Phi(0,t) = 0 = \Phi(1,t), \; t \geq 0$$

and the initial conditions

$$\Phi(x,0) = \sin \pi x, \qquad 0 < x < 1,$$
$$\Phi_t(x,0) = 0, \qquad 0 < x < 1.$$

150   Numerical Techniques in Electromagnetics

Table 3.5 Solution of the Wave Equation in Example 3.2

x	0	0.1	0.2	0.3	0.4	0.5	0.6	...
t								
0.0	0	0.3090	0.5879	0.8990	0.9511	1.0	0.9511	
0.1	0	0.2939	0.5590	0.7694	0.9045	0.9511	0.9045	
0.2	0	0.2500	0.4755	0.6545	0.7694	0.8090	0.7694	
0.3	0	0.1816	0.3455	0.4755	0.5590	0.5878	0.5590	
0.4	0	0.0955	0.1816	0.2500	0.2939	0.3090	0.2939	
0.5	0	0	0	0	0	0	0	
0.6	0	-0.0955	-0.1816	-0.2500	-0.2939	-0.3090	-0.2939	
0.7	0	-0.1816	-0.3455	-0.4755	-0.5590	-0.5878	-0.5590	
⋮	⋮	⋮	⋮	⋮	⋮	⋮	⋮	⋮

Solution

The analytical solution is easily obtained as

$$\Phi(x,t) = \sin \pi x \, \cos \pi t. \quad (3.2.1)$$

Using the explicit finite difference scheme of Eq. (3.18) with $r = 1$, we obtain

$$\Phi(i, j+1) = \Phi(i-1, j) + \Phi(i+1, j) - \Phi(i, j-1), \quad j \geq 1. \quad (3.2.2)$$

For $j = 0$, substituting

$$\Phi_t = \frac{\Phi(i,1) - \Phi(i,-1)}{2\Delta t} = 0$$

or

$$\Phi(i,1) = \Phi(i,-1)$$

into Eq. (3.2.1) gives the starting formula

$$\Phi(i,1) = \frac{1}{2}[\Phi(i-1,0) + \Phi(i+1,0)]. \quad (3.2.3)$$

Since $u = 1$, and $r = 1$, $\Delta t = \Delta x$. Also, since the problem is symmetric with respect to $x = 0.5$, we solve for Φ using Eqs. (3.2.2) and (3.2.3) within $0 < x < 0.5$, $t \geq 0$. We can either calculate the values by hand or write a simple computer program. With the FORTRAN code in Fig. 3.8, the result shown in Table 3.5 is obtained for $\Delta t = \Delta x = 0.1$. The finite difference solution agrees with the exact solution in Eq. (3.2.1) to six decimal places. The accuracy of the FD solution can be increased by choosing a smaller spatial increment Δx and a smaller time increment Δt.

```
0001
0002      C ****************************************************
0003      C FORTRAN CODE FOR EXAMPLE 3.2
0004      C ON ONE-DIMENSIONAL WAVE EQUATION
0005      C SOLVED USING AN EXPLICIT FINITE DIFFERENCE SCHEME
0006      C ****************************************************
0007
0008            DIMENSION PHI(0:50,0:200), PHIEX(0:50,0:200)
0009            DATA  PIE/3.141592654/
0010      C SET SPATIAL AND TIME INCREMENTS
0011            DX = 0.1
0012            DT = 0.1
0013      C INITIALIZE - THIS ALSO TAKES CARE OF BOUNDARY CONDITIONS
0014            DO 10 I = 0,10
0015            DO 10 J = 0,20
0016            PHI(I,J) = 0.0
0017       10   CONTINUE
0018      C INSERT THE INITIAL CONDITIONS
0019            DO 20 I = 0,10
0020            X = DX*FLOAT(I)
0021            PHI(I,0) = SIN(PIE*X)
0022       20   CONTINUE
0023            DO 30 I=1,9
0024            PHI(I,1) = ( PHI(I-1,0) + PHI(I+1,0) )/2.0
0025       30   CONTINUE
0026      C NOW APPLY THE EXPLICIT FD SCHEME
0027            DO 40 J = 1,10
0028            DO 40 I = 1,9
0029              PHI(I,J+1) = PHI(I-1,J) + PHI(I+1,J) - PHI(I,J-1)
0030       40   CONTINUE
0031      C CALCULATE EXACT RESULT
0032            DO 50 J = 0,10
0033            T = DT*FLOAT(J)
0034            CT = COS(PIE*T)
0035            DO 50 I=0,10
0036            X = DX*FLOAT(I)
0037            PHIEX(I,J) = SIN(PIE*X)*CT
0038       50   CONTINUE
0039      C OUTPUT THE FD APPROXIMATE AND EXACT RESULTS
0040            DO 70 J = 0,7
0041            DO 70 I = 0,7
0042            WRITE(6,60) J,I,PHI(I,J),PHIEX(I,J)
0043       60   FORMAT(3X,'J=',I3,3X,'I=',I3,3X,2(F14.10,3X),/)
0044       70   CONTINUE
0045          STOP
0046        END
```

Figure 3.8 FORTRAN code for Example 3.2.

3.5 Finite Differencing of Elliptic PDEs

A typical elliptic PDE is Poisson's equation, which in two dimensions is given by

$$\nabla^2 \Phi = \frac{\partial^2 \Phi}{\partial x^2} + \frac{\partial^2 \Phi}{\partial y^2} = g(x,y). \tag{3.23}$$

We can use the central difference approximation for the partial derivatives of which the simplest forms are

$$\frac{\partial^2 \Phi}{\partial x^2} = \frac{\Phi(i+1,j) - 2\Phi(i,j) + \Phi(i-1,j)}{(\Delta x)^2} + O(\Delta x)^2 \tag{3.24a}$$

$$\frac{\partial^2 \Phi}{\partial y^2} = \frac{\Phi(i,j+1) - 2\Phi(i,j) + \Phi(i,j-1)}{(\Delta y)^2} + O(\Delta y)^2 \tag{3.24b}$$

where $x = i\Delta x, y = j\Delta y$, and $i,j = 0,1,2,\cdots$. If we assume that $\Delta x = \Delta y = h$, to simplify calculations, substituting Eq. (3.24) into Eq. (3.23) gives

$$\left[\Phi(i+1,j) + \Phi(i-1,j) + \Phi(i,j+1) + \Phi(i,j-1)\right] - 4\Phi(i,j) = h^2 g(i,j)$$

or

$$\boxed{\Phi(i,j) = \frac{1}{4}\left[\Phi(i+1,j) + \Phi(i-1,j) + \Phi(i,j+1) + \Phi(i,j-1) - h^2 g(i,j)\right]} \tag{3.25}$$

at every point (i,j) in the mesh for Poisson's equation. The spatial increment h is called the *mesh size*. A special case of Eq. (3.23) is when the source term vanishes, i.e., $g(x,y) = 0$. This leads to Laplace's equation. Thus for Laplace's equation, Eq. (3.25) becomes

$$\boxed{\Phi(i,j) = \frac{1}{4}\left[\Phi(i+1,j) + \Phi(i-1,j) + \Phi(i,j+1) + \Phi(i,j-1)\right]} \tag{3.26}$$

It is worth noting that Eq. (3.26) states that the value of Φ for each point is the average of those at the four surrounding points. The five-point computational molecule for the difference scheme in Eq. (3.26) is illustrated in Fig. 3.9(a) where values of the coefficients are shown. This is

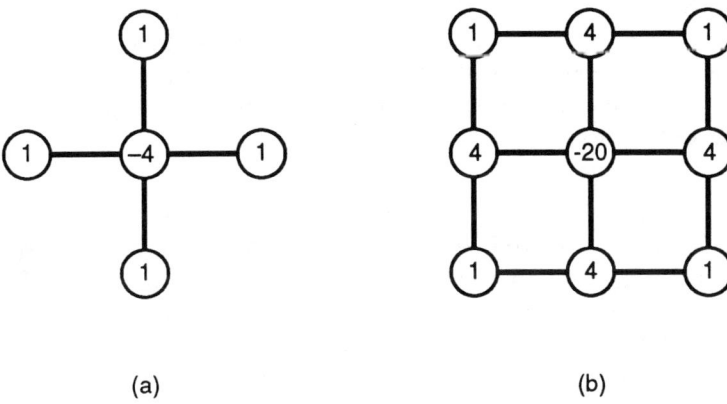

Figure 3.9 Computational molecules for Laplace's equation based on: (a) second order approximation, (b) fourth order approximation.

a convenient way of displaying finite difference algorithms for elliptic PDEs. The molecule in Fig. 3.9(a) is the second order approximation of Laplace's equation. This is obviously not the only way to approximate Laplace's equation, but it is the most popular choice. An alternative fourth order difference is

$$-20\Phi(i,j) + 4\Big[\Phi(i+1,j) + \Phi(i-1,j) + \Phi(i,j+1) + \Phi(i,j-1)\Big]$$
$$+ \Phi(i+1,j-1) + \Phi(i-1,j-1) + \Phi(i-1,j+1)$$
$$+ \Phi(i+1,j+1) = 0. \qquad (3.27)$$

The corresponding computational molecule is shown in Fig. 3.9(b).

The application of the finite difference method to elliptic PDEs often leads to a large system of algebraic equations, and their solution is a major problem in itself. Two commonly used methods of solving the system of equations are band matrix and iterative methods.

3.5.1 Band Matrix Method

From Eqs. (3.25) to (3.27), we notice that only nearest neighboring nodes affect the value of Φ at each node. Hence application of any of Eqs. (3.25) to (3.27) to all free nodes in the solution region results in a set of simultaneous equations of the form

$$[A][X] = [B] \qquad (3.28)$$

where $[A]$ is a *sparse* matrix (it has many zero elements), $[X]$ is a column matrix consisting of the unknown values of Φ at the free nodes, and $[B]$ is a column matrix containing the known values of Φ at fixed nodes. Matrix $[A]$ is also banded in that its nonzero terms appear clustered near the main diagonal. Matrix $[X]$, containing the unknown elements, can be obtained from

$$[X] = [A]^{-1}[B] \qquad (3.29)$$

or by solving Eq. (3.28) using the Gauss elimination discussed in Appendix D.1.

3.5.2 Iterative Methods

The iterative methods are generally used to solve a large system of simultaneous equations. An iterative method for solving equations is one in which a first approximation is used to calculate a second approximation, which in turn is used to calculate the third approximation, and so on. The three common iterative methods (Jacobi, Gauss-Seidel, and successive over-relaxation (SOR)) are discussed in Appendix D.2. We will apply only SOR here.

To apply the method of SOR to Eq. (3.25), for example, we first define the *residual* $R(i,j)$ at node (i,j) as the amount by which the value of $\Phi(i,j)$ does not satisfy Eq. (3.25), i.e.,

$$R(i,j) = \Phi(i+1,j) + \Phi(i-1,j) + \Phi(i,j+1) + \Phi(i,j-1) - 4\Phi(i,j) - h^2 g(i,j). \qquad (3.30)$$

The value of the residual at kth iteration, denoted by $R^k(i,j)$, may be regarded as a correction which must be added to $\Phi(i,j)$ to make it nearer to the correct value. As convergence to the correct value is approached,

$R^k(i,j)$ tends to zero. Hence to improve the rate of convergence, we multiply the residual by a number ω and add that to $\Phi(i,j)$ at the kth iteration to get $\Phi(i,j)$ at $(k+1)$th iteration. Thus

$$\Phi^{k+1}(i,j) = \Phi^k(i,j) + \frac{\omega}{4} R^k(i,j)$$

or

$$\Phi^{k+1}(i,j) = \Phi^k(i,j) + \frac{\omega}{4}\left[\Phi^k(i+1,j) + \Phi^{k+1}(i-1,j) + \Phi^{k+1}(i,j-1)\right.$$
$$\left. - 4\,\Phi^k(i,j) - h^2 g(i,j)\right]. \quad (3.31)$$

The parameter ω is called the *relaxation factor* while the technique is known as the method of *successive over-relaxation* (SOR). The value of ω lies between 1 and 2. (When $\omega = 1$, the method is simply called successive relaxation.) Its optimum value ω_{opt} must be found by trial-and-error. In order to start Eq. (3.31), an initial guess, $\Phi^0(i,j)$, is made at every free node. Typically, we may choose $\Phi^0(i,j) = 0$ or the average of Φ at the fixed nodes.

Example 3.3

Solve Laplace's equation

$$\nabla^2 V = 0, \qquad 0 \le x,\ y \le 1$$

with $V(x,1) = 45x(1-x), V(x,0) = 0 = V(0,y) = V(1,y)$.

Solution

Let $h = 1/3$ so that the solution region is as in Fig. 3.10. Applying Eq. (3.26) to each of the four points leads to

$$4V_1 - V_2 - V_3 - 0 = 10$$
$$-V_1 + 4V_2 - 0 - V_4 = 10$$
$$-V_1 - 0 + 4V_3 - V_4 = 0$$
$$-0 - V_2 - V_3 + 4V_4 = 0.$$

This can be written as

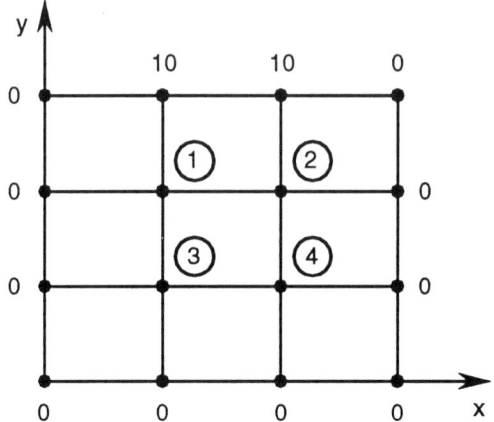

Figure 3.10 Finite difference grid for Example 3.3.

$$\begin{bmatrix} 4 & -1 & -1 & 0 \\ -1 & 4 & 0 & -1 \\ -1 & 0 & 4 & -1 \\ 0 & -1 & -1 & 4 \end{bmatrix} \begin{bmatrix} V_1 \\ V_2 \\ V_3 \\ V_4 \end{bmatrix} = \begin{bmatrix} 10 \\ 10 \\ 0 \\ 0 \end{bmatrix}$$

or

$$[A][V] = [B]$$

where $[A]$ is the band matrix, $[V]$ is the column matrix containing the unknown potentials at the free nodes, and $[B]$ is the column matrix of potentials at the fixed nodes. Solving the equations either by matrix inversion or by Gauss elimination, we obtain

$$V_1 = 3.75, \quad V_2 = 3.75, \quad V_3 = 1.25, \quad V_4 = 1.25.$$

Example 3.4

Solve Poisson's equation

$$\nabla^2 V = -\frac{\rho_S}{\epsilon}, \qquad 0 \le x, y \le 1$$

and obtain the potential at the grid points shown in Fig. 3.11. Assume $\rho_S = x(y-1)$ nC/m² and $\epsilon_r = 1.0$. Use the method of successive overrelaxation.

Finite Difference Methods 157

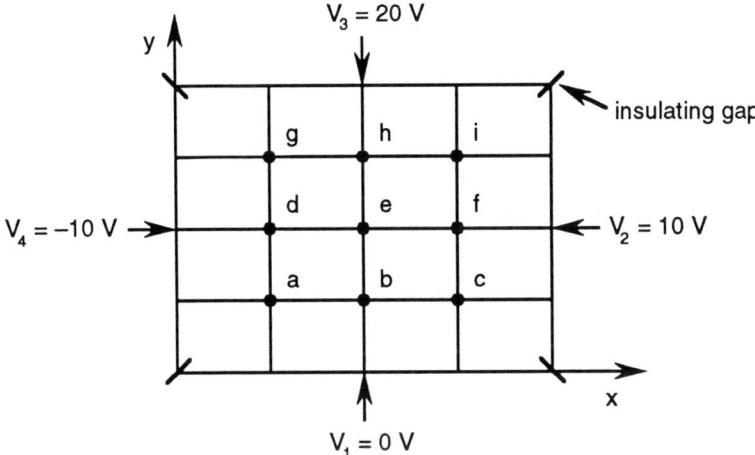

Figure 3.11 Solution region for the problem in Example 3.4.

Solution

This problem has an exact analytical solution and is deliberately chosen so that we can verify the numerical results with exact ones and also see how a problem with a complicated analytical solution is easily solved using finite difference method. For the exact solution, we use the superposition theorem and let

$$V = V_1 + V_2$$

where V_1 is the solution to Laplace's equation $\nabla^2 V_1 = 0$ with the inhomogeneous boundary conditions shown in Fig. 3.11 and V_2 is the solution to Poisson's equation $\nabla^2 V_2 = g = -\rho_S/\epsilon$ subject to homogeneous boundary conditions. From Example 2.1, it is evident that

$$V_1 = V_I + V_{II} + V_{III} + V_{IV}$$

where V_I to V_{IV} are defined by Eqs. (2.1.12) to (2.1.15). V_2 can be obtained by the series expansion method of Section 2.7. From example 2.12,

$$V_2 = \sum_{m=1}^{\infty} \sum_{n=1}^{\infty} A_{mn} \sin \frac{m\pi x}{a} \sin \frac{n\pi y}{b}$$

where

158 Numerical Techniques in Electromagnetics

$$A_{mn} = \int_0^a \int_0^b g(x,y) \sin\frac{m\pi x}{a} \sin\frac{n\pi y}{b}\, dx\, dy$$

$$= \frac{\left[1.0 - \frac{1}{b}[1 - (-1)^n]\right]}{[(m\pi/a)^2 + (n\pi/b)^2]} \cdot \frac{(-1)^{m+n}\, 144ab}{mn\pi},$$

$a = b = 1$, and $g(x,y) = -x(y-1).10^{-9}/\epsilon_o$.

For the finite difference solution, it can be shown that in a rectagular region, the optimum over-relaxation factor is given by the smaller root of the quadratic equation [10]

$$t^2\omega^2 - 16\omega + 16 = 0$$

where $t = \cos(\pi/N_x) + \cos(\pi/N_y)$ and N_x and N_y are the number of intervals along x and y axes, respectively. Hence

$$\omega = \frac{8 - \sqrt{64 - 16t^2}}{t^2}.$$

We try three cases of $N_x = N_y = 4, 12$, and 20 so that $\Delta x = \Delta y = h = 1/4, 1/12$, and $1/20$, respectively. Also we set

$$g(x,y) = -\frac{\rho_S}{\epsilon} = -\frac{x(y-1).10^{-9}}{10^{-9}/36\pi} = -36\pi x(y-1).$$

Figure 3.12 presents the FORTRAN code for the solution of this problem. The potentials at the free nodes for the different cases of h are shown in Table 3.6. Notice that as the mesh size h reduces, the solution becomes more accurate, but it takes more iterations for the same tolerance.

```
0001      C*************************************************************
0002      C     FINITE DIFFERENCE SOLUTION OF POISSON'S EQUATION
0003      C     Vxx + Vyy = G
0004      C     USING THE METHOD OF SUCCESSIVE OVER-RELAXATION
0005      C
0006      C     NX    :  NO. OF INTERVALS ALONG X-AXIS
0007      C     NY    :  NO. OF INTERVALS ALONG Y-AXIS
0008      C     A X B :  DIMENSION OF THE SOLUTION REGION
0009      C     V(I,J):  POTENTIAL AT GRID POINT (X,Y) = H*(I,J)
0010      C              WHERE 1 = 0,1,...,NX, J = 0,1,....,NY
0011      C     H     :  MESH SIZE
0012      C *************************************************************
0013
0014            DIMENSION V(0:20,0:20)
0015            DATA PIE/3.14159/
```

Table 3.6 Successive Over-relaxation Solution of Example 3.4.

Node	$h = 1/4$ $\omega_{opt} = 1.171$ 8 iterations	$h = 1/2$ $\omega_{opt} = 1.729$ 26 iterations	$h = 1/20$ $\omega_{opt} = 1.729$ 43 iterations	Exact Solution
a	-3.247	-3.409	-3.424	-3.429
b	-1.703	-1.982	-2.012	-2.029
c	4.305	4.279	4.277	4.277
d	-0.0393	-0.0961	-0.1087	-0.1182
e	3.012	2.928	2.921	2.913
f	9.368	9.556	9.578	9.593
g	3.044	2.921	2.909	2.902
h	6.111	6.072	6.069	6.065
i	11.04	11.12	11.23	11.13

```
0016            DATA A,B/1.0,1.0/
0017            DATA V1,V2,V3,V4/0.0,10.0,20.0,-10.0/
0018
0019      C     SPECIFY BOUNDARY VALUES AND NECESSARY PARAMETERS
0020            NX= 4
0021            NY= 4
0022            H = A/FLOAT(NX)
0023      C     SET INITIAL GUESS EQUAL TO ZEROS OR TO AVERAGE OF
0024      C     FIXED VALUES
0025            DO 10 I=1,NX-1
0026            DO 10 J=1,NY-1
0027            V(I,J)=(V1 + V2 + V3 + V4)/4.0
0028      10    CONTINUE
0029      C     SET POTENTIALS AT FIXED NODES
0030            DO 20 I = 1, NX-1
0031            V(I,0)=V1
0032            V(I,NY)=V3
0033      20    CONTINUE
0034            DO 30 J=1,NY-1
0035            V(0,J)=V4
0036            V(NX,J)=V2
0037      30    CONTINUE
0038            V(0,0)=(V1 + V4)/2.0
0039            V(NX,0)=(V1 + V2)/2.0
0040            V(0,NY)=(V3 + V4)/2.0
0041            V(NX,NY)=(V2 + V3)/2.0
0042      C     FIND THE OPTIMUM OVER-RELAXATION FACTOR
0043            T = COS(PIE/NX) + COS(PIE/NY)
0044            W = ( 8.0 - SQRT(64.0 - 16.0*T*T))/(T*T)
0045            WRITE(6,40) W
0046      40    FORMAT(2X,'SOR FACTOR OMEGA=',F10.6)
0047            W4 = W/4.0
0048      C     ITERATION BEGINS
0049            NCOUNT = 0
```

```
0050        50      RMIN = 0.0
0051                DO 70 I =1,NX-1
0052                X = H*FLOAT(I)
0053                DO 70 J = 1,NY-1
0054                Y = H*FLOAT(J)
0055                G = -36.0*PIE*X*(Y - 1.0)
0056                R = W4*( V(I+1,J) + V(I-1,J) + V(I,J+1) + V(I,J-1)
0057              1       -4.0*V(I,J) - G*H*H  )
0058                RMIN = RMIN + ABS(R)
0059                V(I,J) =  V(I,J) + R
0060        70      CONTINUE
0061                RMIN = RMIN/FLOAT(NX*NY)
0062                IF(RMIN.GE.0.0001) THEN
0063        C       SOLUTION HAS CONVERGED
0064                NCOUNT = NCOUNT + 1
0065                IF(NCOUNT.LT.100) THEN
0066                GO TO 50
0067                ELSE
0068                WRITE(6,80)
0069        80      FORMAT(2X,'SOLUTION DOES NOT CONVERGE IN 100 ITERATIONS')
0070                GO TO 100
0071                ENDIF
0072                ENDIF
0073        C       SOLUTION HAS CONVERGED
0074                WRITE(6,90) NCOUNT
0075        90      FORMAT(2X,'SOLUTION CONVERGES IN',2X,I3,2X,'ITERATIONS',/)
0076        100     CONTINUE
0077        C       OUTPUT THE FINITE DIFFERENCE APPROX. RESULTS
0078                DO 120 I = 1,NX-1
0079                DO 120 J = 1,NY-1
0080                WRITE(6,110) I,J,V(I,J)
0081        110     FORMAT(2X,'I=',I3,2X,'J=',I3,2X,'V =',E12.6,/)
0082        120     CONTINUE
0083        C       -----------------------------------------------------------
0084        C       CALCULATE THE EXACT SOLUTION
0085        C
0086        C       POISSON'S EQUATION WITH HOMOGENEOUS BOUNDARY CONDITIONS
0087        C       SOLVED BY SERIES EXPANSION
0088        C
0089                DO 150 I =1,NX-1
0090                X = H*FLOAT(I)
0091                DO 150 J = 1,NY-1
0092                Y = H*FLOAT(J)
0093                SUM = 0.0
0094                DO 130 M = 1,10     ! TAKE ONLY 10 TERMS OF THE SERIES
0095                FM = FLOAT(M)
0096                DO 130 N = 1,10
0097                FN = FLOAT(N)
0098                FACTOR1 = (FM*PIE/A)**2  +  (FN*PIE/B)**2
0099                FACTOR2 = ( (-1.0)**(M+N) )*144.0*A*B/(PIE*FM*FN)
0100                FACTOR3 = 1.0 - (1.0 - (-1.0)**N)/B
0101                FACTOR = FACTOR2*FACTOR3/FACTOR1
0102                SUM = SUM + FACTOR*SIN(FM*PIE*X/A)*SIN(FN*PIE*Y/B)
0103        130     CONTINUE
0104                VH = SUM
0105        C
0106        C       LAPLACE'S EQUATION WITH INHOMOGENEOUS BOUNDARY CONDITIONS
0107        C       SOLVED USING THE METHOD OF SEPARATION OF VARIABLES
0108        C
0109                C1=4.0*V1/PIE
0110                C2=4.0*V2/PIE
0111                C3=4.0*V3/PIE
0112                C4=4.0*V4/PIE
```

```
0113            SUM=0.0
0114            DO 140 K =1,10   ! TAKE ONLY 10 TERMS OF THE SERIES
0115            N=2*K-1
0116            AN=FLOAT(N)
0117            A1=SIN(AN*PIE*X/B)
0118            A2=SINH(AN*PIE*(A-Y)/B)
0119            A3=AN*SINH(AN*PIE*A/B)
0120            TERM1=C1*A1*A2/A3
0121            B1=SINH(AN*PIE*X/A)
0122            B2=SIN(AN*PIE*Y/A)
0123            B3=AN*SINH(AN*PIE*B/A)
0124            TERM2=C2*B1*B2/B3
0125            D1=SIN(AN*PIE*X/B)
0126            D2=SINH(AN*PIE*Y/B)
0127            D3=AN*SINH(AN*PIE*A/B)
0128            TERM3=C3*D1*D2/D3
0129            E1=SINH(AN*PIE*(B-X)/A)
0130            E2=SIN(AN*PIE*Y/A)
0131            E3=AN*SINH(AN*PIE*B/A)
0132            TERM4=C4*E1*E2/E3
0133            TERM = TERM1 + TERM2 + TERM3 + TERM4
0134            SUM=SUM + TERM
0135   140      CONTINUE
0136            VI = SUM
0137            V(I,J) = VH + VI
0138   150      CONTINUE
0139   C
0140   C    OUTPUT EXACT RESULTS
0141   C
0142            DO 160 I = 1,NX-1
0143            DO 160 J = 1,NY-1
0144            WRITE(6,110) I,J,V(I,J)
0145   160      CONTINUE
0146            STOP
0147            END
```

Figure 3.12 FORTRAN code for Example 3.4.

3.6 Accuracy and Stability of FD Solutions

The question of accuracy and stability of numerical methods is extremely important if our solution is to be reliable and useful. Accuracy has to do with the closeness of the approximate solution to exact solutions (assuming they exist). Stability is the requirement that the scheme does not increase the magnitude of the solution with increase in time.

There are three sources of errors that are nearly unavoidable in numerical solution of physical problems [8]:

(1) modeling errors,
(2) truncation (or discretization) errors,
(3) roundoff errors.

Each of these error types will affect accuracy and therefore degrade the solution.

The modeling errors are due to several assumptions made in arriving at the mathematical model. For example, a nonlinear system may be represented by a linear PDE. Truncation errors arise from the fact that in numerical analysis, we can deal only with a finite number of terms from processes which are usually described by infinite series. For example, in deriving finite difference schemes, some higher-order terms in the Taylor series expansion were neglected, thereby introducing truncation error. Truncation errors may be reduced by using finer meshes, that is, by reducing the mesh size h and time increment Δt. Alternatively, truncation errors may be reduced by using a large number of terms in the series expansion of derivatives, that is, by using higher-order approximations. However, care must be exercised in applying higher-order approximations. Instability may result if we apply a difference equation of an order higher than the PDE being examined. These higher-order difference equations may introduce "spurious solutions."

Roundoff errors reflect the fact that computations can be done only with a finite precision on a computer. This unavoidable source of errors is due to the limited size of registers in the arithmetric unit of the computer. Roundoff errors can be minimized by the use of double-precision arithmetic. The only way to avoid roundoff errors completely is to code all operations using integer arithmetic. This is hardly possible in most practical situations.

Although it has been noted that reducing the mesh size h will increase accuracy, it is not possible to indefinitely reduce h. Decreasing the truncation error by using a finer mesh may result in increasing the roundoff error due to the increased number of arithmetic operations. A point is reached where the minimum total error occurs for any particular algorithm using any given word length [9]. This is illustrated in Fig. 3.13. The concern about accuracy leads us to question whether the finite difference solution can grow unbounded, a property termed the instability of the difference scheme. A numerical algorithm is said to be stable if a small error at any stage produces a smaller cumulative error. It is unstable otherwise. The consequence of instability (producing unbounded solution) is disastrous. To determine whether a finite difference scheme is stable, we define an error, ϵ^n, which occurs at time step n, assuming that there is one independent variable. We define the amplification of this error at time step $n+1$ as

$$\epsilon^{n+1} = g\epsilon^n \tag{3.32}$$

where g is known as the *amplification factor*. In more complex situations, we have two or more independent variables, and Eq. (3.32) becomes

$$[\epsilon]^{n+1} = [G][\epsilon]^n \tag{3.33}$$

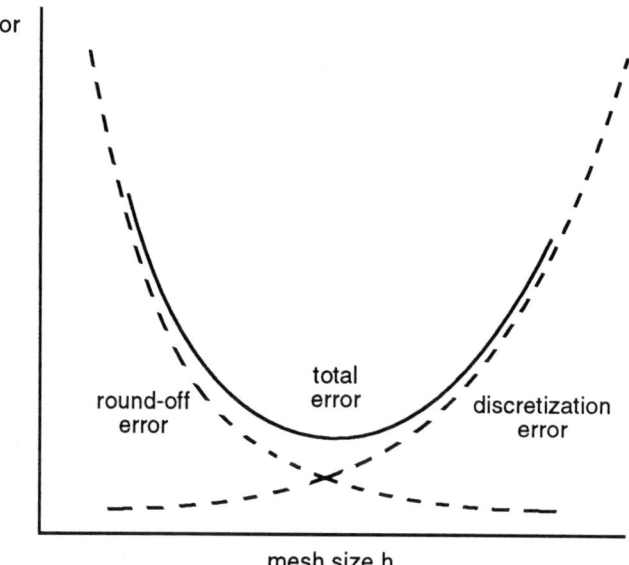

Figure 3.13 Error as a function of the mesh size.

where $[G]$ is the amplification matrix. For the stability of the difference scheme, it is required that Eq. (3.32) satisfy

$$\left|\epsilon^{n+1}\right| \leq \left|\epsilon^{n}\right|$$

or

$$|g| \leq 1. \tag{3.34a}$$

For the case in Eq. (3.33), the determinant of $[G]$ must vanish, i.e.,

$$\det[G] = 0. \tag{3.34b}$$

One useful and simple method of finding a stability criterion for a difference scheme is to construct a Fourier analysis of the difference equation and thereby derive the amplification factor. We illustrate this technique, known as *von Neumann's method* [4,5,7,10], by considering the explicit scheme of Eq. (3.13):

$$\Phi_i^{n+1} = (1 - 2r)\Phi_i^n + r\left(\Phi_{i+1}^n + \Phi_{i-1}^n\right) \tag{3.35}$$

where $r = \Delta t / k(\Delta x)^2$. We have changed our usual notation so that we can use $j = \sqrt{-1}$ in the Fourier series. Let the solution be

$$\Phi_i^n = \sum A_n(t)e^{jkix}, \quad 0 \le x \le 1 \tag{3.36a}$$

where k is the wave number. Since the differential equation (3.10) approximated by Eq. (3.13) is linear, we need consider only a Fourier mode, i.e.,

$$\Phi_i^n = A(t)e^{jkix}. \tag{3.36b}$$

Substituting Eq. (3.36b) into Eq. (3.35) gives

$$A^{n+1}e^{jkix} = (1-2r)A^n e^{jkix} + r\left(e^{jkx} + e^{-jkx}\right)A^n e^{jkix}$$

or

$$A^{n+1} = A^n[1 - 2r + 2\cos kx]. \tag{3.37}$$

Hence the amplification factor is

$$g = \frac{A^{n+1}}{A^n} = 1 - 2r + 2\cos kx$$
$$= 1 - 4r \sin^2 \frac{kx}{2}. \tag{3.38}$$

In order to satisfy Eq. (3.34a),

$$\left| 1 - 4r \sin^2 \frac{kx}{2} \right| \le 1.$$

Since this condition must hold for every wave number k, we take the maximum value of the sine function so that

$$1 - 4r \ge -1 \quad \text{and} \quad r \ge 0$$

or

$$r \ge \frac{1}{2} \quad \text{and} \quad r \ge 0.$$

Of course, $r = 0$ implies $\Delta t = 0$, which is impractical. Thus

$$0 < r \le \frac{1}{2}. \tag{3.39}$$

Example 3.5

For the finite difference scheme of Eq. (3.18), use the von Neumann approach to determine the stability condition.

Solution

We assume a trial solution of the form

$$\Phi_i^n = A^n e^{jkix}.$$

Substituting this into Eq. (3.18) results in

$$A^{n+1}e^{jkix} = 2(1-r)A^n e^{jkix} + r(e^{jkx} + e^{-jkx})A^n e^{jkix} - A^{n-1}e^{jkix}$$

or

$$A^{n+1} = A^n \left[2(1-r) + 2r\cos kx\right] - A^{n-1}. \tag{3.5.1}$$

In terms of $g = A^{n+1}/A^n$, Eq. (3.5.1) becomes

$$g^2 - 2pg + 1 = 0 \tag{3.5.2}$$

where $p = 1 - 2r\sin^2 \frac{kx}{2}$. The quadratic equation (3.5.2) has solutions

$$g_1 = p + [p^2 - 1]^{1/2}, \qquad g_2 = p - [p^2 - 1]^{1/2}.$$

For $|g_i| \leq 1$, where $i = 1, 2$, p must lie between 1 and -1, i.e., $-1 \leq p \leq 1$ or

$$-1 \leq 1 - 2r\sin^2 \frac{kx}{2} \leq 1$$

which implies that $r \leq 1$ or $u\Delta t \leq \Delta x$ for stability. This idea can be extended to show that the stability condition for two-dimensional wave equation is $u\Delta t/h < \frac{1}{\sqrt{2}}$, where $h = \Delta x = \Delta y$.

3.7 Practical Applications I—Guided Structures

The finite difference method has been applied successfully to solve many EM-related problems. Besides those simple examples we have considered earlier in this chapter, the method has been applied to diverse problems [11] including:

- transmission-line problems [12–21],
- waveguides [21–26],
- microwave circuit [27–30],
- EM penetration and scattering problems [31,32],

166   Numerical Techniques in Electromagnetics

- EM pulse (EMP) problems [33],
- EM exploration of minerals [34], and
- EM energy deposition in human bodies [35,36].

It is practically impossible to cover all those applications within the limited scope of this text. In this section, we consider the relatively easier problems of transmission lines and waveguides while the problems of penetration and scattering of EM waves will be treated in the next section. Other applications utilize basically similar techniques.

3.7.1 Transmission Lines

The finite difference techniques are suited for computing the characteristic impedance, phase velocity, and attenuation of several transmission lines—polygonal lines, shielded strip lines, coupled strip lines, microstrip lines, coaxial lines, and rectangular lines [12–19]. The knowledge of the basic parameters of these lines is of paramount importance in the design of microwave circuits.

For concreteness, consider the microstrip line shown in Fig. 3.14(a). The geometry in Fig. 3.14(a) is deliberately selected to be able to illustrate how one accounts for discrete inhomogeneities (i.e., homogeneous media separated by interfaces) and lines of symmetry using finite difference technique. The techniques presented are equally applicable to other lines. Due to the fact that the mode is TEM, having components of neither **E** nor **H** fields in the direction of propagation, the fields obey Laplace's equation over the line cross section. The TEM mode assumption provides good approximations if the line dimensions are much smaller than half a wavelength, which means that the operating frequency is far below cutoff frequency for all higher order modes [16]. Also owing to biaxial symmetry about the two axes only one quarter of the cross section need be considered as shown in Fig. 3.14(b).

The finite difference approximation of Laplace's equation, $\nabla^2 V = 0$, has been derived in Eq. (3.26), namely,

$$V(i,j) = \frac{1}{4}\Big[V(i+1,j) + V(i-1,j) + V(i,j+1) + V(i,j-1)\Big]. \quad (3.40)$$

For the sake of conciseness, let us denote

$$\begin{aligned}
V_o &= V(i,j) \\
V_1 &= V(i,j+1) \\
V_2 &= V(i-1,j) \\
V_3 &= V(i,j-1) \\
V_4 &= V(i+1,j)
\end{aligned} \quad (3.41)$$

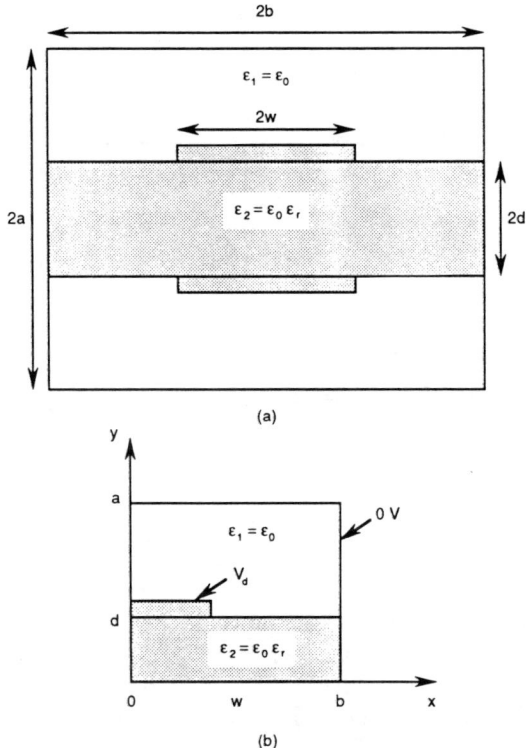

Figure 3.14 (a) Shielded double strip line with partial dielectric support; (b) problem in (a) simplified by making full use of symmetry.

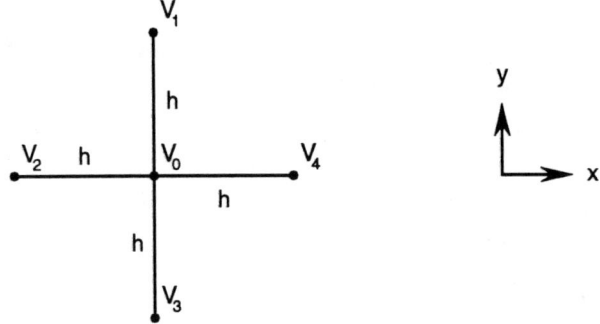

Figure 3.15 Computation molecule for Laplace's equation.

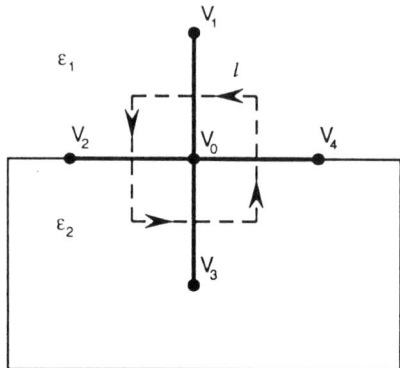

Figure 3.16 Interface between media of dielectric permittivities ϵ_1 and ϵ_2.

so that Eq. (3.40) becomes

$$V_o = \frac{1}{4}\left[V_1 + V_2 + V_3 + V_4\right] \quad (3.42)$$

with the computation molecule shown in Fig. 3.15. Equation (3.42) is the general formula to be applied to all free nodes in the free space and dielectric region of Fig. 3.14(b).

On the dielectric boundary, the boundary condition,

$$D_{1n} = D_{2n}, \quad (3.43)$$

must be imposed. We recall that this condition is based on Gauss's law for the electric field, i.e.,

$$\oint_\ell \mathbf{D} \cdot d\mathbf{l} = \oint_\ell \epsilon \mathbf{E} \cdot d\mathbf{l} = Q_{\text{enc}} = 0 \quad (3.44)$$

since no free charge is deliberately placed on the dielectric boundary. Substituting $\mathbf{E} = -\nabla V$ in Eq. (3.44) gives

$$0 = \oint_\ell \epsilon \nabla V \cdot d\mathbf{l} = \oint_\ell \epsilon \frac{\partial V}{\partial n} dl \quad (3.45)$$

where $\partial V/\partial n$ denotes the derivative of V normal to the contour ℓ. Applying Eq. (3.45) to the interface in Fig. 3.16 yields

$$0 = \epsilon_1 \frac{(V_1 - V_0)}{h}h + \epsilon_1 \frac{(V_2 - V_0)}{h}\frac{h}{2} + \epsilon_2 \frac{(V_2 - V_0)}{h}\frac{h}{2}$$
$$+ \epsilon_2 \frac{(V_3 - V_0)}{h}h + \epsilon_2 \frac{(V_4 - V_0)}{h}\frac{h}{2} + \epsilon_1 \frac{(V_4 - V_0)}{h}\frac{h}{2}.$$

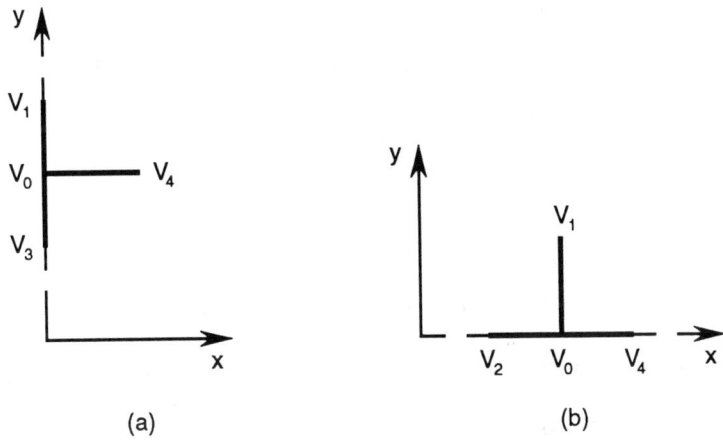

Figure 3.17 Computation molecule used for satisfying symmetry conditions: (a) $\partial V/\partial x = 0$, (b) $\partial V/\partial y = 0$.

Rearranging the terms,

$$2(\epsilon_1 + \epsilon_2)V_0 = \epsilon_1 V_1 + \epsilon_2 V_3 + \frac{(\epsilon_1 + \epsilon_2)}{2}(V_1 + V_4)$$

or

$$\boxed{V_0 = \frac{\epsilon_1}{2(\epsilon_1 + \epsilon_2)}V_1 + \frac{\epsilon_2}{2(\epsilon_1 + \epsilon_2)}V_3 + \frac{1}{4}V_2 + \frac{1}{4}V_4} \quad (3.46)$$

This is the finite difference equivalent of the boundary condition in Eq. (3.43). Notice that the discrete inhomogeneity does not affect points 2 and 4 on the boundary but affects points 1 and 3 in proportion to their corresponding permittivities. Also note that when $\epsilon_1 = \epsilon_2$, Eq. (3.46) reduces to Eq. (3.42).

On the line of symmetry, we impose the condition

$$\frac{\partial V}{\partial n} = 0. \quad (3.47)$$

This implies that on the line of symmetry along the y-axis, ($x = 0$ or $i = 0$) $\frac{\partial V}{\partial x} = (V_4 - V_2)/h = 0$ or $V_2 = V_4$ so that Eq. (3.42) becomes

$$\boxed{V_o = \frac{1}{4}\left[V_1 + V_3 + 2V_4\right]} \quad (3.48a)$$

or
$$V(0,j) = \frac{1}{4}\Big[V(0,j+1) + V(0,j-1) + 2V(1,j)\Big]. \qquad (3.48b)$$

On the line of symmetry along the x-axis ($y=0$ or $j=0$), $\dfrac{\partial V}{\partial y} = (V_1 - V_3)/h = 0$ or $V_3 = V_1$ so that

$$\boxed{V_o = \frac{1}{4}\Big[2V_1 + V_2 + V_4\Big]} \qquad (3.49a)$$

or
$$V(i,0) = \frac{1}{4}\Big[2V(i,1) + V(i-1,0) + V(i+1,0)\Big]. \qquad (3.49b)$$

The computation molecules for Eqs. (3.48) and (3.49) are displayed in Fig. 3.17.

By setting the potential at the fixed nodes equal to their prescribed values and applying Eqs. (3.42), (3.46), (3.48), and (3.49) to the free nodes according to the band matrix or iterative methods discussed in Section 3.5, the potential at the free nodes can be determined. Once this is accomplished, the quantities of interest can be calculated.

The characteristic impedance Z_o and phase velocity u of the line are defined as

$$Z_o = \sqrt{\frac{L}{C}} \qquad (3.50a)$$

$$u = \frac{1}{\sqrt{LC}} \qquad (3.50b)$$

where L and C are the inductance and capacitance per unit length, respectively. If the dielectric medium is nonmagnetic ($\mu = \mu_o$), the characteristic impedance Z_{oo} and phase velocity u_o with the dielectric removed (i.e., the line is air-filled) are given by

$$Z_{oo} = \sqrt{\frac{L}{C_o}} \qquad (3.51a)$$

$$u_o = \frac{1}{\sqrt{LC_o}} \qquad (3.51b)$$

where C_o is the capacitance per unit length without the dielectric. Combining Eqs. (3.50) and (3.51) yields

$$Z_o = \frac{1}{u_o\sqrt{CC_o}} = \frac{1}{uC} \qquad (3.52a)$$

$$u = u_o\sqrt{\frac{C_o}{C}} = \frac{u_o}{\sqrt{\epsilon_{\text{eff}}}} \qquad (3.52b)$$

$$\epsilon_{\text{eff}} = \frac{C}{C_o} \qquad (3.52c)$$

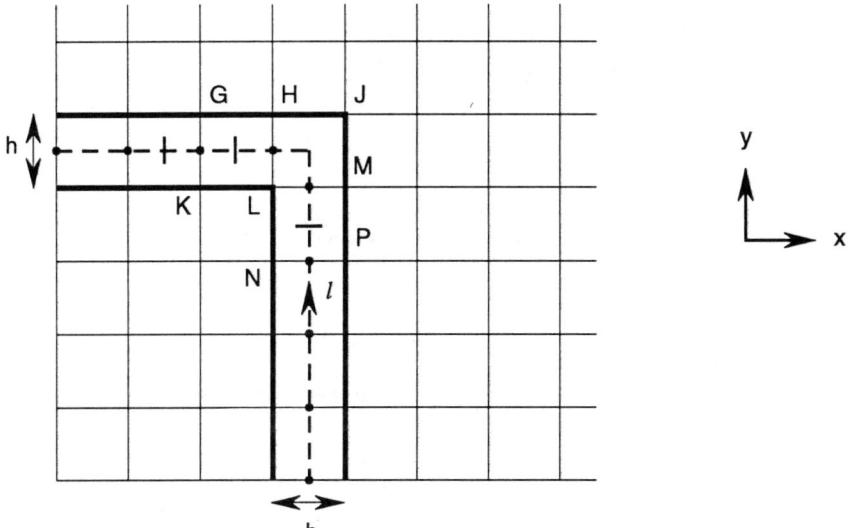

Figure 3.18 The rectangular path ℓ used in calculating charge enclosed.

where $u_o = c = 3 \times 10^8$ m/s, the speed of light in free space, and ϵ_{eff} is the effective dielectric constant. Thus to find Z_o and u for an inhomogeneous medium requires calculating the capacitance per unit length of the structure, with and without the dielectric substrate.

If V_d is the potential difference between the inner and the outer conductors,

$$C = \frac{4Q}{V_d}, \quad (3.53)$$

so that the problem is reduced to finding the charge per unit length Q. (The factor 4 is needed since we are working on only one quarter of the cross section.) To find Q, we apply Gauss's law to a closed path ℓ enclosing the inner conductor. We may select ℓ as the rectangular path between two adjacent rectangles as shown in Fig. 3.18.

$$Q = \oint_\ell \mathbf{D} \cdot d\mathbf{l} = \oint_\ell \epsilon \frac{\partial V}{\partial n} d\ell$$

$$= \epsilon \left(\frac{V_P - V_N}{\Delta x} \right) \Delta y + \epsilon \left(\frac{V_M - V_L}{\Delta x} \right) \Delta y + \epsilon \left(\frac{V_H - V_L}{\Delta y} \right) \Delta x$$

$$+ \epsilon \left(\frac{V_G - V_K}{\Delta y} \right) \Delta x + \cdots . \quad (3.54)$$

Since $\Delta x = \Delta y = h$,

$$Q = \left(\epsilon V_P + \epsilon V_M + \epsilon V_H + \epsilon V_G + \cdots\right) - \left(\epsilon V_N + 2\epsilon V_L + \epsilon V_K + \cdots\right)$$

or

$$Q = \epsilon_o \left[\sum \epsilon_{ri} V_i \text{ for nodes } i \text{ on external rectangle GHJMP} \right.$$
$$\left. \text{with corners (such as J) not counted}\right]$$
$$-\epsilon_o \left[\sum \epsilon_{ri} V_i \text{ for nodes } i \text{ on inner rectangle KLN} \right.$$
$$\left. \text{with corners (such as L) counted twice}\right], \quad (3.55)$$

where V_i and ϵ_{ri} are the potential and dielectric constant at the ith node. If i is on the dielectric interface, $\epsilon_{ri} = (\epsilon_{r1} + \epsilon_{r2})/2$. Also if i is on the line of symmetry, we use $V_i/2$ instead of V_i to avoid including V_i twice in Eq. (3.53), where factor 4 is applied. We also find

$$C_o = 4Q_o/V_d \quad (3.56)$$

where Q_o is obtained by removing the dielectric, finding V_i at the free nodes and then using Eq. (3.55) with $\epsilon_{r1} = 1$ at all nodes. Once Q and Q_o are calculated, we obtain C and C_o from Eqs. (3.53) and (3.56) and Z_o and u from Eq. (3.52).

An outline of the procedure is given below:

(1) Calculate V (with the dielectric space replaced by free space) using Eqs. (3.42), (3.46), (3.48), and (3.49).
(2) Determine Q using Eq. (3.55).
(3) Find $C_o = \dfrac{4Q}{V_d}$.
(4) Repeat steps (1) and (2) (with the dielectric space) and find $C = \dfrac{4Q}{V_d}$.
(5) Finally, calculate $Z_o = \dfrac{1}{c\sqrt{C\,C_o}}$, $c = 3 \times 10^8$ m/s.

The attenuation of the line can be calculated by following similar procedure outlined in [14,20,21]. The procedure for handling boundaries at infinity and that for boundary singularities in finite difference analysis are discussed in [37,38].

3.7.2 Waveguides

The solution of waveguide problems is well suited for finite difference schemes because the solution region is closed. This amounts to solving the Helmholtz or wave equation

$$\nabla^2 \Phi + k^2 \Phi = 0 \tag{3.57}$$

where $\Phi = E_z$ for TM modes or $\Phi = H_z$ for TE modes, while k is the wave number given by

$$k^2 = \omega^2 \mu \epsilon - \beta^2. \tag{3.58}$$

The permittivity ϵ of the dielectric medium can be real for a lossless medium or complex for a lossy medium. We consider all fields to vary with time and axial distance as $\exp j(\omega t - \beta z)$. In the eigenvalue problem of Eq. (3.57), both k and Φ are to be determined. The cutoff wavelength is $\lambda_c = 2\pi/k_c$. For each value of the cutoff wave number k_c, there is a solution for the eigenfunction Φ_i, which represents the field configuration of a propagating mode.

To apply the finite difference method, we discretize the cross section of the waveguide by a suitable square mesh. Applying Eq. (3.24) to Eq. (3.57) gives

$$\boxed{\Phi(i+1,j) + \Phi(i-1,j) + \Phi(i,j+1) + \Phi(i,j-1) - (4 - h^2 k^2)\Phi(i,j) = 0} \tag{3.59}$$

where $\Delta x = \Delta y = h$ is the mesh size. Equation (3.59) applies to all the free or interior nodes. At the boundary points, we apply Dirichlet condition ($\Phi = 0$) for the TM modes and Neumann condition ($\partial \Phi / \partial n = 0$) for the TE modes. This implies that at point A in Fig. 3.19, for example,

$$\Phi_A = 0 \tag{3.60}$$

for TM modes. At point A, $\partial \Phi / \partial n = 0$ implies that $\Phi_D = \Phi_E$ so that Eq. (3.57) becomes

$$\Phi_B + \Phi_C + 2\Phi_D - (4 - h^2 k^2)\Phi_A = 0 \tag{3.61}$$

for TE modes. By applying Eq. (3.59) and either Eq. (3.60) or (3.61) to all mesh points in the waveguide cross section, we obtain m simultaneous equations involving the m unknowns ($\Phi_1, \Phi_2, \cdots, \Phi_m$). These simultaneous equations may be conveniently cast into the matrix equation

$$(A - \lambda I)\Phi = 0 \tag{3.62a}$$

174 Numerical Techniques in Electromagnetics

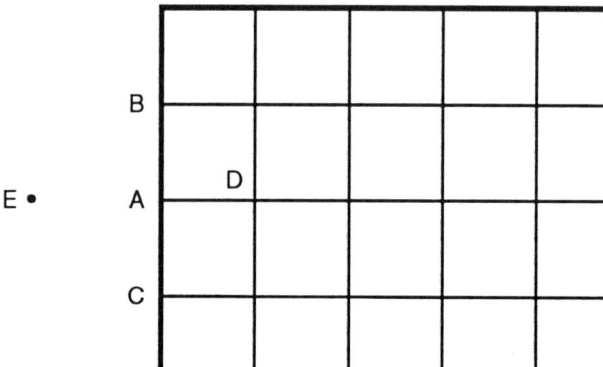

Figure 3.19 Finite difference mesh for a waveguide.

or
$$A\Phi = \lambda\Phi \tag{3.62b}$$

where A is an $m \times m$ band matrix of known integer elements, I is an identity matrix, $\Phi = (\Phi_1, \Phi_2, \cdots, \Phi_m)$ is the eigenvector, and

$$\lambda = (kh)^2 = \left(\frac{2\pi h}{\lambda_c}\right)^2 \tag{3.63}$$

is the eigenvalue. There are several ways of determining λ and the corresponding Φ. We consider two of these options.

The first option is the *direct method*. Equation (3.62) can be satisfied only if the determinant of $(A - \lambda I)$ vanishes, i.e.,

$$|A - \lambda I| = 0. \tag{3.64}$$

This results in a polynomial in λ, which can be solved [39] for the various eigenvalues λ. For each λ, we obtain the corresponding Φ from Eq. (3.59). This method requires storing the relevant matrix elements and does not take advantage of the fact that matrix A is sparse. In favor of the method is the fact that a computer subroutine usually exists [40] that solves the eigenvalue problem in Eq. (3.64) and that determines all the eigenvalues of the matrix. These eigenvalues give the dominant and higher modes of the waveguide, although accuracy deteriorates rapidly with mode number.

The second option is the *iterative method*. In this case, the matrix elements are usually generated rather than stored. We begin with $\Phi_1 = \Phi_2 = \cdots = \Phi_m = 1$ and a guessed value for k. The field Φ_{ij}^{k+1} at the (i,j)th node in the $(k+1)$th iteration is obtained from its known value in the kth iteration using

$$\Phi^{k+1}(i,j) = \Phi^k(i,j) + \frac{\omega R_{ij}}{(4 - h^2 k^2)} \tag{3.65}$$

where ω is the acceleration factor, $1 < \omega < 2$, and R_{ij} is the residual at the (i,j)th node given by

$$R_{ij} = \Phi(i,j) + \Phi(i,j-1) + \Phi(i+1,j) + \Phi(i-1,j) - (4 - h^2 k^2)\Phi(i,j). \quad (3.66)$$

After three or four scans of the complete mesh using Eq. (3.66), the value of $\lambda = h^2 k^2$ should be updated using Raleigh formula

$$k^2 = \frac{\int_S \Phi \nabla^2 \Phi \, dS}{\int_S \Phi^2 \, dS}. \quad (3.67)$$

The finite difference equivalent of Eq. (3.67) is

$$k^2 = \frac{\sum_{i=1}\sum_{j=1} \Phi(i,j)\Big[\Phi(i+1,j) + \Phi(i-1,j) + \Phi(i,j+1) + \Phi(i,j-1) - 4\Phi(i,j)\Big]}{h^2 \sum_{i=1}\sum_{j=1} \Phi^2(i,j)} \quad (3.68)$$

where Φs are the latest field values after three or four scans of the mesh and the summation is carried out over all points in the mesh. The new value of k obtained from Eq. (3.68) is now used in applying Eq. (3.65) over the mesh for another three or four times to give more accurate field values, which are again substituted into Eq. (3.68) to update k. This process is continued until the difference between consecutive values of k is within a specified acceptable tolerance.

If the first option is to be applied, matrix A must first be found. To obtain matrix A is not easy. Assuming TM modes, one way of calculating A is to number the free nodes from left to right, bottom to top, starting from the left-hand corner as shown typically in Fig. 3.20. If there are n_x and n_y divisions along the x and y directions, the number of free nodes is

$$n_f = (n_x - 1)(n_y - 1). \quad (3.69)$$

Each free node must be assigned two sets of numbers, one to correspond to m in Φ_m and the other to correspond to (i,j) in $\Phi(i,j)$. An array $NL(i,j) = m$, $i = 1, 2, \cdots, n_x - 1$, $j = 1, 2, \cdots, n_y - 1$ is easily developed to relate the two numbering schemes. To determine the value of element A_{mn}, we search $NL(i,j)$ to find (i_m, j_m) and (i_n, j_n), which are the values

176 Numerical Techniques in Electromagnetics

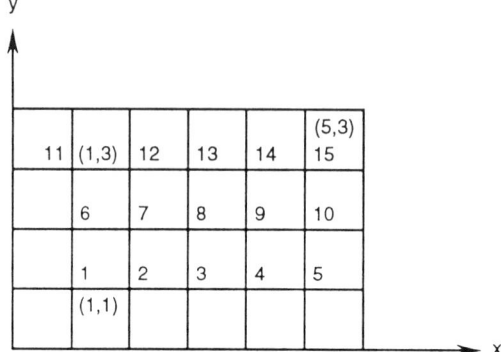

Figure 3.20 Relating node numbering schemes for $n_x = 6$, $n_y = 4$.

of (i,j) corresponding to nodes m and n, respectively. With these ideas, we obtain

$$A_{mn} = \begin{cases} 4, & m = n \\ -1, & i_m = i_n, \quad j_m = j_n + 1 \\ -1, & i_m = i_n, \quad j_m = j_n - 1 \\ -1, & i_m = i_n + 1, \quad j_m = j_n \\ -1, & i_m = i_n - 1, \quad j_m = j_n \\ 0, & \text{otherwise.} \end{cases} \qquad (3.70)$$

Example 3.6

Calculate Z_o for the microstrip transmission line in Fig. 3.14 with

$$a = b = 2.5 \text{ cm}, \quad d = 0.5 \text{ cm}, \quad w = 1 \text{ cm}$$

$$t = 0.001 \text{ cm}, \quad \epsilon_1 = \epsilon_o, \quad \epsilon_2 = 2.35\epsilon_o.$$

Solution

This problem is representative of the various types of problems that can be solved using the concepts developed in Section 3.7.1. The computer program in Fig. 3.21 was developed based on the five-step procedure outlined above. By specifying the step size h and the number of iterations, the program first sets the potential at all nodes equal to zero. The potential on the outer conductor is set equal to zero, while that on the inner conductor is set to 100 volts so that $V_d = 100$. The program finds C_o when the dielectric

Finite Difference Methods

Table 3.7 Characteristic Impedance of a Microstrip Line for Example 3.6

h	Number of iterations	Z_o
0.25	700	49.05
0.1	500	57.85
0.05	500	65.82
0.05	700	63.10
0.05	1000	61.53
Other method [41]: $Z_o = 62.50$		

slab is removed and C when the slab is in place and finally determines Z_o. For a selected h, the number of iterations must be large enough and greater than the number of divisions along x or y direction. Table 3.7 shows some typical results.

```
0001    C**********************************************************
0002    C USING THE FINITE DIFFERENCE METHOD
0003    C THIS PROGRAM CALCULATES THE CHARACTERISTIC IMPEDANCE
0004    C OF A MICROSTRIP LINE
0005    C**********************************************************
0006
0007          DIMENSION V(-1:200,-1:200), SV(2),Q(2)
0008          DATA A,B,D,W/2.5,2.5,0.5,1.0/ ! line data
0009          DATA ER,E0,U/2.35,8.81E-12,3.0E+8/
0010
0011          H = 0.025
0012          NT = 1000    ! NO. OF ITERATIONS
0013          NX = A/H
0014          NY = B/H
0015          ND = D/H
0016          NW = W/H
0017          VD = 100.0
0018    C
0019    C CALCULATE CHARGE WITH AND WITHOUT DIELECTRIC
0020    C
0021          ERR = 1.0
0022          DO 90 L=1,2
0023          E1 = E0
0024          E2 = E0*ERR
0025    C
0026    C INITIALIZATION
0027    C
0028          DO 10 I=0,NX
0029          DO 10 J=0,NY
0030          V(I,J) = 0.0
0031    10    CONTINUE
0032    C
0033    C SET POTENTIAL ON INNER CONDUCTOR (FIXED NODES) EQUAL
          TO VD
0034    C
```

```
0035              DO 20 I=0,NW
0036              V(I,ND) = VD
0037       20     CONTINUE
0038       C
0039       C  CALCULATE POTENTIAL AT FREE NODES
0040       C
0041              P1 = E1/(2.0*(E1 + E2))
0042              P2 = E2/(2.0*(E1 + E2))
0043              DO 50 K=1,NT
0044              DO 40 I=0,NX-1
0045              DO 40 J=0,NX-1
0046              IF( (J.EQ.ND).AND.(I.LE.NW) ) GO TO 40
0047              IF(J.EQ.ND) THEN
0048       C IMPOSE BOUNDARY CONDITION AT THE INTERFACE
0049              V(I,J) = 0.25*(V(I+1,J) + V(I-1,J)) +
0050         1             P1*V(I,J+1) + P2*V(I,J-1)
0051              GO TO 40
0052              ENDIF
0053              IF(I.EQ.0) THEN
0054       C IMPOSE SYMMETRY CONDITION ALONG Y-AXIS
0055              V(I,J) = (2.0*V(I+1,J) + V(I,J+1) + V(I,J-1) )/4.0
0056              GO TO 40
0057              ENDIF
0058              IF(J.EQ.0) THEN
0059       C IMPOSE SYMMETRY CONDITION ALONG X-AXIS
0060              V(I,J) = (V(I+1,J) + V(I-1,J) + 2.0*V(I,J+1) )/4.0
0061              GO TO 40
0062              ENDIF
0063       30     V(I,J) =(V(I+1,J)+V(I-1,J)+V(I,J+1)+V(I,J-1))/4.0
0064       40     CONTINUE
0065       50     CONTINUE
0066       C
0067       C  NOW, CALCULATE THE TOTAL CHARGE ENCLOSED IN A
0068       C  RECTANGULAR PATH SURROUNDING THE INNER CONDUCTOR
0069       C
0070              IOUT = (NX + NW)/2
0071              JOUT = (NY + ND)/2
0072       C SUM POTENTIAL ON INNER AND OUTER LOOPS
0073              DO 80 K=1,2
0074              SUM = 0.0
0075              DO 60 I=1,IOUT-1
0076              SUM = SUM + E1*V(I,JOUT)
0077       60     CONTINUE
0078              SUM = SUM + E1*V(0,JOUT)/2.0 ! SYMMETRY POINT
0079              DO 70 J=1,JOUT-1
0080              IF(J.LT.ND) SUM = SUM + E2*V(IOUT,J)
0081              IF(J.EQ.ND) SUM = SUM + (E1+E2)*V(IOUT,J)/2.0
0082              IF(J.GT.ND) SUM = SUM + E1*V(IOUT,J)
0083       70     CONTINUE
0084              SUM = SUM + E2*V(IOUT,0)/2.0 ! SYMMETRY POINT
0085              IF(K.EQ.1) SV(1) = SUM
0086              IOUT = IOUT -1   ! FOR INNER LOOP
0087              JOUT = JOUT -1
0088       80     CONTINUE
0089              SUM = SUM + 2.0*E1*V(IOUT,JOUT) ! CORNER POINT
0090              SV(2) = SUM
0091              Q(L) = ABS( SV(1) - SV(2) )
0092              ERR = ER
0093       90     CONTINUE
0094       C
0095       C  FINALLY, CALCULATE Zo
0096       C
0097              CO = 4.0*Q(1)/VD
```

```
0098            C1 = 4.0*Q(2)/VD
0099            Z0 = 1.0/( U*SQRT(C0*C1) )
0100            WRITE(6,*) H,NT,Z0
0101            PRINT *,H,NT,Z0
0102            STOP
0103            END
```

Figure 3.21 Computer program for Example 6.3.

3.8 Practical Applications II—Wave Scattering (FD-TD)

The finite-difference time-domain (FD-TD) formulation of EM field problems is a convenient tool for solving scattering problems. The FD-TD method, first introduced by Yee [42] in 1966 and later developed by Taflove and others [31,32,35,43–46], is a direct solution of Maxwell's time-dependent curl equations. The scheme treats the irradiation of the scatterer as an initial value problem. Our discussion on the FD-TD method will cover:

- Yee's finite difference algorithm,
- accuracy and stability,
- lattice (boundary) truncation conditions,
- initial fields, and
- programming aspects.

Some model examples with FORTRAN codes will be provided to illustrate the method.

3.8.1 Yee's Finite Difference Algorithm

In an isotropic medium, Maxwell's equations can be written as

$$\nabla \times \mathbf{E} = -\mu \frac{\partial \mathbf{H}}{\partial t} \qquad (3.71a)$$

$$\nabla \times \mathbf{H} = \sigma \mathbf{E} + \epsilon \frac{\partial \mathbf{E}}{\partial t}. \qquad (3.71b)$$

The vector Eq. (3.71) represents a system of six scalar equations, which can be expressed in rectangular coordinate system (x, y, z) as:

$$\frac{\partial H_x}{\partial t} = \frac{1}{\mu}\left(\frac{\partial E_y}{\partial z} - \frac{\partial E_z}{\partial y}\right), \qquad (3.72a)$$

$$\frac{\partial H_y}{\partial t} = \frac{1}{\mu}\left(\frac{\partial E_z}{\partial x} - \frac{\partial E_x}{\partial z}\right), \qquad (3.72b)$$

$$\frac{\partial H_z}{\partial t} = \frac{1}{\mu}\left(\frac{\partial E_x}{\partial y} - \frac{\partial E_y}{\partial x}\right), \quad (3.72\text{c})$$

$$\frac{\partial E_x}{\partial t} = \frac{1}{\epsilon}\left(\frac{\partial H_z}{\partial y} - \frac{\partial H_y}{\partial z} - \sigma E_x\right), \quad (3.72\text{d})$$

$$\frac{\partial E_y}{\partial t} = \frac{1}{\epsilon}\left(\frac{\partial H_x}{\partial z} - \frac{\partial H_z}{\partial x} - \sigma E_y\right), \quad (3.72\text{e})$$

$$\frac{\partial E_z}{\partial t} = \frac{1}{\epsilon}\left(\frac{\partial H_y}{\partial x} - \frac{\partial H_x}{\partial y} - \sigma E_z\right). \quad (3.72\text{f})$$

Following Yee's notation, we define a grid point in the solution region as

$$(i,j,k) = (i\Delta x, j\Delta y, k\Delta z) \quad (3.73)$$

and any function of space and time as

$$F^n(i,j,k) = F(i\delta, j\delta, k\delta, n\Delta t) \quad (3.74)$$

where $\delta = \Delta x = \Delta y = \Delta z$ is the space increment, Δt is the time increment, while i, j, k, and n are integers. Using central finite difference approximation for space and time derivatives that are second-order accurate,

$$\frac{\partial F^n(i,j,k)}{\partial x} = \frac{F^n(i+1/2,j,k) - F^n(i-1/2,j,k)}{\delta} + O(\delta^2) \quad (3.75)$$

$$\frac{\partial F^n(i,j,k)}{\partial t} = \frac{F^{n+1/2}(i,j,k) - F^{n-1/2}(i,j,k)}{\Delta t} + O(\Delta t^2). \quad (3.76)$$

In applying Eq. (3.75) to all the space derivatives in Eq. (3.72), Yee positions the components of **E** and **H** about a unit cell of the lattice as shown in Fig. 3.22. To incorporate Eq. (3.76), the components of **E** and **H** are evaluated at alternate half-time steps. Thus we obtain the explicit finite difference approximation of Eq. (3.72) as:

$$\boxed{\begin{aligned}H_x^{n+1/2}(i,j+1/2,k+1/2) &= H_x^{n-1/2}(i,j+1/2,k+1/2) \\ &+ \frac{\delta t}{\mu(i,j+1/2,k+1/2)\delta}\Big[E_y^n(i,j+1/2,k+1) \\ &- E_y^n(i,j+1/2,k) \\ &+ E_z^n(i,j,k+1/2) - E_z^n(i,j+1,k+1/2)\Big],\end{aligned}} \quad (3.77\text{a})$$

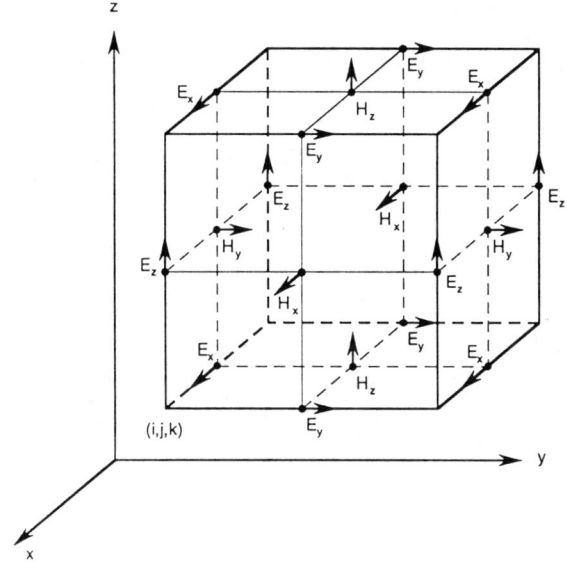

Figure 3.22 Positions of the field components in a unit cell of the Yee's lattice.

$$\begin{aligned}H_y^{n+1/2}(i+1/2,j,k+1/2) &= H_y^{n-1/2}(i+1/2,j,k+1/2) \\ &+ \frac{\delta t}{\mu(i+1/2,j,k+1/2)\delta}\Big[E_z^n(i+1,j,k+1/2) \\ &- E_z^n(i,j,k+1/2) \\ &+ E_x^n(i+1/2,j,k) - E_x^n(i+1/2,j,k+1)\Big],\end{aligned} \qquad (3.77b)$$

$$H_z^{n+1/2}(i+1/2,j+1/2,k) = H_z^{n-1/2}(i+1/2,j+1/2,k)$$

$$+ \frac{\delta t}{\mu(i+1/2,j+1/2,k)\delta} \Big[E_x^n(i+1/2,j+1,k)$$

$$- E_x^n(i+1/2,j,k)$$

$$+ E_y^n(i,j+1/2,k) - E_y^n(i+1,j+1/2,k) \Big],$$

(3.77c)

$$E_x^{n+1}(i+1/2,j,k) = \left(1 - \frac{\sigma(i+1/2,j,k)\delta t}{\epsilon(i+1/2,j,k)}\right) E_x^n(i+1/2,j,k)$$

$$+ \frac{\delta t}{\epsilon(i+1/2,j,k)\delta} \Big[H_z^{n+1/2}(i+1/2,j+1/2,k)$$

$$- H_z^{n+1/2}(i+1/2,j-1/2,k)$$

$$+ H_y^{n+1/2}(i+1/2,j,k-1/2)$$

$$- H_y^{n+1/2}(i+1/2,j,k+1/2) \Big],$$

(3.77d)

$$E_y^{n+1}(i,j+1/2,k) = \left(1 - \frac{\sigma(i,j+1/2,k)\delta t}{\epsilon(i,j+1/2,k)}\right) E_y^n(i,j+1/2,k)$$

$$+ \frac{\delta t}{\epsilon(i,j+1/2,k)\delta} \Big[H_x^{n+1/2}(i,j+1/2,k+1/2)$$

$$- H_x^{n+1/2}(i,j+1/2,k-1/2)$$

$$+ H_z^{n+1/2}(i-1/2,j+1/2,k)$$

$$- H_z^{n+1/2}(i+1/2,j+1/2,k) \Big],$$

(3.77e)

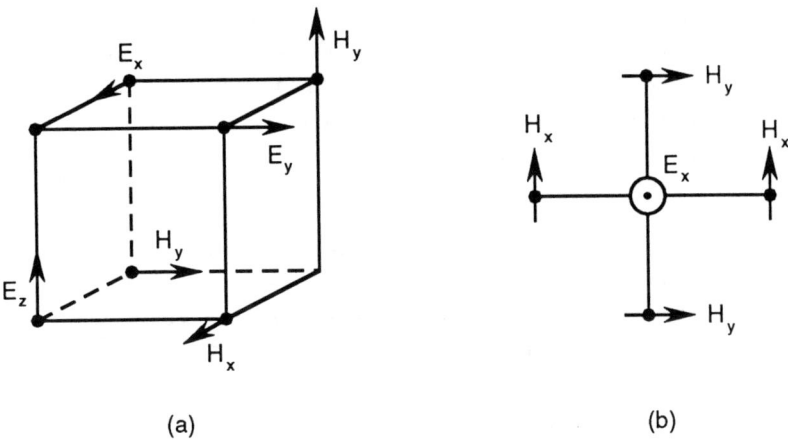

Figure 3.23 Typical relations between field components: (a) within a quarter of a unit cell, (b) in a plane.

$$E_z^{n+1}(i,j,k+1/2) = \left(1 - \frac{\sigma(i,j,k+1/2)\delta t}{\epsilon(i,j,k+1/2)}\right) E_z^n(i,j,k+1/2)$$

$$+ \frac{\delta t}{\epsilon(i,j,k+1/2)\delta} \Big[H_y^{n+1/2}(i+1/2,j,k+1/2)$$

$$- H_y^{n+1/2}(i-1/2,j,k+1/2) \quad (3.77f)$$

$$+ H_x^{n+1/2}(i,j-1/2,k+1/2)$$

$$- H_x^{n+1/2}(i,j+1/2,k+1/2) \Big].$$

Notice from Eq. (3.77) and Fig. 3.22 that the components of **E** and **H** are interlaced within the unit cell and are evaluated at alternate half-time steps. All the field components are present in a quarter of a unit cell as shown typically in Fig. 3.23(a). Figure 3.23(b) illustrates typical relations between field components on a plane; this is particularly useful when incorporating boundary conditions. The figure can be inferred from Eq. (3.72d) or Eq. (3.77d). In translating the hyperbolic system of Eq. (3.77) into a computer code, one must make sure that, within the same time loop, one type of field components is calculated first and the results obtained are used in calculating another type.

3.8.2 Accuracy and Stability

To ensure the accuracy of the computed results, the spatial increment δ must be small compared to the wavelength (usually $\leq \lambda/10$) or minimum dimension of the scatterer. To ensure the stability of the finite difference scheme of Eq. (3.77), the time increment Δt must satisfy the following stability condition [43,47]:

$$u_{max} \Delta t \leq \left[\frac{1}{\Delta x^2} + \frac{1}{\Delta y^2} + \frac{1}{\Delta z^2} \right]^{-1/2} \qquad (3.78)$$

where u_{max} is the maximum wave phase velocity within the model. Since we are using a cubic cell with $\Delta x = \Delta y = \Delta z = \delta$, Eq. (3.78) becomes

$$\boxed{\frac{u_{max} \Delta t}{\delta} \leq \frac{1}{\sqrt{n}}} \qquad (3.79)$$

where n is the number of space dimensions. For practical reasons, it is best to choose the ratio of the time increment to spatial increment as large as possible yet satisfying Eq. (3.79).

3.8.3 Lattice Truncation Conditions

A basic difficulty encountered in applying the FD-TD method to scattering problems is that the domain in which the field is to be computed is open or unbounded (see Fig. 1.3). Since no computer can store an unlimited amount of data, a finite difference scheme over the whole domain is impractical. We must limit the extent of our solution region. In other words, an artificial boundary must be enforced, as in Fig. 3.24, to create the numerical illusion of an infinite space. The solution region must be large enough to enclose the scatterer, and suitable boundary conditions on the artificial boundary must be used to simulate the extension of the solution region to infinity. Outer boundary conditions of this type have been called either *radiation conditions, absorbing boundary conditions*, or *lattice truncation conditions*. Although several types of boundary conditions have been proposed [48,49], we will only consider those developed by Taflove et al. [43,44].

Finite Difference Methods 185

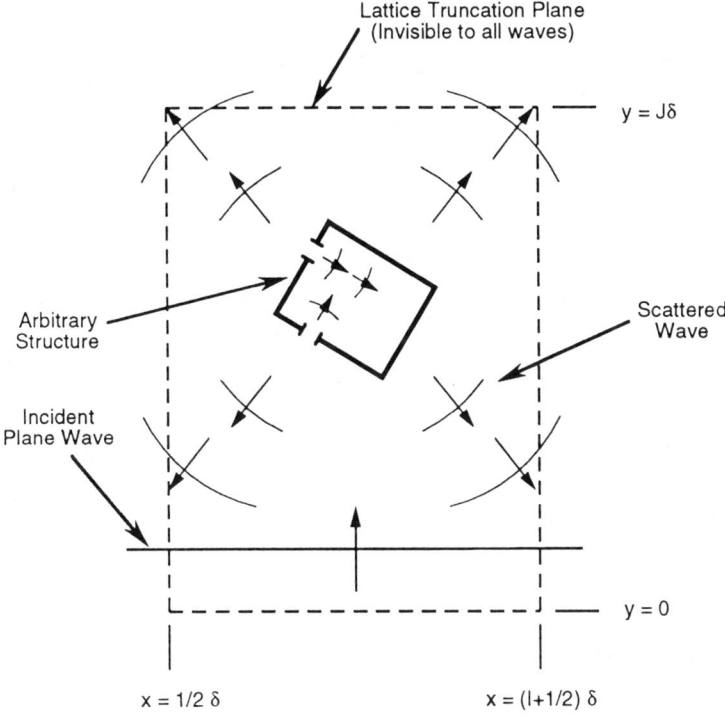

Figure 3.24 Solution region with lattice truncation.

The lattice truncation conditions developed by Taflove et al. allow excellent overall accuracy and numerical stability even when the lattice truncation planes are positioned no more than 5δ from the surface of the scatterer. The conditions relate in a simple way the values of the field components at the truncation planes to the field components at points one or more δ within the lattice (or solution region).

For simplicity, we first consider one-dimensional wave propagation. Assume waves have only E_z and H_x components and propagate in the $\pm y$ directions. Also assume a time step of $\delta t = \delta y/c$, the maximum allowed by the stability condition of Eq. (3.79). If the lattice extends from $y = 0$ to $y = J\Delta y$, with E_z component at the end points, the truncation conditions are:

$$E_z^n(0) = E_z^{n-1}(1) \tag{3.80a}$$

$$E_z^n(J) = E_z^{n-1}(J-1). \tag{3.80b}$$

With these lattice conditions, all possible $\pm y$-directed waves are absorbed at $y = 0$ and $J\Delta y$ without reflection. Equation (3.80) assumes free-space propagation. If we wish to simulate the lattice truncation in a dielectric medium of refractive index m, Eq. (3.80) is modified to

$$E_z^n(0) = E_z^{n-m}(1) \tag{3.81a}$$

$$E_z^n(J) = E_z^{n-m}(J-1). \tag{3.81b}$$

For the three-dimensional case, we consider scattered waves having all six field components and propagating in all possible directions. Assume a time-step of $\delta t = \delta/2c$, a value which is about 13% lower than the maximum allowed ($\delta t = \delta/\sqrt{3}c$) by Eq. (3.79). If the lattice occupies $\frac{1}{2}\delta < x < (I_{max} + \frac{1}{2})\delta, 0 < y < J_{max}\delta, 0 < z < K_{max}$, the truncation conditions are [36,44]:

(a) plane $i = 1/2$

$$H_y^n(1/2, j, k+1/2) = \frac{1}{3}\big[H_y^{n-2}(3/2, j, k-1/2) + H_y^{n-2}(3/2, j, k+1/2)$$
$$+ H_y^{n-2}(3/2, j, k+3/2)\big], \tag{3.82a}$$

$$H_z^n(1/2, j+1/2, k) = \frac{1}{3}\big[H_z^{n-2}(3/2, j+1/2, k-1) + H_z^{n-2}(3/2, j+1/2, k)$$
$$+ H_z^{n-2}(3/2, j+1/2, k+1)\big], \tag{3.82b}$$

(b) plane $i = I_{max} + 1/2$

$$H_y^n(I_{max}+1/2, j, k+1/2) = \frac{1}{3}\big[H_y^{n-2}(I_{max}-1/2, j, k-1/2)$$
$$+ H_y^{n-2}(I_{max}-1/2, j, k+1/2)$$
$$+ H_y^{n-2}(I_{max}-1/2, j, k+3/2)\big], \tag{3.82c}$$

$$H_z^n(I_{max}+1/2, j+1/2, k) = \frac{1}{3}\big[H_z^{n-2}(I_{max}-1/2, j+1/2, k-1)$$
$$+ H_z^{n-2}(I_{max}-1/2, j+1/2, k)$$
$$+ H_z^{n-2}(I_{max}-1/2, j+1/2, k+1)\big], \tag{3.82d}$$

(c) plane $j = 0$,

$$E_x^n(i+1/2, 0, k) = E_x^{n-2}(i+1/2, 1, k), \tag{3.82e}$$

$$E_z^n(i,0,k+1/2) = E_z^{n-2}(i,1,k+1/2), \tag{3.82f}$$

(d) plane $j = J_{max}$

$$E_x^n(i+1/2, J_{max}, k) = E_x^{n-2}(i+1/2, J_{max}-1, k) \tag{3.82g}$$

$$E_z^n(i, J_{max}, k+1/2) = E_z^{n-2}(i, J_{max}-1, k+1/2), \tag{3.82h}$$

(e) plane $k = 0$,

$$\begin{aligned}E_x^n(i+1/2, j, 0) = \frac{1}{3}&[E_x^{n-2}(i-1/2, j, 1) \\&+ E_x^{n-2}(i+1/2, j, 1) \\&+ E_x^{n-2}(i+3/2, j, 1)],\end{aligned} \tag{3.82i}$$

$$\begin{aligned}E_y^n(i, j+1/2, 0) = \frac{1}{3}&[E_y^{n-2}(i-1, j+1/2, 1) \\&+ E_y^{n-2}(i, j+1/2, 1) \\&+ E_y^{n-2}(i+1, j+1/2, 1)],\end{aligned} \tag{3.82j}$$

(f) plane $k = K_{max}$,

$$\begin{aligned}E_x^n(i+1/2, j, K_{max}) = \frac{1}{3}&[E_x^{n-2}(i-1/2, j, K_{max}-1) \\&+ E_x^{n-2}(i+1/2, j, K_{max}-1) \\&+ E_x^{n-2}(i+3/2, j, K_{max}-1)],\end{aligned} \tag{3.82k}$$

$$\begin{aligned}E_y^n(i, j+1/2, K_{max}) = \frac{1}{3}&[E_y^{n-2}(i-1, j+1/2, K_{max}-1) \\&+ E_y^{n-2}(i, j+1/2, K_{max}-1) \\&+ E_y^{n-2}(i+1, j+1/2, K_{max}-1)].\end{aligned} \tag{3.82l}$$

These boundary conditions minimize the reflection of any outgoing waves by simulating the propagation of the wave from the lattice plane adjacent to the lattice truncation plane in a number of time steps corresponding to the propagation delay. The averaging process is used to take into account all possible local angles of incidence of the outgoing wave at the lattice boundary and possible multiple incidences [43]. If the solution region is a dielectric medium of refractive index m rather than free space, we replace the superscript $n-2$ in Eq. (3.82) by $n-m$.

3.8.4 Initial Fields

The initial field components are obtained by simulating either an incident plane wave pulse or single-frequency plane wave. The simulation should not take excessive storage nor cause spurious wave reflections. A desirable plane wave source condition takes into account the scattered fields at the source plane. For the three-dimensional case, a typical wave source condition at plane $y = j_s$ (near $y = 0$) is

$$E_z^n(i, j_s, k + 1/2) \leftarrow 1000\sin(2\pi f n\delta t) + E_z^n(i, j_s, k + 1/2) \qquad (3.83)$$

where f is the irradiation frequency. Equation (3.83) is a modification of the algorithm for all points on plane $y = j_s$; the value of the sinusoid is added to the value of E_z^n obtained from Eq. (3.77).

Thus, at $t = 0$, the plane wave source of frequency f is assumed to be turned on. The propagation of waves from this source is simulated by time stepping, that is, repeatedly implementing Yee's finite difference algorithm on a lattice of points. The incident wave is tracked as it first propagates to the scatterer and then interacts with it via surface-current excitation, diffusion, penetration, and diffraction. Time stepping is continued until the sinusoidal steady state is achieved at each point. The field envelope, or maximum absolute value, during the final half-wave cycle of time stepping is taken as the magnitude of the phasor of the steady-state field [32,43].

From experience, the number of time steps needed to reach the sinusoidal steady state can be greatly reduced by introducing a small isotropic conductivity σ_{ext} within the solution region exterior to the scatterer. This causes the fields to converge more rapidly to the expected steady state condition.

3.8.5 Programming Aspects

Since most EM scattering problems involve nonmagnetic media ($\mu_r = 1$), the quantity $\delta t/\mu(i,j,k)\delta$ can be assumed constant for all (i,j,k). The nine multiplications per unit cell per time required by Yee's algorithm of Eq. (3.77) can be reduced to six multiplications, thereby reducing computer time. Following Taflove et al. [31,35,44], we define the following constants:

$$R = \delta t/2\epsilon_o, \qquad (3.84a)$$

$$R_a = (c\delta t/\delta)^2, \qquad (3.84b)$$

$$R_b = \delta t/\mu_o \delta, \qquad (3.84c)$$

$$C_a = \frac{1 - R\sigma(m)/\epsilon_r(m)}{1 + R\sigma(m)/\epsilon_r(m)}, \qquad (3.84d)$$

Finite Difference Methods

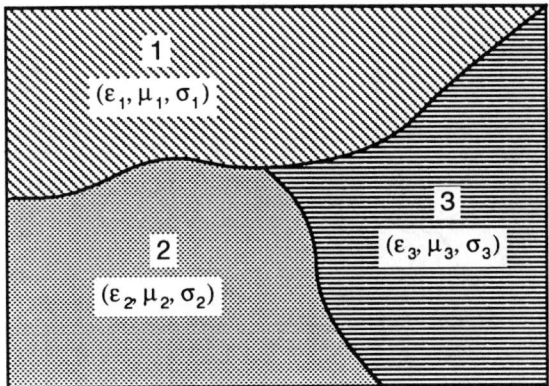

Figure 3.25 A typical inhomogeneous solution region with integer m assigned to each medium.

$$C_b = \frac{R_a}{\epsilon_r(M) + R\sigma(m)}. \quad (3.84e)$$

where $m = \text{MEDIA}(i,j,k)$ is an integer referring to the dielectric or conducting medium type at location (i,j,k). For example, for a solution region comprising of three different homogeneous media shown in Fig. 3.25, m is assumed to be 1 to 3. (This m should not be confused with the refractive index of the medium, mentioned earlier.) In addition to the constants in Eq. (3.84), we define the proportional electric field

$$\widetilde{\mathbf{E}} = R_b \mathbf{E}. \quad (3.85)$$

Thus Yee's algorithm is modified and simplified for easy programming as [50,51]:

$$H_x^n(i,j,k) = H_x^{n-1}(i,j,k) + \widetilde{E}_y^{n-1}(i,j,k+1) \\ - \widetilde{E}_y^{n-1}(i,j,k) - \widetilde{E}_z^{n-1}(i,j+1,k) + \widetilde{E}_z^{n-1}(i,j,k), \quad (3.86a)$$

$$H_y^n(i,j,k) = H_y^{n-1}(i,j,k) + \widetilde{E}_z^{n-1}(i+1,j,k) - \widetilde{E}_z^{n-1}(i,j,k) \\ - \widetilde{E}_x^{n-1}(i,j,k+1) + \widetilde{E}_x^{n-1}(i,j,k), \quad (3.86b)$$

$$H_z^n(i,j,k) = H_z^{n-1}(i,j,k) + \widetilde{E}_x^{n-1}(i,j+1,k) - \widetilde{E}_x^{n-1}(i,j,k) \\ - \widetilde{E}_y^{n-1}(i+1,j,k) + \widetilde{E}_y^{n-1}(i,j,k), \quad (3.86c)$$

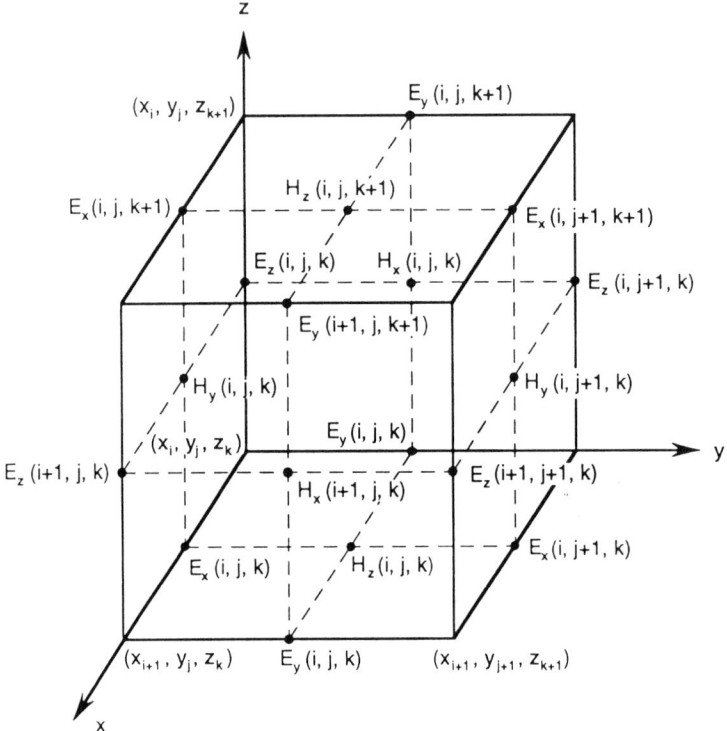

Figure 3.26 Modified node numbering.

$$\widetilde{E}_x^n(i,j,k) = C_a(m)\widetilde{E}_x^{n-1}(i,j,k) + C_b(m)\Big[H_z^{n-1}(i,j,k) - H_z^{n-1}(i,j-1,k)$$
$$- H_y^{n-1}(i,j,k) + H_y^{n-1}(i,j,k-1)\Big], \qquad (3.86\text{d})$$

$$\widetilde{E}_y^n(i,j,k) = C_a(m)\widetilde{E}_y^{n-1}(i,j,k) + C_b(m)\Big[H_x^{n-1}(i,j,k) - H_x^{n-1}(i,j,k-1)$$
$$- H_z^{n-1}(i,j,k) + H_z^{n-1}(i-1,j,k)\Big], \qquad (3.86\text{e})$$

$$\widetilde{E}_z^n(i,j,k) = C_a(m)\widetilde{E}_z^{n-1}(i,j,k) + C_b(m)\Big[H_y^{n-1}(i,j,k) - H_y^{n-1}(i-1,j,k)$$
$$- H_x^{n-1}(i,j,k) + H_x^{n-1}(i,j-1,k)\Big]. \qquad (3.86\text{f})$$

The relationship between the original and modified algorithms is shown in Table 3.8 and illustrated in Fig. 3.26. Needless to say, the truncation

Table 3.8 Relationship Between Original and Modified Field Components (lattice size = $I_{max}\delta \times J_{max}\delta \times K_{max}\delta$)

Original	Modified	Limits on modified (i,j,k)
$H_x^{n+1/2}(x_i, y_{j+1/2}, z_{k+1/2})$	$H_x^n(i,j,k)$	$i = 0, ..., I_{max}$
		$j = 0, ..., J_{max} - 1$
		$k = 0, ..., K_{max} - 1$
$H_y^{n+1/2}(x_{i+1/2}, y_j, z_{k+1/2})$	$H_y^n(i,j,k)$	$i = 0, ..., I_{max} - 1$
		$j = 0, ..., J_{max}$
		$k = 0, ..., K_{max} - 1$
$H_z^{n+1/2}(x_{i+1/2}, y_{j+1/2}, z_k)$	$H_z^n(i,j,k)$	$i = 0, ..., I_{max} - 1$
		$j = 0, ..., J_{max} - 1$
		$k = 0, ..., K_{max}$
$E_x^n(x_{i+1/2}, y_j, z_k)$	$E_x^n(i,j,k)$	$i = 0, ..., I_{max} - 1$
		$j = 0, ..., J_{max}$
		$k = 0, ..., K_{max}$
$E_y^n(x_i, y_{j+1/2}, z_k)$	$E_y^n(i,j,k)$	$i = 0, ..., I_{max}$
		$j = 0, ..., J_{max} - 1$
		$k = 0, ..., K_{max}$
$E_z^n(x_i, y_j, z_{k+1/2})$	$E_z^n(i,j,k)$	$i = 0, ..., I_{max}$
		$j = 0, ..., J_{max}$
		$k = 0, ..., K_{max} - 1$

conditions in Eq. (3.82) must be modified accordingly. This modification eliminates the need for computer storage of separate ϵ and σ arrays; only a MEDIA array which specifies the type-integer of the dielectric or conducting medium at the location of each electric field component in the lattice need be stored. Also the programming problem of handling half integral values of i, j, k has been eliminated.

With the modified algorithm, we determine the scattered fields as follows. Let the solution region, completely enclosing the scatterer, be defined by $0 < i < I_{max}, 0 < j < J_{max}, 0 < k < K_{max}$. At $t \leq 0$, the program is started by setting all field components at the grip points equal to zero:

$$\widetilde{E}_x^0(i,j,k) = \widetilde{E}_y^0(i,j,k) = \widetilde{E}_z^0(i,j,k) = 0 \tag{3.87a}$$

$$H_x^0(i,j,k) = H_y^0(i,j,k) = H_z^0(i,j,k) = 0 \tag{3.87b}$$

for $0 < i < I_{max}, 0 < j < J_{max}, 0 < k < K_{max}$. If we know

$$H_x^{n-1}(i,j,k), E_z^{n-1}(i,j,k),$$

and

$$E_y^{n-1}(i,j,k)$$

at all grid points in the solution region, we can determine new $H_x^n(i,j,k)$ everywhere from Eq. (3.86a). The same applies for finding other field components except that the lattice truncation conditions of Eq. (3.82) must be applied when necessary. The plane wave source is activated at $t = \delta t$, the first time step, and left on during the entire run. The field components are advanced by Yee's finite difference formulas in Eq. (3.86) and by the lattice truncation condition in Eq. (3.82). The time stepping is continued for $t = N_{max}\delta t$, where N_{max} is chosen large enough that the sinusoidal steady state is achieved. In obtaining the steady state solutions, the program must not be left for too long (i.e., N_{max} should not be too large), otherwise the imperfection of the boundary conditions causes the model to become unstable.

The FD-TD method has the following inherent advantages over other modeling techniques, such as the moment method:

(1) It is conceptually simple.
(2) The algorithm does not require the formulation of integral equations, and relatively complex scatterers can be treated without the inversion of large matrices.
(3) It is simple to implement for complicated, inhomogeneous conducting or dielectric structures because constitutive parameters (σ, μ, ϵ) can be assigned to each lattice point.
(4) Its computer memory requirement is not prohibitive for many complex structures of interest.

The method has the following disadvantages:

(1) Its implementation necessitates modeling object as well as its surroundings. Thus, the required program execution time may be excessive.

(2) Its accuracy is at least one order of magnitude worse than that of the method of moments, for example.
(3) Since the computational meshes are rectangular in shape, they do not conform to scatterers with curved surfaces, as is the case of the cylindrical or spherical boundary.
(4) As in all finite difference algorithms, the field quantities are only known at grid nodes.

Time-domain modeling in three dimensions involves a number of issues which are yet to be resolved even for frequency-domain modeling. Among these are whether it is best to reduce Maxwell's equations to a second-order equation for the electric (or magnetic) field or to work directly with the coupled first-order equation. The former approach is used in [34], for example, for solving the problem of EM exploration for minerals. The latter approach has been used with great success in computing EM scattering from objects as demonstrated in this section. In spite of these unresolved issues, the FD-TD algorithm has been applied to solve scattering and other problems including:

- aperture penetration [44,52],
- microwave circuits [53–58],
- eigenvalue problems [59],
- EM absorption in human tissues [35,36,60], and
- other areas [61,62].

The following two examples are taken from the work of Taflove et al. [32,43,44]. The problems whose exact solutions exist will be used to illustrate the applications and accuracy of FD-TD algorithm.

Example 3.7

Consider the scattering of a $+y$-directed plane wave of frequency 2.5 GHz by a uniform, circular, dielectric cylinder of radius 6 cm. We assume that the cylinder is infinite in the z direction and that the incident fields do not vary along z. Thus $\partial/\partial z = 0$ and the problem is reduced to the two-dimensional scattering of the incident wave with only E_z, H_x, and H_y components. Our objective is to compute one of the components, say E_z, at points within the cylinder.

Assuming a lossless dielectric with

$$\epsilon_d = 4\epsilon_o, \qquad \mu_d = \mu_o, \qquad \sigma_d = 0, \qquad (3.7.1)$$

the speed of the wave in the cylinder is

$$u_d = \frac{c}{\sqrt{\epsilon_r}} = 1.5 \times 10^8 \text{ m/s}. \qquad (3.7.2)$$

194 Numerical Techniques in Electromagnetics

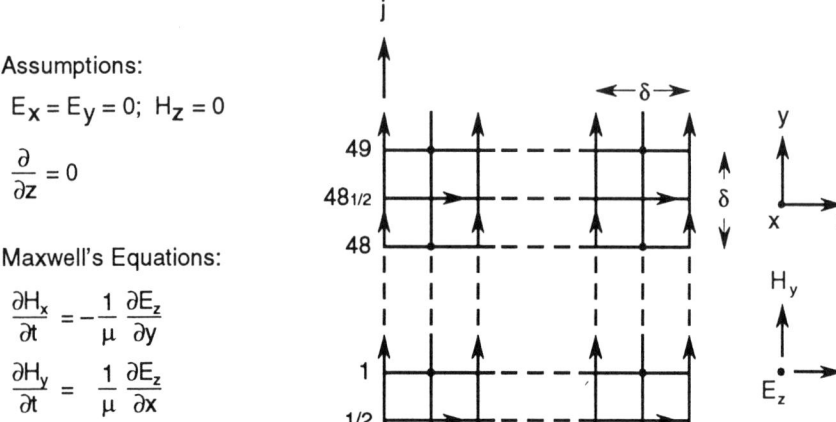

Figure 3.27 Two-dimensional lattice for Example 3.7.

Hence $\lambda_d = u_d/f = 6$ cm. We may select $\delta = \Delta x = \Delta y = \Delta z = \lambda_d/20 = 0.3$ cm and $\delta t = \delta/2c = 5$ ps. Thus we use the two-dimensional grid of Fig. 3.27 as the solution domain. Due to the symmetry of the scatterer, the domain can be reduced relative to Fig. 3.27 to the 25 × 49 subdomain of Fig. 3.28. Choosing the cylinder axis as passing through point $(i,j) = (25.5, 24.5)$ allows the *symmetry condition* to be imposed at line $i = 26$, i.e.,

$$\widetilde{E}_z^n(26,j) = \widetilde{E}_z^n(25,j). \qquad (3.7.3)$$

Soft grid truncation conditions are applied at $j = 0, 49$ and $i = 1/2$, i.e.,

$$\widetilde{E}_z^n(i,0) = \frac{1}{3}\left[\widetilde{E}_z^{n-2}(i-1,1) + \widetilde{E}_z^{n-2}(i,1) + \widetilde{E}_z^{n-2}(i+1,1)\right], \qquad (3.7.4)$$

$$\widetilde{E}_z^n(i,49) = \frac{1}{3}\left[\widetilde{E}_z^{n-2}(i-1,48) + \widetilde{E}_z^{n-2}(i,48) + \widetilde{E}_z^{n-2}(i+1,48)\right], \qquad (3.7.5)$$

$$H_y^n(0.5,49) = \frac{1}{3}\left[H_y^{n-2}(1.5,j) + H_y^{n-2}(1.5,j+1) + H_y^{n-2}(1.5,j+1)\right], \qquad (3.7.6)$$

where $n - 2$ is due to the fact that $\delta = 2c\delta t$ is selected. Notice that the actual values of (i, j, k) are used here, while the modified values for easy programming are used in the program; the relationship between the two types of values is in Table 3.8.

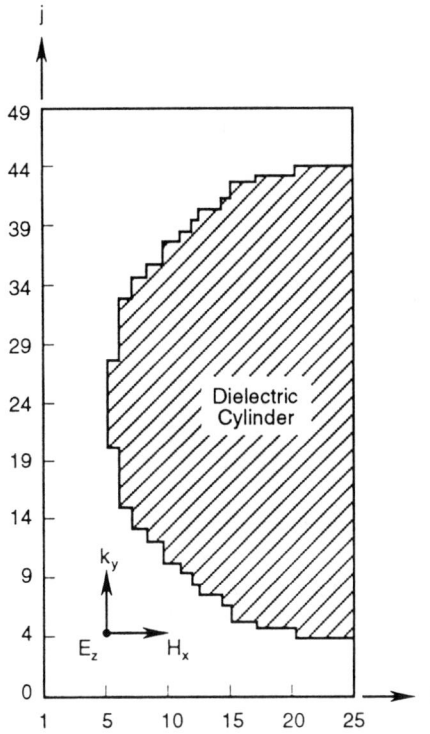

Figure 3.28 Finite difference model of cylindrical dielectric scatterer relative to the grid of Figure 3.27.

Grid points (i, j) internal to the cylinder, determined by

$$[(i - 25.5)^2 + (j - 24.5)^2]^{1/2} \leq 20, \quad (3.7.7)$$

are assigned the constitutive parameters ϵ_d, μ_o, and ϵ_d, while grid points external to the cylinder are assigned parameters of free space ($\epsilon = \epsilon_o, \mu = \mu_o, \sigma = 0$).

A FORTRAN program has been developed by Bemmel [63] based on the ideas expounded above. A similar but more general code is THREDE developed by Holland [50]. The program starts by setting all field components at grid points equal to zero. A plane wave source

$$\widetilde{E}_z^n(i, 2) \leftarrow 1000 \sin(2\pi f n \delta t) + \widetilde{E}_z^n(i, 2) \quad (3.7.8)$$

is used to generate the incident wave at $j = 2$ and $n = 1$, the first time step, and left on during the entire run. The program is time stepped to

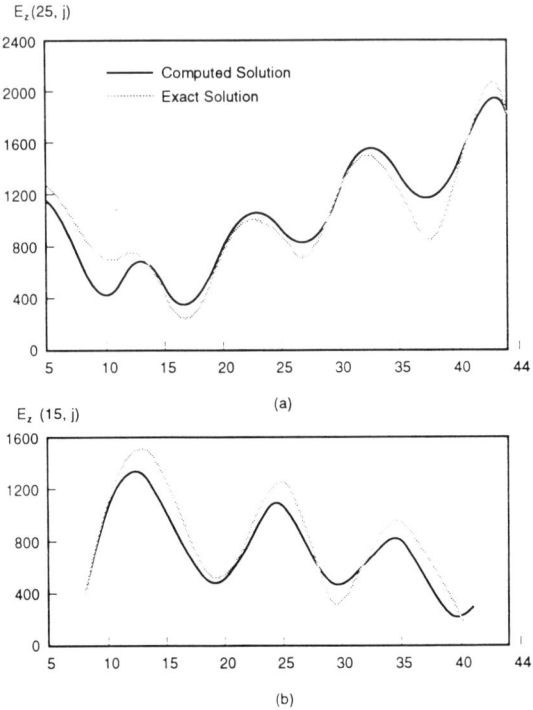

Figure 3.29 Computed internal E_z on line: (a) $i = 25$, (b) $i = 15$.

$t = N_{max}\delta t$, where N_{max} is large enough that sinusoidal steady state is achieved. Since $f = 2.5$ GHz, the wave period $T = 1/f = 400$ ps $= 80\delta t$. Hence $N_{max} = 500 = 6.25T$ is sufficient to reach steady state. Thus the process is terminated after 500 timesteps. Typical results are portrayed in Fig. 3.29 for the envelope of $E_z^n(15, j)$ for $460 \leq n \leq 500$. Figure 3.29 also shows the exact solution using series expansion [64]. Bemmel's code has both the numerical and exact solutions. By simply changing the constitutive parameters of the media and specifying the boundary of the scatterer (through a look-up table for complex objects), the program can be applied to almost any two-dimensional scattering or penetration problem.

Example 3.8

Consider the penetration of a $+y$-directed plane wave of frequency 2.5 GHz by a uniform, dielectric sphere of radius 4.5 cm. The problem is similar to

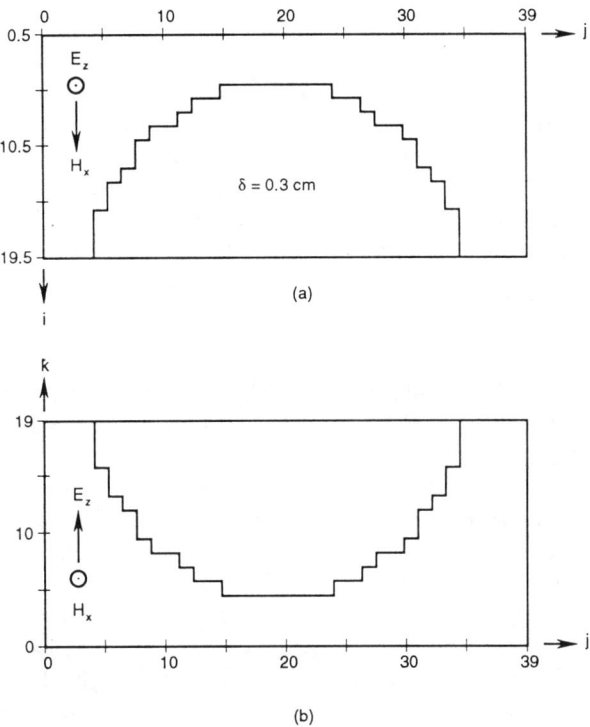

Figure 3.30 FD-TD model of dielectric sphere.

the previous example except that it is three-dimensional and more general. We assume that the incident wave has only E_z and H_x components.

Like in the previous example, we assume that internal to the lossless dielectric sphere,

$$\epsilon_d = 4\epsilon_o, \qquad \mu_d = \mu_o, \qquad \sigma_d = 0. \qquad (3.8.1)$$

We select

$$\delta = \lambda_d/20 = 0.3\,\text{cm} \qquad (3.8.2)$$

and

$$\delta t = \delta/2c = 5\,\text{ps}. \qquad (3.8.3)$$

This choice of the grid size implies that the radius of the sphere is $4.5/0.3 = 15$ units. The sphere model centered at grid point $(19.5, 20, 19)$ in a $19 \times 39 \times 19$ lattice is portrayed in Fig. 3.30 at two lattice symmetry planes $k = 19$ and $i = 19.5$. Grid points (i, j, k) internal to the sphere are determined by

198 Numerical Techniques in Electromagnetics

$$[(i-19.5)^2 + (j-20)^2 + (k-19)^2]^{1/2} \le 15. \qquad (3.8.4)$$

Rather than assigning $\sigma = 0$ to points external to the sphere, a value $\sigma = 0.1$ mho/m is assumed to reduce spurious wave reflections. The FORTRAN code shown in Fig. 3.31, a modified version of Bemmel's [63], is used to generate field components E_y and E_z near the sphere irradiation axis. With the dimensions and constitutive parameters of the sphere specified as input data, the program is developed based on the following steps:

(1) Compute the parameters of each medium using Eq. (3.84) where $m = 1, 2$.
(2) Initialize field components.
(3) Use the FD-TD algorithm in Eq. (3.86) to generate field components. This is the heart of the program. It entails taking the following steps:
  (i) Calculate actual values of grid point (x, y, z) using the relationship in Table 3.8. This will be needed later to identify the constitutive parameters of the medium at that point using subroutine MEDIA.
  (ii) Apply soft lattice truncation conditions in Eq. (3.82) at appropriate boundaries, i.e., at $x = \delta/2, y = 0, y_{max}$, and $z = 0$. Notice that some of the conditions in Eq. (3.82) are not necessary in this case because we restrict the solution to one fourth of the sphere due to geometrical symmetry. At other boundaries ($x = x_{max}$ and $z = z_{max}$), the symmetry conditions are imposed. For example, at $k = 19$,

$$\tilde{E}_x^n(i, j, 20) = \tilde{E}_x^n(i, j, 18).$$

  (iii) Apply FD-TD algorithm in Eq. (3.86).
  (iv) Activate the plane wave source, i.e.,

$$\tilde{E}_z^n(i, j, k) \leftarrow \sin(2\pi f n \delta t) + \tilde{E}_z^n(i, j_s, k)$$

  where $j_s = 3$ or any plane near $y = 0$. Time step until steady state is reached.
(4) Obtain the maximum absolute values (envelopes) of field components in the last half-wave and output the results.

Figure 3.32 illustrates the results of the program. The values of $|E_y|$ and $|E_z|$ near the sphere axis are plotted against j for observation period $460 \le n \le 500$. The computed results are compared with Mie's exact solution [65] covered in Section 2.8. The code for calculating the exact solution is also found in Bemmel's work [63].

```
0001      C************************************************************
0002      C   APPLICATION OF THE FINITE DIFFERENCE METHOD
0003      C   This program involves the penetration of a
0004      C   lossless dielectric SPHERE by a plane wave.
0005      C   The program provides in the maximum absolute value of
0006      C   Ey and Ez during the final half-wave of time-stepping
0007      C   Assumption:
0008      C   +y-directed incident wave with components Ez and Hx.
0009      C   I,J,K,NN correspond to X,Y,Z, and Time.
0010      C   IMAX,JMAX,KMAX are the maximum values of x,y,z
0011      C   NNMAX is the total number of timesteps.
0012      C   NHW represents one half-wave cycle.
0013      C   MED is the number of different uniform media
              sections.
0014      C   JS is the j-position of the plane wave front.
0015      C
0016      C   THIS PROGRAM WAS DEVELOPED BY V. BEMMEL [63]
0017      C   AND LATER IMPROVED BY D. TERRY
0018      C************************************************************
0019
0020            PARAMETER( IMAX=19, JMAX=39, KMAX=19)
0021            PARAMETER( NMAX=2, NNMAX=500, NHW=40, MED=2, JS=3 )
0022            PARAMETER( DELTA=3E-3, CL=3.0E8, F=2.5E9 )
0023            PARAMETER( PIE=3.141592654)
0024
0025      C Define scatterer dimensions
0026            PARAMETER( OI=19.5, OJ=20.0, OK=19.0, RADIUS=15.0)
0027
0028            DIMENSION EX(0:IMAX+1, 0:JMAX+1, 0:KMAX+1, 0:NMAX),
0029          &     EY(0:IMAX+1, 0:JMAX+1, 0:KMAX+1, 0:NMAX),
0030          &     EZ(0:IMAX+1, 0:JMAX+1, 0:KMAX+1, 0:NMAX),
0031          &     HX(0:IMAX+1, 0:JMAX+1, 0:KMAX+1, 0:NMAX),
0032          &     HY(0:IMAX+1, 0:JMAX+1, 0:KMAX+1, 0:NMAX),
0033          &     HZ(0:IMAX+1, 0:JMAX+1, 0:KMAX+1, 0:NMAX),
0034          &     EY1(0:JMAX+1), EZ1(0:JMAX+1),
0035          &     ER(MED), SIG(MED), CA(MED), CB(MED)
0036
0037            DIMENSION IXMED(0:IMAX+1, 0:JMAX+1, 0:KMAX+1)
0038            DIMENSION IYMED(0:IMAX+1, 0:JMAX+1, 0:KMAX+1)
0039            DIMENSION IZMED(0:IMAX+1, 0:JMAX+1, 0:KMAX+1)
0040
0041            DIMENSION CBMRB( 2 )   ! STORE CB(M)/RB HERE
0042            DATA ER/1.0,4.0/       ! CONSTITUTIVE PARAMETERS
0043            DATA SIG/0.1,0.0/
0044      C
0045      C Statement function to compute position w.r.t. center
              of sphere
0046      C
0047            position(RI,RJ,RK)=SQRT((RI-OI)**2+(RJ-OJ)
                                        **2+(RK-OK)**2)
0048            EO=(1E-9)/(36*PIE)
0049            UO=(1E-7)*4*PIE
0050            DT=DELTA/(2*CL)
0051            R=DT/EO
0052            RA=(DT**2)/(UO*EO*(DELTA**2))
0053            RB=DT/(UO*DELTA)
0054            TPIFDT = 2.0*PIE*F*DT
```

```
0055      C*********************************************************
0056      C   STEP # 1 - COMPUTE MEDIA PARAMETERS
0057      C*********************************************************
0058             DO 1 M=1,MED
0059             CA(M)=1.0-R*SIG(M)/ER(M)
0060             CB(M)=RA/ER(M)
0061             CBMRB(M) = CB(M)/RB
0062      1      CONTINUE
0063      C
0064      C   (i) CALCULATE THE REAL/ACTUAL GRID POINTS
0065      C
0066      C Initialize the media arrays.Index (M) determines which
0067      C medium each point is actually located in and is used to
0068      C index into arrays which determine the constitutive
0069      C parameters of the medium.There are separate M determining
0070      C arrays for EX, EY, and EZ. These arrays correlate the
0071      C integer values of I,J,K to the actual position within
0072      C the lattice.Computing these values now and storing them in these
0073      C arrays as opposed to computing them each time they are
0074      C needed saves a large amount of computation time.
0075      C
0076             DO 3 I=0, IMAX+1
0077             DO 3 J=0, JMAX+1
0078             DO 3 K=0, KMAX+1
0079             IF(position(float(I)+0.5,float(J),float(K))
                 .LE.RADIUS) then
0080             IXMED( I, J, K ) = 2
0081             else
0082             IXMED( I, J, K ) = 1
0083             endif
0084             IF(position(float(I),float(J)+0.5,float(K))
                 .LE.RADIUS) then
0085             IYMED( I, J, K ) = 2
0086             else
0087             IYMED( I, J, K ) = 1
0088             endif
0089             IF(position(float(I),float(J),float(K)+0.5)
                 .LE.RADIUS) then
0090             IZMED( I, J, K ) = 2
0091             else
0092             IZMED( I, J, K ) = 1
0093             endif
0094      3      CONTINUE

0095      C*************************************************
0096      C   STEP # 2 - INITIALIZE FIELD COMPONENTS
0097      C*************************************************
0098      C components for output
0099             DO 4 J=0,JMAX+1
0100             EY1(J)=0.
0101             EZ1(J)=0.
0102      4      CONTINUE
0103
0104             DO 5 I=0,IMAX+1
0105             DO 5 J=0,JMAX+1
0106             DO 5 K=0,KMAX+1
0107             DO 5 N = 0, NMAX
0108             EX(I,J,K,N)=0.
0109             EY(I,J,K,N)=0.
```

```
0110          EZ(I,J,K,N)=0.
0111          HX(I,J,K,N)=0.
0112          HY(I,J,K,N)=0.
0113          HZ(I,J,K,N)=0.
0114    5   CONTINUE
0115          print *, 'initialization complete'

0116   C**************************************************************
0117   C STEP # 3 - USE FD/TD ALGORITHM TO GENERATE
0118   C FIELD COMPONENTS
0119   C**************************************************************
0120   C   SINCE ONLY FIELD COMPONENTS AT CURRENT TIME (t) AND
           PREVIOUS
0121   C   TWO TIME STEPS ( t-1 AND t-2) ARE REQUIRED FOR
           COMPUTATION,
0122   C   WE SAVE MEMORY SPACE BY USING THE FOLLOWING INDICES
0123   C       NCUR is index in for time t
0124   C       NPR1 is index in for t-1
0125   C       NPR2 is index in for t-2
0126   C
0127          NCUR = 2
0128          NPR1 = 1
0129          NPR2 = 0
0130          DO 15 NN = 1, NNMAX   ! TIME LOOP
0131          IF (MOD( NN,10).EQ.0) THEN
0132          print *,'NN=',NN   ! DISPLAY PROGRESS
0133          ENDIF
0134   C   Next time step - move indices up a notch.
0135          NPR2 = NPR1
0136          NPR1 = NCUR
0137          NCUR = MOD( NCUR+1, 3)
0138          DO 20 K=0,KMAX   ! Z LOOP
0139          DO 25 J=0,JMAX   ! Y LOOP
0140          DO 30 I=0,IMAX   ! X LOOP
0141   C
0142   C   (ii)-APPLY SOFT LATTICE TRUNCATION CONDITIONS
0143   C
0144   C---x=delta/2
0145          IF(I.EQ.0) THEN
0146            IF ((K.NE.KMAX).AND.(K.NE.0)) THEN
0147          HY(0,J,K,NCUR) = (HY(1,J,K-1,NPR2)
0148        &         + HY(1,J,K,NPR2)+HY(1,J,K+1,NPR2))/3.
0149          HZ(0,J,K,NCUR)=(HZ(1,J,K-1,NPR2)
0150        &         + HZ(1,J,K,NPR2)+HZ(1,J,K+1,NPR2))/3.
0151          else
0152          IF (K.EQ.KMAX) THEN
0153          HY(0,J,KMAX,NCUR) = (HY(1,J,KMAX-1,NPR2)
0154        &         + HY(1,J,KMAX,NPR2))/2.
0155          HZ(0,J,K,NCUR)=( HZ(1,J,K-1,NPR2)
0156        &         + HZ(1,J,K,NPR2) )/2.
0157          ELSE
0158          HY(0,J,K,NCUR) = ( HY(1,J,K,NPR2)
0159        &         + HY(1,J,K+1,NPR2))/2.
0160          HZ(0,J,0,NPR2)=(HZ(1,J,0,NPR2)
0161        &         + HZ(1,J,1,NPR2))/2.
0162          ENDIF
0163          ENDIF
0164          ENDIF
0165   C---y=0
0166          IF(J.EQ.0) THEN
0167          EX(I,0,K,NCUR)=EX(I,1,K,NPR2)
0168          EZ(I,0,K,NCUR)=EZ(I,1,K,NPR2)
```

```
0169                ELSE
0170       C---y=ymax
0171             IF(J.EQ.JMAX) THEN
0172             EX(I,JMAX,K,NCUR)=EX(I,JMAX-1,K,NPR2)
0173             EZ(I,JMAX,K,NCUR)=EZ(I,JMAX-1,K,NPR2)
0174             ENDIF
0175                ENDIF
0176       C---z=0
0177             IF(K.EQ.0) THEN
0178             IF ((I.NE.0).AND.(I.NE.IMAX)) THEN
0179             EX(I,J,0,NCUR) = (EX(I-1,J,1,NPR2)
0180           &              + EX(I,J,1,NPR2)+EX(I+1,J,1,NPR2))/3.
0181             EY(I,J,0,NCUR) = (EY(I-1,J,1,NPR2)
0182           &              + EY(I,J,1,NPR2)+EY(I+1,J,1,NPR2))/3.
0183             ELSE
0184             IF(I.EQ.0)  THEN
0185             EX(0,J,0,NCUR)=(EX(0,J,1,NPR2)+EX(1,J,1,NPR2))/2.
0186             EY(I,J,0,NCUR)=(EY(I,J,1,NPR2)+EY(I+1,J,1,NPR2))/2.
0187             ELSE
0188             EX(I,J,0,NCUR)=(EX(I-1,J,1,NPR2)+EX(I,J,1,NPR2))/2.
0189             EY(I,J,0,NCUR)=(EY(I-1,J,1,NPR2)+EY(I,J,1,NPR2))/2.
0190             ENDIF
0191             ENDIF
0192             ENDIF
0193       C
0194       C  (iii)  APPLY FD/TD ALGORITHM
0195       C
0196       C-----a. HX  generation:
0197             HX(I,J,K,NCUR)=HX(I,J,K,NPR1)+RB*(EY(I,J,K+1,NPR1)
0198           & - EY(I,J,K,NPR1)+EZ(I,J,K,NPR1)-EZ(I,J+1,K,NPR1))
0199       C-----b. HY  generation:
0200             HY(I,J,K,NCUR)=HY(I,J,K,NPR1)+RB*(EZ(I+1,J,K,NPR1)
0201           & - EZ(I,J,K,NPR1)+EX(I,J,K,NPR1)-EX(I,J,K+1,NPR1))
0202       C-----c. HZ  generation:
0203             HZ(I,J,K,NCUR)=HZ(I,J,K,NPR1)+RB*(EX(I,J+1,K,NPR1)
0204           & - EX(I,J,K,NPR1)+EY(I,J,K,NPR1)-EY(I+1,J,K,NPR1))
0205       C---k=kmax  ! SYMMETRY
0206             IF(K.EQ.KMAX) THEN
0207             HX(I,J,KMAX,NCUR)=HX(I,J,KMAX-1,NCUR)
0208             HY(I,J,KMAX,NCUR)=HY(I,J,KMAX-1,NCUR)
0209             ENDIF
0210       C-----d. EX  generation:
0211             IF((J.NE.0).AND.(J.NE.JMAX).AND.(K.NE.0)) THEN
0212             M = IXMED( I, J, K )
0213             EX(I,J,K,NCUR) = CA(M)*EX(I,J,K,NPR1) + CBMRB(M)*
0214           & (HZ(I,J,K,NCUR)-HZ(I,J-1,K,NCUR)+HY(I,J,K-1,NCUR)
0215           &                 -HY(I,J,K,NCUR))
0216             ENDIF
0217       C-----e. EY  generation:
0218             IF(K.NE.0) THEN
0219             M = IYMED( I, J, K )
0220             EY(I,J,K,NCUR)=CA(M)*EY(I,J,K,NPR1) + CBMRB(M)*
0221           &  HX(I,J,K,NCUR)-HX(I,J,K-1,NCUR)+HZ(I-1,J,K,NCUR)
0222           &                 -HZ(I,J,K,NCUR))
0223             ENDIF
0224       C-----f. EZ  generation:
0225             IF ((J.NE.0).AND.(J.NE.JMAX)) THEN
0226             M = IZMED( I, J, K )
0227       C  sig(ext)=0 for Ez only from Taflove[14]
0228             IF(M.EQ.1) THEN
0229             CAM=1.0
0230             ELSE
```

```
0231            CAM=CA(M)
0232          ENDIF
0233          EZ(I,J,K,NCUR)=CAM*EZ(I,J,K,NPR1)+CBMRB(M)*
0234        &   (HY(I,J,K,NCUR)-HY(I-1,J,K,NCUR)+HX(I,J-1,K,NCUR)
0235        &                  -HX(I,J,K,NCUR))
0236     C
0237     C  (iv) APPLY THE PLANE-WAVE SOURCE
0238     C
0239          IF(J.EQ.JS) THEN
0240          EZ(I,JS,K,NCUR) = EZ(I,JS,K,NCUR)+SIN( TPIFDT*NN )
0241          ENDIF
0242          ENDIF
0243     C---i=imax+1/2  ! SYMMETRY
0244          IF(I.EQ.IMAX) THEN
0245          EY(IMAX+1,J,K,NCUR)=EY(IMAX,J,K,NCUR)
0246          EZ(IMAX+1,J,K,NCUR)=EZ(IMAX,J,K,NCUR)
0247          ENDIF
0248     c---k=kmax
0249          IF(K.EQ.KMAX) THEN
0250          EX(I,J,KMAX+1,NCUR)=EX(I,J,KMAX-1,NCUR)
0251          EY(I,J,KMAX+1,NCUR)=EY(I,J,KMAX-1,NCUR)
0252          ENDIF
0253    30    CONTINUE !  I LOOP
0254     C*************************************************************
0255     C   STEP # 4 - RETAIN THE MAXIMUM ABSOLUTE VALUES DURING
0256     C                 THE LAST HALF-WAVE
0257     C*************************************************************
0258          IF ( (K.EQ.KMAX).AND.(NN.GT.(NNMAX-NHW)) ) THEN
0259          TEMP = ABS( EY(IMAX,J,KMAX-1,NCUR) )
0260          IF (TEMP .GT. EY1(J) ) THEN
0261          EY1(J) = TEMP
0262          ENDIF
0263          TEMP = ABS( EZ(IMAX,J,KMAX,NCUR) )
0264          IF (TEMP .GT. EZ1(J) ) THEN
0265          EZ1(J) = TEMP
0266          ENDIF
0267          ENDIF
0268    25    CONTINUE !  J LOOP
0269    20    CONTINUE     !  K LOOP
0270    15    CONTINUE        !  NN (time) LOOP
0271     C*************************************************************
0272     C-----Output E as a function of j
0273          DO 130 J = 0, JMAX-1
0274          WRITE(6,135) J+1, float(J), EY1(J)
0275   130    CONTINUE
0276          DO 140 J = 0, JMAX-1
0277          WRITE(6,135) J+1, float(J), EZ1(J)
0278   140    CONTINUE
0279   135    FORMAT(I5,2F15.8)
0280          STOP
0281          END
```

Figure 3.31 Computer program for FD-TD three-dimensional scattering problem.

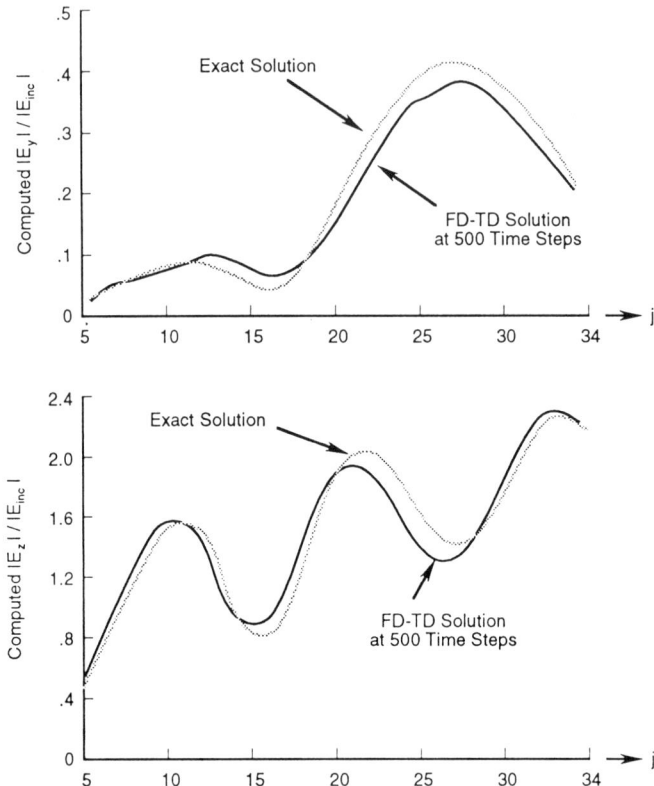

Figure 3.32 Computed $E_y(19.5, j, 18)$ amd $E_z(19, j, 18.5)$ within the lossless dielectric sphere.

3.9 Finite Differencing for Nonrectangular Systems

So far in this chapter, we have considered only rectangular solution regions within which a rectangular grid can be readily placed. Although we can always replace a nonrectangular solution region by an approximate rectangular one, our discussion in this chapter would be incomplete if we failed to apply the method to nonrectangular coordinates since it is sometimes preferable to use these coordinates. We will demonstrate the finite differencing technique in cylindrical coordinates (ρ, ϕ, z) and spherical coordinates (r, θ, ϕ) by solving Laplace's equation $\nabla^2 V$. The idea is readily extended to other PDEs.

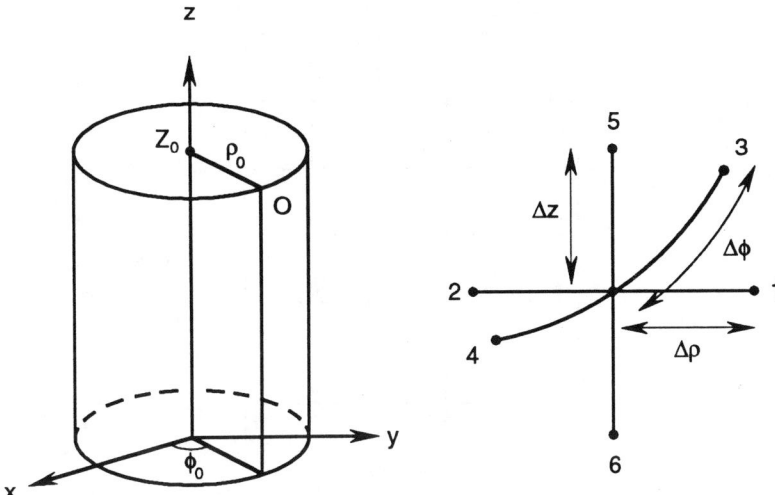

Figure 3.33 Typical node in cylindrical coordinate.

3.9.1 Cylindrical Coordinates

Laplace's equation in cylindrical coordinates can be written as

$$\nabla^2 V = \frac{\partial^2 V}{\partial \rho^2} + \frac{1}{\rho}\frac{\partial V}{\partial \rho} + \frac{1}{\rho^2}\frac{\partial^2 V}{\partial \phi^2} + \frac{\partial^2 V}{\partial z^2}. \tag{3.88}$$

Refer to the cylindrical system and finite difference molecule shown in Fig. 3.33. At point $O(\rho_o, \phi_o, z_o)$, the equivalent finite difference approximation is

$$\frac{V_1 - 2V_o + V_2}{(\Delta\rho)^2} + \frac{1}{\rho_o}\frac{V_1 - V_2}{2\Delta\rho} + \frac{V_3 - 2V_o + V_4}{(\rho_o\Delta\phi)^2} + \frac{V_5 - 2V_o + V_6}{(\Delta z)^2} = 0 \tag{3.89}$$

where $\Delta\rho, \Delta\phi$ and Δz are the step sizes along ρ, ϕ, and z, respectively, and

$$\begin{aligned} V_o &= V(\rho_o, \phi_o, z_o), \quad V_1 = V(\rho_o + \Delta\rho, \phi_o, z_o), \quad V_2 = V(\rho_o - \Delta\rho, \phi_o, z_o), \\ V_3 &= V(\rho_o, \phi_o + \rho_o\Delta\phi, z_o), \quad V_4 = V(\rho_o, \phi_o - \rho_o\Delta\phi, z_o), \\ V_5 &= V(\rho_o, \phi_o, z_o + \Delta x), \quad V_6 = V(\rho_o, \phi_o, z_o - \Delta z). \end{aligned} \tag{3.90}$$

We now consider a special case of Eq. (3.89) for an axisymmetric system [66]. In this case, there is no dependence of ϕ so that $V = V(\rho, z)$. If we

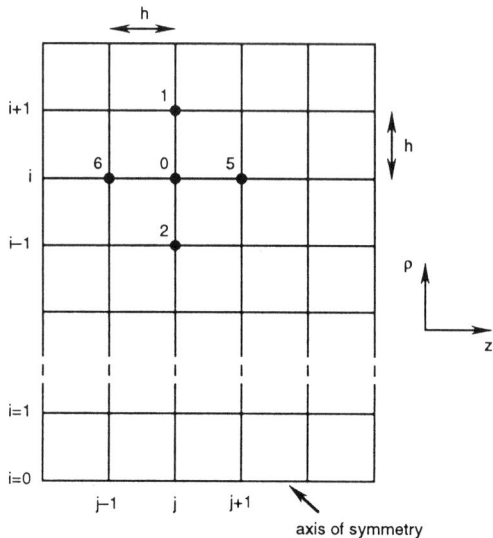

Figure 3.34 Finite difference grid for an axisymmetric system.

assume square nets so that $\Delta\rho = \Delta z = h$, the solution region is discretized as in Fig. 3.34 and Eq. (3.89) becomes

$$\left(1 + \frac{h}{2\rho_o}\right)V_1 + \left(1 - \frac{h}{2\rho_o}\right)V_2 + V_5 + V_6 - 4V_o = 0. \quad (3.91)$$

If point O is at $(\rho_o, z_o) = (ih, jh)$, then

$$1 + \frac{h}{2\rho_o} = \frac{2i+1}{2i}, \qquad 1 - \frac{h}{2\rho_o} = \frac{2i-1}{2i}$$

so that Eq. (3.91) becomes

$$\boxed{V(i,j) = \frac{1}{4}\left[V(i,j-1) + V(i,j+1) + \left(\frac{2i-1}{2i}\right)V(i-1,j) + \left(\frac{2i+1}{2i}\right)V(i+1,j)\right]}$$
(3.92)

Notice that in Eq. (3.91), it appears we have a singularity for $\rho_o = 0$. However, by symmetry, all odd order derivatives must be zero. Hence

$$\left.\frac{\partial V}{\partial \rho}\right|_{\rho=0} = 0 \quad (3.93)$$

since
$$V(\Delta\rho, z_o) = V(-\Delta\rho, z_o). \tag{3.94}$$

Therefore by L'Hopital's rule,
$$\lim_{\rho_o \to 0} \frac{1}{\rho_o} \frac{\partial V}{\partial \rho}\bigg|_{\rho_o} = \frac{\partial^2 V}{\partial \rho^2}\bigg|_{\rho_o}. \tag{3.95}$$

Thus, at $\rho = 0$, Laplace's equation becomes
$$2\frac{\partial^2 V}{\partial \rho^2} + \frac{\partial^2 V}{\partial z^2} = 0. \tag{3.96}$$

The finite difference equivalent to Eq. (3.96) is
$$V_o = \frac{1}{6}(4V_1 + V_5 + V_6)$$

or
$$\boxed{V(0,j) = \frac{1}{6}\left[V(0,j-1) + V(0,j+1) + 4V(1,j)\right]} \tag{3.97}$$

which is used at $\rho = 0$.

To solve Poisson's equation $\nabla^2 V = -\rho_v/\epsilon$ in cylindrical coordinates, we obtain the finite difference form by replacing zero on the right-hand side of Eq. (3.89) with $g = -\rho_v/\epsilon$. We obtain

$$\boxed{V(i,j) = \frac{1}{4}\left[V(i,j+1) + V(i,j-1) + \frac{2i-1}{2i}V(i-1,j) + \frac{2i+1}{2i}V(i+1,j) - gh^2\right]} \tag{3.98}$$

where h is the step size.

As in Section 3.7.1, the boundary condition $D_{1n} = D_{2n}$ must be imposed at the interface between two media. As an alternative to applying Gauss's law as in Section 3.7.1, we will apply Taylor series expansion [67]. Applying the series expansion to point 1, 2, 5 in medium 1 in Fig. 3.35, we obtain

$$\begin{aligned} V_1 &= V_o + \frac{\partial V_o^{(1)}}{\partial \rho}h + \frac{\partial^2 V_o^{(1)}}{\partial \rho^2}\frac{h^2}{2} + \cdots \\ V_2 &= V_o - \frac{\partial V_o^{(1)}}{\partial \rho}h + \frac{\partial^2 V_o^{(1)}}{\partial \rho^2}\frac{h^2}{2} - \cdots \\ V_5 &= V_o + \frac{\partial V_o^{(1)}}{\partial z}h + \frac{\partial^2 V_o^{(1)}}{\partial z^2}\frac{h^2}{2} + \cdots \end{aligned} \tag{3.99}$$

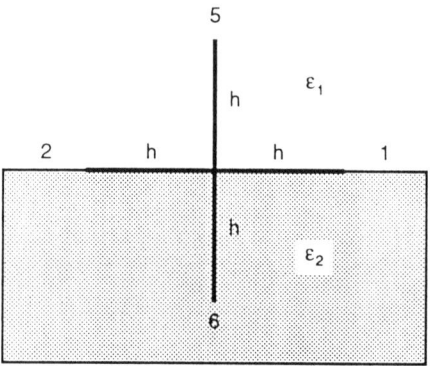

Figure 3.35 Interface between two dielectric media.

where superscript (1) denotes medium 1. Combining Eqs. (3.88) and (3.99) results in

$$h^2 \nabla^2 V = V_1 + V_2 + 2V_5 - 4V_o - 2h\frac{\partial V_o^{(1)}}{\partial z} + \frac{h(V_1 - V_2)}{2\rho_o} = 0$$

or

$$\frac{\partial V_o^{(1)}}{\partial z} = \frac{V_1 + V_2 + 2V_5 - 4V_o + \dfrac{h(V_1 - V_2)}{2\rho_o}}{2h}. \qquad (3.100)$$

Similarly, applying Taylor series to points 1, 2, and 6 in medium 2, we get

$$\begin{aligned}
V_1 &= V_o + \frac{\partial V_o^{(1)}}{\partial \rho}h + \frac{\partial^2 V_o^{(1)}}{\partial \rho^2}\frac{h^2}{2} + \cdots \\
V_2 &= V_o - \frac{\partial V_o^{(1)}}{\partial \rho}h + \frac{\partial^2 V_o^{(1)}}{\partial \rho^2}\frac{h^2}{2} - \cdots \\
V_6 &= V_o - \frac{\partial V_o^{(1)}}{\partial z}h + \frac{\partial^2 V_o^{(1)}}{\partial z^2}\frac{h^2}{2} - \cdots.
\end{aligned} \qquad (3.101)$$

Combining Eqs. (3.88) and (3.101) leads to

$$h^2 \nabla^2 V = V_1 + V_2 + 2V_6 - 4V_o + 2h\frac{\partial V_o^{(2)}}{\partial z} + \frac{h(V_1 - V_2)}{2\rho_o} = 0$$

or

$$-\frac{\partial V_o^{(2)}}{\partial z} = \frac{V_1 + V_2 + 2V_6 - 4V_o + \dfrac{h(V_1 - V_2)}{2\rho_o}}{2h}. \qquad (3.102)$$

Finite Difference Methods 209

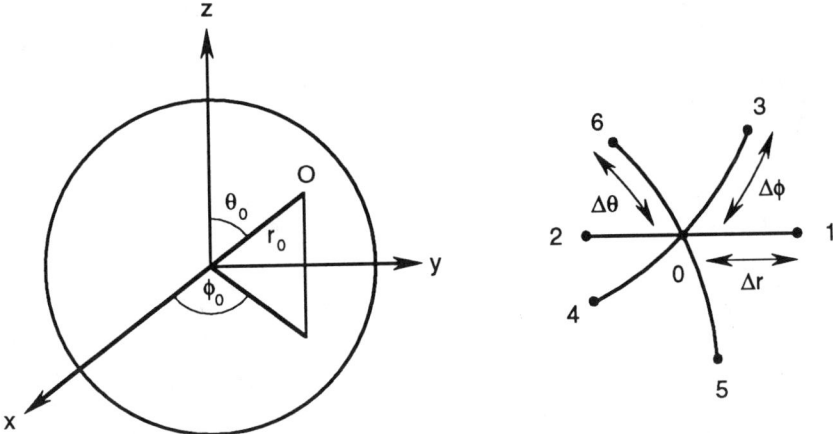

Figure 3.36 Typical node in spherical coordinates.

But $D_{1n} = D_{2n}$ or

$$\epsilon_1 \frac{\partial V_o^{(1)}}{\partial z} = \epsilon_2 \frac{\partial V_o^{(2)}}{\partial z}. \tag{3.103}$$

Substituting Eqs. (3.100) and (3.102) into Eq. (3.103) and solving for V_o yields

$$V_o = \frac{1}{4}\left(1 + \frac{h}{2\rho_o}\right)V_1 + \frac{1}{4}\left(1 - \frac{h}{2\rho_o}\right)V_2 + \frac{\epsilon_1}{2(\epsilon_1 + \epsilon_2)}V_5 + \frac{\epsilon_2}{2(\epsilon_1 + \epsilon_2)}V_6. \tag{3.104}$$

Equation (3.104) is only applicable to interface points. Notice that Eq. (3.104) becomes Eq. (3.91) if $\epsilon_1 = \epsilon_2$.

Typical examples of finite difference approximations for boundary points, written for square nets in rectangular and cylindrical systems, are tabulated in Table 3.9. For more examples, see [12, 68]. The FD-TD has also been applied in solving time-varying axisymmetric problems. See [69], for example.

3.9.2 Spherical Coordinates

In spherical coordinates, Laplace's equation can be written as

$$\nabla^2 V = \frac{\partial^2 V}{\partial r^2} + \frac{2}{r}\frac{\partial V}{\partial r} + \frac{1}{r^2}\frac{\partial^2 V}{\partial \theta^2} + \frac{\cot\theta}{r^2}\frac{\partial V}{\partial \theta} + \frac{1}{r^2 \sin^2\theta}\frac{\partial^2 V}{\partial \phi^2}. \tag{3.105}$$

At a grid point $O(r_o, \theta_o, \phi_o)$ shown in Fig. 3.36, the finite difference approximation to Eq. (3.105) is

210   Numerical Techniques in Electromagnetics

Table 3.9   Finite Difference Approximations at Boundary Points

Description	Figure	Cartesian Equation	Cylindrical Equation
1. Bottom edge		$4V_0 = V_1 + V_2 + 2V_3$	$4V_0 = V_1 + V_2 + 4V_3$
2. Top edge		$4V_0 = V_1 + V_2 + 2V_4$	$4V_0 = V_1 + V_2 + 2V_3$
3. Left edge		$4V_0 = 2V_1 + V_3 + V_4$	$8V_0 = 4V_1 + (\frac{2i+1}{i})V_3$ $+ (\frac{2i-1}{i})V_4$
4. Right edge		$4V_0 = 2V_1 + V_3 + V_4$	$8V_0 = 4V_2 + (\frac{2i+1}{i})V_3$ $+ (\frac{2i-1}{i})V_4$
5. Bottom left corner point		$2V_0 = V_1 + V_3$	$3V_0 = V_1 + 2V_3$
6. Bottom right corner point		$2V_0 = V_2 + V_3$	$3V_0 = V_2 + 2V_3$
7. Top left corner point		$2V_0 = V_1 + V_4$	$3V_0 = V_1 + 2V_4$
8. Top right corner point		$2V_0 = V_2 + V_4$	$3V_0 = V_2 + 2V_4$

$$\frac{V_1 - 2V_0 + V_2}{(\Delta r)^2} + \frac{2}{r_o}\left(\frac{V_1 - V_2}{2\Delta r}\right) + \frac{V_6 - 2V_0 + V_5}{(r_o \Delta\theta)^2}$$
$$+ \frac{\cot\theta_o}{r_o^2}\left(\frac{V_5 - V_6}{2\Delta\theta}\right) + \frac{V_3 - 2V_0 + V_4}{(r_o \Delta\phi \sin\theta_o)^2} = 0. \quad (3.106)$$

Note that θ increases from node 6 to 5 and hence we have $V_5 - V_6$ and not $V_6 - V_5$ in Eq. (3.106).

Example 3.9

Consider an earthed metal cylindrical tank partly filled with a charge liquid, such as hydrocarbons, as illustrated in Fig. 3.37(a). The electrostatic potential can become hazardous. Using the finite difference method, determine the potential distribution in the entire domain. Plot the potential

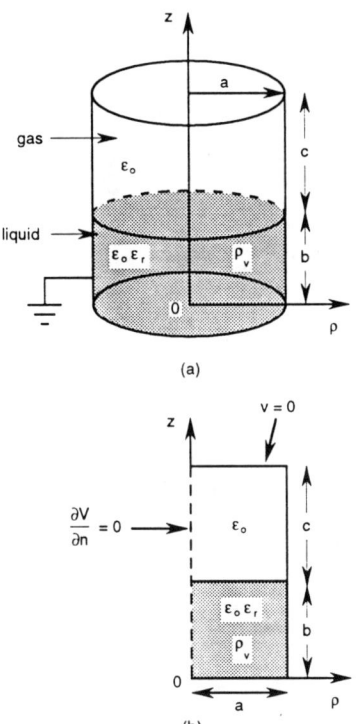

Figure 3.37 For Example 3.9: (a) earth cylindrical tank, (b) solution region.

along $\rho = 0.5$, $0 < z < 2$m and on the surface of the liquid. Take

$$a = b = c = 1.0 \text{ m},$$
$$\epsilon_r = 2.0 \quad \text{(hydrocarbons)},$$
$$\rho_v = 10^{-5} \text{ C/m}^3.$$

Solution

The exact analytic solution to this problem involves solving

$$\nabla^2 V_1 = 0 \quad \text{for gas space},$$
$$\nabla^2 V_2 = -\frac{\rho_v}{\epsilon} \quad \text{for liquid space}.$$

By using series expansion technique, it can be shown that [70]

$$V_1 = 2\rho_v \sum_{n=1}^{N} \frac{1}{C_n K_n} \left[\cosh(\lambda_n b) - 1\right] J_0(\lambda_n \rho) \sinh\left[\lambda_n(b+c-z)\right] \quad (3.9.1)$$

$$V_2 = \frac{2\rho_v}{\epsilon_r} \sum_{n=1}^{N} \frac{J_0(\lambda_n \rho)}{C_n} \left(\frac{\sinh(\lambda_n z)}{K_n}\left[\cosh(\lambda_n b)\cosh(\lambda_n c)\right.\right.$$

$$+ \epsilon_r \sinh(\lambda_n b)\sinh(\lambda_n c)$$

$$\left.\left. - \cosh(\lambda_n c)\right] + 1 - \cosh(\lambda_n z)\right) \quad (3.9.2)$$

where subscripts 1 and 2 denote gas and liquid space, respectively,

$$C_n = \epsilon_o a \lambda_n^3 J_1(\lambda_n a), \quad (3.9.3a)$$
$$K_n = \sinh(\lambda_n b)\cosh(\lambda_n c) + \epsilon_r \cosh(\lambda_n b)\sinh(\lambda_n c), \quad (3.9.3b)$$

and λ_n are the roots of $J_0(\lambda_n a) = 0$.

It is apparent from Fig. 3.37(a) and from the fact that ρ_v is uniform that $V = V(\rho, z)$ (i.e., the problem is two-dimensional) and the domain of the problem is symmetrical about the z-axis. Therefore, it is only necessary to investigate the solution region in Fig. 3.37(b) and impose the condition that the z-axis is a flux line, i.e., $\partial V/\partial n = \partial V/\partial \rho = 0$.

The finite difference grid of Fig. 3.34 is used with $0 \leq i \leq I_{max}$ and $0 \leq j \leq J_{max}$. Choosing $\Delta \rho = \Delta z = h = 0.05$ m makes $I_{max} = 20$ and $J_{max} = 40$. Equation (3.92) is applied for gas space, and Eq. (3.98) for liquid space. Along the z-axis, i.e., $i = 0$, we impose the Neumann condition in Eq. (3.97). To account for the fact that the gas has dielectric constant ϵ_{r1} while the liquid has ϵ_{r2}, we impose the boundary condition in Eq. (3.104) on the liquid-gas interface.

Based on these ideas, the computer program shown in Fig. 3.38 was developed to determine the potential distribution in the entire domain. The values of the potential along $\rho = 0.5$, $0 < z < 2$ and along the gas-liquid interface are plotted in Fig. 3.39. It is evident from the figure that the finite difference solution compares well with the exact solution of Eqs. (3.9.1) to (3.9.3). It is the simplicity in concept and ease of programming finite difference schemes that make them very attractive for solving problems such as this.

```
0001      C ****************************************************
0002      C FORTRAN CODE FOR EXAMPLE 3.9
0003      C AN AXISYMMETRIC PROBLEM OF AN EARTHED CYLINDER
0004      C PARTIALLY FILLED WITH A CHARGED LIQUID
0005      C SOLVED USING FINITE DIFFERENCE SCHEME
0006      C ****************************************************
0007
0008            DIMENSION V(0:50,0:50)
0009            DATA A,B,C/1.0,1.0,1.0/
0010            DATA ER1,ER2,E0/1.0,2.0,8.854E-12/
0011
0012            H = 0.05
0013            NA = A/H  ! same as Imax
0014            NB = B/H
0015            NC = C/H
0016            NBC = NB + NC  ! same as Jmax
0017            NMAX = 500  ! NO. OF ITERATIONS
0018            RHOV = 1.E-5
0019            G = -RHOV/(ER2*E0)
0020            GH2 = G*H*H
0021      C
0022      C INITIALIZE - THIS ALSO TAKES CARE OF DIRICHLET
0023      C    CONDITIONS
0024            DO 10 I = 0,NA
0025            DO 10 J = 0,NBC
0026            V(I,J) = 0.0
0027      10    CONTINUE
0028      C
0029      C NOW, APPLY FINITE DIFFERENCE SCHEME
0030      C
0031            DO 60 N = 1,NMAX
0032            DO 40 I=1,NA-1
0033            FM = FLOAT(2*I - 1)/FLOAT(2*I)
0034            FP = FLOAT(2*I + 1)/FLOAT(2*I)
0035      C IN LIQUID
0036            DO 20 J = 1,NB-1
0037            V(I,J) = 0.25*( V(I,J-1) + V(I,J+1) + FM*V(I-1,J)
0038           &              + FP*V(I+1,J) - GH2 )
0039      20    CONTINUE
0040      C IN GAS
0041            DO 30 J = NB+1,NBC-1
0042            V(I,J) = 0.25*( V(I,J-1) + V(I,J+1) + FM*V(I-1,J)
0043           &              + FP*V(I+1,J) )
0044      30    CONTINUE
0045      C ALONG THE GAS-LIQUID INTERFACE
0046            V(I,NB)=V(I,NB+1)*ER1/(ER1+ER2)
0047                  +V(I,NB-1)*ER2/(ER1+ER2)
0047      40    CONTINUE
0048      C IMPOSE NEUMANN CONDITION ALONG THE Z-AXIS
0049            DO 50 J = 1,NBC-1
0050            V(0,J) = ( 4.0*V(1,J) + V(0,J-1) + V(0,J+1) )/6.0
0051      50    CONTINUE
0052      60    CONTINUE
0053      C
0054      C OUTPUT THE POTENTIAL ALONG RHO = 0.5, 0 < Z < 1.0
0055      C
0056            NR5=0.5/H
0057            DO 80 J = 0,NBC
0058            WRITE(6,90) J, V(NR5,J)/1000.0  ! IN kV
```

```
0059    80          CONTINUE
0060    90          FORMAT(2X,'J=',I3,3X,'V(0.5,J)=',E12.5,/)
0061    C
0062    C   OUTPUT THE POTENTIAL ON THE INTERFACE BETWEEN
                GAS-LIQUID
0063    C
0064                DO 100 I = 0,NA
0065                WRITE(6,110) I, V(I,NB)/1000.0 ! IN kV
0066    100         CONTINUE
0067    110         FORMAT(2X,'I=',I3,3X,'V(I,NB)=',E12.5,/)
0068                STOP
0069                END
```

Figure 3.38   FORTRAN code for Example 3.9.

3.10 Numerical Integration

Numerical integration (also called *numerical quadrature*) is used in science and engineering whenever a function cannot easily be integrated in closed form or when the function is described in the form of discrete data. Integration is a more stable and reliable process than differentiation. The term quadrature or integration rule will be used to indicate any formula that yields an integral approximation. Several integration rules have been developed over the years. The common ones include:

- Euler's rule,
- Trapezoidal rule,
- Simpson's rule,
- Newton-Cotes rules, and
- Gaussian (quadrature) rules.

The first three are very simple and will be considered first to help build up background for other rules which are more general and accurate. A discussion on the subject of numerical integration with diverse FORTRAN codes can be found in Davis and Rabinowitz [71]. A program package called QUADPACK for automatic integration covering a wide variety of problems and various degrees of difficulty is presented in Piessens et al. [72]. Our discussion will be brief but sufficient for the purpose of this text.

3.10.1 Euler's Rule

To apply the Euler or rectangular rule in evaluating the integral

$$I = \int_a^b f(x)\,dx, \tag{3.107}$$

Finite Difference Methods 215

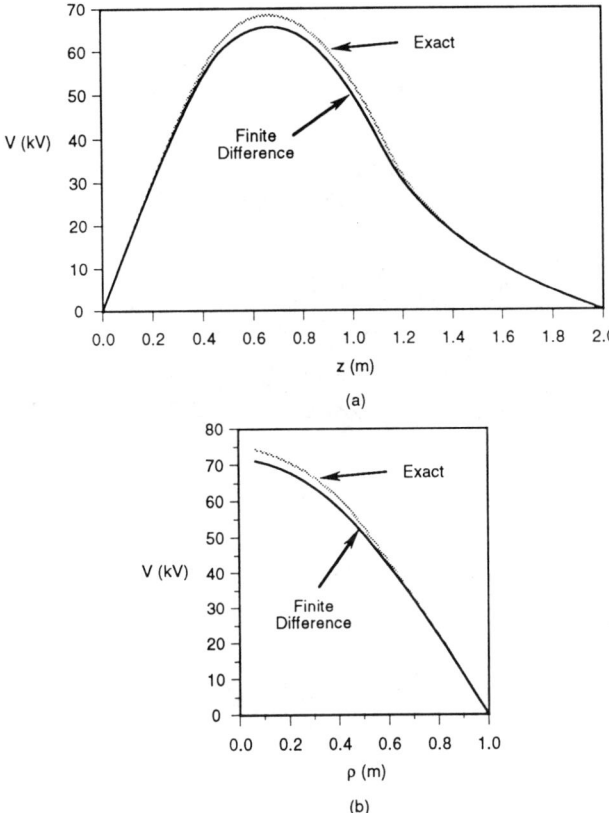

Figure 3.39 Potential distribution in the tank of Fig. 3.37: (a) along $\rho = 0.5\ m$, $0 \le z \le 2\ m$: (b) along the gas-liquid interface.

where $f(x)$ is shown in Fig. 3.40, we seek an approximation for the area under the curve. We divide the curve into n equal intervals as shown. The subarea under the curve within $x_{i-1} < x < x_i$ is

$$A_i = \int_{x_{i-1}}^{x_i} f(x)\ dx \simeq h f_i \tag{3.108}$$

where $f_i = f(x_i)$. The total area under the curve is

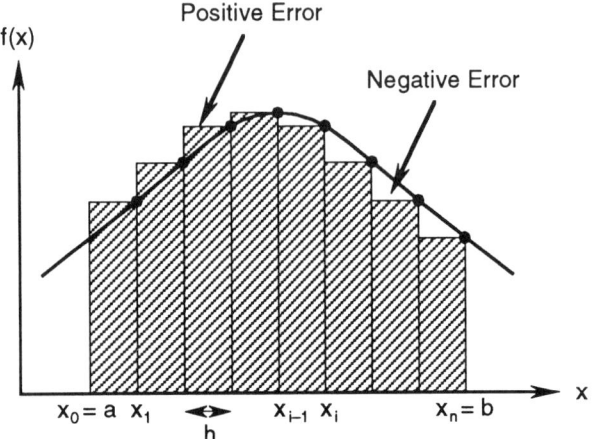

Figure 3.40 Integration using Euler's rule.

$$I = \int_a^b f(x) \, dx \simeq \sum_{i=1}^{n} A_i$$
$$= h\left[f_1 + f_2 + \ldots\ldots + f_n\right]$$

or

$$\boxed{I = h \sum_{i=1}^{n} f_i} \quad (3.109)$$

It is clear from Fig. 3.40 that this quadrature method gives an inaccurate result since each A_i is less or greater than the true area introducing negative or positive error, respectively.

3.10.2 Trapezoidal Rule

To evaluate the same integral in Eq. (3.107) using the trapezoidal rule, the subareas are chosen as shown in Fig. 3.41. For the interval $x_{i-1} < x < x_i$,

$$A_i = \int_{x_{i-1}}^{x_i} f(x) \, dx \simeq \left(\frac{f_{i-1} + f_i}{2}\right)h. \quad (3.110)$$

Finite Difference Methods

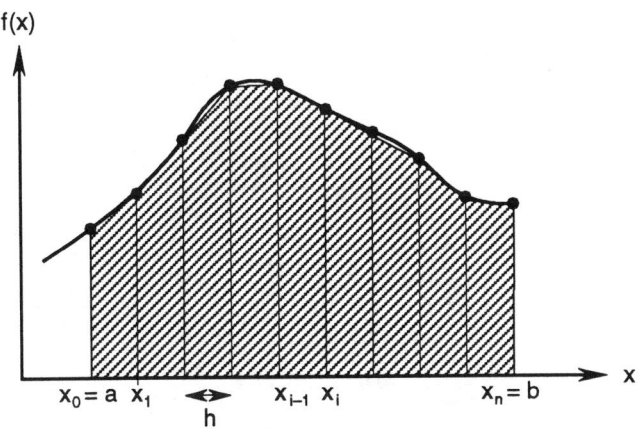

Figure 3.41 Integration using the trapezoidal rule.

Hence

$$I = \int_a^b f(x)\, dx \simeq \sum_{i=1}^n A_i$$

$$= h\left[\frac{f_o + f_1}{2} + \frac{f_1 + f_2}{2} + \cdots + \frac{f_{n-2} + f_{n-1}}{2} + \frac{f_{n-1} + f_n}{2}\right]$$

$$= \frac{h}{2}\left[f_o + 2f_1 + 2f_2 + \ldots\ldots + 2f_{n-1} + f_n\right]$$

or

$$\boxed{I = h\sum_{i=1}^{n-1} f_i + \frac{h}{2}(f_o + f_n)} \tag{3.111}$$

3.10.3 Simpson's Rule

Simpson's rule gives a still more accurate result than the trapezoidal rule. While the trapezoidal rule approximates the curve by connecting successive points on the curve by straight lines, Simpson's rule connects successive groups of three points on the curve by a second-degree polynomial (i.e., a parabola). Thus

$$A_i = \int_{x_{i-1}}^{x_i} f(x)\,dx \simeq \frac{h}{3}(f_{i-1} + f_i + f_{i+1}). \tag{3.112}$$

Therefore

$$\begin{aligned}I &= \int_a^b f(x)\,dx \simeq \sum_{i=1}^n A_i \\ &= \frac{h}{3}\Big[f_0 + 4f_1 + 2f_2 + 4f_3 + \ldots\ldots + 2f_{n-2} + 4f_{n-1} + f_n\Big]\end{aligned} \tag{3.113}$$

where n is even.

The computational molecules for Euler's, trapezoidal, and Simpson's rules are shown in Fig. 3.42. Now that we have considered simple quadrature rules to help build up background, we now consider more general, accurate methods.

3.10.4 Newton-Cotes Rules

To apply a Newton-Cotes rule to evaluate the integral in Eq. (3.107), we divide the interval $a < x < b$ into m equal intervals so that

$$h = \frac{b-a}{m} \tag{3.114}$$

where m is a multiple of n, and n is the number of intervals covered at a time or the order of the approximating polynomial. The subarea in the interval $x_{n(i-1)} < x < x_{ni}$ is

$$A_i = \int_{x_{n(i-1)}}^{x_{ni}} f(x)\,dx \simeq \frac{nh}{N}\sum_{k=0}^n C_k^n f(x_{n(i-1)+k}). \tag{3.115}$$

The coefficients C_k^n, $0 \leq k \leq n$, are called Newton-Cotes numbers and tabulated in Table 3.10. The numbers are obtained from

$$C_k^n = \frac{1}{n}\int_0^N L_k(s)\,ds \tag{3.116}$$

where

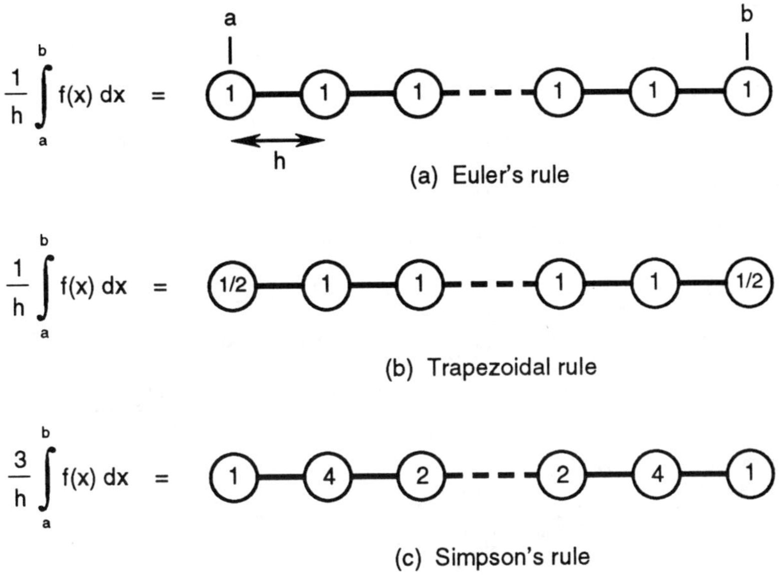

Figure 3.42 Computational molecules for integration.

$$L_k(s) = \prod_{j=0, \neq k}^{n} \frac{s-j}{k-j}. \qquad (3.117)$$

It is easily shown that the coefficients are symmetric, i.e.,

$$C_k^n = C_{n-k}^n \qquad (3.118a)$$

and they sum up to unity, i.e.,

$$\sum_{k=0}^{n} C_k^n = 1. \qquad (3.118b)$$

Table 3.10 Newton-Cotes Numbers

n	N	NC_0^n	NC_1^n	NC_2^n	NC_3^n	NC_4^n	NC_5^n	NC_6^n	NC_7^n	NC_8^n
1	2	1	1							
2	6	1	4	1						
3	8	1	3	3	1					
4	90	7	32	12	32	7				
5	288	19	75	50	50	75	19			
6	840	41	216	27	272	27	216	41		
7	17280	751	3577	1323	2989	2989	1323	3577	751	
8	24350	989	5888	-928	10946	-4540	10946	-928	5888	989

For example, for $n = 2$,

$$C_0^2 = \frac{1}{2} \int_0^6 \frac{(s-1)(s-2)}{(-1)(-2)} ds = \frac{1}{6},$$

$$C_1^2 = \frac{1}{2} \int_0^6 \frac{s(s-2)}{1(-1)} ds = \frac{4}{6},$$

$$C_2^2 = \frac{1}{2} \int_0^6 \frac{s(s-1)}{2(1)} ds = \frac{1}{6}.$$

Once the subareas are found using Eq. (3.115), then

$$\boxed{I = \int_a^b f(x)\, dx \simeq \sum_{i=1}^{m/n} A_i} \qquad (3.119)$$

The most widely known Newton-Cotes formulas are:
$n = 1$ (2-point; trapezoidal rule)

$$A_i \simeq \frac{h}{2}(f_{i+1} + f_i), \qquad (3.120)$$

$n = 2$ (3-point; Simpson's 1/3 rule)

$$A_i \simeq \frac{h}{3}(f_{i-1} + 4f_i + f_{i+1}), \qquad (3.121)$$

$n = 3$ (4-point; Newton's rule)

$$A_i \simeq \frac{3h}{8}(f_i + 3f_{i+1} + 3f_{i+2} + f_{i+3}). \qquad (3.122)$$

3.10.5 Gaussian Rules

The integration rules considered so far involve the use of equally spaced abscissa points. The idea of integration rules using unequally spaced abscissa points stems from Gauss. The Gaussian rules are more complicated but more accurate than the Newton-Cotes rules. A Gaussian rule has the general form

$$\boxed{\int_a^b f(x)\, dx \simeq \sum_{i=1}^n w_i f(x_i)} \qquad (3.123)$$

where (a, b) is the interval for which a sequence of orthogonal polynomials $\{w_i(x)\}$ exists, x_i are the zeros of $w_i(x)$, and the weights w_i are such that Eq. (3.123) is of degree of precision $2n - 1$. Any of the orthogonal polynomials discussed in Chapter 2 can be used to give a particular Gaussian rule. Commonly used rules are Gauss-Legendre, Gauss-Chebyshev, etc., since the sample point x_i are the roots of the Legendre, Chebyshev, etc., of degree n. For the Legendre ($n = 1$ to 16) and Laguerre ($n = 1$ to 16) polynomials, the zeros x_i and weights w_i have been tabulated in [73].

Using Gauss-Legendre rule,

$$\int_a^b f(x)\, dx \simeq \frac{b-a}{2} \sum_{i=1}^n w_i f(u_i) \qquad (3.124)$$

where $u_i = \frac{(b-a)}{2} x_i + \frac{(b+a)}{2}$ are the transformation of the roots x_i of Legendre polynomials from limits $(-1, 1)$ to finite limits (a, b). The values of the abscissas x_i and weights w_i for n up to 7 are presented in Table 3.11; for higher values of n, the interested reader is referred to [73,74]. Note that $-1 < x_i < 1$ and $\sum_{i=1}^n w_i = 2$.

The Gauss-Chebyshev rule is similar to Gauss-Legendre rule. We use Eq. (3.124) except that the sample points x_i the roots of Chebyshev polynomial $T_n(x)$, are

$$x_i = \cos\frac{(2i-1)}{2n}, \qquad i = 1, 2, ..., n \qquad (3.125)$$

and the weights are all equal [75], i.e.,

$$w_i = \frac{\pi}{n}. \qquad (3.126)$$

When either of the limits of integration a or b or both are $\pm\infty$, we use Gauss-Laguerre or Gauss-Hermite rule. For the Gauss-Laguerre rule,

$$\int_0^\infty f(x)\, dx \simeq \sum_{i=1}^n w_i f(x_i) \qquad (3.127)$$

Table 3.11 Abscissas (Roots of Legendre Polynomials) and Weights for Gauss-Legendre Integration

$\pm x_i$	w_i
$n = 2$	
0.57735 02691 89626	1.00000 00000 00000
$n = 3$	
0.00000 00000 00000	0.88888 88888 88889
0.77459 66692 41483	0.55555 55555 55556
$n = 4$	
0.33998 10435 84856	0.65214 51548 62546
0.86113 63115 94053	0.34785 48451 37454
$n = 5$	
0.00000 00000 00000	0.56888 88888 88889
0.53846 93101 05683	0.47862 86704 99366
0.90617 98459 38664	0.23692 68850 56189
$n = 6$	
0.23861 91860 83197	0.46791 39345 72691
0.66120 93864 66265	0.36076 15730 48139
0.93246 95142 03152	0.17132 44923 79170
$n = 7$	
0.00000 00000 00000	0.41795 91836 73469
0.40584 51513 77397	0.38183 00505 05119
0.74153 11855 99394	0.27970 53914 89277
0.94910 79123 42759	0.12948 49661 68870

where the appropriate abscissas x_i, the roots of Laguerre polynomials, and weights w_i are listed for n up to 7 in Table 3.12. For the Gauss-Hermite rule,

$$\int_{-\infty}^{\infty} f(x)\, dx \simeq \sum_{i=1}^{n} w_i f(x_i) \tag{3.128}$$

where the abscissas x_i, the roots of the Hermite polynomials, and weights w_i are listed for n up to 7 in Table 3.13. An integral over (a, ∞) is taken care of by a change of variable so that

$$\int_{a}^{\infty} f(x)\, dx = \int_{0}^{\infty} f(y+a)\, dy. \tag{3.129}$$

Table 3.12 Abscissas (Roots of Laguerre Polynomials) and Weights for Gauss-Laguerre Integration

$\pm x_i$	w_i
$n = 2$	
0.58578 64376 27	1.53332 603312
3.41421 35623 73	4.45095 733505
$n = 3$	
0.41577 45567 83	1.07769 285927
2.29428 03602 79	2.76214 296190
6.28994 50829 37	5.60109 462543
$n = 4$	
0.32254 76896 19	0.83273 912383
1.74576 11011 58	2.04810 243845
4.53662 02969 21	3.63114 630582
9.39507 09123 01	6.48714 508441
$n = 5$	
0.26356 03197 18	0.67909 404220
1.41340 30591 07	1.63848 787360
3.59642 57710 41	2.76944 324237
12.64080 08442 76	7.21918 635435
$n = 6$	
0.22284 66041 79	0.57353 550742
1.18893 21016 73	1.36925 259071
2.99273 63260 59	2.26068 459338
5.77514 35691 05	3.35052 458236
9.83746 74183 83	4.88682 680021
15.98287 39806 02	7.84901 594560
$n = 7$	
0.19304 36765 60	0.49647 759754
1.02666 48953 39	1.17764 306086
2.56787 67449 51	1.91824 978166
4.90035 30845 26	2.77184 863623
8.18215 34445 63	3.84124 912249
12.73418 02917 98	5.38067 820792
19.39572 78622 63	8.40543 248683

We apply Eq. (3.127) with $f(x)$ evaluated at points $x_i + a$, $i = 1, 2, \cdots, n$ and x_is are tabulated in Table 3.12.

A major drawback with Gaussian rules is that if one wishes to improve the accuracy, one must increase n which means that the values of w_i and x_i must be included in the program for each value of n. Another disadvantage is that the function $f(x)$ must be explicit since the sample points x_i are unassigned.

3.10.6 Multiple Integration

This is an extension of one-dimensional (1D) integration discussed so far. A double integral is evaluated by means of two successive applications of the rules presented above for single integral [76]. To evaluate the integral using the Newton Cotes or Simpson's 1/3 rule ($n = 2$), for example,

$$I = \int_a^b \int_c^d f(x,y) \, dx \, dy \tag{3.130}$$

over a rectangular region $a < x < b, c < y < d$, we divide the region into $m \cdot l$ smaller rectangles with sides

$$h_x = \frac{b-a}{m} \tag{3.131a}$$

$$h_y = \frac{d-c}{l} \tag{3.131b}$$

where m and l are multiplies of $n = 2$. The subarea

$$A_{ij} = \int_{y_n(j-1)}^{y_n(j+1)} dy \int_{x_{(i-1)}}^{x_n(i+1)} f(x,y) \, dx \tag{3.132}$$

is evaluated by integrating along x and then along y:

$$A_{ij} \simeq \frac{h_x}{3}(g_{j-1} + 4g_j + g_{j+1}) \tag{3.133}$$

where

$$g_j \simeq \frac{h_y}{3}\left(f_{i-1,j} + 4f_{i,j} + f_{i+1,j}\right). \tag{3.134}$$

Substitution of Eq. (3.134) into Eq. (3.133) yields

$$A_{ij} = \frac{h_x h_y}{9}\Big[\left(f_{i+1,j+1} + f_{i+1,j-1} + f_{i-1,j+1} + f_{i-1,j-1}\right)$$

$$+ 4\left(f_{i,j+1} + f_{i,j-1} + f_{i+1,j} + f_{i-1,j}\right) + 16 f_{i,j}\Big]. \tag{3.135}$$

The corresponding schematic or integration molecule is shown in Fig. 3.43. Summing the value of A_{ij} for all subareas yields

$$I = \sum_{i=1}^{m/n} \sum_{j=1}^{l/n} A_{ij}. \tag{3.136}$$

Table 3.13 Abscissas (Roots of Hermite Polynomials) and Weights for Gauss-Hermite Integration.

$\pm x_i$	w_i
$n = 2$	
0.70710 67811 86548	1.46114 11826 611
$n = 3$	
0.00000 00000 00000	1.18163 59006 037
1.22474 48713 91589	1.32393 11752 136
$n = 4$	
0.52464 76232 75290	1.05996 44828 950
1.65068 01238 85785	1.24022 58176 958
$n = 5$	
0.00000 00000 00000	0.94530 87204 829
0.95857 24646 13819	0.98658 09967 514
2.02018 28704 56086	1.18148 86255 360
$n = 6$	
0.43607 74119 27617	0.87640 13344 362
1.33584 90740 13697	0.93558 05576 312
2.35060 49736 74492	1.13690 83326 745
$n = 7$	
0.00000 00000 00000	0.81026 46175 568
0.81628 78828 58965	0.82868 73032 836
1.67355 16287 67471	0.89718 46002 252
2.65196 13568 35233	1.10133 07296 103

The procedure applied in the 2D integral can be extended to a 3D integral. To evaluate

$$I = \int_a^b \int_c^d \int_e^f f(x,y,z) \, dx \, dy \, dz \tag{3.137}$$

using the $n = 2$ rule, the cuboid $a < x < b$, $c < y < d$, $e < z < f$ is divided

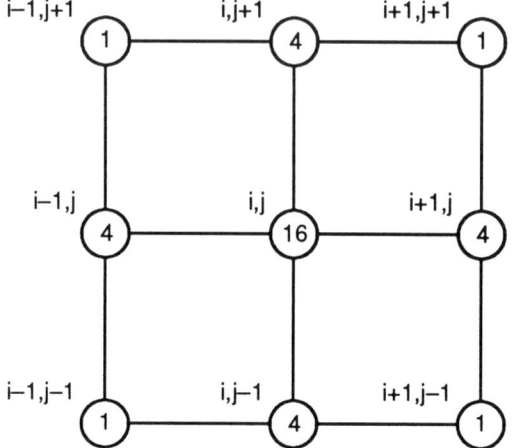

Figure 3.43 Double integration molecule for Simpson's 1/3 rule.

into $m \cdot l \cdot p$ smaller cuboids of sides

$$h_x = \frac{b-a}{m} \tag{3.138a}$$

$$h_y = \frac{d-c}{l} \tag{3.138b}$$

$$h_z = \frac{f-e}{p} \tag{3.138c}$$

where m, l, and p are multiples of $n = 2$. The subvolume A_{ijk} is evaluated by integrating along x according to Eq. (3.121) to obtain

$$g_{j,k} = \frac{h_x}{3}\left(f_{i+1,j,k} + 4f_{i,j,k} + f_{i-1,j,k}\right), \tag{3.139}$$

then along y

$$g_k = \frac{h_y}{3}\left(g_{j+1,k} + 4g_{j,k} + g_{j-1,k}\right), \tag{3.140}$$

and finally along z to obtain

$$A_{ijk} = \frac{h_z}{3}\left(g_{k+1} + 4g_k + g_{k-1}\right). \tag{3.141}$$

Substituting Eqs. (3.139) and (3.140) into Eq. (3.141) results in [76]

$$A_{ijk} = \frac{h_x h_y h_z}{27} \Big[\left(f_{i-1,j-1,k+1} + 4f_{i-1,j,k+1} + f_{i-1,j+1,k+1} \right)$$

$$+ \left(4f_{i,j-1,k+1} + 16f_{i,j,k+1} + 4f_{i,j+1,k+1} \right)$$

$$+ \left(f_{i+1,j-1,k+1} + 4f_{i+1,j,k+1} + f_{i+1,j+1,k+1} \right)$$

$$+ \left(4f_{i-1,j-1,k} + 16f_{i-1,j,k} + 4f_{i-1,j+1,k} \right)$$

$$+ \left(16f_{i,j-1,k} + 64f_{i,j,k} + 16f_{i,j+1,k} \right)$$

$$+ \left(4f_{i+1,j-1,k} + 16f_{i+1,j,k} + 4f_{i+1,j+1,k} \right)$$

$$+ \left(f_{i-1,j-1,k-1} + 4f_{i-1,j,k-1} + f_{i-1,j+1,k-1} \right)$$

$$+ \left(4f_{i,j-1,k-1} + 16f_{i,j,k-1} + 4f_{i,j+1,k-1} \right)$$

$$+ \left(f_{i+1,j-1,k-1} + 4f_{i+1,j,k-1} + f_{i+1,j+1,k-1} \right) \Big]. \qquad (3.142)$$

The integration molecule is portrayed in Fig. 3.44. Observe that the molecule is symmetric with respect to all planes that cut the molecule in half.

Example 3.10

Write a program that uses the Newton-Cotes rule ($n=6$) to evaluate Bessel function of order m, i.e.,

$$J_m(x) = \frac{1}{\pi} \int_0^\pi \cos(x \sin \theta - m\theta) \, d\theta.$$

Run the program for $m = 0$ and $x = 0.1, 0.2, ..., 2.0$.

Solution

The computer program is shown in Fig. 3.45. The program is based on Eqs. (3.115) and (3.119). It evaluates the integral within a subinterval $\theta_{n(i-1)} <$

228 Numerical Techniques in Electromagnetics

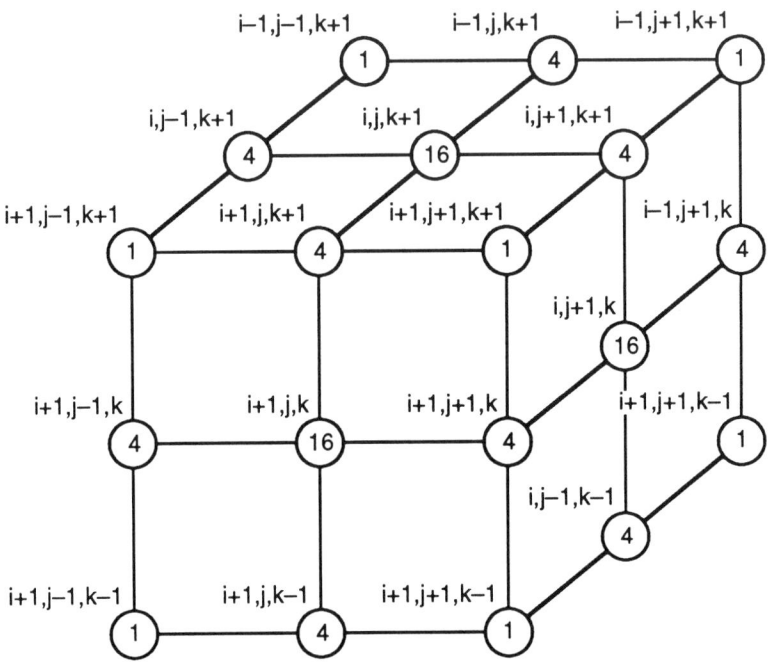

Figure 3.44 Triple integration molecule for Simpson's 1/3 rule.

$\theta < \theta_{ni}$. The summation over all the subintervals gives the required integral. The result for $m = 0$ and $0.1 < x < 2.0$ is shown in Table 3.14; the values agree up to six significant figures with those in standard tables [74, p.390]. The program is intentionally made general so that n, the corresponding Newton-Cotes numbers, and the integrand can be changed easily. Although the integrand in Fig. 3.45 is real, the program can be modified for complex integrand by simply declaring complex the affected variables.

```
0001
0002    C   **************************************************
0003    C   INTEGRATION BY NEWTON-COTES RULE (N=6)
0004    C
0005    C   A, B = ARE THE INTEGRATION LIMITS
0006    C   SUM  = THE RESULT OF THE INTEGRATION
0007    C   THE FUNCTION TO BE INTEGRATED MUST BE DEFINED IN
0008    C   A SUBPROGRAM FUNCTION FUN(ARGUMENTS)
0009    C
0010    C   SPECIFICALLY, THIS PROGRAM EVALUATES BESSEL FUNCTION
```

```
0011      C     JN(X), OF ORDER N AND ARGUMENT X
0012      C     ************************************************************
0013
0014            DATA  PIE/3.1415927/
0015
0016            A=0.0
0017            B=PIE
0018            N=4
0019            DO 20 I=1,20
0020            X=0.1*FLOAT(I)
0021            CALL INTEGRATION(A,B,N,X,SUM)
0022            WRITE(6,10) X,SUM
0023            PRINT *,X,SUM
0024      10    FORMAT(2X,'X = ',F6.3,2X,'JO = ',E20.12,/)
0025      20    CONTINUE
0026            STOP
0027            END

0001      C     ****************************************************************
0002      C           SUBROUTINE FOR INTEGRATION USING NEWTON-COTES RULE (N=6)
0003      C           A IS THE LOWER LIMIT OF INTEGRATION
0004      C           B IS THE UPPER LIMIT OF INTEGRATION
0005      C           H IS THE INTERVAL SIZE
0006      C     ****************************************************************
0007            SUBROUTINE INTEGRATION(AA,BB,NO,X,SUM)
0008            IMPLICIT REAL(A-H,O-Z)
0009            DIMENSION C(0:6)
0010            DATA NC /840/
0011            DATA  ( C(I),I=0,6 )/41.,216.,27.,272.,27.,216.,41./
0012
0013      C     START COMPUTATION
0014
0015            N=6
0016            M=240    ! M MUST BE A MULTIPLE OF N
0017            H=(BB-AA)/FLOAT(M)
0018            SUM=0.0
0019            NN=M/N
0020            T=AA
0021            DO 20 I=1,NN
0022            DO 10 J=0,N
0023            A=C(J)*FUN(T,NO,X)
0024            SUM=SUM + A
0025            T=T + H
0026      10    CONTINUE
0027            T=T - H
0028      20    CONTINUE
0029            SUM=SUM*H*FLOAT(N)/FLOAT(NC)
0030            RETURN
0031            END

0001      C     ***********************************
0002      C           THE FUNCTION TO BE INTEGRATED
0003      C     ***********************************
0004            FUNCTION FUN(THETA,N,X)
0005            DATA  PIE/3.1415927/
0006
0007            FUN = COS( X*SIN(THETA) - FLOAT(N)*THETA )/PIE
0008            RETURN
0009            END
```

Figure 3.45 Computer program for Example 3.10.

Table 3.14 Result of the Program in Fig. 3.45 for $m = 0$

x	$J_0(x)$
0.1	0.9975015
0.2	0.9900251
0.3	0.9776263
0.4	0.9603984
0.5	0.9384694
⋮	⋮
1.5	0.5118274
1.6	0.4554018
1.7	0.3979859
1.8	0.3399859
1.9	0.2818182
2.0	0.2238902

3.11 Concluding Remarks

Only a brief treatment of the finite difference analysis of PDEs is given here. There are many valuable references on the subject which answer many of the questions left unanswered here [3–8,10]. The book by Smith [5] gives an excellent exposition with numerous examples. The problems of stability and convergence of finite difference solutions are further discussed in [77,78], while the error estimates in [79].

As noted in Section 3.8, the finite difference method has some inherent advantages and disadvantages. It is conceptually simple and easy to program. The finite difference approximation to a given PDE is by no means unique; more accurate expressions can be obtained by employing more elaborate and complicated formulas. However, the relatively simple approximations may be employed to yield solutions of any specified accuracy simply by reducing the mesh size provided that the criteria for stability and convergence are met.

A very important difficulty in finite differencing of PDEs, especially parabolic and hyperbolic types, is that if one value of Φ is not calculated and therefore set equal to zero by mistake, the solution may become unstable. For example, in finding the difference between $\Phi_i = 1000$ and $\Phi_{i+1} = 1002$, if Φ_{i+1} is set equal to zero by mistake, the difference of 1000 instead of 2 may cause instability. To guard against such error, care must be taken to ensure that Φ is calculated at every point, particularly at boundary points.

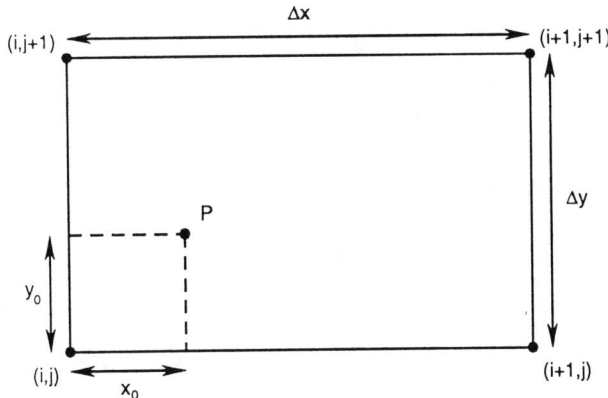

Figure 3.46 Evaluating Φ at a point P not on the grid.

A serious limitation of the finite difference method is that interpolation of some kind must be used to determine solutions at points not on the grid. Suppose we want to find Φ at a point P which is not on the grid, as in Fig. 3.46. Assuming Φ is known at the four grid points surrounding P, at a distance x_o along the bottom edge of the rectangle in Fig. 3.46,

$$\Phi_b = \frac{x_o}{\Delta x}[\Phi(i+1,j) - \Phi(i,j)] + \Phi(i,j). \tag{3.143}$$

At a distance x along the top edge,

$$\Phi_t = \frac{x_o}{\Delta x}[\Phi(i+1,j+1) - \Phi(i,j+1)] + \Phi(i,j+1). \tag{3.144}$$

The value of Φ at P is estimated by combining Eqs. (3.143) and (3.144), i.e.,

$$\Phi_p = \frac{y_o}{\Delta y}(\Phi_t - \Phi_b) + \Phi_b. \tag{3.145}$$

One obvious way to avoid interpolation is to use a finer grid if possible.

References

[1] L. D. Kovach, *Boundary-value Problems.* Reading, MA: Addison-Wesley, 1984, pp. 355–379.

[2] A. Thom and C. J. Apelt, *Field Computations in Engineering and Physics.* London: D. Van Nostrand, 1961, p. v.

[3] R. D. Richtmyer and K. W. Morton, *Difference Methods for Initial-value Problems,* 2nd ed. New York: Interscience Publ., 1976, pp. 185–193.

[4] D. Potter, *Computational Physics*. London: John Wiley, 1973, pp. 40–79.

[5] G. D. Smith, *Numerical Solution of Partial Differential Equations: Finite Difference Methods*, 2nd ed. Oxford: Clarendon, 1978.

[6] J. H. Ferziger, *Numerical Methods for Engineering Application*. New York: John Wiley, 1981.

[7] L. Lapidus and G. F. Pinder, *Numerical Solution of Partial Differential Equations in Science and Engineering*. New York: John Wiley, 1982, pp. 166–185.

[8] V. Vemuri and W. J. Karplus, *Digital Computer Treatment of Partial Differential Equations*. Englewood Cliffs, NJ: Prentice-Hall, 1981, pp. 88–92.

[9] A. Wexler, "Computation of electromagnetic fields," *IEEE Trans. Micro. Theo. Tech.*, vol. MTT-17, no. 8, Aug. 1969, pp. 416–439.

[10] G. de Vahl Davis, *Numerical Methods in Engineering and Science*. London: Allen & Unwin, 1986.

[11] Special issue of IEEE Transactions on Microwave Theory and Techniques, vol. MTT-17, no. 8, Aug. 1969 on "Computer-oriented microwave practices" covers various applications of finite difference methods to EM problems.

[12] H. E. Green, "The numerical solution of some important transmission-line problems," *IEEE Trans. Micro. Theo. Tech.*, vol. MTT-13, no. 5, Sept. 1965, pp. 676–692.

[13] M. N. O. Sadiku, "Finite difference solution of electrodynamic problems," *Int. Jour. Elect. Engr. Educ.*, vol. 28, April 1991, pp. 107–122.

[14] M. F. Iskander, "A new course on computational methods in electromagnetics," *IEEE Trans. Educ.*, vol. 31, no. 2, May 1988, pp. 101–115.

[15] W. S. Metcalf, "Characteristic impedance of rectangular transmission lines," *Proc. IEE*, vol. 112, no. 11, Nov. 1965, pp. 2033–2039.

[16] M. V. Schneider, "Computation of impedance and attenuation of TEM-lines by finite difference methods," *IEEE Micro. Theo. Tech.*, vol. MTT-13, no. 6, Nov. 1965, pp. 793–800.

[17] M. Sendaula, M. Sadiku, and R. Heiman, "Crosstalk computation in coupled transmission lines," *Proc. IEEE Southeastcon*, April 1991, pp. 790–795.

[18] A. R. Djordjevic, et al. "Time-domain response of multiconductor

transmission lines," *Proc. IEEE*, vol. 75, no. 6, June 1987, pp. 743–764.

[19] R. R. Gupta, "Accurate impedance determination of coupled TEM conductors," *IEEE Trans. Micro. Theo. Tech.*, vol. MTT-17, no. 12, Aug. 1969, pp. 479–489.

[20] E. Yamashita, et al., "Characterization method and simple design formulas of MCS lines proposed for MMIC's," *IEEE Trans. Micro. Theo. Tech.*, vol. MTT-35, no. 12, Dec. 1987, pp. 1355–1362.

[21] J. R. Molberg and D. K. Reynolds, "Iterative solutions of the scalar Helmholtz equations in lossy regions," *IEEE Trans. Micro. Theo. Tech.*, vol. MTT-17, no. 8, Aug. 1969, pp. 460–477.

[22] J. B. Davies and C. A. Muilwyk, "Numerical solution of uniform hollow waveguides with boundaries of arbitrary shape," *Proc. IEE*, vol. 113, no. 2, Feb. 1966, pp. 277–284.

[23] J. S. Hornsby and A. Gopinath, "Numerical analysis of a dielectric-loaded waveguide with a microstrip line—finite-difference methods," *IEEE Trans. Micro. Theo. Tech.*, vol. MTT-17, no. 9, Sept. 1969, pp. 684–690.

[24] M. J. Beubien and A. Wexler, "An accurate finite-difference method for higher-order waveguide modes," *IEEE Trans. Micro. Theo. Tech.*, vol. MTT-16, no. 12, Dec. 1968, pp. 1007–1017.

[25] C. A. Muilwyk and J. B. Davies, "The numerical solution of rectangular waveguide junctions and discontinuities of arbitrary cross section," *IEEE Trans. Micro. Theo. Tech.*, vol. MTT-15, no. 8, Aug. 1967, pp. 450–455.

[26] J. H. Collins and P. Daly, "Calculations for guided electromagnetic waves using finite-difference methods," *J. Electronics & Control*, vol. 14, 1963, pp. 361–380.

[27] T. Itoh (ed.), *Numerical Techniques for Microwaves and Millimeter-wave Passive Structures*. New York: John Wiley, 1989.

[28] D. H. Sinnott, et al., "The finite difference solution of microwave circuit problems," *IEEE Trans. Micro. Theo. Tech.*, vol. MTT-17, no. 8, Aug. 1969, pp. 464–478.

[29] W. K. Gwarek, "Analysis of an arbitrarily-shaped planar circuit—a time-domain approach," *IEEE Trans. Micro. Theo. Tech.*, vol. MTT-33, no. 10, Oct. 1985, pp. 1067–1072.

[30] M. De Pourceq, "Field and power-density calculations in closed microwave systems by three-dimensional finite differences," *IEE Proc.*, vol. 132, Pt. H, no. 6, Oct. 1985, pp. 360–368.

[31] A. Taflove and K. R. Umashankar, "Solution of complex electromagnetic penetration and scattering problems in unbounded regions," in A. J. Kalinowski (ed.), *Computational Methods for Infinite Domain Media-structure Interaction.* Washington, DC: ASME, vol. 46, 1981, pp. 83–113.

[32] A. Taflove, "Application of the finite-difference time-domain method to sinusoidal steady-state electromagnetic-penetration problems," *IEEE Trans. EM Comp.*, vol. EMC-22, no. 3, Aug. 1980, pp. 191–202.

[33] K. S. Kunz and K. M. Lee, "A three-dimensional finite-difference solution of the external response of an aircraft to a complex transient EM environment" (2 parts), *IEEE Trans. EM Comp.*, vol. EMC-20, no. 2, May 1978, pp. 328–341.

[34] M. L. Oristaglio and G. W. Hohman, "Diffusion of electromagnetic fields into a two-dimensional earth: a finite-difference approach," *Geophysics*, vol. 49, no. 7, July 1984, pp. 870–894.

[35] A. Taflove and M. E. Brodwin, "Computation of the electromagnetic fields and induced temperature within a model of the microwave-irradiated human eye," *IEEE Trans. Micro. Theo. Tech.*, vol. MTT-23, no. 11, Nov. 1975, pp. 888–896.

[36] R. W. M. Lau and R. J. Sheppard, "The modelling of biological systems in three dimensions using the time domain finite-difference method" (2 parts), *Phys. Med. Biol.*, vol. 31, no. 11, 1986, pp. 1247–1266.

[37] F. Sandy and J. Sage, "Use of finite difference approximations to partial differential equations for problems having boundaries at infinity," *IEEE Trans. Micro. Theo. Tech.*, May 1971, pp. 484–486.

[38] K. B. Whiting, "A treatment for boundary singularities in finite difference solutions of Laplace's equation," *IEEE Trans. Micro. Theo. Tech.*, vol. MTT-16, no. 10, Oct. 1968, pp. 889–891.

[39] G. E. Forsythe and W. R. Wasow, *Finite Difference Methods for Partial Differential Equations.* New York: John Wiley, 1960.

[40] M. L. James et al., *Applied Numerical Methods for Digital Computation,* 3rd ed. New York: Harper & Row, 1985, pp. 203–274.

[41] Y. Naiheng and R. F. Harrington, "Characteristic impedance of transmission lines with arbitrary dielectrics under the TEM approximation," *IEEE Trans. Micro. Theo. Tech.*, vol. MTT-34, no. 4, April 1986, pp. 472–475.

[42] K. S. Yee, "Numerical solution of initial boundary-value problems

involving Maxwell's equations in isotropic media," *IEEE Trans. Ant. Prop.*, vol. AP-14, May 1966, pp. 302–307.

[43] A. Taflove and M. E. Brodwin, "Numerical solution of steady-state electromagnetic scattering problems using the time-dependent Maxwell's equations," *IEEE Micro. Theo. Tech.*, vol. MTT-23, no. 8, Aug. 1975, pp. 623–630.

[44] A. Taflove and K. Umashankar, "A hybrid moment method/finite-difference time-domain approach to electromagnetic coupling and aperture penetration into complex geometries," *IEEE Trans. Ant. Prop.*, vol. AP-30, no. 4, July 1982, pp. 617–627. Also in B. J. Strait (ed.), *Applications of the Method of Moments to Electromagnetic Fields*. Orlando, FL.: SCEE Press, Feb. 1980, pp. 361–426.

[45] K. Umashankar and A. Taflove, "A novel method to analyze electromagnetic scattering of complex objects," *IEEE Trans. EM Comp.*, vol. EMC-24, no.4, Nov. 1982, pp. 397–405.

[46] A. Taflove and K. R. Umashankar, "The finite-difference time-domain method for numerical modeling of electromagnetic wave interactions," *Electromagnetics*, vol. 10, 1990, pp. 105–126.

[47] M. N. O. Sadiku, V. Bemmel, and S. Agbo, "Stability criteria for finite-difference time-domain algorithm," *Proc. IEEE Southeastcon*, April 1990, pp. 48–50.

[48] J. G. Blaschak and G. A. Kriegsmann, "A comparative study of absorbing boundary conditions," *J. Comp. Phys.*, vol. 77, 1988, pp. 109–139.

[49] G. Mux, "Absorbing boundary conditions for the finite-difference approximation of the time-domain electromagnetic-field equations," *IEEE Trans. EM Comp.*, vol. EMC-23, no. 4, Nov. 1981, pp. 377–382.

[50] R. Holland, "THREDE: A free-field EMP coupling and scattering code," *IEEE Trans. Nucl. Sci.*, vol. NS-24, no. 6, Dec. 1977, pp. 2416–2421.

[51] R. W. Ziolkowski et al., "Three-dimensional computer modeling of electromagnetic fields: a global lookback lattice truncation scheme," *J. Comp. Phy.*, vol. 50, 1983, pp. 360–408.

[52] E. R. Demarest, "A finite difference—time domain technique for modeling narrow apertures in conducting scatterers," *IEEE Trans. Ant. Prog.*, vol. AP-35, no. 7, July 1987, pp. 826–831.

[53] T. Shibata et al., "Analysis of microstrip circuits using three-dimensional full-wave electromagnetic field analysis in the time domain,"

IEEE Trans. Micro. Theo. Tech., vol. 36, no. 6, June 1988, pp. 1064–1070.

[54] W. K. Gwarek, "Analysis of arbitrarily shaped two-dimensional microwave circuits by finite-difference time-domain method," IEEE Trans. Micro. Theo. Tech., vol. 36, no. 4, April 1988, pp. 738–744.

[55] X. Zhang, et al., "Calculation of the dispersive characteristics of microstrips by the time-domain finite-difference method," IEEE Trans. Micro. Theo. Tech., vol. 36, no. 2, Feb. 1988, pp. 263–267.

[56] X. Zhang and K. K. Mei, "Time-domain finite-difference approach to the calculation of the frequency-dependent characteristics of microstrip discontinuities," IEEE Trans. Micro. Theo. Tech., vol. 36, no. 12, Dec. 1988, pp. 1775–1787.

[57] R. W. Larson, "Special purpose computers for the time domain advance of Maxwell's equations," IEEE Trans. Magnetics, vol. 25, no. 4, July 1989, pp. 2913–2915.

[58] K. K. Mei, et al., "Conformal time domain finite difference method," Radio Sci., vol. 19, no. 5, Sept./Oct. 1984, pp. 1145–1147.

[59] D. H. Choi and W. J. R. Hoefer, "The finite-difference time-domain method and its application to eigenvalue problems," IEEE Trans. Micro. Theo. Tech., vol. MTT-36, no. 12, Dec. 1986, pp. 1464–1470.

[60] D. M. Sullivan, et al., "Use of the finite-difference time-domain method in calculating EM absorption in human tissues," IEEE Trans. Biomed. Engr., vol. BME-34, no. 2, Feb. 1987, pp. 148–157.

[61] A. Christ and H. L. Hartnagel, "Three-dimensional finite-difference method for the analysis of microwave-device embedding," IEEE Trans. Micro. Theo. Tech., vol. MTT-35, no. 8, Aug. 1987, pp. 688–696.

[62] J. H. Whealton, "A 3D analysis of Maxwell's equations for cavities of arbitrary shape," J. Comp. Phys., vol. 75, 1988, pp. 168–189.

[63] V. Bemmel, "Time-domain finite-difference analysis of electromagnetic scattering and penetration problems," M. Sc. Thesis, Dept. of Electrical and Computer Engr., Florida Atlantic University, Boca Raton, Dec. 1987.

[64] D. S. Jones, The Theory of Electromagnetism. New York: MacMillan, 1964, pp. 450–452.

[65] J. A. Stratton, Electromagnetic Theory. New York: McGraw-Hill, 1941, pp. 563–573.

[66] M. DiStasio and W. C. McHarris, "Electrostatic problems? Relax!," *Am. J. Phy.*, vol. 47, no. 5, May 1979, pp. 440–444.

[67] M. N. O. Sadiku, "Finite difference solution of axisymmetric potential problems," *Int. J. Appl. Engr. Educ.*, vol. 6, no. 4, 1990, pp. 479–485.

[68] H. E. Green, "The numerical solution of transmission line problems," in L. Young (ed.), *Advances in Microwaves*, vol. 2. New York: Academic Press, 1967, pp. 327–393.

[69] C. D. Taylor, et al., "Electromagnetic pulse scattering in time varying inhomogeneous media," *IEEE Trans. Ant. Prog.*, vol. AP-17, no. 5, Sept. 1969, pp. 585–589.

[70] K. Asano, "Electrostatic potential and field in a cylindrical tank containing charged liquid," *Proc. IEE*, vol. 124, no. 12, Dec. 1977, pp. 1277–1281.

[71] P. J. Davis and P. Rabinowitz, *Methods of Numerical Integration*. New York: Academic Press, 1975.

[72] R. Piessens, et al., *QUADPACK : A Subroutine Package for Automatic Integration*. Berlin: Springer-Verlag, 1980.

[73] *Tables of Functions and Zeros of Functions*. Washington, DC: National Bureau of Standards, Applied Mathematical Series no. 37, 1954.

[74] M. Abramwitz and I. A. Stegun (eds.), *Handbook of Mathematical Functions*. Washington, DC: National Bureau of Standards, Applied Mathematical Series no. 55, 1964 .

[75] L. G. Kelly, *Handbook of Numerical Methods and Applications*. Reading, MA: Addison-Wesley, 1967, pp. 57–61.

[76] M. N. O. Sadiku and J. Jongakiem, "Newton-Cotes rules for triple integrals," *Proc. IEEE Southeastcon*, April 1990, pp. 471–475.

[77] B. P. Rynne, "Instabilities in time marching methods for scattering problems," *Electromagnetics*, vol. 6, no. 2, 1986, pp. 129–144.

[78] J. I. Steger and R. F. Warming, "On the convergence of certain finite-difference schemes by an inverse-matrix method," *J. Comp. Phys.*, vol. 17, 1975, pp. 103–121.

[79] D. W. Kelly, et al., "A posteriori error estimates in finite difference techniques," *J. Comp. Phys.*, vol. 74, 1988, pp. 214–232.

Problems

3.1 Show that the following finite difference approximations for Φ_x are valid:

(a) forward difference,
$$\frac{-\Phi_{i+2} + 4\Phi_{i+1} - 3\Phi_i}{2\Delta x}$$

(b) backward difference,
$$\frac{3\Phi_i - 4\Phi_{i-1} + \Phi_{i-2}}{2\Delta x}$$

(c) central difference,
$$\frac{-\Phi_{i+2} + 8\Phi_{i+1} - 8\Phi_{i-1} + \Phi_{i-2}}{12\Delta x}.$$

3.2 Solve the equation $\Phi_t = \Phi_{xx}$, $0 \le x \le 1$, subject to initial and boundary conditions
$$\Phi(x,0) = \sin \pi x, \quad 0 \le x \le 1,$$
$$\Phi(0,t) = 0 = \Phi(1,t), \quad t > 0.$$
Use $\Delta x = 0.25$ and $r = 0.5$.

3.3 Derive the Crank-Nicholson implicit algorithm for the hyperbolic equation $\Phi_{xx} = a^2 \Phi_{yy}$, $a^2 =$ constant. Let $\Delta x = \Delta y = \Delta$.

3.4 Prove that the fourth-order approximation of Laplace's equation $\Phi_{xx} + \Phi_{yy} = 0$ is
$$60\Phi(i,j) - 16\Big[\Phi(i+1,j) + \Phi(i-1,j) + \Phi(i,j+1) + \Phi(i,j-1)\Big]$$
$$+ \Phi(i+2,j) + \Phi(i-2,j) + \Phi(i,j+2) + \Phi(i,j-2) = 0.$$
Draw the computational molecule for the finite difference scheme.

3.5 It is desired to solve

$$\frac{\partial^2 \Phi}{\partial x^2} + \frac{\partial^2 \Phi}{\partial y^2} + 50 = 0$$

in the square region $0 \le x, y \le 1$ subject to the boundary conditions $\Phi = 10$ at $x = 0, 1$, $\Phi_y = 40$ at $y = 0$, $\Phi_y = -20$ at $y = 1$.

(a) Set up a system of finite difference equations which will allow the solution to be found at $x = y = 0.25$ using $\Delta x = \Delta y = h = 0.25$. Perform three iterations.

(b) Develop a program to solve the same problem using $h = 0.05, 0.1$, and 0.2.

3.6 (a) If $\Delta x \ne \Delta y$, show that for the computational molecule in Fig. 3.47(a), Eq. (3.42) becomes

$$V_o = \frac{V_1}{2(1+\alpha)} + \frac{V_2}{2(1+\alpha)} + \frac{V_3}{2(1+1/\alpha)} + \frac{V_4}{2(1+1/\alpha)}$$

where $\alpha = (\Delta x/\Delta y)^2$.

(b) Show that for the molecule in Fig. 3.47(b), Eq. (3.42) becomes

$$V_o = \frac{V_1}{(1+\Delta x_1/\Delta x_2)(1+\Delta x_1 \Delta x_2/\Delta y_3 \Delta y_4)}$$

$$+ \frac{V_2}{(1+\Delta x_2/\Delta x_1)(1+\Delta x_1 \Delta x_2/\Delta y_3 \Delta y_4)}$$

$$+ \frac{V_3}{(1+\Delta y_3/\Delta y_4)(1+\Delta y_3 \Delta y_4/\Delta x_1 \Delta x_2)}$$

$$+ \frac{V_4}{(1+\Delta y_4/\Delta y_3)(1+\Delta y_3 \Delta y_4/\Delta x_1 \Delta x_2)}.$$

The molecule in Fig. 3.47 (b) is useful in treating irregular boundaries.

(c) For the nine-point molecule in Fig. 3.47(c), show that

$$V_o = \frac{1}{8}\sum_{i=1}^{8} V_i.$$

240   Numerical Techniques in Electromagnetics

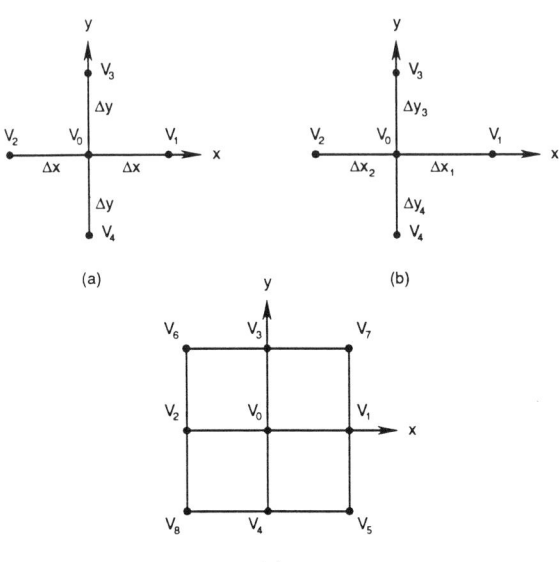

Figure 3.47   For Problem 3.6.

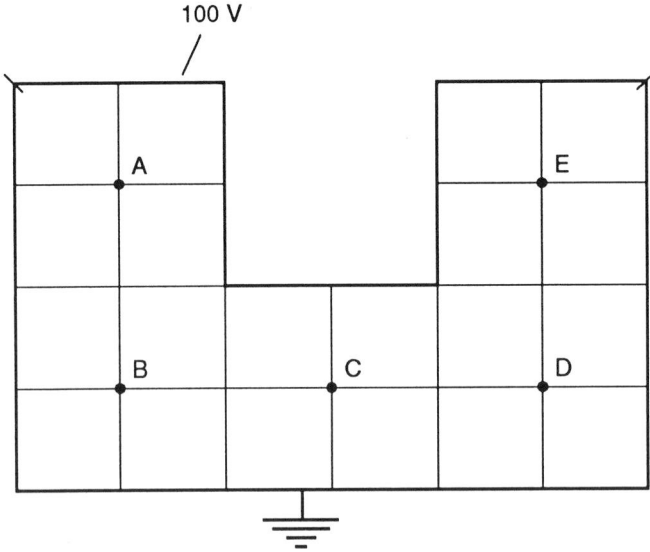

Figure 3.48   For Problem 3.7.

Finite Difference Methods 241

This is a more accurate difference equation than Eq. (3.42).

3.7 For a long hollow conductor with a uniform U-shape cross-section shown in Fig. 3.48, find the potential at points A, B, C, D, and E.

3.8 Modify the FORTRAN code of Fig. 3.12 to solve the following three-dimensional problem:

$$\nabla^2 V = -\rho_v/\epsilon, \qquad 0 \le x, y, z \le 1 \text{ meter},$$

where $\rho_v = xyz^2$ nC/m^2 and $\epsilon = 2\epsilon_o$ subject to the boundary conditions

$$V(0, y, z) = 0 = V(1, y, z)$$
$$V(x, 0, z) = 0 = V(x, 1, z)$$
$$V(x, y, 0) = 0, \ V(x, y, 1) = V_o.$$

Find the potential at the center of the cube and compare your result with the analytic solution. Take $V_o = 100$ volts.

3.9 (a) Find the exact solution of the elliptic equation

$$\frac{\partial^2 V}{\partial x^2} + \frac{\partial^2 V}{\partial y^2} = k\frac{\partial V}{\partial x}, \quad k > 0, \quad 0 \le x, y \le 1$$

subject to the boundary conditions

$$V(0, y) = \sin \pi y, \quad 0 \le y \le 1,$$
$$V(1, y) = 0 = V(x, 0) = V(x, 1).$$

(b) Show that the finite difference equations

(i) $4V(m, n) - (1 - kh/2)V(m+1, n) + (1 + kh/2)V(m-1, n) - V(m, n+1) - V(m, n-1) = 0$,

(ii) $(4 + kh)V(m, n) - V(m+1, n) - (1 + kh)V(m-1, n) - V(m, n-1) - V(m, n-1) = 0$,

(iii) $(2 + 2\cosh \frac{kh}{2})V(m, n) - e^{-kh/2}V(m+1, n) + e^{kh/2}V(m-1, n) - V(m, n+1) - V(m, n-1) = 0$, where $h = \Delta x = \Delta y$ is the mesh size, are all consistent replacements of the elliptic equation in part (a).

(c) Taking $k = 40$ and $h = 1/4$ and applying the boundary conditions in part (a), find the exact value of V at $(x,y) = (1/2, 1/2)$.

(d) Taking $k = 40$ and $h = 1/4$ and applying the boundary conditions in part (a), find the value of V at $(x,y) = (1/2, 1/2)$ using the three schemes in part (b).

3.10 Show that the leapfrog method applied to the parabolic equation (3.10) is unstable, whereas applying the DuFort-Frankel scheme yields an unconditionally stable solution.

3.11 The advective equation

$$\frac{\partial \Phi}{\partial t} + u \frac{\partial \Phi}{\partial x} = 0, \qquad u > 0$$

can be discretized as

$$\Phi_i^{n+1} = \Phi_i^n - r(\Phi_{i+1}^n - \Phi_{i-1}^n),$$

where $r = u\Delta t/2\Delta x$. Show that the difference scheme is unstable. An alternative scheme is:

$$\Phi_{n+1} = \frac{1}{2}(\Phi_{i+1}^n + \Phi_{i-1}^n) - r(\Phi_{i+1}^n - \Phi_{i-1}^n).$$

Find the condition on r for which this scheme is stable.

3.12 The two-dimensional parabolic equation

$$\frac{\partial U}{\partial t} = \frac{\partial^2 U}{\partial x^2} + \frac{\partial^2 U}{\partial y^2}, \qquad 0 \le x, y \le 1, \quad t > 0$$

is approximated by the finite difference methods:

(i) $U_{i,j}^{n+1} = \left[1 + r(\delta_x^2 + \delta_y^2)\right] U_{ij}^n$

(ii) $U_{i,j}^{n+1} = (1 + r\delta_x^2)(1 + r\delta_y^2) U_{i,j}^n$ where $r = \Delta t/h^2$, $h = \Delta x = \Delta y$

and

$$\delta_x^2 U_{i,j}^n = U_{i-1,j}^n - 2U_{i,j}^n + U_{i+1,j}^n$$

$$\delta_y^2 U_{i,j}^n = U_{i,j-1}^n - 2U_{i,j}^n + U_{i,j+1}^n.$$

Show that (i) is stable for $r \leq 1/4$ and (ii) is stable for $r \leq 1/2$.

3.13 (a) The constitutive parameter of the earth allows the displacement currents to be negligibly small. In this type of medium, show that Maxwell's equations for two-dimensional TE mode, where

$$\mathbf{E}(x,y,t) = E_z \mathbf{a}_z$$

and

$$\mathbf{E}(x,y,t) = H_x \mathbf{a}_x + H_y \mathbf{a}_y,$$

reduce to the diffusion equation

$$\frac{\partial^2 E}{\partial x^2} + \frac{\partial^2 E}{\partial y^2} - \mu\sigma\frac{\partial E}{\partial t} = \mu\frac{\partial J_s}{\partial t}$$

where $E = E_z$ and J_s is the source current density in the z direction.

(b) Taking $J_s = 0$, $\Delta x = \Delta y = \Delta$ and

$$\sum E_{i,j} = E_{i+1,j}^n + E_{i-1,j}^n + E_{i,j+1}^n + E_{i,j-1}^n,$$

show that applying Euler, leapfrog, and DuFort-Frankel difference methods to the diffusion equation gives:
Euler:
$$E_{i,j}^{n+1} = (1-4r)E_{i,j}^n + r\sum E_{i,j}^n,$$

Leapfrog:
$$E_{i,j}^{n+1} = E_{i,j}^{n-1} + 2r(\sum E_{i,j}^n - 4E_{i,j}^n),$$

DuFort-Frankel:
$$E_{i,j}^{n+1} = \frac{1-4r}{1+4r}E_{i,j}^{n-1} + \frac{2r}{1+4r}\sum E_{i,j}^n$$

where $r = \Delta t/(\sigma\mu\Delta^2)$.

(c) Analyze the stability of these finite difference schemes by substituting for $E_{i,j}^n$ a Fourier mode of the form

$$E_{i,j}^n = E(x=i\Delta, y=j\Delta, t=n\Delta t) = A_n \cos(k_x i\Delta)\,\cos(k_y j\Delta).$$

3.14 Yee's FD-TD algorithm for one-dimensional wave problems is given by

$$H_y^{n+1/2}(k+1/2) = H_y^{n-1/2}(k+1/2) + \frac{\delta t}{\mu \delta}\left[E_x^n(k) - E_x^n(k+1)\right].$$

Determine the stability criterion for the scheme by letting

$$E_x^n(k) = A^n e^{j\beta k \delta}, \qquad H_y^n(k) = \frac{A^n}{\eta} e^{j\beta k \delta},$$

where $\eta = (\mu/\epsilon)^{1/2}$ is the intrinsic impedance of the medium.

3.15 (a) The potential system in Fig. 3.49(a) is symmetric about the y-axis. Set the initial values at free nodes equal to zero and calculate (by hand) the potential at nodes 1 to 5 for 5 or more iterations.

(b) Consider the square mesh in Fig. 3.49(b). By setting initial values at the free nodes equal to zero, find (by hand calculation) the potential at nodes 1 to 4 for 5 or more iterations.

3.16 Solve Prob. 6.8 using the finite difference method.

3.17 Modify the program in Fig. 3.21 or write your own program to calculate Z_o for the microstrip line shown in Fig. 3.50. Take $a = 2.02$, $b = 7.0$, $h = 1.0 = w$, $t = 0.01$, $\epsilon_1 = \epsilon_0$, $\epsilon_2 = 9.6\epsilon_0$.

3.18 Use the FDM to calculate the characteristic impedance of the high-frequency, air-filled rectangular transmission line shown in Fig. 3.51. Take advantage of the symmetry of the problem and consider cases for which:

(a) $B/A = 1.0$, $a/A = 1/2$, $b/B = 1/2$, $a = 1$,

(b) $B/A = 1/2$, $a/A = 1/3$, $b/B = 1/3$, $a = 1$.

3.19 Use the FDM to determine the lowest (or dominant) cut-off wavenumber k_c of the TM_{11} mode in waveguides with square ($a \times a$) and rectangular ($a \times b$, $b = 2a$) crosssections. Compare your results with the exact solution

$$k_c = \sqrt{(m\pi/a)^2 + (n\pi/a)^2}$$

Finite Difference Methods 245

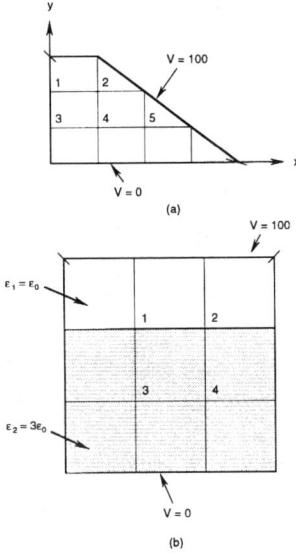

Figure 3.49 For Problem 3.15.

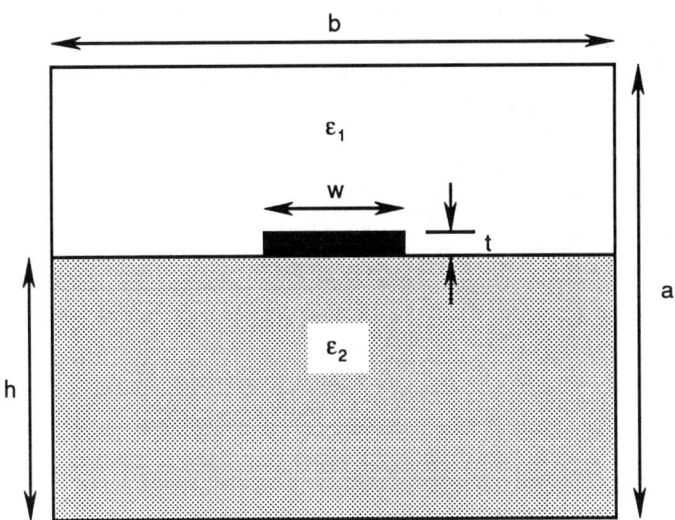

Figure 3.50 For Problem 3.17.

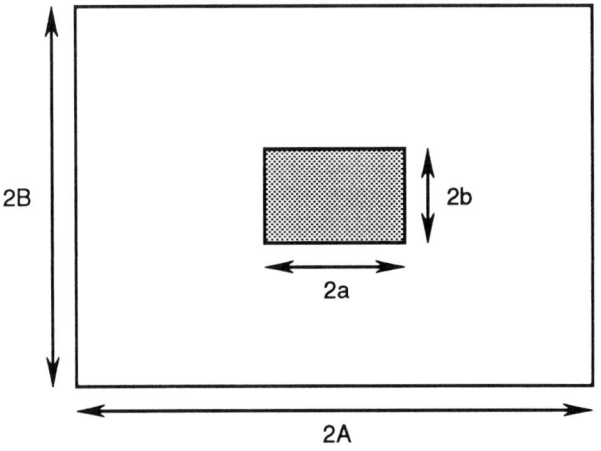

Figure 3.51 For Problem 3.18.

where $m = n = 1$. Take $a = 1$.

3.20 Modify the program shown in Fig. 3.38 to determine the electric field intensity in the entire domain. Plot E_z along the z-axis in the tank and E_ρ on the gas-liquid interface. Compare your result with the exact solution for E_z and E_ρ, namely [70]

$$\mathbf{E}_g = \mathbf{a}_\rho \, \rho_v \sum_{n=1}^{\infty} \frac{2\lambda_n J_1(\lambda_n \rho)}{C_n K_n} \big[\cosh(\lambda_n b) - 1\big] \sinh\big[\lambda_n(b+c-z)\big]$$

$$+ \mathbf{a}_z \, \rho_v \sum_{n=1}^{\infty} \frac{2\lambda_n J_0(\lambda_n \rho)}{\epsilon_r C_n K_n} \big[\cosh(\lambda_n b) - 1\big] \cosh\big[\lambda_n(b+c-z)\big],$$

$$\mathbf{E}_l = \mathbf{a}_\rho \, \rho_v \sum_{n=1}^{\infty} \frac{2\lambda_n J_1(\lambda_n \rho)}{\epsilon_r C_n} \left[\frac{\sinh(\lambda_n z)}{K_n} \Big(\cosh(\lambda_n b) \cosh(\lambda_n c) \right.$$

$$\left. + \epsilon_r \sinh(\lambda_n b) \sinh(\lambda_n c) - \cosh(\lambda_n c) \Big) - \cosh(\lambda_n z) + 1 \right]$$

$$- \mathbf{a}_z \, \rho_v \sum_{n=1}^{\infty} \frac{2\lambda_n J_0(\lambda_n \rho)}{\epsilon_r C_n} \left[\frac{\cosh(\lambda_n z)}{K_n} \Big(\cosh(\lambda_n b) \cosh(\lambda_n c) \right.$$

$$\left. + \epsilon_r \sinh(\lambda_n b) \sinh(\lambda_n c) - \cosh(\lambda_n c) \Big) - \sinh(\lambda_n z) \right]$$

where $\mathbf{E}_g$ and $\mathbf{E}_l$ are the electric fields in gas and liquid, respectively, and C_n and K_n are as defined in Example 3.9.

3.21 (a) Instead of the 5-point scheme of Eq. (3.92), use a more accurate 5-point formula

$$2i(8i^2 - 5)V(i,j) = (4i^3 + 2i^2 - 4i + 1)V(i+1,j)$$
$$+ (4i^3 - 2i^2 - 4i - 1)V(i-1,j)$$
$$i(4i^2 - 1)V(i,j+1) + i(4i^2 - 1)V(i,j-1)$$

in Example 3.9 while other things remain the same.

(b) Repeat part (a) with the following 9-point scheme:

$$\left(20 - \frac{14}{20i^2 - 5}\right) V(i,j) =$$

$$\left(1 + \frac{1}{2i} - \frac{3}{40i^2 + 20i}\right) \Big[V(i+1,j-1) + V(i+1,j-1)\Big]$$

$$+ \left(1 - \frac{1}{2i} - \frac{3}{40i^2 - 20i}\right) \Big[V(i-1,j-1) + V(i-1,j+1)\Big]$$

$$+ \left(4 + \frac{2}{i} - \frac{7}{20i^2 + 10i}\right) V(i+1,j) + \left(4 - \frac{2}{i} - \frac{7}{20i^2 - 10i}\right) V(i-1,j)$$

$$+ \left(4 + \frac{3}{20i^2 - 5}\right) \Big[V(i,j-1) + V(i,j+1)\Big].$$

3.22 For two-dimensional problems in which the field components do not vary with z coordinate ($\partial/\partial z = 0$), show that Yee's algorithm of Eq. (3.77) becomes:

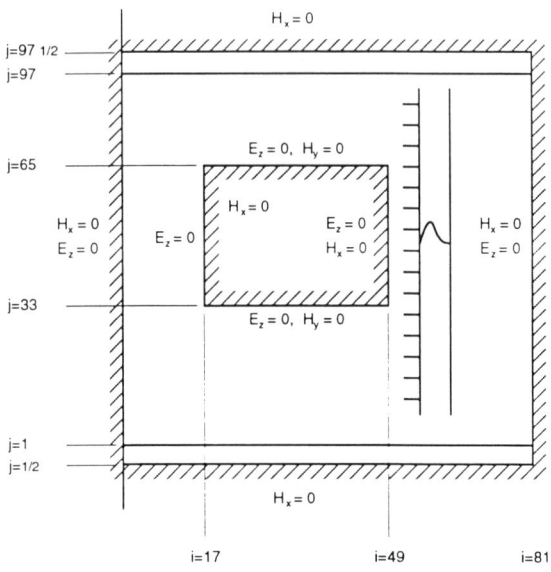

Figure 3.52 For Problem 3.23.

(a) for TE waves ($E_z = 0$)

$$H_z^{n+1/2}(i+1/2, j+1/2) = H_z^{n-1/2}(i+1/2, j+1/2)$$
$$- \alpha \left[E_y^n(i+1, j+1/2) - E_y^n(i, j+1/2) \right]$$
$$+ \alpha \left[E_x^n(i+1/2, j+1) - E_x^n(i+1/2, j) \right],$$

$$E_x^{n+1}(i+1/2, j) = E_x^n(i+1/2, j) + \beta \left[H_z^{n+1/2}(i+1/2, j+1/2) - H_z^{n+1/2}(i+1/2, j-1/2) \right],$$

$$E_y^{n+1}(i, j+1/2) = \gamma E_y^n(i, j+1/2) - \beta \left[H_z^{n+1/2}(i+1/2, j+1/2) - H_z^{n+1/2}(i-1/2, j+1/2) \right];$$

(b) for TM waves ($H_z = 0$)

$$E_z^{n+1}(i, j) = \gamma E_z^n(i, j) + \beta \left[H_y^{n+1/2}(i+1/2, j) - H_y^{n+1/2}(i-1/2, j) \right]$$
$$- \beta \left[H_x^{n+1/2}(i, j+1/2) - H_x^{n+1/2}(i, j-1/2) \right],$$

$$H_x^{n+1/2}(i,j+1/2) = H_x^{n-1/2}(i,j+1/2) - \alpha[E_z^n(i,j+1) - E_z^n(i,j)],$$

$$H_y^{n+1/2}(i+1/2,j) = H_y^{n-1/2}(i+1/2,j) + \alpha[E_z^n(i+1,j) - E_z^n(i,j)],$$

where

$$\alpha = \frac{\delta t}{\mu \delta}, \quad \beta = \frac{\delta t}{\epsilon \delta}, \quad \gamma = 1 - \frac{\sigma \delta t}{\epsilon},$$

and $\delta = \Delta x = \Delta y$.

3.23 Consider the diffraction/scattering of an incident TM wave by a perfectly conducting square of side $4a$. The conducting obstacle occupies $17 < i < 49, 33 < j < 65$, while artificial boundaries are placed at $i = 1, 81, j = 0.5, 97.5$ as shown in Fig. 3.52. Assume an incident wave with only E_z and H_y components given by

$$E_z = \begin{cases} \sin \pi \theta, & 0 < \theta < 1 \\ 0, & \text{otherwise} \end{cases}$$

$$H_y = \frac{1}{\eta_o} E_z$$

where $\eta_o = 120\pi \ \Omega$, $\theta = \dfrac{(x - 50a + ct)}{8a}$, $\Delta x = \Delta y = a/8$, $\Delta t = c\Delta x = a/16$. Write a program that applies the algorithm in Prob. 3.22(b). Assume "hard lattice truncation conditions" at the artificial boundaries shown in Fig. 3.52 and reproduce Yee's result [42] in his figures 3 to 6.

3.24 Repeat the previous problem but assume "soft lattice truncation conditions" of Eqs. (3.80) to (3.82) at the artificial boundaries.

3.25 Consider the finite cylindrical conductor held at $V = 100$ volts and enclosed in a larger grounded cylinder as in Fig. 3.53. Such a deceptively simple looking problem is beyond closed form solution, but by employing finite difference techniques, the problem can be solved without much effort. Using the finite difference method, write a program that determines the potential distribution in the axisymmetric solution region. Output the potential at $(\rho, z) = (2,10), (5,10), (8,10), (5,2),$ and $(5,18)$.

3.26 The problem in Fig. 3.54 is a prototype of an electrostatic particle focusing system which is employed in a recoil-mass time-of-flight

250 Numerical Techniques in Electromagnetics

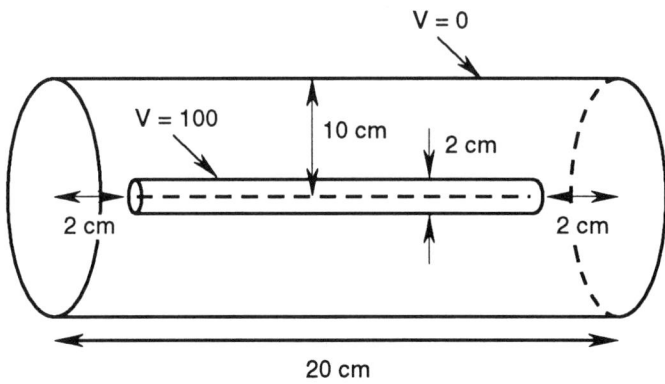

Figure 3.53 For Problem 3.25.

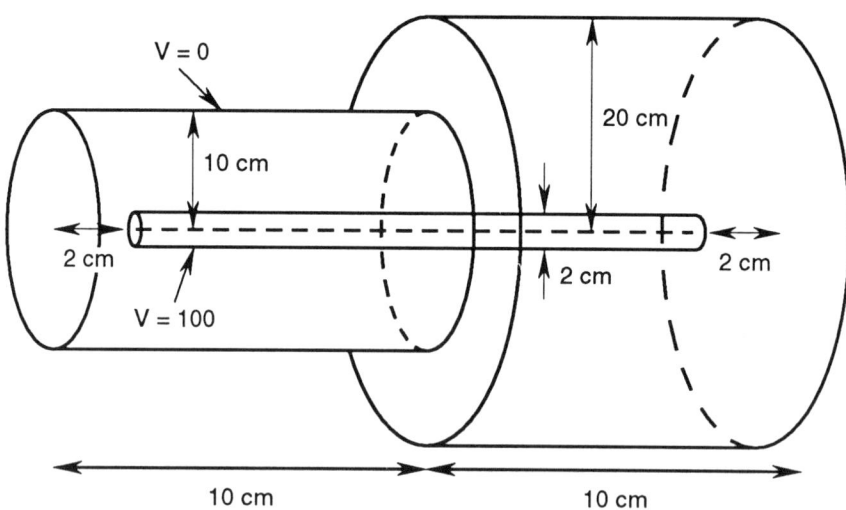

Figure 3.54 For Problem 3.26.

spectrometer. Write a program to determine the potential distribution in the system. The problem is similar to the previous problem except that the outer conductor abruptly expands radius by a factor of 2. Output the potential at $(\rho, z) = (5, 18)$, $(5, 10)$, $(5, 2)$, $(10, 2)$, and $(15, 2)$.

3.27 For axisymmetric problems (no variation with respect to ϕ), show that Yee's algorithm for TM waves can be written as

$$H_\phi^{n+1}(i,j) = H_\phi^n(i,j) + \alpha \left[E_z^{n+1/2}(i, j+1/2) - E_z^{n+1/2}(i, j-1/2) \right]$$

$$- \alpha \left[E_\rho^{n+1/2}(i+1/2, j) - E_\rho^{n+1/2}(i-1/2, j) \right].$$

$$E_z^{n+3/2}(i, j+1/2) = \gamma E_z^{n+1/2}(i, j+1/2)$$
$$+ \beta \left[\frac{1}{j} H_\phi^{n+1}(i, j+1/2) \right.$$
$$\left. + H_\phi^{n+1}(i, j+1) - H_\phi^{n+1}(i, j) \right],$$

$$E_\rho^{n+3/2}(i+1/2, j) = \gamma E_\rho^{n+1/2}(i+1/2, j) - \beta \left[H_\phi^{n+1}(i+1, j) - H_\phi^{n+1}(i, j) \right],$$

where

$$\alpha = \frac{\delta t}{\mu \delta}, \beta = \frac{\delta t}{\epsilon \delta}, \gamma = 1 - \frac{\sigma \delta t}{\epsilon}, \delta = \Delta \rho = \Delta z,$$

and $H_\phi(z, \rho, t) = H_\phi(z = i\Delta z, \rho = (j-1/2)\Delta\rho, t = n\delta t) = H_\phi^n(i, j)$.

3.28 Given the tabulated values of $y = \sin x$ for $x = 0.4$ to 0.52 radians in intervals of $\Delta x = 0.02$, find: (a) $\frac{dy}{dx}$ at $x = 0.44$, (b) $\int_{0.4}^{0.52} y\, dx$ using Simpson's rule .

x	sin
0.40	0.38942
0.42	0.40776
0.44	0.42594
0.46	0.44395
0.48	0.46178
0.50	0.47943
0.52	0.49688

3.29 (a) Use a pocket calculator to determine the approximate area under the curve $f(x) = 4 - x^2, 0 < x < 1$ by the trapezoidal rule with $h = 0.2$.

(b) Repeat part (a) using the Newton-Cotes rule with $n = 3$.

3.30 For a half-wave dipole, evaluating the integral

$$\int_0^\infty \frac{\cos^2(\frac{\pi}{2}\cos\theta)}{\sin\theta} d\theta$$

is usually required. Evaluate this integral numerically using any quadrature rule of your choice.

3.31 Compute

$$\int_0^1 e^{-x} dx$$

using the Newton-Cotes rule for cases $n = 2, 4$, and 6. Compare your results with exact values.

3.32 (a) Write a program which evaluates numerically the integral defining the error function

$$erf(x) = \frac{2}{\sqrt{\pi}} \int_0^x e^{-t^2} dt.$$

Print a table of $erf(x)$ for $x = 0.0, 0.1, 0.2, ..., 2.0$.

(b) Write a program to evaluate the $erf(x)$ using the Taylor series

$$erf(x) = \frac{2}{\sqrt{\pi}} \sum_{n=0}^{\infty} \frac{(-1)^n x^{2n+1}}{n!(2n+1)}.$$

Use as many terms in the series as you deem necessary. Print a table of $erf(x)$ for the same values of x as in part (a).

(c) Compare your results in (a) and (b) with those in standard tables such as in [74, pp. 310–312].

3.33 The criterion for accuracy of the numerical approximation of an integral

$$I = \int_a^b f(x)\,dx \simeq \sum_{i=0}^{\infty} a_i\, f(x_i)$$

is that the formula is exact for all polynomials of degree less than or equal to n. If $a = 0, b = 4$, and the values of $f(x)$ are available at points $x_0 = 0$, $x_1 = 1$, $x_2 = 3, x_4 = 4$, find the values of the coefficients a_i for which the above requirement of accuracy is met.

3.34 The arc length AB of an ellipse shown in Fig. 3.55 is given by

$$F(k,\phi) = a \int_0^{\phi} \left[1 - k^2 \sin^2 \lambda\right]^{1/2} d\lambda, \qquad 0 \le k \le 1$$

where $k^2 = \frac{a^2 - b^2}{a^2}$, a and b are the semimajor and semiminor axes of the ellipse. This elliptic integral of the first type cannot be evaluated in a closed form. Write a general FORTRAN subprogram **ELLIPSE (A, B, PHI, F)** using Simpson's rule to determine $F(k,\phi)$. Verify your subprogram for an ellipse described by $9x^2 + 16y^2 = 36$, $\phi = 90°$.

3.35 Evaluate the following double integral using the trapezoidal rule:

(a) $\int_0^{\pi/2} \int_0^{\pi/2} \sin(\sqrt{2xy})\,dx\,dy,$

(b) $\int_1^5 \int_1^5 \left[x^2 + y^2\right]^{-1/2} dx\,dy,$

(c) $\int_2^4 \left[\int_4^6 \ln(xy^2)dx\right] dy.$

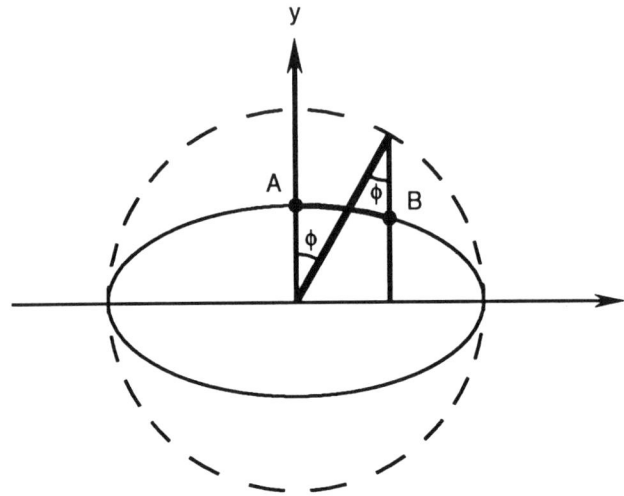

Figure 3.55 For Problem 3.34.

Chapter 4

Variational Methods

"The best thing to give to your enemy is forgiveness; to an opponent, tolerance; to a friend, your heart; to your child, a good example; to a father, deference; to your mother, conduct that will make her proud of you; to yourself, respect; to all men, charity." Arthur J. Balfour

4.1 Introduction

In solving problems arising from mathematical physics and engineering, we find that it is often possible to replace the problem of integrating a differential equation by the equivalent problem of seeking a function that gives a minimum value of some integral. Problems of this type are called *variational problems.* The methods that allow us to reduce the problem of integrating a differential equation to the equivalent variational problem are usually called *variational methods* [1]. The variational methods form a common base for both the method of moments (MOM) and the finite element method (FEM). Therefore, it is appropriate that we study the variational methods before MOM and FEM. Besides, it is relatively easy to formulate the solution of certain differential and integral equations in variational terms. Also, variational methods give accurate results without making excessive demands on computer storage and time.

Variational methods can be classified into two groups: direct and indirect methods. The direct method is the classical Rayleigh-Ritz method, while the indirect methods are collectively referred to as the method of weighted residuals: collocation (or point-matching), subdomain, Galerkin, and least square methods. The variational solution of a given PDE using an indirect method usually involves two basic steps [2]:

- cast the PDE into variational form, and

- determine the approximate solution using one of the methods.

The literature on the theory and applications of variational methods to EM problems is quite extensive, and no attempt will be made to provide an exhaustive list of references. Numerous additional references may be found in those cited in this chapter. Owing to a lack of space, we only can hint at some of the topics usually covered in an introduction to this subject.

4.2 Operators in Linear Spaces

In this section, we will review some principles of operators in linear spaces and establish notation [2–5]. We define the *inner (dot or scalar) product* of functions u and v as

$$\langle u, v \rangle = \int_\Omega uv^* d\Omega \tag{4.1}$$

where * denotes the complex conjugate and the integration is performed over Ω, which may be one-, two-, or three-dimensional physical space depending on the problem. In a sense, the inner product $\langle u, v \rangle$ gives the component or projection of function u in the direction of v. If **u** and **v** are vector fields, we modify Eq. (4.1) slightly to include a dot between them, i.e.,

$$\langle \mathbf{u}, \mathbf{v} \rangle = \int_\Omega \mathbf{u} \cdot \mathbf{v}^* d\Omega. \tag{4.2}$$

However, we shall consider u and v to be complex-valued scalar functions. For each pair of u and v belonging to the linear space, a number $\langle u, v \rangle$ is obtained that satisfies:

$$
\begin{aligned}
(1) \quad & \langle u, v \rangle = \langle v, u \rangle^*, & (4.3a) \\
(2) \quad & \langle \alpha u_1 + \beta u_2, v \rangle = \alpha \langle u_1, v \rangle + \beta \langle u_2, v \rangle, & (4.3b) \\
(3) \quad & \langle u, v \rangle > 0 \quad \text{if } u \neq 0, & (4.3c) \\
(4) \quad & \langle u, v \rangle = 0 \quad \text{if } u = 0. & (4.3d)
\end{aligned}
$$

If $\langle u, v \rangle = 0$, u and v are said to be *orthogonal*. Notice that these properties mimic familiar properties of the dot product in three-dimensional space. Equation (4.3) is easily derived from Eq. (4.1). Note that from Eq. (4.3a,b),

$$\langle u, \alpha v \rangle = \alpha^* \langle v, u \rangle^* = \alpha^* \langle u, v \rangle$$

where α is a complex scalar.

Equation (4.1) is called an *unweighted* or *standard inner product*. A *weighted inner product* is given by

$$\langle u, v \rangle = \int_\Omega uv^* w \, d\Omega \tag{4.4}$$

where w is a suitable weight function.

We define the norm of the function u as

$$\|u\| = \sqrt{\langle u, u \rangle}. \tag{4.5}$$

The norm is a measure of the "length" or "magnitude" of the function. (As far as a field is concerned, the norm is its rms value.) A vector is said to be *normal* if its norm is 1. Since the *Schwarz inequality*

$$|\langle u, v \rangle| \leq \|u\| \, \|v\| \tag{4.6}$$

holds for any inner product space, the angle θ between two nonzero vectors **u** and **v** can be obtained as

$$\theta = \cos^{-1} \frac{\langle \mathbf{u}, \mathbf{v} \rangle}{\|\mathbf{u}\| \, \|\mathbf{v}\|}. \tag{4.7}$$

We now consider the operator equation

$$\boxed{L \, \Phi = g} \tag{4.8}$$

where L is any linear operator, Φ is the unknown function, and g is the source function. The space spanned by all functions resulting from the operator L is

$$\langle L \, \Phi, g \rangle = \langle \Phi, L^a g \rangle. \tag{4.9}$$

The operator L is said to be:

(1) self-adjoint if $L = L^a$, i.e., $\langle L \, \Phi, g \rangle = \langle \Phi, Lg \rangle$,
(2) positive definite if $\langle L \, \Phi, g \rangle > 0$ for any $\Phi = 0$ in the domain of L,
(3) negative definite if $\langle L \, \Phi, g \rangle < 0$ for any $\Phi = 0$ in the domain of L.

The properties of the solution of Eq. (4.8) depend strongly on the properties of the operator L. If, for example, L is positive definite, we can easily show that the solution of Φ in Eq. (4.8) is unique, i.e., Eq. (4.8) cannot have more than one solution. To do this, suppose that Φ and Ψ are two solutions to Eq. (4.8) such that $L\Phi = g$ and $L\Psi = g$. Then, by virtue of linearity of L, $f = \Phi - \Psi$ is also is solution. Therefore, $Lf = 0$. Since L is positive definite, $f = 0$ implying that $\Phi = \Psi$ and confirming the uniqueness of the solution Φ.

Example 4.1

Find the inner product of $u(x) = 1 - x$ and $v(x) = 2x$ in the interval $(0, 1)$.

Solution

In this case, both u and v are real functions. Hence

$$\langle u, v \rangle = \langle v, u \rangle = \int_0^1 (1 - x)\, 2x\, dx$$

$$= 2\left(\frac{x^2}{2} - \frac{x^3}{3}\right)\bigg|_0^1 = 0.333.$$

Example 4.2

Show that the operator

$$L = -\nabla^2 = -\frac{\partial^2}{\partial x^2} - \frac{\partial^2}{\partial y^2}$$

is self-adjoint.

Solution

$$\langle Lu, v \rangle = -\int_S v \nabla^2 u \, dS.$$

Taking u and v to be real functions (for convenience) and applying the Green's identity

$$\oint_\ell v \frac{\partial u}{\partial n} \, dl = \int_S \nabla u \cdot \nabla v \, dS + \int_S v \nabla^2 u \, dS$$

yields

$$\langle Lu, v \rangle = \int_S \nabla u \cdot \nabla v \, dS - \oint_\ell v \frac{\partial u}{\partial n} \, dl \qquad (4.2.1)$$

where S is bounded by ℓ and **n** is the outward normal. Similarly

$$\langle u, Lv \rangle = \int_S \nabla u \cdot \nabla v \, dS - \oint_\ell u \frac{\partial v}{\partial n} \, dl. \qquad (4.2.2)$$

The line integrals in Eqs. (4.2.1) and (4.2.2) vanish under either the homogeneous Dirichlet or Neumann boundary conditions. Under the homogeneous mixed boundary conditions, they become equal. Thus, L is self-adjoint under any one of these boundary conditions.

4.3 Calculus of Variations

The calculus of variations, an extension of ordinary calculus, is a discipline that is concerned primarily with the theory of maxima and minima. Here we are concerned with seeking the extremum (minima or maxima) of an integral expression involving a function of functions or *functionals*. Whereas a function produces a number as a result of giving values to one or more independent variables, a functional produces a number that depends on the entire form of one or more functions between prescribed limits. In a sense, a functional is a measure of the function. A simple example is the inner product $\langle u, v \rangle$.

In the calculus of variation, we are interested in the necessary condition for a functional to achieve a stationary value. This necessary condition on the functional is generally in the form of a differential equation with boundary conditions on the required function.

Consider the problem of finding a function $y(x)$ such that the function

$$I(y) = \int_a^b F(x, y, y') \, dx, \qquad (4.10a)$$

subject to the boundary conditions

$$y(a) = A, \quad y(b) = B, \qquad (4.10b)$$

is rendered stationary. The integrand $F(x, y, y')$ is a given function of x, y, and $y' = dy/dx$. In Eq. (4.10a), $I(y)$ is called a *functional* or *variational* (or *stationary*) *principle*. The problem here is finding an extremizing function $y(x)$ for which the functional $I(y)$ has an extremum. Before attacking this problem, it is necessary that we introduce the operator δ, called the *variational symbol.*

The variation δy of a function $y(x)$ is an infinitesimal change in y' for a fixed value of the independent variable x, i.e., for $\delta x = 0$. The variation δy of y vanishes at points where y is prescribed (since the prescribed value cannot be varied) and it is arbitrary elsewhere (see Fig. 4.1). Due to the change in y (i.e., $y \to y + \delta y$), there is a corresponding change in F. The first variation of F at y is defined by

$$\delta F = \frac{\partial F}{\partial y} \delta y + \frac{\partial F}{\partial y'} \delta y'. \qquad (4.11)$$

This is analogous to the total differential of F,

$$dF = \frac{\partial F}{\partial x} dx + \frac{\partial F}{\partial y} dy + \frac{\partial F}{\partial y'} dy' \qquad (4.12)$$

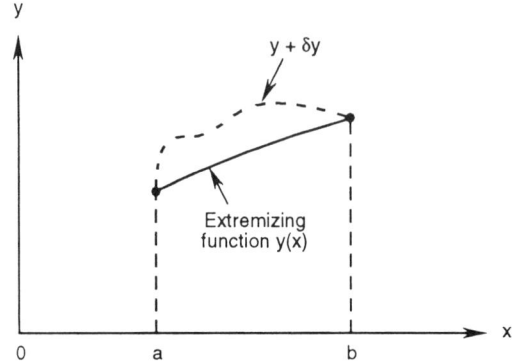

Figure 4.1 Variation of extremizing function with fixed ends.

where $dx = 0$ since x does not change as y changes to $y + \delta y$. Thus, we note that the operator δ is similar to the differential operator. Therefore, if $F_1 = F_1(y)$ and $F_2 = F_2(y)$, then

(i) $\delta(F_1 \pm F_2) = \delta F_1 \pm \delta F_2,$ \hfill (4.13a)

(ii) $\delta(F_1 F_2) = F_2 \delta F_1 + F_1 \delta F_2,$ \hfill (4.13b)

(iii) $\delta(F_1 F_2) = \dfrac{F_2 \delta F_1 - F_1 \delta F_2}{F_2^2},$ \hfill (4.13c)

(iv) $\delta(F_1)^n = n\,(F_1)^{n-1} \delta F_1,$ \hfill (4.13d)

(v) $\dfrac{d}{dx}(\delta y) = \delta(\dfrac{dy}{dx}),$ \hfill (4.13e)

(vi) $\delta \displaystyle\int_a^b y(x)\,dx = \int_a^b \delta y(x)\,dx.$ \hfill (4.13f)

A necessary condition for the function $I(y)$ in Eq. (4.10a) to have an extremum is that the variation vanishes, i.e.,

$$\boxed{\delta I = 0} \tag{4.14}$$

To apply this condition, we must be able to find the variation δI of I in Eq. (4.10a). To this end, let $h(x)$ be an increment in $y(x)$. For Eq. (4.10b) to be satisfied by $y(x) + h(x)$,

$$h(a) = h(b) = 0. \tag{4.15}$$

The corresponding increment in I in Eq. (4.10a) is

$$\Delta I = I(y+h) - I(y)$$
$$= \int_a^b \left[F(x, y+h, y'+h') - F(x, y, y') \right] dx.$$

Variational Methods

On applying Taylor's expansion,

$$\Delta I = \int_a^b \left[F_y(x,y,y')h - F_{y'}(x,y,y')h' \right] dx$$
$$+ \text{higher order terms}$$
$$= \delta I + O(h^2)$$

where

$$\delta I = \int_a^b \left[F_y(x,y,y')h - F_{y'}(x,y,y')h' \right] dx.$$

Integration by parts leads to

$$\delta I = \int_a^b \left[\frac{\partial F}{\partial y} - \frac{d}{dx}\left(\frac{\partial F}{\partial y'}\right) \right] h \, dx + \frac{\partial F}{\partial y'} h \bigg|_{x=0}^{x=b}.$$

The last term vanishes since $h(b) = h(a) = 0$ according to Eq. (4.15). In order that $\delta I = 0$, the integrand must vanish, i.e.,

$$\frac{\partial F}{\partial y} - \frac{d}{dx}\left(\frac{\partial F}{\partial y'}\right) = 0$$

or

$$\boxed{F_y - \frac{d}{dx} F_{y'} = 0} \qquad (4.16)$$

This is called *Euler's (or Euler-Lagrange) equation*. Thus a necessary condition for $I(y)$ to have an extremum for a given function $y(x)$ is that $y(x)$ satisfies Euler's equation.

This idea can be extended to more general cases. In the case considered so far, we have one dependent variable y and one independent variable x, i.e., $y = y(x)$. If we have one dependent variable u and two independent variables x and y, i.e., $u = u(x,y)$, then

$$I(u) = \int_S F(x,y,u,u_x,u_y) \, dS \qquad (4.17)$$

where $u_x = \partial u/\partial x$, $u_y = \partial u/\partial y$, and $dS = dxdy$. The functional in Eq. (4.17) is stationary when $\delta I = 0$, and it is easily shown that the corresponding Euler's equation is [6]

$$\boxed{\frac{\partial F}{\partial u} - \frac{\partial}{\partial x}\left(\frac{\partial F}{\partial u_x}\right) - \frac{\partial}{\partial y}\left(\frac{\partial F}{\partial u_y}\right) = 0} \qquad (4.18)$$

262 Numerical Techniques in Electromagnetics

Next we consider the case of two independent variables x and y and two dependent variables $u(x,y)$ and $v(x,y)$. The functional to be minimized is

$$I(u,v) = \int_S F(x,y,u,v,u_x,u_y,v_x,v_y)\,dS. \tag{4.19}$$

The corresponding Euler's equation is

$$\boxed{\begin{aligned}\frac{\partial F}{\partial u} - \frac{\partial}{\partial x}\left(\frac{\partial F}{\partial u_x}\right) - \frac{\partial}{\partial y}\left(\frac{\partial F}{\partial u_y}\right) = 0 \\ \frac{\partial F}{\partial v} - \frac{\partial}{\partial x}\left(\frac{\partial F}{\partial v_x}\right) - \frac{\partial}{\partial y}\left(\frac{\partial F}{\partial v_y}\right) = 0\end{aligned}} \quad \begin{aligned}(4.20a)\\(4.20b)\end{aligned}$$

Another case is when the functional depends on second- or higher-order derivatives. For example,

$$I(y) = \int_a^b F(x,y,y',y'',\ldots,y^{(n)})\,dx. \tag{4.21}$$

In this case, the corresponding Euler's equation is

$$\boxed{F_y - \frac{d}{dx}F_{y'} + \frac{d^2}{dx^2}F_{y''} - \frac{d^3}{dx^3}F_{y'''} + \cdots + (-1)^n \frac{d^n}{dx^n} F_{y^{(n)}} = 0} \tag{4.22}$$

Note that each of Euler's equations (4.16), (4.18), (4.20), and (4.22) is a differential equation.

Example 4.3

Given the functional

$$I(\Phi) = \int_S \left[\frac{1}{2}(\Phi_x^2 + \Phi_y^2) - f(x,y)\Phi\right]dxdy,$$

obtain the relevant Euler's equation.

Solution

Let

$$F(x,y,\Phi,\Phi_x,\Phi_y) = \frac{1}{2}(\Phi_x^2 + \Phi_y^2) - f(x,y)\Phi$$

showing that we have two independent variables x and y and one dependent variable Φ. Hence, Euler's equation (4.18) becomes

$$-f(x,y) - \frac{\partial}{\partial x}\Phi_x - \frac{\partial}{\partial y}\Phi_y = 0$$

or

$$\Phi_{xx} + \Phi_{yy} = -f(x,y),$$

i.e.,

$$\nabla^2 \Phi = -f(x,y)$$

which is Poisson's equation. Thus, solving Poisson's equation is equivalent to finding Φ that extremizes the given functional $I(\Phi)$.

4.4 Construction of Functionals from PDEs

In the previous section, we noticed that Euler's equation produces the governing differential equation corresponding to a given functional or variational principle. Here we seek the inverse procedure of constructing a variational principle for a given differential equation. To be specific, suppose we have one dependent variable Φ and two independent variables x and y. The procedure for finding the functional associated with the differential equation involves four basic steps [2,7]:

(1) Multiply the operator equation $L\Phi = g$ (Euler's equation) with the variation $\delta\Phi$ of the dependent variable Φ and integrate over the domain of the problem.
(2) Use the divergence theorem to transfer the derivatives to variation $\delta\Phi$.
(3) Express the boundary integrals in terms of the specified boundary conditions.
(4) Bring the variational operator δ outside the integrals.

The procedure is best illustrated with an example. Suppose we are interested in finding the variational principle associated with the Poisson's equation

$$\nabla^2 \Phi = -f(x,y) \qquad (4.23)$$

which is the exact opposite of what we did in Example 4.3. After taking step 1, we have

$$\delta I = \iint [-\nabla^2 \Phi - f]\, \delta\Phi\, dxdy = 0$$
$$= -\iint \nabla^2 \Phi\, \delta\Phi\, dxdy - \iint f\delta\Phi\, dxdy.$$

This can be evaluated by applying divergence theorem or integrating by parts. To integrate by parts, let $u = \delta\Phi$, $dv = \frac{\partial}{\partial x}(\frac{\partial \Phi}{\partial x})dx$ so that $du = \frac{\partial}{\partial x}\delta\Phi\, dx$, $v = \frac{\partial \Phi}{\partial x}$ and

$$-\int \left[\int \frac{\partial}{\partial x}(\frac{\partial \Phi}{\partial x})\delta\Phi\, dx\right] dy = -\int \left[\delta\Phi \frac{\partial \Phi}{\partial x} - \int \frac{\partial \Phi}{\partial x}\frac{\partial}{\partial x}\delta\Phi\, dx\right] dy.$$

Thus

$$\delta I = \iint \left[\frac{\partial \Phi}{\partial x}\frac{\partial}{\partial x}\delta\Phi + \frac{\partial \Phi}{\partial y}\frac{\partial}{\partial y}\delta\Phi - \delta f\Phi\right] dxdy$$

$$- \int \delta\Phi \frac{\partial \Phi}{\partial x} dy - \int \delta\Phi \frac{\partial \Phi}{\partial y} dx$$

$$\delta I = \frac{\delta}{2} \iint \left[(\frac{\partial \Phi}{\partial x})^2 + (\frac{\partial \Phi}{\partial y})^2 - 2f\Phi\right] dxdy$$

$$- \delta \int \Phi \frac{\partial \Phi}{\partial x} dy - \delta \int \Phi \frac{\partial \Phi}{\partial y} dx. \qquad (4.24)$$

The last two terms vanish if we assume either the homogeneous Dirichlet or Neuman conditions at the boundaries. Hence

$$\delta I = \delta \iint \frac{1}{2}\left[\Phi_x^2 + \Phi_y^2 - 2\Phi f\right] dxdy,$$

i.e.,

$$I(\Phi) = \frac{1}{2} \iint \left[\Phi_x^2 + \Phi_y^2 - 2\Phi f\right] dxdy \qquad (4.25)$$

as expected.

Rather than following the four steps listed above to find the function $I(\Phi)$ corresponding to the operator equation (4.8), an alternative approach is provided by Mikhlin [1, pp. 74–78]. According to Mikhlin, if L in Eq. (4.8) is real, self-adjoint, and positive definite, the solution of Eq. (4.8) minimizes the functional

$$\boxed{I(\Phi) = \langle L\Phi, \Phi \rangle - 2\langle \Phi, g \rangle} \qquad (4.26)$$

(See Prob. 4.5 for a proof.) Thus Eq. (4.25), for example, can be obtained from Eq. (4.23) by applying Eq. (4.26). This approach has been applied to derive variational solutions of integral equations [8].

Other systematic approaches for the derivation of variational principles for EM problems include Hamilton's principle or the principle of least action [9,10], Lagrange multipliers [10–14], and a technique described as variational electromagnetics [15,16]. The method of Lagrange undetermined multipliers is particularly useful for deriving a functional for a PDE whose arguments are constrained. Table 4.1 provides the variational principles for some differential equations commonly found in EM-related problems.

Variational Methods

Table 4.1 Variational Principle Associated with Common PDEs in EM[1]

Name of equation	Partial Differential Equation (PDE)	Variational principle
Inhomogeneous wave equation	$\nabla^2\Phi + k^2\Phi = g$	$I(\Phi) = \dfrac{1}{2}\int_v \left[|\nabla\Phi|^2 - k^2\Phi^2 + 2g\Phi\right] dv$
Homogeneous wave equation	$\nabla^2\Phi + k^2\Phi = 0$	$I(\Phi) = \dfrac{1}{2}\int_v \left[|\nabla\Phi|^2 - k^2\Phi^2\right] dv$
or		
	$\nabla^2\Phi - \dfrac{1}{u^2}\Phi_{tt} = 0$	$I(\Phi) = \dfrac{1}{2}\int^{t_o}\int_v \left[|\nabla\Phi|^2 - \dfrac{1}{u^2}\Phi_t^2\right] dv\, dt$
Diffusion equation	$\nabla^2\Phi - k\Phi_t = 0$	$I(\Phi) = \dfrac{1}{2}\int^{t_o}\int_v \left[|\nabla\Phi|^2 - k\Phi\Phi_t\right] dv\, dt$
Poisson's equation	$\nabla^2\Phi = g$	$I(\Phi) = \dfrac{1}{2}\int_v \left[|\nabla\Phi|^2 + 2g\Phi\right] dv$
Laplace's equation	$\nabla^2\Phi = 0$	$I(\Phi) = \dfrac{1}{2}\int_v \left[|\nabla\Phi|^2\right] dv$

[1]Note that $|\nabla\Phi|^2 = \nabla\Phi\cdot\nabla\Phi = \Phi_x^2 + \Phi_y^2 + \Phi_z^2$.

Example 4.4

Find the functional for the ordinary differential equation

$$y'' + y + x = 0, \quad 0 < x < 1$$

subject to $y(0) = y(1) = 0$.

Solution

Given that

$$\frac{d^2y}{dx^2} + y + x = 0, \quad 0 < x < 1,$$

we obtain

$$\delta I = \int_0^1 \left(\frac{d^2y}{dx^2} + y + x\right)\delta y\, dx = 0$$

$$= \int_0^1 \frac{d^2y}{dx^2}\delta y\, dx + \int_0^1 y\, \delta y\, dx + \int_0^1 x\, \delta y\, dx.$$

Integrating the first term by parts,

$$\delta I = \delta y \left.\frac{dy}{dx}\right|_{x=0}^{x=1} - \int_0^1 \frac{dy}{dx}\frac{d}{dx}\,\delta y + \int_0^1 \frac{1}{2}\delta(y^2)\,dx + \delta\int_0^1 xy\,dx.$$

Since y is fixed at $x = 0, 1$, $\delta y(1) = \delta y(0) = 0$. Hence

$$\delta I = -\delta\int_0^1 \frac{1}{2}\left(\frac{dy}{dx}\right)^2 + \frac{1}{2}\delta\int y^2\,dx + \delta\int_0^1 xy\,dx$$

$$= \frac{\delta}{2}\int_0^1 \left[-y'^2 + y^2 + 2xy\right]\,dx$$

or

$$I(y) = \frac{1}{2}\int_0^1 \left[-y'^2 + y^2 + 2xy\right]\,dx.$$

Check: Taking $F(x,y,y') = y'^2 - y^2 - 2xy$, Euler's equation $F_y - \frac{d}{dx}F_{y'} = 0$ gives the differential equation

$$y'' + y + x = 0.$$

4.5 Rayleigh-Ritz Method

The Rayleigh-Ritz method is the direct variational method for minimizing a given functional. It is direct in that it yields a solution to the variational problem without recourse to the associated differential equation [17]. In other words, it is the direct application of variational principles discussed in the previous sections. The method was first presented by Rayleigh in 1877 and extended by Ritz in 1909. Without loss of generality, let the associated variational principle be

$$I(\Phi) = \int_S F(x, y, \Phi, \Phi_x, \Phi_y)\,dS. \qquad (4.27)$$

Our objective is to minimize this integral. In the Rayleigh-Ritz method, we select a linearly independent set of functions called *expansion functions* (or *basis functions*) u_n and construct an approximate solution to Eq. (4.27), satisfying some prescribed boundary conditions. The solution is in the form of a finite series

$$\tilde{\Phi} \simeq \sum_{n=1}^{N} a_n u_n + u_o \qquad (4.28)$$

where a_n are expansion coefficients to be determined and $\tilde{\Phi}$ is an approximate solution to Φ (the exact solution). We substitute Eq. (4.28) into Eq. (4.27) and convert the integral $I(\Phi)$ into a function of N coefficients $a_1, a_2, \cdots, a_N$, i.e.,

$$I(\Phi) = I(a_1, a_2, \cdots, a_N).$$

The minimum of this function is obtained when its partial derivatives with respect to each coefficient is zero:

$$\frac{\partial I}{\partial a_1} = 0, \quad \frac{\partial I}{\partial a_2} = 0, \cdots, \frac{\partial I}{\partial a_N} = 0$$

or

$$\frac{\partial I}{\partial a_n} = 0, \quad n = 1, 2, \cdots, N. \tag{4.29}$$

Thus we obtain a set of N simultaneous equations. The system of linear algebraic equations obtained is solved to get a_n, which are finally substituted into the approximate solution of Eq. (4.28). In the approximate solution of Eq. (4.28), if $\tilde{\Phi} \to \Phi$ as $N \to \infty$ in some sense, then the procedure is said to *converge* to the exact solution.

An alternative, perhaps easier, procedure to determine the expansion coefficients a_n is by solving a system of simultaneous equations obtained as follows [4,18]. Substituting Eq. (4.28) (ignoring u_o since it can be lumped with the right-hand side of the equation) into Eq. (4.26) yields

$$I = \langle \sum_{m=1}^{N} a_m L u_m, \sum_{n=1}^{N} a_n u_n \rangle - 2 \langle \sum_{m=1}^{N} a_m u_m, g \rangle$$

$$= \sum_{m=1}^{N} \sum_{n=1}^{N} \langle L u_m, u_n \rangle a_n a_m - 2 \sum_{m=1}^{N} \langle u_m, g \rangle a_m.$$

Expanding this into powers of a_m results in

$$I = \langle L u_m, u_m \rangle a_m^2 + \sum_{n \neq m}^{N} \langle L u_m, u_n \rangle a_m a_n + \sum_{k \neq m}^{N} \langle L u_k, u_m \rangle a_k a_m$$

$$- 2 \langle g, u_m \rangle a_m + \text{terms not containing } a_m. \tag{4.30}$$

Assuming L is self-adjoint and replacing k with n in the second summation,

$$I = \langle L u_m, u_m \rangle a_m^2 + 2 \sum_{n \neq m}^{N} \langle L u_m, u_n \rangle a_n a_m - 2 \langle g, u_m \rangle a_m + \cdots. \tag{4.31}$$

Since we are interested in selecting a_m such that I is minimized, Eq. (4.31) must satisfy Eq. (4.29). Thus differentiating Eq. (4.31) with respect to a_m and setting the result equal to zero leads to

$$\sum_{n=1}^{N} \langle Lu_m, u_n \rangle a_n = \langle g, u_m \rangle, \quad m = 1, 2, \cdots, N \quad (4.32)$$

which can be put in matrix form as

$$\begin{bmatrix} \langle Lu_1, u_1 \rangle & \langle Lu_1, u_2 \rangle & \cdots & \langle Lu_1, u_N \rangle \\ \vdots & & & \vdots \\ \langle Lu_N, u_1 \rangle & \langle Lu_N, u_2 \rangle & \cdots & \langle Lu_N, u_N \rangle \end{bmatrix} \begin{bmatrix} a_1 \\ \vdots \\ a_N \end{bmatrix} = \begin{bmatrix} \langle g, u_1 \rangle \\ \vdots \\ \langle g, u_N \rangle \end{bmatrix} \quad (4.33a)$$

or

$$[A][X] = [B] \quad (4.33b)$$

where $A_{mn} = \langle Lu_m, u_n \rangle$, $B_m = \langle g, u_m \rangle$, $X_n = a_n$. Solving for $[X]$ in Eq. (4.33) and substituting a_m in Eq. (4.28) gives the approximate solution $\tilde{\Phi}$. Equation (4.33) is called the *Rayleigh-Ritz system*.

We are yet to know how the expansion functions are selected. They are selected to satisfy the prescribed boundary conditions of the problem. u_o is chosen to satisfy the inhomogeneous boundary conditions, while u_n $(n = 1, 2, \cdots, N)$ are selected to satisfy the homogeneous boundary conditions. If the prescribed boundary conditions are all homogeneous (Dirichlet conditions), $u_o = 0$. The next section will discuss more on the selection of the expansion functions.

The Rayleigh-Ritz method has two major limitations. First, the variational principle in Eq. (4.27) may not exist in some problems such as in nonself-adjoint equations (odd order derivatives). Second, it is difficult, if not impossible, to find the functions u_o satisfing the global boundary conditions for the domains with complicated geometries [19].

Example 4.5

Use the Rayleigh-Ritz method to solve the ordinary differential equation:

$$\Phi'' + 4\Phi - x^2 = 0, \quad 0 < x < 1$$

subject to $\Phi(0) = 0 = \Phi(1)$.

Solution

The exact solution is

$$\Phi(x) = \frac{\sin 2(1-x) - \sin 2x}{8 \sin 2} + \frac{x^2}{4} - \frac{1}{8}.$$

The variational principle associated with $\Phi'' + 4\Phi - x^2 = 0$ is

$$I(\Phi) = \int_0^1 \left[(\Phi')^2 - 4\Phi^2 + 2x^2\Phi\right] dx. \tag{4.5.1}$$

This is readily verified using Euler's equation. We let the approximate solution be

$$\widetilde{\Phi} = u_o + \sum_{n=1}^{N} a_n u_n \tag{4.5.2}$$

where $u_o = 0$, $u_n = x^n(1-x)$ since $\Phi(0) = 0 = \Phi(0)$ must be satisfied. (This choice of u_n is not unique. Other possible choices are $u_n = x(1 - x^n)$ and $u_n = \sin n\pi$. Note that each choice satisfies the prescribed boundary conditions.) Let us try different values of N, the number of expansion coefficients. We can find the expansion coefficients a_n in two ways: using the functional directly as in Eq. (4.29) or using the Rayleigh-Ritz system of Eq. (4.33).

Method 1

For $N = 1$, $\widetilde{\Phi} = a_1 u_1 = a_1 x(1-x)$. Substituting this into Eq. (4.5.1) gives

$$I(a_1) = \int_0^1 \left[a_1^2(1-2x)^2 - 4a_1^2(x-x^2)^2 + 2a_1 x^3(1-x)\right] dx$$

$$= \frac{1}{5}a_1^2 + \frac{1}{10}a_1.$$

$I(a_1)$ is minimum when

$$\frac{\partial I}{\partial a_1} = 0 \quad \rightarrow \quad \frac{2}{5}a_1 + \frac{1}{10} = 0 \text{ or } a_1 = -\frac{1}{4}.$$

Hence the quadratic approximate solution is

$$\widetilde{\Phi} = -\frac{1}{4}x(1-x). \tag{4.5.3}$$

For $N = 2$, $\tilde{\Phi} = a_1 u_1 + a_2 u_2 = a_1 x(1 - x) + a_2 x^2(1 - x)$. Substituting $\tilde{\Phi}$ into Eq. (4.5.1),

$$I(a_1, a_2) = \int_0^1 \Big[[a_1(1 - 2x) + a_2(2x - 3x^2)]^2 - 4[a_1(x - x^2) + a_2(x^2 - x^3)]^2$$
$$+ 2a_1 x^2(x - x^2) + 2a_2 x^2(x^2 - x^3)\Big] dx$$
$$= \frac{1}{5}a_1^2 + \frac{2}{21}a_2^2 + \frac{1}{5}a_1 a_2 + \frac{1}{10}a_1 + \frac{1}{15}a_2.$$

$$\frac{\partial I}{\partial a_1} = 0 \quad \rightarrow \quad \frac{2}{5}a_1 + \frac{1}{5}a_2 + \frac{1}{10} = 0$$

or

$$4a_1 + 2a_2 = -1. \qquad (4.5.4a)$$

$$\frac{\partial I}{\partial a_2} = 0 \quad \rightarrow \quad \frac{4}{21}a_1 + \frac{1}{5}a_2 + \frac{1}{15} = 0$$

or

$$21a_1 + 20a_2 = -7. \qquad (4.5.4b)$$

Solving Eq. (4.5.4) gives

$$a_1 = -\frac{6}{38}, \quad a_2 = -\frac{7}{38}$$

and hence the cubic approximate solution is

$$\tilde{\Phi} = -\frac{6}{38}x(1 - x) - \frac{7}{38}x^2(1 - x)$$

or

$$\tilde{\Phi} = \frac{x}{38}(7x^2 - x - 6).$$

Method 2

We now determine a_m using Eq. (4.33). From the given differential equation,

$$L = \frac{d^2}{dx^2} + 4, \quad g = x^2.$$

Hence

$$A_{mn} = \langle L u_m, u_n \rangle = \langle u_m, L u_n \rangle$$
$$= \int_0^1 x^m(1 - x)\Big[(\frac{d^2}{dx^2} + 4)x^n(1 - x)\Big] dx,$$

$$A_{mn} = \frac{n(n-1)}{m+n-1} - \frac{2n^2}{m+n} + \frac{n(n+1)+4}{m+n+1} - \frac{8}{m+n+2} + \frac{4}{m+n+3},$$

$$B_n = \langle g, u_n \rangle = \int_0^1 x^2 x^n (1-x) dx = \frac{1}{n+3} - \frac{1}{n+4}.$$

When $N = 1$, $A_{11} = -\frac{1}{5}$, $B_1 = \frac{1}{20}$, i.e.,

$$-\frac{1}{5} a_1 = \frac{1}{20} \to a_1 = -\frac{1}{4}$$

as before. When $N = 2$,

$$A_{11} = -\frac{1}{5}, \quad A_{12} = A_{21} = -\frac{1}{10}, \quad A_{22} = -\frac{2}{21}, \quad B_1 = \frac{1}{20}, \quad B_2 = \frac{1}{30}.$$

Hence

$$\begin{bmatrix} -\frac{1}{5} & -\frac{1}{10} \\ -\frac{1}{10} & -\frac{2}{21} \end{bmatrix} \begin{bmatrix} a_1 \\ a_2 \end{bmatrix} = \begin{bmatrix} \frac{1}{20} \\ \frac{1}{30} \end{bmatrix}$$

which gives $a_1 = -\frac{6}{38}$, $a_2 = -\frac{7}{38}$ as obtained previously. When $N = 3$,

$$A_{13} = A_{31} = -\frac{13}{210}, \quad A_{23} = A_{32} = -\frac{28}{105}, \quad A_{33} = -\frac{22}{315}, \quad B_3 = \frac{1}{42},$$

i.e.,

$$\begin{bmatrix} -\frac{1}{5} & -\frac{1}{10} & -\frac{13}{210} \\ -\frac{1}{10} & -\frac{2}{21} & -\frac{28}{105} \\ -\frac{13}{210} & -\frac{28}{105} & -\frac{22}{315} \end{bmatrix} \begin{bmatrix} a_1 \\ a_2 \\ a_3 \end{bmatrix} = \begin{bmatrix} \frac{1}{20} \\ \frac{1}{30} \\ \frac{1}{42} \end{bmatrix}$$

from which we obtain

$$a_1 = -\frac{6}{38}, \quad a_2 = -\frac{7}{38}, \quad a_3 = 0$$

showing that we obtain the same solution as for the case $N = 2$. For different values of x, $0 < x < 1$, the Rayleigh-Ritz solution is compared with the exact solution in Table 4.2.

Example 4.6

Using the Rayleigh-Ritz method, solve Poisson's equation:

$$\nabla^2 V = -\rho_o, \qquad \rho_o = \text{constant}$$

in a square $-1 \le x, y \le 1$, subject to the homogeneous boundary conditions $V(x, \pm 1) = 0 = V(\pm 1, y)$.

272   Numerical Techniques in Electromagnetics

Table 4.2 Comparison of Exact Solution with the Rayleigh-Ritz Solution of $\Phi'' + 4\Phi - x^2 = 0$, $\Phi(0) = 0 = \Phi(1)$

x	Exact solution	Rayleigh-Ritz	Solution
		$N = 1$	$N = 2$
0.0	0.0	0.0	0.0
0.2	-0.0301	-0.0400	-0.0312
0.4	-0.0555	-0.0600	-0.0556
0.6	-0.0625	-0.0625	-0.0644
0.8	-0.0489	-0.0400	-0.0488
1.0	0.0	0.0	0.0

Solution

Due to the symmetry of the problem, we choose the basis functions as

$$u_{mn} = (1 - x^2)(1 - y^2)(x^{2m} y^{2n} + x^{2n} y^{2m}), \quad m, n = 0, 1, 2, \cdots.$$

Hence

$$\tilde{\Phi} = (1 - x^2)(1 - y^2)\left[a_1 + a_2(x^2 + y^2) + a_3 x^2 y^2 + a_4(x^4 + y^4) + \cdots\right].$$

Case 1: When $m = n = 0$, we obtain the first approximation ($N = 1$) as

$$\tilde{\Phi} = a_1 u_1$$

where $u_1 = (1 - x^2)(1 - y^2)$.

$$A_{11} = \langle Lu_1, u_1 \rangle = \int_{-1}^{1}\int_{-1}^{1} \left(\frac{\partial^2 u_1}{\partial x^2} + \frac{\partial^2 u_1}{\partial y^2}\right) u_1 \, dx dy$$

$$= -8 \int_0^1 \int_0^1 (2 - x^2 - y^2)(1 - x^2)(1 - y^2) \, dx dy$$

$$= -\frac{256}{45},$$

$$B_1 = \langle g, u_1 \rangle = -\int_{-1}^{1}\int_{-1}^{1} (1 - x^2)(1 - y^2) \rho_o \, dx dy = -\frac{16 \rho_o}{9}.$$

Hence
$$-\frac{256}{45}a_1 = -\frac{16}{9}\rho_o \quad \to \quad a_1 = \frac{5}{16}\rho_o$$
and
$$\widetilde{\Phi} = \frac{5}{16}\rho_o(1-x^2)(1-y^2).$$

Case 2: When $m = n = 1$, we obtain the second order approximation ($N = 2$) as
$$\widetilde{\Phi} = a_1 u_1 + a_2 u_2$$
where $u_1 = (1-x^2)(1-y^2)$, $u_2 = (1-x^2)(1-y^2)(x^2+y^2)$. A_{11} and B_1 are the same as in case 1.

$$A_{12} = A_{21} = \langle Lu_1, u_2 \rangle = -\frac{1024}{525},$$
$$A_{22} = \langle Lu_2, u_2 \rangle = -\frac{11264}{4725},$$
$$B_2 = \langle g, u_2 \rangle = -\frac{32}{45}\rho_o.$$

Hence
$$\begin{bmatrix} -\frac{256}{45} & -\frac{1024}{525} \\ -\frac{1024}{525} & -\frac{11264}{4725} \end{bmatrix} \begin{bmatrix} a_1 \\ a_2 \end{bmatrix} = \begin{bmatrix} -\frac{16}{9}\rho_o \\ -\frac{32}{45}\rho_o \end{bmatrix}.$$

Solving this yields
$$a_1 = \frac{1295}{4432}\rho_o = 0.2922\rho_o, \quad a_2 = \frac{525}{8864}\rho_o = 0.0592\rho_o$$
and
$$\widetilde{\Phi} = (1-x^2)(1-y^2)\Big(0.2922 + 0.0592(x^2+y^2)\Big)\rho_o.$$

4.6 Weighted Residual Method

As noted earlier, the Rayleigh-Ritz method is applicable when a suitable functional exists. In cases where such a functional cannot be found, we apply one of the techniques collectively referred to as the *method of weighted residuals*. The method is more general and has wider application than the Rayleigh-Ritz method because it is not limited to a class of variational problems.

Consider the operator equation

$$L\Phi = g. \tag{4.34}$$

In the weighted residual method, the solution to Eq. (4.34) is approximated, in the same manner as in the Rayleigh-Ritz method, using the expansion functions, u_n, i.e.,

$$\tilde{\Phi} = \sum_{n=1}^{N} a_n u_n \tag{4.35}$$

where a_n the expansion coefficients. If a set of *weighting functions* $\{w_m\}$ (also known as *testing functions*) are chosen and the inner product of Eq. (4.35) is taken for each w_m, we obtain

$$\sum_{n=1}^{N} a_n \langle w_m, L u_n \rangle = \langle w_m, g \rangle, \quad m = 1, 2, \cdots, N. \tag{4.36}$$

The system of linear equations (4.36) can be cast into matrix form as

$$[A][X] = [B] \tag{4.37}$$

where $A_{mn} = \langle w_m, L u_n \rangle$, $B_m = \langle w_m, g \rangle$, $X_n = a_n$. Solving for $[X]$ in Eq. (4.37) and substituting for a_n in Eq. (4.35) gives the approximate solution to Eq. (4.34). However, substitution of the approximate solution in the operator equation results in a *residual* R (an error in the equation), i.e.,

$$\boxed{R = L(\tilde{\Phi} - \Phi) = L\tilde{\Phi} - g} \tag{4.38}$$

In the weighted residual method, the weighting functions w_m (which, in general, are not the same as the expansion functions u_n) are chosen such that the integral of a weighted residual of the approximation is zero, i.e.,

$$\int w_m R \, dv = 0$$

or

$$\boxed{\langle w_m, R \rangle = 0} \tag{4.39}$$

There are different ways of choosing the weighting functions w_m leading to:

- collocation (or point-matching method),
- subdomain method,
- Galerkin method,
- least squares method.

4.6.1 Collocation Method

We select the Dirac delta function as the weighting function, i.e.,

$$w_m(\mathbf{r}) = \delta(\mathbf{r} - \mathbf{r}_m) = \begin{cases} 1, & \mathbf{r} = \mathbf{r}_m \\ 0, & \mathbf{r} \neq \mathbf{r}_m. \end{cases} \quad (4.40)$$

Substituting Eq. (4.40) into Eq. (4.39) results in

$$R(\mathbf{r}) = 0. \quad (4.41)$$

Thus we select as many collocation (or matching) points in the interval as there are unknown coefficients a_n and make the residual zero at those points. This is equivalent to enforcing

$$\sum_{n=1}^{N} L\, a_n u_n = g \quad (4.42)$$

at discrete points in the region of interest, generally where boundary conditions must be met. Although the point-matching method is the simplest specialization for computation, it is not possible to determine in advance for a particular operator equation what collocation points would be suitable. An accurate result is ensured only if judicious choice of the match points is taken. (This will be illustrated in Example 4.7.) It is important to note that the finite difference method is a particular case of collocation with locally defined expansion functions [20]. The validity and legitimacy of the point-matching technique are discussed in [21,22].

4.6.2 Subdomain Method

We select weighting functions w_m, each of which exists only over subdomains of the domain of Φ. Typical examples of such functions for one-dimensional problems are illustrated in Fig. 4.2 and defined as follows.

(1) piecewise uniform (or pulse) function:

$$w_m(x) = \begin{cases} 1, & x_{m-1} < x < x_{m+1} \\ 0, & \text{otherwise} \end{cases} \quad (4.43a)$$

(2) piecewise linear (or triangular) function:

$$w_m(x) = \begin{cases} \dfrac{\Delta - |x - x_m|}{\Delta}, & x_{m-1} < x < x_{m+1} \\ 0, & \text{otherwise} \end{cases} \quad (4.43b)$$

276 Numerical Techniques in Electromagnetics

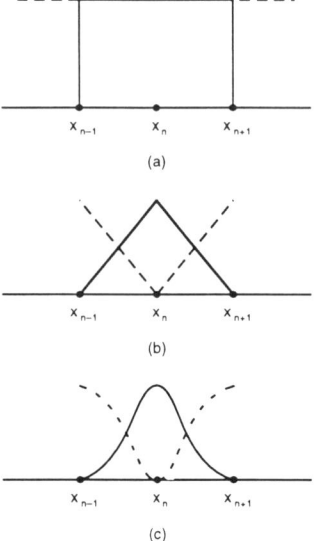

Figure 4.2 Typical subdomain weighting functions: (a) piecewise uniform function, (b) piecewise linear function, (c) piecewise sinusoidal function.

(3) piecewise sinusoidal function:

$$w_m(x) = \begin{cases} \dfrac{\sin k(x - |x - x_m|)}{\Delta}, & x_{m-1} < x < x_{m+1} \\ 0, & \text{otherwise} \end{cases} \quad (4.43c)$$

Using the unit pulse function, for example, is equivalent to dividing the domain of Φ into as many subdomains as there are unknown terms and letting the average value of R over such subdomains vanish.

4.6.3 Galerkin Method

We select basis functions as the weighting function, i.e., $w_m = u_m$. When the operator is a linear differential operator of even order, the Galerkin method reduces to the Rayleigh-Ritz method. This is due to the fact that the differentiation can be transferred to the weighting functions and the resulting coefficient matrix $[A]$ will be symmetric [7]. In order for the Galerkin method to be applicable, the operator must be of a certain specific type. Also, the expansion function u_n must span both the domain and the range of the operator.

4.6.4 Least Squares Method

This involves minimizing the integral of the square of the residual, i.e.,

$$\frac{\partial}{\partial a_m} \int R^2 \, dv = 0$$

or

$$\int \frac{\partial R}{\partial a_m} R \, dv = 0. \tag{4.44}$$

Comparing Eq. (4.44) with Eq. (4.39) shows that we must choose

$$w_m = \frac{\partial R}{\partial a_m} = L u_m. \tag{4.45}$$

This may be viewed as requiring that

$$\frac{1}{2} \int R^2 \, dv$$

be minimum. In other words, the choice of w_m corresponds to minimizing the mean square residual. It should be noted that the least squares method involves higher order derivatives which will, in general, lead to a better convergence than the Rayleigh-Ritz method or Galerkin method, but it has the disadvantage of requiring higher order weighting functions [19].

The concept of convergence discussed in the previous section applies here as well. That is, if the approximate solution $\tilde{\Phi}$ were to converge to the exact solution Φ as $N \to \infty$, the residual must approach zero as $N \to \infty$. Otherwise, the sequence of approximate solutions may not converge to any meaningful result.

The inner product involved in applying a weighted residual method can sometimes be evaluated analytically, but in most practical situations it is evaluated numerically. Due to a careless evaluation of the inner product, one may think that the least squares technique is being used when the resulting solution is identical to a point-matching solution. To avoid such erroneous results or conclusions, one must be certain that the overall number of points involved in the numerical integration is not smaller than the number of unknowns, N, involved in the weighted residual method [23].

The accuracy and efficiency of a weighted residual method is largely dependent on the selection of expansion functions. The solution may be exact or approximate depending on how we select the expansion and weighting functions [17]. The criteria for selecting expansion and weighting functions in a weighted residual method are provided in the work of Sarkar and others

[24-27]. We summarize their results here. The expansion functions u_n are selected to satisfy the following requirements [27]:

(1) The expansion functions should be in the domain of the operator L in some sense, i.e., they should satisfy the differentiability criterion and they must satisfy the boundary conditions for the operator. It is not necessary for each expansion function to satisfy exactly the boundary conditions. What is required is that the total solution must satisfy the boundary conditions at least in some distributional sense. The same holds for the differentiability conditions.

(2) The expansion functions must be such that $L\, u_n$ form a complete set for the range of the operator. It really does not matter whether the expansion functions are complete in the domain of the operator. What is important is that u_n must be chosen in such a way that $L\, u_n$ is complete. This will be illustrated in Example 4.8.

From a mathematical point of view, the choice of expansion functions does not depend on the choice of weighting functions. It is required that the weighting functions w_n must make the difference $\Phi - \tilde{\Phi}$ small. For the Galerkin method to be applicable, the expansion functions u_n must span both the domain and the range of the operator. For the least squares method, the weighting functions are already presented and defined by $L\, u_n$. It is necessary that $L\, u_n$ form a complete set. The least squares technique mathematically and numerically is one of the safest techniques to utilize when very little is known about the nature of the operator and the exact solution.

Example 4.7

Find an approximate solution to

$$\Phi'' + 4\Phi - x^2 = 0, \quad 0 < x < 1,$$

with $\Phi(0) = 0$, $\Phi'(1) = 1$, using the method of weighted residuals.

Solution

The exact solution is

$$\Phi(x) = \frac{\cos 2(x-1) + 2\sin 2x}{8\cos 2} - \frac{x^2}{4} - \frac{1}{8}. \qquad (4.7.1)$$

Let the approximate solution be

$$\tilde{\Phi} = u_0 + \sum_{n=1}^{N} a_n u_n. \qquad (4.7.2)$$

The boundary conditions can be decomposed into two parts:
(1) homogeneous part → $\Phi(0) = 0, \Phi'(0) = 0$,
(2) inhomogeneous part → $\Phi'(1) = 1$.

We choose u_0 to satisfy the inhomogeneous boundary conditions. A reasonable choice is

$$u_0 = x. \tag{4.7.3a}$$

We choose u_n $(n = 1, 2, \cdots, N)$ to satisfy the homogeneous boundary conditions. Suppose we select

$$u_n(x) = x^n(x - \frac{n+1}{n}). \tag{4.7.3b}$$

Thus, if we take $N = 2$, the approximate solution is

$$\begin{aligned}\widetilde{\Phi} &= u_0 + a_1 u_1 + a_2 u_2 \\ &= x + a_1 x(x - 2) + a_2 x^2(x - 3/2)\end{aligned} \tag{4.7.4}$$

where the expansion coefficients, a_1 and a_2, are to be determined. We find the residual R using Eq. (4.38), namely,

$$\begin{aligned}R &= L\widetilde{\Phi} - g \\ &= \left(\frac{d^2}{dx^2} + 4\right)\widetilde{\Phi} - x^2 \\ &= a_1(4x^2 - 8x + 2) + a_2(4x^3 - 6x^2 + 6x - 3) - x^2 + 4x.\end{aligned} \tag{4.7.5}$$

We now apply each of the four techniques discussed and compare the solutions.

Method 1: (collocation or point-matching method)

Since we have two unknowns a_1 and a_2, we select two match points, at $x = \frac{1}{3}$ and $x = \frac{2}{3}$, and set the residual equal to zero at those points, i.e.,

$$R(\frac{1}{3}) = 0 \quad \rightarrow \quad 6a_1 + 41a_2 = 33$$

$$R(\frac{2}{3}) = 0 \quad \rightarrow \quad 42a_1 + 13a_2 = 60.$$

Solving these equations,

$$a_1 = \frac{677}{548}, \quad a_2 = \frac{342}{548}.$$

Substituting a_1 and a_2 into Eq. (4.7.4) gives

$$\widetilde{\Phi}(x) = -1.471x + 0.2993x^2 + 0.6241x^3. \quad (4.7.6)$$

To illustrate the dependence of the solution on the match points, suppose we select $x = \frac{1}{4}$ and $x = \frac{3}{4}$ as the match points. Then

$$R(\frac{1}{4}) = 0 \quad \rightarrow \quad -4a_1 + 29a_2 = 15$$

$$R(\frac{3}{4}) = 0 \quad \rightarrow \quad 28a_1 + 3a_2 = 39.$$

Solving for a_1 and a_2, we obtain

$$a_1 = \frac{543}{412}, \quad a_2 = \frac{288}{412}$$

with the approximate solution

$$\widetilde{\Phi}(x) = -1.636x + 0.2694x^2 + 0.699x^3. \quad (4.7.7)$$

We will refer to the solutions in Eqs. (4.7.6) and (4.7.7) as collocation 1 and collocation 2, respectively. It is evident from Table 4.3 that collocation 2 is more accurate than collocation 1.

Method 2: (subdomain method)

Divide the interval $0 < x < 1$ into two segments since we have two unknowns a_1 and a_2. We select pulse functions as weighting functions:

$$w_1 = 1, \quad 0 < x < \frac{1}{2},$$

$$w_2 = 1, \quad \frac{1}{2} < x < 1$$

so that

$$\int_0^{1/2} w_1 R \, dx = 0 \quad \rightarrow \quad -8a_1 + 45a_2 = 22$$

$$\int_{1/2}^1 w_2 R \, dx = 0 \quad \rightarrow \quad 40a_1 + 3a_2 = 58.$$

Solving the two equations gives

$$a_1 = \frac{53}{38}, \quad a_2 = \frac{28}{38}$$

and hence Eq. (4.7.4) becomes
$$\widetilde{\Phi}(x) = -1.789x + 0.2895x^2 + 0.7368x^3. \tag{4.7.8}$$

Method 3: (Galerkin method)

In this case, we select $w_m = u_m$, i.e.,
$$w_1 = x(x-2), \quad w_2 = x^2(x - 3/2).$$
We now apply Eq. (4.39), namely, $\int w_m R \, dx = 0$. We obtain
$$\int_0^1 (x^2 - 2x) R \, dx = 0 \quad \rightarrow \quad 24a_1 + 11a_2 = 41$$
$$\int_0^1 (x^3 - \frac{3}{2}x^2) R \, dx = 0 \quad \rightarrow \quad 77a_1 + 15a_2 = 119.$$
Solving these leads to
$$a_1 = \frac{694}{487}, \quad a_2 = \frac{301}{487}.$$
Substituting a_1 and a_2 into Eq. (4.7.4) gives
$$\widetilde{\Phi}(x) = -1.85x + 0.4979x^2 + 0.6181x^3. \tag{4.7.9}$$

Method 4: (least squares method)

For this method, we select $w_m = \dfrac{\partial R}{\partial a_m}$, i.e.,
$$w_1 = 4x^2 - 8x + 2, \quad w_2 = 4x^3 - 6x^2 + 6x - 3.$$
Applying Eq. (4.39)
$$\int_0^1 w_1 R \, dx = 0 \quad \rightarrow \quad 7a_1 - 2a_2 = 8$$
$$\int_0^1 w_2 R \, dx = 0 \quad \rightarrow \quad -112a_1 + 438a_2 = 161.$$
Thus
$$a_1 = \frac{3826}{2842}, \quad a_2 = \frac{2023}{2842}$$
and Eq. (4.7.4) becomes
$$\widetilde{\Phi}(x) = -1.6925x + 0.2785x^2 + 0.7118x^3. \tag{4.7.10}$$
Notice that the approximate solutions in Eqs. (4.7.6) to (4.7.10) all satisfy the boundary conditions $\Phi(0) = 0$ and $\Phi'(1) = 1$. The five approximate solutions are compared in Table 4.3.

282 Numerical Techniques in Electromagnetics

Table 4.3 Comparison of the Weighted Residual Solutions of the Problem in Example 4.7 with the Exact Solution in Eq. (4.7.1)

x	Exact solution	Collocation 1	Collocation 2	Subdomain	Galerkin	Least squares
0.0	0.0000	0.0000	0.0000	0.0000	0.0000	0.0000
0.1	-0.1736	-0.1435	-0.1602	-0.1753	-0.1794	-0.1657
0.2	-0.3402	-0.2772	-0.3108	-0.3403	-0.3451	-0.3217
0.3	-0.4928	-0.3975	-0.4477	-0.4907	-0.4935	-0.4635
0.4	-0.6248	-0.5006	-0.5666	-0.6221	-0.6208	-0.5869
0.5	-0.7303	-0.5827	-0.6633	-0.7300	-0.7233	-0.6877
0.6	-0.8042	-0.6400	-0.7336	-0.8100	-0.7972	-0.7615
0.7	-0.8424	-0.6690	-0.7734	-0.8577	-0.8390	-0.8041
0.8	-0.8422	-0.6657	-0.7785	-0.8687	-0.8449	-0.8113
0.9	-0.8019	-0.6264	-0.7446	-0.8385	-0.8111	-0.7788
1.0	-0.7216	-0.5476	-0.6676	-0.7627	-0.7340	-0.7022

Example 4.8

This example illustrates the fact that expansion functions u_n must be selected such that $L\,u_n$ form a complete set for the range of the operator L. Consider the differential equation

$$-\Phi'' = 2 + \sin x, \qquad 0 \leq x \leq 2\pi \qquad (4.8.1)$$

subject to

$$\Phi(0) = \Phi(2\pi) = 0. \qquad (4.8.2)$$

Suppose we carelessly select

$$u_n = \sin nx, \qquad n = 1, 2, \cdots \qquad (4.8.3)$$

as the expansion functions, the approximate solution is

$$\widetilde{\Phi} = \sum_{n=1}^{N} a_n \sin nx. \qquad (4.8.4)$$

If we apply the Galerkin method, we obtain

$$\widetilde{\Phi} = \sin x. \qquad (4.8.5)$$

Although u_n satisfy both the differentiability and boundary conditions, Eq. (4.8.5) does not satisfy Eq. (4.8.1). Hence Eq. (4.8.5) is an incorrect solution. The problem is that the set $\{\sin nx\}$ does not form a complete set. If we add constant and cosine terms to the expansion functions in Eq. (4.8.4), then

$$\tilde{\Phi} = a_0 + \sum_{n=1}^{N}[a_n \sin nx + b_n \cos nx]. \qquad (4.8.6)$$

As $N \to \infty$, Eq. (4.8.6) is the classical Fourier series solution. Applying the Galerkin method leads to

$$\tilde{\Phi} = \sin nx \qquad (4.8.7)$$

which is again an incorrect solution. The problem is that even though u_n form a complete set, $L\,u_n$ do not. In order for $L\,u_n$ to form a complete set, $\tilde{\Phi}$ must be of the form

$$\tilde{\Phi} = \sum_{n=1}^{n}[a_n \sin nx + b_n \cos nx] + a_0 + cx + dx^2. \qquad (4.8.8)$$

Notice that the expansion functions $\{1, x, x^2, \sin nx, \cos nx\}$ in the interval $[0, 2\pi]$ form a linearly dependent set. This is because any function such as x or x^2 can be represented in the interval $[0, 2\pi]$ by the set $\{\sin nx, \cos nx\}$. Applying the Galerkin method, Eq. (4.8.8) leads to

$$\tilde{\Phi} = \sin x + x(2\pi - x) \qquad (4.8.9)$$

which is the exact solution Φ.

4.7 Eigenvalue Problems

As mentioned in Section 1.3.2, eigenvalue (nondeterministic) problems are described by equations of the type

$$L\Phi = \lambda M\Phi \qquad (4.46)$$

where L and M are differential or integral, scalar or vector operators. The problem here is the determination of the eigenvalues λ and the corresponding eigenfunctions Φ. It can be shown [11] that the variational principle for λ takes the form

$$\lambda = \frac{\langle \Phi, L\Phi \rangle}{\langle \Phi, M\Phi \rangle} = \frac{\int \Phi L\Phi \, dv}{\int \Phi M\Phi \, dv}. \qquad (4.47)$$

We may apply Eq. (4.47) to the Helmholtz equation for scalar waves, for example,
$$\nabla^2 \Phi + k^2 \Phi = 0. \tag{4.48}$$
Comparing Eq. (4.48) with Eq. (4.46), we obtain $L = -\nabla^2$, $M = 1$ (the identity operator), $\lambda = k^2$ so that
$$k^2 = -\frac{\int \Phi \nabla^2 \Phi \, dv}{\int \Phi^2 \, dv}. \tag{4.49}$$

Applying Green's identity (see Example 1.1),
$$\int_v (U \nabla^2 V + \nabla U \cdot \nabla V) \, dv = \oint U \frac{\partial V}{\partial n} \, dS,$$
to Eq. (4.49) yields
$$\boxed{k^2 = \frac{\int_v |\nabla \Phi|^2 \, dv - \oint \Phi \frac{\partial \Phi}{\partial n} \, dS}{\int_v \Phi^2 \, dv}} \tag{4.50}$$

Consider the following special cases.

(a) For homogeneous boundary conditions of the Dirichlet type ($\Phi = 0$) or Neumann type ($\frac{\partial \Phi}{\partial n} = 0$). Equation (4.50) reduces to
$$k^2 = \frac{\int_v |\nabla \Phi|^2 \, dv}{\int_v \Phi^2 \, dv}. \tag{4.51}$$

(b) For mixed boundary conditions ($\frac{\partial \Phi}{\partial n} + h\Phi = 0$), Eq. (4.50) becomes
$$k^2 = \frac{\int_v |\nabla \Phi|^2 \, dv + \oint h \Phi^2 \, dS}{\int_v \Phi^2 \, dv}. \tag{4.52}$$

It is usually possible to represent the ratio in each of Eqs. (4.47) to (4.52) in a rather different form. We choose the basis functions $u_1, u_2, \cdots, u_N$ which satisfy the boundary conditions and assume the approximate solution
$$\tilde{\Phi} = a_1 u_1 + a_2 u_2 + \cdots + a_N u_N$$

or
$$\tilde{\Phi} = \sum_{n=1}^{N} a_n u_n. \tag{4.53}$$

Substituting this into Eq. (4.46) gives

$$\sum_{n=1}^{N} a_n L u_n = \lambda \sum_{n=1}^{N} a_n M u_n. \tag{4.54}$$

Choosing the weighting functions w_m and taking the inner product of Eq. (4.54) with each w_m, we obtain

$$\sum_{n=1}^{N} \left[\langle w_m, L u_n \rangle - \lambda \langle w_m, M u_n \rangle \right] a_n = 0, \quad m = 1, 2, \cdots, N. \tag{4.55}$$

This can be cast into matrix form as

$$\sum_{n=1}^{N} (A_{mn} - \lambda B_{mn}) X_n = 0 \tag{4.56}$$

where $A_{mn} = \langle w_m, L u_n \rangle$, $B_{mn} = \langle w_m, M u_n \rangle$, $X_n = a_n$. Thus we have a set of homogeneous equations. In order for $\tilde{\Phi}$ in Eq. (4.53) not to vanish, it is necessary that the expansion coefficients a_n not all be equal to zero. This implies that the determinant of simultaneous equations (4.56) vanish, i.e.,

$$\begin{vmatrix} A_{11} - \lambda B_{11} & A_{12} - \lambda B_{12} & \cdots & A_{1N} - \lambda B_{1N} \\ \vdots & & & \vdots \\ A_{N1} - \lambda B_{N1} & A_{N2} - \lambda B_{N2} & \cdots & A_{NN} - \lambda B_{NN} \end{vmatrix} = 0$$

or
$$\left| [A] - \lambda [B] \right| = 0. \tag{4.57}$$

Solving this gives N approximate eigenvalues $\lambda_1, \cdots, \lambda_N$. The various ways of choosing w_m leads to different weighted residual techniques as discussed in the previous section.

Examples of eigenvalue problems for which variational methods have been applied include [28–37]:

- the cutoff frequency of a waveguide,
- the propagation constant of a waveguide, and
- the resonant frequency of a resonator.

Example 4.9

Solve the eigenvalue problem

$$\Phi'' + \lambda \Phi = 0, \qquad 0 < x < 1$$

with boundary conditions $\Phi(0) = 0 = \Phi(1)$.

Solution

The exact eigenvalues are

$$\lambda_n = (n\pi^2), \quad n = 1, 2, 3, \cdots \tag{4.9.1}$$

and the corresponding (normalized) eigenfunctions are

$$\Phi_n = \sqrt{2}\sin(n\pi x) \tag{4.9.2}$$

where Φ_n has been normalized to unity, i.e., $\langle \Phi_n, \Phi_n \rangle = 1$.

The approximate eigenvalues and eigenfunctions can be obtained by either using Eq. (4.47) or Eq. (4.57). Let the approximate solution be

$$\widetilde{\Phi}(x) = \sum_{k=0}^{N} a_k u_k, \qquad u_k = x(1 - x^k). \tag{4.9.3}$$

From the given problem, $L = -\frac{d^2}{dx^2}$, $M = 1$ (identity operator). Using the Galerkin method, $w_m = u_m$.

$$A_{mn} = \langle u_m, L u_n \rangle = \int_0^1 (x - x^{m+1}) \left[-\frac{d^2}{dx^2}(x - x^{n+1}) \right] dx$$

$$= \frac{mn}{m+n+1}, \tag{4.9.4}$$

$$B_{mn} = \langle u_m, M u_n \rangle = \int_0^1 (x - x^{m+1})(x - x^{n+1}) dx$$

$$= \frac{mn(m+n+6)}{3(m+3)(n+3)(m+n+3)}. \tag{4.9.5}$$

The eigenvalues are obtained from

$$\big|[A] - \lambda[B]\big| = 0. \tag{4.9.6}$$

For $N = 1$,
$$A_{11} = \frac{1}{3}, \quad B_1 = \frac{1}{30},$$
giving
$$\frac{1}{3} - \lambda \frac{1}{30} = 0 \quad \rightarrow \quad \lambda = 10.$$

The first approximate eigenvalue is $\lambda = 10$, a good approximation to the exact value of $\pi^2 = 9.8696$. The corresponding eigenfunction $\tilde{\Phi} = a_1(x - x^2)$ can be normalized to unity so that

$$\tilde{\Phi} = \sqrt{30}(x - x^2).$$

For $N = 2$, evaluating Eqs. (4.9.4) and (4.9.5), we obtain

$$\begin{bmatrix} \frac{1}{3} & \frac{1}{2} \\ \frac{1}{2} & \frac{4}{5} \end{bmatrix} \begin{bmatrix} a_1 \\ a_2 \end{bmatrix} = \lambda \begin{bmatrix} \frac{1}{30} & \frac{1}{20} \\ \frac{1}{20} & \frac{8}{105} \end{bmatrix} \begin{bmatrix} a_1 \\ a_2 \end{bmatrix}$$

or

$$\begin{vmatrix} 10 - \lambda & 0 \\ 0 & 42 - \lambda \end{vmatrix} = 0$$

giving eigenvalues $\lambda_1 = 10$, $\lambda_2 = 42$, compared with the exact values $\lambda_1 = \pi^2 = 9.8696$, $\lambda_2 = 4\pi^2 = 39.4784$, and the corresponding normalized eigenfunctions are

$$\tilde{\Phi}_1 = \sqrt{30}(x - x^2)$$
$$\tilde{\Phi}_2 = 2\sqrt{210}(x - x^2) - 2\sqrt{210}(x - x^3).$$

Continuing this way for higher N, the approximate eigenvalues shown in Table 4.4 are obtained. Unfortunately, the labor of computation increases as more u_k are included in $\tilde{\Phi}$. Notice from Table 4.4 that the approximate eigenvalues are always greater than the exact values. This is always true for a self-adjoint, positive definite operator [17]. Figure 4.3 shows the comparison between the approximate and exact eigenfunctions.

Example 4.10

Calculate the cutoff frequency of the inhomogeneous rectangular waveguide shown in Fig. 4.4. Take $\epsilon = 4\epsilon_o$ and $s = a/3$.

288 Numerical Techniques in Electromagnetics

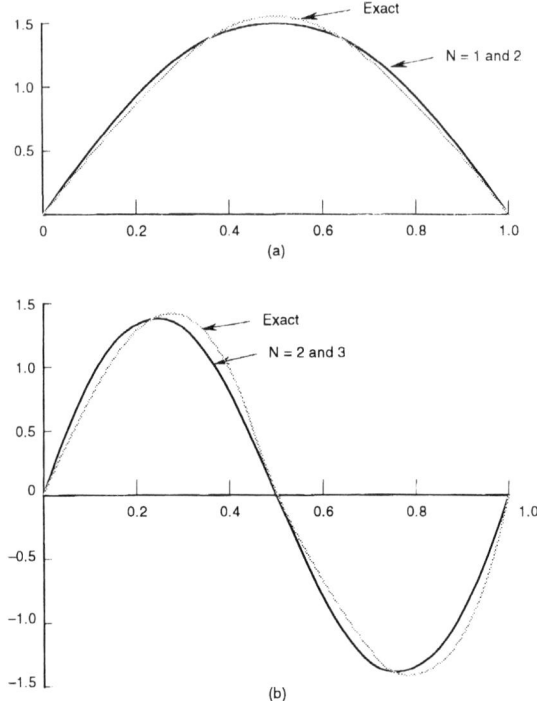

Figure 4.3 Comparison of approximate eigenfunctions with the exact solutions: (a) first eigenfunction, (b) second eigenfunction. (After Harrington [17]; with permission of Krieger Publishing Co.)

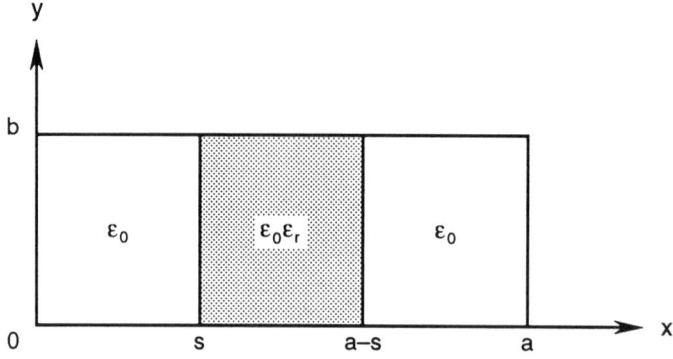

Figure 4.4 A symmetrically loaded rectangular waveguide.

Table 4.4 Comparison Between Approximate and Exact Eigenvalues for Example 4.9

Exact	Approximate			
	$N=1$	$N=2$	$N=3$	$N=4$
9.870	10.0	10.0	9.8697	9.8697
39.478		42.0	39.497	39.478
88.826			102.133	102.133
157.914				200.583

Solution

We will find the lowest mode having $\frac{\partial}{\partial y} \equiv 0$. It is this dominant mode that is of most practical value. Since the dielectric constant varies from one region to another, it is reasonable to choose Φ to be an electric field, i.e., $\Phi = E_y$. Also, since $k^2 = \frac{\omega^2}{u^2} = \omega^2 \mu \epsilon$, Eq. (4.49) becomes

$$\omega^2 \mu_0 \epsilon_0 \int_0^s E_y^2 \, dx + \omega^2 \mu_0 \epsilon_0 \epsilon_r \int_s^{a-s} E_y^2 \, dx + \omega^2 \mu_0 \epsilon_0 \int_{a-s}^a E_y^2 \, dx$$
$$= -\int_0^a E_y \frac{d^2 E_y}{dx^2} \, dx. \qquad (4.10.1)$$

This can be written as

$$\omega^2 \mu_0 \epsilon_0 \int_0^a E_y^2 \, dx + \omega^2 \mu_0 \epsilon_0 (\epsilon_r - 1) \int_s^{a-s} E_y^2 \, dx = -\int_0^a E_y \frac{d^2 E_y}{dx^2} \, dx. \qquad (4.10.2)$$

We now choose the trial function for E_y. It must be chosen to satisfy the boundary conditions, namely, $E_y = 0$ at $x = 0, a$. Since $E_y \sim \sin \frac{n\pi x}{a}$ for the empty waveguide, it makes sense to choose the trial function of the form

$$E_y = \sum_{n=1,3,5}^{\infty} c_n \sin \frac{n\pi x}{a}. \qquad (4.10.3)$$

We choose the odd values of n because the dielectric is symmetrically placed; otherwise we would have both odd and even terms.
Let us consider the trial function

$$E_y = \sin \frac{\pi x}{a}. \qquad (4.10.4)$$

Substituting Eq. (4.10.4) into Eq. (4.10.2) yields

$$\omega^2 \mu_0 \epsilon_0 \int_0^a \sin^2 \frac{\pi x}{a} \, dx + \omega^2 \mu_0 \epsilon_0 (\epsilon_r - 1) \int_s^{a-s} \sin^2 \frac{\pi x}{a} \, dx +$$
$$= \frac{\pi^2}{a^2} \int_0^a \sin^2 \frac{\pi x}{a} \, dx \qquad (4.10.5)$$

which leads to

$$\omega^2 \mu_0 \epsilon_0 \left[1 + (\epsilon_r - 1)\left[(1 - \frac{2s}{a}) + \frac{1}{\pi} \sin \frac{2\pi s}{a}\right] \right] = \frac{\pi^2}{a^2}.$$

But $k_0^2 = \omega^2 \mu_0 \epsilon_0 = \frac{4\pi^2}{\lambda_c^2}$, where λ_c is the cutoff wavelength of the empty waveguide. Hence

$$\frac{4\pi^2}{\lambda_c^2} = \frac{(\pi/a)^2}{1 + (\epsilon_r - 1)\left[(1 - \frac{2s}{a}) + \frac{1}{\pi} \sin \frac{2\pi s}{a}\right]}.$$

Taking $\epsilon_r = 4$ and $s = a/3$ gives

$$\frac{4\pi^2}{\lambda_c^2} = \frac{(\pi/a)^2}{2 + \frac{3\sqrt{3}}{2\pi}}$$

or

$$\frac{a}{\lambda_c} = 0.2974.$$

This is a considerable reduction in a/λ_c compared with the value of $a/\lambda_c = 0.5$ for the empty guide. The accuracy of the result may be improved by choosing more terms in Eq. (4.10.3).

4.8 Practical Applications

The various techniques discussed in this chapter have been applied to solve a considerable number of EM problems. We select a simple example for illustration [38,39]. This example illustrates the conventional use of the least squares method.

Consider a strip transmission line enclosed in a box containing a homogeneous medium as shown in Fig. 4.5. If a TEM mode of propagation is assumed, Laplace's equation

$$\nabla^2 V = 0 \qquad (4.58)$$

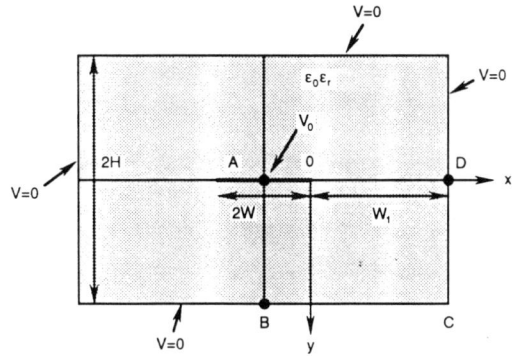

Figure 4.5 The strip line enclosed in a shielded box.

is obeyed. Due to symmetry, we will consider only one quarter section of the line as in Fig. 4.6 and adopt a boundary condition $\frac{\partial V}{\partial x} = 0$ at $x = -W$. We allow for the singularity at the edge of the strip. The variation of the potential in the vicinity of such a singularity is approximated, in terms of trigonometric basis functions, as

$$V = V_o + \sum_{k=1,3,5}^{\infty} c_k \, \rho^{k/2} \cos \frac{k\phi}{2}, \qquad (4.59)$$

where V_o is the potential on the trip conductor and the expansion coefficients c_k are to be determined. If we truncate the infinite series in Eq. (4.59) so that we have N unknown coefficients, we determine the coefficients by requiring that Eq. (4.59) be satisfied at $M(\geq N)$ points on the boundary. If $M = N$, we are applying the collocation method. If $M > N$, we obtain an overdetermined system of equations which can be solved by the method of least squares. Enforcing Eq. (4.59) at M boundary points, we obtain M simultaneous equations

$$\begin{bmatrix} V_1 \\ V_2 \\ \vdots \\ V_M \end{bmatrix} = \begin{bmatrix} A_{11} & A_{12} & \cdots & A_{1N} \\ A_{21} & A_{22} & \cdots & A_{2N} \\ \vdots & & & \vdots \\ A_{M1} & A_{M2} & \cdots & A_{MN} \end{bmatrix} \begin{bmatrix} c_1 \\ c_2 \\ \vdots \\ c_M \end{bmatrix}$$

i.e.,

$$[V] = [A][X] \qquad (4.60)$$

where $[X]$ is an $N \times 1$ matrix containing the unknown expansion coefficients, $[V]$ is an $M \times 1$ column matrix containing the boundary conditions, and $[A]$

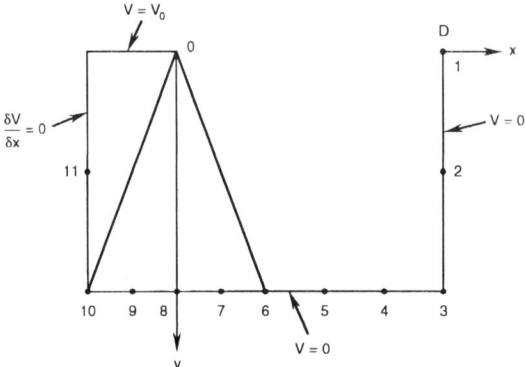

Figure 4.6 A quarter-section of the strip line.

is the $M \times N$ coefficient matrix. Due to redundancy, $[X]$ cannot be uniquely determined from Eq. (4.60) if $M > N$. To solve this redundant system of equations by the method of least squares, we define the residual matrix $[R]$ as

$$[R] = [A][X] - [V]. \tag{4.61}$$

We seek for $[X]$, which minimizes $[R]^2$. Consider

$$[I] = [R]^t[R] = \left[[A][X] - [V]\right]^t \left[[A][X] - [V]\right].$$

$$\frac{\partial [I]}{\partial [X]} = 0 \quad \rightarrow \quad [A]^t[A][X] - [A]^t[V] = 0$$

or

$$[X] = \left[[A]^t[A]\right]^{-1} [A]^t[V] \tag{4.62}$$

where the superscript t denotes the transposition of the relevant matrix. Thus we have reduced the original redundant system of equations to a determinate set of N simultaneous equations in N unknown coefficients c_1, $c_2, \cdots, c_N$.

Once $[X] = [c_1, c_2, \cdots, c_N]$ is determined from Eq. (4.62), the approximate solution in Eq. (4.59) is completely determined. We can now determine the capacitance and consequently the characteristic impedance of the line for a given value of width-to-height ratio. The capacitance is determined from

$$C = \frac{Q}{V_o} = Q \tag{4.63}$$

if $V_o = 1$ V. The characteristic impedance is found from [40]

$$Z_o = \frac{\sqrt{\epsilon_r}}{cC} \quad (4.64)$$

where $c = 3 \times 10^8$ m/s, the speed of light in vacuum. The major problem here is finding Q in Eq. (4.63). If we divide the boundary BCD into segments,

$$Q = \int \rho_L \, dl = 4 \sum_{BCD} \rho_L \Delta l$$

where the charge density $\rho_L = \mathbf{D} \cdot \mathbf{a}_n = \epsilon \mathbf{E} \cdot \mathbf{a}_n$ and $\mathbf{E} = -\nabla V$. But

$$\nabla V = \frac{\partial V}{\partial \rho} \mathbf{a}_\rho + \frac{1}{\rho} \frac{\partial V}{\partial \phi} \mathbf{a}_\phi,$$

$$\mathbf{E} = -\sum_{k=\text{odd}} \frac{k}{2} c_k \, \rho^{k/2-1} \left(\cos \frac{k\phi}{2} \mathbf{a}_\rho - \sin \frac{k\phi}{2} \mathbf{a}_\phi \right).$$

Since $\mathbf{a}_x = \cos\phi \, \mathbf{a}_\rho - \sin\phi \, \mathbf{a}_\phi$ and $\mathbf{a}_y = \sin\phi \, \mathbf{a}_\rho + \cos\phi \, \mathbf{a}_\phi$,

$$\rho_L|_{CD} = \epsilon \mathbf{E} \cdot \mathbf{a}_x$$

$$= -\epsilon \sum_{k=\text{odd}} \frac{k}{2} c_k \, \rho^{k/2-1} \left(\cos \frac{k\phi}{2} \cos\phi + \sin \frac{k\phi}{2} \sin\phi \right) \quad (4.65a)$$

and

$$\rho_L|_{BC} = \epsilon \mathbf{E} \cdot \mathbf{a}_y$$

$$= -\epsilon \sum_{k=\text{odd}} \frac{k}{2} c_k \, \rho^{k/2-1} \left(\cos \frac{k\phi}{2} \sin\phi - \sin \frac{k\phi}{2} \cos\phi \right). \quad (4.65b)$$

Example 4.11

Using the collocation (or point matching) method, write a computer program to calculate the characteristics impedance of the line shown in Fig. 4.5. Take:

(a) $W = H = 1.0$ m, $W_1 = 5.0$ m, $\epsilon_r = 1$, $V_0 = 1$ V,
(b) $W = H = 0.5$ m, $W_1 = 5.0$ m, $\epsilon_r = 1$, $V_0 = 1$ V.

Solution

The computer program is presented in Fig. 4.7. For the first run, we take the number of matching points $N = 11$; the points are selected as illustrated

294 Numerical Techniques in Electromagnetics

in Fig. 4.6. The selection of the points is based on our prior knowledge of the fact that the flux lines are concentrated on the side of the strip line numbered 6 to 10; hence more points are chosen on that side.

The first step is to determine the potential distribution within the strip line using Eq. (4.59). In order to determine the expansion coefficients c_k in Eq. (4.59), we let Eq. (4.59) be satisfied at the matching points. On points 1 to 10 in Fig. 4.6, $V = 0$ so that Eq. (4.59) can be written as

$$-V_o = \sum_{k=1,3,5}^{\infty} c_k \rho^{k/2} \cos \frac{k\phi}{2}. \quad (4.11.1)$$

The infinite series is terminated at $k = 19$ so that 10 points are selected on the sides of the strip line. The 11th point is selected such that $\frac{\partial V}{\partial x} = 0$ is satisfied at the point. Hence at point 11,

$$0 = \frac{\partial V}{\partial x} = \cos\phi \frac{\partial V}{\partial \rho} - \frac{\sin\phi}{\rho} \frac{\partial V}{\partial \phi}$$

or

$$0 = \sum_{k=1,3,5} \frac{k}{2} c_k \, \rho^{k/2-1} \left(\cos \frac{k\phi}{2} \cos\phi + \sin \frac{k\phi}{2} \sin\phi \right). \quad (4.11.2)$$

With Eqs. (4.11.1) and (4.11.2), we set up a matrix equation of the form

$$[B] = [F][A] \quad (4.11.3)$$

where

$$B_k = \begin{cases} -V_o, & k \neq N \\ 0, & k = N, \end{cases}$$

$F_{ki} =$

$$\begin{cases} \rho_i^{k/2} \cos k\phi_i/2, & i = 1, \cdots, N-1, \; k = 1, \cdots, N \\ \frac{k}{2}\rho_i^{k/2-1}\Big(\cos(k\phi_i/2)\cos\phi_i + \sin(k\phi_i/2)\sin\phi_i\Big), & i = N, \; k = 1, \cdots, N \end{cases}$$

where (ρ_i, ϕ_i) are the cylindrical coordinates of the ith point. Matrix $[A]$ consists of the unknown expansion coefficients c_k. By matrix inversion using subroutine INVERSE in Appendix D, we obtain $[A]$ as

$$[A] = [F]^{-1}[B]. \quad (4.11.4)$$

Table 4.5 Characteristic Impedance of the Strip Transmission Line of Fig. 4.5; for Example 4.11 with $W_1 = 5.0$

$W = H$	N	N_x	N_y	c_1	Calculated $Z_o(\Omega)$	Exact [39] $Z_o(\Omega)$
1.0	7	5	2	−1.1549	67.846	65.16
	11	8	3	−1.1266	65.16	
0.5	7	5	2	−1.1549	96.92	100.57
	11	8	3	−1.1266	99.60	

Once the expansion coefficients are determined, we now calculate the total charge on the sides of the strip line using Eq. (4.65) and

$$Q = 4 \sum_{BDC} \rho_L \Delta l.$$

Finally, we obtain Z_o from Eqs. (4.63) and (4.64). Table 4.5 shows the results obtained using the program in Fig. 4.7 for different cases. In Table 4.5, $N = N_x + N_y$, where N_x and N_y are the number of matching points selected along the x and y axes, respectively. By comparing Fig. 4.5 with Fig. 2.12, one may be tempted to apply Eq. (2.155) to obtain the exact solution of part (a) as 61.1 Ω. But we must recall that Eq. (2.155) was derived based on the assumption that $w \gg b$ in Fig. 2.12 or $W \gg H$ in Fig. 4.5. The assumption is not valid in this case, the exact solution given in [39] is more appropriate.

```
0001      C    ***********************************************
0002      C    THIS PROGRAM CALCULATES THE CHARACTERISTIC IMPEDANCE
0003      C    Zo  OF A BOXED MICROSTRIP LINE USING COLLOCATION
0004      C    POINT-MATCHING METHOD
0005      C    ***********************************************
0006
0007           DIMENSION X(30), Y(30), R(30), PHI(30), F(30,30)
0008           DIMENSION A(30), B(30)
0009           DATA (X(I),I=1,11)/3*5.0,4.0,2.0,1.0,0.5,
                    0.0,-0.5,2*-1.0/
0010           DATA ( Y(J),J=1,11 )/0.0,0.5,8*1.0,0.5/
0011           DATA EO,PIE,VL/8.8541E-12,3.141592,3.0E+8/
0012
0013      C    INPUT DATA
0014           ER = 1.0
0015           EPS = EO*ER
0016           W = 1.0
0017           H = 1.0
0018           W1 = 5.0
0019           VO = 1.0
0020           NMAX = 11
```

```
0021            NBC = 7    ! NO. OF POINTS ON BC OR X-AXIS
0022            NCD = NMAX - NBC  ! NO. OF POINTS ON DC OR Y-AXIS
0023            DX = (W + W1)/FLOAT(NBC)
0024            DY = H/FLOAT(NCD)
0025       C
0026       C    CALCULATE (R,PHI) FOR EACH POINT (X,Y)
0027       C
0028            DO 40 N = 1,NMAX
0029            R(N) = SQRT( X(N)**2 + Y(N)**2 )
0030            IF( X(N) ) 10,20,30
0031       10   PHI(N) = PIE - ATAN( Y(N)/ABS(X(N)) )
0032            GO TO 40
0033       20   PHI(N) = PIE/2.0
0034            GO TO 40
0035       30   PHI(N) = ATAN( Y(N)/X(N) )
0036       40   CONTINUE
0037       C
0038       C    CALCULATE MATRICES F(I,J) AND B(I)
0039       C
0040            DO 60 I = 1,NMAX
0041            B(I) = -VO
0042            M = 0
0043            DO 60 J = 1,NMAX
0044            M = 2*J - 1
0045            FM = FLOAT(M)/2.0
0046            IF(I.EQ.NMAX) GO TO 50
0047            F(I,J) = ( R(I)**(FM) )*COS(FM*PHI(I))
0048            GO TO 60
0049       50   CC = COS(PHI(I))*COS(PHI(I)*FM)
0050            SS = SIN(PHI(I))*SIN(PHI(I)*FM)
0051            F(I,J) = ( R(I)**(FM-1.) )*(CC + SS)*FM
0052            B(I) = 0.0
0053       60   CONTINUE
0054       C
0055       C    DETERMINE THE EXPANSION COEFFICIENTS A(I)
0056       C
0057            IDM = 30
0058            CALL INVERSE(F,NMAX,IDM)
0059            DO 80 I = 1, NMAX
0060            A(I) = 0.0
0061            DO 70 J=1,NMAX
0062            A(I) = A(I) + F(I,J)*B(J)
0063       70   CONTINUE
0064       80   CONTINUE
0065       C
0066       C    NOW, CALCULATE CHARGE ON THE X-SIDE, i.e. BC
0067       C
0068            RHO = 0.0
0069            YC = H
0070            XC = - W - DX/2.0
0071            DO 120 I=1,NBC
0072            XC = XC + DX
0073            RC = SQRT(XC**2 + YC**2 )
0074            IF(XC) 90,90,100
0075       90   PC = PIE - ATAN(YC/ABS(XC))
0076            GO TO 110
0077       100  PC = ATAN(YC/XC)
0078       110  CONTINUE
0079            DO 120 K=1,NMAX
0080            FK = FLOAT(2*K - 1)/2.0
0081            RRO = SIN(PC)*COS(FK*PC) - COS(PC)*SIN(FK*PC)
0082            RHO = RHO - A(K)*FK*(RC**(FK-1))*RRO*DX*EPS
0083       120  CONTINUE
```

```
0084      C
0085      C         NEXT, CALCULATE THE CHARGE ON THE Y-SIDE, i.e. DC
0086      C
0087                XC = W1
0088                YC = - DY/2.0
0089                DO 130 I=1,NCD
0090                YC = YC + DY
0091                RC = SQRT(XC**2 + YC**2)
0092                PC = ATAN(YC/XC)
0093                DO 130 K =1, NMAX
0094                FK = FLOAT(2*K - 1)/2.0
0095                RRO = COS(PC)*COS(FK*PC) + SIN(PC)*SIN(FK*PC)
0096                RHO = RHO - A(K)*FK*(RC**(FK-1))*RRO*DY*EPS
0097      130       CONTINUE
0098      C
0099      C         CALCULATE THE CHARACTERISTIC IMPEDANCE Zo
0100      C
0101                Q = 4.0*RHO
0102                C = ABS(Q)/VO
0103                ZO = SQRT(ER)/( C*VL )
0104                PRINT *,C,ZO
0105                WRITE(6,140) ZO
0106      140       FORMAT(2X,'Zo = ',3X,E12.6)
0107                STOP
0108                END
```

Figure 4.7 Computer program for Example 4.11.

4.9 Concluding Remarks

This chapter has provided an elementary introduction to the basic idea of variational techniques. The variational methods provide simple but powerful solutions to physical problems provided we can find approximate basis functions. A prominent feature of the variational method lies in the ability to achieve high accuracy with few terms in the approximate solution. A major drawback is the difficulty encountered in selecting the basis functions. In spite of the drawback, the variational methods have been very useful and provide basis for both the method of moments and the finite element method to be discussed in the forthcoming chapters.

Needless to say, our discussion on variational techniques in this chapter has only been introductory. An exhaustive treatment of the subject can be found in [1,6,10,11,41–43]. Wexler's paper [4] provides a review on computational techniques in EM including variational methods. Various applications of variational methods to EM-related problems include:

- waveguides and resonators [28–37]
- transmission lines [38,39,44–47]
- acoustic radiation [48]
- wave propagation [49–51]

- transient problems [52]
- scattering problems [53–59].

The problem of variational principles for EM waves in inhomogeneous media is discussed in [60].

References

[1] S. G. Mikhlin, *Variational Methods in Mathematical Physics*. New York: Macmillan, 1964, pp. xv, 4–78.

[2] J. N. Reddy, *An Introduction to the Finite Element Method*. New York: McGraw-Hill, 1984, pp. 13–63.

[3] R. B. Guenther and J. W. Lee, *Partial Differential Equations of Mathematical Physics and Integral Equations*. Englewood Cliffs, NJ: Prentice-Hall, 1988, pp. 434–485.

[4] A. Wexler, "Computation of electromagnetic fields," *IEEE Micro. Theo. Tech.*, vol. MTT-17, no. 8, Aug. 1969, pp. 416–439.

[5] M. M. Ney, "Method of moments as applied to electromagnetic problems," *IEEE Trans. Micro. Theo. Tech.*, vol. MTT-33, no. 10, Oct. 1985, pp. 972–980.

[6] I. M. Gelfand and S. V. Fomin, *Calculus of Variations* (translated from Russian by R. A. Silverman). Englewood Cliffs, NJ: Prentice-Hall, 1963.

[7] J. N. Reddy and M. L. Rasmussen, *Advanced Engineering Analysis*. New York: John Wiley, 1982, pp. 377–386.

[8] B. H. McDonald, et al., "Variational solution of integral equations," *IEEE Trans. Micro. Theo. Tech.*, vol. MTT-22, no. 3, Mar. 1974, pp. 237–248. See also vol. MTT-23, no. 2, Feb. 1975, pp. 265, 266 for correction to the paper.

[9] K. Morishita and N. Kumagai, "Unified approach to the derivation of variational expression for electromagnetic fields," *IEEE Trans. Micro. Theo. Tech.*, vol. MTT-25, no. 1, Jan. 1977, pp. 34–39.

[10] B. L. Moiseiwitch, *Variational Principles*. London: Interscience Pub., 1966.

[11] L. Cairo and T. Kahan, *Variational Techniques in Electromagnetics*. New York: Gordon & Breach, 1965, pp. 48–65.

[12] K. Kalikstein, "Formulation of variational principles via Lagrange multipliers," *J. Math. Phys.*, vol. 22, no. 7, July 1981, pp. 1433–1437.

[13] K. Kalikstein and A. Sepulveda, "Variational principles and variational functions in electromagnetic scattering," *IEEE Trans. Ant. Prog.*, vol. AP-29, no. 5, Sept. 1981, pp. 811–815.

[14] P. Hammond, "Equilibrium and duality in electromagnetic field problems," *J. Frank. Inst.*, vol. 306, no. 1, July 1978, pp. 133–157.

[15] S. K. Jeng and C. H. Chen, "On variational electromagnetics," *IEEE Trans. Ant. Prog.*, vol. AP-32, no. 9, Sept. 1984, pp. 902–907.

[16] S. J. Chung and C. H. Chen, "Partial variational principle for electromagnetic field problems: theory and applications," *IEEE Trans. Micro. Theo. Tech.*, vol. 36, no. 3, Mar. 1988, pp. 473–479.

[17] R. F. Harrington, *Field Computation by Moment Methods*. Malabar, FL: R. E. Krieger, 1968, pp. 19, 126–131.

[18] S. G. Mikhlin and K. I. Smolitskiy, *Approximate Methods for Solution of Differential and Integral Equations*. New York: Elsevier, 1967, pp. 147–270.

[19] T. J. Chung, *Finite Element Analysis in Fluid Dynamics*. New York: McGraw-Hill, 1978, pp. 36–43.

[20] O. C. Zienkiewicz and R. L. Taylor, *The Finite Element Method*. London: McGraw-Hill, 1989, vol. 1, 4th ed., pp. 206–259.

[21] L. Lewin, "On the restricted validity of point-matching techniques," *IEEE Trans. Micro. Theo. Tech.*, vol. MTT-18, no. 12, Dec. 1970, pp. 1041–1047.

[22] R. F. Muller, "On the legitimacy of an assumption underlying the point-matching method," *IEEE Trans. Micro. Theo. Tech.*, vol. MTT-18, June 1970, pp. 325–327.

[23] A. R. Djordjevic and T. K. Sakar, "A theorem on the moment methods," *IEEE Trans. Ant. Prop.*, vol. AP-35, no. 3, Mar. 1987, pp. 353–355.

[24] T. K. Sarkar, "A study of the various methods for computing electromagnetic field utilizing thin wire integral equations," *Radio Sci.*, vol. 18, no. 1, Jan./Feb., pp. 29–38.

[25] T. K. Sarkar, "A note on the variational method (Rayleigh-Ritz), Galerkin's method, and the method of least squares," *Radio Sci.*, vol. 18, no. 6, Nov./Dec. 1983, pp. 1207–1224.

[26] T. K. Sarkar, "A note on the choice of weighting functions in the method of moments," *IEEE Trans. Ant. Prog.*, vol. AP-33, no. 4, April 1985, pp. 436–441.

[27] T. K. Sarkar, et al., "On the choice of expansion and weighting functions in the numerical solution of operator equations," *IEEE Trans*

Ant. Prog., vol. AP- 33, no. 9, Sept. 1985, pp. 988–996.

[28] A. D. Berk, "Variational principles for electromagnetic resonators and waveguides," *IRE Trans. Ant. Prop.*, vol. AP-4, April 1956, pp. 104–111.

[29] G. J. Gabriel and M. E. Brodwin, "The solution of guided waves in inhomogeneous anisotropic media by perturbation and variation methods," *IEEE Trans. Micro. Theo. Tech.*, vol. MTT-13, May 1965, pp. 364–370.

[30] W. English and F. Young, "An E vector variational formulation of the Maxwell equations for cylindrical waveguide problems," *IEEE Trans. Micro. Theo. Tech.*, vol. MTT-19, Jan. 1971, pp. 40–46.

[31] H. Y. Yee and N. F. Audeh, "Uniform waveguides with arbitrary cross-section considered by the point-matching method," *IEEE Trans. Micro. Theo. Tech.*, vol. MTT-13, Nov. 1965, pp. 847–851.

[32] J. A. Fuller and N. F. Audeh, "The point-matching solution of uniform nonsymmetric waveguides," *IEEE Trans. Micro. Theo. Tech.*, vol. MTT-17, no. 2, Feb. 1969.

[33] R. B. Wu and C. H. Chen, "On the variational reaction theory for dielectric waveguides," *IEEE Trans. Micro. Theo. Tech.*, no.6, June 1985, pp. 477–483.

[34] T. E. Rozzi, "The variational treatment of thick interacting inductive irises," *IEEE Trans. Micro. Theo. Tech*, vol. MTT-21, no. 2, Feb. 1973, pp. 82–88.

[35] A. D. McAulay, "Variational finite-element solution of dissipative waveguide and transportation application," *IEEE Trans. Micro. Theo. Tech.*, vol. MTT-25, no. 5, May 1977, pp. 382–392.

[36] L. V. Lindell, "A variational method for nonstandard eigenvalue problems in waveguides and resonator analysis," *IEEE Trans. Micro. Theo. Tech.*, vol. MTT-30, no. 8, Aug. 1982, pp. 1194–1204. See comment on this paper in vol. MTT-31, no. 9, Sept. 1983, pp. 786–789.

[37] K. Chang, "Variational solutions on two opposite narrow resonant strips in waveguide," *IEEE Trans. Micro. Theo. Tech.*, vol. MTT-35, no. 2, Feb. 1987, pp. 151–158.

[38] T. K. Seshadri, et al., "Application of 'corner function approach' to strip line problems," *Int. Jour. Electron.*, vol. 44, no. 5, May 1978, pp. 525–528.

[39] T. K. Seshadri, et al., "Least squares collocation as applied to the analysis of strip transmission lines," *Proc. IEEE*, vol. 67. no. 2,

Feb. 1979, pp. 314–315.

[40] M. N. 0. Sadiku, *Elements of Electromagnetics*. New York: Holt, Rinehart and Winston, Chap. 11, 1989.

[41] P. M. Morse and H. Feshback, *Methods of Theoretical Physics*. New York: McGraw-Hill, 2 volumes, 1953.

[42] R. E. Collin, *Field Theory of Guided Waves*. New York: McGraw-Hill, 1960, pp. 148–164, 314–367.

[43] D. G. Bodner and D. T. Paris, "New variational principle in electromagnetics," *IEEE Trans. Ant. Prog.*, vol. AP-18, no. 2, March 1970, pp. 216–223.

[44] T. D. Tsiboukis, "Estimation of the characteristic impedance of a transmission line by variational methods," *IEE Proc.*, vol. 132, Pt. H, no. 3, June 1985, pp. 171–175.

[45] E. Yamashita and R. Mittra, "Variational method for the analysis of microstrip lines," *IEEE Trans. Micro. Theo. Tech.*, vol. MTT-16, no. 4, Apr. 1968, pp. 251–256.

[46] E. Yamashita, "Variational method for the analysis of microstrip-like transmission lines," *IEEE Trans. Micro. Theo. Tech.*, vol. MTT-16, no. 8, Aug. 1968, pp. 529–535.

[47] F. Medina and M. Horno, "Capacitance and inductance matrices for multistrip structures in multilayered anisotropic dielectrics," *IEEE Trans. Micro. Theo. Tech.*, vol. MTT-35, no. 11, Nov. 1987, pp. 1002–1008.

[48] F. H. Fenlon, "Calculation of the acoustic radiation of field at the surface of a finite cylinder by the method of weighted residuals," *Proc. IEEE*, vol. 57, no. 3, March 1969, pp. 291–306.

[49] C. H. Chen and Y. W. Kiang, "A variational theory for wave propagation in a one-dimensional inhomogeneous medium," *IEEE Trans. Ant. Prog.*, vol. AP-28, no. 6, Nov. 1980, pp. 762–769.

[50] S. K. Jeng and C. H. Chen, "Variational finite element solution of electromagnetic wave propagation in a one-dimensional inhomogeneous anisotropic medium," *J. Appl. Phys.*, vol. 55, no. 3, Feb. 1984, pp. 630–636.

[51] J. A. Bennett, "On the application of variation techniques to the ray theory of radio propagation," *Radio Sci.*, vol. 4, no. 8, Aug. 1969, pp. 667–678.

[52] J. T. Kuo and D. H. Cho, "Transient time-domain electromagnetics," *Geophys.*, vol. 45, no. 2, Feb. 1980, pp. 271–291.

[53] R. D. Kodis, "An introduction to variational methods in electromagnetic scattering," *J. Soc. Industr. Appl. Math.*, vol. 2, no. 2, June 1954, pp. 89–112.

[54] D. S. Jones, "A critique of the variational method in scattering problems," *IRE Trans.*, vol. AP-4, no. 3, 1965, pp. 297–301.

[55] R. J. Wagner, "Variational principles for electromagnetic potential scattering," *Phys. Rev.*, vol. 131, no. 1, July 1963, pp. 423–434.

[56] J. A. Krill and R. A. Farrell, "Comparison between variational, perturbational, and exact solutions for scattering from a random rough-surface model," *J. Opt. Soc. Am.*, vol. 68, June 1978, pp. 768–774.

[57] R. B. Wu and C. H. Chen, "Variational reaction formulation of scattering problem for anisotropic dielectric cylinders," *IEEE Trans. Ant. Prog.*, vol. 34, no. 5, May 1986, pp. 640–645.

[58] J. A. Krill and R. H. Andreo, "Vector stochastic variational principles for electromagnetic wave scattering," *IEEE Trans. Ant. Prog.*, vol. AP-28, no. 6, Nov. 1980, pp. 770–776.

[59] R. W. Hart and R. A. Farrell, "A variational principle for scattering from rough surfaces," *IEEE Trans. Ant. Prog.*, vol. AP-25, no. 5, Sept. 1977, pp. 708–713.

[60] J. R. Willis, "Variational principles and operator equations for electromagnetic waves in inhomogeneous media," *Wave Motion*, vol. 6, no. 2, 1984, pp. 127–139.

Problems

4.1 Find $\langle u, v \rangle$ if:

(a) $u = x^2$, $v = 2 - x$ in the interval $-1 < x < 1$,

(b) $u = 1$, $v = x^2 - 2y^2$ in the rectangular region $0 < x < 1$, $1 < y < 2$,

(c) $u = x + y$, $v = xz$ in the cylindrical region $x^2 + y^2 \leq 4$, $0 < z < 5$.

4.2 Show that:

(a) $\langle h(x), f(x) \rangle = \langle h(x), f(-x) \rangle$,

(b) $\langle h(ax), f(x) \rangle = \langle h(x), \frac{1}{a}f(\frac{x}{a}) \rangle$,

(c) $\langle \frac{df}{dx}, h(x) \rangle = -\langle f(x), \frac{dh}{dx} \rangle,$

(d) $\langle \frac{d^n f}{dx^n}, h(x) \rangle = (-1)^n \langle f(x), \frac{d^n h}{dx^n} \rangle.$

Note from (d) that $L = \frac{d}{dx}, \frac{d^3}{dx^3}$, etc., are not self-adjoint, whereas $L = \frac{d^2}{dx^2}, \frac{d^4}{dx^4}$, etc., are.

4.3 Find the Euler partial differential equation for each of the following functionals:

(a) $\quad \int_a^b \sqrt{1+y'}\, dx,$

(b) $\quad \int_a^b y\sqrt{1+y'^2}\, dx,$

(c) $\quad \int_a^b \cos(xy')\, dx,$

(d) $\quad \int_a^b (y'^2 - y^2)\, dx,$

(e) $\quad \int_a^b [5y^2 - (y'')^2 + 10x]\, dx,$

(f) $\quad \int_a^b (3uv - u^2 + u'^2 - v'^2)\, dx.$

4.4 Determine the extremal $y(x)$ for each of the following variational problems:

(a) $$\int_0^1 (2y'^2 + yy' + y' + y)\, dx, \quad y(0) = 0,\ y(1) = 1,$$

(b) $$\int (y'^2 - y^2)\, dx, \quad y(0) = 1,\ y(\pi/2).$$

4.5 If L is a positive definite, self-adjoint operator and $L\Phi = g$ has a solution Φ_o, show that the function

$$I = \langle L\Phi, \Phi \rangle - 2\langle \Phi, g \rangle,$$

where Φ and g are real functions, is minimized by the solution Φ_o.

4.6 Show that a function that minimizes the functional

$$I(\Phi) = \frac{1}{2} \int_S \left[|\nabla \Phi|^2 - k^2 \Phi^2 + 2g\Phi \right] dS$$

is the solution to the inhomogeneous Helmholtz equation

$$\nabla^2 \Phi + k^2 \Phi = g.$$

4.7 A real field $\Phi(x, y, z, t)$ obeys the variational

$$\delta \int F\, dxdydzdt,$$

where $F(x, y, z, t) = \frac{1}{2}\left[\left(\frac{\partial \Phi}{\partial t}\right)^2 - |\nabla \Phi|^2 - \lambda \Phi^2\right]$; find the differential equation of motion obeyed by Φ.

4.8 Show that the variational principle for the boundary value problem

$$\nabla^2 \Phi = f(x, y, z)$$

subject to the mixed boundary condition

is

$$\frac{\partial \Phi}{\partial n} + g\Phi = h \quad \text{on } S$$

$$I(\Phi) = \int_v \left[|\nabla \Phi|^2 - 2fg\right] dv + \oint \left[g\Phi^2 - 2h\Phi\right] dS.$$

4.9 Obtain the variational principle for the differential equation

$$-\frac{d^2y}{dx^2} + y = \sin \pi x, \qquad 0 < x < 1$$

subject to $y(0) = 0 = y(1)$.

4.10 Determine the variational principle for

$$\Phi'' = \Phi - 4xe^x, \qquad 0 < x < 1$$

subject to $\Phi'(0) = \Phi(0) + 1$, $\Phi'(1) = \Phi(1) - e$.

4.11 For the boundary value problem

$$-\Phi'' = x, \quad 0 < x < 1,$$
$$\Phi(0) = 0, \quad \Phi(1) = 2,$$

determine the approximate solution using the Rayleigh-Ritz method with basis functions

$$u_k = x^k(x-1), \quad k = 0, 1, 2, ..., M.$$

Try cases when $M = 1$, 2, and 3.

4.12 Rework Example 4.5 using

(a) $u_m = x(1 - x^m)$, (b) $u_m = \sin m\pi x$, $m = 1, 2, 3, \cdots, M$. Try cases when $M = 1$, 2, and 3.

4.13 Solve the differential equation

$$-\Phi''(x) + 0.1\Phi(x) = 1, \qquad 0 \le x \le 10$$

subject to the boundary conditions $\Phi'(0) = 0 = \Phi(0)$ using the trial function

$$\widetilde{\Phi}(x) = a_1 \cos \frac{\pi x}{20} + a_2 \cos \frac{3\pi x}{20} + a_3 \cos \frac{5\pi x}{20}.$$

Determine the expansion coefficients using: (a) collocation method, (b) subdomain method, (c) Galerkin method, (d) least squares method.

4.14 For the boundary value problem

$$\Phi'' + \Phi + x = 0, \qquad 0 < x < 1$$

with homogeneous boundary conditions $\Phi = 0$ at $x = 0$ and $x = 1$, determine the coefficients of the approximate solution function

$$\widetilde{\Phi}(x) = x(1-x)(a_1 + a_2 x)$$

using: (a) collocation method (choose $x = 1/4$, $x = 1/2$ as collocation points), (b) Galerkin method, (c) least squares method.

4.15 Given the boundary value problem

$$y'' + (1 + x^2)y + 1 = 0, \qquad -1 < x < 1,$$

$$y(-1) = 0 = y(1),$$

find the expansion coefficients of the approximate solution

$$\widetilde{y} = a_1(1 - x^2) + a_2 x^2 (1 - x^2)$$

by using: (i) the various weighted residual methods, (ii) the Rayleigh-Ritz method.

4.16 Rework the previous problem using the approximate solution

$$\widetilde{y} = a_1(1 - x^2)(1 - 4x^2) + a_2 x^2 (1 - x^2).$$

Use the Galerkin and the least squares methods to determine a_1 and a_2.

4.17 Consider the problem

$$\Phi'' + x\Phi' + \Phi = 2x, \qquad 0 < x < 1$$

subject to $\Phi(0) = 1, \Phi(1) = 0$. Find the approximate solution using the Galerkin method. Use $u_k = x^k(1-x)$, $k = 0, 1, \cdots, N$. Try $N = 3$.

4.18 Determine the first three eigenvalues of the equation

$$y'' + \lambda y = 0, \qquad 0 < x < 1,$$

$y(0) = 0 = y(1)$ using collocation at $x = 1/4, 1/2, 3/4$.

4.19 Determine the fundamental eigenvalue of the problem

$$-\Phi''(x) + 0.1\Phi(x) = \lambda \Phi(x), \qquad 0 < x < 10$$

subject to $\Phi(0) = 0 = \Phi(10)$. Use the trial function

$$\tilde{\Phi}(x) = x(x - 10).$$

4.20 Obtain the lowest eigenvalue of the problem

$$\nabla^2 \Phi + \lambda \Phi = 0, \qquad 0 < \rho < 1$$

with $\Phi = 0$ at $\rho = 1$.

4.21 Rework Example 4.10 using the trial function

$$E_y = \sin\frac{\pi x}{a} + c_1 \sin\frac{3\pi x}{a}$$

where c_1 is a coefficient to be chosen such that $\omega^2 \epsilon_o \mu_o$ is minimized.

4.22 Consider the waveguide in Fig. 4.4 as homogeneous. To determine the cutoff frequency, we may use the polynomial trial function

$$H_z = Ax^3 + Bx^2 + Cx + D.$$

By applying the conditions

$H_z = 1$ at $x = a$, $\qquad H_z = -1$ at $x = a$,

$\dfrac{\partial H_z}{\partial x} = 0$ at $x = 0, a$,

determine A, B, C, and D. Using the trial function, calculate the cutoff frequency.

Chapter 5

Moment Methods

"Planning is doing today to make us better tomorrow because the future belongs to those who make the hard decisions today." Business Week

5.1 Introduction

In Section 1.3.2, it was mentioned that most EM problems can be stated in terms of an inhomogeneous equation

$$L\Phi = g \tag{5.1}$$

where L is an operator which may be differential, integral or integro-differential, g is the known excitation or source function, and Φ is the unknown function to be determined. So far, we have limited our discussion to cases for which L is differential. In this chapter, we will treat L as an integral or integro-differential operator.

The *method of moments* (MOM) is a general procedure for solving Eq. (5.1). It is essentially the method of weighted residuals discussed in Section 4.6. Therefore, the method is applicable for solving both differential and integral equations.

What we call the method of moments is called by different names by different users. It is sometimes called the *method of projections* or the *Petrov-Galerkin method* [1,2]. It is similar in form to Rumsey's reaction concept [3]. The name "method of moments" has its origin in Russian literature [4,5]. In western literature, the first use of the name is usually attributed to Harrington [6]. Harrington's choice of the name is based on the following statement:

"Certainly others have used it [the method] in the past, and I didn't want to introduce a new jargon. After a search of the literature, I decided that the exposition most closely analogous to what I was using was that given by Kantorovich and Krylov [4]. They called it the 'method of moments,' and hence that is the name I chose." [7]

The method owes its name to the process of taking moments by multiplying by appropriate weighting functions and integrating, as discussed in Section 4.6.

The use of MOM in EM has become popular since the work of Richmond [8] in 1965 and Harrington [9] in 1967. The method has been successfully applied to a wide variety of EM problems of practical interest such as radiation due to thin-wire elements and arrays, scattering problems, analysis of microstrips and lossy structures, propagation over an inhomogeneous earth, and antenna beam pattern, to mention but a few. An updated review of the method is found in a paper by Ney [10]. The literature on MOM is already so large as to prohibit a comprehensive bibliography. A partial bibliography is provided by Adams [11].

The procedure for applying MOM to solve Eq. (5.1) usually involves four steps:

(1) derivation of the appropriate integral equation (IE),
(2) conversion (discretization) of the IE into a matrix equation using basis (or expansions) functions and weighting (or testing) functions,
(3) evaluation of the matrix elements, and
(4) solving the matrix equation and obtaining the parameters of interest.

The basic tools for step (2) have already been mastered in Section 4.6; in this chapter we will apply them to IEs rather than PDEs. Just as we studied PDEs themselves in Section 1.3.2, we will first study IEs.

5.2 Integral Equations

An integral equation is any equation involving unknown function Φ under the integral sign. Simple examples of integral equations are Fourier, Laplace, and Hankel transforms.

5.2.1 Classification of Integral Equations

Linear integral equations that are most frequently studied fall into two categories named after Fredholm and Volterra. One class is the Fredholm equations of the first, second, and third kind, namely,

$$f(x) = \int_a^b K(x,t)\Phi(t)dt, \tag{5.2}$$

$$f(x) = \Phi(x) - \lambda \int_a^b K(x,t)\Phi(t)dt, \tag{5.3}$$

$$f(x) = a(x)\Phi(x) - \lambda \int_a^b K(x,t)\Phi(t)dt, \tag{5.4}$$

where λ is a scalar (or possibly complex) parameter. Functions $K(x,t)$ and $f(x)$ and the limits a and b are known, while $\Phi(x)$ is unknown. The function $K(x,t)$ is known as the *kernel* of the integral equation. The parameter λ is sometimes equal to unity.

The second class of integral equations are the Volterra equations of the first, second, and third kind, namely,

$$f(x) = \int_a^x K(x,t)\Phi(t)dt, \tag{5.5}$$

$$f(x) = \Phi(x) - \lambda \int_a^x K(x,t)\Phi(t)dt, \tag{5.6}$$

$$f(x) = a(x)\Phi(x) - \lambda \int_a^x K(x,t)\Phi(t)dt, \tag{5.7}$$

with a variable upper limit of integration. If $f(x) = 0$, the integral equations (5.2) to (5.7) become homogeneous. Note that Eqs. (5.2) to (5.7) are all linear equations in that Φ enters the equations in a linear manner. An integral equation is nonlinear if Φ appears in the power of $n > 1$ under the integral sign. For example, the integral equation

$$f(x) = \Phi(x) - \int_a^b K(x,t)\Phi^2(t)dt \tag{5.8}$$

is nonlinear. Also, if limit a or b or the kernel $K(x,t)$ becomes infinite, an integral equation is said to be singular. Finally, a kernel $K(x,t)$ is said to be symmetric if $K(x,t) = K(t,x)$.

5.2.2 Connection Between Differential and Integral Equations

The above classification of one-dimensional integral equations arises naturally from the theory of differential equations, thereby showing a close connection between the integral and differential formulation of a given problem.

Most ordinary differential equations can be expressed as integral equations, but the reverse is not true. While boundary conditions are imposed externally in differential equations, they are incorporated within an integral equation.

For example, consider the first order ordinary differential equation

$$\frac{d\Phi}{dx} = F(x, \Phi), \quad a \le x \le b \tag{5.9}$$

subject to $\Phi(a) =$ constant. This can be written as the Volterra integral of the second kind. Integrating Eq. (5.9) gives

$$\Phi(x) = \int_a^x F(x, \Phi(t))dt + c_1$$

where $c_1 = \Phi(a)$. Hence Eq. (5.9) is the same as

$$\Phi(x) = \Phi(a) + \int_a^x F(t, \Phi)dt. \tag{5.10}$$

Any solution of Eq. (5.10) satisfies both Eq. (5.9) and the boundary condition. Thus an integral equation formulation includes both the differential equation and the boundary conditions.

Similarly, consider the second order ordinary differential equation

$$\frac{d^2\Phi}{dx^2} = F(x, \Phi), \quad a \le x \le b. \tag{5.11}$$

Integrating once yields

$$\frac{d\Phi}{dx} = \int_a^x F(x, \Phi(t))dt + c_1 \tag{5.12}$$

where $c_1 = \Phi'(a)$. Integrating Eq. (5.12),

$$\Phi(x) = c_2 + c_1 x + \int_a^x (x-t) F(x, \Phi(t))dt$$

where $c_2 = \Phi(a) - \Phi'(a)a$. Hence

$$\Phi(x) = \Phi(a) + (x-a)\Phi'(a) + \int_a^x (x-t) F(x, \Phi)dt. \tag{5.13}$$

Again, we notice that the integral equation (5.13) represents both the differential equation (5.11) and the boundary conditions. We have only considered one-dimensional integral equations. Integral equations involving unknown functions in two or more space dimensions will be discussed later.

Example 5.1

Solve the Volterra integral equation

$$\Phi(x) = 1 + \int_0^x \Phi(t)dt.$$

Solution

This can be solved directly or indirectly by finding the solution of the corresponding differential equation. To solve it directly, we differentiate both sides of the given integral equation. In general, given an integral

$$g(x) = \int_{\alpha(x)}^{\beta(x)} f(x,t)dt \qquad (5.1.1)$$

with variable limits, we differentiate this using the Leibnitz rule, namely,

$$g'(x) = \int_{\alpha(x)}^{\beta(x)} \frac{\partial f(x,t)}{\partial x} dt + f(x,\beta)\beta' - f(x,\alpha)\alpha'. \qquad (5.1.2)$$

Differentiating the given integral equation, we obtain

$$\frac{d\Phi}{dx} = \Phi(x) \qquad (5.1.3a)$$

or

$$\frac{d\Phi}{\Phi} = dx. \qquad (5.1.3b)$$

Integrating gives

$$\ln \Phi = x + \ln c_o$$

or

$$\Phi = c_o e^x$$

where $\ln c_o$ is the integration constant. From the given integral equation

$$\Phi(0) = 1 = c_o.$$

Hence

$$\Phi(x) = e^x \qquad (5.1.4)$$

is the required solution. This can be checked by substituting it into the given integral equation.

314   Numerical Techniques in Electromagnetics

An indirect way of solving the integral equation is by comparing it with Eq. (5.10) and noting that

$$a = 0, \Phi(a) = \Phi(0) = 1$$

and that $F(x,\Phi) = \Phi(x)$. Hence the corresponding first order differential equation is

$$\frac{d\Phi}{dx} = \Phi, \quad \Phi(0) = 1$$

which is the same as Eq. (5.1.3), and the solution in Eq. (5.1.4) follows.

Example 5.2

Find the integral equation corresponding to the differential equation

$$\Phi''' - 3\Phi'' - 6\Phi' + 8\Phi = 0$$

subject to $\Phi''(0) = \Phi'(0) = \Phi(0) = 1$.

Solution

Let $\Phi''' = F(\Phi, \Phi, \phi, x) = 3\Phi'' + 6\Phi' - 8\Phi$. Integrating both sides,

$$\Phi'' = 3\Phi' + 6\Phi - 8\int_0^x \Phi(t)dt + c_1 \qquad (5.2.1)$$

where c_1 is determined from the initial values, i.e.,

$$1 = 3 + 6 + c_1 \quad \rightarrow \quad c_1 = -8.$$

Integrating both sides of Eq. (5.2.1) gives

$$\Phi' = 3\Phi + 6\int_0^x \Phi(t)dt - 8\int_0^x (x-t)\Phi(t)dt - 8x + c_2 \qquad (5.2.2)$$

where

$$1 = 3 + c_2 \quad \rightarrow \quad c_2 = -2.$$

Finally, we integrate both sides of Eq. (5.2.2) to get

$$\Phi = 3\int_0^x \Phi(t)dt + 6\int_0^x (x-t)\Phi(t)dt - 4\int_0^x (x-t)^2\Phi(t)dt - 4x^2 - 2x + c_3$$

where $1 = c_3$. Thus the integral equation equivalent to the given differential equation is

$$\Phi(x) = 1 - 2x - 4x^2 + \int_0^x [3 + 6(x-t) - 4(x-t)^2]\Phi(t)dt.$$

5.3 Green's Functions

A more systematic means of obtaining an IE from a PDE is by constructing an auxiliary function known as the *Green's function*[1] of that problem. The Green's function, also known as the *source function* or *influence function*, is the kernel function obtained from a linear boundary value problem and forms the essential link between the differential and integral formulations. Green's function also provides a method of dealing with the source term (g in $L\Phi = g$) in a PDE. In other words, it provides an alternative approach to the series expansion method of Section 2.7 for solving inhomogeneous boundary-value problems by reducing the inhomogeneous problem to a homogeneous one.

To obtain the field caused by a distributed source by the Green's function technique, we find the effects of each elementary portion of source and sum them up. If $G(\mathbf{r}, \mathbf{r}')$ is the field at the observation point (or field point) $\mathbf{r}$ caused by a unit point source at the source point $\mathbf{r}'$, then the field at $\mathbf{r}$ by a source distribution $g(\mathbf{r}')$ is the integral of $g(\mathbf{r}')G(\mathbf{r},\mathbf{r}')$ over the range of $\mathbf{r}'$ occupied by the source. The function G is the Green's function [12]. Thus, physically, the Green's function $G(\mathbf{r},\mathbf{r}')$ represents the potential at $\mathbf{r}$ due to a unit point charge at $\mathbf{r}'$. For example, the solution to the Dirichlet problem

$$\nabla^2 \Phi = g \quad \text{in} \quad R$$
$$\Phi = f \quad \text{on} \quad B \qquad (5.14)$$

is given by

$$\Phi = \int_R g(\mathbf{r}')G(\mathbf{r},\mathbf{r}')dv' + \oint_B f\frac{\partial G}{\partial n}dS \qquad (5.15)$$

where n denotes the outward normal to the boundary B of the solution region R. It is obvious from Eq. (5.15) that the solution Φ can be determined provided the Green's function G is known. So the real problem is not that of finding the solution but that of constructing the Green's function for the problem.

Consider the linear second order PDE

$$L\Phi = g. \qquad (5.16)$$

We define the Green's function corresponding to the differential operator L as a solution of the point source inhomogeneous equation

$$L\, G(\mathbf{r},\mathbf{r}') = \delta(\mathbf{r},\mathbf{r}') \qquad (5.17)$$

---

[1] Named after George Green (1793–1841), an English mathematician.

where **r** and **r**' are the position vectors of the field point (x, y, z) and source point (x', y', z'), respectively, and $\delta(\mathbf{r}, \mathbf{r}')$ is the Dirac delta function, which vanishes for $\mathbf{r} \neq \mathbf{r}'$ and satisfies

$$\int \delta(\mathbf{r}, \mathbf{r}')g(\mathbf{r}')dv' = g(\mathbf{r}). \tag{5.18}$$

From Eq. (5.17), we notice that the Green function $G(\mathbf{r}, \mathbf{r}')$ can be interpreted as the solution to the given boundary value problem with the source term g replaced by the unit impulse function. Thus $G(\mathbf{r}, \mathbf{r}')$ physically represents the response of the linear system to a unit impulse applied at the point $\mathbf{r} = \mathbf{r}'$.

The Green's function has the following properties [13]:

(a) G satisfies the equation $L G = 0$ except at the source point $\mathbf{r}'$, i.e.,

$$LG(\mathbf{r}, \mathbf{r}') = \delta(\mathbf{r}, \mathbf{r}'). \tag{5.17}$$

(b) G is symmetric in the sense that

$$G(\mathbf{r}, \mathbf{r}') = G(\mathbf{r}', \mathbf{r}). \tag{5.19}$$

(c) G satisfies that prescribed boundary value f on B, i.e.,

$$G = f \text{ on } B. \tag{5.20}$$

d) G is continuous in $\mathbf{r}$, $\mathbf{r}'$, but the directional derivative $\partial G/\partial n$ has a discontinuity at $\mathbf{r}'$ which is specified by the equation

$$\lim_{\epsilon \to 0} \oint_S \frac{\partial G}{\partial n} dS = 1 \tag{5.21}$$

where n is the outward normal to the sphere of radius ϵ as shown in Fig 5.1, i.e.,

$$|\mathbf{r} - \mathbf{r}'| = \epsilon^2.$$

Property (b) expresses the *principle of reciprocity*; it implies that an exchange of source and observer does not affect G. The property is proved by Myint-U [13] by applying Green's second identity in conjunction with Eq. (5.17) while property (d) is proved by applying divergence theorem along with Eq. (5.17).

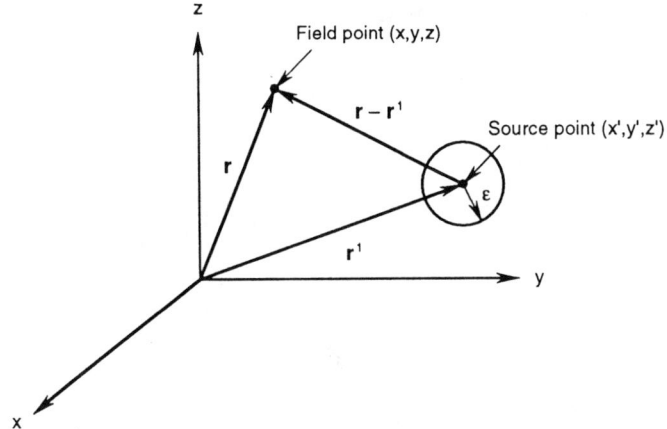

Figure 5.1 Illustration of the field point (x, y, z) amd source point (x', y', z').

5.3.1 For Free Space

We now illustrate how to construct the free space Green's function G corresponding to a PDE. It is usually convenient to let G be the sum of a particular integral of the inhomogeneous equation $LG = g$ and the solution of the associated homogeneous equation $L\,G = 0$. In other words, we let

$$G(\mathbf{r},\mathbf{r}') = F(\mathbf{r},\mathbf{r}') + U(\mathbf{r},\mathbf{r}') \tag{5.22}$$

where F, known as the free-space Green's function or fundamental solution, satisfies

$$L\,F = \delta(\mathbf{r},\mathbf{r}') \text{ in } R \tag{5.23}$$

and U satisfies

$$L\,U = 0 \text{ in } R \tag{5.24}$$

so that by superposition $G = F + U$ satisfies Eq. (5.17). Also $G = F$ on the boundary B requires that

$$U = -F + f \quad \text{on} \quad B. \tag{5.25}$$

Notice that F need not satisfy the boundary condition.

We apply this to two specific examples. First, consider the two-dimensional problem for which

$$L = \frac{\partial^2}{\partial x^2} + \frac{\partial^2}{\partial y^2} = \nabla^2. \tag{5.26}$$

The corresponding Green's function $G(x, y; x', y')$ satisfies

$$\nabla^2 G(x, y; x', y') = \delta(x - x')\delta(y - y'). \tag{5.27}$$

Hence, F must satisfy

$$\nabla^2 F = \delta(x - x')\delta(y - y').$$

For $\rho = [(x - x')^2 + (y - y')^2]^{1/2} > 0$, i.e., for $x \neq x', y \neq y'$,

$$\nabla^2 F = \frac{1}{\rho}\frac{\partial}{\partial \rho}\left(\rho\frac{\partial F}{\partial \rho}\right) = 0 \tag{5.28}$$

which is integrated twice to give

$$F = A \ln \rho + B. \tag{5.29}$$

Applying the property in Eq. (5.21)

$$\lim_{\epsilon \to 0} \oint \frac{dF}{d\rho} dl = \lim_{\epsilon \to 0} \int_0^{2\pi} \frac{A}{\rho} \rho d\phi = 2\pi A = 1$$

or $A = \dfrac{1}{2\pi}$.

Since B is arbitrary, we may choose $B = 0$. Thus

$$F = \frac{1}{2\pi} \ln \rho$$

and

$$G = F + U = \frac{1}{2\pi} \ln \rho + U. \tag{5.30}$$

We choose U so that G satisfies prescribed boundary conditions.

For the three-dimensional problem,

$$L = \nabla^2 = \frac{\partial^2}{\partial x^2} + \frac{\partial^2}{\partial y^2} + \frac{\partial^2}{\partial z^2} \tag{5.31}$$

and the corresponding Green's function $G(x, y, z; x', y', z')$ satisfies

$$LG(x, y, z; x', y', z') = \delta(x - x')\delta(y - y')\delta(z - z'). \tag{5.32}$$

Hence, F must satisfy

$$\nabla^2 F = \delta(x-x')\delta(y-y')\delta(z-z')$$
$$= \delta(\mathbf{r}-\mathbf{r}').$$

For $\mathbf{r} \neq \mathbf{r}'$,
$$\nabla^2 F = \frac{1}{r^2}\frac{d}{dr}\left(r^2 \frac{dF}{dr}\right) = 0 \tag{5.33}$$

which is integrated twice to yield
$$F = -\frac{A}{r} + B. \tag{5.34}$$

Applying Eq. (5.21),
$$1 = \lim_{\epsilon \to 0} \oint \frac{dF}{dr} dS = \lim_{\epsilon \to 0} \int_0^{2\pi}\int_0^{\pi} \frac{A}{r^2} r^2 \sin\phi \, d\theta d\phi = 4\pi A$$

or $A = \frac{1}{4\pi}$. Choosing $B = 0$ leads to
$$F = -\frac{1}{4\pi r}$$

and
$$G = F + U = -\frac{1}{4\pi r} + U \tag{5.35}$$

where U is chosen so that G satisfies prescribed boundary conditions.

Table 5.1 lists some Green functions that are commonly used in the solution of EM-related problems. It should be observed from Table 5.1 that the form of the three-dimensional Green's function for the steady-state wave equation tends to the Green's function for Laplace's equation as the wave number k approaches zero. It is also worthy of remark that each of the Green's functions in closed form as in Table 5.1 can be expressed in series form. For example, the Green's function
$$\begin{aligned}F &= -\frac{j}{4} H_0^{(1)}(k|\boldsymbol{\rho}-\boldsymbol{\rho}'|) \\ &= -\frac{j}{4} H_0^{(1)}(k[\rho^2 + \rho'^2 - 2\rho\rho'\cos(\phi-\phi')]^{1/2})\end{aligned} \tag{5.36}$$

can be written in series form as
$$F = \begin{cases} -\dfrac{j}{4} \displaystyle\sum_{n=-\infty}^{\infty} H_n^{(1)}(k\rho') J_n(k\rho) e^{-jn(\phi-\phi')}, & \rho < \rho' \\ -\dfrac{j}{4} \displaystyle\sum_{n=-\infty}^{\infty} H_n^{(1)}(k\rho) J_n(k\rho') e^{-jn(\phi-\phi')}, & \rho > \rho' \end{cases} \tag{5.37}$$

Table 5.1 Free-space Green's Functions

Operator equation	Laplace's equation	Steady-state Helmholtz's (or wave) equation	Modified steady-state Helmholtz's (or wave) equation
Solution Region	$\nabla^2 G = \delta(\mathbf{r}, \mathbf{r}')$	$\nabla^2 G + k^2 G = \delta(\mathbf{r}, \mathbf{r}')$	$\nabla^2 G - k^2 G = \delta(\mathbf{r}, \mathbf{r}')$
1-dimensional	no solution for $(-\infty, \infty)$	$-\dfrac{j}{2k}\exp(ik\|x - x'\|)$	$-\dfrac{1}{2k}\exp(-k\|x - x'\|)$
2-dimensional	$\dfrac{1}{2\pi}\ln\|\rho - \rho'\|$	$-\dfrac{j}{4}H_0^{(1)}(k\|\rho - \rho\|)$	$-\dfrac{1}{2\rho}K_o(k\|\rho - \rho'\|)$
3-dimensional	$-\dfrac{1}{4\pi\|\rho - \rho'\|}$	$\dfrac{\exp(jk\|\rho - \rho'\|)}{4\pi\|\rho - \rho'\|}$	$\dfrac{\exp(-k\|\rho - \rho'\|)}{4\pi\|\rho - \rho'\|}$

This is obtained from addition theorem for Hankel functions [14]. It should be noted that Green's functions are very difficult to construct in an explicit form except for the simplest shapes of domain and variety of operators.

With the aid of the Green's function, we can construct the integral equation corresponding to Poisson's equation in three dimensions

$$\nabla^2 V = -\frac{\rho_v}{\epsilon} \tag{5.38}$$

as

$$V = \int \frac{\rho_v}{\epsilon} G(\mathbf{r}, \mathbf{r}') dv'$$

or

$$V = \int \frac{\rho_v \, dv'}{4\pi\epsilon r}. \tag{5.39}$$

Similarly, the integral equation corresponding to Helmholtz's equation in three dimensions

$$\nabla^2 \Phi + k^2 \Phi = g \tag{5.40}$$

as

$$\Phi = \int g G(\mathbf{r}, \mathbf{r}') dv'$$

or

$$\Phi = \int \frac{g e^{jkr} dv'}{4\pi r} \tag{5.41}$$

where an outgoing wave is assumed.

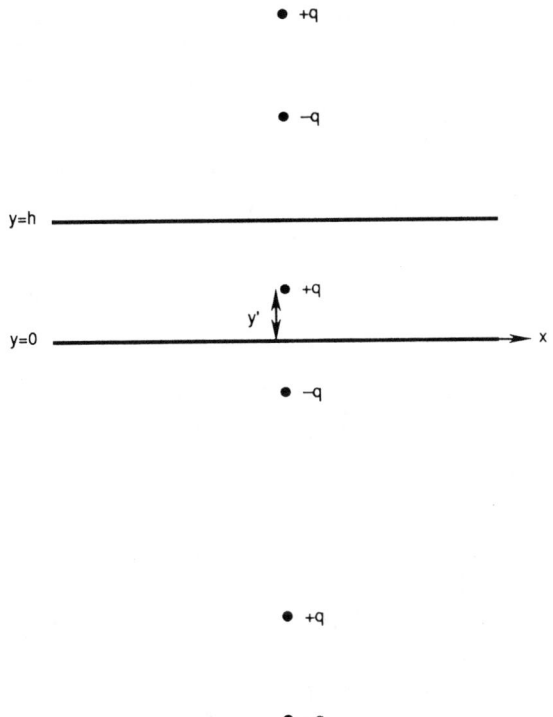

Figure 5.2 A single charge placed between two conducting planes produces the same potential as does the system of image charges when no conducting planes are present.

5.3.2 For Domain with Conducting Boundaries

The Green's functions derived so far are useful if the domain is free space. When the domain is bounded by one or more grounded planes, there are two ways to obtain the Green's function:

(a) the method of images [12,15–22] and
(b) the eigenfunction expansion [12,16,17,22–30].

(a) Method of Images

The method of images is a powerful technique for obtaining the field due to one or more sources with conducting boundary planes. If a point charge q is at some distance h from a grounded conducting plane, the boundary condition imposed by the plane on the resulting potential field may be satisfied by replacing the plane with an "image charge" $-q$ located at a position which is the mirror location of q. Using this idea to obtain the Green's function is perhaps best illustrated with an example.

Consider the region between the ground planes at $y = 0$ and $y = h$ as shown in Fig. 5.2. The Green's function $G(x,y;x',y')$ is the potential at the point (x,y), which results when a unit line charge of 1 C/m is placed at the point (x',y'). If no ground planes were present, the potential at distance ρ from a unit line charge would be

$$V(\rho) = \frac{1}{4\pi\epsilon} \ln \rho^2. \tag{5.42}$$

In order to satisfy the boundary conditions on the ground planes, an infinite set of images is derived as shown in Fig. 5.2. The potential due to such a sequence of line charges (including the original) within the strip is the superposition of an infinite series of images:

$$\begin{aligned}G(x,y;x',y') =& \frac{1}{4\pi\epsilon} \Big(\ln\big[(x-x')^2 + (y+y')^2\big] - \ln\big[(x-x')^2 + (y+y')^2\big] \\ &+ \sum_{n=1}^{\infty} (-1)^n \Big[\ln\big[(x-x')^2 + (y+y'-2nh)^2\big] \\ &- \ln\big[(x-x')^2 + (y-y'-2nh)^2\big] \\ &+ \ln\big[(x-x')^2 + (y+y'+2nh)^2\big] \\ &- \ln\big[(x-x')^2 + (y-y'+2nh)^2\big] \Big] \Big) \\ =& \frac{1}{4\pi\epsilon} \sum_{n=-\infty}^{\infty} \ln\left[\frac{(x-x')^2 + (y+y'-2nh)^2}{(x-x')^2 + (y-y'-2nh)^2}\right]. \end{aligned} \tag{5.43}$$

This series converges slowly and is awkward for numerical computation. It can be summed to give [15]

$$G(x,y;x',y') = \frac{1}{4\pi\epsilon} \ln\left[\frac{\sinh^2\left(\frac{\pi(x-x')}{2h}\right) + \sin^2\left(\frac{\pi(y+y')}{2h}\right)}{\sinh^2\left(\frac{\pi(x-x')}{2h}\right) + \sin^2\left(\frac{\pi(y-y')}{2h}\right)}\right]. \tag{5.44}$$

This expression can be shown to satisfy the appropriate boundary conditions along the ground plane, i.e., $G(x,y;x',y') = 0$ at $y = 0$ or $y = h$. Note that G has exactly one singularity at $x = x'$, $y = y'$ in the region $0 \leq y \leq h$.

In order to evaluate an integral involving $G(x,y;x',y')$ in Eq. (5.44), it is convenient to take out the singular portion of the unit source function. We rewrite Eq. (5.44) as

$$G(x,y;x',y') = -\frac{1}{4\pi\epsilon}\ln\left[(x-x')^2 + (y+y')^2\right] + g(x,y;x',y') \qquad (5.45)$$

where

$$g(x,y;x',y') =$$

$$\frac{1}{4\pi\epsilon}\ln\left[\frac{[(x-x')^2 + (y-y')^2]\left[\sinh^2\left(\frac{\pi(x-x')}{2h}\right) + \sin^2\left(\frac{\pi(y+y')}{2h}\right)\right]}{\sinh^2\left(\frac{\pi(x-x')}{2h}\right) + \sin^2\left(\frac{\pi(y-y')}{2h}\right)}\right]. \qquad (5.46)$$

Note that $g(x,y;x',y')$ is finite everywhere in $0 \leq y \leq h$. The integral involving g is evaluated numerically, while the one involving the singular logarithmic term is evaluated analytically with the aid of integral tables.

The method of images has been applied in deriving the Green's functions for multiconductor transmission lines [18–20] and planar microwave circuits [16,17,21]. The method is restricted to the shapes enclosed by boundaries that are straight lines.

(b) Eigenfunction Expansion

This method is suitable for deriving the Green's function for differential equations whose homogeneous solution is known. The Green's function is represented in terms of a series of orthonormal functions that satisfy the boundary conditions associated with the differential equation. To illustrate the eigenfunction expansion procedure, suppose we are interested in the Green's function for the wave equation

$$\frac{\partial^2 \Psi}{\partial x^2} + \frac{\partial^2 \Psi}{\partial y^2} + k^2 \Psi = 0 \qquad (5.47)$$

subject to

$$\frac{\partial \Psi}{\partial n} = 0 \quad \text{or} \quad \Psi = 0. \qquad (5.48)$$

Let the eigenfunctions and eigenvalues of Eq. (5.47) that satisfy Eq. (5.48) be Ψ_j and k_j, respectively, i.e.,

$$\nabla^2 \Psi_j + k_j^2 \Psi_j = 0. \qquad (5.49)$$

Since Ψ_j form an orthonormal set,

$$\int_S \Psi_j^* \Psi_i dx\, dy = \begin{cases} 1, & j=i \\ 0, & j \neq i \end{cases} \qquad (5.50)$$

where the asterisk (*) denotes complex conjugation. Assuming that Ψ_j form a complete set of orthonormal functions, $G(x,y;x',y')$ can be expanded in terms of Ψ_j, i.e.,

$$G(x,y;x',y') = \sum_{j=1}^{\infty} a_j \Psi_j(x,y). \qquad (5.51)$$

Since the Green's function must satisfy

$$(\nabla^2 + k^2)G(x,y;x',y') = \delta(x-x')\,\delta(y-y'), \qquad (5.52)$$

substituting Eqs. (5.49) and (5.51) into Eq. (5.52), we obtain

$$\sum_{j=1}^{\infty} a_j(k^2 - k_j^2)\Psi_j = \delta(x-x')\,\delta(y-y'). \qquad (5.53)$$

Multiplying both sides by Ψ_i^* and integrating over the region S gives

$$\sum_{j=1}^{\infty} a_j(k^2 - k_j^2) \int_S \Psi_j \Psi_i^*\, dx\, dy = \Psi_i^*(x',y'). \qquad (5.54)$$

Imposing the orthonormal property in Eq. (5.50) leads to

$$a_i(k^2 - k_i^2) = \Psi_i^*(x',y')$$

or

$$a_i = \frac{\Psi_i^*(x',y')}{(k^2 - k_i^2)}. \qquad (5.55)$$

Thus

$$G(x,y;x',y') = \sum_{j=1}^{\infty} \frac{\Psi_j(x,y)\,\Psi_j^*(x',y')}{(k^2 - k_j^2)}. \qquad (5.56)$$

The eigenfunction expansion approach has been applied to derive the Green's functions for plane conducting boundaries such as rectangular box and prism [22], planar microwave circuits [16,17,25], multilayered dielectric structures [23,24], waveguides [28], and surfaces of revolution [27]. The approach is limited to separable coordinate systems since the requisite eigenfunctions can be determined for only these cases.

Moment Methods 325

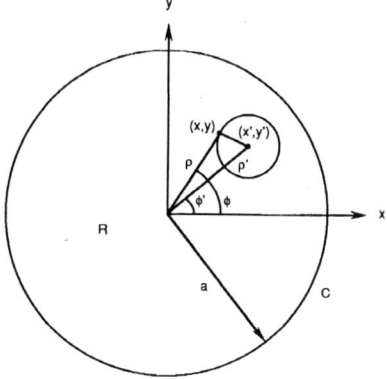

Figure 5.3 A disk of radius a.

Example 5.3

Construct a Green's function for

$$\nabla^2 V = 0$$

subject $V(a,\phi) = f(\phi)$ within as a circular disk $\rho \leq a$.

Solution

Since $g = 0$, the solution is obtained from Eq. (5.15) as

$$V = \oint_C f \frac{\partial G}{\partial n} \, dl \tag{5.3.1}$$

where the circle C is the boundary of the disk as shown in Fig. 5.3. Let

$$G = F + U,$$

where F is already found to be

$$F = \frac{1}{2\pi} \ln |\boldsymbol{\rho} - \boldsymbol{\rho}'|,$$

i.e.,

$$F(\rho,\phi;\rho',\phi') = \frac{1}{4\pi} \ln[\rho'^2 + \rho'^2 - 2\rho\rho' \cos(\phi - \phi')]. \tag{5.3.2}$$

The major problem is finding U. But

$$\nabla^2 U = 0 \quad \text{in } R \tag{5.3.3a}$$

with
$$U = -F \quad \text{on} \quad C$$

or
$$U(a,\phi;\rho',\phi') = -\frac{1}{4\pi}\ln[a^2 + \rho'^2 - 2a\rho'\cos(\phi-\phi')]. \tag{5.3.3b}$$

Thus U can be found by solving the PDE in Eq. (5.3.3a) subject to the condition in Eq. (5.3.3b). Applying the separation of variables method,

$$U = \frac{A_0}{2} + \sum_{n=1}^{\infty}\rho^n[A_n\cos n\phi + B_n\sin n\phi]. \tag{5.3.4}$$

The term ρ^{-n} is not included since U must be bounded at $\rho = 0$. To impose the boundary condition in Eq. (5.3.3b) on the solution in Eq. (5.3.4), we first express Eq. (5.3.3b) in Fourier series using the identity

$$\sum_{n=1}^{\infty}\frac{z^n}{n}\cos n\theta = \int_0^z \frac{\cos\theta - \lambda}{1+\lambda^2 - 2\lambda\cos\theta}d\lambda = -\frac{1}{2}\ln[1 + z^2 - 2z\cos\theta]. \tag{5.3.5}$$

Hence Eq. (5.3.3b) becomes

$$\begin{aligned}U(a,\phi;\rho',\phi') &= -\frac{1}{4\pi}\ln a^2[1 + (\rho'/a)^2 - \frac{2\rho'}{a}\cos(\phi-\phi')] \\ &= -\frac{1}{2\pi}\ln a + \frac{1}{2\pi}\sum_{n=1}^{\infty}[\frac{\rho'}{a}]^n\frac{\cos n(\phi-\phi')}{n} \\ &= -\frac{1}{2\pi}\ln a + \frac{1}{2\pi}\sum_{n=1}^{\infty}[\frac{\rho'}{a}]^n\frac{(\cos n\phi\cos n\phi' + \sin n\phi\sin n\phi')}{n}.\end{aligned} \tag{5.3.6}$$

Comparing Eq. (5.3.4) with Eq. (5.3.6) at $\rho = a$, we obtain the coefficients A_n and B_n as
$$\frac{A_0}{2} = -\frac{1}{2\pi}\ln a$$
$$a^n A_n = \frac{1}{2\pi n}[\frac{\rho'}{a}]^n\cos n\phi'$$
$$a^n B_n = \frac{1}{2\pi n}[\frac{\rho'}{a}]^n\sin n\phi'.$$

Thus Eq. (5.3.4) becomes

$$\begin{aligned}U(\rho,\phi;\rho',\phi') &= -\frac{1}{2\pi}\ln a + \frac{1}{2\pi}\sum_{n=1}^{\infty}[\frac{\rho'}{a}]^n[\frac{\rho}{a}]^n\frac{\cos n(\phi-\phi')}{n} \\ &= -\frac{1}{2\pi}\ln a - \frac{1}{4\pi}\ln\left[1 + [\frac{\rho\rho'}{a^2}]^2 - \frac{2\rho\rho'}{a^2}\cos(\phi-\phi')\right].\end{aligned} \tag{5.3.7}$$

From Eqs. (5.3.2) and (5.3.7), we obtain the Green's function as

$$G = \frac{1}{4\pi}\ln\left[\rho^2 + \rho'^2 - 2\rho\rho'\cos(\phi-\phi')\right] - \frac{1}{4\pi}\ln\left[a^2 + \frac{\rho'^2\rho^2}{a^2} - 2\rho\rho'\cos(\phi-\phi')\right]. \tag{5.3.8}$$

An alternative means of constructing the Green's function is the method of images. Let us obtain Eq. (5.3.8) using the method of images. Let

$$G(P,P') = \frac{1}{2\pi}\ln r + U.$$

The problem reduces to finding the induced field U, which is harmonic within the disk and is equal to $-\frac{1}{2\pi}\ln r$ on C. Let P' be the singular point of Green's function and let P_o be the image of P' with respect to the circle C as shown in Fig. 5.4. The triangles OQP' and OQP_0 are similar because the angle at O is common and the sides adjacent to it are proportional. Thus

$$\frac{\rho'}{a} = \frac{a}{\rho_o} \rightarrow \rho'\rho_o = a^2. \tag{5.3.9}$$

That is, the product of OP' and OP_o is equal to the square of the radius OQ. At a point Q on C, it is evident from Fig. 5.4 that

$$r_{QP'} = \frac{\rho'}{a}r_{QP_o}.$$

Therefore,

$$U = -\frac{1}{2\pi}\ln\frac{\rho' r_{PP_o}}{a} \tag{5.3.10}$$

and

$$G = \frac{1}{2\pi}\ln r_{PP'} - \frac{1}{2\pi}\ln\frac{\rho'}{a}r_{PP_o}. \tag{5.3.11}$$

Since $r_{PP'}$ is the distance between $P(\rho,\phi)$ and $P'(\rho',\phi')$ while r_{PP_o} is the distance between $P(\rho,\phi)$ and $P_o(\rho'_o,\phi) = P_o(a^2/\rho',\phi)$,

$$r_{PP'}^2 = \rho^2 + \rho'^2 - 2\rho\rho'\cos(\phi-\phi'),$$

$$r_{PP_o}^2 = \rho^2 + \frac{a^4}{\rho'^2} - 2\rho\frac{a^2}{\rho'}\cos(\phi-\phi').$$

Substituting these in Eq. (5.3.11), we obtain

$$G = \frac{1}{4\pi}\ln\left[\rho^2 + \rho'^2 - 2\rho\rho'\cos(\phi-\phi')\right] - \frac{1}{4\pi}\ln\left[a^2 + \frac{\rho'^2\rho^2}{a^2} - 2\rho\rho'\cos(\phi-\phi')\right] \tag{5.3.12}$$

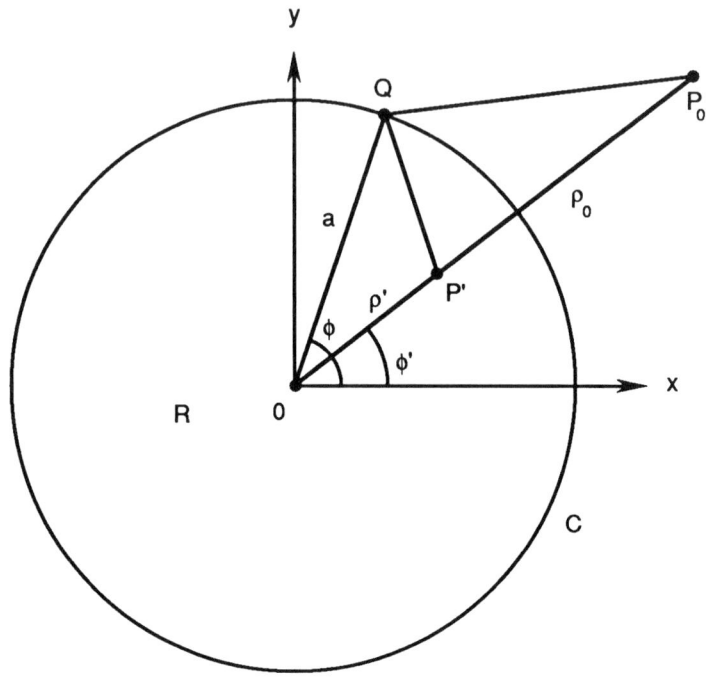

Figure 5.4 The image point P_o of P' with respect to circle C so that $OP'.OP_o = OQ = a^2$ and OQP' and OP_oQ are similar triangles.

which is the same as Eq. (5.3.8). From Eq. (5.3.8) or (5.3.12), the directional derivative $\partial G/\partial n (= \nabla G \cdot \mathbf{a}_n)$ on C is given by

$$\left.\frac{\partial G}{\partial \rho'}\right|_{\rho'=a} = \frac{2a - 2\rho\cos(\phi - \phi')}{4\pi[a^2 + \rho^2 - 2a\rho\cos(\phi - \phi')]}$$

$$- \frac{\frac{2\rho^2}{a} - 2\rho\cos(\phi - \phi')}{4\pi[a^2 + \rho^2 - 2a\rho\cos(\phi - \phi')]},$$

$$= \frac{a^2 - \rho^2}{2\pi a[a^2 + \rho^2 - 2a\rho\cos(\phi - \phi')]}.$$

Hence the solution in Eq. (5.3.1) becomes (with $dl = a d\phi'$)

$$V(\rho, \phi) = \frac{1}{2\pi} \int_0^{2\pi} \frac{(a^2 - \rho^2) f(\phi') d\phi'}{[a^2 + \rho^2 - 2a\rho\cos(\phi - \phi')]} \qquad (5.3.13)$$

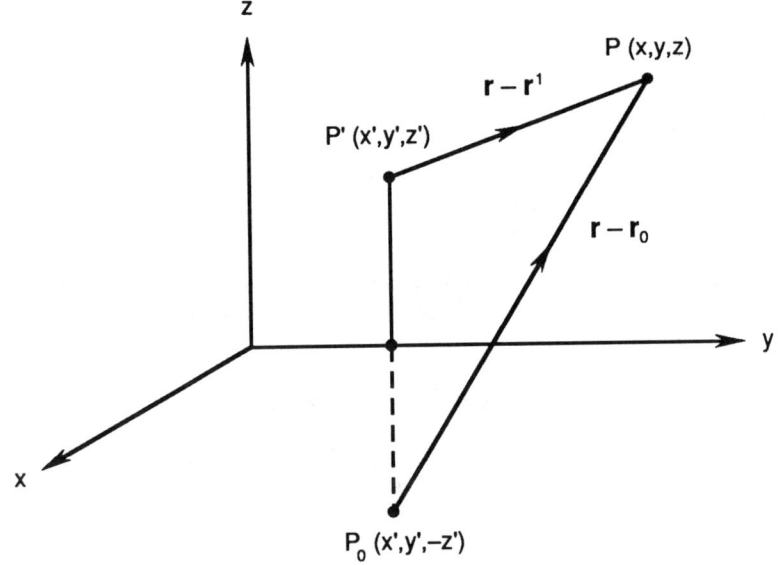

Figure 5.5 Half-space problem of Example 5.4.

which is known as *Poisson's integral formula*.

Example 5.4

Obtain the solution for the Laplace operator on unbounded half-space, $z \leq 0$ with the condition $V(z=0) = f$.

Solution

Again the solution is

$$V = \oint_S f \frac{\partial G}{\partial n} dS$$

where S is the plane $z = 0$. We let

$$G = \frac{1}{4\pi|\mathbf{r} - \mathbf{r}'|} + U$$

so that the major problem is reduced to finding U. Using the method of images, it is easy to see that the image point of $P'(x', y', z')$ is $P_0(x', y', -z')$ as shown in Fig. 5.5. Hence

$$U = -\frac{1}{4\pi|\mathbf{r} - \mathbf{r}_0|}$$

and
$$G = \frac{1}{4\pi|\mathbf{r}-\mathbf{r}'|} - \frac{1}{4\pi|\mathbf{r}-\mathbf{r}_o|},$$
where
$$|\mathbf{r}-\mathbf{r}'| = [(x-x')^2 + (y-y')^2 + (z-z')^2]^{1/2}$$
$$|\mathbf{r}-\mathbf{r}_o| = [(x-x')^2 + (y-y')^2 + (z+z')^2]^{1/2}.$$

Notice that G reduces to zero at $z=0$ and has the required singularity at $P'(x', y', z')$. The directional derivative $\partial G/\partial n$ on plane $z=0$ is

$$\left.\frac{\partial G}{\partial z'}\right|_{z'=0} = \frac{1}{4\pi}\left[\frac{(z-z')}{|\mathbf{r}-\mathbf{r}'|^3} + \frac{(z+z')}{|\mathbf{r}-\mathbf{r}_o|^3}\right]\bigg|_{z'=0}$$

$$= \frac{z}{2\pi\left[(x-x')^2 + (y-y')^2 + z^2\right]^{3/2}}.$$

Hence
$$V(x,y,z) = \frac{1}{2\pi}\int_{-\infty}^{\infty}\int_{-\infty}^{\infty}\frac{zf(x',y')dx'dy'}{\left[(x-x')^2 + (y-y')^2 + z^2\right]^{3/2}}$$

Example 5.5

Using Green's function, construct the solution for Poisson's equation
$$\frac{\partial^2 V}{\partial x^2} + \frac{\partial^2 V}{\partial y^2} = f(x,y)$$
subject to the boundary conditions
$$V(0,y) = V(a,y) = V(x,0) = V(x,b) = 0.$$

Solution

According to Eq. (5.15), the solution is
$$V(x,y) = \int_0^b \int_0^a f(x',y')G(x,y;x',y')\, dx'dy' \qquad (5.5.1)$$
so that our problem is essentially that of obtaining the Green's function $G(x,y;x',y')$. The Green's function satisfies

$$\frac{\partial^2 G}{\partial x^2} + \frac{\partial^2 G}{\partial y^2} = \delta(x-x')\delta(y-y'). \tag{5.5.2}$$

To apply the series expansion method of finding G, we must first determine eigenfunctions $\Psi(x,y)$ of Laplace's equation, i.e.,

$$\nabla^2 \Psi = \lambda \Psi$$

where Ψ satisfies the boundary conditions. It is evident that the normalized eigenfunctions are

$$\Psi_{mn} = \frac{2}{\sqrt{ab}} \sin\frac{m\pi x}{a} \sin\frac{n\pi y}{b}$$

with the corresponding eigenvalues

$$\lambda_{mn} = -\left(\frac{m^2\pi^2}{a^2} + \frac{n^2\pi^2}{b^2}\right).$$

Thus,

$$G(x,y;x',y') = \frac{2}{\sqrt{ab}} \sum_{m=1}^{\infty}\sum_{n=1}^{\infty} A_{mn}(x',y') \sin\frac{m\pi x}{a} \sin\frac{n\pi y}{b}. \tag{5.5.3}$$

The expansion coefficients A_{mn} are determined by substituting Eq. (5.5.3) into Eq. (5.5.2), multiplying both sides by $\sin\frac{m\pi x}{a} \sin\frac{n\pi y}{b}$ and integrating over $0 < x < a$, $0 < y < b$. Using the orthonormality property of the eigenfunctions and the shifting property of the delta function results in

$$-\left(\frac{m^2\pi^2}{a^2} + \frac{n^2\pi^2}{b^2}\right) A_{mn} = \frac{2}{\sqrt{ab}} \sin\frac{m\pi x'}{a} \sin\frac{n\pi y'}{b}.$$

Obtaining A_{mn} from this and substituting in Eq. (5.5.3) gives

$$G(x,y;x',y') = -\frac{4}{ab} \sum_{m=1}^{\infty}\sum_{n=1}^{\infty} \frac{\sin\frac{m\pi x}{a} \sin\frac{m\pi x'}{a} \sin\frac{n\pi y}{b} \sin\frac{n\pi y'}{b}}{m^2\pi^2/a^2 + n^2\pi^2/b^2}. \tag{5.5.4}$$

Another way of obtaining Green's function is by means of a single series rather than a double summation in Eq. (5.5.4). It can be shown that [28,29]

$$G(x,y;x',y') =$$

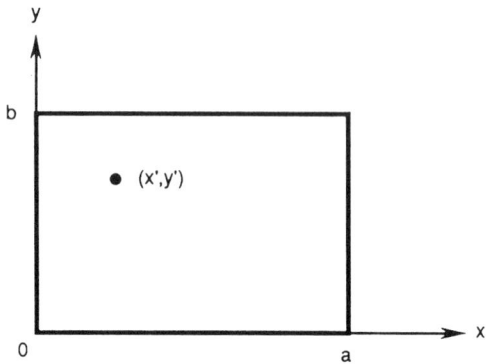

Figure 5.6 Line source in a rectangular region.

$$\begin{cases} -\dfrac{2}{\pi} \displaystyle\sum_{n=1}^{\infty} \dfrac{\sin \dfrac{n\pi x}{b} \, \sinh \dfrac{n\pi(a-x')}{b} \, \sin \dfrac{n\pi y}{b} \, \sinh \dfrac{n\pi y'}{b}}{n \sinh \dfrac{n\pi a}{b}}, & x < x' \\[2ex] -\dfrac{2}{\pi} \displaystyle\sum_{n=1}^{\infty} \dfrac{\sinh \dfrac{n\pi x'}{b} \, \sinh \dfrac{n\pi(a-x)}{b} \, \sin \dfrac{n\pi y}{b} \, \sinh \dfrac{n\pi y'}{b}}{n \sinh \dfrac{n\pi a}{b}}, & x > x' \end{cases} \quad (5.5.5)$$

By Fourier series expansion, it can be verified that the expressions in Eqs. (5.5.4) and (5.5.5) are identical. Besides the factor $\dfrac{1}{\epsilon}$, the Green's function in Eq. (5.5.4) or (5.5.5) gives the potential V due to a unit line source at (x', y') in the region $0 < x < a$, $0 < y < b$ as shown in Fig. 5.6.

Example 5.6

An infinite line source I_z is located at (ρ', ϕ') in a wedge waveguide shown in Fig. 5.7. Derive the electric field due to the line.

Solution

Assuming the time factor $e^{j\omega t}$, the z-component of $\mathbf{E}$ for the TE mode satisfies the wave equation

$$\nabla^2 E_z + k^2 E_z = j\omega\mu I_z \qquad (5.6.1)$$

with

$$\frac{\partial E_z}{\partial R} = 0$$

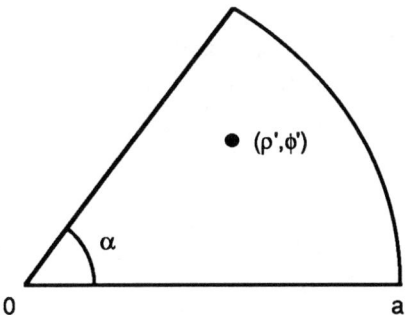

Figure 5.7 Line source in a waveguide.

where $k = \omega\sqrt{\mu\epsilon}$ and R is the outward unit normal at any point on the periphery of the cross section. The Green's function for this problem satisfies

$$\nabla^2 G + k^2 G = j\omega\mu\delta(\rho - \rho') \tag{5.6.2}$$

with

$$\frac{\partial G}{\partial R} = 0$$

so that the solution to Eq. (5.6.1) is

$$E_z = j\omega\mu \int_S I_z(\rho', \phi') G(\rho, \phi; \rho', \phi') \, dS. \tag{5.6.3}$$

To determine the Green's function $G(\rho, \phi; \rho', \phi')$, we find Ψ_i so that Eq. (5.56) can be applied. The boundary condition $\frac{\partial G}{\partial R} = 0$ implies that

$$\left.\frac{1}{\rho}\frac{\partial G}{\partial \phi}\right|_{\phi=0} = 0 = \left.\frac{1}{\rho}\frac{\partial G}{\partial \phi}\right|_{\phi=\alpha} = \left.\frac{\partial G}{\partial \rho}\right|_{\rho=a}. \tag{5.6.4}$$

The set of functions which satisfy the boundary conditions are

$$\Psi_{m\nu}(\rho, \phi) = J_\nu(k_{m\nu}\rho)\cos\nu\phi \tag{5.6.5}$$

where

$$\nu = n\pi/\alpha, \quad n = 0, 1, 2, \cdots, \tag{5.6.6a}$$

$k_{m\nu}$ are chosen to satisfy

$$\left.\frac{\partial}{\partial \rho} J_\nu(k_{m\nu}\rho)\right|_{\rho=a} = 0, \tag{5.6.6b}$$

and the subscript m is used to denote the mth root of Eq. (5.6.6b); m can take the value zero for $n = 0$. The functions $\Psi_{m\nu}$ are orthogonal if and only if ν is an integer which implies that ν is an integral multiple of α. Let $\alpha = \pi/\ell$, where ℓ is a positive integer, so that $\Phi_{m\nu}$ are mutually orthogonal. To obtain the Green's function using Eq. (5.56), these eigenfunctions must be normalized over the region, i.e.,

$$\int_0^a J_\nu^2(k_{m\nu}\rho)\, d\rho = \begin{cases} a^2/2, & m = \nu \\ \frac{1}{2}[a^2 - (\nu^2/k_{m\nu}^2)]J_\nu^2(k_{m\nu}a), & \text{otherwise} \end{cases} \quad (5.6.7a)$$

$$\int_0^\alpha \cos^2 \nu\phi\, d\phi = \begin{cases} \dfrac{\pi}{\ell}, & \nu = 0 \\ \dfrac{\pi}{2\ell}, & \text{otherwise} \end{cases} \quad (5.6.7b)$$

where $\nu = n\ell$. Using the normalized eigenfunctions, we obtain

$$G(\rho,\phi;\rho',\phi') = \frac{j2\ell}{\omega\epsilon\pi a^2} - 4j\ell\omega\mu \sum_{n=1}^\infty \sum_{m=1}^\infty \frac{J_\nu(k_{m\nu}\rho)J_\nu(k_{m\nu}\rho')\cos\nu\phi\cos\nu\phi'}{\epsilon_\nu \pi \left(a^2 - \dfrac{\nu^2}{k_{m\nu}^2}\right) J_\nu^2(k_{m\nu}a)(k^2 - k_{m\nu}^2)} \quad (5.6.8)$$

where

$$\epsilon_\nu = \begin{cases} 2, & \nu = 0 \\ 1, & \nu \neq 0 \end{cases}. \quad (5.6.9)$$

We have employed the fact that $\omega\mu/k^2 = \dfrac{1}{\omega\epsilon}$ to obtain the first term on the right-hand side of Eq. (5.6.8).

5.4 Applications I—Quasi-Static Problems

The method of moments has been applied to so many EM problems that covering all of them is practically impossible. We will only consider the relatively easy ones to illustrate the techniques involved. Once the basic approach has been mastered, it will be easy for the reader to extend the idea to attack more complicated problems.

We will apply MOM to a static problem in this section; more involved application will be considered in the sections to follow. We will consider the problem of determining the characteristic impedance Z_o of a strip transmission line [31].

Consider the strip transmission of Fig. 5.8(a). If the line is assumed to be infinitely long, the problem is reduced to a two-dimensional TEM

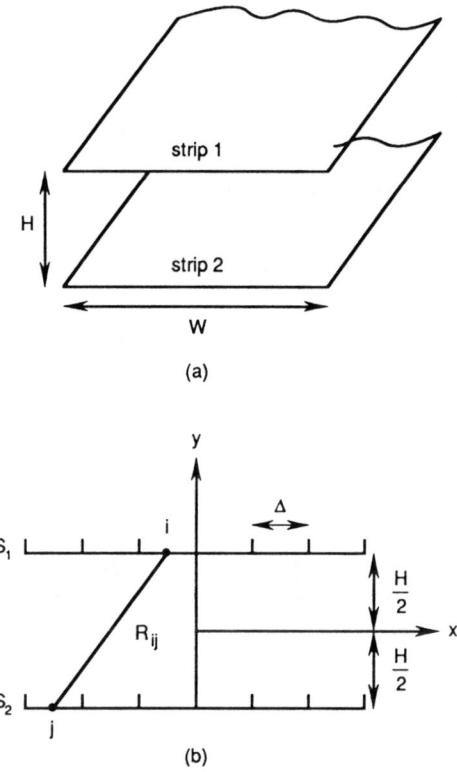

Figure 5.8 (a) The trip transmission line. (b) The two-dimensional view.

problem of line sources in a plane as in Fig. 5.8(b). Let the potential difference of the strips be $V_d = 2V$ so that strip 1 is maintained at $+1V$ while strip 2 is at $-1V$. Our objective is to find the surface charge density $\rho(x,y)$ on the strips so that the total charge per unit length on one strip can be found as

$$Q_\ell = \int \rho\, dl. \tag{5.57}$$

(Q_ℓ is charge per unit length as distinct from the total charge on the strip because we are treating a three-dimensional problem as a two-dimensional one.) Once Q is known, the capacitance per unit length C_ℓ can be found from

$$C_\ell = \frac{Q_\ell}{V_d}. \tag{5.58}$$

Finally, the line characteristic impedance is obtained:

$$Z_o = \frac{(\mu\epsilon)^{1/2}}{C_\ell} = \frac{1}{uC_\ell} \tag{5.59}$$

where $u = 1/\sqrt{\mu\epsilon}$ is the speed of the wave in the (lossless) dielectric medium between the strips. Everything is straightforward once the charge density $\rho(x,y)$ in Eq. (5.57) is known. To find ρ using MOM, we divide each strip into n subareas of equal width Δ so that subareas in strip 1 are numbered $1, 2, \ldots, n$, while those in strip 2 are numbered $n+1, n+2, n+3, \ldots, 2n$. The potential at an arbitrary field point is

$$V(x,y) = \frac{1}{2\pi\epsilon} \int \rho(x',y') \ln \frac{R}{r_o} \, dx' dy' \tag{5.60}$$

where R is the distance between source and field points, i.e.,

$$R = [(x-x')^2 + (y-y')^2]^{1/2}. \tag{5.61}$$

Since the integral in Eq. (5.60) may be regarded as rectangular subareas in a numerical sense, the potential at the center of a typical subarea S_i is

$$V_i = \frac{1}{2\pi\epsilon} \sum_{j=1}^{2n} \rho_j \int_{S_i} \ln \frac{R_{ij}}{r_o} \, dx'$$

or

$$V_i = \sum_{j=1}^{2n} A_{ij} \rho_j \tag{5.62}$$

where

$$A_{ij} = \frac{1}{2\pi\epsilon} \int_{S_i} \ln \frac{R_{ij}}{r_o} \, dx', \tag{5.63}$$

R_{ij} is the distance between ith and jth subareas, and $A_{ij}\rho_j$ represents the potential at point i due to subarea j. In Eq. (5.62), we have assumed that the charge density ρ is constant within each subarea. For all the subareas S_i, $i = 1, 2, \ldots, 2n$, we have

$$V_1 = \sum_{j=1}^{2n} \rho_j A_{1j} = 1$$

$$V_2 = \sum_{j=1}^{2n} \rho_j A_{2j} = 1$$

$$\vdots$$

$$V_n = \sum_{j=1}^{2n} \rho_j A_{nj} = 1$$

$$V_{n+1} = \sum_{j=1}^{2n} \rho_j A_{n+1,j} = -1$$

$$\vdots$$

$$V_{2n} = \sum_{j=1}^{2n} \rho_j A_{2n,j} = -1.$$

Thus we obtain a set of $2n$ simultaneous equations with $2n$ unknown charge densities ρ_i. In matrix form,

$$\begin{bmatrix} A_{11} & A_{12} & \cdots & A_{1,2n} \\ A_{21} & A_{22} & \cdots & A_{2,2n} \\ \vdots & & & \\ A_{2n,1} & A_{2n,2} & \cdots & A_{2n,2n} \end{bmatrix} \begin{bmatrix} \rho_1 \\ \rho_2 \\ \vdots \\ \rho_{2n} \end{bmatrix} = \begin{bmatrix} 1 \\ 1 \\ \vdots \\ -1 \\ -1 \end{bmatrix}$$

or simply

$$[A][\rho] = [B]. \tag{5.64}$$

It can be shown that [32] the elements of matrix $[A]$ expressed in Eq. (5.63) can be reduced to

$$A_{ij} = \begin{cases} \dfrac{\Delta}{2\pi\epsilon} \ln \dfrac{R_{ij}}{r_o}, & i \neq j \\[2ex] \dfrac{\Delta}{2\pi\epsilon} [\ln \dfrac{\Delta}{r_o} - 1.5], & i = j \end{cases} \tag{5.65}$$

where r_o is a constant scale factor (commonly taken as unity). From Eq. (5.64), we obtain $[\rho]$ either by solving the simultaneous equation or by matrix inversion, i.e.,

$$[\rho] = [A]^{-1}[B]. \tag{5.66}$$

Once $[\rho]$ is known, we determine C_ℓ from Eqs. (5.57) and (5.58) as

$$C_\ell = \sum_{j=1}^{n} \rho_j \Delta / V_d \tag{5.67}$$

where $V_d = 2V$. Obtaining Z_o follows from Eqs. (5.59) and (5.67).

338 Numerical Techniques in Electromagnetics

Table 5.2 Characteristic Impedance of a Strip Transmission Line

n	Z_o(in Ω)
3	53.02
7	51.07
11	50.49
18	50.39
39	49.71
59	49.61

Example 5.7

Write a program to find the characteristic impedance Z_o of a strip line with $H = 2$m, $W = 5$m, $\epsilon = \epsilon_o, \mu_o = \mu_o$, and $V_d = 2V$.

Solution

The FORTRAN program is shown in Fig. 5.9. With the given data, the program calculates the elements of matrices [A] and [B] and determines $[\rho]$ by matrix inversion. With the computed charge densities the capacitance per unit length is calculated using Eq. (5.67) and the characteristic impedance from Eq. (5.59). Table 5.2 presents the computed values of Z_o for a different number of segments per strip, n. The results agree well with $Z_o = 50\Omega$ from Wheeler's curve [33].

```
0001      C***********************************************************
0002      C      USING MOMENTS METHOD,
0003      C      THIS PROGRAM DETERMINES THE CHARACTERISTIC IMPEDANCE
0004      C      OF A STRIP TRANSMISSION LINE
0005      C      WITH CROSS-SECTION  W X H
0006      C      THE STRIPS MAINTAINED AT 1 VOLT AND - 1 VOLT.
0007      C
0008      C      ONE STRIP IS LOCATED ON THE Y= H/2 PLANE WHILE THE OTHER
0009      C      IS LOCATED ON THE  Y= -H/2  PLANE.
0010      C
0011      C      ALL DIMENSIONS ARE IN S.I. UNITS
0012      C
0013      C      N IS THE NUMBER OF SUBSECTIONS INTO WHICH
                 EACH STRIP IS DIVIDED
0014      C***********************************************************
0015
0016             DIMENSION A(100,100), B(100), X(100), Y(100), R0(100)
0017             DATA  PIE/3.14159/
```

Moment Methods 339

```
0018
0019        C       FIRST, SPECIFY THE PARAMETERS
0020
0021                CL=3.E+8   ! SPEED OF LIGHT IN FREE SPACE
0022                ER=1.0
0023                EO=8.8541E-12
0024                H=2.0
0025                W=5.0
0026                N=10
0027                NT=2N
0028                DELTA = W/FLOAT(N)
0029        C
0030        C       SECOND, CALCULATE THE MATCH POINTS
0031        C       AND THE COEFFICIENT MATRIX [A]
0032        C
0033                DO 10 K=1,N
0034                X(K) = DELTA*( FLOAT(K) -0.5 )   ! FOR LOWER STRIP
0035                Y(K) = -H/2.0
0036                X(K+N) = X(K)                              ! FOR UPPER STRIP
0037                Y(K+N) = H/2.0
0038        10      CONTINUE
0039
0040                FACTOR = DELTA/(2.0*PIE*EO)
0041
0042                DO 20 I=1,NT
0043                DO 20 J=1,NT
0044                IF(I-J) 12,11,12
0045
0046        11      A(I,J)= -( ALOG(DELTA) - 1.5)*FACTOR
0047                GO TO 13
0048        12      R=SQRT( (X(I) - X(J))**2 + (Y(I) - Y(J))**2 )
0049                A(I,J)= - ALOG(R)*FACTOR
0050        13      CONTINUE
0051        20      CONTINUE
0052        C
0053        C       NOW DETERMINE THE MATRIX OF CONSTANT VECTOR [B]
0054        C
0055                DO 30 K=1,N
0056                B(K)=1.0
0057        C       B(K+N)= -1.0
0058        30      CONTINUE
0059        C
0060        C       INVERT MATRIX  A(I,J) AND CALCULATE MATRIX RO(N)
0061        C       CONSISTING OF THE UNKNOWN ELEMENTS
0062        C       ALSO CALCULATE THE TOTAL CHARGE Q,
0063        C       THE CAPACITANCE C, AND THE CHARACTERISTIC IMPEDANCE Zo.
0064
0065                NIV=NT
0066                NMAX=100
0067                CALL INVERSE(A,NIV,NMAX) !  APPENDIX D
0068                DO 50 I=1,NT
0069                RO(I)=0.0
0070                DO 40 M=1,NT
0071                RO(I)=RO(I) + A(I,M)*B(M)
0072        40      CONTINUE
0073        50      CONTINUE
0074                SUM=0.0
0075                DO 60 I=1,N
0076                SUM= SUM + RO(I)
0077        60      CONTINUE
0078                Q=SUM*DELTA
0079                VD=2.0
0080                C=ABS(Q)/VD
```

```
0081        Z0 = SQRT(ER)/CL*C
0082        PRINT *,Z0
0083        WRITE(6,70) Z0
0084    70  FORMAT(2X,'Z0=',E14.6,/)
0085        STOP
0086        END
```

Figure 5.9 FORTRAN program for Example 5.7.

5.5 Applications II—Scattering Problems

The purpose of this section is to illustrate with two examples how the method of moments can be applied to solve electromagnetic scattering problems. The first example is on scattering of a plane wave by a perfectly conducting cylinder [6], while the second is on scattering of a plane wave by an arbitrary array of parallel wires [34].

5.5.1 Scattering by Conducting Cylinder

Consider an infinitely long, perfectly conducting cylinder located at a far distance from a radiating source. Assuming a time-harmonic field with time factor $e^{j\omega t}$, Maxwell's equations can be written in phasor form as

$$\nabla \cdot \mathbf{E}_s = 0 \qquad (5.68a)$$

$$\nabla \cdot \mathbf{H}_s = 0 \qquad (5.68b)$$

$$\nabla \times \mathbf{E}_s = -j\omega\mu \mathbf{H}_s \qquad (5.68c)$$

$$\nabla \times \mathbf{H}_s = \mathbf{J}_s + j\omega\epsilon \mathbf{E}_s \qquad (5.68d)$$

where the subscript s denotes phasor or complex quantities. Henceforth, we will drop subscript s for simplicity and use the same symbols for the frequency-domain quantities and time-domain quantities. It is assumed that the reader can differentiate between the two quantities. Taking the curl of Eq. (5.68c) and applying Eq. (5.68d), we obtain

$$\nabla \times \nabla \times \mathbf{E} = -j\omega\mu \nabla \times \mathbf{H} = -j\omega\mu(\mathbf{J} + j\omega\epsilon\mathbf{E}). \qquad (5.69)$$

Introducing the vector identity

$$\nabla \times \nabla \times \mathbf{A} = \nabla(\nabla \cdot \mathbf{A}) - \nabla^2 \mathbf{A}$$

into Eq. (5.69) gives

$$\nabla(\nabla \cdot \mathbf{E}) - \nabla^2 \mathbf{E} = -j\omega\mu(\mathbf{J} + j\omega\epsilon\mathbf{E}).$$

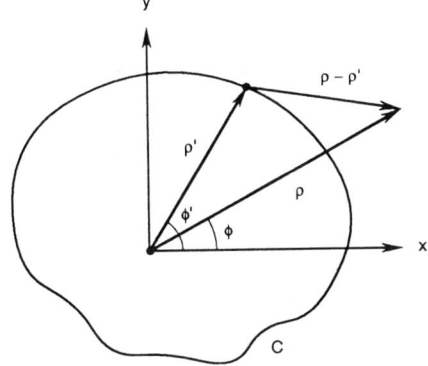

Figure 5.10 Cross section of the cylinder.

In view of Eq. (5.68a), $\nabla(\nabla \cdot \mathbf{E}) = 0$ so that

$$\nabla^2 \mathbf{E} + k^2 \mathbf{E} = j\omega\mu\mathbf{J} \qquad (5.70)$$

where $k = \omega(\mu\epsilon)^{1/2} = 2\pi/\lambda$ is the wave number and λ is the wavelength. Equation (5.70) is the vector form of the Helmholtz wave equation. If we assume a TM wave ($H_z = 0$) with $\mathbf{E} = E_z(x,y)\mathbf{a}_z$, the vector equation (5.70) becomes a scalar equation, namely,

$$\nabla^2 E_z + k^2 E_z = j\omega\mu J_z \qquad (5.71)$$

where $\mathbf{J} = J_z\mathbf{a}_z$ is the source current density. The integral solution to Eq. (5.71) is

$$E_z(x,y) = E_z(\rho) = -\frac{k\eta_o}{4}\int_S J_z(\rho')H_0^{(2)}(k|\rho - \rho'|)dS' \qquad (5.72)$$

where $\rho = x\mathbf{a}_x + y\mathbf{a}_y$ is the field point, $\rho' = x'\mathbf{a}_x + y'\mathbf{a}_y$ is the source point, $\eta_o = (\mu_o/\epsilon_o)^{1/2} \simeq 377\ \Omega$ is the intrinsic impedance of free space, and $H_0^{(2)} =$ Hankel function of the second kind of zero order since an outward-traveling wave is assumed and there is no ϕ dependence. The integration in Eq. (5.72) is over the cross section of the cylinder shown in Fig. 5.10.

If field E_z^i is incident on a perfectly conducting cylinder, it induces surface current J_z on the conducting cylinder, which in turn produces a scattered field E_z^s. The scattered field E_z^s due to J_z is expressed by Eq. (5.72). On the boundary C, the tangential component of the total field must vanish. Thus

$$E_z^i + E_z^s = 0 \text{ on } C. \qquad (5.73)$$

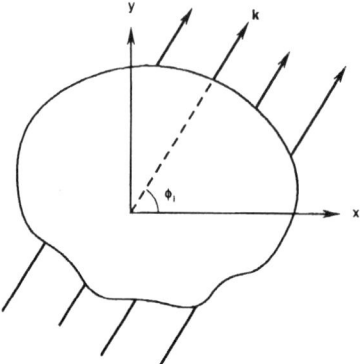

Figure 5.11 Typical propogation of vector **k**.

Substitution of Eq. (5.72) into Eq. (5.73) yields

$$E_z^i(\rho) = \frac{k\eta_o}{4} \int_C J_z(\rho) H_0^{(2)}(k|\boldsymbol{\rho} - \boldsymbol{\rho}'|) \, dl'. \tag{5.74}$$

In the integral equation (5.74), the induced surface current density J_z is the only unknown. We determine J_z using the moment method.

We divide the boundary C into N segments and apply the point matching technique. On a segment ΔC_n, Eq. (5.74) becomes

$$E_z^i(\rho_n) = \frac{k\eta_o}{4} \sum_{m=1}^{N} J_z(\rho_m) H_0^{(2)}(k|\boldsymbol{\rho}_n - \boldsymbol{\rho}_m|) \Delta C_m \tag{5.75}$$

where the integration in Eq. (5.74) has been replaced by summation. On applying Eq. (5.75) to all segments, a system of simultaneous equations results. The system of equations can be cast in matrix form as

$$\begin{bmatrix} E_z^i(\rho_1) \\ E_z^i(\rho_2) \\ \vdots \\ E_z^i(\rho_N) \end{bmatrix} = \begin{bmatrix} A_{11} & A_{12} & \ldots & A_{1N} \\ A_{21} & A_{22} & \ldots & A_{2N} \\ \vdots & & \vdots & \\ A_{N1} & A_{N2} & \ldots & A_{NN} \end{bmatrix} \begin{bmatrix} J_z(\rho_1) \\ J_z(\rho_2) \\ \vdots \\ J_z(\rho_N) \end{bmatrix} \tag{5.76a}$$

or

$$[E] = [A][J]. \tag{5.76b}$$

Hence

$$[J] = [A]^{-1}[E]. \tag{5.77}$$

To determine the exact values of elements of matrix [A] may be difficult. Approximately [6],

$$A_{mn} \simeq \begin{cases} \frac{\eta_o k}{4} \Delta C_n H_o^{(2)}\{k(x_n - x_m)^2 + (y_n - y_m)^2]^{1/2}\}, & m \neq n \\ \frac{\eta_o k}{4}[1 - j\frac{2}{\pi}\log_{10}(\frac{\gamma k \Delta C_n}{4e})], & m = n \end{cases} \quad (5.78)$$

where (x_n, y_n) is the midpoint of ΔC_n, $e = 2.718\ldots$, and $\gamma = 1.781\ldots$. Thus for a given cross section and specified incident field E_z^i, the induced surface current density J_z can be found from Eq. (5.77). To be specific, assume the propagation vector **k** is directed as shown in Fig. 5.11 so that

$$E_z^i = E_o e^{j\mathbf{k}\cdot\mathbf{r}}$$

where $\mathbf{r} = x\mathbf{a}_x + y\mathbf{a}_y$, $\mathbf{k} = k(\cos\phi_i \mathbf{a}_x + \sin\phi_i \mathbf{a}_y)$, $k = 2\pi/\lambda$, and ϕ_i is the incidence angle. Taking $E_o = 1$ so that $|E_z^i| = 1$,

$$E_z^i = e^{jk(x\cos\phi_i + y\sin\phi_i)}. \quad (5.79)$$

Given any C (dictated by the cross section of the cylinder), we can substitute Eqs. (5.78) and (5.79) into Eq. (5.76) and determine [J] from Eq. (5.77). Once J_z, the induced current density, is known, we calculate the *scattering cross section* σ defined by

$$\sigma(\phi, \phi_i) = 2\pi\rho \left|\frac{E_z^s(\phi)}{E_z^s(\phi_i)}\right|^2$$

$$= \frac{k\eta_o^2}{4}\left|\int_C J_z(x', y') e^{jk(x'\cos\phi + y'\sin\phi)} \, dl'\right|^2 \quad (5.80)$$

where ϕ is the angle at the observation point, the point at which σ is evaluated. In matrix form,

$$\sigma(\phi_i, \phi) = \frac{k\eta^2}{4}\left|[V_n^s][Z_{nm}]^{-1}[V_m^i]\right|^2 \quad (5.81)$$

where

$$V_m^i = \Delta C_m e^{jk(x_m\cos\phi_i + y_m\cos\phi_i)}, \quad (5.82a)$$

$$V_n^s = \Delta C_n e^{jk(x_n\cos\phi + y_n\cos\phi)}, \quad (5.82b)$$

and

$$Z_{mn} = \Delta C_m \, A_{mn}. \quad (5.82c)$$

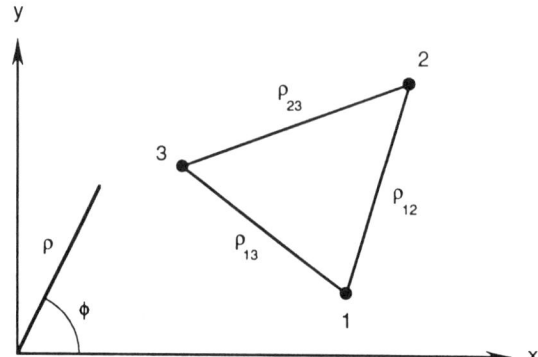

Figure 5.12 An array of three wires parallel to the z-axis.

5.5.2 Scattering by an Arbitrary Array of Parallel Wires

This problem is of more general nature than the one just described. As a matter of fact, any infinitely long, perfectly conducting, thin metal can be modeled as an array of parallel wires. It will be shown that the scattering pattern due to an arbitrary array of line sources approaches that of a solid conducting cylinder of the same cross-sectional geometry if a sufficiently large number of wires are present and they are arrayed on a closed curve. Hence the problem of scattering by a conducting cylinder presented above can also be modeled with the techniques to be described here (see Problems 5.15 and 5.17).

Consider an arbitrary array of N parallel, infinitely long wires placed parallel to the z-axis. Three of such wires are illustrated in Fig. 5.12. Let a harmonic TM wave be incident on the wires. Assuming a time factor $e^{j\omega t}$, the incident wave in phasor form is given by

$$E_z^i = E_i(x,y)e^{-jhz} \tag{5.83}$$

where

$$E_i(x,y) = E_o e^{-jk(x \sin\theta_i \cos\phi_i + y \sin\theta_i \sin\phi_i)} \tag{5.84a}$$

$$h = k\cos\theta_i, \tag{5.84b}$$

$$k = \frac{2\pi}{\lambda} = \omega(\mu\epsilon)^{1/2}, \tag{5.84c}$$

and θ_i and ϕ_i define the axis of propagation as illustrated in Fig. 5.13. The incident wave induces current on the surface of wire n. The induced current density has only z component.

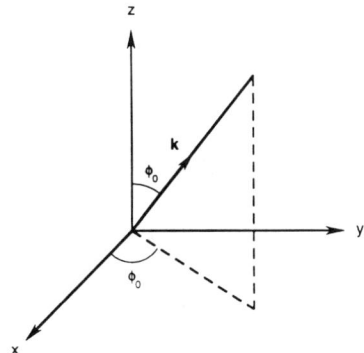

Figure 5.13 Propagation vector **k**.

It can be shown that the field due to a harmonic current I_n uniformly distributed on a circular cylinder of radius a_n has a z component given by

$$E_n = -I'_n H_0^{(2)}(g\rho_n)e^{-jhz}, \quad \rho_n > a_n \tag{5.85}$$

where

$$I'_n = \frac{\omega\mu g^2}{4k^2} I_n J_0(ga_n), \tag{5.86}$$

$$g^2 + h^2 = k^2, \tag{5.87}$$

J_0 is Bessel function of order zero, and H_0 is Hankel function of the second kind of order zero. By induction theorem, if I_n is regarded as the induced current, Eq. (5.85) may be considered as the scattered field, i.e.,

$$E_z^s = -\sum_{n=1}^{N} I'_n H_0^{(2)}(g\rho_n)e^{-jhz} \tag{5.88}$$

where the summation is taken over all the N wires. On the surface of each wire (assumed perfectly conducting),

$$E_z^i + E_z^s = 0$$

or

$$E_z^i = -E_z^s, \quad \rho = \rho_n. \tag{5.89}$$

Substitution of Eqs. (5.83) and (5.88) into Eq. (5.89) leads to

$$\sum_{n=1}^{N} I'_n H_0^{(2)}(g\rho_{mn}) = E_i(x_m, y_m) \tag{5.90}$$

where

$$\rho_{mn} = \begin{cases} \sqrt{(x_m - x_n)^2 + (y_m - y_n)^2} & , m \neq n \\ a_m & , m = n \end{cases} \quad (5.91)$$

and a_m is the radius of the mth wire. In matrix form, Eq. (5.90) can be written as

$$[A][I] = [B]$$

or

$$[I] = [A]^{-1}[B] \quad (5.92)$$

where

$$I_n = I'_n, \quad (5.93a)$$
$$A_{mn} = H_0^{(2)}(g\rho_{mn}), \quad (5.93b)$$
$$B_m = E_o e^{-jk(x_m \sin\theta_i \cos\phi_i + y_m \sin\theta_i \sin\phi_i)}. \quad (5.93c)$$

Once I'_n is calculated from Eq. (5.92), the scattered field can be obtained as

$$E_z^s = -\sum_{n=1}^N I'_n H_0^{(2)}(g\rho_n)e^{-jhz}. \quad (5.94)$$

Finally, we may calculate the "distant scattering pattern," defined as

$$E(\phi) = \sum_{n=1}^N I'_n e^{jg(x_n \cos\phi + y_n \sin\phi)}. \quad (5.95)$$

The following example, taken from Richmond's work [34], will be used to illustrate the techniques discussed in the latter half of this section.

Example 5.8

Consider the two arrays shown in Fig. 5.14. For Fig. 5.14(a), take

no. of wires, $N = 15$
wire radius, $ka = 0.05$
wire spacing $ks = 1.0$
$\theta_o = 90°$, $\phi_o = 40°$, $270° < \phi < 90°$

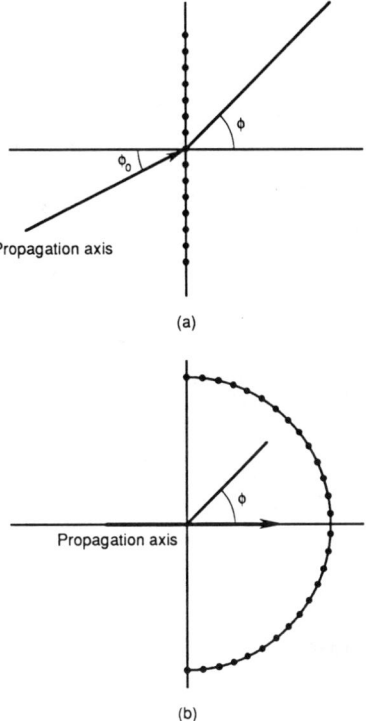

Figure 5.14 For Example 5.8: (a) A plane array of 15 parallel wires, (b) a semicircular array of 30 parallel wires.

and for Fig. 5.14(b), take

$$\text{no. of wires, } N = 30$$
$$\text{wire radius, } ka = 0.05$$
$$\text{cylinder radius, } R = 1.12\lambda$$
$$\theta_o = 90°, \phi_o = 0, \; 0 < \phi < 180°.$$

For the two arrays, calculate and plot the scattering pattern as a function of ϕ.

Solution

The FORTRAN code for calculating the scattering pattern $E(\phi)$ based on Eq. (5.95) is shown in Fig. 5.15. The same program can be used for the two arrays in Fig. 5.14, except that the input data on N, ka, ks, and the

348 Numerical Techniques in Electromagnetics

locations $(x_n, y_n), n = 1, 2, \ldots, N$ of the wires are different. The program essentially calculates I_n required in Eq. (5.95) using Eqs. (5.92) and (5.93). The plots of $E(\phi)$ against ϕ are portrayed in Figs. 5.16 and 5.17 for the arrays in Figs. 5.14a and 5.14b, respectively.

```
0001      C ==============================================
0002      C   THIS PROGRAM CALCULATES THE SCATTERING PATTERN OF AN
0003      C   ARRAY OF PARALLEL WIRES
0004      C   REFERENCE - RICHMOND'S WORK [8]
0005      C ==============================================
0006
0007          PARAMETER (IDM=50)
0008          DIMENSION X(IDM),Y(IDM),RHO(IDM,IDM),E(IDM)
0009          REAL LAMBDA,K
0010          COMPLEX J,H0,SUM,A(IDM,IDM),B(IDM),I(IDM)
0011          DATA PIE,LAMBDA/3.14159,1.0/
0012
0013          J=(0.0,1.0)
0014      C
0015      C   SPECIFY NECESSARY DATA AND CALCULATE
              RELEVANT PARAMETERS
0016      C
0017          NN = 30
0018          THETA0 = PIE/2.0
0019          PHI0 = 0.0
0020          E0 = 1.0
0021          R = 1.125*LAMBDA
0022          K = 2.0*PIE/LAMBDA
0023          AA = 0.05/K
0024          S = 1.0/K
0025          H = K*COS(THETA0)
0026          G = SQRT( K**2 - H**2 )
0027      C
0028      C   DEFINE WIRE LOCATIONS AND CALCULATE RHO
0029      C
0030          DO 10 N=1,NN
0031          PHI = PIE*( FLOAT(N-1)/FLOAT(NN-1) ) - PIE/2.0
0032          X(N) = R*COS(PHI)
0033          Y(N) = R*SIN(PHI)
0034      10  CONTINUE
0035          DO 40 M=1,NN
0036          DO 40 N=1,NN
0037          IF(M-N) 20,30,20
0038      20  RHO(M,N) = SQRT( (X(M) - X(N))**2 + (Y(M) - Y(N))**2 )
0039          GO TO 40
0040      30  RHO(M,N) = AA
0041      40  CONTINUE
0042      C
0043      C CONSTRUCT MATRIX [A]
0044      C
0045          DO 50 M=1,NN
0046          DO 50 N=1,NN
0047          ARGU = G*RHO(M,N)
0048          CALL HANKEL(ARGU,H0)
0049          A(M,N) = H0
0050      50  CONTINUE
0051      C
0052      C CONSTRUCT MATRIX [B]
0053      C
```

```
0054            DO 60 M=1,NN
0055            ALPHA = X(M)*SIN(THETAO)*COS(PHIO)
0056           1      + Y(M)*SIN(THETAO)*SIN(PHIO)
0057            B(M) = EO*CEXP( -J*K*ALPHA )
0058        60  CONTINUE
0059       C
0060       C SOLVE FOR MATRIX [I] CONSISTING OF "MODIFIED CURRENT"OR
0061       C CURRENT COEFFICIENTS
0062       C
0063            CALL INVERSE(A,NN,IDM)
0064            DO 70 N=1,NN
0065            I(N) = (0.0,0.0)
0066            DO 70 M=1,NN
0067            I(N) = I(N) + A(N,M)*B(M)
0068        70  CONTINUE
0069       C
0070       C FINALLY, CALCULATE THE SCATTERING PATTERN E(PHI)
0071       C
0072            DPHI = 5.0
0073            DO 90 L=1,37
0074            PHI = DPHI*FLOAT(L-1)
0075            SUM = (0.0,0.0)
0076            DO 80 N=1,NN
0077            ALP = X(N)*COSD(PHI) + Y(N)*SIND(PHI)
0078            SUM = SUM + I(N)*CEXP( J*G*ALP )
0079        80  CONTINUE
0080            E(L) = CABS(SUM)
0081            WRITE(6,100) PHI,E(L)
0082        90  CONTINUE
0083       100  FORMAT(2X,'PHI=',F8.2,3X,'E(PHI)=',F12.6,/)
0084            CALL PLOT(E,37,0.0,DPHI,1)
0085            STOP
0086            END

0001       C**************************************************************
0002       C THIS PROGRAM EVALUATES HANKEL FUNCTION OF THE SECOND KIND
0003       C AND ORDER ZERO USING NEWTON-COTES RULE (N=5)
0004       C**************************************************************
0005       C
0006       C INITIALIZATION OF VARIABLES
0007       C
0008            SUBROUTINE HANKEL(X,H)
0009            DIMENSION O1(150),C(10),A1(10),A2(100),O2(150)
0010            DATA ( C(I), I=1,6 )/19.0,75.0,50.0,50.0,75.0,19.0/
0011            COMPLEX H
0012            A=0.
0013            B=3.141592654
0014            M=30.
0015            H1=(B-A)/M
0016            H2=H1/2
0017            N=5
0018            NX=288
0019            E=.577216
0020       C
0021       C SET VALUES OF ANGLES
0022       C
0023            DO 5 I=1,M+1
0024            O1(I)=(I-1)*H1
0025            O2(I)=(I-1)*H2
0026         5  CONTINUE
0027       C
0028       C COMPUTE VALUES OF XJO &YO
0029       C
```

```
0030              L=M/N
0031              SUM1=0.
0032              SUM3=0.
0033              DO 500 I=1,L
0034              SUM2=0.
0035              SUM4=0.
0036       C
0037       C SET UP COMPUTATIONS OF AIS
0038       C
0039              DO 100 II=1,N+1
0040              J=(I-1)*N+II
0041              SUM2=SUM2+C(II)*COS(X*COS(O1(J)))
0042              IF(J.EQ.1)O2(J)=.00405
0043              SUM4=SUM4+C(II)*COS(X*COS(O2(J)))*(E+
0044           2 ALOG(2*X*(SIN(O2(J)))**2))
0045       100    CONTINUE
0046       C
0047       C COMPUTE THE SUM A1S
0048       C
0049              A1(I)=SUM2
0050              A2(I)=SUM4
0051              SUM1=SUM1+A1(I)
0052              SUM3=SUM3+A2(I)
0053       500    CONTINUE
0054       C
0055       C COMPUTE HO(X)=XJO(X)+I*YO(X)
0056       C
0057              XJO=(N*H1*SUM1)/(NX*B)
0058              YO=(N*H2*4*SUM3)/(NX*B**2)
0059              H=CMPLX( XJO,-YO )
0060              RETURN
0061              END
```

Figure 5.15 Computer program for Example 5.8.

5.6 Applications III—Radiation Problems

In this section, we consider the application of MOM to wires or cylindrical antennas. The distinction between scatterers considered in the previous section and antennas to be treated here is primarily that of the location of the source. An object acts as a scatterer if it is far from the source; it acts as an antenna if the source is on it [6].

Consider a perfectly conducting cylindrical antenna of radius a, extending from $z = -\ell/2$ to $z = \ell/2$ as shown in Fig. 5.18. Let the antenna be situated in a lossless homogeneous dielectric medium ($\sigma = 0$). Assuming a z-directed current on the cylinder ($\mathbf{J} = J_z a_z$), only axial electric field E_z is produced due to axial symmetry. The electric field can be expressed in terms of the retarded potentials of Eq. (1.38) as

$$E_z = -j\omega A_z - \frac{\partial V}{\partial z}. \tag{5.96}$$

Moment Methods 351

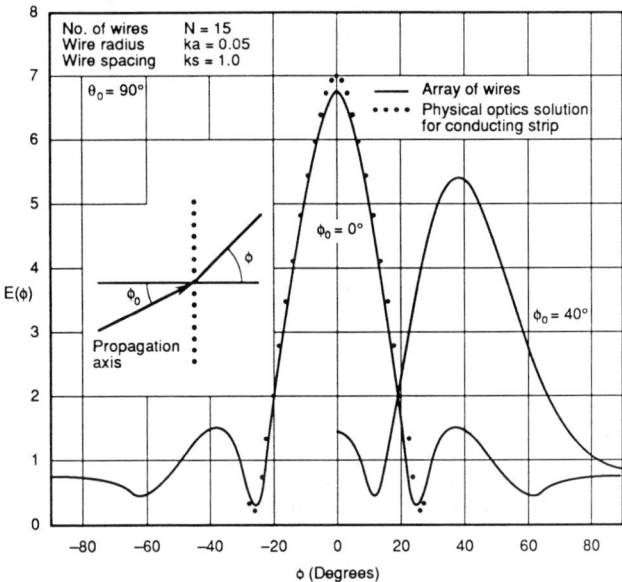

Figure 5.16 Scattering pattern for the plane array of Fig. 5.14(a).

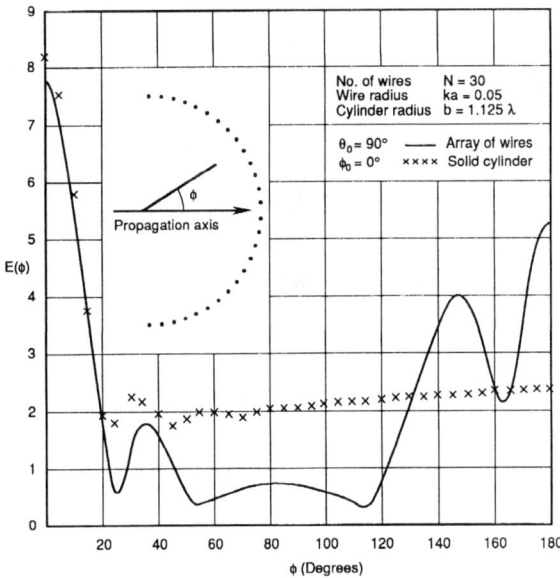

Figure 5.17 Scattering pattern for the semicircular array of Fig. 5.14(b).

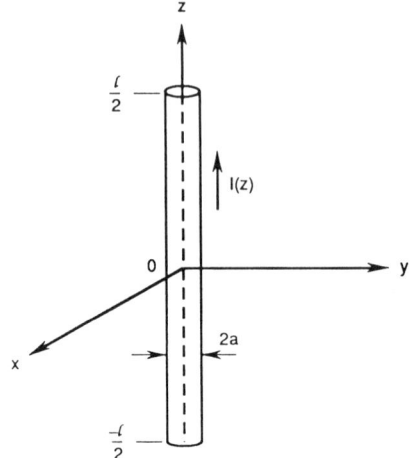

Figure 5.18 Cylindrical antenna of length l and radius a.

Applying the Lorentz condition of Eq. (1.41), namely,

$$\frac{\partial A_z}{\partial z} = -j\omega\mu\epsilon V, \qquad (5.97)$$

Eq. (5.96) becomes

$$E_z = -j\omega\left(1 + \frac{1}{k^2}\frac{d^2}{dz^2}\right)A_z \qquad (5.98)$$

where $k = \omega(\mu\epsilon)^{1/2} = 2\pi/\lambda$, ω is the angular frequency of the suppressed harmonic time variation $e^{j\omega t}$. From Eq. (1.44)

$$A_z = \mu \int_{-l/2}^{l/2} I(z')G(z,z')dz' \qquad (5.99)$$

where $G(z,z')$ is the free space Green's function, i.e.,

$$G(z,z') = \frac{e^{-jkR}}{4\pi R} \qquad (5.100)$$

and R is the distance between observation point (x, y, z) and source point (x', y', z') or

$$R = [(x - x')^2 + (y - y')^2 + (z - z')^2]^{1/2}. \qquad (5.101)$$

Combining Eqs. (5.98) and (5.99) gives

$$E_z = -j\omega\mu\left(1 + \frac{1}{k^2}\frac{d^2}{dz^2}\right)\int_{-l/2}^{l/2} I(z')G(z,z')dz'. \qquad (5.102)$$

This integro-differential equation is not convenient for numerical analysis because it requires evaluation of the second derivative with respect to z of the integral. We will now consider two types of modification of Eq. (5.102) leading to Hallen's (magnetic vector potential) and Pocklington's (electric field) integral equations. Either of these integral equations can be used to determine the current distribution on a cylindrical antenna or scatterer and subsequently calculate all other quantities of interest.

5.6.1 Hallen's Integral Equation

We can rewrite Eq. (5.102) in a compact form as

$$\left(\frac{d^2}{dz^2} + k^2\right)F(z) = k^2 S(z), \quad -\ell/2 < z < \ell/2 \qquad (5.103)$$

where

$$F(z) = \int_{-\ell/2}^{\ell/2} I(z')G(z,z')dz', \qquad (5.104\text{a})$$

$$S(z) = -\frac{E_z}{j\omega\mu}. \qquad (5.104\text{b})$$

Equation (5.103) is a second-order linear ordinary differential equation. The solution to the homogeneous equation

$$\left(\frac{d^2}{dz^2} + k^2\right)F(z) = 0,$$

which is consistent with the boundary condition that the current must be zero at the wire ends ($z = \pm\ell/2$), is

$$F_h(z) = c_1 \cos kz + c_2 \sin kz \qquad (5.105)$$

where c_1 and c_2 are integration constants. The particular solution of Eq. (5.103) can be obtained, for example, by the Lagrange method of varying constants [35] as

$$F_p(z) = k \int_{-\ell/2}^{\ell/2} S(z') \sin k(z - z')dz'. \qquad (5.106)$$

Thus from Eqs. (5.104) to (5.106), the solution to Eq. (5.103) is

$$\int_{-\ell/2}^{\ell/2} I(z')G(z,z')dz' = c_1 \cos kz + c_2 \sin kz - \frac{j}{\eta}\int_{-\ell/2}^{\ell/2} E_z(z') \sin k(z - z')dz' \qquad (5.107)$$

where $\eta = \sqrt{\mu/\epsilon}$ is the intrinsic impedance of the surrounding medium. Equation (5.107) is referred to as *Hallen's integral equation* [36] for a perfectly conducting cylindrical antenna or scatterer. The equation has been generalized by Mei [37] to perfectly conducting wires of arbitrary shape. Hallen's IE is computationally convenient since its kernel contains only ℓ/r terms. Its major advantage is the ease with which a converged solution may be obtained, while its major drawback lies in the additional work required in finding the integration constants c_1 and c_2 [35,38].

5.6.2 Pocklington's Integral Equation

We can also rewrite Eq. (5.102) by introducing the operator in parentheses under the integral sign so that

$$\int_{-\ell/2}^{\ell/2} I(z')\left(\frac{\partial^2}{\partial z^2} + k^2\right)G(z,z')dz' = j\omega\epsilon E_z. \tag{5.108}$$

This is known as Pocklington's integral equation [39]. Note that Pocklington's IE has E_z, which represents the field from the source on the right-hand side. Both Pocklington's and Hallen's IEs can be used to treat wire antennas. The third type of IE derivable from Eq. (5.102) is the Schelkunoff's IE, found in [35].

5.6.3 Expansion and Weighting Functions

Having derived suitable integral equations, we can now find solutions for a variety of wire antennas or scatterers. This usually entails reducing the integral equations to a set of simultaneous linear equations using the method of moments. The unknown current $I(z)$ along the wire is approximated by a finite set $u_n(z)$ of basis (or expansion) functions with unknown amplitudes as discussed in the last chapter. That is, we let

$$I(z) = \sum_{n=1}^{N} I_n u_n(z), \tag{5.109}$$

where N is the number of basis functions needed to cover the wire and the expansion coefficients I_n are to be determined. The functions u_n are chosen to be linearly independent. The basis functions commonly used in solving antenna or scattering problems are of two types: entire domain functions and subdomain functions. The entire domain basis functions exist over the full domain $-\ell/2 < z < \ell/2$. Typical examples are [8,40]:

(1) Fourier:

$$u_n(z) = \cos(n-1)\nu/2, \tag{5.110a}$$

(2) Chebychev:
$$u_n(z) = T_{2n-2}(\nu), \tag{5.110b}$$

(3) Mauclaurin:
$$u_n(z) = \nu^{2n-2}, \tag{5.110c}$$

(4) Legendre:
$$u_n(z) = P_{2n-2}(\nu), \tag{5.110d}$$

(5) Hermite:
$$u_n(z) = H_{2n-2}(\nu), \tag{5.110e}$$

where $\nu = 2z/\ell$ and $n = 1, 2, \ldots, N$. The subdomain basis functions exist only on one of the N nonoverlapping segments into which the domain is divided. Typical examples are [41,42]:

(1) piecewise constant (pulse) function:
$$u_n(z) = \begin{cases} 1, & z_{n-1/2} < z < z_{n+1/2} \\ 0, & \text{otherwise,} \end{cases} \tag{5.111a}$$

(2) piecewise linear (triangular) function:
$$u_n(z) = \begin{cases} \dfrac{\Delta - |z - z_n|}{\Delta}, & z_{n-1} < z < z_{n+1} \\ 0, & \text{otherwise,} \end{cases} \tag{5.111b}$$

(3) piecewise sinusoidal function:
$$u_n(z) = \begin{cases} \dfrac{\sin k(z - |z - z_n|)}{\sin k\Delta}, & z_{n-1} < z < z_{n+1} \\ 0, & \text{otherwise,} \end{cases} \tag{5.111c}$$

where $\Delta = \ell/N$, assuming equal subintervals although this is unnecessary. Figure 5.19 illustrates these subdomain functions. The entire domain basis functions are of limited applications since they require a prior knowledge of the nature of the function to be represented. The subdomain functions are the most commonly used, particularly in developing general-purpose user-oriented computer codes for treating wire problems. For this reason, we will focus on using subdomain functions as basis functions.

Substitution of the approximate representation of current $I(z)$ in Eq. (5.109) into Pocklington's IE of Eq. (5.108) gives

$$\int_{-\ell/2}^{\ell/2} \sum_{n=1}^{N} I_n u_n(z') K(z_m, z') dz' \simeq E_z(z_m) \tag{5.112}$$

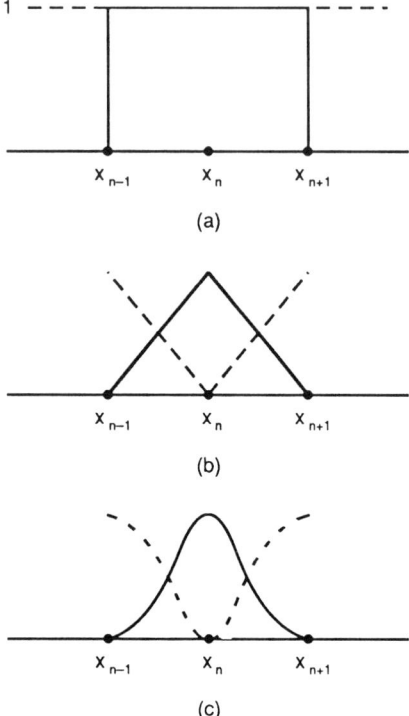

Figure 5.19 Typical subdomain weighting functions: (a) Piecewise uniform function, (b) piecewise linear function, (c) piecewise sinusoidal function.

where

$$K(z_m, z') = \frac{1}{j\omega\epsilon}\left(\frac{\partial^2}{\partial z^2} + k^2\right)G(z_m, z')$$

is the kernel, $z = z_m$ on segment m is the point on the wire at which the IE is being enforced. Equation (5.112) may be written as

$$\sum_{n=1}^{N} I_n \int_{\Delta z_n} K(z_m, z') u_n(z') dz' \simeq E_z(z_m)$$

or

$$\sum_{n=1}^{N} I_n g_m = E_z(z_m) \qquad (5.113)$$

where
$$g_m = \int_{\Delta z'_n} K(z_m, z') u_n(z') dz'. \tag{5.114}$$

In order to solve for the unknown current amplitudes I_n ($n = 1, 2, \ldots, N$), N equations need to be derived from Eq. (5.113). We achieve this by multiplying Eq. (5.113) by weighting (or testing) functions $w_n (n = 1, 2, \ldots, n)$ and integrating over the wire length. In other words, we let Eq. (5.113) be satisfied in an average sense over the entire domain. This leads to forming the inner product between each of the weighting functions and g_m so that Eq. (5.113) is reduced to

$$\sum_{n=1}^{N} I_n \langle w_n, g_m \rangle = \langle w_n, E_z \rangle, \quad m = 1, 2, \ldots, N. \tag{5.115}$$

Thus we have a set of N simultaneous equations which can be written in matrix form as

$$\begin{bmatrix} \langle w_1, g_1 \rangle & \cdots & \langle w_1, g_N \rangle \\ \langle w_2, g_1 \rangle & \cdots & \langle w_2, g_N \rangle \\ \vdots & & \vdots \\ \langle w_N, g_1 \rangle & \cdots & \langle w_N, g_N \rangle \end{bmatrix} \begin{bmatrix} I_1 \\ I_2 \\ \vdots \\ I_N \end{bmatrix} = \begin{bmatrix} \langle w_1, E_{z1} \rangle \\ \langle w_2, E_{z2} \rangle \\ \vdots \\ \langle w_N, E_{zN} \rangle \end{bmatrix}$$

or
$$[Z][I] = [V] \tag{5.116}$$

where $z_{mn} = \langle w_n, g_m \rangle$ and $V_m = \langle w_m, E_z \rangle$. The desired solution for the current is then obtained by solving the simultaneous equations (5.116) or by matrix inversion, i.e.,

$$[I] = [Z]^{-1}[V]. \tag{5.117}$$

Because of the similarity of Eq. (5.116) to the network equations, the matrices $[Z], [V]$, and $[I]$ are referred to as *generalized* impedance, voltage, and current matrices, respectively [6]. Once the current distribution $I(z')$ is determined from Eq. (5.116) or (5.117), parameters of practical interest such as input impedance and radiation patterns are readily obtained.

The weighting functions $\{w_n\}$ must be chosen so that each of Eq. (5.115) is linearly independent and computation of the necessary numerical integration is minimized. Evaluation of the integrals in Eq. (5.116) is often the most time-consuming portion of scattering or radiation problems. Sometimes we select similar types of functions for both weighting and expansion. As discussed in the previous chapter, choosing $w_n = u_n$ leads to Galerkin's

358 Numerical Techniques in Electromagnetics

method, while choosing $w_n = \delta(z - z_n)$ results in point matching (or collocation) method. The point matching method is simpler than Galerkin's method and is sufficiently adequate for many EM problems. However, it tends to be a slower converging method. The general rules that should be followed in selecting the weighting functions are addressed in [43]. The following examples are taken from [41,44–46].

Example 5.9

Solve the Hallen's integral equation

$$\int_{-\ell/2}^{\ell/2} I(z')G(z,z')dz' = -\frac{j}{\eta_o}(A\cos kz + B\sin k|z|)$$

where $k = 2\pi/\lambda$ is the phase constant and $\eta_o = 377\Omega$ is the intrinsic impedance of free space. Consider a straight wire dipole with length $\lambda = 0.5\lambda$ and radius $a = 0.005\lambda$.

Solution

The integral equation has the form

$$\int_{-\ell/2}^{\ell/2} I(z')K(z,z')dz' = D(z) \qquad (5.9.1)$$

which is a Fredholm integral equation of the first kind. In Eq. (5.9.1),

$$K(z,z') = G(z,z') = \frac{e^{-jkR}}{4\pi R}, \qquad (5.9.2a)$$

$$R = \sqrt{a^2 + (z-z')^2}, \qquad (5.9.2b)$$

and

$$D(z) = -\frac{j}{\eta_o}\left[A\cos(kz) + B\sin(k|z|)\right]. \qquad (5.9.2c)$$

If the terminal voltage of the wire antenna is V_T, the constant $B = V_T/2$. The absolute value in $\sin k|z|$ expresses the assumption of antenna symmetry, i.e., $I(-z') = I(z')$. Thus

$$\int_{-\ell/2}^{\ell/2} I(z')\frac{e^{-jkR}}{4\pi R}dz' = -\frac{j}{\eta_o}[A\cos kz + \frac{V_T}{2}\sin k|z|]. \qquad (5.9.3)$$

If we let

$$I(z) = \sum_{n=1}^{N} I_n\, u_n(z), \qquad (5.9.4)$$

Eq. (5.9.3) will contain N unknown variables I_n and the unknown constant A. To determine the $N+1$ unknowns, we divide the wire into N segments. For the sake of simplicity, we choose segments of equal lengths $\Delta z = \ell/N$ and select $N+1$ matching points such as:

$$z = -\ell/2, -\ell/2 + \Delta z, \ldots, 0, \ldots, \ell/2 - \Delta z, \ell/2.$$

At each match point $z = z_m$,

$$\int_{-\ell/2}^{\ell/2} \sum_{n=1}^{N} I_n u_n(z') K(z_m, z') dz' = D(z_m). \tag{5.9.5}$$

Taking the inner products (moments) by multiplying either side with a weighting function $w_m(z)$ and integrating both sides,

$$\int_{-\ell/2}^{\ell/2} \int_{-\ell/2}^{\ell/2} \sum_{n=1}^{N} I_n\, u_n(z') K(z_m, z')\, dz' w_m(z)\, dz = \int_{-\ell/2}^{\ell/2} D(z_m) w_m(z)\, dz. \tag{5.9.6}$$

By reversing the order of the summation and integration,

$$\sum_{n=1}^{N} I_n \int_{-\ell/2}^{\ell/2} u_n(z') \int_{-\ell/2}^{\ell/2} K(z_m, z') w_m(z)\, dz dz' = \int_{-\ell/2}^{\ell/2} D(z_m) w_m(z)\, dz. \tag{5.9.7}$$

The integration on either side of Eq. (5.9.7) can be carried out numerically or analytically if possible. If we use the point matching method by selecting the weighting function as delta function, then

$$w_m(z) = \delta(z - z_m).$$

Since the integral of any function multiplied by $\delta(z - z_m)$ gives the value of the function at $z = z_m$, Eq. (5.9.7) becomes

$$\sum_{n=1}^{N} I_n \int_{-\ell/2}^{\ell/2} u_n(z') K(z_m, z') dz' = D(z_m), \tag{5.9.8}$$

where $m = 1, 2, \ldots, N+1$. Also, if we choose pulse function as the basis or expansion function,

$$u_n(z) = \begin{cases} 1, & z_n - \Delta z/2 < z < z_n + \Delta z/2 \\ 0, & \text{elsewhere}, \end{cases}$$

and Eq. (5.9.8) yields

$$\sum_{n=1}^{N} I_n \int_{z_n-\Delta z/2}^{z_n+\Delta z/2} K(z_m, z') dz' = D(z_m). \qquad (5.9.9)$$

Substitution of Eq. (5.9.2) into Eq. (5.9.9) gives

$$\sum_{n=1}^{N} I_n \int_{z_n-\Delta z/2}^{z_n+\Delta z/2} \frac{e^{jkR_m}}{4\pi R_m} dz' = -\frac{j}{\eta_o}[A\cos kz_m + \frac{V_T}{2}\sin k|z_m|] \qquad (5.9.10)$$

where $m = 1, 2, \ldots, N+1$ and $R_m = [a^2 + (z_m - z')^2]^{1/2}$. Thus we have a set of $N+1$ simultaneous equations, which can be cast in matrix form as

$$\begin{bmatrix} F_{11} & F_{12} & \cdots & F_{1,N} & \frac{j}{\eta}\cos(kz_1) \\ F_{21} & F_{22} & \cdots & F_{2,N} & \frac{j}{\eta}\cos(kz_2) \\ \vdots & & & & \vdots \\ F_{N+1,1} & F_{N+1,2} & \cdots & F_{N+1,N} & \frac{j}{\eta}\cos(kz_{N+1}) \end{bmatrix} \begin{bmatrix} I_1 \\ I_2 \\ \vdots \\ A \end{bmatrix}$$

$$= \begin{bmatrix} -\frac{j}{2\eta}V_T \sin k|z_1| \\ -\frac{j}{2\eta}V_T \sin k|z_2| \\ \vdots \\ -\frac{j}{2\eta}V_T \sin k|z_{N+1}| \end{bmatrix} \qquad (5.9.11a)$$

or

$$[F][X] = [Q] \qquad (5.9.11b)$$

where

$$F_{mn} = \int_{z_n-\Delta z/2}^{z_n+\Delta z/2} \frac{e^{-jkR_m}}{4\pi R_m} dz'. \qquad (5.9.12)$$

The $N+1$ unknowns are determined by solving Eq. (5.9.11) in the usual manner. To evaluate F_{mn} analytically rather than numerically, let the integrand in Eq. (5.9.12) be separated into its real (RE) and imaginary (IM) parts,

$$\frac{e^{-jkR_m}}{R_m} = \text{RE} + j\,\text{IM}$$
$$= \frac{\cos kR_m}{R_m} - j\frac{\sin kR_m}{R_m}. \qquad (5.9.13)$$

IM as a function of z' is a smooth curve so that

$$\int_{z_n-\Delta z/2}^{z_n+\Delta z/2} \text{IM}(z')dz' = -\int_{z_n-\Delta z/2}^{z_n+\Delta z/2} \frac{\sin k[a^2 + (z_m - z')^2]^{1/2}}{[a^2 + (z_m - z')^2]^{1/2}} dz'$$

$$\simeq -\frac{\Delta z \sin k[a^2 + (z_m - z_n)^2]^{1/2}}{[a^2 + (z_m - z_n)^2]^{1/2}}. \tag{5.9.14}$$

The approximation is accurate as long as $\Delta z < 0.05\lambda$. On the other hand, RE changes rapidly as $z' \to z_m$ due to R_m. Hence

$$\int_{z_n-\Delta z/2}^{z_n+\Delta z/2} \text{RE}(z')dz' = -\int_{z_n-\Delta z/2}^{z_n+\Delta z/2} \frac{\cos k[a^2 + (z_m - z')^2]^{1/2}}{[a^2 + (z_m - z')^2]^{1/2}} dz'$$

$$\simeq \cos k[a^2 + (z_m - z_n)^2]^{1/2} \int_{z_n-\Delta z/2}^{z_n+\Delta z/2} \frac{dz'}{[a^2 + (z_m - z')^2]^{1/2}}$$

$$= \cos k[a^2 + (z_m - z_n)^2]^{1/2} \cdot$$

$$\ln\left[\frac{z_m + \Delta z/2 - z_n + [a^2 + (z_m - z_n + \Delta z/2)^2]^{1/2}}{z_m - \Delta z/2 - z_n + [a^2 + (z_m - z_n - \Delta z/2)^2]^{1/2}}\right]. \tag{5.9.15}$$

Thus

$$F_{mn} \simeq \frac{1}{4\pi} \cos k[a^2 + (z_m - z_n)^2]^{1/2} \times$$

$$\ln\left[\frac{z_m + \Delta z/2 - z_n + [a^2 + (z_m - z_n + \Delta z/2)^2]^{1/2}}{z_m - \Delta z/2 - z_n + [a^2 + (z_m - z_n - \Delta z/2)^2]^{1/2}}\right]$$

$$- \frac{j\Delta z \sin k[a^2 + (z_m - z_n)^2]^{1/2}}{4\pi[a^2 + (z_m - z_n)^2]^{1/2}}. \tag{5.9.16}$$

A typical example of the current distribution obtained for $\ell = \lambda, a = 0.01\lambda$ is shown in Fig. 5.20, where the sinusoidal distribution commonly assumed for wire antennas is also shown for comparison. Notice the remarkable difference between the two near the dipole center.

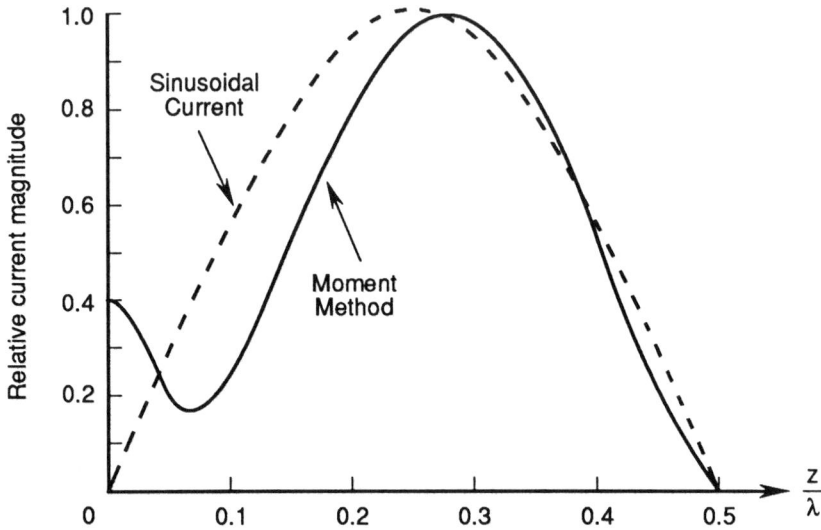

Figure 5.20 Current distribution of straight center-fed dipole.

Example 5.10

Consider a perfectly conducting scatterer or antenna of cylindrical cross section shown in Fig. 5.21. Determine the axial current $I(z)$ on the structure by solving the electric field integral equation (EFIE)

$$\frac{j\eta}{4\pi k}\left(\frac{d^2}{dz^2}+k^2\right)\int_{-h}^{h} I(z')G(z,z')\,dz' = E_z^i(z) \qquad (5.10.1)$$

where

$$G(z,z') = \frac{1}{2\pi}\int_0^{2\pi}\frac{e^{-jkR}}{R}d\phi',$$

$$R = [(z-z')^2 + 4a^2\sin^2\frac{\phi'}{2}]^{1/2},$$

$$\eta = \sqrt{\frac{\mu}{\epsilon}}, \quad \text{and} \quad k = \frac{2\pi}{\lambda}.$$

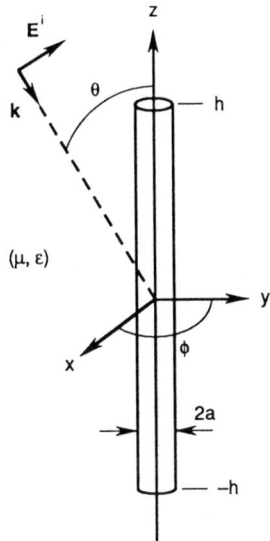

Figure 5.21 Cylindrical scatterer or antenna.

Solution

If the radius $a \ll \lambda$ (the wavelength) and $a \ll 2h$ (the length of the wire), the structure can be regarded as a "thin-wire" antenna or scatterer. As a scatterer, we may consider a plane wave excitation

$$E_z^i(z) = E_o \sin\theta e^{jkz\cos\theta} \qquad (5.10.2a)$$

where θ is the angle of incidence. As an antenna, we may assume a delta-gap generator

$$E_z^i = V\,\delta(z - z_g) \qquad (5.10.2b)$$

where V is the generator voltage and $z = z_g$ is the location of the generator.

In order to apply the method of moments to the given integral equation (5.10.1), we expand the currents in terms of pulse basis functions as

$$I(z) = \sum_{n=1}^{N} I_n u_n(z) \qquad (5.10.3)$$

where

$$u_n(z) = \begin{cases} 1, & z_{n-1/2} < z < z_{n+1/2} \\ 0, & \text{elsewhere.} \end{cases}$$

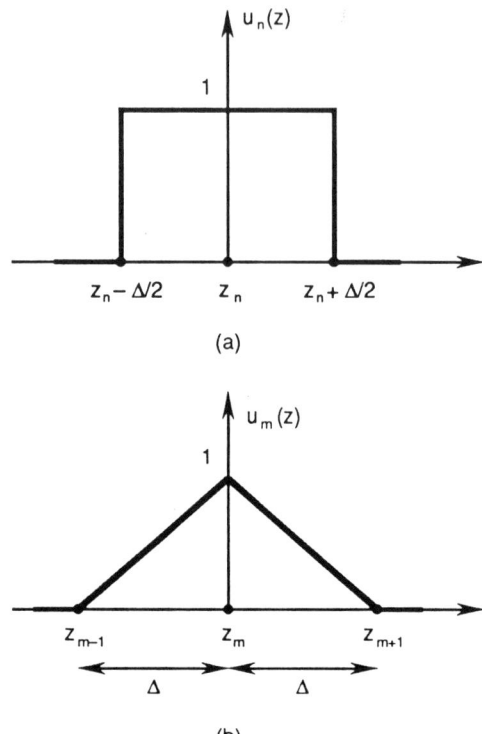

Figure 5.22 For example 5.10: (a) Pulse basis function, (b) triangular weighting function.

Substituting Eq. (5.10.3) into Eq. (5.10.1) and weighting the result with triangular functions

$$w_m(z) = \begin{cases} \dfrac{z - z_{m-1}}{\Delta}, & z_{m-1} < z < z_m \\ -\dfrac{z - z_{m+1}}{\Delta}, & z_{m-1} < z < z_{m+1} \\ 0, & \text{elsewhere,} \end{cases} \quad (5.10.4)$$

where $\Delta = 2h/N$, leads to

$$\sum_{n=1}^{N} Z_{mn} I_n = V_m, \quad m = 1, 2, \ldots, N. \quad (5.10.5)$$

Figure 5.22 illustrates $u_n(z)$ and $w_m(z)$. Equation (5.10.5) can be cast in matrix form as

Moment Methods   365

$$[Z][I] = [V] \tag{5.10.6}$$

where $[I]$ can be solved using any standard method. For the impedance matrix $[Z]$, the elements are given by

$$Z_{mn} = \frac{j\eta}{4\pi k}\frac{2}{\Delta}\left[\frac{1}{2}G_{m-1,n} - (1 - \frac{k^2\Delta^2}{2})G_{m,n} + \frac{1}{2}G_{m+1,n}\right] \tag{5.10.7}$$

where

$$G_{m,n} = \int_{z_n-\Delta/2}^{z_n+\Delta/2} G(z_m, z')\, dz'. \tag{5.10.8}$$

To obtain Eq. (5.10.7), we have used the approximation

$$\int_{z_{m-1}}^{z_{m+1}} w_m(z) f(z)\, dz = \Delta f(z_m).$$

For the plane wave excitation, the elements of the forcing vector $[V]$ are

$$V_m = \Delta E_0 e^{jkz_m \cos\theta}. \tag{5.10.9a}$$

For delta-gap generator,

$$V_m = V\,\delta_{mg} \tag{5.10.9b}$$

where g is the index of the feed zone pulse.

To solve Eq. (5.10.6) requires that we incorporate a method to perform numerically the double integration in Eq. (5.10.8). The kernel $G(z,z')$ exhibits a logarithmic singularity as $|z - z'| \to 0$, and therefore care must be exercised. To circumvent the difficulty, we let

$$G(z,z') = \frac{1}{2\pi}\int_0^{2\pi}\frac{e^{-jkR}}{R}d\phi' = G_o(z,z') + G_1(z,z') \tag{5.10.10}$$

where

$$G_o(z,z') = \frac{1}{2\pi}\int_0^{2\pi}\frac{d\phi'}{R} \tag{5.10.11}$$

and

$$G_1(z,z') = \frac{1}{2\pi}\int_0^{2\pi}\frac{e^{-jkR}-1}{R}d\phi'. \tag{5.10.12}$$

We note that

$$G_o(z,z') \xrightarrow{(\frac{z-z'}{2a})\to 0} -\frac{1}{\pi a}\ln\frac{|z-z'|}{8a}$$

and hence we replace $G_o(z,z')$ by

$$\left[G_o(z,z') + \frac{1}{\pi a}\ln\frac{|z-z'|}{8a}\right] - \frac{1}{\pi a}\ln\frac{|z-z'|}{8a}. \qquad (5.10.13)$$

The term $G_1(z,z')$ is nonsingular, while the singularity of $G_o(z,z')$ can be avoided by using Eq. (5.10.13). Thus the double integral involved in evaluating Z_{mn} is easily done numerically. It is interesting to note that Z_{mn} would remain the same if we had chosen triangular basis function and pulse weighting function [46].

5.7 Applications IV—EM Absorption in the Human Body

The interest in hyperthermia (or electromagnetic heating of deep-seated tumors) and in the assessment of possible health hazards due to EM radiation have prompted the development of analytical and numerical techniques for evaluating EM power deposition in the interior of the human body or a biological system [47]. The overall need is to provide a scientific basis for the establishment of an EM radiation safety standard. Since human experimentation is not possible, irradiation experiments must be performed on animals. Theoretical models are required to interpret, confirm the experiment, develop an extrapolation process, and thereby develop a radiation safety standard for man [48].

The mathematical complexity of the problem has led researchers to investigate simple models of tissue structures such as plane slab, dielectric cylinder homogeneous and layered spheres, and prolate spheroid. A review of these earlier efforts is given in [49,50]. Although spherical models are still being used to study the power deposition characteristics of the head of man and animals, realistic block model composed of cubical cells is being used to simulate the whole body.

The key issue in this bioelectromagnetic effort is how much EM energy is absorbed by a biological body and where is it deposited. This is usually quantified in terms of the specific absorption rate (SAR), which is the mass-normalized rate of energy absorbed by the body. At a specific location, SAR may be defined by

$$\text{SAR} = \frac{\sigma}{\rho}|E|^2 \qquad (5.118)$$

where σ = tissue conductivity, ρ = tissue mass density, E = RMS value of the internal field strength. Thus the localized SAR is directly related to the internal electric field and the major effort involves the determination of the electric field distribution within the biological body. The method of moments has been extensively utilized to calculate localized SARs in block model representation of humans and animals.

As mentioned in Section 5.1, an application of MOM to EM problems usually involved four steps:

(1) deriving the appropriate IE,
(2) transforming the IE into a matrix equation (discretization),
(3) evaluating the matrix elements, and
(4) solving the resulting set of simultaneous equations.

We will apply these steps for calculating the electric field induced in an arbitrary human body or a biological system illuminated by an incident EM wave.

5.7.1 Derivation of Integral Equations

In general, the induced electric field inside a biological body was found to be quite complicated even for the simple case of assuming the plane wave as the incident field. The complexity is due to the irregularity of the body geometry, and the fact that the body is finitely conducting. To handle the complexity, the so-called *tensor integral-equation* (TIE) was developed by Livesay and Chen [51]. Only the essential steps will be provided here; the interested reader is referred to [51–53].

Consider a biological body of an arbitrary shape, with constitutive parameters ϵ, μ, σ illuminated by an incident (or impressed) plane EM wave as shown in Fig. 5.23. The induced current in the body gives rise to a scattered field $\mathbf{E}^s$, which may be accounted for by replacing the body with an equivalent free-space current density $\mathbf{J}_{eq}$ given by

$$\mathbf{J}_{eq}(\mathbf{r}) = \Big(\sigma(\mathbf{r}) + j\omega\big[\epsilon(\mathbf{r}) - \epsilon_o\big]\Big)\mathbf{E}(\mathbf{r}) = \tau(\mathbf{r})\mathbf{E}(\mathbf{r}) \qquad (5.119)$$

where a time factor $e^{j\omega t}$ is assumed. The first term in Eq. (5.119) is the conduction current density, while the second term is the polarization current density. With these equivalent current density $\mathbf{J}_{eq}$, we can obtain the scattered fields $\mathbf{E}^s$ and $\mathbf{H}^s$ by solving Maxwell's equations

$$\nabla \times \mathbf{E}^s = -\mathbf{J}_{eq} - j\omega \mathbf{H}^s \qquad (5.120a)$$

$$\nabla \times \mathbf{H}^s = j\omega \mathbf{E}^s \qquad (5.120b)$$

where $\mathbf{E}^s$, $\mathbf{H}^s$, and $\mathbf{J}_{eq}$ are all in phasor (complex) form. Elimination of $\mathbf{E}^s$ or $\mathbf{H}^s$ in Eq. (5.120) leads to

$$\nabla \times \nabla \times \mathbf{E}^s - k_o^2 \mathbf{E}^s = -j\omega\mu_o \mathbf{J}_{eq} \qquad (5.121a)$$

$$\nabla \times \nabla \times \mathbf{H}^s - k_o^2 \mathbf{H}^s = \nabla \times \mathbf{J}_{eq} \qquad (5.121b)$$

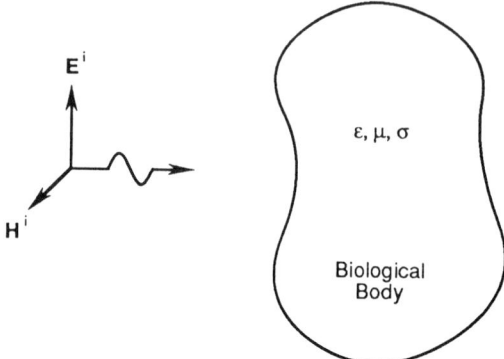

Figure 5.23 A biological body illuminated by a plane EM wave.

where $k_o^2 = \omega^2 \mu_o \epsilon_o$. The solutions to Eq. (5.121) are

$$\mathbf{E}^s = -j\omega \left[1 + \frac{1}{k_o^2}\nabla\nabla\cdot\right]\mathbf{A} \qquad (5.122a)$$

$$\mathbf{H}^s = \frac{1}{\mu_o}\nabla \times \mathbf{A} \qquad (5.122b)$$

where

$$\mathbf{A} = \mu_o \int_v G_o(\mathbf{r},\mathbf{r}')\mathbf{J}_{eq}(\mathbf{r}')\,dv' \qquad (5.123)$$

and

$$G_o(\mathbf{r},\mathbf{r}') = \frac{e^{-jk_o(\mathbf{r}-\mathbf{r}')}}{4\pi|\mathbf{r}-\mathbf{r}'|} \qquad (5.124)$$

is the free-space scalar Green's function. By the operator $\nabla\nabla\cdot$, we mean that $\nabla\nabla\cdot\mathbf{A} = \nabla(\nabla\cdot\mathbf{A})$. It is evident from Eq. (5.122) that $\mathbf{E}^s$ and $\mathbf{H}^s$ depend on $\mathbf{J}_{eq}$. Suppose $\mathbf{J}_{eq}$ is an infinitesimal, elementary source at $\mathbf{r}'$ pointed in the x direction so that

$$\mathbf{J}_{eq} = \delta(\mathbf{r}-\mathbf{r}')\mathbf{a}_x, \qquad (5.125)$$

the corresponding vector potential is obtained from Eq. (5.123) as

$$\mathbf{A} = \mu_o G_o(\mathbf{r},\mathbf{r}')\mathbf{a}_x. \qquad (5.126)$$

If $\mathbf{G}_{ox}(\mathbf{r},\mathbf{r}')$ is the electric field produced by the elementary source, then $\mathbf{G}_{ox}(\mathbf{r},\mathbf{r}')$ must satisfy

$$\nabla\times\nabla\times\mathbf{G}_{ox}(\mathbf{r},\mathbf{r}') - k_o^2\mathbf{G}_{ox}(\mathbf{r},\mathbf{r}') = -j\omega\mu_o\delta(\mathbf{r},\mathbf{r}') \qquad (5.127)$$

with solution

$$\mathbf{G}_{ox}(\mathbf{r},\mathbf{r}') = -j\omega\mu_o\left(1 + \frac{1}{k^2}\nabla\nabla\cdot\right)G_o(\mathbf{r},\mathbf{r}'). \quad (5.128)$$

$\mathbf{G}_{ox}(\mathbf{r},\mathbf{r}')$ is referred to as a free-space vector Green's function with a source pointed in the x direction. We could also have $\mathbf{G}_{oy}(\mathbf{r},\mathbf{r}')$ and $\mathbf{G}_{oz}(\mathbf{r},\mathbf{r}')$ corresponding to infinitesimal, elementary sources pointed in the y and z direction, respectively. We now introduce a dyadic function[2] which can store the three vector Green functions $\mathbf{G}_{ox}(\mathbf{r},\mathbf{r}')$, $\mathbf{G}_{oy}(\mathbf{r},\mathbf{r}')$, and $\mathbf{G}_{oz}(\mathbf{r},\mathbf{r}')$, i.e.,

$$\mathbf{G}_o(\mathbf{r},\mathbf{r}') = \mathbf{G}_{ox}(\mathbf{r},\mathbf{r}')\mathbf{a}_x + \mathbf{G}_{oy}(\mathbf{r},\mathbf{r}')\mathbf{a}_y + \mathbf{G}_{oz}(\mathbf{r},\mathbf{r}')\mathbf{a}_z. \quad (5.129)$$

This is called free-space dyadic Green's function [53]. It is a solution to the dyadic differential equation

$$\nabla \times \nabla \times \mathbf{G}_o(\mathbf{r},\mathbf{r}') - k_o^2\mathbf{G}_o(\mathbf{r},\mathbf{r}') = \widetilde{I}\delta(\mathbf{r}-\mathbf{r}') \quad (5.130)$$

where $\widetilde{I}$ denotes the unit dyad (or idem factor) defined by

$$\widetilde{I} = \mathbf{a}_x\mathbf{a}_x + \mathbf{a}_y\mathbf{a}_y + \mathbf{a}_z\mathbf{a}_z. \quad (5.131)$$

The physical meaning of $\mathbf{G}_o(\mathbf{r},\mathbf{r}')$ is rather obvious. $\mathbf{G}_o(\mathbf{r},\mathbf{r}')$ is the electric field at a field point $\mathbf{r}$ due to an infinitesimal source at $\mathbf{r}'$.

From Eqs. (5.121a) and (5.130), the solution of $\mathbf{E}$ is evidently

$$\mathbf{E}^s(\mathbf{r}) = -j\omega\mu_o \int \mathbf{G}_o(\mathbf{r},\mathbf{r}') \cdot \mathbf{J}_{eq}(\mathbf{r}') \, dv'. \quad (5.132)$$

Since $\mathbf{G}_o(\mathbf{r},\mathbf{r}')$ has a singularity of the order $|\mathbf{r}-\mathbf{r}'|^3$, the integral in Eq. (5.132) diverges if the field point $\mathbf{r}$ is inside the volume v of the body (or source region). This difficulty is overcome by excluding a small volume surrounding the field point first and then letting the small volume approach zero. The process entails defining the principal value (PV) and adding a correction term needed to yield the correct solution. Thus

$$\mathbf{E}^s(\mathbf{r}) = PV \int_v \mathbf{J}_{eq}(\mathbf{r}) \cdot \mathbf{G}(\mathbf{r},\mathbf{r}') \, dv' + \Big[\mathbf{E}^s(\mathbf{r})\Big]_{\text{correction}}. \quad (5.133)$$

---

[2] A dyad is a group of two or a pair of quantities. A dyadic function, denoted by $\widetilde{D}$, is formed by two functions, i.e., $\widetilde{D} = \mathbf{AB}$. See Tai [53] or Balanis [28] for an exposition on dyadic functions.

370 Numerical Techniques in Electromagnetics

The correction term has been evaluated [51,52] to be $-\mathbf{J}_{eq}/j3\omega\epsilon_o$ so that

$$\mathbf{E}^s(\mathbf{r}) = PV \int_v \mathbf{J}_{eq}(\mathbf{r}) \cdot \mathbf{G}(\mathbf{r},\mathbf{r}') \, dv' - \frac{\mathbf{J}_{eq}(\mathbf{r})}{3j\omega\epsilon_o}. \qquad (5.134)$$

The total electric field inside the body is the sum of the incident field $\mathbf{E}^i$ and scattered field $\mathbf{E}^s$, i.e.,

$$\mathbf{E}(\mathbf{r}) = \mathbf{E}^i(\mathbf{r}) + \mathbf{E}^s(\mathbf{r}). \qquad (5.135)$$

Combining Eqs. (5.119), (5.134), and (5.135) gives the desired tensor integral equation for $\mathbf{E}(\mathbf{r})$:

$$\left[1 + \frac{\tau(\mathbf{r})}{3j\omega\epsilon_o}\right]\mathbf{E}(\mathbf{r}) - PV \int_v \tau(\mathbf{r}')\mathbf{E}(\mathbf{r}) \cdot \mathbf{G}(\mathbf{r},\mathbf{r}') \, dv' = \mathbf{E}^i(\mathbf{r}). \qquad (5.136)$$

In Eq. (5.136), $\tau(\mathbf{r}) = \sigma(\mathbf{r}) + j\omega[\epsilon(\mathbf{r}) - \epsilon_o]$ and the incident electric field $\mathbf{E}$ are known quantities; the total electric field $\mathbf{E}$ inside the body is unknown and is to be determined by MOM.

5.7.2 Transformation to Matrix Equation (discretization)

The inner product $\mathbf{E}(\mathbf{r}) \cdot \mathbf{G}(\mathbf{r},\mathbf{r}')$ in Eq. (5.136) may be represented as

$$\mathbf{E}(\mathbf{r}) \cdot \mathbf{G}(\mathbf{r},\mathbf{r}') = \begin{bmatrix} G_{xx}(\mathbf{r},\mathbf{r}') & G_{xy}(\mathbf{r},\mathbf{r}') & G_{xz}(\mathbf{r},\mathbf{r}') \\ G_{yx}(\mathbf{r},\mathbf{r}') & G_{yy}(\mathbf{r},\mathbf{r}') & G_{yz}(\mathbf{r},\mathbf{r}') \\ G_{zx}(\mathbf{r},\mathbf{r}') & G_{zy}(\mathbf{r},\mathbf{r}') & G_{zz}(\mathbf{r},\mathbf{r}') \end{bmatrix} \begin{bmatrix} E_x(\mathbf{r}') \\ E_y(\mathbf{r}') \\ E_z(\mathbf{r}') \end{bmatrix} \qquad (5.137)$$

showing that $\mathbf{G}(\mathbf{r},\mathbf{r}')$ is a symmetric dyad. If we let

$$x_1 = x, \quad x_2 = y, \quad x_3 = z,$$

then $G_{x_p x_q}(\mathbf{r},\mathbf{r}')$ can be written as

$$G_{x_p x_q}(\mathbf{r},\mathbf{r}') = -j\omega\mu_o\left[\delta_{pq} + \frac{1}{k_o^2}\frac{\partial^2}{\partial x_q \partial x_p}\right]G_o(\mathbf{r},\mathbf{r}'), \quad p,q = 1,2,3. \qquad (5.138)$$

We now apply MOM to transform Eq. (5.136) into a matrix equation. We partition the body into N subvolumes or cells, each denoted by v_m ($m = 1, 2, \cdots, N$), and assume that $\mathbf{E}(\mathbf{r})$ and $\tau(\mathbf{r})$ are constant within each cell. If $\mathbf{r}_m$ is the center of the mth cell, requiring that each scalar component of Eq. (5.136) be satisfied at $\mathbf{r}_m$, this leads to

$$\left[1 + \frac{\tau(\mathbf{r})}{3j\omega\epsilon_o}\right]E_{x_p}(\mathbf{r}_m) - \sum_{q=1}^{3}\left[\sum_{q=1}^{3}\tau(\mathbf{r}_n)\,PV\int_{v_m}G_{x_p x_q}(\mathbf{r}_m,\mathbf{r}')\,dv'\right]E_{x_q}(\mathbf{r}_n)$$

$$= E^i_{x_p}(\mathbf{r}_m). \qquad (5.139)$$

If we let $[G_{x_p x_q}]$ be an $N \times N$ matrix with elements

$$G^{mn}_{x_p x_q} = \tau(\mathbf{r}_n)PV\int_{v_n}G_{x_p x_q}(\mathbf{r}_m,\mathbf{r}')\,dv' - \delta_{pq}\delta_{mn}\left[1 + \frac{\tau(\mathbf{r})}{3j\omega\epsilon_o}\right], \qquad (5.140)$$

where $m,n = 1,2,\cdots,N$, $p,q = 1,2,3$, and let $[E_{x_p}]$ and $[E^i_{x_p}]$ be column matrices with elements

$$E_{x_p} = \begin{bmatrix} E_{x_p}(\mathbf{r}_1) \\ \vdots \\ E_{x_p}(\mathbf{r}_N) \end{bmatrix}, \qquad E^i_{x_p} = \begin{bmatrix} E^i_{x_p}(\mathbf{r}_1) \\ \vdots \\ E^i_{x_p}(\mathbf{r}_N) \end{bmatrix}, \qquad (5.141)$$

then from Eqs. (5.136) and (5.139), we obtain $3N$ simultaneous equations for E_x, E_y and E_z at the centers of N cells by the point matching technique. These simultaneous equations can be written in matrix form as

$$\begin{bmatrix} [G_{xx}] & [G_{xy}] & [G_{xz}] \\ --- & --- & --- \\ [G_{yx}] & [G_{yy}] & [G_{yz}] \\ --- & --- & --- \\ [G_{zx}] & [G_{zy}] & [G_{zz}] \end{bmatrix} \begin{bmatrix} [E_x] \\ --- \\ [E_y] \\ --- \\ [E_z] \end{bmatrix} = \begin{bmatrix} [E^i_x] \\ --- \\ [E^i_y] \\ --- \\ [E^i_z] \end{bmatrix} \qquad (5.142a)$$

or simply

$$[G][E] = -[E^i] \qquad (5.142b)$$

where $[G]$ is $3N \times 3N$ matrix and $[E]$ and $[E^i]$ are $3N$ column matrices.

372   Numerical Techniques in Electromagnetics

5.7.3 Evaluation of Matrix Elements

Although the matrix $[E^i]$ in Eq. (5.142) is known, while the matrix $[E]$ is to be determined, the elements of the matrix $[G]$, defined in Eq. (5.140), are yet to be calculated. For the off-diagonal elements of $[G_{x_p x_q}]$, $\mathbf{r}_m$ is not in the nth cell ($\mathbf{r}_m$ is not in v_n) so that $G_{x_p x_q}(\mathbf{r}_m, \mathbf{r}')$ is continuous in v_n and the principal value operation can be dropped. Equation (5.140) becomes

$$G^{mn}_{x_p x_q} = \tau(\mathbf{r}_n) \int_{v_n} G_{x_p x_q}(\mathbf{r}_m, \mathbf{r}') \, dv', \quad m \neq n. \tag{5.143}$$

As a first approximation,

$$G^{mn}_{x_p x_q} = \tau(\mathbf{r}_n) G_{x_p x_q}(\mathbf{r}_m, \mathbf{r}') \Delta v_n, \quad m \neq n \tag{5.144}$$

where Δv_n is the volume of cell v_n. Incorporating Eqs. (5.128) and (5.138) into Eq. (5.144) yields

$$G^{mn}_{x_p x_q} = \frac{-j\omega\mu k_o \Delta v_n \tau(\mathbf{r}_n) \exp(-j\alpha_{mn})}{4\pi\alpha^3_{mn}} \Big[(\alpha_{mn} - 1 - j\alpha_{mn})\delta_{pq}$$
$$+ \cos\theta^{mn}_{x_p} \cos\theta^{mn}_{x_q}(3 - \alpha^2_{mn} + 3j\alpha_{mn})\Big], \quad m \neq n \tag{5.145}$$

where

$$\alpha_{mn} = k_o R_{mn}, \quad R_{mn} = |\mathbf{r}_m - \mathbf{r}_n|,$$

$$\cos\theta^{mn}_{x_p} = \frac{x^m_p - x^n_p}{R_{mn}}, \quad \cos\theta^{mn}_{x_q} = \frac{x^m_q - x^n_q}{R_{mn}},$$

$$\mathbf{r}_m = (x^m_1, x^m_2, x^m_3), \quad \mathbf{r}_n = (x^n_1, x^n_2, x^n_3).$$

The approximation in Eq. (5.145) yields adequate results provided N is large. If greater accuracy is desired, the integral in Eq. (5.143) must be evaluated numerically.

For the diagonal terms ($m = n$), Eq. (5.140) becomes

$$G^{nn}_{x_p x_q} = \tau(\mathbf{r}_n) PV \int_{v_n} G_{x_p x_q}(\mathbf{r}_n, \mathbf{r}') \, dv' - \delta_{pq}\left[1 + \frac{\tau(\mathbf{r})}{3j\omega\epsilon_o}\right]. \tag{5.146}$$

To evaluate this integral, we approximate cell v_n by an equivolumic sphere of radius a_n centered at $\mathbf{r}_n$, i.e.,

$$\Delta v = \frac{4}{3}\pi a^3_n$$

or
$$a_n = \left(\frac{3\Delta v}{4\pi}\right)^{1/3}. \tag{5.147}$$

After a lengthy calculation, we obtain [51]

$$G^{nn}_{x_p x_q} = \delta_{pq}\left[\frac{-2j\omega\mu_o \tau(r_n)}{3k_o^3}\left(\exp(-jk_o a_n)(1+jk_o a_n) - 1\right) - \left(1 + \frac{\tau(r_n)}{3j\omega\epsilon_o}\right)\right]. \tag{5.148}$$

In case the shape of cell v_n differs considerably from that of a sphere, the approximation in Eq. (5.148) may yield poor results. To have a greater accuracy, a small cube, cylinder, or sphere is created around r_n to evaluate the correction term, and the integration through the remainder of v_n is performed numerically.

5.7.4 Solution of the Matrix Equation

Once the elements of matrix $[G]$ are evaluated, we are ready to solve Eq. (5.142), namely,

$$[G][E] = -[E^i]. \tag{5.142}$$

With the known incident electric field represented by $[E^i]$, the total induced electric field represented by $[E]$ can be obtained from Eq. (5.142) by inverting $[G]$ or by employing a Gauss-Jordan elimination method. If matrix inversion is used, the total induced electric field inside the biological body is obtained from

$$[E] = -[G]^{-1}[E^i]. \tag{5.149}$$

Guru and Chen [55] have developed computer programs that yield accurate results on the induced electric field and the absorption power density in various biological bodies irradiated by various EM waves. The validity and accuracy of their numerical results were verified by experiments.

In the following examples, we illustrate the accuracy of the numerical procedure with one simple elementary shape and one advanced shape of biological bodies. The examples are taken from the works of Chen and others [52,56–58].

Example 5.11

Determine the distribution of the energy absorption rate or EM heating induced by plane EM waves of 918 MHz in spherical models of animal brain having radius 3 cm. Assume the $\mathbf{E}^i$ field expressed as

$$\mathbf{E}^i = E_o e^{-jk_o z}\mathbf{a}_x = \mathbf{a}_x E_o(\cos k_o z - j\sin k_o z) \text{ V/m} \tag{5.11.1}$$

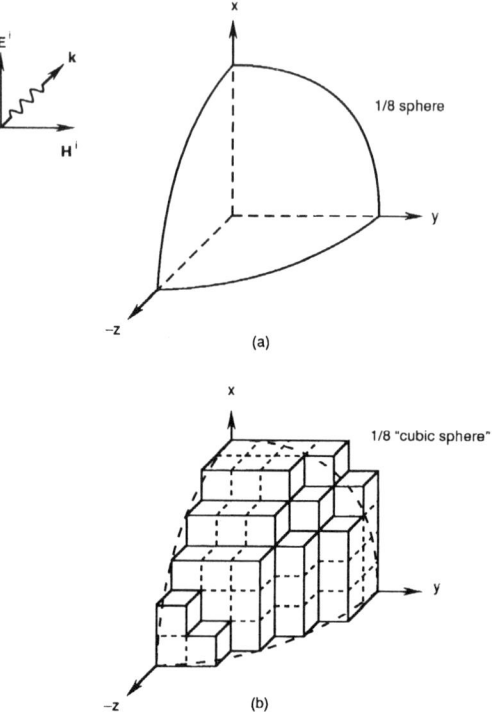

Figure 5.24 For Example 5.11: (a) One eighth of a sphere, (b) a "cubic sphere" constructed from 73 cubic cells.

where $k_o = 2\pi/\lambda = 2\pi f/c$, $E_o = \sqrt{2\eta_o P_i}$, P_i is the incident power in mW/cm² and $\eta_o = 377\Omega$ is the intrinsic impedance of free space. Take $P_i = 1$ mW/cm² ($E_o = 86.83$ V/m), $\epsilon_r = 35$, $\sigma = 0.7$ mhos/m.

Solution

In order to apply MOM, we first approximate the spherical model by a "cubic sphere." Figure 5.24 portrays an example in which one eighth of a sphere is approximated by 40 or 73 cubic cells. The center of each cell, for the case of 40 cells, is determined from Fig. 5.25. $\mathbf{E}^i$ at the center of each cell can be calculated using Eq. (5.11.1). With the computed $\mathbf{E}^i$ and the elements of the matrix $[G_{r_p r_q}]$ calculated using Eqs. (5.145) and (5.148), the induced electric field $\mathbf{E}$ in each cell is computed from Eq. (5.149). Once $\mathbf{E}$ is obtained, the absorption rate of the EM energy is determined using

$$P = \frac{\sigma}{2}|\mathbf{E}|^2. \qquad (5.11.2)$$

Moment Methods 375

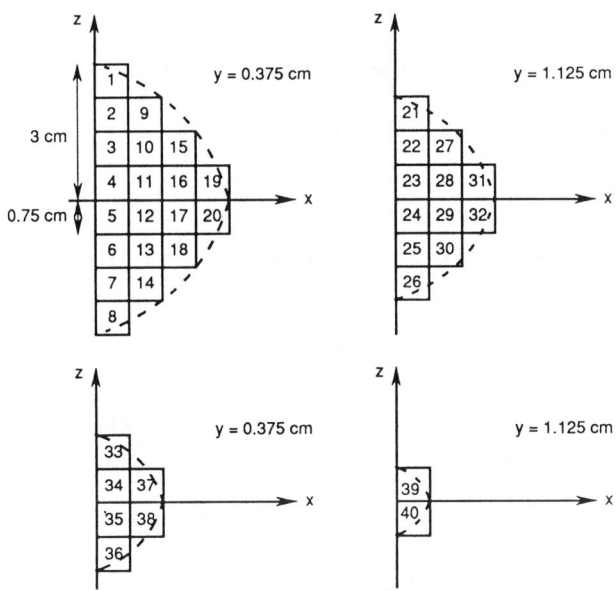

Figure 5.25 Geometry and dimensions of one half of the spherical model of the brain constructed from 40 cells. The cell numbering is used in the program of Fig. 5.26.

The average heating is obtained by averaging P in the brain. The curve showing relative heating as a function of location is obtained by normalizing the distribution of P with respect to the maximum value of P at a certain location in the brain.

The computer program for the computation is shown in Fig. 5.26. It is a modified version of the one developed by Jongakiem [58]. The numerical results are shown in Fig. 5.27(a), where relative heating along the x-, y-, and z-axes in the brain is plotted. The three curves identified by X, Y, and Z correspond with the distributions of the relative heating along x-, y-, and z-axes, respectively. Observe the strong standing wave patterns with peak heating located somewhere near the center of the brain. The average and maximum heating are found to be 0.3202 and 0.885 in mW/cm^3. The exact solution obtained from Mie theory (see Section 2.8) is shown in Fig. 5.27(b). The average and maximum heating from exact solution are 0.295 and 0.814 mW/cm^3, respectively. A comparison of Figs. 5.27(a) and (b) confirms the accuracy of the numerical procedure.

```
0001      C***********************************************************
0002      C   THIS PROGRAM DETERMINED EM POWER ABSORPTION
             OF BIOLOGICAL BODIES
0003      C   PROGRAMMER: JONG A KIEM RAYMOND
0004      C   PROGRAM LIMITATIONS:
0005      C
0006      C     BODIES ARE ASSUMED TO HAVE A PERMEABILITY EQUAL
                TO FREE SPACE
0007      C     MAX NUMBER OF CELLS = 150
0008      C     MAX NUMBER OF CELLS IS SUFFICIENTLY LARGE ELSE G-MATRIX
0009      C     NEED TO BE OBTAINED BY INTEGRATION METHODS
0010      C     THE SHAPE OF A CELL CAN BE APPROXIMATED BY A SPHERE
0011      C
0012      C     DIST = R(M,N) = DISTANCE BETWEEN CENTER OF CELL M AND N
0013      C     E(N) = CONCANTINATED MATRIX OF ALL FIELD INTENSITIES
0014      C     EI(N) = CONCANTINATED MATRIX OF ALL INPUT FIELD INTENSITIES
0015      C
0016      C     NOTE: IN THE CONCANTINATED MATRICES, THE ORDER IN WHICH
                THE DATA IS
0017      C     STORED IS X,Y,Z I.E. ALL X FIRST, THEN ALL Y,
                FINALLY ALL Z
0018      C     COORDINATES IN SEQUENTIAL ORDER
0019      C
0020      C         EPS(N) = EPSO*EPSR(N) = PERMITTIVITY OF CELL N
0021      C         EPSO = FREE SPACE PERMITTIVITY
0022      C         F = FREQUENCY
0023      C         G = COEFFICIENT MATRIX
0024      C         J = COMPLEX VALUE
0025      C         KO = FREE SPACE WAVE NUMBER
0026      C         MXCLL = MAXIMUM NUMBER OF CELLS
0027      C         MXCLL3 = MAXIMUM DIMENSION COEFFICIENT
                    MATRIX = 3*MXCLL
0028      C         PZ(I) = POWER ALONG THE Z DIRECTION
0029      C         NCELL = NUMBER OF CELLS
0030      C         SIGMA(N) = CONDUCTIVITY OF CELL N
0031      C         SAR = SPECIFIC ABSORPTION RATE
0032      C         TAU(N) = ARRAY STORING VALUES OF CELL N = E(N)/Jeq(N)
0033      C         UO = FREE SPACE PERMEABILITY
0034      C         V(N) = VOLUME OF CELL N
0035      C         W = RADIAN FREQUENCY
0036      C         X1 = ROW LOCATION IN G-MATRIX
0037      C         X2 = COLUMN LOCATION IN G-MATRIX
0038      C         X(1,N) = X COMPONENT OF CENTER OF CELL N
0039      C         X(2,N) = Y COMPONENT OF CENTER OF CELL N
0040      C         X(3,N) = Z COMPONENT OF CENTER OF CELL N
0041      C
0042      C ------ INPUT DATA MUST OBEY THE FOLLOWING FORMAT ----------- C
0043      C
0044      C    **** DATA ****              ***** FORMAT *****
0045      C
0046      C    NCELL,F                     FORMAT(1X,I4,E10.4)
0047      C    V(N),X(1,N),X(2,N),X(3,N),
                SIGMA(N),EPS(N)            FORMAT(1X,E10.5,4F10.5,E10.4)
0048      C    EI(I),EI(I+NCELL),EI(I+2*NCELL) FORMAT(1X,6E12.5)
0049
0050      C ----------- SUBROUTINES AND FUNCTIONS --------------------- C
0051
0052      C         DELTA = DELTA FUNCTION
0053      C         MAGSQ(C) = MAGNITUDE SQUARED OF COMPLEX NUMBER
0054      C         R(M,N) = DISTANCE BETWEEN CENTERS OF VOLUME M
                    AND VOLUME N
0055      C
```

```
0056
0057              IMPLICIT COMPLEX (C), INTEGER (K-N)
0058
0059              PARAMETER (MXCLL=600,MXCLL3=1800,MXCL31=1801)
0060              COMPLEX J,TAU,G,E,EI,G1
0061              REAL KO,COSANGLE,MAGSQ,PZ(10)
0062              INTEGER P,Q,X2,X1
0063
0064              DIMENSION G1(MXCLL3,MXCL31),X(3,MXCLL),SIGMA(MXCLL)
0065     C        DIMENSION EPS(MXCLL)
0066              DIMENSION V(MXCLL),E(MXCLL3),EI(MXCLL3),TAU(MXCLL)
0067
0068
0069     C ---------- STATEMENT FUNCTIONS ------------------------ C
0070
0071              R2(M,N) = (X(1,M)-X(1,N))**2+(X(2,M)-X(2,N))
0072                        **2+(X(3,M)-X(3,N))**2
                  R(M,N) = SQRT(R2(M,N))
0073              COSANGLE(P,M,N,DIST)= ( X(P,M) - X(P,N) )/DIST
0074              MAGSQ(C) = (CABS(C))**2
0075
0076              J = (0.,1.)
0077              PI = 4.*ATAN(1.)
0078              EPSO = 1.E-9/(36.*PI)
0079              UO = 4.E-7*PI
0080
0081     C ------------- READ AND GENERATE DATA ---------------- C
0082
0083              READ(5,1) NCELL,F
0084              WRITE(6,1) NCELL,F
0085     1        FORMAT(1X,I4,E10.4)
0086
0087              W = 2.*PI*F
0088              KO = W*SQRT(UO*EPSO)
0089
0090              DO 10 N = 1,NCELL
0091                 READ(5,15)  V(N),X(1,N),X(2,N),X(3,N),SIGMA(N),EPS
0092                 WRITE(6,15) V(N),X(1,N),X(2,N),X(3,N),SIGMA(N),EPS
0093                 TAU(N) = SIGMA(N) + J*W*( EPS - EPSO)
0094     10      CONTINUE
0095     15      FORMAT(1X,E10.5,4F10.5,E10.4)
0096
0097              DO 20 I=1,NCELL
0098                 READ (5,25) EI(I),EI(I+NCELL),EI(I+2*NCELL)
0099                 G1(I,3*NCELL+1) = -EI(I)
0100                 G1(I+NCELL,3*NCELL+1)= -EI(I+NCELL)
0101                 G1(I+2*NCELL,3*NCELL+1)= -EI(I+2*NCELL)
0102     20         WRITE(6,25) EI(I),EI(I+NCELL),EI(I+2*NCELL)
0103     25      FORMAT(1X,6E12.5)
0104
0105     C -------- GENERATION OF COEFFICIENT MATRIX  G -------------- C
0106
0107     C ------ M CONTROLS THE NUMBER OF VOLUMES
0108     C ------ P
0109     C ------ Q IS DUE TO DOTPRODUCT
0110     C ------ N IS ALSO DUE TO DOTPRODUCT
0111
0112              C = -J*W*UO*KO
0113              C1 = -2.*J*W*UO/(3.*KO**2)
0114              C2 = 3.*J*W*EPSO
0115
0116              DO 50 P = 1,3
0117                 DO 50 Q = 1,3
```

```
0118                X1 = (P-1) * NCELL
0119             DO 50 M = 1,NCELL
0120                X1 = X1 + 1
0121                X2 = (Q-1) * NCELL
0122             DO 50 N = 1,NCELL
0123             X2= X2 + 1
0124             IF (M .EQ. N) THEN
0125                AN = ( 3.*V(N)/(4.*PI) )**(1./3.)
0126                C3 = J*KO*AN
0127                C4 = TAU(N)/C2
0128
0129                G1(X1,X2)=DELTA(P,Q)*( C1*TAU(N)*((1.+C3)
0130                                     *EXP(-C3)-1.)-(1.+C4) )
0131             ELSE
0132                DIST = R(M,N)
0133                IF ( DIST .EQ. 0. ) THEN
0134                   PRINT *, 'CELLS HAVE SAME CENTER ',M,N
0135                END IF
0136                AMN = KO*DIST
0137                C3 = ( C*TAU(N)*V(N)*EXP(-J*AMN) )/(4.*PI*(AMN**3))
0138                G1(X1,X2) = C3*( (AMN**2 - 1. - J*AMN)*DELTA(P,Q)
0139           &    +COSANGLE(P,M,N,DIST)*COSANGLE(Q,M,N,DIST)
                                     *(3.-AMN**2+3.*J*AMN))
0140             END IF
0141
0142       50    CONTINUE
0143
0144             CALL GAUSSC(MXCLL3,MXCLL3,MXCL31,3*NCELL,3*NCELL+1,G1,E)
0145
0146             WRITE (6,105)
0147       105   FORMAT (1X,'OUTPUT DATA')
0148             WRITE (6,106)
0149       106   FORMAT(1X,'VOLUME  EX    EY   EZ SAR')
0150
0151             DO 110 I = 1,NCELL
0152             SAR = SQRT(MAGSQ(E(I))+MAGSQ(E(I+NCELL))
                                     +MAGSQ(E(I+2*NCELL)) )
0153             SAR = 0.5 * SIGMA(I) * SAR
0154       110   WRITE(6,115) I,E(I),E(I+NCELL),E(I+2*NCELL),SAR
0155       115   FORMAT(I4,7E10.4)
0156
0157             CLOSE (6)
0158             OPEN (6,FILE='PLHEAD.DAT',STATUS='NEW')
0159       C
0160       C NOTE: THERE ARE 8 CELLS ALONG THE Z-DIRECTION
0161       C     SEE FIG. 5.22 DR. SADIKU'S NOTES
0162       C
0163             PZMAX = -1.E-10
0164             DO 140 I=1,8
0165                PZ(I) = MAGSQ(E(I))+MAGSQ(E(I+NCELL))+MAGSQ(E(I+2*NCELL))
0166                PZ(I) = (SIGMA(I)/2.)*PZ(I)
0167       140   PZMAX = AMAX1(PZ(I),PZMAX)
0168
0169       145   FORMAT(1X,'MAXIMUM POWER FOR CELLS
                                     ALONG THE Z-AXIS IS ',F10.5)
0170       146   FORMAT(1X,'NORMALIZED POWER ALONG THE Z-AXIS')
0171             WRITE(6,145) PZMAX
0172             WRITE(6,146)
0173             DO 150 I=1,8
0174                PZ(I) = PZ(I)/PZMAX
0175       150   WRITE(6,155) I,PZ(I)
0176       155   FORMAT(1X,I4,F10.6)
```

```
0177       C
0178       C AVERAGE POWER CALCULATION
0179       C
0180             SPOW = 0.
0181             SVOL = 0.
0182             PMAX = -1.E-10
0183             DO 160 I=1,NCELL
0184               POWER = MAGSQ(E(I))+MAGSQ(E(I+NCELL))
                                       +MAGSQ(E(I+2*NCELL))
0185               POWER = (SIGMA(I)/2.)*POWER*VOL(I)
0186               PMAX = AMAX1(PMAX,POWER)
0187               IF ( ABS( PMAX - POWER ) .LT. 1.E-6) IVPMAX = I
0188               SVOL = SVOL + V(I)
0189       160     SPOW = POWER + SPOW
0190
0191             WRITE(6,165) SPOW
0192             WRITE(6,170) PMAX/V(IVPMAX)
0193       165   FORMAT(1X,'AVERAGE HEATING IN W/M**3 IS ',E12.6)
0194       170   FORMAT(1X,'MAX HEATING IN W/M**3 IS ',E12.6)
0195             CLOSE (6)
0196             STOP
0197             END

0001       C****************************************************
0002       C    SUBROUTINE FOR GAUSSIAN ELIMINATION
0003       C  THIS SUBROUTINE SOLVES COMPLEX SIMULTANEOUS EQUATIONS
0004       C  USING GAUSS ELIMIANATION WITH PARTIAL PIVOTING
0005       C  THE ORIGINAL PROGRAM WAS WRITTEN TO HANDLE REAL MATRICES
0006       C  AND CAN BE FOUND IN:
0007       C         APPLIED NUMERICAL METHODS FOR DIGITAL COMPUTATION
0008       C         WITH FORTRAN AND CSMP - SECOND EDITION
0009       C         BY M. L. JAMES, G. M. SMITH, AND J. C. WOLFORD
0010       C  A(I,J) = AUGMENTED MATRIX
0011       C  JJ = PIVOT PURPOSES
0012       C  BIG = VALUE THAT MAY BE USED AS PIVOT ELEMENT
0013       C  CMAX = MAXIMUM DIMENSION OF MATRIX CONTAINING RESULTS
0014       C  CMAX1 = MAXIMUM DIMENSION OF AUGMENTED MATRIX
0015       C  M = NUMBER OF COLUMNS IN AUGMENTED MATRIX = N+1
0016       C  N = NUMBER OF SIMULTANEOUS EQUATIONS
0017       C  RMAX = MAXIMUM ROW DIMENSION OF AUGMENTED MATRIX
0018       C  X(N) = RESULT
0019       C
0020
0021             SUBROUTINE GAUSSC(RMAX,CMAX,CMAX1,N,M,A,X)
0022
0023
0024             COMPLEX A,AB,X,TEMP,QUOT,SUM,BIG
0025             INTEGER RMAX,CMAX1,CMAX
0026             DIMENSION A(RMAX,CMAX1),X(CMAX)
0027
0028             L = N-1
0029             DO 12 K = 1,L
0030             JJ = K
0031             BIG = ABS( A(K,K) )
0032             KP1 = K + 1
0033       C
0034       C  SEARCH FOR THE LARGEST POSSIBLE PIVOT ELEMENT
0035       C
0036             DO 7 I = KP1,N
0037             AB = ABS( A(I,K) )
0038             IF ( ABS(BIG - AB) ) 6,7,7
0039       6     BIG = AB
0040             JJ = I
```

```
0041        7     CONTINUE
0042        C
0043        C DECISION ON NECESSITY OF ROW INTERCHANGE
0044        C
0045              IF (JJ-K) 8,10,8
0046        C
0047        C ROW INTERCHANGE
0048        C
0049        8     DO 9 J=K,M
0050              TEMP = A(JJ,J)
0051              A(JJ,J) = A(K,J)
0052        9     A(K,J) = TEMP
0053        C
0054        C CALCULATION OF ELEMENTS OF NEW MATRIX
0055        C
0056        10    DO 11 I = KP1,N
0057              QUOT = A(I,K)/A(K,K)
0058              DO 11 J=KP1,M
0059        11    A(I,J) = A(I,J) - QUOT*A(K,J)
0060              DO 12 I = KP1,N
0061        12    A(I,K) = (0.,0.)
0062        C
0063        C FIRST STEP IN BACKSUBSTITUTION
0064        C
0065              X(N) = A(N,M)/A(N,N)
0066        C
0067        C REMAINDER IN BACKSUBSTITUTION
0068        C
0069              DO 14 NN=1,L
0070              SUM = (0.,0.)
0071              I = N-NN
0072              IP1 = I + 1
0073              DO 13 J = IP1,N
0074        13    SUM = SUM + A(I,J)*X(J)
0075        14    X(I) = (A(I,M)-SUM)/A(I,I)
0076              RETURN
0077              END

0001        C**********************************
0002        C     SUBPROGRAM DELTA
0003              FUNCTION DELTA(M,N)
0004
0005              IF (M .EQ. N) THEN
0006                   DELTA = 1.
0007              ELSE
0008                   DELTA = 0.
0009              END IF
0010              RETURN
0011              END
```

Figure 5.26   Computer program for Example 5.11.

Example 5.12

Having validated the accuracy of the tensor-integral-equation (TIE) method, determine the induced electric field and specific absorption rate (SAR) of EM energy inside a model of typical human body irradiation (Fig. 5.28), by EM wave at 80 MHz with vertical polarization, i.e.,

$$\mathbf{E} = \mathbf{a}_x \quad \text{V/m}$$

at normal incidence. Assume the body at 80 MHz is that of a high-water content tissue with $\epsilon = 80\epsilon_o$, $\mu = \mu_o$, $\sigma = 0.84$ mhos/m.

Solution

The body is partitioned into 108 cubic cells of various sizes ranging from 5 cm^3 to 12 cm^3. To ensure accurate results, the cell size is kept smaller than a quarter-wavelength (of the medium). With the coordinates of the center of each cell figured out from Fig. 5.28, the program in Fig. 5.26 can be used to find induced electric field components E_x, E_y, and E_z at the centers of the cells due to an incident electric field 1 V/m (maximum value) at normal incidence. The SAR is calculated from $(\sigma/2)(E_x^2 + E_y^2 + E_z^2)$. Figures 5.29 to 5.31 illustrate E_x, E_y, and E_z at the center of each cell. Observe that E_y and E_z are much smaller than E_x at this frequency due to the polarization of the incident wave.

As mentioned earlier, the model of the human body shown in Fig. 5.28 is due to Chen et al. [52]. An improved, more realistic model due to Gandhi et al. [59–61] is shown in Fig. 5.32.

5.8 Concluding Remarks

The method of moments is a powerful numerical method capable of applying weighted residual techniques to reduce an integral equation to a matrix equation. The solution of the matrix equation is usually carried out via inversion, elimination, or iterative techniques. Although MOM is commonly applied to open problems such as those involving radiation and scattering, it has been successfully applied to closed problems such as waveguides and cavities.

Needless to say, the issues on MOM covered in this chapter have been carefully selected. We have only attempted to cover the background and reference material upon which the reader can easily build. The interested reader is referred to the literature for more indept treatment of each subject. General concepts on MOM are covered in [10] and [62]. Clear and elementary discussions on IEs and Green's functions may be found in [12,28–30,62–65]. For further study on the theory of the method of moments, one should see [6,9,10,28,40].

The number of problems that can be treated by MOM is endless, and the examples given in this chapter just scratch the surface. The following problems represent typical EM-related application areas:

- electrostatic problems [31,66–69]
- wire antennas and scatterers [34,37,42,44,70,78]

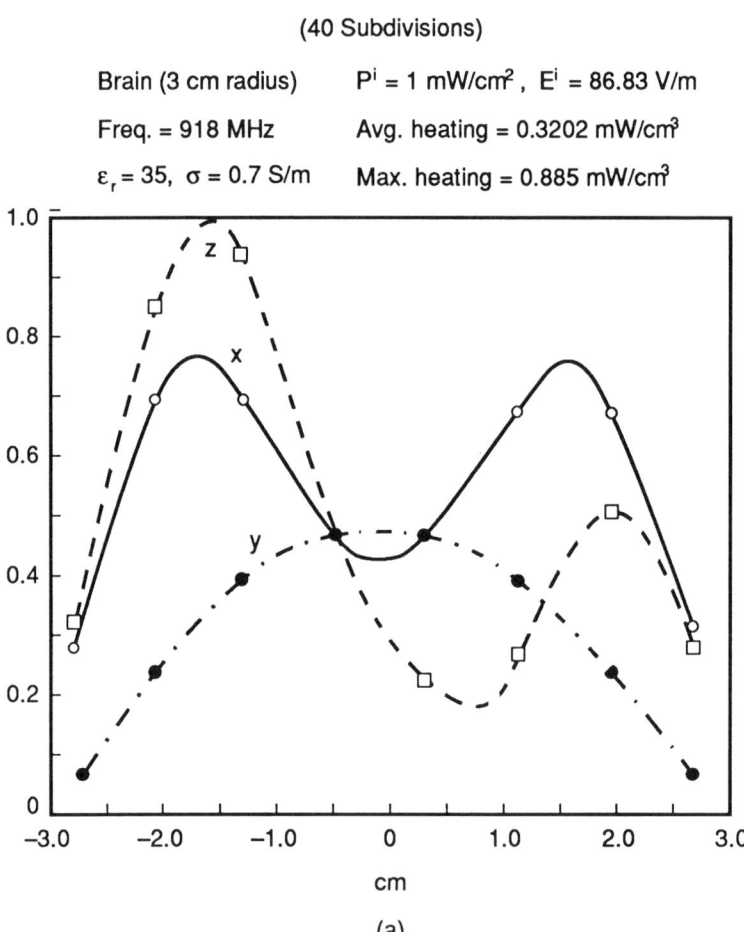

Figure 5.27 Distributions of heating along the x-, y-, and z-axis of a spherical model of an animal brain [57]: (a) MOM solution, (b) exact solution.

Moment Methods 383

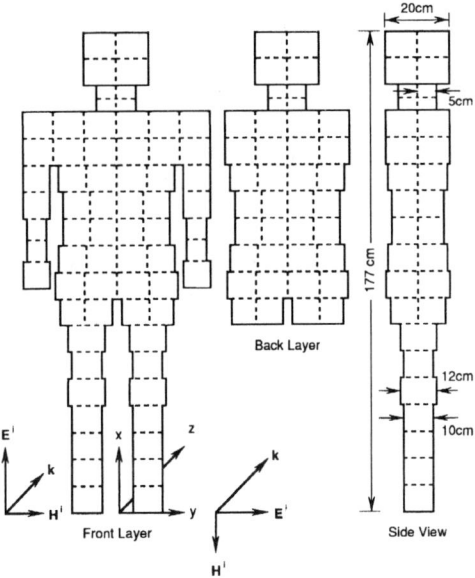

Figure 5.28 Geometry and dimensions of a model of typical human body of height 1.77 m [52].

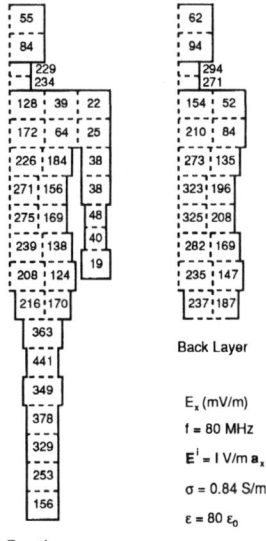

Figure 5.29 Induced E_x (in mV/m) at the center of each cell due to E_x^i of 1 V/m [52].

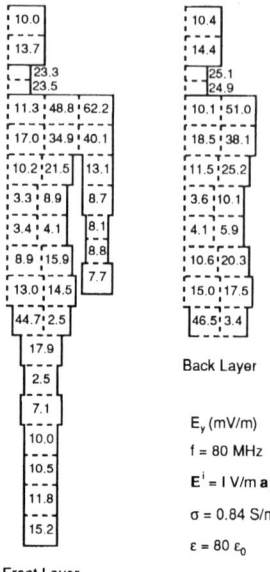

Figure 5.30 Induced E_y (in mV/m) at the center of each cell due to E_x^i of 1 V/m [52].

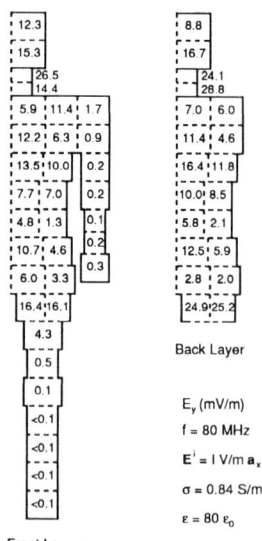

Figure 5.31 Induced E_z (in mV/m) at the center of each cell due to E_x^i of 1 V/m [52].

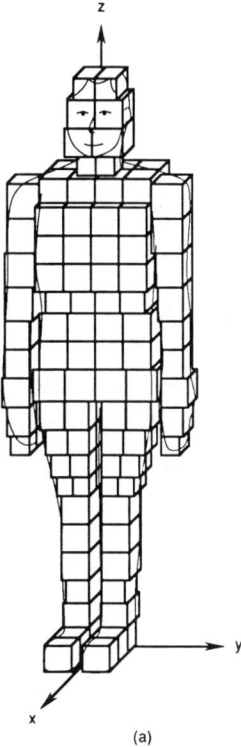

(a)

Figure 5.32 A more realistic block model of the human body [59]: (a) In three dimensions, (b) front and side views.

- scattering and radiation from bodies of revolution [79,80]
- scattering and radiation from bodies of arbitrary shape [38,81,82]
- transmission lines [18-20,23,24,83-86]
- aperture problems [87-89]
- biomagnetic problems [47-52,90-92].

A number of user-oriented computer programs have evolved over the years to solve electromagnetic integral equations by the method of moments. These codes can handle radiation and scattering problems in both the frequency and time domains. Reviews of the codes may be found in [7,38,93]. The most popular of these codes is the Numerical Electromagnetic Code (NEC) developed at the Lawrence Livermore National Laboratory [7,94]. NEC is a frequency domain antenna modeling FORTRAN IV code applying the MOM to IEs for wire and surface structures. Its most notable features are probably that it is user oriented, includes documen-

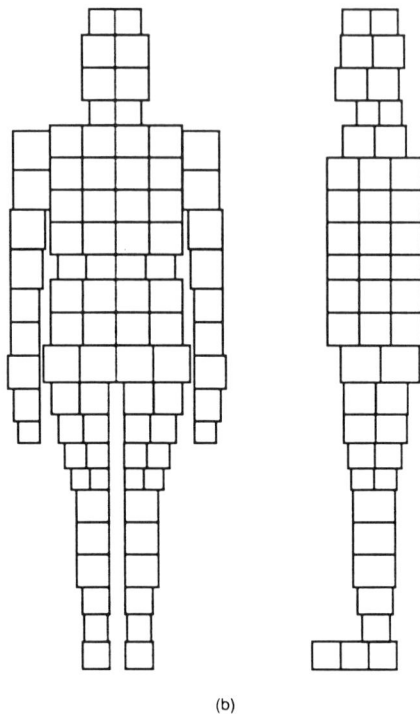

(b)

Figure 5.32 (continued)

tation, and is available; for these reasons, it is being used in public and private institutions. A compact version of NEC is the mini-numerical electromagnetic code (MININEC) [95], which is intended to be used in personal computers.

It is important that we recognize the fact that MOM is limited in application to radiation and scattering from bodies that are electrically large. The size of the scatterer or radiator must be of the order λ^3. This is because the cost of storing, inverting, and computing matrix elements becomes prohibitively large. At high frequencies, asymptotic techniques such as the geometrical theory of diffraction (GTD) are usually employed to derive approximate but accurate solutions [46,96,97].

References

[1] R. F. Harrington, "The method of moments in electromagnetics," *J. Elect. Waves Appl.*, vol. 1, no. 3, 1987, pp. 181–200.

[2] R. F. Harrington and T. K. Sarkar, "Boundary elements and the method of moments" in *Boundary Elements* (eds. C. A. Brebbia et al.). Southampton: CML Publ., 1983, pp. 31–40.

[3] V. H. Rumsey, "The reaction concept in electromagnetic theory," *Phys. Rev.*, Series 2, vol. 94, June 1954, pp. 1483–1491.

[4] L. V. Kantorovich and V. I. Krylov, *Approximate Methods of Higher Analysis* (translated from Russian by C. D. Benster). New York: John Wiley, 1964.

[5] Y. U. Vorobev, *Method of Moments in Applied Mathematics* (translated from Russian by Seckler). New York: Gordon & Breach, 1965.

[6] R. F. Harrington, *Field Computation by Moment Methods*. Malabar, FL: Krieger, 1968.

[7] B. J. Strait, *Applications of the Method of Moments to Electromagnetics*. St. Cloud, FL: SCEEE Press, 1980.

[8] J. H. Richmond, "Digital computer solutions of the rigorous equations for scattering problems," *Proc. IEEE*, vol. 53, Aug. 1965, pp. 796–804.

[9] R. F. Harrington, "Matrix methods for field problems," *Proc. IEEE*, vol. 55, no. 2, Feb. 1967, pp. 136–149.

[10] M. M. Ney, "Method of moments as applied to electromagnetics problems," *IEEE Trans. Micro. Theo. Tech.*, vol. MTT-33, no. 10, Oct. 1985, pp. 972–980.

[11] A. T. Adams, "An introduction to the method of moments," Syracuse Univ., *Report RADC* TR-73-217, vol. 1, Aug. 1974.

[12] P. M. Morse and H. Feshbach, *Methods of Theoretical Physics*. New York: McGraw-Hill, Part I, Chap. 7, 1953, pp. 791–895.

[13] T. Myint-U, *Partial Differential Equations of Mathematical Physics*. New York: North-Holland, 1980, 2nd ed., Chap. 10, pp. 282–305.

[14] R. F. Harrington, *Time-harmonic Electromagnetic Fields*. New York: McGraw-Hill, 1961, p. 232.

[15] I. V. Bewley, *Two-dimensional Fields in Electrical Engineering*. New York: Dover Publ., 1963, pp. 151–166.

[16] K. C. Gupta, et al., *Computer-aided Design of Microwave Circuits*. Dedham: Artech House, 1981, pp. 237–261.

[17] T. Itoh (ed.), *Numerical Techniques for Microwaves and Millimeter-wave Passive Structures.* New York: John Wiley & Sons, 1989, pp. 221–250.

[18] D. W. Kammler, "Calculation of characteristic admittance and coupling coefficients for strip transmission lines," *IEEE Trans. Micro. Tech.*, vol. MTT-16, no. 11, Nov. 1968, pp. 925–937.

[19] Y. M. Hill, et al., "A general method for obtaining impedance and coupling characteristics of practical microstrip and triplate transmission line configurations," *IBM J. Res. Dev.*, vol. 13, May 1969, pp. 314–322.

[20] W. T. Weeks, "Calculation of coefficients of capacitance of multiconductor transmission lines in the presence of a dielectric interface," *IEEE Trans. Micro. Theo. Tech.*, vol. MTT-18, no. 1, Jan. 1970, pp. 35–43.

[21] R. Chadha and K. C. Gupta, "Green's functions for triangular segments in planar microwave circuits," *IEEE Trans. Micro. Theo. Tech.*, vol. MTT-28, no. 10, Oct. 1980, pp. 1139–1143.

[22] R. Terras and R. Swanson, "Image methods for constructing Green's functions and eigenfunctions for domains with plane boundaries," *J. Math. Phys.*, vol. 21, no. 8, Aug. 1980, pp. 2140–2153.

[23] E. Yamashita and K. Atsuki, "Stripline with rectangular outer conductor and three dielectric layers," *IEEE Trans. Micro. Theo. Tech.*, vol. MTT-18, no. 5, May 1970, pp. 238–244.

[24] R. Crampagne, et al., "A simple method for determining the Green's function for a large class of MIC lines having multilayered dielectric structures," *IEEE Trans. Micro. Theo. Tech.*, vol. MTT-26, no. 2, Feb. 1978, pp. 82–87.

[25] R. Chadha and K. C. Gupta, "Green's functions for circular sectors, annular rings, and annular sectors in planar microwave circuits," *IEEE Trans. Micro. Theo. Tech.*, vol. MTT-29, no. 1, Jan. 1981, pp. 68–71.

[26] P. H. Pathak, "On the eigenfunction expansion of electromagnetic dydadic Green's functions," *IEEE Trans. Ant. Prog.*, vol. AP-31, no. 6, Nov. 1983.

[27] R. F. Harrington and J. R. Mautz, "Green's functions for surfaces of revolution," *Radio Sci.*, vol. 7, no. 5, May 1972, pp. 603–611.

[28] C. A. Balanis, *Advanced Engineering Electromagnetics.* New York: John Wiley, 1989, pp. 670–742, 851–916.

[29] E. Butkov, *Mathematical Physics*. New York: Addison-Wesley, 1968, pp. 503–552.

[30] J. D. Jackson, *Classical Electrodynamics*, 2nd ed. New York: John Wiley, 1975, pp. 119–135.

[31] L. L. Tsai and C. E. Smith, "Moment Methods in electromagnetics undergraduates," *IEEE Trans. Educ.*, vol. E-21, no. 1, Feb. 1978, pp. 14–22.

[32] P. P. Silvester and R. L. Ferrari, *Finite Elements for Electrical Engineers*. Cambridge, UK: Cambridge University Press, 1983, pp. 103–105.

[33] H. A. Wheeler, "Transmission-line properties of parallel strips separated by a dielectric sheet," *IEEE Trans. Micro. Theo. Tech.*, vol. MTT-13, Mar. 1965, pp. 172–185.

[34] J. H. Richmond, "Scattering by an arbitrary array of parallel wires," *IEEE Trans. Micro. Theo. Tech.*, vol. MTT-13, no. 4, July 1965, pp. 408–412.

[35] B. D. Popovic, et al., *Analysis and Synthesis of Wire Antennas*. Chichester, UK: Research Studies Press, 1982, pp. 3–21.

[36] E. Hallen, "Theoretical investigations into the transmitting and receiving qualities of antennae," *Nova Acta Regiae Soc. Sci. Upsaliensis*, Ser. IV, no. 11, 1938, pp. 1–44.

[37] K. K. Mei, "On the integral equations of thin wire antennas," *IEEE Trans. Ant. Prog.*, vol. AP-13, 1965, pp. 374–378.

[38] J. Moore and P. Pizer (eds.), *Moment Methods in Electromagnetics: Techniques and Applications*. Letchworth, UK: Research Studies Press, 1984.

[39] H. C. Pocklington, "Electrical oscillations in wire," *Cambridge Phil. Soc. Proc.*, vol. 9, 1897, pp. 324–332.

[40] R. Mittra (ed.), *Computer Techniques for Electromagnetics*. Oxford: Pergamon Press, 1973, pp. 7–95.

[41] C. A. Balanis, *Antenna Theory: Analysis and Design*. New York: Harper & Row, 1982, pp. 283–321.

[42] C. M. Butler and D. R. Wilton, "Analysis of various numerical techniques applied to thin-wire scatterers," *IEEE Trans. Ant. Prog.*, vol. AP-23, no. 4, July 1975, pp. 524–540. Also in [46, pp. 46–52].

[43] T. K. Sarkar, "A note on the choice of weighting functions in the method of moments," *IEEE Trans. Ant. Prog.*, vol. AP-33, no. 4, April 1985, pp. 436–441.

[44] F. M. Landstorfer and R. F. Sacher, *Optimisation of Wire Antenna.* Letchworth, UK: Research Studies Press, 1985, pp. 18–33.

[45] K. A. Michalski and C. M. Butler, "An efficient technique for solving the wire integral equation with non-uniform sampling," *Conf. Proc. IEEE Southeastcon.*, April 1983, pp. 507–510.

[46] R. F. Harrington, et al. (eds.), *Lectures on Computational Methods in Electromagnetics.* St. Cloud, FL: SCEEE Press, 1981.

[47] R. Kastner and R. Mittra, "A new stacked two-dimensional spectral iterative technique (SIT) for analyzing microwave power deposition in biological media," *IEEE Trans. Micro. Theo. Tech.*, vol. MTT-31, no. 1, Nov. 1983, pp. 898–904.

[48] P. W. Barber, "Electromagnetic power deposition in prolate spheroidal models of man and animals at resonance," *IEEE Trans. Biomed. Engr.*, vol. BME-24, no. 6, Nov. 1977, pp. 513–521.

[49] J. M. Osepchuk (ed.), *Biological Effects of Electromagnetic Radiation.* New York: IEEE Press, 1983.

[50] R. J. Spiegel, "A review of numerical models for predicting the energy deposition and resultant thermal response of humans exposed to electromagnetic fields," *IEEE Trans. Micro. Theo. Tech.*, vol. MTT-32, no. 8, Aug. 1984, pp. 730–746.

[51] D. E. Livesay and K. M. Chen, "Electromagnetic fields induced inside arbitrary shaped biological bodies," *IEEE Trans. Micro. Theo. Tech.*, vol. MTT-22, no. 12, Dec. 1974, pp. 1273–1280.

[52] J. A. Kong (ed.), *Research Topics in Electromagnetic Theory.* New York: John Wiley, 1981, pp. 290–355.

[53] C. T. Tai, *Dyadic Green's Functions in Electromagnetic Theory.* Scranton, PA: Intext Educational Pub., 1971, pp. 46–54.

[54] J. Van Bladel, "Some remarks on Green's dyadic infinite space," *IRE Trans. Ant. Prog.*, vol. AP-9, Nov. 1961, pp. 563–566.

[55] B. S. Guru and K. M. Chen, "A computer program for calculating the induced EM field inside an irradiated body," Dept. of Electrical Engineering and System Science, Michigan State Univ., East Lansing, MI, 48824, 1976.

[56] K. M. Chen and B. S. Guru, "Internal EM field and absorbed power density in human torsos induced by 1-500 MHz EM waves," *IEEE Micro. Theo. Tech.*, vol. MTT-25, no. 9., Sept. 1977, pp. 746–756.

[57] R. Rukspollmuang and K. M. Chen, "Heating of spherical versus realistic models of human and infrahuman heads by electromagnetic waves," *Radio Sci.*, vol. 14, no. 6S, Nov.-Dec., 1979, pp. 51–62.

[58] R. Jongakiem, "Electromagnetic absorption in biological bodies," *M. S. Thesis*, Dept. of Electrical and Computer Engr., Florida Atlantic Univ., Boca Raton, Aug. 1988.

[59] O. P. Gandhi, "Electromagnetic absorption in an inhomogeneous model of man for realistic exposure conditions," *Bioelectromagnetics*, vol. 3, 1982, pp. 81–90.

[60] O. P. Gandhi, et al., "Part-body and multibody effects on absorption of radio-frequency electromagnetic energy by animals and by models of man," *Radio Sci.*, vol. 14, no. 6S, Nov.-Dec., 1979, pp. 15–21.

[61] M. J. Hagmann, O. P. Gandhi, and C. H. Durney, "Numerical calculation of electromagnetic energy deposition for a realistic model of man," *IEEE Trans. Micro. Theo. Tech.*, vol. MTT-27, no. 9, Sept. 1979, pp. 804–809.

[62] J. J. Wang, "Generalized moment methods in electromagnetics," *IEE Proc.*, vol. 137, Pt. H, no. 2, April 1990, pp. 127–132.

[63] G. Goertzel and N. Tralli, *Some Mathematical Methods of Physics.* New York: McGraw-Hill, 1960.

[64] W. V. Lovitt, *Linear Integral Equations.* New York: Dover Publ., 1950.

[65] C. D. Green, *Integral Equation Methods.* New York: Barnes & Nobles, 1969.

[66] R. F. Harrington, et al., "Computation of Laplacian potentials by an equivalent source method," *Proc. IEE*, vol. 116, no. 10, Oct. 1969, pp. 1715–1720.

[67] J. R. Mautz and R. F. Harrington, "Computation of rotationally symmetric Laplacian," *Proc. IEE*, vol. 117, no. 4, April 1970, pp. 850–852.

[68] S. M. Rao, et al., "A simple numerical solution procedure for static problems involving arbitrary-shaped surfaces," *IEEE Trans. Ant. Prog.*, vol. AP-27, no. 5, Sept. 1979, pp. 604–608.

[69] K. Adamiak, "Application of integral equations for solving inverse problems in stationary electromagnetic fields," *Int. J. Num. Meth. Engr.*, vol. 21, 1985, pp. 1447–1485.

[70] A. W. Glisson, "An integral equation for electromagnetic scattering from homogeneous dielectric bodies," *IEEE Trans. Ant. Prog.*, vol. AP-33, no. 2, Sept. 1984, pp. 172–175.

[71] E. Max, "Integral equation for scattering by a dielectric," *IEEE Trans. Ant. Prog.*, vol. AP-32, no. 2, Feb. 1984, pp. 166–172.

[72] A. T. Adams, et al., "Near fields of wire antennas by matrix methods," *IEEE Trans. Ant. Prog.*, vol. AP-21, no. 5, Sept. 1973, pp. 602–610.

[73] R. F. Harrington and J. R. Mautz, "Electromagnetic behavior of circular wire loops with arbitrary excitation and loading," *Proc. IEE*, vol. 115, Jan. 1969, pp. 68–77.

[74] S. A. Adekola and O. U. Okereke, "Analysis of a circular loop antenna using moment methods," *Int. J. Elect.*, vol. 66, no. 5, 1989, pp. 821–834.

[75] E. K. Miller, et al., "Computer evaluation of large low-frequency antennas," *IEEE Trans. Ant. Prog.*, vol. AP-21, no. 3, May 1973, pp. 386–389.

[76] K. S. H. Lee, et al., "Limitations of wire-grid modeling of a closed surface," *IEEE Trans. Elect. Comp.*, vol. EMC-18, no. 3, Aug. 1976, pp. 123–129.

[77] E. H. Newman and D. M. Pozar, "Considerations for efficient wire/surface modeling," *IEEE Trans. Ant. Prog.*, vol. AP-28, no. 1, Jan. 1980, pp. 121–125.

[78] J. Perini and D. J. Buchanan, "Assessment of MOM techniques for shipboard applications," *IEEE Trans. Elect. Comp.*, vol. EMC-24, no. 1, Feb. 1982, pp. 32–39.

[79] J. R. Mautz and R. F. Harrington, "Radiation and scattering from bodies of revolution," *Appl. Sci. Res.*, vol. 20, June 1969, pp. 405–435.

[80] A. W. Glisson and C. M. Butler, "Analysis of a wire antenna in the presence of a body of revolution," *IEEE Trans. Ant. Prog.*, vol. AP-28, Sept. 1980, pp. 604–609.

[81] J. H. Richmond, "A wire-grid model for scattering by conducting bodies," *IEEE Trans. Ant. Prog.*, vol. AP-14, Nov. 1966, pp. 782–786.

[82] S. M. Rao, et al., "Electromagnetic scattering by surfaces of arbitrary shape," *IEEE Trans. Ant. Prog.*, vol. AP-30, May 1966, pp. 409–418.

[83] A. Farrar and A. T. Adams, "Matrix methods for microstrip three-dimensional problems," *IEEE Trans. Micro. Theo. Tech.*, vol. MTT-20, no. 8, Aug. 1972, pp. 497–505.

[84] A. Farrar and A. T. Adams, "Computation of propagation constants for the fundamental and higher-order modes in microstrip," *IEEE*

Trans. Micro. Theo. Tech., vol. MTT-24, no. 7, July 1972, pp. 456–460.

[85] A. Farrar and A. T. Adams, "Characteristic impedance of microstrip by the method of moments," *IEEE Trans. Micro. Theo. Tech.*, vol. MTT-18, no. 1, Jan 1970, pp. 68, 69.

[86] A. Farrar and A. T. Adams, "Computation of lumped microstrip capacitance by matrix methods—rectangular sections and end effects," *IEEE Trans. Micro. Theo. Tech.*, vol. MTT-19, no. 5, May 1971, pp. 495, 496.

[87] R. F. Harrington and J. R. Mautz, "A generalized network formulation for aperture problems," *IEEE Trans. Ant. Prog.*, vol. AP-24, Nov. 1976, pp. 870–873.

[88] R. H. Harrington and D. T. Auckland, "Electromagnetic transmission through narrow slots in thick conducting screens," *IEEE Trans. Ant. Prog.*, vol. AP-28, Sept. 1980, pp. 616–622.

[89] C. M. Butler and K. R. Umashankar, "Electromagnetic excitation of a scatterer coupled to an aperture in a conducting screen," *Proc. IEE*, vol. 127, Pt. H, June 1980, pp. 161–169.

[90] Special issue on electromagnetic wave interactions with biological systems, *IEEE Trans. Micro. Theo. Tech.*, vol. MTT-32, no. 8, Aug. 1984.

[91] Special issue on effects of electromagnetic radiation, *IEEE Engr. Med. Biol. Mag.*, March 1987.

[92] Helsinki symposium on biological effects of electromagnetic radiation, *Radio Sci.*, vol. 17, no. 5S, Sept.-Oct. 1982.

[93] R. M. Bevensee, "Computer codes for EMP interaction and coupling," *IEEE Trans. Ant. Prog.*, vol. AP-26, no. 1, Jan. 1978, pp. 156–165.

[94] G. J. Burke and A. J. Poggio, *Numerical Electromagnetic Code (NEC)—Method of Moments*. Lawrence Livermore National Lab., Jan. 1981.

[95] J. W. Rockway, et al., *The MININEC System: Microcomputer Analysis of Wire Antennas*. Norwood, MA: Artech House, 1988.

[96] R. Mittra (ed.), *Numerical and Asymptotic Techniques in Electromagnetics*. New York: Springer-Verlag, 1975.

[97] G. L. James, *Geometrical Theory of Diffraction for Electromagnetic Waves*, 3rd ed. London: Peregrinus, 1986.

Problems

5.1 Classify the following integral equations and show that they have the stated solutions:

(a) $\Phi(x) = \dfrac{5x}{b} + \dfrac{1}{2}\displaystyle\int_0^1 xt\Phi(t)\,dt$ [solution $\Phi(x) = x$],

(b) $\Phi(x) = \cos x - \sin x + 2\displaystyle\int_0^x \sin(x-t)\Phi(t)\,dt$ [solution $\Phi(x) = e^{-x}$],

(c) $\Phi(x) = -\cosh x + \lambda\int_{-1}^1 \cosh(x+t)\Phi(t)\,dt$ [solution $\Phi(x) = \dfrac{\cosh x}{\frac{\lambda}{2}\sinh 2 + \lambda - 1}$].

5.2 Solve the following Volterra integral equations:

(a) $\Phi(x) = 5 + 2\int_0^x t\Phi(t)\,dt$,

(b) $\Phi(x) = x + \int_0^x (t-x)\Phi(t)\,dt$.

5.3 Find the integral equation corresponding to each of the following differential equations:

(a) $y'' = -y$, $y(0) = 0$, $y'(1) = 1$,

(b) $y'' + y = \cos x$, $y(0) = 0$, $y'(0) = 1$.

5.4 Construct the Green's function for the differential equation

$$G_{xx} + k^2 G = -\delta(x - x'), \quad 0 < x < a$$

subject to $G(0) = G(a) = 0$.

5.5 Show that

$$G(x, z; x', z') = \dfrac{j}{a}\sum_{n=1}^\infty \dfrac{\sin(n\pi x/a)\sin(n\pi x'/a)}{k_n} e^{jk_n(z-z')},$$

where $k_n^2 = k^2 - (n\pi/a)^2$ is the Green's function for Helmholtz's equation.

5.6 Find the Green's function in the semi-infinite plane $0 < x < \infty$, $0 < y < \infty$ for the problem

$$\nabla^2 \Phi = f \quad \text{in} \quad R$$
$$\Phi = g \quad \text{on} \quad x = 0.$$

5.7 Find the Green's function satisfying

$$G_{xx} + G_{yy} + 2G_x = \delta(x - x')\delta(y - y'), \quad 0 < x < a, 0 < y < b$$

and

$$G(0, y) = G(a, y) = G(x, 0) = G(x, b) = 0.$$

5.8 (a) Verify by Fourier expansion that Eqs. (5.5.4) and (5.5.5) in Example 5.5 are equivalent.

(b) Show that another form of expressing Eq. (5.5.4) is

$$G(x, y; x', y') = \begin{cases} -\dfrac{2}{\pi} \sum_{m=1}^{\infty} \sinh \dfrac{m\pi(b - y')}{a} \sin \dfrac{m\pi y}{a} \dfrac{m\pi x}{a} \dfrac{m\pi x'}{a}, & y < y' \\ -\dfrac{2}{\pi} \sum_{m=1}^{\infty} \sinh \dfrac{m\pi y'}{a} \sin \dfrac{m\pi(b - y)}{a} \dfrac{m\pi x}{a} \dfrac{m\pi x'}{a}, & y > y' \end{cases}$$

5.9 The two-dimensional delta function expressed in cylindrical coordinates reads

$$\delta(\boldsymbol{\rho} - \boldsymbol{\rho}') = \frac{1}{\rho}\delta(\rho - \rho')\delta(\phi - \phi').$$

Obtain the Green's function for the potential problem

$$\nabla^2 G = \frac{1}{\rho}\delta(\rho - \rho')\delta(\phi - \phi')$$

with the region defined in Fig. 5.33. Assume homogeneous Dirichlet boundary conditions.

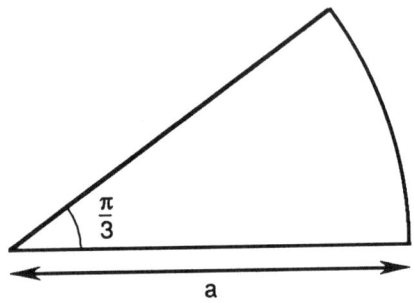

Figure 5.33 For Problem 5.9.

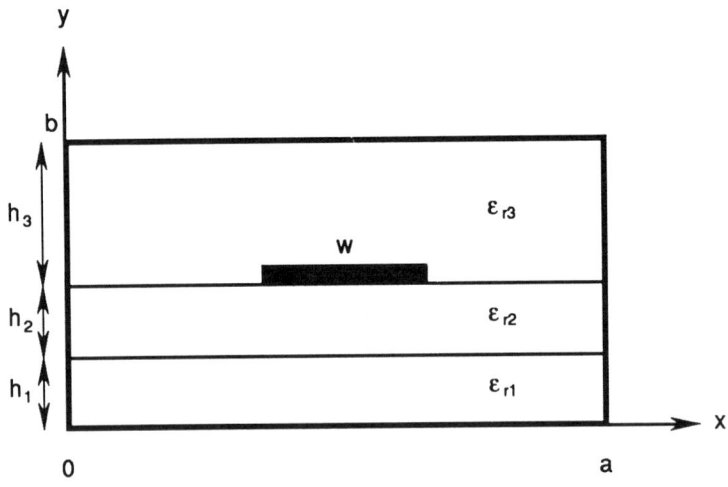

Figure 5.34 For Problem 5.10.

5.10 Consider the transmission line with cross section as shown in Fig. 5.34. In a TEM wave approximation, the potential distribution satisfies Poisson's equation

$$\nabla^2 V = -\frac{\rho_s}{\epsilon}$$

subject to the following continuity and boundary conditions:

$$\frac{\partial}{\partial x}V(x, h_1 - 0) = \frac{\partial}{\partial x}V(x, h_1 + 0)$$
$$\frac{\partial}{\partial x}V(x, h_1 + h_2 - 0) = \frac{\partial}{\partial x}V(x, h_1 + h_2 + 0)$$
$$\epsilon_1 \frac{\partial}{\partial y}V(x, h_1 - 0) = \epsilon_2 \frac{\partial}{\partial y}V(x, h_1 + 0)$$
$$\epsilon_2 \frac{\partial}{\partial y}V(x, h_1 + h_2 - 0) = \epsilon_3 \frac{\partial}{\partial y}V(x, h_1 + h_2 + 0) - \rho_s(x, h_1 + h_2)$$
$$V(0, y) = V(a, y) = V(x, 0) = V(x, b) = 0.$$

Using series expansion method, evaluate the Green's function at $y = h_1 + h_2$, i.e., $G(x, y; x', h_1 + h_2)$.

5.11 Show that the free-space Green's function for $L = \nabla^2 + k^2$ in two-dimensional space is $-\frac{j}{4}H_0^{(1)}(k\rho)$.

5.12 The spherical Green's function $h_0^{(2)}(|\mathbf{r} - \mathbf{r}'|)$ can be expanded in terms of spherical Bessel functions and Legendre polynomials. Show that

$$h_0^{(2)}(|\mathbf{r} - \mathbf{r}'|) =$$

$$\frac{j\exp(-j|\mathbf{r} - \mathbf{r}'|)}{(|\mathbf{r} - \mathbf{r}'|)} = \begin{cases} \sum_{n=0}^{\infty}(2n+1)h_n^{(2)}(r')j_n(r)P_n(\cos\alpha), & r < r' \\ \sum_{n=0}^{\infty}(2n+1)h_n^{(2)}(r)j_n(r')P_n(\cos\alpha), & r > r' \end{cases}$$

where $\cos\alpha = \cos\theta\cos\theta' + \sin\theta\sin\theta'\cos(\phi - \phi')$. From this, derive the plane wave expansion

$$e^{-j\mathbf{k}\cdot\mathbf{r}} = \sum_{n=0}^{\infty}(-j)^n(2n+1)j_n(kr)P_n(\cos\alpha).$$

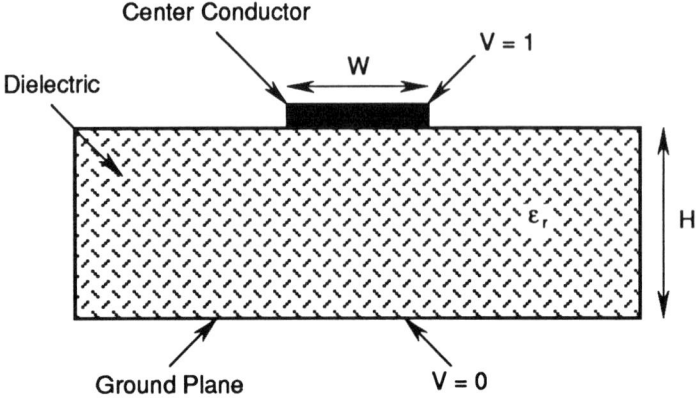

Figure 5.35 For Problem 5.13.

5.13 Consider the cross section of a microstrip transmission line shown in Fig. 5.35. Let $G_{ij}\rho_j$ be the potential at the field point i on the center conductor due to the charge on subsection j. (It is assumed that the charge is concentrated in the filament along the center of the subsection.) G_{ij} is the Green's function for this problem and is given by

$$G_{ij} = \frac{1}{4\pi\epsilon_r} \sum_{n=1}^{\infty} \left[k^{2(n-1)} \ln \frac{A_{ij}^2 + (4n-2)^2}{A_{ij}^2 + (4n-4)^2} \right.$$
$$\left. + k^{2n-1} \ln \frac{A_{ij}^2 + (4n-2)^2}{A_{ij}^2 + (4n)^2} \right]$$

where

$$A_{ij} = \frac{\Delta}{H} \left| 2(i-1) - 2(j-1) - 1 \right|, \quad k = \frac{\epsilon_r - 1}{\epsilon_r + 1},$$

$\Delta = W/N$, and N is the number of equal subsections into which the center conductor is divided. By setting the potential equal to unity on the center conductor, one can find

$$C = \sum_{j=1}^{\infty} \rho_j \quad \text{(farads/m)}$$

and

$$Z_o = \frac{1}{c\sqrt{C_o C}}$$

where $c = 3 \times 10^8$ m/s and C_o is the capacitance per unit length for an air-filled transmission line (i.e., set $k = 1$ in G_{ij}). Find Z_o for $N = 30$ and:

(a) $\epsilon_r = 6.0$, $W = 4$ cm, $H = 4$ cm

(b) $\epsilon_r = 16.0$, $W = 8$ cm, $H = 4$ cm.

5.14 A rectangular section of microstrip transmission line of length L, width W, and height H above the ground plane is shown in Fig. 5.36. The section is subdivided into N subsections. A typical subsection ΔS_j, of sides Δx_j and Δy_j, is assumed to bear a uniform surface charge density ρ_j. The potential V_i at ΔS_i due to a uniform charge density ρ_j on ΔS_j ($j = 1, 2, \cdots, N$) is

$$V_i = \sum_{j=1}^{N} G_{ij} \rho_j$$

where

$$G_{ij} = \sum_{n=1}^{\infty} \frac{k^{n-1}(-1)^{n+1}}{2\pi\epsilon_0(\epsilon_r + 1)} \cdot$$

$$\left[(x_j - x_i) \ln \frac{(y_j - y_i) + \sqrt{(x_j - x_i)^2 + (y_j - y_i)^2 + (2n-2)^2 H^2}}{(y_j + \Delta y_j - y_i) + \sqrt{(x_j - x_i)^2 + (y_j + \Delta y_j - y_i)^2 + (2n-2)^2 H^2}} \right.$$

$$+ (x_j + \Delta x_j - x_i) \cdot$$

$$\ln \frac{(y_j + \Delta y_j - y_i) + \sqrt{(x_j + \Delta x_j - x_i)^2 + (y_j + \Delta y_j - y_i)^2 + (2n-2)^2 H^2}}{(y_j - y_i) + \sqrt{(x_j + \Delta x_j - x_i)^2 + (y_j - y_i)^2 + (2n-2)^2 H^2}}$$

$$+ (y_j - y_i) \ln \frac{(x_j - x_i) + \sqrt{(x_j - x_i)^2 + (y_j - y_i)^2 + (2n-2)^2 H^2}}{(x_j + \Delta x_j - x_i) + \sqrt{(x_j + \Delta x_j - x_i)^2 + (y_j - y_i)^2 + (2n-2)^2 H^2}}$$

$$+ (y_j + \Delta y_j - y_i) \cdot$$

$$\ln \frac{(x_j + \Delta x_j - x_i) + \sqrt{(x_j + \Delta x_j - x_i)^2 + (y_j + \Delta y_j - y_i)^2 + (2n-2)^2 H^2}}{(x_j - x_i) + \sqrt{(x_j - x_i)^2 + (y_j + \Delta y_j - y_i)^2 + (2n-2)^2 H^2}}$$

$$- (2n - 2) H \tan^{-1} \frac{(x_j - x_i)(y_j - y_i)}{(2n-2) H \sqrt{(x_j - x_i)^2 + (y_j - y_i)^2 + (2n-2)^2 H^2}}$$

$$- (2n - 2) \cdot$$

$$H \tan^{-1} \frac{(x_j + \Delta x_j - x_i)(y_j + \Delta y_j - y_i)}{(2n-2) H \sqrt{(x_j + \Delta x_j - x_i)^2 + (y_j + \Delta y_j - y_i)^2 + (2n-2)^2 H^2}}$$

$$+ (2n - 2) H \tan^{-1} \frac{(x_j - x_i)(y_j + \Delta y_j - y_i)}{(2n-2) H \sqrt{(x_j - x_i)^2 + (y_j + \Delta y_j - y_i)^2 + (2n-2)^2 H^2}}$$

$$+ (2n - 2) H \tan^{-1} \frac{(x_j + \Delta x_j - x_i)(y_j - y_i)}{(2n-2) H \sqrt{(x_j + \Delta x_j - x_i)^2 + (y_j - y_i)^2 + (2n-2)^2 H^2}} \Bigg]$$

400 Numerical Techniques in Electromagnetics

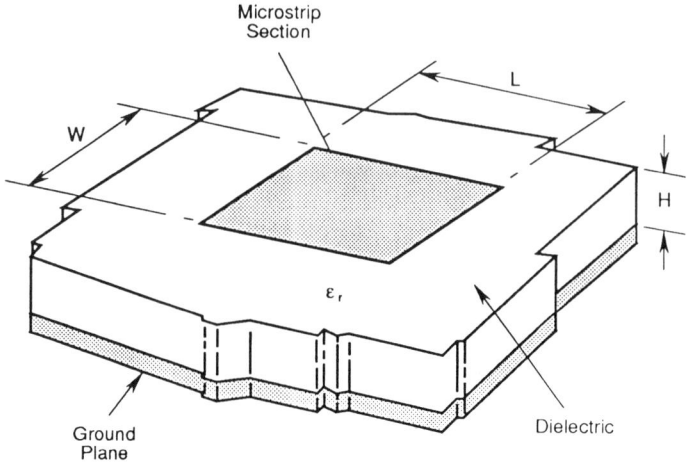

Figure 5.36 For Problem 5.14.

and $k = \dfrac{\epsilon_r - 1}{\epsilon_r + 1}$. If the ground plane is assumed to be at zero potential while the conducting strip at 1 V potential, we can find

$$C = \sum_{j=1}^{N} \rho_j.$$

Find C for:

(a) $\epsilon_r = 9.6$, $W = L = H = 2$ cm,

(b) $\epsilon_r = 9.6$, $W = H = 2$ cm, $L = 1$ cm.

5.15 For a conducting elliptic cylinder with cross section in Fig. 5.37(a), write a program to determine the scattering cross section $\sigma(\phi_i, \phi)$ due to a plane TM wave. Consider $\phi = 0°$, $10°$, $\cdots$, $180°$ and cases $\phi_i = 0°$, $30°$, and $90°$. Plot $\sigma(\phi_i, \phi)$ against ϕ for each ϕ_i. Take $\lambda = 1$m, $2a = \lambda/2$, $2b = \lambda$, $N = 18$.

Hint: Due to symmetry, consider only one half of the cross section as in Fig. 5.37(b). An ellipse is described by

$$\frac{x^2}{a^2} + \frac{y^2}{b^2} = 1.$$

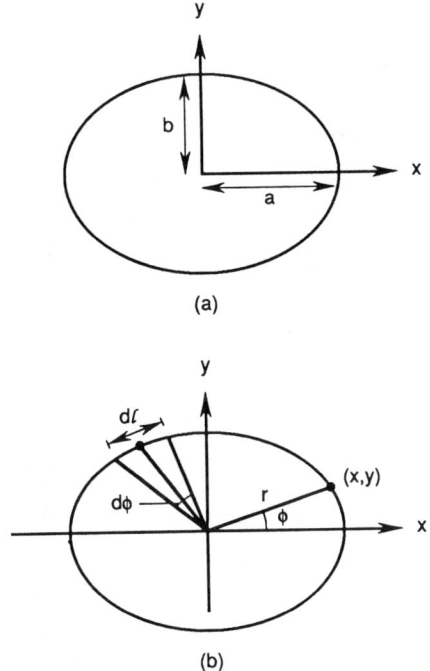

Figure 5.37 For Problem 5.15.

With $x = r\cos\phi$, $y = r\sin\phi$, it is readily shown that

$$r = \frac{a}{\sqrt{\cos^2\phi + \nu^2 \sin^2\phi}}, \quad \nu = a/b, \; dl = r\,d\phi.$$

5.16 Use the program in Fig. 5.15 (or develop your own program) to calculate the scattering pattern for each array of parallel wires shown in Fig. 5.38.

5.17 Repeat Prob. 5.15 using the techniques of Section 5.5.2. That is, consider the cylinder in Fig. 5.37(a) as an array of parallel wires.

5.18 The integral equation

$$-\frac{1}{2\pi}\int_{-w}^{w} I(z')\ln|z - z'|dz' = f(z), \quad -w < z < w$$

can be cast into matrix equation

$$[S][I] = [F]$$

402 Numerical Techniques in Electromagnetics

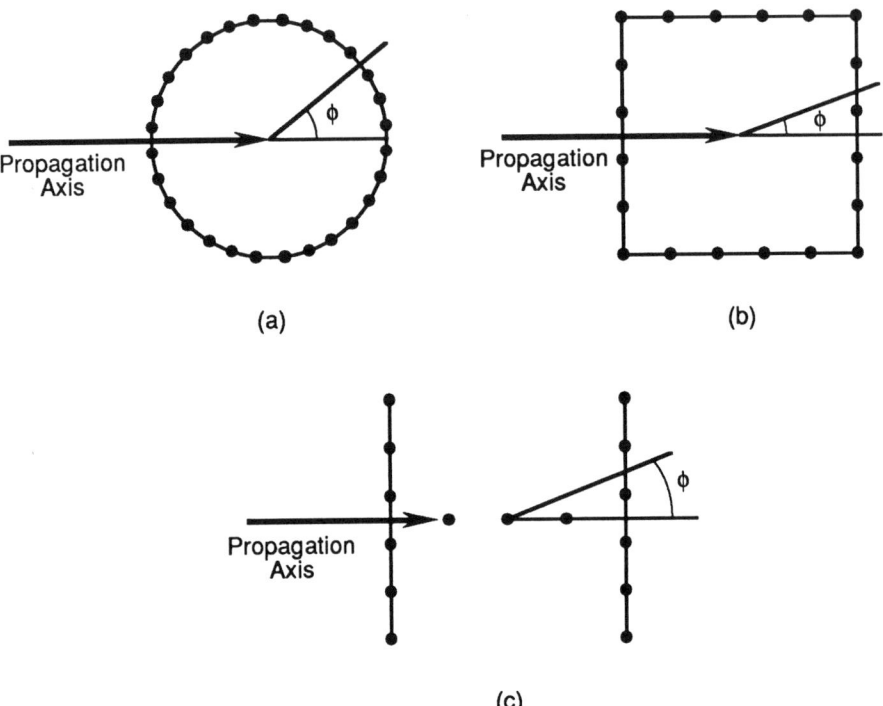

Figure 5.38 Arrays of parallel wires: (a) cylinder, (b) square, (c) I-beam, for Problem 5.16.

using pulse basis function and delta expansion function (point matching).

(a) Show that

$$S_{mn} = \frac{\Delta}{2\pi}\left[1 - \ln\Delta - \frac{1}{2}\ln\left|(m-n)^2 - \frac{1}{4}\right| - (m-n)\ln\frac{|m-n+1/2|}{|m-n-1/2|}\right]$$
$$F_m = f(z_m)$$

where $z_n = -w + \Delta(n - 1/2)$, $n = 1, 2, \cdots, N$, $\Delta = 2w/N$. Note that $[S]$ is a Toepliz matrix having only N distinct elements.

(b) Determine the unknowns $\{I_m\}$ with $f(z) = 1$, $N = 10$, $2w = 1$.

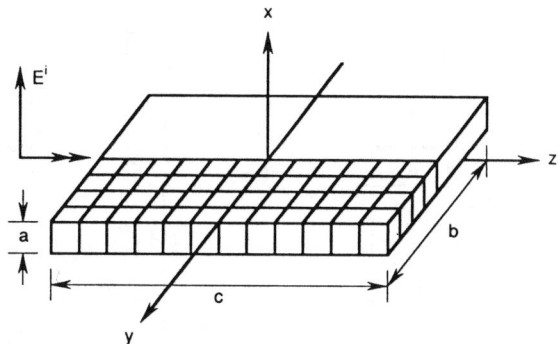

Figure 5.39 For Problem 5.22.

(c) Repeat part (b) with $f(z) = z$, $N = 10$, $2w = 1$.

5.19 A two-term representation of the current distribution on a thin, center-fed half-wavelength dipole antenna is given by

$$I(z) = \sum_{n=1}^{2} B_n \sin\left(\frac{2\pi n}{\lambda}(\lambda/4 - |z|)\right).$$

Substituting this into Hallen's integral equation gives

$$\sum_{n=1}^{2} B_n \int_{-\lambda/4}^{\lambda/4} \sin\left[\frac{2\pi n}{\lambda}(\lambda/4 - |z'|)\right] G(z,z')\,dz' + \frac{j4\pi}{\eta_o} \cos k_o z$$
$$= -\frac{j2\pi}{\eta_o} V_T \sin k_o |z|$$

where $\eta_o = 120\pi$, $k_o = \dfrac{2\pi}{\lambda} = \dfrac{2\pi f}{c}$, and $G(z,z')$ is given by Eq. (5.9.2). Taking $V_T = 1$ volt, $\lambda = 1$m, $a/\lambda = 7.022 \times 10^{-3}$, and match points at $z = 0, \lambda/8, \lambda/4$, determine the constants B_1, B_2, and C_1. Plot the real and imaginary parts of $I(z)$ against z.

5.20 Using Hallen's IE, determine the current distribution $I(z)$ on a straight dipole of length ℓ. Plot $|I| = |I_r + jI_i|$ against z. Assume excitation by a unit voltage, $N = 51$, $\Omega = 2\ln\dfrac{\ell}{a} = 12.5$, and consider cases: (a) $\ell = \lambda/2$, (b) $\ell = 1.5\lambda$.

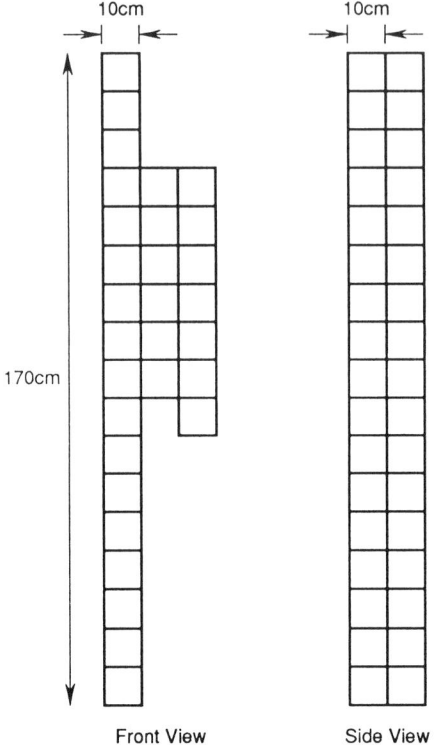

Figure 5.40 An adult torso; for Problem 5.23.

5.21 (a) Show that Pocklington integral equation (5.108) can be written as

$$-E_z^i = \frac{\lambda\sqrt{\mu/\epsilon}}{8j\pi a^2} \int_{-\ell/2}^{\ell/2} \frac{I(z')e^{-jkR}}{R^5}\left[(1+jkR)(2R^2-3a^2)+k^2a^2R^2\right]dz'.$$

(b) By changing variables, $z' - z = a\tan\theta$, show that

$$-E_z^i = \frac{\lambda\sqrt{\mu/\epsilon}}{8j\pi^2 a^2}\int_{\theta_1}^{\theta_2} I(\theta')e^{-jka/\cos\theta'}$$
$$\cdot\left[(jka+\cos\theta')(2-2\cos^2\theta')+k^2a^2\cos\theta'\right]d\theta'$$

where $\theta_1 = -\tan^{-1}\frac{\ell/2+z}{a}$, $\theta_2 = \tan^{-1}\frac{\ell/2-z}{a}$.

5.22 Using the program in Fig. 5.26 (or your own self-developed program), calculate the electric field inside a thin conducting layer ($\mu = \mu_o$, $\epsilon = 70\epsilon_o$, $\sigma = 1$ mho/m) shown in Fig. 5.39. Assume plane wave with electric field perpendicular to the plane of the layer, i.e.,

$$\mathbf{E}^i = e^{-jk_o z}\mathbf{a}_x \quad \text{V/m}$$

where $k_o = 2\pi f/c$. Consider only one half of the layer. Calculate $|E_x|/|E^i|$ and neglect E_y and E_z at the center of the cells since they are very small compared with E_x. Take $a = 0.5$ cm, $b = 4$ cm, $c = 6$ cm.

5.23 Consider an adult torso with a height 1.7 m and a shape shown in Fig. 5.40. If the torso is illuminated by a vertically polarized EM wave of 80 MHz with an incident electric field of 1 V/m, calculate the absorbed power density given by

$$\frac{\sigma}{2}(E_x^2 + E_y^2 + E_z^2)$$

at the center of each cell. Take $\mu = \mu_o$, $\epsilon = 80\epsilon_o$, $\sigma = 0.84$ mhos/m.

Chapter 6

Finite Element Method

"Remember—the value of time, the success of perseverance, the pleasure of working, the dignity of simplicity, the worth of character, the power of kindness, the influence of example, the obligation of duty, the wisdom of economy, the virtue of patience, the improvement of talent, the joy of originating."
                                                                    Bulletin

6.1 Introduction

The finite element method (FEM) has its origin in the field of structural analysis. Although the earlier mathematical treatment of the method was provided by Courant [1] in 1943, the method was not applied to electromagnetic (EM) problems until 1968. Since then the method has been employed in diverse areas such as waveguide problems, electric machines, semiconductor devices, microstrips, and absorption of EM radiation by biological bodies. For a review of the historical development of FEM, the interested reader is referred to textbooks that are exclusively devoted to the FEM (see references at the end of this chapter).

Although the finite difference method (FDM) and the method of moments (MOM) are conceptually simpler and easier to program than the finite element method (FEM), FEM is a more powerful and versatile numerical technique for handling problems involving complex geometries and inhomogeneous media. The systematic generality of the method makes it possible to construct general-purpose computer programs for solving a wide range of problems. Consequently, programs developed for a particular discipline have been applied successfully to solve problems in a different field with little or no modification [2].

408   Numerical Techniques in Electromagnetics

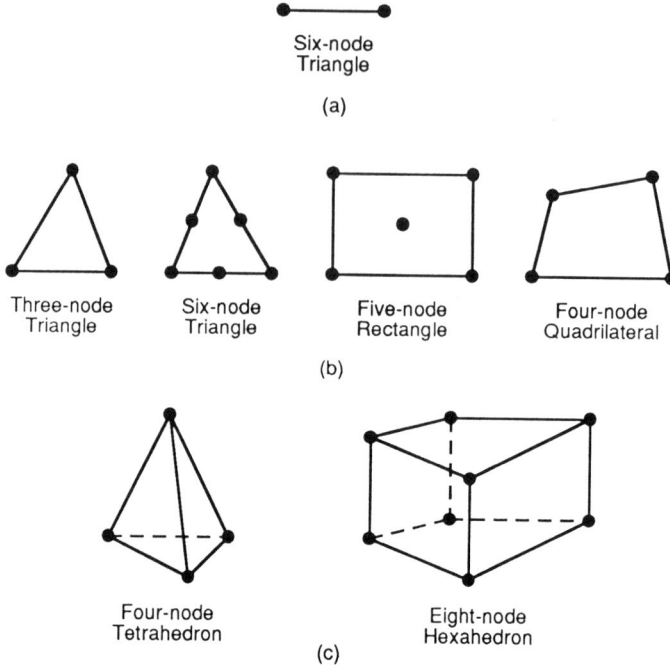

Figure 6.1 Typical finite elements: (a) One-dimensional, (b) two-dimensional, (c) three-dimensional.

The finite element analysis of any problem involves basically four steps [3]:

- discretizing the solution region into finite number of *subregions* or *elements*,
- deriving governing equations for a typical element,
- assembling of all elements in the solution region, and
- solving the system of equations obtained.

Discretization of the continuum involves dividing up the solution region into subdomains, called *finite elements*. Figure 6.1 shows some typical elements for one-, two-, and three-dimensional problems. The problem of discretization will be fully treated in Sections 6.5 and 6.6. The other three steps will be described in detail in the subsequent sections.

6.2 Solution of Laplace's Equation

As an application of FEM to electrostatic problems, let us apply the four steps mentioned above to solve Laplace's equation, $\nabla^2 V = 0$. For the

Finite Element Method

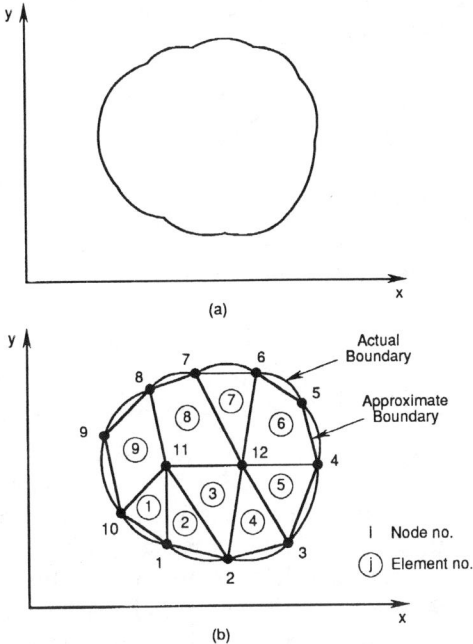

Figure 6.2 (a) The solution region; (b) its finite element discretization.

purpose of illustration, we will strictly follow the four steps mentioned above.

6.2.1 Finite Element Discretization

To find the potential distribution $V(x,y)$ for the two-dimensional solution region shown in Fig. 6.2(a), we divide the region into a number of finite elements as illustrated in Fig. 6.2(b). In Fig. 6.2(b), the solution region is subdivided into nine nonoverlapping *finite elements*; elements 6, 8, and 9 are four-node quadrilaterals, while other elements are three-node triangles. In practical situations, however, it is preferred for ease of computation to have elements of the same type throughout the region. That is, in Fig. 6.2(b), we could have split each quadrilateral into two triangles so that we have 12 triangular elements altogether. The subdivision of the solution region into elements is usually done by hand, but in situations where a large number of elements is required, automatic schemes to be discussed in Sections 6.5 and 6.6 are used.

We seek an approximation for the potential V_e within an element e and then interrelate the potential distribution in various elements such that the potential is continous across interelement boundaries. The approximate solution for the whole region is

$$V(x,y) \simeq \sum_{e=1}^{N} V_e(x,y), \tag{6.1}$$

where N is the number of triangular elements into which the solution region is divided. The most common form of approximation for V within an element is polynomial approximation, namely,

$$V_e(x,y) = a + bx + cy \tag{6.2}$$

for a triangular element and

$$V_e(x,y) = a + bx + cy + dxy \tag{6.3}$$

for a quadrilateral element. The constants $a, b, c,$ and d are to be determined. The potential V_e in general is nonzero within element e but zero outside e. In view of the fact that quadrilateral elements do not conform to curved boundary as easily as triangular elements, we prefer to use triangular elements throughout our analysis in this chapter. Notice that our assumption of linear variation of potential within the triangular element as in Eq. (6.2) is the same as assuming that the electric field is uniform within the element, i.e.,

$$\mathbf{E}_e = -\nabla V_e = -(b\mathbf{a}_x + c\mathbf{a}_y). \tag{6.4}$$

6.2.2 Element Governing Equations

Consider a typical triangular element shown in Fig. 6.3. The potential $V_{e1}, V_{e2},$ and V_{e3} at nodes 1, 2, and 3, respectively, are obtained using Eq. (6.2), i.e.,

$$\begin{bmatrix} V_{e1} \\ V_{e2} \\ V_{e3} \end{bmatrix} = \begin{bmatrix} 1 & x_1 & y_1 \\ 1 & x_2 & y_2 \\ 1 & x_3 & y_3 \end{bmatrix} \begin{bmatrix} a \\ b \\ c \end{bmatrix}. \tag{6.5}$$

The coefficients $a, b,$ and c are determined from Eq. (6.5) as

$$\begin{bmatrix} a \\ b \\ c \end{bmatrix} = \begin{bmatrix} 1 & x_1 & y_1 \\ 1 & x_2 & y_2 \\ 1 & x_3 & y_3 \end{bmatrix}^{-1} \begin{bmatrix} V_{e1} \\ V_{e2} \\ V_{e3} \end{bmatrix}. \tag{6.6}$$

Substituting this into Eq. (6.2) gives

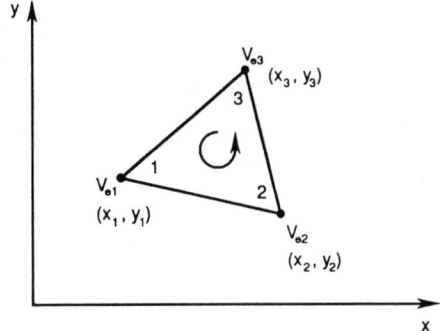

Figure 6.3 Typical triangular element; local node numbering 1-2-3 must proceed counterclockwise as indicated by the arrow.

$V_e =$

$$[1 \quad x \quad y] \frac{1}{2A} \begin{bmatrix} (x_2 y_3 - x_3 y_2) & (x_3 y_1 - x_1 y_3) & (x_1 y_2 - x_2 y_1) \\ (y_2 - y_3) & (y_3 - y_1) & (y_1 - y_2) \\ (x_3 - x_2) & (x_1 - x_3) & (x_2 - x_1) \end{bmatrix} \begin{bmatrix} V_{e1} \\ V_{e2} \\ V_{e3} \end{bmatrix}$$

or

$$V_e = \sum_{i=1}^{3} \alpha_i(x,y) V_{ei} \qquad (6.7)$$

where

$$\alpha_1 = \frac{1}{2A}\Big[(x_2 y_3 - x_3 y_2) + (y_2 - y_3)x + (x_3 - x_2)y\Big], \qquad (6.8a)$$

$$\alpha_2 = \frac{1}{2A}\Big[(x_3 y_1 - x_1 y_3) + (y_3 - y_1)x + (x_1 - x_3)y\Big], \qquad (6.8b)$$

$$\alpha_3 = \frac{1}{2A}\Big[(x_1 y_2 - x_2 y_1) + (y_1 - y_2)x + (x_2 - x_1)y\Big], \qquad (6.8c)$$

and A is the area of the element e, i.e.,

$$2A = \begin{vmatrix} 1 & x_1 & y_1 \\ 1 & x_2 & y_2 \\ 1 & x_3 & y_3 \end{vmatrix}$$
$$= (x_1 y_2 - x_2 y_1) + (x_3 y_1 - x_1 y_3) + (x_2 y_3 - x_3 y_2)$$

or

$$A = \frac{1}{2}\Big[(x_2 - x_1)(y_3 - y_1) - (x_3 - x_1)(y_2 - y_1)\Big]. \qquad (6.9)$$

412 Numerical Techniques in Electromagnetics

The value of A is positive if the nodes are numbered counterclockwise (starting from any node) as shown by the arrow in Fig. 6.3. Note that Eq. (6.7) gives the potential at any point (x, y) within the element provided that the potentials at the vertices are known. This is unlike finite difference analysis, where the potential is known at the grid points only. Also note that α_i are linear interpolation functions. They are called the *element shape functions* and they have the following properties [4]:

$$\alpha_i = \begin{cases} 1, & i = j \\ 0, & i \neq j \end{cases} \tag{6.10a}$$

$$\sum_{i=1}^{3} \alpha_i(x, y) = 1. \tag{6.10b}$$

The shape functions α_1, α_2, and α_3 are illustrated in Fig. 6.4.

The functional corresponding to Laplace's equation, $\nabla^2 V = 0$, is given by

$$W_e = \frac{1}{2} \int \epsilon |\mathbf{E}|^2 \, dS = \frac{1}{2} \int \epsilon |\nabla V_e|^2 \, dS. \tag{6.11}$$

(Physically, the functional W_e is the energy per unit length associated with the element e.) From Eq. (6.7),

$$\nabla V_e = \sum_{i=1}^{3} V_{ei} \nabla \alpha_i. \tag{6.12}$$

Substituting Eq. (6.12) into Eq. (6.11) gives

$$W_e = \frac{1}{2} \sum_{i=1}^{3} \sum_{j=1}^{3} \epsilon V_{ei} \left[\int \nabla \alpha_i \cdot \nabla \alpha_j \, dS \right] V_{ej}. \tag{6.13}$$

If we define the term in brackets as

$$C_{ij}^{(e)} = \int \nabla \alpha_i \cdot \nabla \alpha_j \, dS, \tag{6.14}$$

we may write Eq. (6.13) in matrix form as

$$W_e = \frac{1}{2} \epsilon \, [V_e]^t \, [C^{(e)}] \, [V_e] \tag{6.15}$$

where the superscript t denotes the transpose of the matrix,

$$[V_e] = \begin{bmatrix} V_{e1} \\ V_{e2} \\ V_{e3} \end{bmatrix} \tag{6.16a}$$

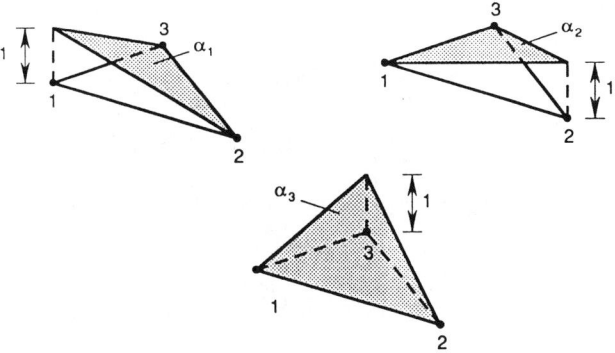

Figure 6.4 Shape functions α_1, α_2, and α_3 for a triangular element.

and

$$[C^{(e)}] = \begin{bmatrix} C^{(e)}_{11} & C^{(e)}_{12} & C^{(e)}_{13} \\ C^{(e)}_{21} & C^{(e)}_{22} & C^{(e)}_{23} \\ C^{(e)}_{31} & C^{(e)}_{32} & C^{(e)}_{33} \end{bmatrix}. \qquad (6.16b)$$

The matrix $[C^{(e)}]$ is usually called the *element coefficient matrix* (or "stiffness matrix" in structural analysis). The matrix element $C^{(e)}_{ij}$ of the coefficient matrix may be regarded as the coupling between nodes i and j; its value is obtained from Eqs. (6.8) and (6.14). For example,

$$\begin{aligned} C^{(e)}_{12} &= \int \nabla \alpha_1 \cdot \nabla \alpha_2 \, dS \\ &= \frac{1}{4A^2}\Big[(y_2 - y_3)(y_3 - y_1) + (x_3 - x_2)(x_1 - x_3)\Big] \int dS \\ &= \frac{1}{4A}\Big[(y_2 - y_3)(y_3 - y_1) + (x_3 - x_2)(x_1 - x_3)\Big]. \qquad (6.17a) \end{aligned}$$

Similarly,

$$C^{(e)}_{13} = \frac{1}{4A}\Big[(y_2 - y_3)(y_1 - y_2) + (x_3 - x_2)(x_2 - x_1)\Big], \qquad (6.17b)$$

$$C^{(e)}_{23} = \frac{1}{4A}\Big[(y_3 - y_1)(y_1 - y_2) + (x_1 - x_3)(x_2 - x_1)\Big], \qquad (6.17c)$$

$$C_{11}^{(e)} = \frac{1}{4A}\left[(y_2 - y_3)^2 + (x_3 - x_2)^2\right], \tag{6.17d}$$

$$C_{22}^{(e)} = \frac{1}{4A}\left[(y_3 - y_1)^2 + (x_1 - x_3)^2\right], \tag{6.17e}$$

$$C_{33}^{(e)} = \frac{1}{4A}\left[(y_1 - y_2)^2 + (x_2 - x_1)^2\right]. \tag{6.17f}$$

Also

$$C_{21}^{(e)} = C_{12}^{(e)}, \quad C_{31}^{(e)} = C_{13}^{(e)}, \quad C_{32}^{(e)} = C_{23}^{(e)}. \tag{6.18}$$

6.2.3 Assembling of All Elements

Having considered a typical element, the next step is to assemble all such elements in the solution region. The energy associated with the assemblage of elements is

$$W = \sum_{e=1}^{N} W_e = \frac{1}{2}\epsilon[V]^t[C][V] \tag{6.19}$$

where

$$[V] = \begin{bmatrix} V_1 \\ V_2 \\ V_3 \\ \vdots \\ V_n \end{bmatrix}, \tag{6.20}$$

n is the number of nodes, N is the number of elements, and $[C]$ is called the overall or *global coefficient matrix*, which is the assemblage of individual element coefficient matrices. Notice that to obtain Eq. (6.19), we have assumed that the whole solution region is homogeneous so that ϵ is constant. For an inhomogeneous solution region such as shown in Fig. 6.5, for example, the region is discretized such that each finite element is homogeneous. In this case, Eq. (6.11) still holds, but Eq. (6.19) does not apply since $\epsilon(=\epsilon_r\epsilon_o)$ or simply ϵ_r varies from element to element. To apply Eq. (6.19), we may replace ϵ by ϵ_o and multiply the integrand in Eq. (6.14) by ϵ_r.

The process by which individual element coefficient matrices are assembled to obtain the global coefficient matrix is best illustrated with an example. Consider the finite element mesh consisting of three finite elements as shown in Fig. 6.6. Observe the numberings of the mesh. The numbering of nodes 1, 2, 3, 4, and 5 is called *global* numbering. The numbering i-j-k is called *local* numbering, and it corresponds with $1 - 2 - 3$ of the element in Fig. 6.3. For example, for element 3 in Fig. 6.6, the global numbering

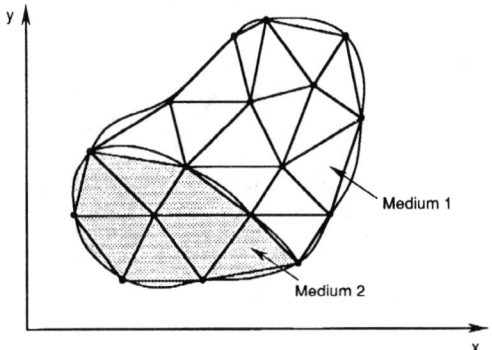

Figure 6.5 Discretization of an inhomogeneous solution region.

3 – 5 – 4 corresponds with local numbering 1 – 2 – 3 of the element in Fig. 6.3. (Note that the local numbering must be in counterclockwise sequence starting from any node of the element.) For element 3, we could choose 4 – 3 – 5 instead of 3 – 5 – 4 to correspond with 1 – 2 – 3 of the element in Fig. 6.3. Thus the numbering in Fig. 6.6 is not unique. Assuming the particular numbering in Fig. 6.6, the global coefficient matrix is expected to have the form

$$[C] = \begin{bmatrix} C_{11} & C_{12} & C_{13} & C_{14} & C_{15} \\ C_{21} & C_{22} & C_{23} & C_{24} & C_{25} \\ C_{31} & C_{32} & C_{33} & C_{34} & C_{35} \\ C_{41} & C_{42} & C_{43} & C_{44} & C_{45} \\ C_{51} & C_{52} & C_{53} & C_{54} & C_{55} \end{bmatrix} \quad (6.21)$$

which is a 5 × 5 matrix since five nodes ($n = 5$) are involved. Again, C_{ij} is the coupling between nodes i and j. We obtain C_{ij} by using the fact that the potential distribution must be continuous across interelement boundaries. The contribution to the i, j position in $[C]$ comes from all elements containing nodes i and j. For example, in Fig. 6.6, elements 1 and 2 have node 1 in common; hence

$$C_{11} = C_{11}^{(1)} + C_{11}^{(2)}. \quad (6.22a)$$

Node 2 belongs to element 1 only; hence

$$C_{33} = C_{33}^{(1)}. \quad (6.22b)$$

Node 4 belongs to elements 1, 2, and 3; consequently

$$C_{44} = C_{22}^{(1)} + C_{33}^{(2)} + C_{33}^{(3)}. \quad (6.22c)$$

416   Numerical Techniques in Electromagnetics

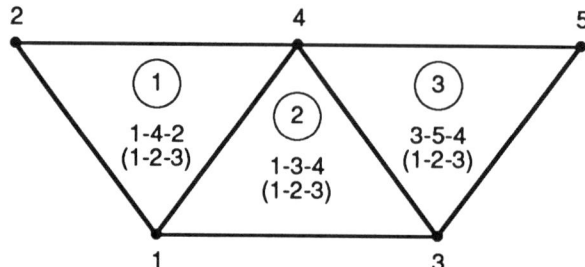

Figure 6.6 Assembly of three elements; i-j-k corresponds to local numbering (1-2-3) of the element in Fig. 6.3.

Nodes 1 and 4 belong simultaneously to elements 1 and 2; hence

$$C_{14} = C_{41} = C_{12}^{(1)} + C_{13}^{(2)}. \qquad (6.22\text{d})$$

Since there is no coupling (or direct link) between nodes 2 and 3,

$$C_{23} = C_{32} = 0. \qquad (6.22\text{e})$$

Continuing in this manner, we obtain all the terms in the global coefficient matrix by inspection of Fig. 6.6 as

$$\begin{bmatrix} C_{11}^{(1)}+C_{11}^{(2)} & C_{13}^{(1)} & C_{12}^{(2)} & C_{12}^{(1)}+C_{13}^{(2)} & 0 \\ C_{31}^{(1)} & C_{33}^{(1)} & 0 & C_{32}^{(1)} & 0 \\ C_{21}^{(2)} & 0 & C_{22}^{(2)}+C_{11}^{(3)} & C_{23}^{(2)}+C_{13}^{(3)} & C_{12}^{(3)} \\ C_{21}^{(1)}+C_{31}^{(2)} & C_{23}^{(1)} & C_{32}^{(2)}+C_{31}^{(3)} & C_{22}^{(1)}+C_{33}^{(2)}+C_{33}^{(3)} & C_{32}^{(3)} \\ 0 & 0 & C_{21}^{(3)} & C_{23}^{(3)} & C_{22}^{(3)} \end{bmatrix}. \qquad (6.23)$$

Note that element coefficient matrices overlap at nodes shared by elements and that there are 27 terms (9 for each of the 3 elements) in the global coefficient matrix $[C]$. Also note the following properties of the matrix $[C]$:

(1) It is symmetric ($C_{ij} = C_{ji}$) just as the element coefficient matrix.

(2) Since $C_{ij} = 0$ if no coupling exists between nodes i and j, it is evident that for a large number of elements $[C]$ becomes sparse. Matrix $[C]$ is also banded if the nodes are carefully numbered. It can be shown using Eq. (6.17) that

$$\sum_{i=1}^{3} C_{ij}^{(e)} = 0 = \sum_{j=1}^{3} C_{ij}^{(e)}$$

(3) It is singular. Although this is not so obvious, it can be shown using the element coefficient matrix of Eq. (6.16b). Adding columns 2 and 3 to column 1 yields zeros in column 1 in Eq. (6.16b).

6.2.4 Solving the Resulting Equations

Using the concepts developed in Chapter 4, it can be shown that Laplace's (or Poisson's) equation is satisfied when the total energy in the solution region is minimum. Thus we require that the partial derivatives of W with respect to each nodal value of the potential be zero, i.e.,

$$\frac{\partial W}{\partial V_1} = \frac{\partial W}{\partial V_2} = \cdots = \frac{\partial W}{\partial V_n} = 0$$

or

$$\frac{\partial W}{\partial V_k} = 0, \quad k = 1, 2, \cdots, n. \tag{6.24}$$

For example, to get $\frac{\partial W}{\partial V_1} = 0$ for the finite element mesh of Fig. 6.6, we substitute Eq. (6.21) into Eq. (6.19) and take the partial derivative of W with respect to V_1. We obtain

$$0 = \frac{\partial W}{\partial V_1} = 2V_1 C_{11} + V_2 C_{12} + V_3 C_{13} + V_4 C_{14} + V_5 C_{15}$$
$$+ V_2 C_{21} + V_3 C_{31} + V_4 C_{41} + V_5 C_{51}$$

or

$$0 = V_1 C_{11} + V_2 C_{12} + V_3 C_{13} + V_4 C_{14} + V_5 C_{15}. \tag{6.25}$$

In general, $\frac{\partial W}{\partial V_k} = 0$ leads to

$$0 = \sum_{i=1}^{n} V_i C_{ik} \tag{6.26}$$

where n is the number of nodes in the mesh. By writing Eq. (6.26) for all nodes $k = 1, 2, \cdots, n$, we obtain a set of simultaneous equations from which the solution of $[V]^t = [V_1, V_2, \cdots, V_n]$ can be found. This can be done in two ways similar to those used in solving finite difference equations obtained from Laplace's (or Poisson's) equation in Section 3.5.

(1) Iteration Method: Suppose node 1 in Fig. 6.6, for example, is a free node. From Eq. (6.25),

$$V_1 = -\frac{1}{C_{11}} \sum_{i=2}^{5} V_i C_{1i}. \tag{6.27}$$

Thus, in general, at node k in a mesh with n nodes

$$\boxed{V_k = -\frac{1}{C_{kk}} \sum_{i=1, i \neq k}^{n} V_i C_{ki}} \tag{6.28}$$

where node k is a free node. Since $C_{ki} = 0$ if node k is not directly connected to node i, only nodes that are directly linked to node k contribute to V_k in Eq. (6.28). Equation (6.28) can be applied iteratively to all the free nodes. The iteration process begins by setting the potentials of fixed nodes (where the potentials are known) to their prescribed values and the potentials at the free nodes (where the potentials are unknown) equal to zero or to the average potential [5]

$$V_{ave} = \frac{1}{2}(V_{min} + V_{max}) \tag{6.29}$$

where V_{min} and V_{max} are the minimum and maximum values of V at the fixed nodes (where the potential V is prescribed or known). With these initial values, the potentials at the free nodes are calculated using Eq. (6.28). At the end of the first iteration, when the new values have been calculated for all the free nodes, they become the old values for the second iteration. The procedure is repeated until the change between subsequent iterations is negligible enough.

(2) Band Matrix Method: If all free nodes are numbered first and the fixed nodes last, Eq. (6.19) can be written such that [4]

$$W = \frac{1}{2}\epsilon \begin{bmatrix} V_f & V_p \end{bmatrix} \begin{bmatrix} C_{ff} & C_{fp} \\ C_{pf} & C_{pp} \end{bmatrix} \begin{bmatrix} V_f \\ V_p \end{bmatrix} \tag{6.30}$$

where subscripts f and p, respectively, refer to nodes with free and fixed (or prescribed) potentials. Since V_p is constant (it consists of known, fixed values), we only differentiate with respect to V_f so that applying Eqs. (6.24) to (6.30) yields

$$[C_{ff} \quad C_{fp}] \begin{bmatrix} V_f \\ V_p \end{bmatrix} = 0$$

or

$$\boxed{[C_{ff}][V_f] = -[C_{fp}][V_p]} \qquad (6.31)$$

This equation can be written as

$$[A][V] = [B] \qquad (6.32\text{a})$$

or

$$[V] = [A]^{-1}[B] \qquad (6.32\text{b})$$

where $[V] = [V_f], [A] = [C_{ff}], [B] = -[C_{fp}][V_p]$. Since $[A]$ is, in general, nonsingular, the potential at the free nodes can be found using Eq. (6.32). We can solve for $[V]$ in Eq. (6.32a) using Gaussian elimination technique. We can also solve for $[V]$ in Eq. (6.32b) using matrix inversion if the size of the matrix to be inverted is not large.

It is sometimes necessary to impose Neumann condition ($\frac{\partial V}{\partial n} = 0$) as a boundary condition or at the line of symmetry when we take advantage of the symmetry of the problem. Suppose, for concreteness, that a solution region is symmetric along the y-axis as in Fig. 6.7. We impose condition $\frac{\partial V}{\partial x} = 0$ along the y-axis by making

$$V_1 = V_2, \quad V_4 = V_5, \quad V_7 = V_8. \qquad (6.33)$$

Notice that as from Eq. (6.11) onward, the solution has been restricted to a two-dimensional problem involving Laplace's equation, $\nabla^2 V = 0$. The basic concepts developed in this section will be extended to finite element analysis of problems involving Poisson's equation ($\nabla^2 V = -\rho_v/\epsilon$, $\nabla^2 \mathbf{A} = -\mu \mathbf{J}$) or wave equation ($\nabla^2 \Phi - \gamma^2 \Phi = 0$) in the next sections.

The following two examples were solved in [3] using the band matrix method; here they are solved using the iterative method.

Example 6.1

Consider the two-element mesh shown in Fig. 6.8(a). Using the finite element method, determine the potentials within the mesh.

420   Numerical Techniques in Electromagnetics

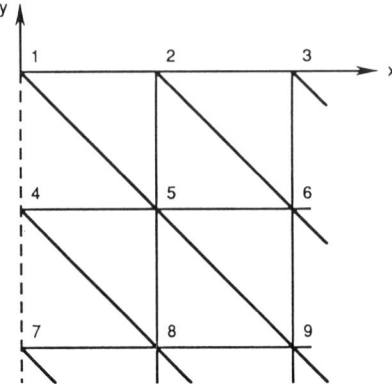

Figure 6.7 A solution region that is symmetric along the y-axis.

Solution

The element coefficient matrices can be calculated using Eqs. (6.17) and (6.18). However, our calculations will be easier if we define

$$P_1 = (y_2 - y_3), \quad P_2 = (y_3 - y_1), \quad P_3 = (y_1 - y_2), \tag{6.1.1}$$

$$Q_1 = (x_3 - x_2), \quad Q_2 = (x_1 - x_3), \quad Q_3 = (x_2 - x_1).$$

With P_i and Q_i ($i = 1, 2, 3$ are the local node numbers), each term in the element coefficient matrix is found as

$$C_{ij}^{(e)} = \frac{1}{4A}(P_i P_j + Q_i Q_j) \tag{6.1.2}$$

where $A = \frac{1}{2}(P_2 Q_3 - P_3 Q_2)$. It is evident that Eq. (6.1.2) is more convenient to use than Eqs. (6.17) and (6.18). For element 1 consisting of nodes 1-2-4 corresponding to the local numbering 1 - 2 - 3 as in Fig. 6.8(b),

$$P_1 = -1.3, \quad P_2 = 0.9, \quad P_3 = 0.4,$$
$$Q_1 = -0.2, \quad Q_2 = -0.4, \quad Q_3 = 0.6,$$
$$A = \frac{1}{2}(0.54 + 0.16) = 0.35.$$

Substituting all of these into Eq. (6.1.2) gives

$$[C^{(1)}] = \begin{bmatrix} 1.2357 & -0.7786 & -0.4571 \\ -0.7786 & 0.6929 & 0.0857 \\ -0.4571 & 0.0857 & 0.3714 \end{bmatrix}. \tag{6.1.3}$$

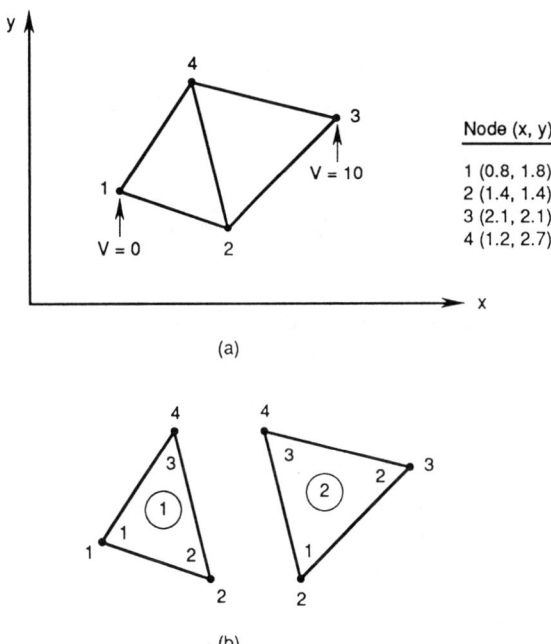

Figure 6.8 For example 6.1: (a) Two-element mesh, (b) local and global numbering at the elements.

Similarly, for element 2 consisting of nodes 2 – 3 – 4 corresponding to local numbering 1 – 2 – 3 as in Fig. 6.8(b),

$$P_1 = -0.6, \quad P_2 = 1.3, \quad P_3 = -0.7,$$
$$Q_1 = -0.9, \quad Q_2 = 0.2, \quad Q_3 = 0.7,$$
$$A = \frac{1}{2}(0.91 + 0.14) = 0.525.$$

Hence

$$[C^{(2)}] = \begin{bmatrix} 0.5571 & -0.4571 & -0.1 \\ -0.4571 & 0.8238 & -0.3667 \\ -0.1 & -0.3667 & 0.4667 \end{bmatrix}. \quad (6.1.4)$$

The terms of the global coefficient matrix are obtained as follows:

$$C_{22} = C_{22}^{(1)} + C_{11}^{(2)} = 0.6929 + 0.5571 = 1.25$$
$$C_{24} = C_{23}^{(1)} + C_{13}^{(2)} = 0.0857 - 0.1 = -0.0143$$
$$C_{44} = C_{33}^{(1)} + C_{33}^{(2)} = 0.3714 + 0.4667 = 0.8381$$

422   Numerical Techniques in Electromagnetics

$$C_{21} = C_{21}^{(1)} = -0.7786$$
$$C_{23} = C_{12}^{(2)} = -0.4571$$
$$C_{41} = C_{31}^{(1)} = -0.4571$$
$$C_{43} = C_{32}^{(2)} = -0.3667.$$

Note that we follow local numbering for the element coefficient matrix and global numbering for the global coefficient matrix. Thus

$$[C] = \begin{bmatrix} C_{11}^{(1)} & C_{12}^{(1)} & 0 & C_{13}^{(1)} \\ C_{21}^{(1)} & C_{22}^{(1)} + C_{11}^{(2)} & C_{12}^{(2)} & C_{23}^{(1)} + C_{12}^{(2)} \\ 0 & C_{21}^{(2)} & C_{22}^{(2)} & C_{23}^{(2)} \\ C_{31}^{(1)} & C_{32}^{(1)} + C_{31}^{(2)} & C_{32}^{(2)} & C_{33}^{(1)} + C_{33}^{(2)} \end{bmatrix}$$

$$= \begin{bmatrix} 1.2357 & -0.7786 & 0 & -0.4571 \\ -0.7786 & 1.25 & -0.4571 & -0.0143 \\ 0 & -0.4571 & 0.8238 & -0.3667 \\ -0.4571 & -0.0143 & -0.3667 & 0.8381 \end{bmatrix}. \qquad (6.1.5)$$

Note that $\sum_{i=1}^{4} C_{ij} = 0 = \sum_{j=1}^{4} C_{ij}$. This may be used to check if C is properly obtained. We now apply Eq. (6.28) to the free nodes 2 and 4, i.e.,

$$V_2 = -\frac{1}{C_{22}}(V_1 C_{12} + V_3 C_{32} + V_4 C_{42})$$
$$V_4 = -\frac{1}{C_{44}}(V_1 C_{14} + V_2 C_{24} + V_3 C_{34})$$

or

$$V_2 = -\frac{1}{1.25}(-4.571 - 0.0143 V_4) \qquad (6.1.6a)$$
$$V_4 = -\frac{1}{0.8381}(-0.143 V_2 - 3.667). \qquad (6.1.6b)$$

By initially setting $V_2 = 0 = V_4$, we apply Eqs. (6.1.6a,b) iteratively. The first iteration gives $V_2 = 3.6568$, $V_4 = 4.4378$ and at the second iteration

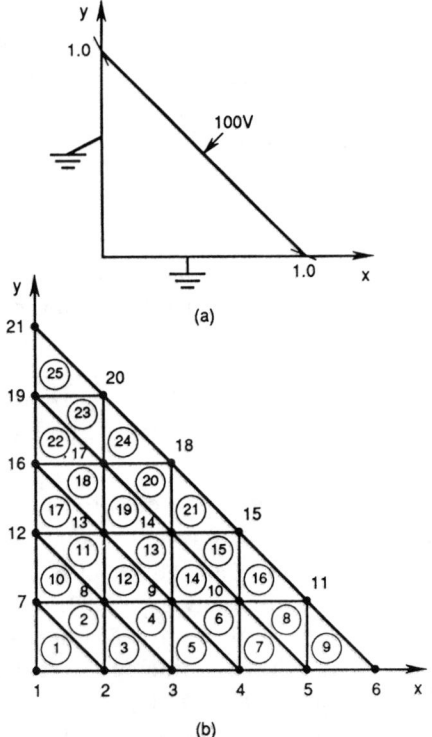

Figure 6.9 For example 6.2: (a) Two-dimensional electrostatic problem, (b) solution region divided into 25 triangular elements.

$V_2 = 3.7075$, $V_4 = 4.4386$. Just after two iterations, we obtain the same results as those from the band matrix method [3]. Thus the iterative technique is faster and is usually preferred for a large number of nodes. Once the values of the potentials at the nodes are known, the potential at any point within the mesh can be determined using Eq. (6.7).

Example 6.2

Write a FORTRAN program to solve Laplace's equation using the finite element method. Apply the program to the two-dimensional problem shown in Fig. 6.9(a).

Solution

The solution region is divided into 25 three-node triangular elements with total number of nodes being 21 as shown in Fig. 6.9(b). This is a necessary step in order to have input data defining the geometry of the problem. Based on the discussions in Section 6.2, a general FORTRAN program for solving problems involving Laplace's equation using three-node triangular elements is developed as shown in Fig. 6.10. The development of the program basically involves four steps indicated in the program and explained as follows.

Step 1: This involves inputting the necessary data defining the problem. This is the only step that depends on the geometry of the problem at hand. Through a data file, we input the number of elements, the number of nodes, the number of fixed nodes, the prescribed values of the potentials at the free nodes, the x and y coordinates of all nodes, and a list identifying the nodes belonging to each element in the order of the local numbering $1-2-3$. For the problem in Fig. 6.9, the three sets of data for coordinates, element-node relationship, and prescribed potentials at fixed nodes are shown in Tables 6.1, 6.2, and 6.3, respectively.

Step 2: This step entails finding the element coefficient matrix $[C^{(e)}]$ for each element and using the terms to form the global matrix $[C]$.

Step 3: At this stage, we first find the list of free nodes from the given list of prescribed nodes. We now apply Eq. (6.28) iteratively to all the free nodes. The solution converges at 50 iterations or less since only 6 nodes are involved in this case. The solution obtained is exactly the same as those obtained using the band matrix method [3].

Step 4: This involves outputting the result of the computation. The output data for the problem in Fig. 6.9 is presented in Table 6.4. The validity of the result in Table 6.4 is checked using the finite difference method. From the finite difference analysis, the potentials at the free nodes are obtained as:

$$V_8 = 15.41, \quad V_9 = 26.74, \quad V_{10} = 56.69,$$
$$V_{13} = 34.88, \quad V_{14} = 65.41, \quad V_{17} = 58.72 V.$$

Although the result obtained using finite difference is considered more accurate in this problem, increased accuracy of finite element analysis can be obtained by dividing the solution region into a greater number of triangular elements, or using higher-order elements to be discussed in Section 6.8. As alluded to earlier, the finite element method has two major advantages over the finite difference method. Field quantities are obtained only at discrete positions in the solution region using FDM; they can be obtained at any point in the solution region in FEM. Also, it is easier to handle complex geometries using FEM than using FDM.

Table 6.1 Nodal Coordinates of the Finite Element Mesh of Fig. 6.9

Node	x	y	Node	x	y
1	0.0	0.0	12	0.0	0.4
2	0.2	0.0	13	0.2	0.4
3	0.4	0.0	14	0.4	0.4
4	0.6	0.0	15	0.6	0.4
5	0.8	0.0	16	0.0	0.6
6	1.0	0.0	17	0.2	0.6
7	0.0	0.2	18	0.4	0.6
8	0.2	0.2	19	0.0	0.8
9	0.4	0.2	20	0.2	0.8
10	0.6	0.2	21	0.0	1.0
11	0.8	0.2			

```
0001      C      FINITE ELEMENT SOLUTION OF LAPLACE'S EQUATION FOR
0002      C      TWO-DIMENSIONAL PROBLEMS
0003      C      TRIANGULAR ELEMENTS ARE USED
0004      C
0005      C      THE UNKNOWN POTENTIALS ARE OBTAINED USING
0006      C      ITERATION METHOD
0007      C
0008      C      ND = NO. OF NODES
0009      C      NE = NO. OF ELEMENTS
0010      C      NP = NO. OF FIXED NODES (WHERE POTENTIAL IS PRESCRIBED)
0011      C      NDP(I) = NODE NO. OF PRESCRIBED POTENTIAL, I = 1,2,...NP
0012      C      VAL(I)) = VALUE OF PRESCRIBED POTENTIAL AT NODE NDP(I)
0013      C      NL(I,J) = LIST OF NODES FOR EACH ELEMENT I, WHERE
0014      c      LF(I) = LIST OF FREE NODES I = 1,2,...,NF=ND-NP
0015      C      J = 1, 2, 3  IS THE LOCAL NODE NUMBER
0016      C      CE(I,J) = ELEMENT COEFFICIENT MATRIX
0017      C      ER(I) = VALUE OF THE RELATIVE PERMITTIVITY FOR ELEMENT I
0018      C      C(I,J) = GLOBAL COEFFICIENT MATRIX
0019      C      X(I), Y(I) = GLOBAL COORDINATES OF NODE I
0020      C      XL(J), YL(J) = LOCAL COORDINATES OF NODE J = 1,2,3
0021      C      V(I) = POTENTIAL AT NODE I
0022      C      MATRICES P(I) AND Q(I) ARE DEFINED IN EQ.(6.1.1)
0023
0024             DIMENSION X(100), Y(100), C(100,100), CE(100,100)
0025             DIMENSION NL(100,3), NDP(100), VAL(100),LF(100)
0026             DIMENSION V(100),P(3),Q(3),XL(3),YL(3),ER(100)
0027
0028      C      ****************************************************
0029      C      FIRST STEP - INPUT DATA DEFINING GEOMETRY AND
```

426 Numerical Techniques in Electromagnetics

Table 6.2 Element-node Identification

Element	Local node no. 1	2	3	Element	Local node no. 1	2	3
1	1	2	7	14	9	10	14
2	2	8	7	15	10	15	14
3	2	3	8	16	10	11	15
4	3	9	8	17	12	13	16
5	3	4	9	18	13	17	16
6	4	10	9	19	13	14	17
7	4	5	10	20	14	18	17
8	5	11	10	21	14	15	18
9	5	6	11	22	16	17	19
10	7	8	12	23	17	20	19
11	8	13	12	24	17	18	20
12	8	9	13	25	19	20	21
13	9	14	13				

```
0030      C                    BOUNDARY CONDITIONS
0031      C     ********************************************************
0032
0033            NI = 50  ! NO. OF ITERATIONS
0034            READ(5,*) NE,ND,NP
0035            READ(5,*)( I, ( NL(I,J), J=1,3),I=1,NE)
0036            READ(5,*) (I, X(I), Y(I), I=1,ND)
0037            READ(5,*) ( NDP(I), VAL(I), I=1,NP)
0038            PIE = 4.0*ATAN(1.0)
0039            E0 = 1.0E-9/(36.0*PIE)
0040            DO 10 I=1,NE
0041            ER(I) = 1.0
0042     10     CONTINUE
0043      C     ********************************************************
0044      C     SECOND STEP - EVALUATE COEFFICIENT MATRIX FOR EACH
0045      C                   ELEMENT AND ASSEMBLE GLOBALLY
0046      C     ********************************************************
0047            DO 20 M =1, ND
0048            DO 20 N=1,ND
0049            C(M,N) = 0.0
0050     20     CONTINUE
0051            DO 70 I = 1, NE
0052      C     FIND LOCAL COORDINATES XL(J), YL(J) FOR ELEMENT I
0053            DO 30 J=1,3
0054            K=NL(I,J)
0055            XL(J) = X(K)
0056            YL(J) = Y(K)
0057     30     CONTINUE
0058            P(1) = YL(2) - YL(3)
```

Finite Element Method 427

Table 6.3 Prescribed Potentials at Fixed Nodes

Node	Prescribed potential	Node	Prescribed potential
1	0.0	18	100.0
2	0.0	20	100.0
3	0.0	21	50.0
4	0.0	19	0.0
5	0.0	16	0.0
6	50.0	12	0.0
11	100.0	7	0.0
15	100.0		

```
0059          P(2) = YL(3) - YL(1)
0060          P(3) = YL(1) - YL(2)
0061          Q(1) = XL(3) - XL(2)
0062          Q(2) = XL(1) - XL(3)
0063          Q(3) = XL(2) - XL(1)
0064          AREA = 0.5*ABS( P(2)*Q(3) - Q(2)*P(3) )
0065    C     DETERMINE COEFFICIENT MATRIX FOR ELEMENT I
0066          DO 40 M=1,3
0067          DO 40 N=1,3
0068          CE(M,N) = ER(I)*( P(M)*P(N) + Q(M)*Q(N) )/(4.0*AREA)
0069    40    CONTINUE
0070    C     ASSEMBLE GLOBALLY - FIND C(I,J)
0071          DO 60 J=1,3
0072          IR = NL(I,J)
0073          DO 50 L=1,3
0074          IC = NL(I,L)
0075          C(IR,IC) = C(IR,IC) + CE(J,L)
0076    50    CONTINUE
0077    60    CONTINUE
0078    70    CONTINUE
0079    C     **************************************************
0080    C     THIRD STEP - SOLVE THE RESULTING SYSTEM
0081    C                  ITERATIVELY
0082    C     **************************************************
0083    C
0084    C     INITIALIZE AND DETERMINE LF(I) - LIST OF FREE NODES I
0085    C
0086          NF = 0
0087          DO 120 I=1,ND
0088          V(I) = 0.0
0089          DO 110 K=1,NP ! CHECK IF NODE I IS A PRESCRIBED NODE
0090          IF(I.EQ.NDP(K)) THEN
0091             V(I) = VAL(K)
0092          print *, i, v(i)
0093             GO TO 120
0094          ENDIF
0095    110   CONTINUE
0096          NF = NF + 1
0097          LF(NF) = I ! IF I IS NOT A PRESCRIBED NODE, IT IS FREE
0098    120   CONTINUE
```

428 Numerical Techniques in Electromagnetics

Table 6.4 Output Data of the Program in Fig. 6.10. No. of Nodes = 21, No. of Elements = 25, No. of Fixed Nodes = 15.

Node	X	Y	Potential
1	0.00	0.00	0.000
2	0.20	0.00	0.000
3	0.40	0.00	0.000
4	0.60	0.00	0.000
5	0.80	0.00	0.000
6	1.00	0.00	50.000
7	0.00	0.20	0.000
8	0.20	0.20	18.182
9	0.40	0.20	36.364
10	0.60	0.20	59.091
11	0.80	0.20	100.000
12	0.00	0.40	0.000
13	0.20	0.40	36.364
14	0.40	0.40	68.182
15	0.60	0.40	100.000
16	0.00	0.60	0.000
17	0.20	0.60	59.091
18	0.40	0.60	100.000
19	0.00	0.80	0.000
20	0.20	0.80	100.00
21	0.00	1.00	50.00

```
0099              PRINT *,NF,ND-NP,'CHECK IF THESE ARE EQUAL'
0100        C
0101        C     NOW, APPLY ITERATIVE METHOD
0102        C
0103              DO 150 N = 1,NI
0104              DO 140 I = 1,NF
0105              SUM = 0.0
0106              K = LF(I)
0107              DO 130 J=1,ND
0108               IF(J.EQ.K) GO TO 130
0109               SUM = SUM + V(J)*C(J,K)
0110        130   CONTINUE
0111              V(K) = - SUM/C(K,K)  ! APPLIES ONLY TO FREE NODES
0112        140   CONTINUE
```

```
0113      150       CONTINUE
0114      C         ****************************************************
0115      C         FOURTH STEP - FINALLY OUTPUT THE RESULTS
0116      C         ****************************************************
0117                WRITE(6,170) ND,NE,NP
0118                DO 160 I=1,ND
0119                WRITE(6,*)I, X(I),Y(I),V(I)
0120      160       CONTINUE
0121      170       FORMAT(2X,'NO. OF NODES = ',I3,2X,'NO. OF ELEMENTS =',
0122            1   I3,2X,'NO. OF FIXED NODES = ',I3,/)
0123                STOP
0124                END
```

Figure 6.10 Computer program for Example 6.2.

6.3 Solution of Poisson's Equation

To solve the two-dimensional Poisson's equation,

$$\nabla^2 V = -\frac{\rho_s}{\epsilon} \tag{6.34}$$

using FEM, we take the same steps as in Section 6.2. Since the steps are essentially the same as in Section 6.2 except that we must include the source term, only the major differences will be highlighted here.

6.3.1 Deriving Element-governing Equations

After the solution region is divided into triangular elements, we approximate the potential distribution $V_e(x,y)$ and the source term ρ_{se} (for two-dimensional problems) over each triangular element by linear combinations of the local interpolation polynomial α_i, i.e.,

$$\boxed{V_e = \sum_{i=1}^{3} V_{ei}\alpha_i(x,y),} \tag{6.35}$$

$$\boxed{\rho_{se} = \sum_{i=1}^{3} \rho_{ei}\alpha_i(x,y).} \tag{6.36}$$

The coefficients V_{ei} and ρ_{ei}, respectively, represent the values of V and ρ_s at vertex i of element e as in Fig. 6.3. The values of ρ_{ei} are known since $\rho_s(x,y)$ is prescribed, while the values of V_{ei} are to be determined.

From Table 4.1, an energy functional whose associated Euler equation is Eq. (6.34) is

$$F(V_e) = \frac{1}{2}\int_S \left[\epsilon|\nabla V_e|^2 - 2\rho_{se}V_e\right] dS. \tag{6.37}$$

$F(V_e)$ represents the total energy per length within element e. The first term under the integral sign, $\frac{1}{2}\mathbf{D}\cdot\mathbf{E} = \frac{1}{2}\epsilon|\nabla V_e|^2$, is the energy density in the electrostatic system, while the second term, $\rho_{se} V_e\, dS$, is the work done in moving the charge $\rho_{se}\, dS$ to its location at potential V_e. Substitution of Eqs. (6.35) and (6.36) into Eq. (6.37) yields

$$F(V_e) = \frac{1}{2}\sum_{i=1}^{3}\sum_{j=1}^{3}\epsilon V_{ei}\left[\int \nabla\alpha_i \cdot \nabla\alpha_j\, dS\right]V_{ej} - \sum_{i=1}^{3}\sum_{j=1}^{3}V_{ei}\left[\int \alpha_i\alpha_j\, dS\right]\rho_{ej}.$$

This can be written in matrix form as

$$F(V_e) = \frac{1}{2}\epsilon[V_e]^t[C^{(e)}][V_e] - [V_e]^t[T^{(e)}][\rho_e] \quad (6.38)$$

where

$$C_{ij}^{(e)} = \int \nabla\alpha_i \cdot \nabla\alpha_j\, dS \quad (6.39)$$

which is already defined in Eq. (6.17) and

$$T_{ij}^{(e)} = \int \alpha_i\alpha_j\, dS. \quad (6.40)$$

It will be shown in Section 6.8 that

$$\boxed{T_{ij}^{(e)} = \begin{cases} A/12, & i \neq j \\ A/6, & i = j \end{cases}} \quad (6.41)$$

where A is the area of the triangular element.

Equation (6.38) can be applied to every element in the solution region. All the resulting equations can be assembled into one global discretized functional. We obtain the discretized functional for the whole solution region (with N elements and n nodes) as the sum of the functionals for the individual elements, i.e., from Eq. (6.38),

$$F(V) = \sum_{e=1}^{N} F(V_e) = \frac{1}{2}\epsilon[V]^t[C][V] - [V]^t[T][\rho] \quad (6.42)$$

where t denotes transposition. In Eq. (6.42), the column matrix $[V]$ consists of the values of V_{ei}, while the column matrix $[\rho]$ contains n values of the source function ρ_s at the nodes. The functional in Eq. (6.42) is now minimized by differentiating with respect to V_{ei} and setting the result equal to zero.

6.3.2 Solving the Resulting Equations

The resulting equations can be solved by either the iteration method or the band matrix method as discussed in Section 6.2.4.

Iteration Method: Consider a solution region in Fig. 6.6 having five nodes so that $n = 5$. From Eq. (6.42),

$$F = \frac{1}{2}\epsilon [V_1 \ V_2 \ \cdots \ V_5] \begin{bmatrix} C_{11} & C_{12} & \cdots & C_{15} \\ C_{21} & C_{22} & \cdots & C_{25} \\ \vdots & & & \vdots \\ C_{51} & C_{52} & \cdots & C_{55} \end{bmatrix} \begin{bmatrix} V_1 \\ V_2 \\ \vdots \\ V_5 \end{bmatrix}$$

$$- [V_1 \ V_2 \ \cdots \ V_5] \begin{bmatrix} T_{11} & T_{12} & \cdots & T_{15} \\ T_{21} & T_{22} & \cdots & T_{25} \\ \vdots & & & \vdots \\ T_{51} & T_{52} & \cdots & T_{55} \end{bmatrix} \begin{bmatrix} \rho_1 \\ \rho_2 \\ \vdots \\ \rho_5 \end{bmatrix}.$$

(6.43)

We minimize the energy by applying

$$\frac{\partial F}{\partial V_k} = 0, \quad k = 1, 2, \cdots, n. \tag{6.44}$$

From Eq. (6.43), we get $\frac{\partial F}{\partial V_1} = 0$, for example, as

$$\frac{\partial F}{\partial V_1} = \epsilon \left[V_1 C_{11} + V_2 C_{21} + \cdots + V_5 C_{51} \right] - \left[T_{11}\rho_1 + T_{21}\rho_2 + \cdots + T_{51}\rho_5 \right] = 0$$

or

$$V_1 = -\frac{1}{C_{11}} \sum_{i=2}^{5} V_i C_{i1} + \frac{1}{\epsilon C_{11}} \sum_{i=1}^{5} T_{i1}\rho_i. \tag{6.45}$$

Thus, in general, for a mesh with n nodes

$$\boxed{V_k = -\frac{1}{C_{kk}} \sum_{i=1, i\neq k}^{n} V_i C_{ki} + \frac{1}{\epsilon C_{kk}} \sum_{i=1}^{n} T_{ki}\rho_i} \tag{6.46}$$

where node k is assumed to be a free node.

By fixing the potential at the prescribed nodes and setting the potential at the free nodes initially equal to zero, we apply Eq. (6.46) iteratively to all free nodes until convergence is reached.

Band Matrix Method: If we choose to solve the problem using the band matrix method, we let the free nodes be numbered first and the prescribed nodes last. By so doing, Eq. (6.42) can be written as

$$F(V) = \frac{1}{2}\epsilon [V_f \ V_p] \begin{bmatrix} C_{ff} & C_{fp} \\ C_{pf} & C_{pp} \end{bmatrix} \begin{bmatrix} V_f \\ V_p \end{bmatrix} - [V_f \ V_p] \begin{bmatrix} T_{ff} & T_{fp} \\ T_{pf} & T_{pp} \end{bmatrix} \begin{bmatrix} \rho_f \\ \rho_p \end{bmatrix}. \quad (6.47)$$

Minimizing $F(V)$ with respect to V_f, i.e.,

$$\frac{\partial F}{\partial V_f} = 0$$

gives

$$0 = \epsilon(C_{ff}V_f + C_{pf}V_p) - (T_{ff}\rho_f + T_{fp}\rho_p)$$

or

$$\boxed{[C_{ff}][V_f] = -[C_{fp}][V_p] + \frac{1}{\epsilon}[T_{ff}][\rho_f] + \frac{1}{\epsilon}[T_{fp}][\rho_p]} \quad (6.48)$$

This can be written as

$$[A][V] = [B] \quad (6.49)$$

where $[A] = [C_{ff}]$, $[V] = [V_f]$ and $[B]$ is the right-hand side of Eq. (6.48). Equation (6.49) can be solved to determine $[V]$ either by matrix inversion or Gaussian elimination technique discussed in Appendix D. There is little point in giving examples on applying FEM to Poisson's problems, especially when it is noted that the difference between Eqs. (6.28) and (6.46) or Eqs. (6.48) and (6.31) is slight. See [19] for an example.

6.4 Solution of the Wave Equation

A typical wave equation is the inhomogeneous scalar Helmholtz's equation

$$\nabla^2 \Phi + k^2 \Phi = g \quad (6.50)$$

where Φ is the field quantity (for waveguide problem, $\Phi = H_z$ for TE mode or E_z for TM mode) to be determined, g is the source function, and $k = \omega\sqrt{\mu\epsilon}$ is the wave number of the medium. The following three distinct special cases of Eq. (6.50) should be noted:

(i) $k = 0 = g$: Laplace's equation;
(ii) $k = 0$: Poisson's equation; and
(iii) k is an unknown, $g = 0$: homogeneous, scalar Helmholtz's equation.

We know from Chapter 4 that the variational solution to the operator equation
$$L\Phi = g \tag{6.51}$$
is obtained by extremizing the functional
$$I(\Phi) = <L, \Phi> - 2<\Phi, g>. \tag{6.52}$$
Hence the solution of Eq. (6.50) is equivalent to satisfying the boundary conditions and minimizing the functional
$$I(\Phi) = \frac{1}{2} \int\int \left[|\nabla\Phi|^2 - k^2\Phi^2 + 2\Phi g\right] dS. \tag{6.53}$$
If other than the natural boundary conditions (i.e., Dirichlet or homogeneous Neumann conditions) must be satisfied, appropriate terms must be added to the functional as discussed in Chapter 4.

We now express potential Φ and source function g in terms of the shape functions α_i over a triangular element as

$$\boxed{\Phi_e(x,y) = \sum_{i=1}^{3} \alpha_i \Phi_{ei}} \tag{6.54}$$

$$\boxed{g_e(x,y) = \sum_{i=1}^{3} \alpha_i g_{ei}} \tag{6.55}$$

where Φ_{ei} and g_{ei} are, respectively, the values of Φ and g at nodal point i of element e.

Substituting Eqs. (6.54) and (6.55) into Eq. (6.53) gives

$$\begin{aligned} I(\Phi_e) &= \frac{1}{2}\sum_{i=1}^{3}\sum_{j=1}^{3}\Phi_{ei}\Phi_{ej}\int\int \nabla\alpha_i\cdot\nabla\alpha_j \, dS \\ &\quad - \frac{k^2}{2}\sum_{i=1}^{3}\sum_{j=1}^{3}\Phi_{ei}\Phi_{ej}\int\int \alpha_i\alpha_j \, dS \\ &\quad + \sum_{i=1}^{3}\sum_{j=1}^{3}\Phi_{ei}g_{ej}\int\int \alpha_i\alpha_j \, dS \\ &= \frac{1}{2}[\Phi_e]^t\,[C^{(e)}]\,[\Phi_e] \\ &\quad - \frac{k^2}{2}[\Phi_e]^t\,[T^{(e)}]\,[\Phi_e] + [\Phi_e]^t\,[T^{(e)}]\,[G_e] \end{aligned} \tag{6.56}$$

where $[\Phi_e] = [\Phi_{e1}, \Phi_{e2}, \Phi_{e3}]^t$, $[G_e] = [g_{e1}, g_{e2}, g_{e3}]^t$, and $[C^{(e)}]$ and $[T^{(e)}]$ are defined in Eqs. (6.17) and (6.41), respectively.

Equation (6.56), derived for a single element, can be applied for all N elements in the solution region. Thus,

$$I(\Phi) = \sum_{e=1}^{N} I(\Phi_e). \tag{6.57}$$

From Eqs. (6.56) and (6.57), $I(\Phi)$ can be expressed in matrix form as

$$I(\Phi) = \frac{1}{2}[\Phi]^t [C][\Phi] - \frac{k^2}{2}[\Phi]^t [T] [\Phi] + [\Phi]^t [T] [G] \tag{6.58}$$

where

$$[\Phi] = [\Phi_1, \Phi_2, \cdots, \Phi_N]^t, \tag{6.59a}$$

$$[G] = [g_1, g_2, \cdots, g_N]^t, \tag{6.59b}$$

$[C]$, and $[T]$ are global matrices consisting of local matrices $[C^{(e)}]$ and $[T^{(e)}]$, respectively.

Consider the special case in which the source function $g = 0$. Again, if free nodes are numbered first and the prescribed nodes last, we may write Eq. (6.58) as

$$I = \frac{1}{2}[\Phi_f \ \Phi_p] \begin{bmatrix} C_{ff} & C_{fp} \\ C_{pf} & C_{pp} \end{bmatrix} \begin{bmatrix} \Phi_f \\ \Phi_p \end{bmatrix} - \frac{k^2}{2}[\Phi_f \ \Phi_p] \begin{bmatrix} T_{ff} & T_{fp} \\ T_{pf} & T_{pp} \end{bmatrix} \begin{bmatrix} \Phi_f \\ \Phi_p \end{bmatrix}. \tag{6.60}$$

Setting $\dfrac{\partial I}{\partial \Phi_f}$ equal to zero gives

$$[C_{ff} \ C_{fp}] \begin{bmatrix} \Phi_f \\ \Phi_p \end{bmatrix} - k^2 [T_{ff} \ T_{fp}] \begin{bmatrix} \Phi_f \\ \Phi_p \end{bmatrix} = 0. \tag{6.61}$$

For TM modes, $\Phi_p = 0$ and hence

$$[C_{ff} - k^2 T_{ff}]\Phi_f = 0. \tag{6.62}$$

Premultiplying by T_{ff}^{-1} gives

$$\boxed{[T_{ff}^{-1} C_{ff} - k^2 I]\Phi_f = 0} \tag{6.63}$$

Letting

$$A = T_{ff}^{-1} C_{ff}, \qquad k^2 = \lambda, \qquad X = \Phi_f \tag{6.64a}$$

we obtain the standard eigenproblem

$$(A - \lambda I)X = 0 \qquad (6.64b)$$

where I is a unit matrix. Any standard procedure [7] (or see Appendix D) may be used to obtain some or all of the eigenvalues $\lambda_1, \lambda_2, \cdots, \lambda_{n_f}$ and eigenvectors $X_1, X_2, \cdots, X_{n_f}$, where n_f is the number of free nodes. The eigenvalues are always real since C and T are symmetric.

Solution of the algebraic eigenvalue problems in Eq. (6.64) furnishes eigenvalues and eigenvectors, which form good approximations to the eigenvalues and eigenfunctions of the Helmholtz problem, i.e., the cutoff wavelengths and field distribution patterns of the various modes possible in a given waveguide.

The solution of the problem presented in this section, as summarized in Eq. (6.63), can be viewed as the finite element solution of homogeneous waveguides. The idea can be extended to handle inhomogeneous waveguide problems [8–11]. However, in applying FEM to inhomogeneous problems, a serious difficulty is the appearance of spurious, nonphysical solutions. Several techniques have been proposed to overcome the difficulty [12–18].

Example 6.3

To apply the ideas presented in this section, we use the finite element analysis to determine the lowest (or dominant) cutoff wavenumber k_c of the TM$_{11}$ mode in waveguides with square ($a \times a$) and rectangular ($a \times b$) cross sections for which the exact results are already known as

$$k_c = \sqrt{(m\pi/a)^2 + (n\pi/b)^2}$$

where $m = n = 1$.

It may be instructive to try with hand calculation the case of a square waveguide with 2 divisions in the x and y directions. In this case, there are 9 nodes, 8 triangular elements, and 1 free node ($n_f = 1$). Equation (6.62) becomes

$$C_{11} - k^2 T_{11} = 0$$

where C_{11} and T_{11} are obtained from Eqs. (6.1.1), (6.1.2), and (6.41) as

$$C_{11} = \frac{a^2}{2A}, \qquad T_{11} = A, \qquad A = \frac{a^2}{8}.$$

Hence

436   Numerical Techniques in Electromagnetics

$$k^2 = \frac{a^2}{2A^2} = \frac{32}{a^2}$$

or

$$ka = 5.656$$

which is about 27% off the exact solution. To improve the accuracy, we must use more elements.

The computer program in Fig. 6.11 applies the ideas in this section to find k_c. The main program calls subroutine GRID (to be discussed in Section 6.5) to generate the necessary input data from a given geometry. If n_x and n_y are the number of divisions in the x and y directions, the total number of elements $n_e = 2n_x n_y$. By simply specifying the values of a, b, n_x, and n_y, the program determines k_c using subroutines GRID, INVERSE, and POWER or EIGEN. Subroutine INVERSE available in Appendix D finds T_{ff}^{-1} required in Eq. (6.64a). Either subroutine POWER or EIGEN calculates the eigenvalues. EIGEN finds all the eigenvalues, while POWER only determines the lowest eigenvalue; both subroutines are available in Appendix D. The results for the square ($a = b$) and rectangular ($b = 2a$) waveguides are presented in Tables 6.5a and 6.5b, respectively.

```
0001      C***********************************************************
0002      C      FINITE ELEMENT SOLUTION OF THE WAVE EQUATION
0003      C      TRIANGULAR ELEMENTS ARE USED
0004      C
0005      C      ND = NO. OF NODES
0006      C      NE = NO. OF ELEMENTS
0007      C      NL(I,J) = LIST OF NODES FOR EACH ELEMENT I, WHERE
0008      C      CE(I,J) = ELEMENT COEFFICIENT MATRIX
0009      C      C(I,J)  = GLOBAL COEFFICIENT MATRIX
0010      C      X(I), Y(I) = GLOBAL COORDINATES OF NODE I
0011      C      XL(J), YL(J) = LOCAL COORDINATES OF NODE J = 1,2,3
0012      C      MATRICES P(I) AND Q(I) ARE DEFINED IN EQ.(9)
0013      C      LF(I) = LIST OF FREE NODES
0014      C      ALAM(I) = CONTAINS EIGENVALUES
0015
0016             DIMENSION C(400,400), CE(400,400),LF(400)
0017             DIMENSION V(400), P(3), Q(3), XL(3), YL(3)
0018             DIMENSION T(400,400), A(400,400), ALAM(400)
0019             COMMON X(400),Y(400),DX(50),DY(50),NL(400,3),NDP(400)
0020
0021      C      ***********************************************************
0022      C      FIRST STEP - INPUT DATA DEFINING GEOMETRY AND
0023      C                   BOUNDARY CONDITIONS (USE SUBROUTINE GRID)
0024      C      ***********************************************************
0025             PRINT *, 'INPUT NX'
0026             READ(5,*)  NX
0027             NY = 2.0*NX
0028             AA = 1.0
0029             BB = 2.0
0030             DELTAX = AA/FLOAT(NX)
0031             DELTAY = BB/FLOAT(NY)
```

Table 6.5a Lowest Wavenumber for a Square Waveguide ($b = a$)

n_x	n_e	$k_c a$	% error
2	8	5.656	27.3
3	18	5.030	13.2
5	50	4.657	4.82
7	98	4.553	2.47
10	200	4.497	1.22

Exact: $k_c a = 4.4429$, $n_y = n_x$

```
0032              DO 10 I=1,NX
0033              DX(I)=DELTAX
0034        10    CONTINUE
0035              DO 20 I=1,NY
0036              DY(I)=DELTAY
0037        20    CONTINUE
0038              CALL GRID(NX,NY,ND,NE,NP)
0039         C    *****************************************************
0040         C    SECOND STEP - EVALUATE COEFFICIENT MATRIX FOR EACH
0041         C                  ELEMENT AND ASSEMBLE GLOBALLY
0042         C    *****************************************************
0043              DO 30 M =1, ND
0044              DO 30 N=1,ND
0045              C(M,N) = 0.0
0046        30    CONTINUE
0047              DO 80 I = 1, NE
0048              DO 40 J=1,3
0049              K=NL(I,J)
0050              XL(J) = X(K)
0051              YL(J) = Y(K)
0052        40    CONTINUE
0053              P(1) = YL(2) - YL(3)
0054              P(2) = YL(3) - YL(1)
0055              P(3) = YL(1) - YL(2)
0056              Q(1) = XL(3) - XL(2)
0057              Q(2) = XL(1) - XL(3)
0058              Q(3) = XL(2) - XL(1)
0059              AREA = 0.5*ABS( P(2)*Q(3) - Q(2)*P(3) )
0060         C    DETERMINE COEFFICIENT MATRIX FOR ELEMENT I
0061              DO 50 M=1,3
0062              DO 50 N=1,3
0063              CE(M,N) = ( P(M)*P(N) + Q(M)*Q(N) )/(4.0*AREA)
0064        50    CONTINUE
0065         C    ASSEMBLE GLOBALLY - FIND C(I,J) AND T(I,J)
0066              DO 70 J=1,3
0067              IR = NL(I,J)
0068              DO 60 L=1,3
0069              IC = NL(I,L)
0070              C(IR,IC) = C(IR,IC) + CE(J,L)
0071              IF(J.EQ.L) THEN
0072              T(IR,IC) = T(IR,IC) + AREA/6.0
0073              GO TO 60
0074              ELSE
```

Table 6.5b Lowest Wavenumber for a Rectangular Waveguide ($b = 2a$)

n_x	n_e	$k_c a$	% error
2	16	4.092	16.5
4	64	3.659	4.17
6	144	3.578	1.87
8	256	3.549	1.04

Exact: $k_c a = 3.5124$, $n_y = 2n_x$

```
0075              T(IR,IC) = T(IR,IC) + AREA/12.0
0076           ENDIF
0077     60    CONTINUE
0078     70    CONTINUE
0079     80    CONTINUE
0080           PRINT *, 'C AND T HAVE BEEN CALCULATED'
0081     C     ******************************************************
0082     C     THIRD STEP - SOLVE THE RESULTING SYSTEM
0083     C     ******************************************************
0084     C     DETERMINE LF(I) - LIST OF FREE NODES
0085           NF = 0
0086           DO 100 I=1,ND
0087           DO 90 K=1,NP ! CHECK IF NODE I IS PRESCRIBED
0088           IF(I.EQ.NDP(K)) GO TO 100
0089     90    CONTINUE
0090           NF = NF + 1
0091           LF(NF) = I   ! NODE I IS FREE
0092     100   CONTINUE
0093           PRINT *,NF,ND-NP,' CHECK IF THESE ARE EQUAL'
0094     C
0095     C     FROM GLOBAL C AND T, FIND C_ff AND T_ff
0096     C
0097           DO 110 I=1,NF
0098           DO 110 J=1,NF
0099           C(I,J) = C(LF(I),LF(J))
0100           T(I,J) = T(LF(I),LF(J))
0101     110   CONTINUE
0102           NMAX = 400
0103           CALL INVERSE(T,NF,NMAX)
0104           DO 120 I = 1,NF
0105           DO 120 J = 1,NF
0106           DO 120 K=1,NF
0107           A(I,J) = A(I,J) + T(I,K)*C(K,J)
0108     120   CONTINUE
0109     C     CALL INVERSE(A,NF,NMAX)
0110     C     CALL POWER(A,ALAMBDA,X,NMAX,NF,IT)
0111           CALL  EIGEN(A,X,NMAX,NF,ALAM)
0112     C     ******************************************************
0113     C     FOURTH STEP - OUTPUT THE RESULTS
0114     C     ******************************************************
0115           WRITE(6,130) ND,NE,NP
0116     130   FORMAT(2X,'NO. OF NODES = ',I3,2X,'NO. OF ELEMENTS =',
0117          1 I3,2X,'NO. OF PRESCRIBED NODES',I3,/)
0118     C     AK = 1.0/SQRT(ALAMBDA)
0119     C     WRITE (6,*)NX,NY,AK,IT
```

```
0120            DO 140 I=1,NF
0121            ALAM(I) = SQRT( ALAM(I) )
0122            PRINT *,I,ALAM(I)
0123            WRITE(6,*) I,ALAM(I)
0124       140  CONTINUE
0125            STOP
0126            END
```

Figure 6.11 Computer program for Example 6.3.

6.5 Automatic Mesh Generation I—Rectangular Domains

One of the major difficulties encountered in the finite element analysis of continuum problems is the tedious and time-consuming effort required in data preparation. Efficient finite element programs must have node and element generating schemes, referred to collectively as *mesh generators*. Automatic mesh generation minimizes the input data required to specify a problem. It not only reduces the time involved in data preparation, it eliminates human errors introduced when data preparation is performed manually. Combining the automatic mesh generation program with computer graphics is particularly valuable since the output can be monitored visually. Since some applications of the FEM to EM problems involve simple rectangular domains, we consider the generation of simple meshes [19] here; automatic mesh generator for arbitrary domains will be discussed in Section 6.6.

Consider a rectangular solution region of size $a \times b$ as in Fig. 6.12. Our goal is to divide the region into rectangular elements, each of which is later divided into two triangular elements. Suppose n_x and n_y are the number of divisions in x and y directions, the total number of elements and nodes are, respectively, given by

$$n_e = 2\, n_x n_y$$
$$n_d = (n_x + 1)(n_y + 1). \qquad (6.65)$$

Thus it is easy to figure out from Fig. 6.12 a systematic way of numbering the elements and nodes. To obtain the global coordinates (x, y) for each node, we need an array containing $\Delta x_i, i = 1, 2, \cdots, n_x$ and $\Delta y_j, j = 1, 2, \cdots, n_y$, which are, respectively, the distances between nodes in the x and y directions. If the order of node numbering is from left to right along horizontal rows and from bottom to top along the vertical rows, then the first node is the origin (0,0). The next node is obtained as $x \to x + \Delta x_1$ while $y = 0$ remains unchanged. The following node has $x \to x + \Delta x_2$, $y = 0$,

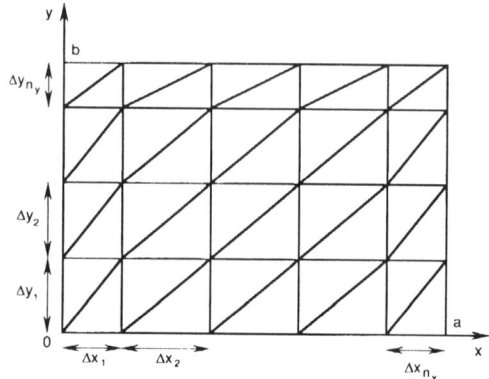

Figure 6.12 Discretization of a rectangular region into a nonuniform mesh.

and so on until Δx_i are exhausted. We start the second next horizontal row by starting with $x = 0$, $y \to y + \Delta y_1$ and increasing x until Δx_i are exhausted. We repeat the process until the last node $(n_x + 1)(n_y + 1)$ is reached, i.e., when Δx_i and Δy_j are exhausted simultaneously.

The procedure presented here allows for generating uniform and nonuniform meshes. A mesh is uniform if all Δx_i are equal and all Δy_i are equal; it is nonuniform otherwise. A nonuniform mesh is preferred if it is known in advance that the parameter of interest varies rapidly in some parts of the solution domain. This allows a concentration of relatively small elements in the regions where the parameter changes rapidly, particularly since these regions are often of greatest interest in the solution. Without the preknowledge of the rapid change in the unknown parameter, a uniform mesh can be used. In that case, we set

$$\Delta x_1 = \Delta x_2 = \cdots = h_x$$
$$\Delta y_1 = \Delta y_2 = \cdots = h_y \qquad (6.66)$$

where $h_x = a/n_x$ and $h_y = b/n_y$.

In some cases, we also need a list of prescribed nodes. If we assume that all boundary points have prescribed potentials, the number n_p of prescribed node is given by

$$n_p = 2(n_x + n_y). \qquad (6.67)$$

A simple way to obtain the list of boundary points is to enumerate points on the bottom, right, top, and left sides of the rectangular region in that order.

The ideas presented here are implemented in the subroutine GRID in Fig. 6.13. The subroutine can be used for generating a uniform or nonuniform mesh out of a given rectangular region. If a uniform mesh is desired, the required input parameters are a, b, n_x, and n_y. If, on the other hand, a nonuniform mesh is required, we need to supply $n_x, n_y, \Delta x_i, i = 1, 2, \cdots, n_x$, and $\Delta y_j, \ j = 1, 2, \cdots, n_y$. The output parameters are n_e, n_d, n_p, connectivity list, the global coordinates (x, y) of each node, and the list of prescribed nodes. It is needless to say that subroutine GRID is not useful for a nonrectangular solution region. See the program in Fig. 6.11 as an example on how to use subroutine GRID. A more general program for discretizing a solution region of any shape will be presented in the next section.

```
0001      C****************************************************
0002      C      SUBROUTINE GRID
0003      C      THIS PROGRAM DIVIDES A RECTANGULAR DOMAIN INTO
0004      C      TRIANGULAR ELEMENTS (NX BY NY NONUNIFORM
0005      C      MESH IN GENERAL)
0006      C      NX & NY ARE THE NO.S OF SUBDIVISION ALONG X & Y AXES
0007      C      NE = NO. OF ELEMENTS IN THE MESH
0008      C      ND = NO. OF NODES IN THE MESH
0009      C      NP = NO. OF BOUNDARY (PRESCRIBED) NODES
0010      C      X(I) & Y(I) ARE GLOBAL COORDINATES OF NODE I
0011      C      DX(I) & DY(I) ARE DISTANCES BETWEEN NODES ALONG X & Y AXES
0012      C      NL(I,J) IS THE LIST OF NODES FOR ELEMENT I, J=1,2,3 ARE
0013      C           LOCAL NUMBERS
0014      C      NDP(I) = LIST OF PRESCRIBED NODES I
0015      C
0016      C      REF: J. N. REDDY, "AN INTRODUCTION TO THE FINITE ELEMENT
0017      C           METHOD", NEW YORK: MCGRAW-HILL, 1984, P. 436.
0018
0019             SUBROUTINE GRID(NX,NY,ND,NE,NP)
0020             COMMON X(400),Y(400),DX(50),DY(50),NL(400,3),NDP(400)
0021
0022      C
0023      C      CALCULATE NE, ND, AND NP
0024      C
0025             NE=2*NX*NY
0026             NP = 2*(NX + NY)
0027             NX1=NX + 1
0028             NY1=NY + 1
0029             NXX1=2*NX
0030             NYY1=2*NY
0031             ND=NX1*NY1
0032      C
0033      C      DETERMINE NL(I,J) STARTING FROM LEFT BOTTOM CORNER
0034      C
0035             NL(1,1)=1
0036             NL(1,2)=NX1 + 2
0037             NL(1,3)=NX1 + 1
0038             NL(2,1)=1
0039             NL(2,2)=2
0040             NL(2,3)=NX1 + 2
0041             K=3
0042             DO 50 IY=1,NY
0043             L=IY*NXX1
0044             M=(IY - 1)*NXX1
0045             IF(NX.EQ.1) GO TO 30
0046             DO 20 N=K,L,2
0047             DO 10 I=1,3
0048             NL(N,I)=NL(N-2,I) + 1
```

```
0049      10       NL(N+1,I)=NL(N-1,I) + 1
0050      20       CONTINUE
0051      30       IF(NY.EQ.1) GO TO 50
0052               DO 40 I=1,3
0053               NL(L+1,I)=NL(M+1,I) + NX1
0054      40       NL(L+2,I)=NL(M+2,I) + NX1
0055      50       K=L + 3
0056      C
0057      C     DETERMINE X(I) AND Y(I)
0058      C
0059      60       L=0
0060               YC=0.0
0061               DO 80 J=1,NY1
0062               XC=0.0
0063               DO 70 I=1,NX1
0064               L=L + 1
0065               X(L)=XC
0066               Y(L)=YC
0067      70       XC=XC + DX(I)
0068      80       YC=YC + DY(J)
0069      C
0070      C     DETERMINE NDP(I)
0071      C
0072               N = 0
0073               DO 90 K=1,NX1
0074               N = N + 1
0075               NDP(N) = N
0076      90       CONTINUE    ! BOTTOM SIDE
0077               DO 100 K=1,NY
0078               N = N+1
0079               NDP(N) = NDP(N-1) + NX1
0080      100      CONTINUE    ! RIGHT SIDE
0081               DO 110 K=1,NX
0082               N = N + 1
0083               NDP(N) = NDP(N-1) - 1
0084      110      CONTINUE    ! TOP SIDE
0085               DO 120 K=1,NY-1
0086               N = N + 1
0087               NDP(N) = NDP(N-1) - NX1
0088      120      CONTINUE    ! LEFT SIDE
0089               WRITE(6,*) NE,ND,NP
0090               DO 130 I=1,NE
0091               WRITE(6,*) I,( NL(I,J),J=1,3)
0092      130      CONTINUE
0093               DO 140 I=1,ND
0094               WRITE(6,*) I,X(I),Y(I)
0095      140      CONTINUE
0096               DO 150 I=1,NP
0097               WRITE(6,*)  NDP(I)
0098      150 CONTINUE
0099               RETURN
0100               END
```

Figure 6.13 Subroutine GRID.

Finite Element Method 443

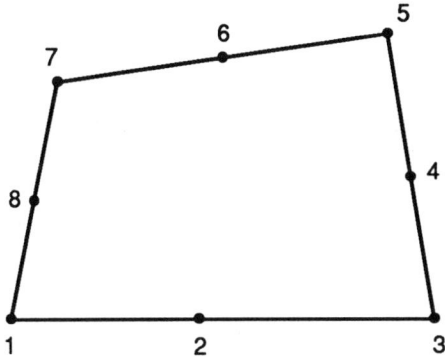

Figure 6.14 Typical quadrilateral block.

6.6 Automatic Mesh Generation II—Arbitrary Domains

As the solution regions become more complex than the ones considered in Section 6.5, the task of developing mesh generators becomes more tedious. A number of mesh generation algorithms [e.g., 21–33] of varying degrees of automation have been proposed for arbitrary solution domains. Reviews of various mesh generation techniques can be found in [34,35].

The basic steps involved in a mesh generation are as follows [36]:

- subdivide solution region into few quadrilateral blocks,
- separately subdivide each block into elements,
- connect individual blocks.

Each step is explained as follows.

6.6.1 Definition of the Blocks

The solution region is subdivided into quadrilateral blocks. Subdomains with different constitutive parameters (σ, μ, ϵ) must be represented by separate blocks. As input data, we specify block topologies and the coordinates at eight points describing each block. Each block is represented by an eight-node quadratic isoparametric element. With natural coordinate system (ζ, η), the x and y coordinates are represented as

$$x(\zeta, \eta) = \sum_{i=1}^{8} \alpha_i(\zeta, \eta) \, x_i \qquad (6.68)$$

$$y(\zeta, \eta) = \sum_{i=1}^{8} \alpha_i(\zeta, \eta) \, y_i \qquad (6.69)$$

444   Numerical Techniques in Electromagnetics

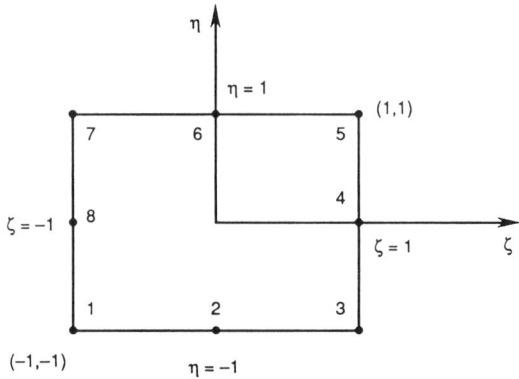

Figure 6.15 Eight-node Serendipity element.

where $\alpha_i(\zeta, \eta)$ is a shape function associated with node i, and (x_i, y_i) are the coordinates of node i defining the boundary of the quadrilateral block as shown in Fig. 6.14. The shape functions are expressed in terms of the quadratic or parabolic isoparametric elements shown in Fig. 6.15. They are given by:

$$\alpha_i = \frac{1}{4}(1+\zeta\zeta_i)(1+\eta\eta_i)(\zeta\zeta_i + \eta\eta_i + 1), \qquad i = 1,3,5,7 \qquad (6.70)$$

for corner nodes,

$$\alpha_i = \frac{1}{2}\zeta_i^2(1+\zeta\zeta_i)(1-\eta^2) + \frac{1}{2}\eta_i^2(1+\eta\eta_i)(1-\zeta^2), \qquad i = 2,4,6,8 \quad (6.71)$$

for midside nodes. Note the following properties of the shape functions:

(1) They satisfy the conditions

$$\sum_{i=1}^{8} \alpha_i(\zeta, \eta) = 1 \qquad (6.72a)$$

$$\alpha_i(\zeta_j, \eta_j) = \begin{cases} 1, & i = j \\ 0, & i \neq j. \end{cases} \qquad (6.72b)$$

(2) They become quadratic along element edges ($\zeta = \pm 1, \eta = \pm 1$).

6.6.2 Subdivision of Each Block

For each block, we specify $NDIVX$ and $NDIVY$, the number of element subdivisions to be made in the ζ and η direction, respectively. Also, we specify the weighting factors $(W_\zeta)_i$ and $(W_\eta)_i$ allowing for graded mesh within a block. In specifying $NDIVX, NDIVY, W_\zeta$, and W_η care must be taken to ensure that the subdivision along block interfaces (for adjacent blocks) are compatible. We initialize ζ and η to a value of -1 so that the natural coordinates are incremented according to

$$\zeta_i = \zeta_i + \frac{2(W_\zeta)_i}{W_\zeta^T \cdot F} \tag{6.73}$$

$$\eta_i = \eta_i + \frac{2(W_\eta)_i}{W_\eta^T \cdot F} \tag{6.74}$$

where

$$W_\zeta^T = \sum_{j=1}^{NDIVX} (W_\zeta)_j \tag{6.75a}$$

$$W_\eta^T = \sum_{j=1}^{NDIVX} (W_\eta)_j \tag{6.75b}$$

and

$$F = \begin{cases} 1, & \text{for linear elements} \\ 2, & \text{for quadratic elements.} \end{cases}$$

Three element types are permitted: (a) linear four-node quadrilateral elements, (b) linear three-node triangular elements, (c) quadratic eight-node isoparametric elements.

6.6.3 Connection of Individual Blocks

After subdividing each block and numbering its nodal points separately, it is necessary to connect the blocks and have each node numbered uniquely. This is accomplished by comparing the coordinates of all nodal points and assigning the same number to all nodes having identical coordinates. That is, we compare the coordinates of node 1 with all other nodes, and then node 2 with other nodes, etc., until all repeated nodes are eliminated. The listing of the FORTRAN code for automatic mesh generation is shown in Fig. 6.16; it is essentially a modified version of the one in Hinton and Owen [36]. The following example taken from [36] illustrates the application of the code.

```
0001      C****************************************************
0002      C THIS PROGRAM PERFORMS A MESH GENERATION OF AN
0003      C ARBITRARY SOLUTION DOMAIN USING A  SYSTEMATIC
0004      C APPROACH. A FEW POINTS ARE  GIVEN TO DETERMINE
0005      C THE  GENERAL  CONFIGURATION  OF  THE REGION.
0006      C THEN THE PROGRAM AUTOMATICALLY GENERATES
0007      C TRIANGULAR OR QUADRILATERAL ELEMENTS
0008      C     REFERENCE: HINTON AND OWEN [36]
0009      C****************************************************
0010            IMPLICIT INTEGER (I-N)
0011            IMPLICIT REAL (A-H,O-Z)
0012            COMMON /MESH1/COORD(1500,2),NL(750,8),MATNO(750),
0013          1 SHAPE(9),NP,NELEM,NTYPE,NDIME,NNODE
0014
0015      C THIS SUBROUTINE ACCEPTS DATA DEFINING THE SOLUTION REGION
0016            CALL INPUT
0017      C THIS SUBROUTINE UNDERTAKES THE MESH SUBDIVISION
0018            CALL GENERATE
0019      C THIS SUBROUTINE SUBDIVIDES INTO TRIANGULAR ELEMENTS
0020            IF(NTYPE.EQ.3)CALL TRIANGLE
0021      C THIS SUBROUTINE OUTPUTS THE GENERATED MESH
0022      C THE SUBROUTINE DOES NOT NEED TO BE CALLED IF A PLOTTING
0023      C SUBROUTINE USED IN DISPLAYING THE OUTPUT
0024            CALL OUTPUT
0025            STOP
0026            END

0001      C****************************************************
0002      C THIS SUBROUTINE ACCEPTS THE DATA WHICH DEFINES THE
0003      C SOLUTION REGION   OUTLINE AND THE MATERIAL ZONES
0004      C NP = NUMBER OF COORDINATE POINTS DEFINING THE
0005      C SOLUTION REGION
0006      C NELEM = NUMBER OF BLOCKS OR ZONES
0007      C NTYPE = THE TYPE OF ELEMENT INTO WHICH THE
0008      C STRUCTURE IS TO BE SUBDIVIDED
0009      C NDIME = THE NUMBER OF COORDINATE DIMENSIONS
0010      C           FOR A PLANE NDIME=2
0011      C NUMEL = BLOCK NUMBER
0012      C ( NL(NUMEL,INODE),INODE=1,NTYPE) )=THE
0013      C BLOCK TOPOLOGY DEFINITION
0014      C MATNO(NUMEL)THE MATERIAL IDENTIFICATION NUMBER:
0015      C INPUT SPECIFICATION FOR EACH BLOCK
0016      C JPOIN = POINT NUMBER
0017      C (COORD(JPOI,IDIME),IDIME=1,NDIME)=X&Y COORDINATES
0018
0019            SUBROUTINE INPUT
0020            COMMON/MESH1/COORD(1500,2),NL(750,8),
0021           1MATNO(750),SHAPE(9),NP,NELEM,NTYPE,NDIME,NNODE
0022            DATA LNODE/8/
0023      C
0024            READ(5,*) NP,NELEM,NTYPE,NDIME
0025            DO 10 IELEM=1,NELEM
0026            READ(5,*) NUMEL,( NL(NUMEL,I), I=1,LNODE ),
0027          1           MATNO(NUMEL)
0028       10   CONTINUE
0029            DO 20 IPOIN=1,NP
0030            READ(5,*)JPOIN, ( COORD(JPOIN,I), I=1,NDIME )
0031       20   CONTINUE
0032            RETURN
0033            END
```

```
0001        C***************************************************
0002        C THIS SUBROUTINE UNDERTAKES THE SUBDIVISION OF EACH
0003        C BLOCK AND ELIMINATES COMMON NODES ALONG BLOCK INTERFACES
0004        C KBLOC = BLOCK NUMBER
0005        C NDIVX/NDIVY = NUMBER OF ELEMENTS IN THE ZETA/ETA
0006        C DIRECTION INTO WHICH THE BLOCK IS TO BE SUBDIVIDED
0007        C WEITX(IDIVX) AND WEITHY = WEIGHTING FACTORS
0008        C
0009              SUBROUTINE GENERATE
0010              DIMENSION WEITX(40),WEITY(40),TCORD(81,2),
                 TNODS(50,8),
0011             1TMATO(50),LREPN(350),LASOC(350),LFINN(350),LFASC(350)
0012              COMMON/MESH1/COORD(1500,2),NL(750,8),MATNO(750),
0013             1SHAPE(9),NP,NELEM,NTYPE,NDIME,MNODE
0014              DATA MREPN/350/,MPOIN/1500/,LNODE/8/
0015        C
0016        C INITIALIZATION SECTION
0017        C
0018              DO 10 IREPN=1,MREPN
0019        10    LREPN(IREPN)=0
0020              NPONT=NP
0021              NBLOC=NELEM
0022              NP=0
0023              NELEM=0
0024              MNODE=4
0025              IF(NTYPE.EQ.8)MNODE=8
0026              KNODE=MNODE/4
0027              FNODE=KNODE
0028              DO 20 IPONT=1,NPONT
0029              DO 20 IDIME=1,NDIME
0030        20    TCORD(IPONT,IDIME)=COORD(IPONT,IDIME)
0031              DO 30 IPOIN=1,MPOIN
0032              DO 30 IDIME=1,NDIME
0033        30    COORD(IPOIN,IDIME)=0.0
0034              DO 40 IBLOC=1,NBLOC
0035              TMATO(IBLOC)=MATNO(IBLOC)
0036              DO 40 INODE=1,LNODE
0037        40    TNODS(IBLOC,INODE)=NL(IBLOC,INODE)
0038        C
0039        C READ AND WRITE BLOCK SUBDIVISION DATA
0040        C
0041              DO 170 IBLOC=1,NBLOC
0042              READ(5,*)KBLOC,NDIVX,NDIVY
0043              READ(5,*) ( WEITX(IDIVX), IDIVX=1,NDIVX )
0044              READ(5,*) ( WEITY(IDIVY), IDIVY=1,NDIVY )
0045        C
0046        C DIVIDE EACH BLOCK INTO ELEMENTS
0047        C
0048              TOTAL=0.0
0049              DO 50 IDIVX=1,NDIVX
0050              IF(WEITX(IDIVX).EQ.0.0)WEITX(IDIVX)=1.0
0051        50    TOTAL=TOTAL+WEITX(IDIVX)
0052              XNORM=2.0/TOTAL
0053              TOTAL=0.0
0054              DO 60 IDIVY=1,NDIVY
0055              IF(WEITY(IDIVY).EQ.0.0)WEITY(IDIVY)=1.0
0056        60    TOTAL=TOTAL+WEITY(IDIVY)
0057              YNORM=2.0/TOTAL
0058              NXTWO=NDIVX*KNODE+1
0059              NYTWO=NDIVY*KNODE+1
0060              IASEY=0
0061              ETASP=-1.0
```

```
0062            KWETY=0
0063            DO 160 IYTWO=1,NYTWO
0064            IASEY=IASEY+1
0065            IF(NTYPE.NE.8.AND.IASEY.EQ.3)IASEY=2
0066            IF(NTYPE.EQ.8.AND.IASEY.EQ.4)IASEY=2
0067            IASEX=0
0068            EXISP=-1.0
0069            KWETX=0
0070            DO 130 IXTWO=1,NXTWO
0071            IASEX=IASEX+1
0072            IF(NTYPE.NE.8.AND.IASEX.EQ.3)IASEX=2
0073            IF(NTYPE.EQ.8.AND.IASEX.EQ.4)IASEX=2
0074            NP=NP+1
0075            CALL SHAPEF(EXISP,ETASP)
0076            DO 70 INODE=1,LNODE
0077            JTEMP=TNODS(IBLOC,INODE)
0078            DO 70 IDIME=1,NDIME
0079      70    COORD(NP,IDIME)=COORD(NP,IDIME) +
0080          1             SHAPE(INODE)*TCORD(JTEMP,IDIME)
0081            GO TO (80,90) KNODE
0082      80    IF(IASEX.NE.2.OR.IASEY.NE.2)GO TO 100
0083            NELEM=NELEM+1
0084            JPOIN=NP-NXTWO
0085            NL(NELEM,1)=JPOIN-1
0086            NL(NELEM,2)=JPOIN
0087            NL(NELEM,3)=NP
0088            NL(NELEM,4)=NP-1
0089            MATNO(NELEM)=TMATO(IBLOC)
0090      90    IF(IASEX.NE.3.OR.IASEY.NE.3)GO TO 100
0091            NELEM=NELEM+1
0092            IPOIN=NP-IXTWO-NDIVX+(IXTWO-1)/2
0093            JPOIN=NP-NXTWO-NDIVX-1
0094            NL(NELEM,1)=JPOIN-2
0095            NL(NELEM,2)=JPOIN-1
0096            NL(NELEM,3)=JPOIN
0097            NL(NELEM,4)=IPOIN
0098            NL(NELEM,5)=NP
0099            NL(NELEM,6)=NP-1
0100            NL(NELEM,7)=NP-2
0101            NL(NELEM,8)=IPOIN-1
0102            MATNO(NELEM)=TMATO(IBLOC)
0103     100    CONTINUE
0104            GO TO (110,120),KNODE
0105     110    KWETX=KWETX+1
0106            GO TO 130
0107     120    IF(KONTX.LT.0) KWETX=KWETX + 1
0108            KONTX=KONTX*(-1)
0109     130    EXISP=EXISP+XNORM*WEITX(KWETX)/FNODE
0110            GO TO (140,150),KNODE
0111     140    KWETY=KWETY+1
0112            GO TO 160
0113     150    IF(KONTY.LT.0) KWETY=KWETY + 1
0114            KONTY=KONTY*(-1)
0115     160    ETASP=ETASP+YNORM*WEITY(KWETY)/FNODE
0116     170    CONTINUE
0117          C
0118          C ELIMINATE REPEATED NODES AT BLOCK INTERFACES
0119          C
0120            NREPN=0
0121            DO 210 IPOIN=1,NP
0122            IF(NREPN.EQ.0)GO TO 190
0123            DO 180 IREPN=1,NREPN
0124            IF(IPOIN.EQ.LREPN(IREPN))GO TO 210
```

```
0125      180   CONTINUE
0126      190   CONTINUE
0127            LPOIN=IPOIN+1
0128            DO 200 JPONT=LPOIN,NP
0129            TOTAL=ABS(COORD(IPOIN,1)-COORD(JPONT,1)) +
0130          1      ABS(COORD(IPOIN,2)-COORD(JPONT,2))
0131            IF(TOTAL.GT.0.00001)GO TO 200
0132            NREPN=NREPN+1
0133            LREPN(NREPN)=JPONT
0134            LASOC(NREPN)=IPOIN
0135      200   CONTINUE
0136      210   CONTINUE
0137            IF(NREPN.EQ.0)GO TO 360
0138            INDEX=0
0139            DO 240 IPOIN=1,NP
0140            DO 220 IREPN=1,NREPN
0141            IF(LREPN(IREPN).EQ.IPOIN)GO TO 230
0142      220   CONTINUE
0143            GO TO 240
0144      230   INDEX=INDEX+1
0145            LFINN(INDEX)=LREPN(IREPN)
0146            LFASC(INDEX)=LASOC(IREPN)
0147      240   CONTINUE
0148            DO 250 IREPN=1,NREPN
0149            LREPN(IREPN)=LFINN(IREPN)
0150      250   LASOC(IREPN)=LFASC(IREPN)
0151            DO 260 IREPN=1,NREPN
0152            DO 260 IELEM=1,NELEM
0153            DO 260 INODE=1,MNODE
0154            IF(NL(IELEM,INODE).EQ.LREPN(IREPN))
0155          1NL(IELEM,INODE)=LASOC(IREPN)
0156      260   CONTINUE
0157            DO 310 IPOIN=1,NP
0158            DO 270 IREPN=1,NREPN
0159            IF(IPOIN.EQ.LREPN(IREPN)) GO TO 310
0160      270   CONTINUE
0161            IF(IPOIN.LT.LREPN(1))GO TO 310
0162            IDIFF=IPOIN-NREPN
0163            IF(IPOIN.GT.LREPN(NREPN))GO TO 290
0164            DO 280 IREPN=1,NREPN
0165            KREPN=NREPN-IREPN+1
0166      280   IF(IPOIN.LT.LREPN(KREPN))IDIFF=IPOIN-KREPN+1
0167      290   DO 300 IDIME=1,NDIME
0168      300   COORD(IDIFF,IDIME)=COORD(IPOIN,IDIME)
0169      310   CONTINUE
0170            DO 350 IELEM=1,NELEM
0171            DO 350 INODE=1,MNODE
0172            NPOSI=NL(IELEM,INODE)
0173            DO 320 IREPN=1,NREPN
0174            IF(NPOSI.EQ.LREPN(IREPN))GO TO 350
0175      320   CONTINUE
0176            IF(NPOSI.LT.LREPN(1))GO TO 350
0177            IDIFF=NPOSI-NREPN
0178            IF(NPOSI.GT.LREPN(NREPN))GO TO 340
0179            DO 330 IREPN=1,NREPN
0180            KREPN=NREPN-IREPN+1
0181      330   IF(NPOSI.LT.LREPN(KREPN))IDIFF=NPOSI-KREPN+1
0182      340   NL(IELEM,INODE)=IDIFF
0183      350   CONTINUE
0184      360   CONTINUE
0185            NP=NP-NREPN
0186            RETURN
0187            END
```

```
0001      C***************************************************
0002      C THIS SUBROUTINE EVALUATES THE SHAPE FUNCTIONS
0003      C
0004            SUBROUTINE SHAPEF(S,T)
0005            COMMON/MESH1/COORD(1500,2),NL(750,8),
0006           1MATNO(750),SHAPE(9),NP,NELEM,NTYPE,NDIME,MNODE
0007
0008            SHAPE(1)=0.25*(1.0-S)*(1.0-T)*(-S-T-1.0)
0009            SHAPE(2)=0.5*(1.0-S*S)*(1.0-T)
0010            SHAPE(3)=0.25*(1.0+S)*(1.0-T)*(S-T-1.0)
0011            SHAPE(4)=0.5*(1.0-T*T)*(1.0+S)
0012            SHAPE(5)=0.25*(1.0+S)*(1.0+T)*(S+T-1.0)
0013            SHAPE(6)=0.5*(1.0-S*S)*(1.0+T)
0014            SHAPE(7)=0.25*(1.0-S)*(1.0+T)*(-S+T-1.0)
0015            SHAPE(8)=0.5*(1.0-T*T)*(1.0-S)
0016            RETURN
0017            END

0001      C***************************************************
0002      C THIS SUBROUTINE SUBDIVIDES EACH 4-NODED
0003      C QUADRILATERAL ELEMENT INTO TWO TRIANGULAR
0004      C ELEMENTS:THE SUBDIVISION IS DONE ACROSS THE
0005      C SHORTER DIAGONAL
0006      C
0007            SUBROUTINE TRIANGLE
0008            DIMENSION CORDE(4,2),LTEMP(4)
0009            COMMON/MESH1/ COORD(1500,2),NL(750,8),
0010           1MATNO(750),SHAPE(9),NP,NELEM,NTYPE,NDIME,MNODE
0011      C
0012            KOUNT=0
0013            DO 10 IELEM=1,NELEM
0014            NOTAL=NELEM+IELEM
0015            MATNO(NOTAL)=MATNO(IELEM)
0016            DO 10 INODE=1,MNODE
0017      10    NL(NOTAL,INODE)=NL(IELEM,INODE)
0018            DO 40 IELEM=1,NELEM
0019            NOTAL=NELEM+IELEM
0020            DO 20 INODE=1,MNODE
0021            INDEX=NL(NOTAL,INODE)
0022            LTEMP(INODE)=INDEX
0023            DO 20 IDIME=1,NDIME
0024      20    CORDE(INODE,IDIME)=COORD(INDEX,IDIME)
0025            DIAG1=SQRT((CORDE(1,1)-CORDE(3,1))**2 +
0026           1      (CORDE(1,2)-CORDE(3,2))**2)
0027            DIAG2=SQRT((CORDE(2,1)-CORDE(4,1))**2 +
0028           1      (CORDE(2,2)-CORDE(4,2))**2)
0029      C
0030      C DIVIDE ACROSS THE SHORTER DIAGONAL
0031      C
0032            DIFER=DIAG1-DIAG2
0033            IF(DIFER.GT.1.0E-9)GO TO 30
0034            KOUNT=KOUNT+1
0035            NL(KOUNT,1)=LTEMP(1)
0036            NL(KOUNT,2)=LTEMP(2)
0037            NL(KOUNT,3)=LTEMP(3)
0038            MATNO(KOUNT)=MATNO(NOTAL)
0039            KOUNT=KOUNT+1
0040            NL(KOUNT,1)=LTEMP(1)
0041            NL(KOUNT,2)=LTEMP(3)
0042            NL(KOUNT,3)=LTEMP(4)
0043            MATNO(KOUNT)=MATNO(NOTAL)
0044            GO TO 40
```

Finite Element Method 451

```
0045   30      KOUNT=KOUNT+1
0046           NL(KOUNT,1)=LTEMP(1)
0047           NL(KOUNT,2)=LTEMP(2)
0048           NL(KOUNT,3)=LTEMP(4)
0049           MATNO(KOUNT)=MATNO(NOTAL)
0050           KOUNT=KOUNT+1
0051           NL(KOUNT,1)=LTEMP(2)
0052           NL(KOUNT,2)=LTEMP(3)
0053           NL(KOUNT,3)=LTEMP(4)
0054           MATNO(KOUNT)=MATNO(NOTAL)
0055   40      CONTINUE
0056           NELEM=2*NELEM
0057           RETURN
0058           END

0001   C***************************************************
0002   C THIS SUBROUTINE OUTPUTS THE COORDINATES AND
0003   C ELEMENT TOPOLOGIES OF THE GENERATED MESH
0004   C
0005           SUBROUTINE OUTPUT
0006           COMMON/MESH1/ COORD(1500,2),NL(750,8),
0007          1 MATNO(750),SHAPE(9),NP,NELEM,NTYPE,NDIME,MNODE
0008   C
0009           WRITE(6,*)NP      ! TOTAL NO. OF POINTS
0010           WRITE(6,*)NELEM   ! TOTAL NO. OF ELEMENTS
0011           DO 10 IPOIN=1,NP
0012   10      WRITE(6,*)IPOIN,( COORD(IPOIN,I), I=1,NDIME )
0013           DO 20 IELEM=1,NELEM
0014   20      WRITE(6,*)IELEM,(NL(IELEM,I),I=1,NTYPE),MATNO(IELEM)
0015           RETURN
0016           END
```

Figure 6.16 FORTRAN code for automatic mesh generation.

Example 6.4

Use the code in Fig. 6.16 to discretize the mesh in Fig. 6.17.

Solution

The input data for the mesh generation is presented in Table 6.6. The subroutine INPUT reads the number of points (NPOIN) defining the mesh, the number of blocks (NELEM), the element type (NNODE), the number of coordinate dimensions (NDIME), the nodes defining each block, and the coordinates of each node in the mesh. The subroutine GENERATE reads the number of divisions and weighting factors along ζ and η directions for each block. It then subdivides the block into quadrilateral elements. At this point, the whole input data shown in Table 6.6 have been read. The subroutine TRIANGLE divides each four-node quadrilateral element across the shorter diagonal. The subroutine OUTPUT provides the coordinates of the nodes, element topologies, and material property numbers of the generated mesh. For the input data in Table 6.6, the generated mesh with 200 nodes and 330 elements is shown in Fig. 6.18.

452 Numerical Techniques in Electromagnetics

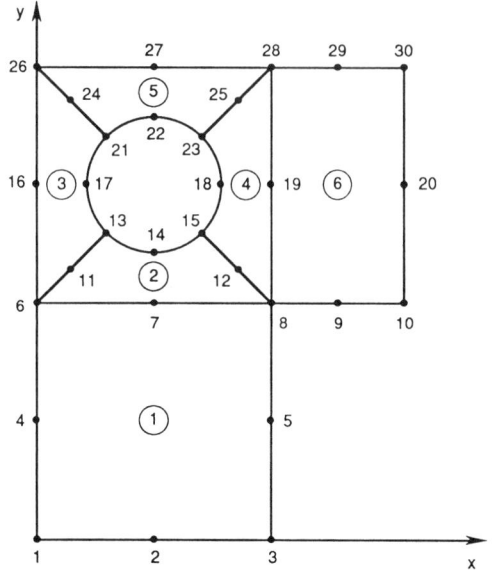

Figure 6.17 Solution region of Example 6.4.

Table 6.6 Input Data for Automatic Mesh Generation for the Solution Region in Fig. 6.17

```
30   6    3    2
1    1    2    3    5    8    7    6    4    1
2    6    7    8    12   15   14   13   11   1
3    6    11   13   17   21   24   26   16   1
4    8    19   28   25   23   18   15   12   1
5    21   22   23   25   28   27   26   24   1
6    8    9    10   20   30   29   28   19   2

     0.0       0.0
2    2.5       0.0
3    5.0       0.0
4    0.0       2.5
5    5.0       2.5
6    0.0       5.0
7    2.5       5.0
8    5.0       5.0
9    6.5       5.0
10   8.0       5.0
11   0.7196    5.7196
12   4.2803    5.7196
13   1.4393    6.4393
14   2.5       6.0
15   3.5607    6.6493
16   0.0       7.5
17   1.0       7.5
18   4.0       7.5
```

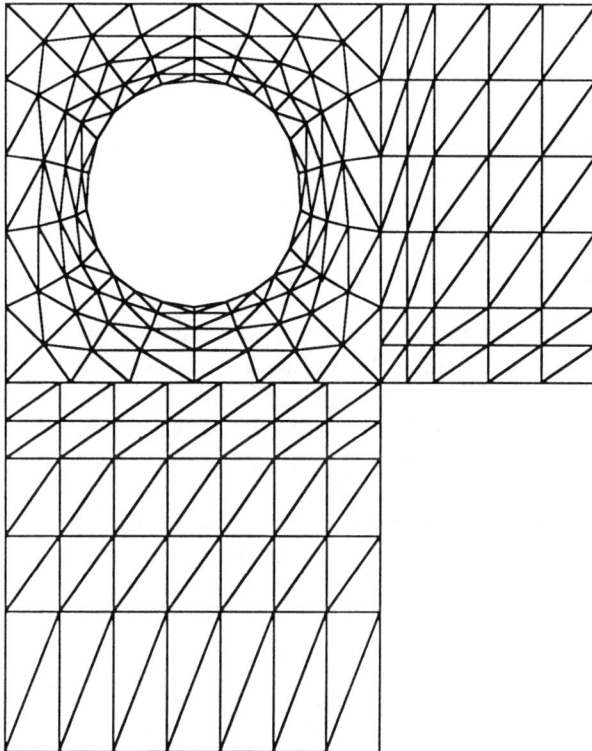

Figure 6.18 The generated mesh corresponding to input data in Table 6.6.

```
19      5.0       7.5
20      8.0       7.5
21      1.4393    8.5607
22      2.5       9.0
23      3.5607    8.5607
24      0.7196    9.2805
25      4.2803    9.2805
26      0.0       10.0
27      2.5       10.0
28      2.5       10.0
29      6.5       10.0
30      8.0       10.0
1       7    5
1.0     1.0       1.0     1.0    1.0    1.0   1.0
2.0     1.0       1.0     0.5    0.5
2       7    4
1.0     1.0       1.0     1.0    1.0    1.0   1.0
1.0     0.75      0.5     0.25
3       4    6
1.0     0.75      0.5     0.25
1.0     1.0       1.0     1.0    1.0    1.0
```

```
4    6    4
1.0    1.0    2.0    2.0    2.0    2.0
1.0    0.75   0.5    0.25
5    6    4 1.0    1.0    1.0    1.0    1.0
0.25   0.5    0.75   1.0
6    5    6
1.0    1.0    2.0    2.0    2.0
1.0    1.0    2.0    2.0    2.0    2.0
```

6.7 Bandwidth Reduction

Since most of the matrices involved in FEM are symmetric, sparse, and banded, we can minimize the storage requirements and the solution time by storing only the elements involved in half bandwidth instead of storing the whole matrix. To take the fullest advantage of the benefits from using a banded matrix solution technique, we must make sure that the matrix bandwidth is as narrow as possible.

If we let d be the maximum difference between the lowest and the highest node numbers of any single element in the mesh, we define the semi-bandwidth B (which includes the diagonal term) of the coefficient matrix $[C]$ as

$$B = (d+1)f \qquad (6.76)$$

where f is the number of degrees of freedom (or number of parameters) at each node. If, for example, we are interested in calculating the electric field intensity $\mathbf{E}$ for a three-dimensional problem, then we need E_x, E_y, and E_z at each node, and $f = 3$ in this case. Assuming that there is only one parameter per node,

$$B = d+1. \qquad (6.77)$$

The semi-bandwidth, which does not include the diagonal term, is obtained from Eq. (6.76) or (6.77) by subtracting one from the right-hand side, i.e., for $f = 1$,

$$B = d. \qquad (6.78)$$

Throughout our discussion in this section, we will stick to the definition of semi-bandwidth in Eq. (6.78). The total bandwidth may be obtained from Eq. (6.78) as $2B + 1$.

The bandwidth of the global coefficient matrix depends on the node numbering. Hence, to minimize the bandwidth, the node numbering should be selected to minimize d. Good node numbering is usually such that nodes with widely different numbers are widely separated. To minimize d, we must number nodes across the narrowest part of the region.

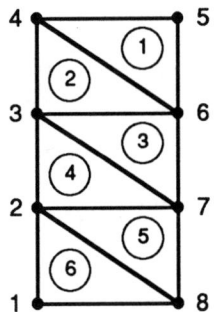

Figure 6.19 Original mesh with $B = 7$.

Consider, for the purpose of illustration, the mesh shown in Fig. 6.19. If the mesh is numbered originally as in Fig. 6.19, we obtain d_e for each element e as

$$d_1 = 2,\ d_2 = 3,\ d_3 = 4,\ d_4 = 5,\ d_5 = 6,\ d_6 = 7. \qquad (6.79)$$

From this, we obtain

$$d = \text{maximum } d_e = 7$$

or

$$B = 7. \qquad (6.80)$$

Alternatively, the semi-bandwidth may be determined from the coefficient matrix, which is obtained by mere inspection of Fig. 6.19 as

$$[C] = \begin{matrix} & \overset{B=7}{\overbrace{\begin{matrix}1 & 2 & 3 & 4 & 5 & 6 & 7 & 8\end{matrix}}} \\ \begin{matrix}1\\2\\3\\4\\5\\6\\7\\8\end{matrix} & \begin{bmatrix} x & x & & & & & & x \\ x & x & x & & & & x & x \\ & x & x & x & & x & & \\ & & x & x & x & x & & \\ & & & x & x & x & & \\ & & x & x & x & x & x & \\ & x & & & & x & x & x \\ x & x & & & & & & x \end{bmatrix}\end{matrix} \qquad (6.81)$$

where x indicates a possible nonzero term and blanks are zeros (i.e., $C_{ij} = 0$, indicating no coupling between nodes i and j). If the mesh is renumbered as in Fig. 6.20(a),

$$d_1 = 4 = d_2 = d_3 = d_4 = d_5 = d_6 \qquad (6.82)$$

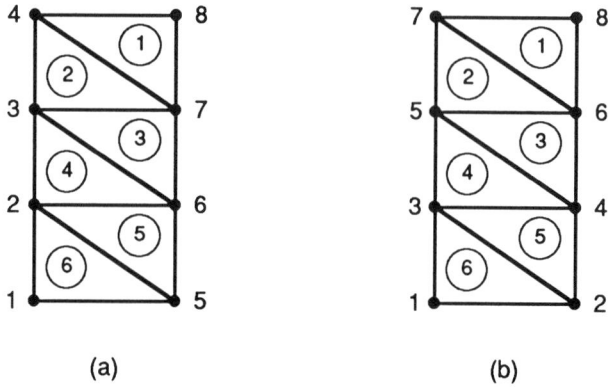

Figure 6.20 Renumbered nodes: (a) $B = 4$, (b) $B = 2$.

and hence
$$d = \text{maximum} \quad d_e = 4$$
or
$$B = 4. \tag{6.83}$$

Finally, we may renumber the mesh as in Fig. 6.20 (b). In this case
$$d_1 = 2 = d_2 = d_3 = d_4 = d_5 = d_6 \tag{6.84}$$
and
$$d = \text{maximum} \quad d_e = 2 \tag{6.85}$$
or
$$B = 2. \tag{6.86}$$

The value $B = 2$ may also be obtained from the coefficient matrix for the mesh in Fig. 6.20 (b), namely,

$$[C] = \begin{array}{c} \\ \\ 1 \\ 2 \\ 3 \\ 4 \\ 5 \\ 6 \\ 7 \\ 8 \end{array} \overset{\overset{B=2}{\overleftrightarrow{P \quad Q}}}{\begin{bmatrix} 1 & 2 & 3 & 4 & 5 & 6 & 7 & 8 \\ x & x & x & & & & & \\ x & x & x & x & & & & \\ x & x & x & x & x & & & \\ & x & x & x & x & & & \\ & & x & x & x & x & x & \\ & & & x & x & x & x & x \\ & & & & x & x & x & x \\ & & & & & x & x & x \end{bmatrix}} \begin{array}{c} \\ \\ \\ \\ \\ \\ R \\ \\ S \end{array} \tag{6.87}$$

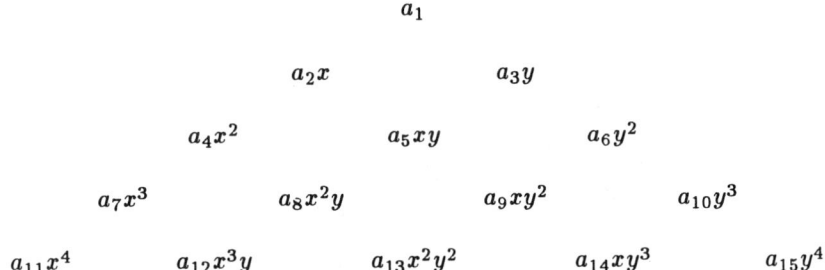

Figure 6.21 The Pascal Triangle. The first row is: (constant, $n = 0$), the second: (linear, $n = 1$), the third: (quadratic, $n = 2$), the fourth: (cubic, $n = 3$), the fifth: (quartic, $n = 4$).

From Eq. (6.86), one immediately notices that $[C]$ is symmetric and that terms are clustered in a band about the diagonal. Hence $[C]$ is sparse and banded so that only the data within the area **PQRS** of the matrix need to be stored—a total of 21 terms out of 64. This illustrates the savings in storage by a careful nodal numbering.

For a simple mesh, hand-labeling coupled with a careful inspection of the mesh (as we have done so far) can lead to a minimum bandwidth. However, for a large mesh, a hand-labeling technique becomes a tedious, time-consuming task, which in most cases may not be successful. It is particularly desirable that an automatic relabeling scheme is implemented within a mesh generation program. A number of algorithms have been proposed for bandwidth reduction by automatic mesh renumbering [37–40]. A simple, efficient algorithm is found in Collins [37].

6.8 Higher Order Elements

The finite elements we have used so far have been the linear type in that the shape function is of the order one. A higher order element is one in which the shape function or interpolation polynomial is of the order two or more.

The accuracy of a finite element solution can be improved by using finer mesh or using higher order elements or both. A discussion on mesh refinement versus higher order elements is given by Desai and Abel [2]; a motivation for using higher order elements is given by Csendes in [41]. In general, fewer higher order elements are needed to achieve the same degree of accuracy in the final results. The higher order elements are particularly useful when the gradient of the field variable is expected to vary rapidly.

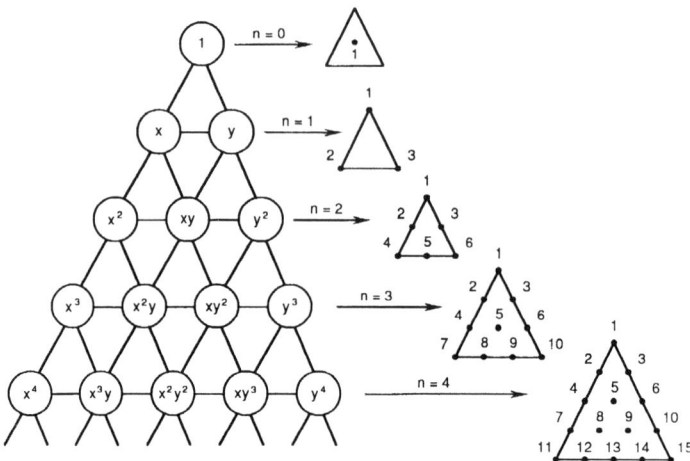

Figure 6.22 The Pascal triangle and the associated polynomial basis function for degree $n = 1$ to 4.

They have been applied with great success in solving EM-related problems [4, 41–46].

6.8.1 Pascal Triangle

Higher order triangular elements can be systematically developed with the aid of the so-called Pascal triangle given in Fig. 6.21. The family of finite elements generated in this manner with the distribution of nodes is illustrated in Fig. 6.22. Note that in higher order elements, some secondary (side and/or interior) nodes are introduced in addition to the primary (corner) nodes so as to produce exactly the right number of nodes required to define the shape function of that order. The Pascal triangle contains terms of the basis functions of various degrees in variables x and y. An arbitrary function $\Phi_i(x, y)$ can be approximated in an element in terms of a complete nth order polynomial as

$$\Phi(x, y) = \sum_{i=1}^{m} \alpha_i \Phi_i \qquad (6.88)$$

where

$$m = \frac{1}{2}(n+1)(n+2) \qquad (6.89)$$

is the number of terms in complete polynomials (also the number of nodes in the triangle). For example, for second order ($n = 2$) or quadratic (six-node)

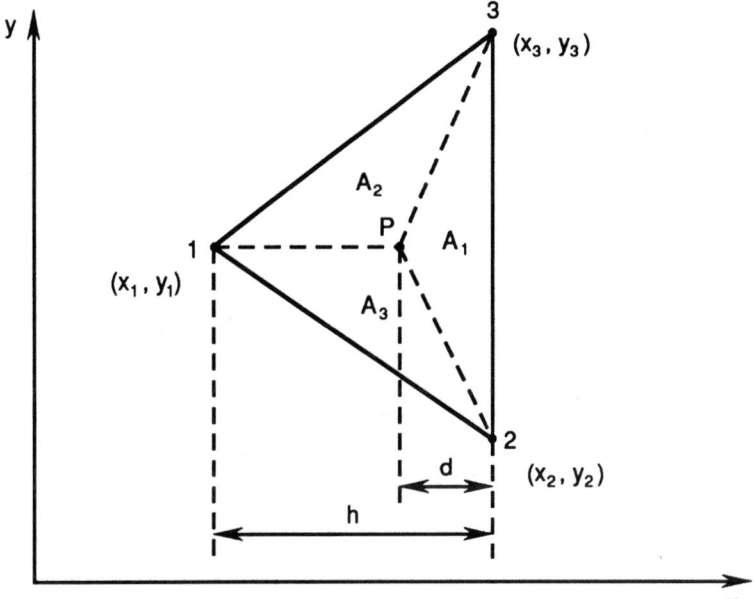

Figure 6.23 Definition of local coordinates.

triangular elements,

$$\Phi_e(x,y) = a_1 + a_2 x + a_3 y + a_4 xy + a_5 x^2 + a_6 y^2. \tag{6.90}$$

This equation has six coefficients, and hence the element must have six nodes. It is also complete through the second order terms. A systematic derivation of the interpolation function α for the higher order elements involves the use of the local coordinates.

6.8.2 Local Coordinates

The triangular local coordinates (ξ_1, ξ_2, ξ_3) are related to Cartesian coordinates (x, y) as

$$x = \xi_1 x_1 + \xi_2 x_2 + \xi_3 x_3 \tag{6.91}$$
$$y = \xi_1 y_1 + \xi_2 y_2 + \xi_3 y_3. \tag{6.92}$$

The local coordinates are dimensionless with values ranging from 0 to 1. By definition, ξ_i at any point within the triangle is the ratio of the perpendicular distance from the point to the side opposite to vertex i to the length of the altitude drawn from vertex i. Thus, from Fig. 6.23 the value

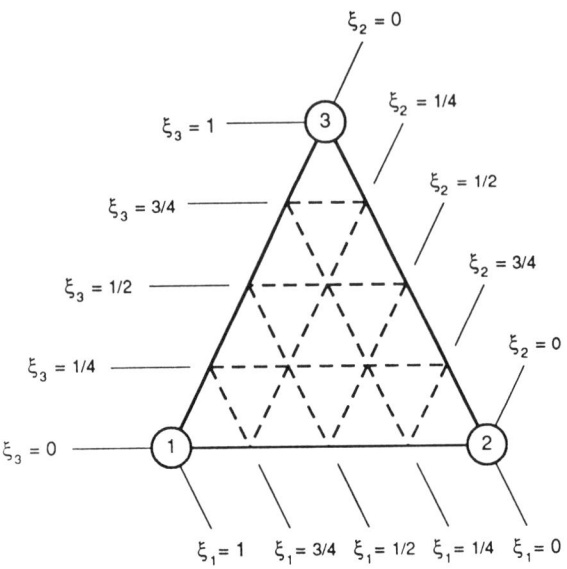

Figure 6.24 Variation of local coordinates.

of ξ_1 at P, for example, is given by the ratio of the perpendicular distance d from the side opposite vertex 1 to the altitude h of that side, i.e.,

$$\xi_1 = \frac{d}{h}. \tag{6.93}$$

Alternatively, from Fig. 6.23, ξ_i at P can be defined as

$$\xi_i = \frac{A_i}{A} \tag{6.94}$$

so that

$$\xi_1 + \xi_2 + \xi_3 = 1 \tag{6.95}$$

since $A_1 + A_2 + A_3 = A$. In view of Eq. (6.94), the local coordinates ξ_i are also called *area coordinates*. The variation of (ξ_1, ξ_2, ξ_3) inside an element is shown in Fig. 6.24. Although the coordinates ξ_1, ξ_2, and ξ_3 are used to define a point P, only two are independent since they must satisfy Eq. (6.95). The inverted form of Eqs. (6.91) and (6.92) is

$$\xi_i = \frac{1}{2A}\left[c_i + b_i x + a_i y\right] \tag{6.96}$$

where

$$a_i = x_k - x_j,$$
$$b_i = y_j - y_k, \qquad (6.97)$$
$$c_i = x_j y_k - x_k y_j$$
$$A = \text{area of the triangle} = \frac{1}{2}(b_1 a_2 - b_2 a_1),$$

and (i, j, k) is an even permutation of (1,2,3). (Notice that a_i and b_i are the same as Q_i and P_i in Eq. (6.1.1).) The differentiation and integration in local coordinates are carried out using [47]:

$$\frac{\partial f}{\partial \xi_1} = a_2 \frac{\partial f}{\partial x} - b_2 \frac{\partial f}{\partial y} \qquad (6.98\text{a})$$

$$\frac{\partial f}{\partial \xi_2} = -a_1 \frac{\partial f}{\partial x} + b_1 \frac{\partial f}{\partial y} \qquad (6.98\text{b})$$

$$\frac{\partial f}{\partial x} = \frac{1}{2A}\left(b_1 \frac{\partial f}{\partial \xi_1} + b_2 \frac{\partial f}{\partial \xi_2}\right) \qquad (6.98\text{c})$$

$$\frac{\partial f}{\partial y} = \frac{1}{2A}\left(a_1 \frac{\partial f}{\partial \xi_1} + a_2 \frac{\partial f}{\partial \xi_2}\right) \qquad (6.98\text{d})$$

$$\iint f \, dS = 2A \int_0^1 \left[\int_0^{1-\xi_2} f(\xi_1, \xi_2) \, d\xi_1 \right] d\xi_2 \qquad (6.98\text{e})$$

$$\iint \xi_1^i \xi_2^j \xi_3^k \, dS = \frac{i! \, j! \, k!}{(1+j+k+2)!} \, 2A \qquad (6.98\text{f})$$

$$dS = 2A \, d\xi_1 d\xi_2. \qquad (6.98\text{g})$$

6.8.3 Shape Functions

We may now express the shape function for higher order elements in terms of local coordinates. Sometimes, it is convenient to label each point in the finite elements in Fig. 6.22 with three integers i, j, and k from which its local coordinates (ξ_1, ξ_2, ξ_3) can be found or vice versa. At each point P_{ijk}

$$(\xi_1, \xi_2, \xi_3) = \left(\frac{i}{n}, \frac{j}{n}, \frac{k}{n}\right). \qquad (6.99)$$

Hence if a value of Φ, say Φ_{ijk}, is prescribed at each point P_{ijk}, Eq. (6.88) can be written as

$$\Phi(\xi_1,\xi_2,\xi_3) = \sum_{i=1}^{m}\sum_{j=1}^{m-i} \alpha_{ijk}(\xi_1,\xi_2,\xi_3)\Phi_{ijk} \qquad (6.100)$$

where

$$\alpha_\ell = \alpha_{ijk} = p_i(\xi)p_j(\xi_2)p_k(\xi_3), \quad \ell = 1, 2, \cdots \qquad (6.101)$$

$$p_r(\xi) = \begin{cases} \dfrac{1}{r!}\displaystyle\prod_{t=0}^{r-1}(n\xi - t), & r > 0 \\ 1, & r = 0 \end{cases} \qquad (6.102)$$

and $r \in (i,j,k)$. $p_r(\xi)$ may also be written as

$$p_r(\xi) = \frac{(n\xi - r + 1)}{r} p_{r-1}(\xi), \quad r > 0 \qquad (6.103)$$

where $p_0(\xi) = 1$.

The relationships between the subscripts $q \in \{1,2,3\}$ on ξ_q, $\ell \in \{1, 2, \cdots, m\}$ on α_ℓ, and $r \in (i,j,k)$ on p_r and P_{ijk} in Eqs. (6.101) to (6.103) are illustrated in Fig. 6.25 for n ranging from 1 to 4. Henceforth point P_{ijk} will be written as P_n for conciseness.

Notice from Eq. (6.102) or Eq. (6.103) that

$$p_0(\xi) = 1$$
$$p_1(\xi) = n\xi$$
$$p_2(\xi) = \frac{1}{2}(n\xi - 1)n\xi$$
$$p_3(\xi) = \frac{1}{6}(n\xi - 2)(n\xi - 1)n\xi$$
$$p_4(\xi) = \frac{1}{24}(n\xi - 3)(n\xi - 2)(n\xi - 1)n\xi, \text{ etc }. \qquad (6.104)$$

Substituting Eq. (6.104) into Eq. (6.101) gives the shape functions α_ℓ for nodes $\ell = 1, 2, \cdots, m$, as shown in Table 6.7 for $n = 1$ to 4. Observe that each α_ℓ takes the value of 1 at node ℓ and value of 0 at all other nodes in the triangle. This is easily verified using Eq. (6.99) in conjunction with Fig. 6.25.

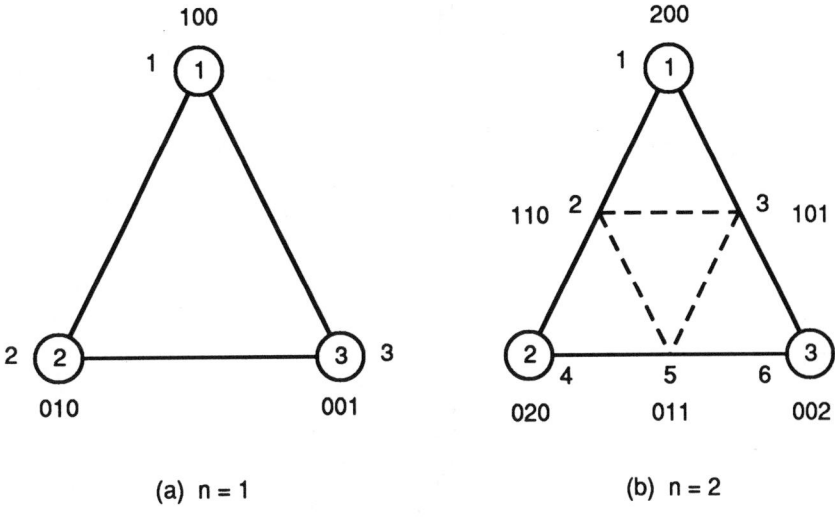

Figure 6.25 Distribution of nodes over triangles for $n = 1$ to 4. The triangles are in standard position.

6.8.4 Fundamental Matrices

The fundamental matrices $[T]$ and $[Q]$ for triangular elements can be derived using the shape functions in Table 6.7. (For simplicity, the brackets [ ] denoting a matrix quantity will be dropped in the remaining part of this section.) From Eq. (6.40), the T matrix is defined as

$$T_{ij} = \iint \alpha_i \alpha_j \, dS. \tag{6.40}$$

From Table 6.7, we substitute α_ℓ in Eq. (6.40) and apply Eqs. (6.98f) and (6.98g) to obtain elements of T. For example, for $n = 1$,

$$T_{ij} = 2A \int_0^1 \int_0^{1-\xi_2} \xi_i \xi_j \, d\xi_1 d\xi_2.$$

When $i \neq j$,

$$T_{ij} = \frac{2A(1!)(1!)(0!)}{4!} = \frac{A}{12}, \tag{6.105a}$$

when $i = j$

$$T_{ij} = \frac{2A(2!)}{4!} = \frac{A}{6}. \tag{6.105b}$$

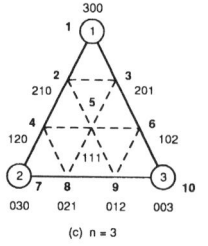

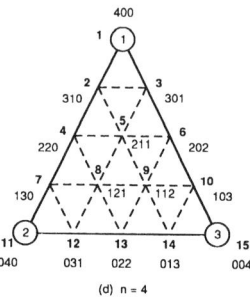

Figure 6.25 (continued)

Hence

$$T = \frac{A}{12} \begin{bmatrix} 2 & 1 & 1 \\ 1 & 2 & 1 \\ 1 & 1 & 2 \end{bmatrix}. \qquad (6.106)$$

By following the same procedure, higher order T matrices can be obtained. The T matrices for orders up to $n = 4$ are tabulated in Table 6.8 where the factor A, the area of the element, has been suppressed. The actual matrix elements are obtained from Table 6.8 by multiplying the tabulated numbers by A and dividing by the indicated common denominator. The following properties of the T matrix are noteworthy:

(a) T is symmetric with positive elements;

(b) elements of T all add up to the area of the triangle, i.e., $\sum_{i}^{m}\sum_{j}^{m} T_{ij} = A$, since by definition $\sum_{\ell=1}^{m} \alpha_\ell = 1$ at any point within the element;

(c) elements for which the two triple subscripts form similar permutations are equal, i.e., $T_{ijk,prq} = T_{ikj,prq} = T_{kij,rpq} = T_{kji,rqp} = T_{jki,qrp} = T_{jik,qpr}$; this should be obvious from Eqs. (6.40) and (6.101).

These properties are not only useful in checking the matrix, they have

Table 6.7 Polynomial Basis Function $\alpha_\ell(\xi_1, \xi_2, \xi_3, \xi_4)$ for First-, Second-, Third-, and Fourth-order

$n = 1$	$n = 2$	$n = 3$	$n = 4$
$\alpha_1 = \xi_1$	$\alpha_1 = \xi_1(2\xi_1 - 1)$	$\alpha_1 = \frac{1}{2}\xi_1(3\xi_1 - 2)(3\xi_1 - 1)$	$\alpha_1 = \frac{1}{6}\xi_1(4\xi_1 - 3)(4\xi_1 - 2)(4\xi_1 - 1)$
$\alpha_2 = \xi_2$	$\alpha_2 = 4\xi_1\xi_2$	$\alpha_2 = \frac{9}{2}\xi_1(3\xi_1 - 1)\xi_2$	$\alpha_2 = \frac{8}{3}\xi_1(4\xi_1 - 2)(4\xi_1 - 1)\xi_2$
$\alpha_3 = \xi_3$	$\alpha_3 = 4\xi_1\xi_3$	$\alpha_3 = \frac{9}{2}\xi_1(3\xi_1 - 1)\xi_3$	$\alpha_3 = \frac{8}{3}\xi_1(4\xi_1 - 2)(4\xi_1 - 1)\xi_3$
	$\alpha_4 = \xi_2(2\xi_2 - 1)$	$\alpha_4 = \frac{9}{2}\xi_1(3\xi_2 - 1)\xi_2$	$\alpha_4 = 4\xi_1(4\xi_1 - 1)(4\xi_2 - 1)\xi_2$
	$\alpha_5 = 4\xi_2\xi_3$	$\alpha_5 = 27\xi_1\xi_2\xi_3$	$\alpha_5 = 32\xi_1(4\xi_1 - 1)\xi_2\xi_3$
	$\alpha_6 = \xi_3(2\xi_3 - 1)$	$\alpha_6 = \frac{9}{2}\xi_1(3\xi_3 - 1)\xi_3$	$\alpha_6 = 4\xi_1(4\xi_1 - 1)(4\xi_3 - 1)\xi_3$
		$\alpha_7 = \frac{1}{2}\xi_2(3\xi_2 - 2)(3\xi_2 - 1)$	$\alpha_7 = \frac{8}{3}\xi_1(4\xi_2 - 2)(4\xi_2 - 1)\xi_2$
		$\alpha_8 = \frac{9}{2}\xi_2(3\xi_2 - 1)\xi_3$	$\alpha_8 = 32\xi_1(4\xi_2 - 1)\xi_2\xi_3$
		$\alpha_9 = \frac{9}{2}\xi_2(3\xi_3 - 1)\xi_3$	$\alpha_9 = 32\xi_1\xi_2(4\xi_3 - 1)\xi_3$
		$\alpha_{10} = \frac{1}{2}\xi_3(3\xi_3 - 2)(3\xi_3 - 1)$	$\alpha_{10} = \frac{8}{3}\xi_1(4\xi_3 - 2)(4\xi_3 - 1)\xi_3$
			$\alpha_{11} = \frac{1}{6}\xi_2(4\xi_2 - 3)(4\xi_2 - 2)(4\xi_2 - 1)$
			$\alpha_{12} = \frac{8}{3}\xi_2(4\xi_2 - 2)(4\xi_2 - 1)\xi_3$
			$\alpha_{13} = 4\xi_2(4\xi_2 - 1)(4\xi_3 - 1)\xi_3$
			$\alpha_{14} = \frac{8}{3}\xi_2(4\xi_3 - 2)(4\xi_3 - 1)\xi_3$
			$\alpha_{15} = \frac{1}{6}\xi_3(4\xi_3 - 3)(4\xi_3 - 2)(4\xi_3 - 1)$

Table 6.8 Table of T Matrix for $n = 1$ to 4

$n = 1$ Common denominator: 12

$$\begin{matrix} 2 & 1 & 1 \\ 1 & 2 & 1 \\ 1 & 1 & 2 \end{matrix}$$

$n = 2$ Common denominator: 180

$$\begin{matrix} 6 & 0 & 0 & -1 & -4 & -1 \\ 0 & 32 & 16 & 0 & 16 & -4 \\ 0 & 16 & 32 & -4 & 16 & 0 \\ -1 & 0 & -4 & 6 & 0 & -1 \\ -4 & 16 & 16 & 0 & 32 & 0 \\ -1 & -4 & 0 & -1 & 0 & 6 \end{matrix}$$

$n = 3$ Common denominator: 6720

$$\begin{matrix}
76 & 18 & 18 & 0 & 36 & 0 & 11 & 27 & 27 & 11 \\
18 & 540 & 270 & -189 & 162 & -135 & 0 & -135 & -54 & 27 \\
18 & 270 & 540 & -135 & 162 & -189 & 27 & -54 & -135 & 0 \\
0 & -189 & -135 & 540 & 162 & -54 & 18 & 270 & -135 & 27 \\
36 & 162 & 162 & 162 & 1944 & 162 & 36 & 162 & 162 & 36 \\
0 & -135 & -189 & -54 & 162 & 540 & 27 & -135 & 270 & 18 \\
11 & 0 & 27 & 18 & 36 & 27 & 76 & 18 & 0 & 11 \\
27 & -135 & -54 & 270 & 162 & -135 & 18 & 540 & -189 & 0 \\
27 & -54 & -135 & -135 & 162 & 270 & 0 & -189 & 540 & 18 \\
11 & 27 & 0 & 27 & 36 & 18 & 11 & 0 & 18 & 76
\end{matrix}$$

proved useful in saving computer time and storage. It is interesting to know that the properties are independent of coordinate system [46].

From Eq. (6.14) or Eq. (6.39), elements of [C] matrix are defined by

$$C_{ij} = \iint \left(\frac{\partial \alpha_i}{\partial x} \frac{\partial \alpha_j}{\partial x} + \frac{\partial \alpha_i}{\partial y} \frac{\partial \alpha_j}{\partial y} \right) dS. \tag{6.107}$$

By applying Eqs. (6.98a) to (6.98d) to Eq. (6.107), it can be shown that [4,43]

$$C_{ij} = \frac{1}{2A} \sum_{q=1}^{3} \cot \theta_q \iint \left(\frac{\partial \alpha_i}{\partial \xi_{q+1}} - \frac{\partial \alpha_i}{\partial \xi_{q-1}} \right) \left(\frac{\partial \alpha_j}{\partial \xi_{q+1}} - \frac{\partial \alpha_j}{\partial \xi_{q-1}} \right) dS$$

or

$n = 4$ Common denominator: 56700

290	160	160	−80	160	−80	0	−160	−160	0	−27	−112	−12	−112	−27
160	2560	1280	−1280	1280	−960	768	256	−256	512	0	512	64	256	−112
160	1280	2560	−960	1280	−1280	512	−256	256	768	−112	256	64	512	0
−80	−1280	−960	3168	384	48	−1280	384	−768	64	−80	−960	48	64	−12
160	1280	1280	384	10752	384	256	−1536	−1536	256	−160	−256	−768	−256	−160
−80	−960	−1280	48	384	3168	64	−768	384	−1280	−12	64	48	−960	−80
0	768	512	−1280	256	64	2560	1280	−256	256	160	1280	−960	512	−112
−160	256	−256	384	−1536	−768	1280	10752	−1536	−256	160	1280	384	256	−160
−160	−256	256	−768	−1536	384	−256	−1536	10752	1280	−160	−160	1280	384	−160
0	512	768	64	256	−1280	256	−256	1280	2560	−112	512	−960	1280	160
−27	0	−112	−80	−160	−12	160	160	−160	−112	290	160	−80	−112	0
−112	512	256	−960	−256	64	1280	1280	256	512	160	2560	−1280	768	0
−12	64	64	48	−768	48	−960	384	384	−960	−80	−1280	3168	−1280	−80
−112	256	512	64	−256	−960	512	256	1280	1280	0	768	−1280	2560	160
−27	−112	0	−12	−160	−80	−112	−160	160	160	−80	0	−80	160	290

Table 6.8 (continued.)

Table 6.9 Table of Q Matrices for $n = 1$ to 4

$n = 1$ Common denominator: 2

$$\begin{pmatrix} 0 & 0 & 0 \\ 0 & 1 & -1 \\ 0 & -1 & 1 \end{pmatrix}$$

$n = 2$ Common denominator: 6

$$\begin{pmatrix} 0 & 0 & 0 & 0 & 0 & 0 \\ 0 & 8 & -8 & 0 & 0 & 0 \\ 0 & -8 & 8 & 0 & 0 & 0 \\ 0 & 0 & 0 & 3 & -4 & 1 \\ 0 & 0 & 0 & -4 & 8 & -4 \\ 0 & 0 & 0 & 1 & -4 & 3 \end{pmatrix}$$

$n = 3$ Common denominator: 80

$$\begin{pmatrix} 0 & 0 & 0 & 0 & 0 & 0 & 0 & 0 & 0 & 0 \\ 0 & 135 & -135 & -27 & 0 & 27 & 3 & 0 & 0 & -3 \\ 0 & -135 & 135 & 27 & 0 & -27 & -3 & 0 & 0 & 3 \\ 0 & -27 & 27 & 135 & -162 & 27 & 3 & 0 & 0 & -3 \\ 0 & 0 & 0 & -162 & 324 & -162 & 0 & 0 & 0 & 0 \\ 0 & 27 & -27 & 27 & -162 & 135 & -3 & 0 & 0 & 3 \\ 0 & 3 & -3 & 3 & 0 & -3 & 34 & -54 & 27 & -7 \\ 0 & 0 & 0 & 0 & 0 & 0 & -54 & 135 & -108 & 27 \\ 0 & 0 & 0 & 0 & 0 & 0 & 27 & -108 & 135 & -54 \\ 0 & -3 & 3 & -3 & 0 & 3 & -7 & 27 & -54 & 34 \end{pmatrix}$$

$$\boxed{C_{ij} = \sum_{q=1}^{3} Q_{ij}^{(q)} \cot \theta_q} \qquad (6.108)$$

where θ_q is the included angle of vertex $q \in \{1, 2, 3\}$ of the triangle and

$$\boxed{Q_{ij}^{(q)} = \iint \left(\frac{\partial \alpha_i}{\partial \xi_{q+1}} - \frac{\partial \alpha_i}{\partial \xi_{q-1}} \right) \left(\frac{\partial \alpha_j}{\partial \xi_{q+1}} - \frac{\partial \alpha_j}{\partial \xi_{q-1}} \right) d\xi_1 d\xi_2} \qquad (6.109)$$

We notice that matrix C depends on the triangle shape, whereas the matrices $Q^{(q)}$ do not. The $Q^{(1)}$ matrices for $n = 1$ to 4 are tabulated in Table 6.9. The following properties of Q matrices should be noted:

$n = 4$ Common denominator: 1890

0	0	0	0	0	0	0	0	0	0	0	0	0	0	0
0	3968	0	−1440	0	0	1440	640	0	0	0	−80	0	0	0	0
0	−3968	3968	1440	0	0	−1440	−640	0	0	0	80	0	0	0	80
0	−1440	1440	4632	−5376	0	744	−1248	768	768	−640	80	0	0	0	−80
0	0	0	−5376	10752	−5376	4632	1536	−1536	−1536	640	−160	−128	96	0	80
0	1440	−1440	744	−5376	1536	−288	−288	768	768	−288	80	256	−192	256	−160
0	640	−640	−1248	1536	−1536	4632	3456	1536	1536	−1248	−160	−128	96	−128	80
0	0	0	768	−1536	768	−288	−4608	10752	10752	−4608	80	−256	192	−256	80
0	0	0	768	−1536	1536	−4608	10752	−7680	−7680	10752	240	256	−192	256	−160
0	−640	640	−288	1536	−7680	1536	−7680	1536	10752	−4608	−160	256	−192	256	−160
0	−80	80	80	−160	−4608	−384	−4608	1536	−4608	3456	−160	−256	192	−256	240
0	0	0	−128	256	−160	240	−160	−160	1536	80	705	−1232	884	−464	107
0	0	0	256	−192	256	−256	256	256	−160	−256	−1232	3456	−3680	1920	−464
0	0	0	−128	256	−192	192	−192	−192	256	192	884	−3680	5592	−3680	884
0	0	0	256	−192	256	−256	256	256	−160	−256	−464	1920	−3680	3456	−1232
0	80	−80	−128	256	−160	80	−160	−160	240	240	107	−464	884	−1232	705

Table 6.9 (continued.)

470   Numerical Techniques in Electromagnetics

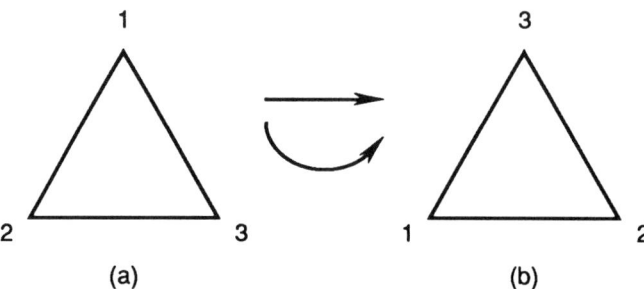

Figure 6.26 One counterclockwise rotation of the triangle in (a) gives the triangle in (b).

(a) they are symmetric;

(b) the row and column sums of any Q matrix are zero, i.e., $\sum_{i=1}^{m} Q_{ij}^{(q)} = 0 = \sum_{j=1}^{m} Q_{ij}^{(q)}$ so that the C matrix is singular.

$Q^{(2)}$ and $Q^{(3)}$ are easily obtained from $Q^{(1)}$ by row and column permutations so that the matrix C for any triangular element is constructed easily if $Q^{(1)}$ is known. One approach [48] involves using a rotation matrix R similar to that in Silvester and Ferrari [4], which is essentially a unit matrix with elements rearranged to correspond to one rotation of the triangle about its centroid in a counterclockwise direction. For example, for $n = 1$, the rotation matrix is basically derived from Fig. 6.26 as

$$R = \begin{bmatrix} 0 & 0 & 1 \\ 1 & 0 & 0 \\ 0 & 1 & 0 \end{bmatrix} \qquad (6.110)$$

where $R_{ij} = 1$ if node i is replaced by node j after one counterclockwise rotation, or $R_{ij} = 0$ otherwise. Table 6.10 presents the R matrices for $n = 1$ to 4. Notice that each row or column of R has only one nonzero element since R is essentially a unit matrix with rearranged elements.

Once the R is known, we obtain

$$Q^{(2)} = R\ Q^{(1)}\ R^t \qquad (6.111\text{a})$$

Table 6.10 R Matrix for $n = 1$ to 4

$n = 1$

$$\begin{bmatrix} 0 & 0 & 1 \\ 1 & 0 & 0 \\ 0 & 1 & 0 \end{bmatrix}$$

$n = 2$

$$\begin{bmatrix} 0 & 0 & 0 & 0 & 0 & 1 \\ 0 & 0 & 1 & 0 & 0 & 0 \\ 0 & 0 & 0 & 0 & 1 & 0 \\ 1 & 0 & 0 & 0 & 0 & 0 \\ 0 & 1 & 0 & 0 & 0 & 0 \\ 0 & 0 & 0 & 1 & 0 & 0 \end{bmatrix}$$

$n = 3$

$$\begin{bmatrix} 0 & 0 & 0 & 0 & 0 & 0 & 0 & 0 & 0 & 1 \\ 0 & 0 & 0 & 0 & 0 & 1 & 0 & 0 & 0 & 0 \\ 0 & 0 & 0 & 0 & 0 & 0 & 0 & 0 & 1 & 0 \\ 0 & 0 & 1 & 0 & 0 & 0 & 0 & 0 & 0 & 0 \\ 0 & 0 & 0 & 0 & 1 & 0 & 0 & 0 & 0 & 0 \\ 0 & 0 & 0 & 0 & 0 & 0 & 0 & 1 & 0 & 0 \\ 1 & 0 & 0 & 0 & 0 & 0 & 0 & 0 & 0 & 0 \\ 0 & 1 & 0 & 0 & 0 & 0 & 0 & 0 & 0 & 0 \\ 0 & 0 & 0 & 1 & 0 & 0 & 0 & 0 & 0 & 0 \\ 0 & 0 & 0 & 0 & 0 & 0 & 1 & 0 & 0 & 0 \end{bmatrix}$$

$n = 4$

$$\begin{bmatrix} 0 & 0 & 0 & 0 & 0 & 0 & 0 & 0 & 0 & 0 & 0 & 0 & 0 & 1 \\ 0 & 0 & 0 & 0 & 0 & 0 & 0 & 0 & 0 & 1 & 0 & 0 & 0 & 0 \\ 0 & 0 & 0 & 0 & 0 & 0 & 0 & 0 & 0 & 0 & 0 & 0 & 1 & 0 \\ 0 & 0 & 0 & 0 & 0 & 1 & 0 & 0 & 0 & 0 & 0 & 0 & 0 & 0 \\ 0 & 0 & 0 & 0 & 0 & 0 & 0 & 1 & 0 & 0 & 0 & 0 & 0 & 0 \\ 0 & 0 & 0 & 0 & 0 & 0 & 0 & 0 & 0 & 0 & 1 & 0 & 0 & 0 \\ 0 & 0 & 1 & 0 & 0 & 0 & 0 & 0 & 0 & 0 & 0 & 0 & 0 & 0 \\ 0 & 0 & 0 & 0 & 1 & 0 & 0 & 0 & 0 & 0 & 0 & 0 & 0 & 0 \\ 0 & 0 & 0 & 0 & 0 & 0 & 1 & 0 & 0 & 0 & 0 & 0 & 0 & 0 \\ 0 & 0 & 0 & 0 & 0 & 0 & 0 & 0 & 0 & 0 & 1 & 0 & 0 & 0 \\ 1 & 0 & 0 & 0 & 0 & 0 & 0 & 0 & 0 & 0 & 0 & 0 & 0 & 0 \\ 0 & 1 & 0 & 0 & 0 & 0 & 0 & 0 & 0 & 0 & 0 & 0 & 0 & 0 \\ 0 & 0 & 0 & 1 & 0 & 0 & 0 & 0 & 0 & 0 & 0 & 0 & 0 & 0 \\ 0 & 0 & 0 & 0 & 0 & 0 & 1 & 0 & 0 & 0 & 0 & 0 & 0 & 0 \\ 0 & 0 & 0 & 0 & 0 & 0 & 0 & 0 & 0 & 1 & 0 & 0 & 0 & 0 \end{bmatrix}$$

472   Numerical Techniques in Electromagnetics

$$\boxed{Q^{(3)} = R \; Q^{(2)} \; R^t} \qquad (6.111b)$$

where R^t is the transpose of R.

Example 6.5

For $n = 2$, calculate $Q^{(1)}$ and obtain $Q^{(2)}$ from $Q^{(1)}$ using Eq. (6.111a).

Solution

By definition,

$$Q_{ij}^{(1)} = \iint \left(\frac{\partial \alpha_i}{\partial \xi_2} - \frac{\partial \alpha_i}{\partial \xi_3}\right)\left(\frac{\partial \alpha_j}{\partial \xi_2} - \frac{\partial \alpha_j}{\partial \xi_3}\right) d\xi_1 d\xi_2.$$

For $n = 2$, $i, j = 1, 2, \cdots, 6$, and α_i are given in terms of the local coordinates in Table 6.7. Since $Q^{(1)}$ is symmetric, only some of the elements need be calculated. Substituting for α_ℓ from Table 6.7 and applying Eqs. (6.98e) and (6.98f), we obtain

$$Q_{1j} = 0, \quad j = 1 \text{ to } 6,$$
$$Q_{i1} = 0, \quad i = 1 \text{ to } 6,$$
$$Q_{22} = \frac{1}{2A}\iint (4\xi_1)^2 d\xi_1 \xi_2 = \frac{8}{6},$$
$$Q_{23} = \frac{1}{2A}\iint (4\xi_1)(-4\xi_1) d\xi_1 \xi_2 = -\frac{8}{6},$$
$$Q_{24} = \frac{1}{2A}\iint (4\xi_1)(4\xi_1 - 1) d\xi_1 \xi_2 = 0 = Q_{26},$$
$$Q_{25} = \frac{1}{2A}\iint (4\xi_1)(4\xi_3 - 4\xi_2) d\xi_1 \xi_2 = 0,$$
$$Q_{33} = \frac{1}{2A}\iint (-4\xi_1)^2 d\xi_1 \xi_2 = \frac{8}{6},$$
$$Q_{34} = \frac{1}{2A}\iint (-4\xi_1)(4\xi_2 - 1) d\xi_1 \xi_2 = 0 = Q_{36},$$
$$Q_{35} = \frac{1}{2A}\iint (-4\xi_1)(4\xi_3 - 4\xi_2) d\xi_1 \xi_2 = 0,$$
$$Q_{44} = \frac{1}{2A}\iint (4\xi_2 - 1)^2 d\xi_1 \xi_2 = \frac{3}{6},$$

Finite Element Method

$$Q_{45} = \frac{1}{2A}\iint (4\xi_2 - 1)(4\xi_3 - 4xi_2)d\xi_1\xi_2 = -\frac{4}{6},$$

$$Q_{46} = \frac{1}{2A}\iint (4\xi_2 - 1)(4\xi_3 - 1)(-1)d\xi_1\xi_2 = \frac{1}{6},$$

$$Q_{55} = \frac{1}{2A}\iint (4\xi_3 - 4\xi_2)^2 d\xi_1\xi_2 = \frac{8}{6},$$

$$Q_{56} = \frac{1}{2A}\iint (4\xi_3 - 4xi_2)(-1)(4\xi_3 - 1)d\xi_1\xi_2 = -\frac{4}{6},$$

$$Q_{66} = \frac{1}{2A}\iint (-1)(4\xi_3 - 1)^2 d\xi_1\xi_2 = \frac{3}{6}.$$

Hence

$$Q^{(1)} = \frac{1}{6}\begin{bmatrix} 0 & 0 & 0 & 0 & 0 & 0 \\ 0 & 8 & -8 & 0 & 0 & 0 \\ 0 & -8 & 8 & 0 & 0 & 0 \\ 0 & 0 & 0 & 3 & -4 & 1 \\ 0 & 0 & 0 & -4 & 8 & -4 \\ 0 & 0 & 0 & 1 & -4 & 3 \end{bmatrix}.$$

We now obtain $Q^{(2)}$ from

$$Q^{(2)} = RQ^{(1)}R^t$$

$$= \frac{1}{6}R\begin{bmatrix} 0 & 0 & 0 & 0 & 0 & 0 \\ 0 & 8 & -8 & 0 & 0 & 0 \\ 0 & -8 & 8 & 0 & 0 & 0 \\ 0 & 0 & 0 & 3 & -4 & 1 \\ 0 & 0 & 0 & -4 & 8 & -4 \\ 0 & 0 & 0 & 1 & -4 & 3 \end{bmatrix}\begin{bmatrix} 0 & 0 & 0 & 1 & 0 & 0 \\ 0 & 0 & 0 & 0 & 1 & 0 \\ 0 & 1 & 0 & 0 & 0 & 0 \\ 0 & 0 & 0 & 0 & 0 & 1 \\ 0 & 0 & 1 & 0 & 0 & 0 \\ 1 & 0 & 0 & 0 & 0 & 0 \end{bmatrix}$$

$$= \frac{1}{6}\begin{bmatrix} 0 & 0 & 0 & 0 & 0 & 1 \\ 0 & 0 & 1 & 0 & 0 & 0 \\ 0 & 0 & 0 & 0 & 1 & 0 \\ 1 & 0 & 0 & 0 & 0 & 0 \\ 0 & 1 & 0 & 0 & 0 & 0 \\ 0 & 0 & 0 & 1 & 0 & 0 \end{bmatrix}\begin{bmatrix} 0 & 0 & 0 & 0 & 0 & 0 \\ 0 & -8 & 0 & 0 & 8 & 0 \\ 0 & 8 & 0 & 0 & -8 & 0 \\ 1 & 0 & -4 & 0 & 0 & 3 \\ -4 & 0 & 8 & 0 & 0 & -4 \\ 3 & 0 & 4 & 0 & 0 & 1 \end{bmatrix}$$

$$Q^{(2)} = \frac{1}{6}\begin{bmatrix} 3 & 0 & -4 & 0 & 0 & 1 \\ 0 & 8 & 0 & 0 & -8 & 0 \\ -4 & 0 & 8 & 0 & 0 & -4 \\ 0 & 0 & 0 & 0 & 0 & 0 \\ 0 & -8 & 0 & 0 & 8 & 0 \\ 1 & 0 & -4 & 0 & 0 & 3 \end{bmatrix}.$$

6.9 Three-dimensional Elements

The finite element techniques developed in the previous sections for two-dimensional elements can be extended to three-dimensional elements. One would expect three-dimensional problems to require a large total number of elements to achieve an accurate result and demand a large storage capacity and computational time. For the sake of completeness, we will discuss the finite element analysis of Helmholtz's equation in three dimensions, namely,

$$\nabla^2 \Phi + k^2 \Phi = g. \tag{6.112}$$

We first divide the solution region into tetrahedral or hexahedral (rectangular prism) elements such as shown in Fig. 6.27. Assuming a four-node tetrahedral element, the function Φ is represented within the element by

$$\Phi_e = a + bx + cy + dz. \tag{6.113}$$

The same applies to the function g. Since Eq. (6.113) must be satisfied at the four nodes of the tetrahedral elements,

$$\Phi_{ei} = a + bx_i + cy_i + dz_i, \qquad i = 1, \cdots, 4. \tag{6.114}$$

Thus we have four simultaneous equations (similar to Eq. (6.5)) from which the coefficients $a, b, c,$ and d can be determined. The determinant of the system of equations is

$$\det = \begin{vmatrix} 1 & x_1 & y_1 & z_1 \\ 1 & x_2 & y_2 & z_2 \\ 1 & x_3 & y_3 & z_3 \\ 1 & x_4 & y_4 & z_4 \end{vmatrix} = 6v, \tag{6.115}$$

where v is the volume of the tetrahedron. By finding $a, b, c,$ and d, we can write

$$\boxed{\Phi_e = \sum_{i=1}^{4} \alpha_i(x,y)\Phi_{ei}} \tag{6.116}$$

where

$$\alpha_1 = \frac{1}{6v} \begin{vmatrix} 1 & x & y & z \\ 1 & x_2 & y_2 & z_2 \\ 1 & x_3 & y_3 & z_3 \\ 1 & x_4 & y_4 & z_4 \end{vmatrix}, \tag{6.117a}$$

$$\alpha_2 = \frac{1}{6v} \begin{vmatrix} 1 & x_1 & y_1 & z_1 \\ 1 & x & y & z \\ 1 & x_3 & y_3 & z_3 \\ 1 & x_4 & y_4 & z_4 \end{vmatrix}, \tag{6.117b}$$

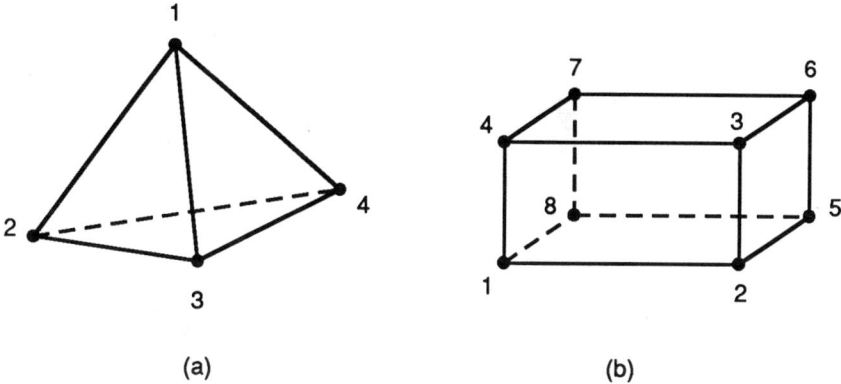

Figure 6.27 Three-dimensional elements: (a) Four-node or linear-order tetrahedral, (b) eight-node or linear-order hexahedral.

with α_3 and α_4 having similar expressions. For higher order approximation, the matrices for αs become large in size and we resort to local coordinates. Another motivation for using local coordinates is the existence of integration equations which simplify the evaluation of the fundamental matrices T and Q.

For the tetrahedral element, the local coordinates are $\xi_1, \xi_2, \xi_3,$ and ξ_4, each perpendicular to a side. They are defined at a given point as the ratio of the distance from that point to the appropriate apex to the perpendicular distance from the side to the opposite apex. They can also be interpreted as volume ratios, i.e., at a point P

$$\xi_i = \frac{v_i}{v} \tag{6.118}$$

where v_i is the volume bound by P and face i. It is evident that

$$\sum_{i=1}^{4} \xi_i = 1 \tag{6.119a}$$

or

$$\xi_4 = 1 - \xi_1 - \xi_2 - \xi_3. \tag{6.119b}$$

476   Numerical Techniques in Electromagnetics

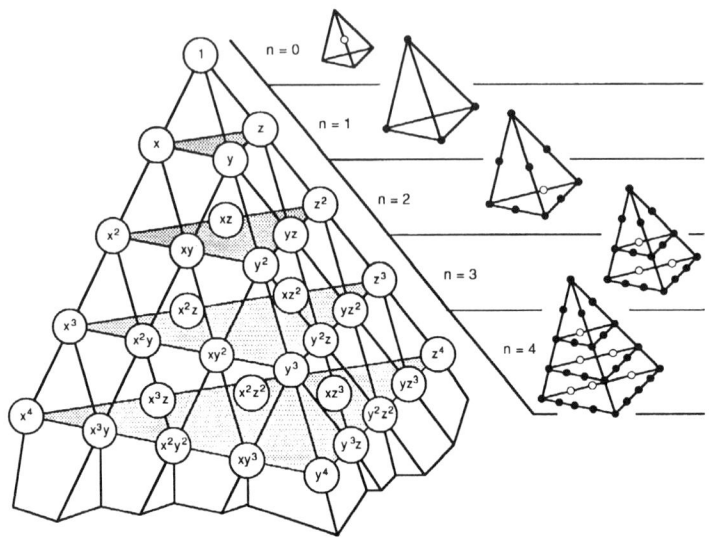

Figure 6.28 Pascal tetrahedron and associated array of terms.

The following properties are useful in calculating integration involving local coordinates [47]:

$$dv = 6v \, d\xi_1 d\xi_2 d\xi_3, \tag{6.120a}$$

$$\iiint f \, dv = 6v \int_0^1 \left[\int_0^{1-\xi_3} \left(\int_0^{1-\xi_2-\xi_3} f \, d\xi_1 \right) d\xi_3 \right] d\xi_3, \tag{6.120b}$$

$$\iiint \xi_1^i \xi_2^j \xi_3^k \xi_4^\ell \, dv = \frac{i! j! k! \ell!}{(i+j+k+\ell+3)} 6v. \tag{6.120c}$$

In terms of the local coordinates, an arbitrary function $\Phi(x, y)$ can be approximated within an element in terms of a complete nth order polynomial as

$$\boxed{\Phi_e(x,y) = \sum_{i=1}^m \alpha_i(x,y) \Phi_{ei}} \tag{6.121}$$

where $m = \frac{1}{6}(n+1)(n+2)(n+3)$ is the number of nodes in the tetrahedron

Finite Element Method 477

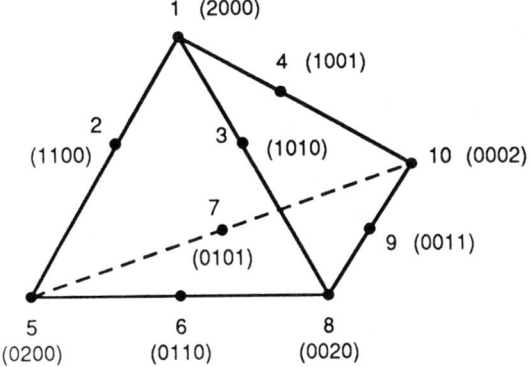

Figure 6.29 Numbering scheme for second-order tetrahedron.

or the number of terms in the polynomial. The terms in a complete three-dimensional polynomial may be arrayed as shown in Fig. 6.28.

Each point in the tetrahedral element is represented by four integers i, j, k, and ℓ which can be used to determine the local coordinates (ξ_1, ξ_2, ξ_3, ξ_4). That is, at $P_{ijk\ell}$,

$$(\xi_1, \xi_2, \xi_3, \xi_4) = (\frac{i}{n}, \frac{j}{n}, \frac{k}{n}, \frac{\ell}{n}). \tag{6.122}$$

Hence at each node,

$$\alpha_q = \alpha_{ijk\ell} = p_i(\xi_1)p_j(\xi_2)p_k(\xi_3)p_\ell(\xi_4), \tag{6.123}$$

where $q = 1, 2, \cdots, m$ and p_r is defined in Eq. (6.102) or (6.103). The relationship between the node numbers q and $ijk\ell$ is illustrated in Fig. 6.29 for the second order tetrahedron ($n = 2$). The shape functions obtained by substituting Eq. (6.102) into Eq. (6.123) are presented in Table 6.11 for $n = 1$ to 3.

The expressions derived from the variational principle for the two-dimensional problems in Sections 6.2 to 6.4 still hold except that the fundamental matrices $[T]$ and $[Q]$ now involve triple integration. For Helmholtz equation (6.50), for example, Eq. (6.62) applies, namely,

$$\left[C_{ff} - k^2 T_{ff}\right]\Phi_f = 0 \tag{6.124}$$

except that

Table 6.11 Shape Functions $\alpha_q(\xi_1, \xi_2, \xi_3, \xi_4)$ for $n = 1$ to 3

$n = 1$	$n = 2$	$n = 3$
$\alpha_1 = \xi_1$	$\alpha_1 = \xi_1(2\xi_2 - 1)$	$\alpha_1 = \frac{1}{2}\xi_1(3\xi_1 - 2)(3\xi_1 - 1)$
$\alpha_2 = \xi_2$	$\alpha_2 = 4\xi_1\xi_2$	$\alpha_2 = \frac{9}{2}\xi_1(3\xi_1 - 1)\xi_2$
$\alpha_3 = \xi_3$	$\alpha_3 = 4\xi_1\xi_3$	$\alpha_3 = \frac{9}{2}\xi_1(3\xi_1 - 1)\xi_3$
$\alpha_4 = \xi_4$	$\alpha_4 = 4\xi_1\xi_4$	$\alpha_4 = \frac{9}{2}\xi_1(3\xi_1 - 1)\xi_4$
	$\alpha_5 = \xi_2(2\xi_2 - 1)$	$\alpha_5 = \frac{9}{2}\xi_1(3\xi_3 - 1)\xi_2$
	$\alpha_6 = 4\xi_2\xi_3$	$\alpha_6 = 27\xi_1\xi_2\xi_3$
	$\alpha_7 = 4\xi_2\xi_4$	$\alpha_7 = 27\xi_1\xi_2\xi_4$
	$\alpha_8 = \xi_2(2\xi_3 - 1)$	$\alpha_8 = \frac{9}{2}\xi_1(3\xi_3 - 1)\xi_3$
	$\alpha_9 = 4\xi_3\xi_4$	$\alpha_9 = 27\xi_1\xi_3\xi_4$
	$\alpha_{10} = \xi_4(2\xi_4 - 1)$	$\alpha_{10} = \frac{9}{2}\xi_1(3\xi_4 - 1)\xi_4$
		$\alpha_{11} = \frac{1}{2}\xi_2(3\xi_2 - 1)(3\xi_2 - 2)$
		$\alpha_{12} = \frac{9}{2}\xi_2(3\xi_2 - 1)\xi_3$
		$\alpha_{13} = \frac{9}{2}\xi_2(3\xi_2 - 1)\xi_4$
		$\alpha_{14} = \frac{9}{2}\xi_2(3\xi_3 - 1)\xi_3$
		$\alpha_{15} = 27\xi_2\xi_3\xi_4$
		$\alpha_{16} = \frac{9}{2}\xi_2(3\xi_3 - 1)\xi_3$
		$\alpha_{17} = \frac{1}{2}\xi_3(3\xi_3 - 1)(3\xi_3 - 2)$
		$\alpha_{18} = \frac{9}{2}\xi_3(3\xi_3 - 1)\xi_4$
		$\alpha_{19} = \frac{9}{2}\xi_3(3\xi_4 - 1)\xi_4$
		$\alpha_{20} = \frac{1}{2}\xi_4(3\xi_4 - 1)(3\xi_4 - 2)$

$$C_{ij}^{(e)} = \int_v \nabla\alpha_i \cdot \nabla\alpha_j \, dv$$
$$= \int_v \left(\frac{\partial\alpha_i}{\partial x}\frac{\partial\alpha_j}{\partial x} + \frac{\partial\alpha_i}{\partial y}\frac{\partial\alpha_j}{\partial y} + \frac{\partial\alpha_i}{\partial z}\frac{\partial\alpha_j}{\partial z} \right) dv, \tag{6.125}$$

$$T_{ij}^{(e)} = \int_v \alpha_i\alpha_j \, dv = v \iiint \alpha_i\alpha_j \, d\xi_1 \, d\xi_2 \, d\xi_3. \tag{6.126}$$

For further discussion on three-dimensional elements, one should consult Silvester and Ferrari [4]. Applications of three-dimensional elements to EM-related problems can be found in [49–52].

6.10 Hybrid Finite Element Methods

Thus far in this chapter, the FEM has been presented for solving interior problems. To apply the FEM to exterior or unbounded problems such as open-type transmission lines (e.g., microstrip), scattering, and radiation problems poses certain difficulties. To overcome these difficulties, several approaches [53–80] have been proposed, all of which have strengths and weaknesses. We will consider two common approaches: the infinite element method and the boundary element method.

6.10.1 Infinite Element Method

Consider the solution region shown in Fig. 6.30(a). We divide the entire domain into a near field (n.f.) region, which is bounded, and a far field (f.f.) region, which is unbounded. The n.f. region is divided into finite triangular elements as usual, while the f.f. region is divided into *infinite elements*. Each infinite element shares two nodes with a finite element. Here we are mainly concerned with the infinite elements.

Consider the infinite element in Fig. 6.30(b) with nodes 1 and 2 and radial sides intersecting at point (x_o, y_o). We relate triangular polar coordinates (ρ, ξ) to the global Cartesian coordinates (x, y) as [62]

$$\begin{aligned} x &= x_o + \rho[(x_1 - x_o) + \xi(x_2 - x_1)] \\ y &= y_o + \rho[(y_1 - y_o) + \xi(y_2 - y_1)] \end{aligned} \tag{6.127}$$

where $1 \leq \rho < \infty$, $0 \leq \xi \leq 1$. The potential distribution within the element is approximated by a linear variation as

$$V = \frac{1}{\rho}[V_1(1-\xi) + V_2\xi]$$

480   Numerical Techniques in Electromagnetics

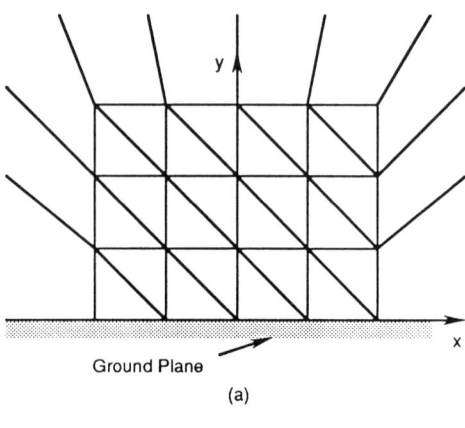

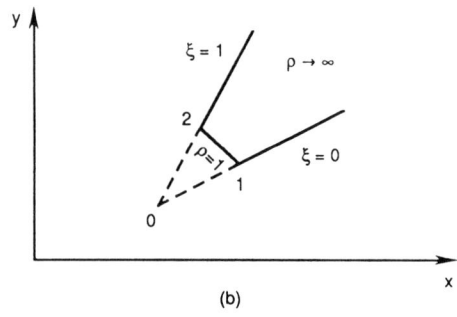

Figure 6.30 (a) Division of solution region into finite and infinite elements; (b) typical infinite element.

or
$$V = \sum_{i=1}^{2} \alpha_i V_i \qquad (6.128)$$

where V_1 and V_2 are potentials at nodes 1 and 2 of the infinite elements, α_1 and α_2 are the interpolation or shape functions, i.e.,

$$\alpha_1 = \frac{1-\xi}{\rho}, \quad \alpha_2 = \frac{\xi}{\rho}. \qquad (6.129)$$

The infinite element is compatible with the ordinary first order finite element and satisfies the boundary condition at infinity. With the shape functions in Eq. (6.129), we can obtain the $[C^{(e)}]$ and $[T^{(e)}]$ matrices. We obtain solution for the exterior problem by using a standard finite element program with the $[C^{(e)}]$ and $[T^{(e)}]$ matrices of the infinite elements added to the $[C]$ and $[T]$ matrices of the n.f. region.

Table 6.12 Comparison Between Method of Moments and Finite Element Method [81]

Method of Moments	Finite Element Method
Conceptually easy	Conceptually involved
Requires problem-dependent Green's functions	Avoids difficulties associated with singularity of Green's functions
Few equations; $O(n)$ for 2-D, $O(n^2)$ for 3-D	Many equations; $O(n^2)$ for 2-D, $O(n^3)$ for 3-D
Only boundary is discretized	Entire domain is discretized
Open boundary easy	Open boundary difficult
Fields by integration	Fields by differentiation
Good representation of far-field condition	Good representation of boundary conditions
Full matrices result	Sparse matrices result
Nonlinearity, inhomogeneity difficult	Nonlinearity, inhomogeneity easy

6.10.2 Boundary Element Method

The finite element method (FEM) and the method of moments (MOM) have become the two most popular methods of numerical analysis of electromagnetic problems. The choice between the two numerical techniques will usually depend on the particular problem, although both are ultimately capable of yielding the same information. The two methods result in a set of simultaneous equations. These equations look quite different from each other, and each set of equations presents its own peculiar problems. Perhaps the best way to compare MOM with FEM is shown in Table 6.12. From the table, it is evident that the two methods have properties that complement each other. In view of this, hybrid methods have been proposed. These methods allow the use of both MOM and FEM with the aim of exploiting the strong points in each method.

One of these hybrid methods involves using the so-called boundary element method (BEM). It is a finite element approach for handling exterior problems [68-80]. It basically involves obtaining the boundary-integral equation from Green's identity and solving this by a discretization procedure similar to that used in regular finite element analysis. Since the BEM is based on the boundary integral equivalent to the governing differential equation, only the surface of the problem domain needs to be modeled. Thus the dimension of the problem is reduced by one as in MOM. For 2-D problems, the boundary elements are taken to be straight line segments, whereas for 3-D problems, they are taken as triangular elements. Thus the shape or interpolation functions corresponding to subsectional bases in the MOM are used in the finite element analysis.

6.11 Concluding Remarks

An introduction to the basic concepts and applications of the finite element method has been presented. It is by no means an exhaustive exposition of the subject. However, we have given the flavor of the way in which the ideas may be developed; the interested reader may build on this by consulting the references. Several introductory texts have been published on FEM. Although most of these texts are written for civil or mechanical engineers, the texts by Silvester and Ferrari [4], Chari and Silvester [41], Steele [82], Hoole [83], and Itoh [84] are for electrical engineers.

The finite element method has been applied with great success to numerous EM-related problems. Such applications are:

- transmission line problems [85–87],
- optical and microwave waveguide problems [8–17,88–92],
- electric machines [41, 93–95],
- semiconductor devices [96–97],
- scattering problems [69, 70, 73, 98, 99],
- human exposition to EM radiation [100–103], and
- others [104–106].

Applications of the FEM to time-dependent phenomena can be found in [99,107–116].

For other issues on FEM not covered in this chapter, one is referred to introductory texts on FEM such as [2, 4, 36, 41, 47, 82–84, 116–123]. Estimating error in finite element solution is discussed in [52, 124, 125]. The reader may benefit from the numerous finite element codes that are commercially available. Among these packages are ABAQUS, ADINA, ASKA, and SAFE. An extensive description of these systems and their capabilities can be found in Brebbia [117]. Although the codes were developed for one field of engineering or the other, they can be applied to problems in a different field with little or no modification.

References

[1] R. Courant, "Variational methods for the solution of problems of equilibrium and vibrations," *Bull. Am. Math. Soc.*, vol. 49, 1943, pp. 1–23.

[2] C. S. Desai and J. F. Abel, *Introduction to the Finite Element Method: A Numerical Approach for Engineering Analysis.* New York: Van Nostrand Reinhold, 1972.

[3] M. N. O. Sadiku, "A simple introduction to finite element analysis of electromagnetic problems," *IEEE Trans. Educ.*, vol. 32, no. 2, May 1989, pp. 85–93.

[4] P. P. Silvester and R. L. Ferrari, *Finite Elements for Electrical Engineers.* Cambridge: Cambridge University Press, 1983, pp. 1–32, 73–88, 143–186.

[5] O. W. Andersen, "Laplacian electrostatic field calculations by finite elements with automatic grid generation," *IEEE Trans. Power App. Syst.*, vol. PAS-92, no. 5, Sept./Oct. 1973, pp. 1485–1492.

[6] S. Nakamura, *Computational Methods in Engineering and Science.* New York: John Wiley, 1977, pp. 446, 447.

[7] B. S. Garbow, *Matrix Eigensystem Routine—EISPACK Guide Extension.* Berlin: Springer-Verlag, 1977.

[8] S. Ahmed and P. Daly, "Finite-element methods for inhomogeneous waveguides," *Proc. IEE,* vol. 116, no. 10, Oct. 1969, pp. 1661–1664.

[9] Z. J. Csendes and P. Silvester, "Numerical solution of dielectric loaded waveguides: I—Finite-element analysis," *IEEE Trans. Micro. Theo. Tech.*, vol. MTT-18, no. 12, Dec. 1970, pp. 1124–1131.

[10] Z. J. Csendes and P. Silvester, "Numerical solution of dielectric loaded waveguides: II—Modal approximation technique," *IEEE Trans. Micro. Theo. Tech.*, vol. MTT-19, no. 6, June 1971, pp. 504–509.

[11] M. Hano, "Finite-element analysis of dielectric-loaded waveguides," *IEEE Trans. Micro. Theo. Tech.*, vol. MTT-32, no. 10, Oct. 1984, pp. 1275–1279.

[12] A. Konrad, "Vector variational formulation of electromagnetic fields in anisotropic media," *IEEE Trans. Micro. Theo. Tech.*, vol. MTT-24, Sept. 1976, pp. 553–559.

[13] M. Koshiba, et al., "Improved finite-element formulation in terms of the magnetic field vector for dielectric waveguides," *IEEE Trans. Micro. Theo. Tech.*, vol. MTT-33, no. 3, March 1985, pp. 227–233.

[14] M. Koshiba, et al., "Finite-element formulation in terms of the electric-field vector for electromagnetic waveguide problems," *IEEE*

Trans. Micro. Theo. Tech., vol. MTT-33, no. 10, Oct. 1985, pp. 900–905.

[15] K. Hayata, et al., "Vectorial finite-element method without any spurious solutions for dielectric waveguiding problems using transverse magnetic-field component," *IEEE Trans. Micro. Theo. Tech.*, vol. MTT-34, no. 11, Nov. 1986.

[16] K. Hayata, et al., "Novel finite-element formulation without any spurious solutions for dielectric waveguides," *Elect. Lett.*, vol. 22, no. 6, March 1986, pp. 295, 296.

[17] S. Dervain, "Finite element analysis of inhomogeneous waveguides," Masters thesis, Department of Electrical and Computer Engineering, Florida Atlantic University, Boca Raton, April 1987.

[18] J. R. Winkler and J. B. Davies, "Elimination of spurious modes in finite element analysis," *J. Comp. Phys.*, vol. 56, no. 1, Oct. 1984, pp. 1–14.

[19] M. N. O. Sadiku, et al., "A further introduction to finite element analysis of electromagnetic problems," *IEEE Trans. Educ.*, vol. 34, no. 4, Nov. 1991, pp. 322–329.

[20] The IMSL Libraries: Problem-solving software systems for numerical FORTRAN programming, IMSL, Houston, TX, 1984.

[21] M. Kono, "A generalized automatic mesh generation scheme for finite element method," *Inter. J. Num. Method Engr.*, vol. 15, 1980, pp. 713–731.

[22] J. C. Cavendish, "Automatic triangulation of arbitrary planar domains for the finite element method," *Inter. J. Num. Meth. Engr.*, vol. 8, 1974, pp. 676–696.

[23] A. O. Moscardini, et al., "AGTHOM—Automatic generation of triangular and higher order meshes," *Inter. J. Num. Meth. Engr.*, vol. 19, 1983, pp. 1331–1353.

[24] C. O. Frederick, et al., "Two-dimensional automatic mesh generation for structured analysis," *Inter. J. Num. Meth. Engr.*, vol. 2 no. 1, 1970, pp. 133–144.

[25] E. A. Heighway, "A mesh generation for automatically subdividing irregular polygon into quadrilaterals," *IEEE Trans. Mag.*, vol. MAG-19, no. 6, Nov. 1983, pp. 2535–2538.

[26] C. Kleinstreuer and J. T. Holdeman, "A triangular finite element mesh generator for fluid dynamic systems of arbitrary geometry," *Inter. J. Num. Meth. Engr.*, vol. 15, 1980, pp. 1325–1334.

[27] A. Bykat, "Automatic generation of triangular grid I—subdivision of a general polygon into convex subregions. II—Triangulation of convex polygons," *Inter. J. Num. Meth. Engr.*, vol. 10, 1976, pp. 1329–1342.

[28] N. V. Phai, "Automatic mesh generator with tetrahedron elements," *Inter. J. Num. Meth. Engr.*, vol. 18, 1982, pp. 273–289.

[29] F. A. Akyuz, "Natural coordinates systems—an automatic input data generation scheme for a finite element method," *Nuclear Engr. Design*, vol. 11, 1970, pp. 195–207.

[30] P. Girdinio, et al., "New developments of grid optimization by the grid iteration method," in Z. J. Csendes (ed.), *Computational Electromagnetism*. New York: North-Holland, 1986, pp. 3–12.

[31] M. Yokoyama, "Automated computer simulation of two-dimensional elastrostatic problems by finite element method," *Inter. J. Num. Meth. Engr.*, vol. 21, 1985, pp. 2273–2287.

[32] G. F. Carey, "A mesh-refinement scheme for finite element computations," *Comp. Meth. Appl. Mech. Engr.*, vol. 7, 1976, pp. 93–105.

[33] K. Preiss, "Checking the topological consistency of a finite element mesh," *Inter. J. Meth. Engr.*, vol. 14, 1979, pp. 1805–1812.

[34] H. Kardestuncer (ed.), *Finite Element Handbook*. New York: McGraw-Hill, 1987, pp. 4.191–4.207.

[35] W. C. Thacker, "A brief review of techniques for generating irregular computational grids," *Inter. J. Num. Meth. Engr.*, vol. 15, 1980, pp. 1335–1341.

[36] E. Hinton and D. R. J. Owen, *An Introduction to Finite Element Computations*. Swansea, UK: Pineridge Press, 1979, pp. 247, 328–346.

[37] R. J. Collins, "Bandwidth reduction by automatic renumbering," *Inter. J. Num. Meth. Engr.*, vol. 6, 1973, pp. 345–356.

[38] E. Cuthill and J. McKee, "Reducing the bandwidth of sparse symmetric matrices," *ACM Nat. Conf.*, San Francisco, 1969, pp. 157–172.

[39] G. A. Akhras and G. Dhatt, "An automatic node relabelling scheme for minimizing a matrix or network bandwidth," *Inter. J. Num. Meth. Engr.*, vol. 10, 1976, pp. 787–797.

[40] F. A. Akyuz and S. Utku, "An automatic node-relabelling scheme for bandwidth minimization of stiffness matrices," *J. Amer. Inst. Aero. Astro.*, vol. 6, no. 4, 1968, pp. 728–730.

[41] M. V. K. Chari and P. P. Silvester (eds.), *Finite Elements for Electrical and Magnetic Field Problems*. Chichester: John Wiley, 1980,

pp. 125-143.

[42] P. Silvester, "Construction of triangular finite element universal matrices," *Inter. J. Num. Meth. Engr.*, vol. 12, 1978, pp. 237–244.

[43] P. Silvester, "High-order polynomial triangular finite elements for potential problems," *Inter. J. Engr. Sci.*, vol. 7, 1969, pp. 849–861.

[44] G. O. Stone, "High-order finite elements for inhomogeneous acoustic guiding structures," *IEEE Trans. Micro. Theory Tech.*, vol. MTT-21, no. 8, Aug. 1973, pp. 538–542.

[45] A. Konrad, "High-order triangular finite elements for electromagnetic waves in anistropic media," *IEEE Trans. Micro. Theory Tech.*, vol. MTT-25, no. 5, May 1977, pp. 353-360.

[46] P. Daly, "Finite elements for field problems in cylindrical coordinates," *Inter. J. Num. Meth. Engr.*, vol. 6, 1973, pp. 169–178.

[47] C. A. Brebbia and J. J. Connor, *Fundamentals of Finite Element Technique*. London: Butterworth, 1973, pp. 114-118, 150–163, 191.

[48] M. Sadiku and L. Agba, "News rules for generating finite elements fundamental matrices," *Proc. IEEE Southeastcon*, 1989, pp. 797–801.

[49] R. L. Ferrari and G. L. Maile, "Three-dimensional finite element method for solving electromagnetic problems," *Elect. Lett.*, vol. 14, no. 15, 1978, pp. 467, 468.

[50] M. de Pourcq, "Field and power-density calculation by three-dimensional finite elements," *IEE Proc.*, vol. 130, Pt. H, no. 6, Oct. 1983, pp. 377–384.

[51] M. V. K. Chari, et al., "Finite element computation of three-dimensional electrostatic and magnetostatic field problems," *IEEE Trans. Mag.*, vol. MAG-19, no. 16, Nov. 1983, pp. 2321–2324.

[52] O. A. Mohammed, et al., "Validity of finite element formulation and solution of three dimensional magnetostatic problems in electrical devices with applications to transformers and reactors," *IEEE Trans. Pow. App. Syst.*, vol. PAS-103, no. 7, July 1984, pp. 1846–1853.

[53] B. H. McDonald and A. Wexler, "Finite-element solution of unbounded field problems," *IEEE Trans. Micro. Theo. Tech.*, vol. MTT-20, no. 12, Dec. 1972, pp. 841–847.

[54] P. P. Silvester, et al., "Exterior finite elements for 2-dimensional field problems with open boundaries," *Proc. IEE*, vol. 124, no. 12, Dec. 1977, pp. 1267–1270.

[55] S. Washisu, et al., "Extension of finite-element method to unbounded

field problems," *Elect. Lett.*, vol. 15, no. 24, Nov. 1979, pp. 772–774.

[56] P. Silvester and M. S. Hsieh, "Finite-element solution of 2-dimensional exterior-field problems," *Proc. IEE*, vol. 118, no. 12, Dec. 1971, pp. 1743–1747.

[57] Z. J. Csendes, "A note on the finite-element solution of exterior-field problems," *IEEE Trans. Micro. Theo. Tech.*, vol. MTT-24, no. 7, July 1976, pp. 468–473.

[58] T. Corzani, et al., "Numerical analysis of surface wave propagation using finite and infinite elements," *Alta Frequenza*, vol. 51, no. 3, June 1982, pp. 127–133.

[59] O. C. Zienkiewicz, et al., "Mapped infinite elements for exterior wave problems," *Inter. J. Num. Meth. Engr.*, vol. 21, 1985.

[60] F. Medina, "An axisymmetric infinite element," *Int. J. Num. Meth. Engr.*, vol. 17, 1981, pp. 1177–1185.

[61] S. Pissanetzky, "A simple infinite element," *Int. J. Comp. Math. Elect. Engr.* (COMPEL), vol. 3, no. 2, 1984, pp. 107–114.

[62] Z. Pantic and R. Mittra, "Quasi-TEM analysis of microwave transmission lines by the finite-element method," *IEEE Trans. Micro. Theo. Tech.*, vol. MTT-34, no. 11, Nov. 1986, pp. 1096–1103.

[63] K. Hayata, et al., "Self-consistent finite/infinite element scheme for unbounded guided wave problems," *IEEE Trans. Micro. Theo. Tech.*, vol. MTT-36, no. 3, Mar. 1988, pp. 614–616.

[64] P. Petre and L. Zombory, "Infinite elements and base functions for rotationally symmetric electromagnetic waves," *IEEE Trans. Ant. Prog.*, vol. 36, no. 10, Oct. 1988, pp. 1490, 1491.

[65] Z. J. Csendes and J. F. Lee, "The transfinite element method for modeling MMIC devices," *IEEE Trans. Micro. Theo. Tech.* vol. 36, no. 12, Dec. 1988, pp. 1639–1649.

[66] K. H. Lee, et al., "A hybrid three-dimensional electromagnetic modeling scheme," *Geophys.*, vol. 46, no. 5, May 1981, pp. 796–805.

[67] S. J. Salon and J. M. Schneider, "A hybrid finite element-boundary integral formulation of Poisson's equation," *IEEE Trans. Mag.*, vol. MAG-17, no. 6, Nov. 1981, pp. 2574–2576.

[68] S. J. Salon and J. Peng, "Hybrid finite-element boundary-element solutions to axisymmetric scalar potential problems," in Z. J. Csendes (ed.), *Computational Electromagnetics*. New York: North-Holland/Elsevier, 1986, pp. 251–261.

[69] J. M. Lin and V. V. Liepa, "Application of hybrid finite element method for electromagnetic scattering from coated cylinders," *IEEE Trans. Ant. Prop.*, vol. 36, no. 1, Jan. 1988, pp. 50–54.

[70] J. M. Lin and V. V. Liepa, "A note on hybrid finite element method for solving scattering problems," *IEEE Trans. Ant. Prop.*, vol. 36, no. 10, Oct. 1988, pp. 1486–1490.

[71] M. H. Lean and A. Wexler, "Accurate field computation with boundary element method," *IEEE Trans. Mag.*, vol. MAG-18, no. 2, Mar. 1982, pp. 331–335.

[72] R. F. Harrington and T. K. Sarkar, "Boundary elements and the method of moments," in C. A. Brebbia, et al. (eds.), *Boundary Elements*. Southampton: CML Publ., 1983, pp. 31–40.

[73] M. A. Morgan, et al., "Finite element-boundary integral formulation for electromagnetic scattering," *Wave Motion*, vol. 6, no. 1, 1984, pp. 91–103.

[74] S. Kagami and I. Fukai, "Application of boundary-element method to electromagnetic field problems," *IEEE Trans. Micro. Theo. Tech.*, vol. 32, no. 4, Apr. 1984, pp. 455–461.

[75] Y. Tanaka, et al., "A boundary-element analysis of TEM cells in three dimensions," *IEEE Trans. Elect. Comp.*, vol. EMC-28, no. 4, Nov. 1986, pp. 179–184.

[76] N. Kishi and T. Okoshi, "Proposal for a boundary-integral method without using Green's function," *IEEE Trans. Micro. Theo. Tech.*, vol. MTT-35, no. 10, Oct. 1987, pp. 887–892.

[77] D. B. Ingham, et al., "Boundary integral equation analysis of transmission-line singularities," *IEEE Trans. Micro. Theo. Tech.*, vol. MTT-29, no. 11, Nov. 1981, pp. 1240–1243.

[78] S. Washiru, et al., "An analysis of unbounded field problems by finite element method," *Electr. Comm. Japan*, vol. 64-B, no. 1, 1981, pp. 60–66.

[79] T. Yamabuchi and Y. Kagawa, "Finite element approach to unbounded Poisson and Helmholtz problems using hybrid-type infinite element," *Electr. Comm. Japan*, Pt. I, vol. 68, no. 3, 1986, pp. 65–74.

[80] K. L. Wu and J. Litva, "Boundary element method for modelling MIC devices," *Elect. Lett.*, vol. 26, no. 8, April 1990, pp. 518–520.

[81] M. N. O. Sadiku and A. F. Peterson, "A comparison of numerical methods for computing electromagnetic fields," *Proc. of IEEE Southeastcon*, April 1990, pp. 42–47.

[82] C. W. Steele, *Numerical Computation of Electric and Magnetic Fields.* New York: Van Nostrand Reinhold, 1987.

[83] S. R. Hoole, *Computer-aided Analysis and Design of Electromagnetic Devices.* New York: Elsevier, 1989.

[84] T. Itoh (ed.), *Numerical Technique for Microwave and Millimeter-wave Passive Structures.* New York: John Wiley, 1989.

[85] R. L. Khan and G. I. Costache, "Finite element method applied to modeling crosstalk problems on printed circuit boards," *IEEE Trans. Elect. Comp.*, vol. 31, no. 1, Feb. 1989, pp. 5–15.

[86] P. Daly, "Upper and lower bounds to the characteristic impedance of transmission lines using the finite element method," *Inter. J. Comp. Math. Elect. Electr. Engr.* (COMPEL), vol. 3, no. 2, 1984, pp. 65–78.

[87] A. Khebir, et al., "An absorbing boundary condition for quasi-TEM analysis of microwave transmission lines via the finite element method," *J. Elect. Waves Appl.*, vol. 4, no. 2, 1990, pp. 145–157.

[88] N. Mabaya, et al., "Finite element analysis of optical waveguides," *IEEE Trans. Micro. Theo. Tech.*, vol. MTT-29, no. 6, June 1981, pp. 600–605.

[89] M. Ikeuchi, et al., "Analysis of open-type dielectric waveguides by the finite-element iterative method," *IEEE Trans. Micro. Theo. Tech.*, vol. MTT-29, no. 3, Mar. 1981, pp. 234–239.

[90] C. Yeh, et al., "Single model optical waveguides," *Appl. Optics*, vol. 18, no. 10, May 1979, pp. 1490–1504.

[91] J. Katz, "Novel solution of 2-D waveguides using the finite element method," *Appl. Optics*, vol. 21, no. 15, Aug. 1982, pp. 2747–2750.

[92] B. A. Rahman and J. B. Davies, "Finite-element analysis of optical and microwave waveguide problems," *IEEE Trans. Micro. Theo. Tech.*, vol. MTT-32, no. 1, Jan. 1984, pp. 20–28.

[93] C. B. Rajanathan, et al., "Finite-element analysis of the Xi-core levitator," *IEE Proc.*, vol. 131, Pt. A, no. 1, Jan. 1984, pp. 62–66.

[94] T. L. Ma and J. D. Lavers, "A finite-element package for the analysis of electromagnetic forces and power in an electric smelting furnace," *IEEE Trans. Indus. Appl.*, vol. IA-22, no. 4, July/Aug. 1986, pp. 578–585.

[95] C. O. Obiozor and M. N. O. Sadiku, "Finite element analysis of a solid rotor induction motor under stator winding effects," *Proc. IEEE Southeastcon*, 1991, pp. 449–453.

[96] J. J. Barnes and R. J. Lomax, "Finite-element methods in semiconductor device simulation," *IEEE Trans. Elect. Dev.*, vol. ED-24, no. 8, Aug. 1977, pp. 1084–1089.

[97] T. Adachi, et al., "Two-dimensional semiconductor analysis using finite-element method," *IEEE Trans. Elect. Dev.*, vol. ED-26, no. 7, July 1979, pp. 1026–1031.

[98] J. L. Mason and W. J. Anderson, "Finite element solution for electromagnetic scattering from two-dimensional bodies," *Inter. J. Num. Meth. Engr.*, vol. 21, 1985, pp. 909–928.

[99] A. C. Cangellaris, et al., "Point-matching time domain finite element methods for electromagnetic radiation and scattering," *IEEE Trans. Ant. Prop.*, vol. AP35, 1987, pp. 1160–1173.

[100] A. Chiba, et al., "Application of finite element method to analysis of induced current densities inside human model exposed to 60 Hz electric field," *IEEE Trans. Power App. Sys.*, vol. PAS-103, no. 7, July 1984, pp. 1895–1902.

[101] Y. Yamashita and T. Takahashi, "Use of the finite element method to determine epicardial from body surface potentials under a realistic torso model," *IEEE Trans. Biomed. Engr.*, vol. BME-31, no. 9, Sept. 1984, pp. 611–621.

[102] M. A. Morgan, "Finite element calculation of microwave absorption by the cranial structure," *IEEE Trans. Biomed. Engr.*, vol. BME-28, no. 10, Oct. 1981, pp. 687–695.

[103] D. R. Lynch, et al., "Finite element solution of Maxwell's equation for hyperthermia treatment planning," *J. Comp. Phys.* vol. 58, 1985, pp. 246–269.

[104] J. R. Brauer, et al., "Dynamic electric fields computed by finite elements," *IEEE Trans. Ind. Appl.*, vol. 25, no. 6, Nov./Dec. 1989, pp. 1088–1092.

[105] C. H. Chen and C. D. Lien, "A finite element solution of the wave propagation problem for an inhomogeneous dielectric slab," *IEEE Trans. Ant. Prop.*, vol. AP-27, no. 6, Nov. 1979, pp. 877–880.

[106] T. L. W. Ma and J. D. Lavers, "A finite-element package for the analysis of electromagnetic forces and power in an electric smelting furnace," *IEEE Trans. Ind. Appl.*, vol. IA-22, no. 4, July/Aug., 1986, pp. 578–585.

[107] V. Shanka, "A time-domain, finite-volume treatment for Maxwell's equations," *Electromagnetics*, vol. 10, 1990, pp. 127–145.

[108] J. H. Argyris and D. W. Scharpf, "Finite elements in time and space," *Nucl. Engr. Space Des.*, vol. 10, no. 4, 1969, pp. 456–464.

[109] I. Fried, "Finite-element analysis of time-dependent phenomena," *AIAA J.*, vol. 7, no. 6, 1969, pp. 1170–1173.

[110] O. C. Zienkiewicz and C. J. Pareth, "Transient field problems: two-dimensional and three-dimensional analysis by isoparametric finite elements," *Inter. J. Num. Meth. Engr.*, vol. 2, 1970, pp. 61–71.

[111] J. H. Argyris and A. S. L. Chan, "Applications of finite elements in space and time," *Ingenieur–Archiv*, vol. 41, 1972, pp. 235–257.

[112] J. C. Bruch and G. Zyvoloski, "Transient two-dimensional heat conduction problems solved by the finite element method," *Inter. J. Num. Meth. Engr.*, vol. 8, 1974, pp. 481–494.

[113] B. Swartz and B. Wendroff, "The relative efficiency of finite difference and finite element methods I: hyperbolic problems and splines," *SIAM J. Numer. Anal.*, vol. 11, no. 5, Oct. 1974.

[114] J. Cushman, "Difference schemes or element schemes," *Int. J. Num. Meth. Engr.*, vol. 14, 1979, pp. 1643–1651.

[115] A. J. Baker and M. O. Soliman, "Utility of a finite element solution algorithm for initial-value problems," *J. Comp. Phys.*, vol. 32, 1979, pp. 289–324.

[116] J. N. Reddy, *An Introduction to the Finite Element Method.* New York: McGraw-Hill, 1984, pp. 299–307.

[117] C. A. Brebbia (ed.), *Applied Numerical Modelling.* New York: John Wiley, 1978, pp. 571–586.

[118] C. A. Brebbia (ed.), *Finite Element Systems: A Handbook.* Berlin: Springer-Verlag, 1985.

[119] O. C. Zienkiewicz, *The Finite Element Method.* New York: McGraw-Hill, 1977.

[120] A. J. Davies, *The Finite Element Method: A First Approach.* Oxford: Clarendon, 1980.

[121] C. Martin and G. F. Carey, *Introduction to Finite Element Analysis: Theory and Application.* New York: McGraw-Hill, 1973.

[122] T. J. Chung, *Finite Element Analysis in Fluid Dynamics.* New York: McGraw-Hill, 1978.

[123] D. H. Norris and G de. Vries, *An Introduction to Finite Element Analysis.* New York: Academic Press, 1978.

[124] R. Thatcher, "Assessing the error in a finite element solution," *IEEE Trans. Micro. Theo. Tech.*, vol. MTT-30, no. 6, June 1982, pp. 911–914.

Problems

6.1 For the triangular elements in Fig. 6.31, determine the element coefficient matrices.

6.2 Find the coefficient matrix for the two-element mesh of Fig. 6.32. Given that $V_2 = 10$ and $V_4 = -10$, determine V_1 and V_3.

6.3 Determine the shape functions $\alpha_1, \alpha_2,$ and α_3 for the element in Fig. 6.33.

6.4 The cross section of an infinitely long rectangular trough is shown in Fig. 6.34; develop a program using FEM to find the potential at the center of the cross section. Take $\epsilon_r = 4.5$.

6.5 Solve the problem in Example 3.3 using the finite element method.

6.6 Modify the program in Fig. 6.10 to calculate the electric field intensity **E** at any point in the solution region.

6.7 The program in Fig. 6.10 applies the iteration method to determine the potential at the free nodes. Modify the program and use the band matrix method to determine the potential. Test the program using the data in Example 6.2.

6.8 A grounded rectangular pipe with the cross section in Fig. 6.35 is half filled with hydrocarbons ($\epsilon = 2.5\epsilon_o, \rho_o = 10^{-5}$ C/m^3). Use FEM to determine the potential along the liquid-air interface. Plot the potential versus x.

6.9 Solve the problem in Example 3.4 using the finite element method.

6.10 The cross section of an isosceles right-triangular waveguide is discretized as in Fig. 6.36. Determine the first 10 TM cutoff wavelengths of the guide.

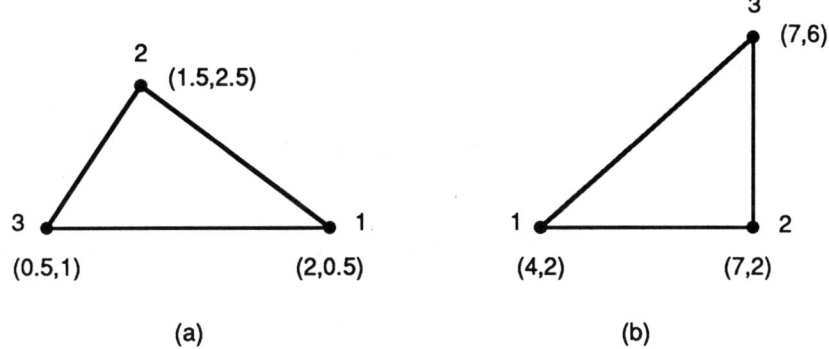

Figure 6.31 For Problem 6.1.

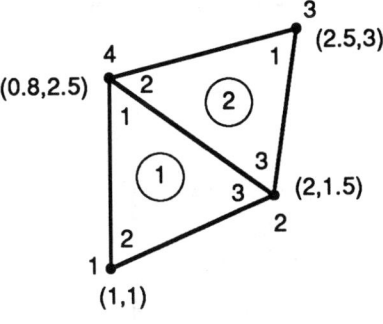

Figure 6.32 For Problem 6.2.

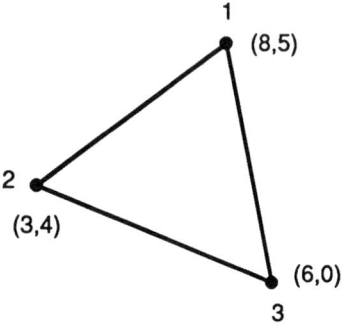

Figure 6.33 For Problem 6.3.

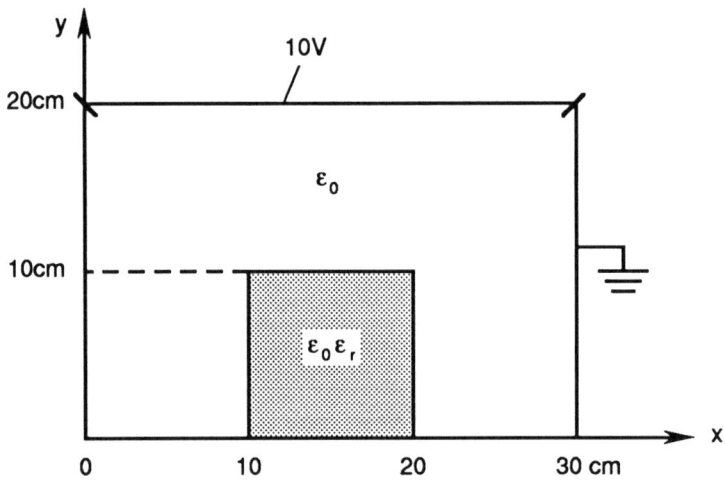

Figure 6.34 For Problem 6.4.

Finite Element Method 495

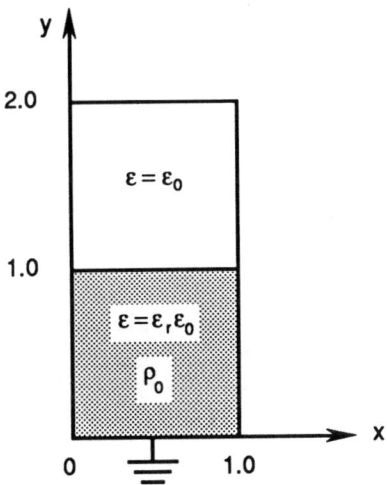

Figure 6.35 For Problem 6.8.

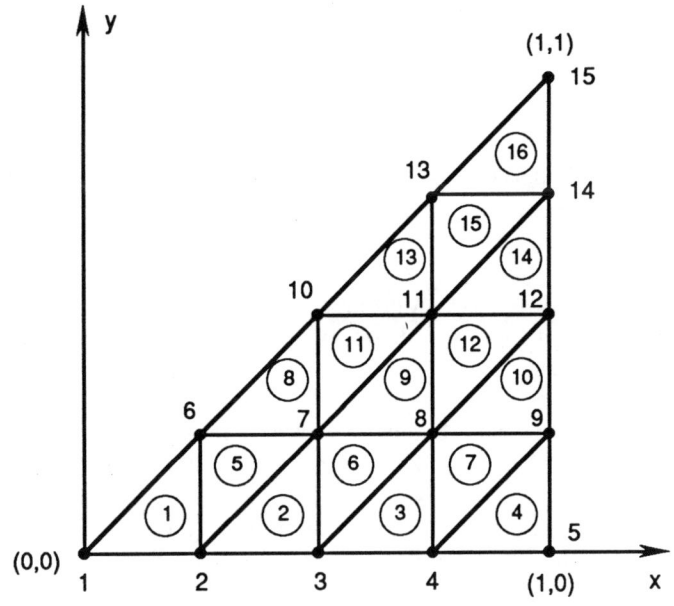

Figure 6.36 For Problem 6.10.

496 Numerical Techniques in Electromagnetics

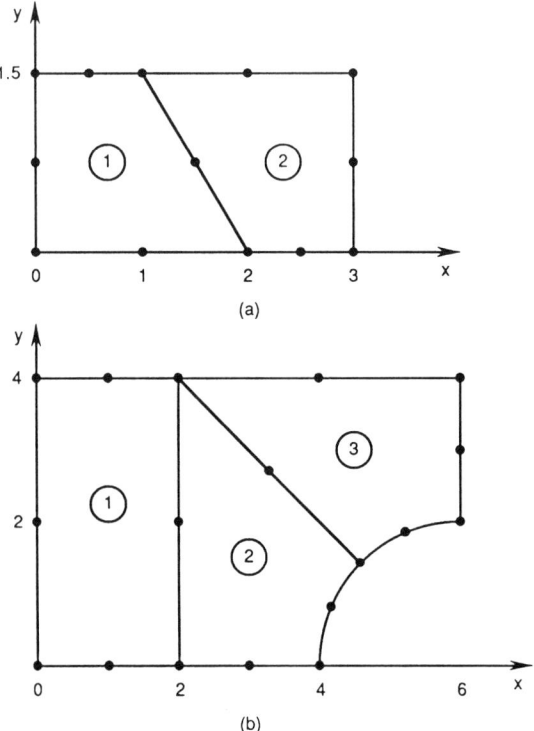

Figure 6.37 For Problem 6.12.

6.11 Using FEM, determine the first 10 cutoff wavelengths of a rectangular waveguide of cross section 2 cm by 1 cm. Compare your results with exact solution. Assume the guide is air-filled.

6.12 Use the mesh generation program in Fig. 6.16 to subdivide the solution region in Fig. 6.37. Subdivide into as many triangular elements as you choose.

6.13 Determine the semi-bandwidth of the mesh shown in Fig. 6.38. Renumber the mesh so as to minimize the bandwidth.

6.14 Find the semi-bandwidth B of the mesh in Fig. 6.39. Renumber the mesh to minimize B and determine the new value of B.

6.15 Rework Prob. 3.17 using the FEM.

Finite Element Method   497

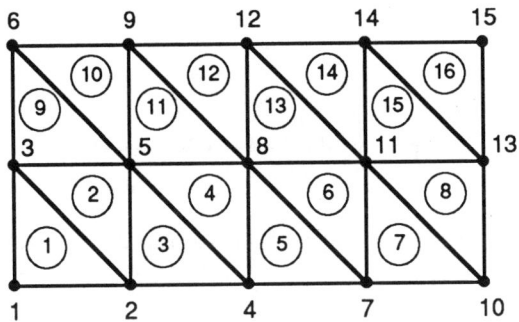

Figure 6.38   For Problem 6.13.

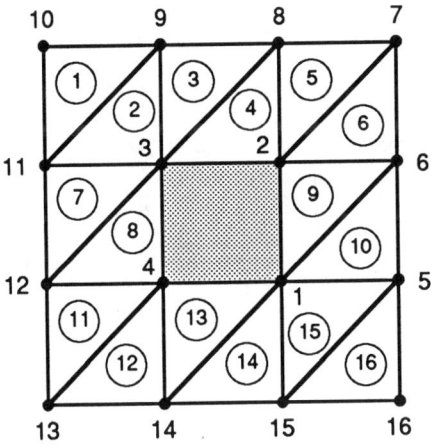

Figure 6.39   For Problem 6.14.

Hint: After calculating V at all free nodes with ϵ lumped with C_{ij}, use Eq. (6.19) to calculate W, i.e.,

$$W = \frac{1}{2}[V]^t[C][V].$$

Then find the capacitance from

$$C = \frac{2W}{V_d^2}$$

where V_d is the potential difference between inner and outer conductors.

6.16 Verify the interpolation functions for the six-node quadratic triangular element.

6.17 Using the area coordinates (ξ_1, ξ_2, ξ_3) for the triangular element in Fig. 6.3, evaluate:

(a) $\int_S x \, dS$,

(b) $\int_S x \, dS$,

(c) $\int_S xy \, dS$.

6.18 Derive matrix T for $n = 2$.

6.19 By hand calculation, obtain $Q^{(2)}$ and $Q^{(3)}$ for $n = 1$ and $n = 2$.

6.20 The $D^{(q)}$ matrix is an auxilliary matrix used along with the T matrix to derive other fundamental matrices. An element of D is defined in [43] as the partial derivative of α_i with respect to ξ_q evaluated at node P_j, i.e.,

$$D_{ij}^{(q)} = \left.\frac{\partial \alpha_i}{\partial \xi_q}\right|_{P_i}, \qquad i,j = 1, 2, \cdots, m$$

where $q \in \{1, 2, 3\}$. For $n = 1$ and 2, derive $D^{(1)}$. From $D^{(1)}$, derive $D^{(2)}$ and $D^{(3)}$.

6.21 (a) The matrix $K^{(pq)}$ can be defined as

$$K_{ij}^{(pq)} = \iint \frac{\partial \alpha_i}{\partial \xi_p} \frac{\partial \alpha_j}{\partial \xi_q} \, dS$$

where $p, q = 1, 2, 3$. Using the $D^{(q)}$ matrix of the previous problem, show that

$$K^{(pq)} = D^{(p)} \, T \, D^{(q)t}$$

where t denotes transposition.

(b) Show that the $Q^{(q)}$ matrix can be written as

$$Q^{(q)} = \left[D^{(q+1)} - D^{(q-1)}\right] T \left[D^{(q+1)} - D^{(q-1)}\right]^t.$$

Use this formula to derive $Q^{(1)}$ for $n = 1$ and 2.

6.22 Verify the interpolation function for the 10-node tetrahedral element.

6.23 Using the volume coordinates for a tetrahedron, evaluate

$$\int z^2 \, dv.$$

Assume that the origin is located at the centroid of the tetrahedron.

Chapter 7

Transmission-line-matrix Method

"Youth is a blunder; manhood a struggle; old age a regret."
                                        Benjamin Disraeli

7.1 Introduction

The link between field theory and circuit theory, the major theories on which electrical engineering is based, has been exploited in developing numerical techniques to solve certain types of partial differential equations arising in field problems with the aid of equivalent electrical networks [1]. There are three ranges in the frequency spectrum for which numerical techniques for field problems in general have been developed. In terms of the wavelength λ and the approximate dimension ℓ of the apparatus, these ranges are [2]:

$$\lambda \gg \ell$$
$$\lambda \approx \ell$$
$$\lambda \ll \ell.$$

In the first range, the special analysis techniques are known as *circuit theory*; in the second, as *microwave theory*; and in the third, as *geometric optics* (frequency independent). Hence the fundamental laws of circuit theory can be obtained from Maxwell's equations by applying an approximation valid when $\lambda \gg \ell$. However, it should be noted that circuit theory was not developed by approximating Maxwell's equations, but rather was developed independently from experimentally obtained laws. The connection between circuit theory and Maxwell's equations (summarizing field theory) is important; it adds to the comprehension of the fundamentals of electromagnetics.

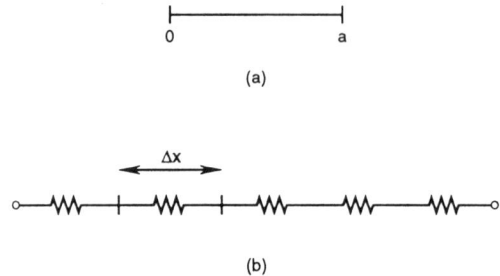

Figure 7.1 (a) One-dimensional conducting system, (b) discretized equivalent.

According to Silvester and Ferrari, circuits are mathematical abstractions of physically real fields; nevertheless, electrical engineers at times feel they understand circuit theory more clearly than fields [3].

The idea of replacing a complicated electrical system by a simple equivalent circuit goes back to Kirchhoff and Helmholtz. As a result of Park's [4], Kron's [5,6] and Schwinger's [7,8] works, the power and flexibility of equivalent circuits become more obvious to engineers. The recent applications of this idea to scattering problems, originally due to Johns [9], has made the method more popular and attractive.

Transmission-line modeling (TLM), otherwise known as the *transmission-line-matrix method*, is a numerical technique for solving field problems using circuit equivalent. It is based on the equivalence between Maxwell's equations and the equations for voltages and currents on a mesh of continuous two-wire transmission lines. The main feature of this method is the simplicity of formulation and programming for a wide range of applications [10]. As compared with the lumped network model, the transmission-line model is more general and performs better at high frequencies where the transmission and reflection properties of geometrical discontinuities cannot be regarded as lumped [7].

Like other numerical techniques, the TLM method is a discretization process. Unlike other methods such as finite difference and finite element methods, which are mathematical discretization approaches, the TLM is a physical discretization approach. In the TLM, the discretization of a field involves replacing a continuous system by a network or array of lumped elements. For example, consider the one-dimensional system (a conducting wire) with no energy storage as in Fig. 7.1(a). The wire can be replaced by a number of lumped resistors providing a discretized equivalent in Fig. 7.1(b). The discretization of the two-dimensional, distributed field is shown in Fig. 7.2. More general systems containing energy-reservoir elements as

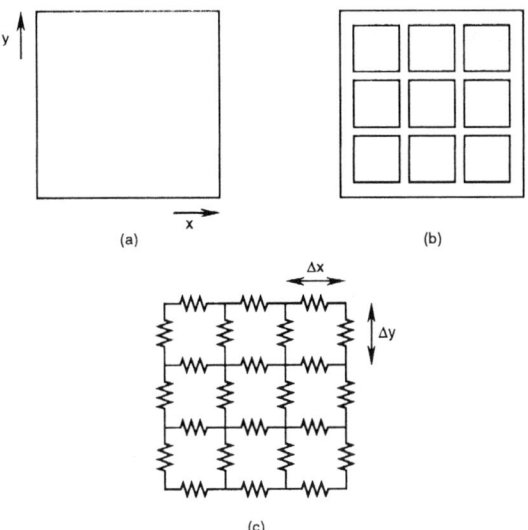

Figure 7.2 (a) Two-dimensional conductive sheet, (b) partially discretized equivalent, (c) fully discretized equivalent.

well as dissipative elements will be considered later.

The TLM method involves dividing the solution region into a rectangular mesh of transmission lines. Junctions are formed where the lines cross forming impedance discontinuities. A comparison between the transmission-line equations and Maxwell's equations allows equivalences to be drawn between voltages and currents on the lines and electromagnetic fields in the solution region. Thus, the TLM method involves two basic steps [11]:

- Replacing the field problem by the equivalent network and deriving analogy between the field and network quantities.
- Solving the equivalent network by iterative methods.

Before we apply the method, it seems fit to briefly review the basic concepts of transmission lines and then show how the TLM method can be applied to a wide range of EM-related problems.

7.2 Transmission-line Equations

Consider an elemental portion of length $\Delta \ell$ of a two-conductor transmission line. We intend to find an equivalent circuit for this line and derive the line equations. The equivalent circuit of a portion of the line is shown in Fig. 7.3, where the line parameters R, L, G, and C are resistance per

504   Numerical Techniques in Electromagnetics

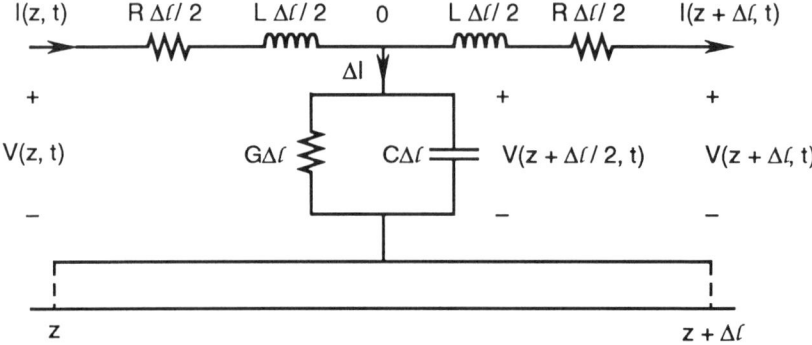

Figure 7.3   T-type equivalent circuit model of a differential length of a two-conductor transmission line.

unit length, inductance per unit length, conductance per unit length, and capacitance per unit length of the line, respectively. The model in Fig. 7.3 may represent any two-conductor line. The model is called the T-type equivalent circuit; other types of equivalent circuits are possible, but we end up with the same set of equations. In the model of Fig. 7.3, we assume without loss of generality that wave propagates in the $+z$ direction, from the generator to the load.

By applying Kirchhoff's voltage law to the left loop of the circuit in Fig. 7.3, we obtain

$$V(z,t) = R\frac{\Delta\ell}{2}I(z,t) + L\frac{\Delta\ell}{2}\frac{\partial I}{\partial t}(z,t) + V(z+\Delta\ell/2,t)$$

or

$$-\frac{V(z+\Delta\ell/2,t) - V(z,t)}{\Delta\ell/2} = R\,I(z,t) + L\frac{\partial I}{\partial t}(z,t). \quad (7.1)$$

Taking the limit of Eq. (7.1) as $\Delta\ell \to 0$ leads to

$$-\frac{\partial V(z,t)}{\partial z} = R\,I(z,t) + L\frac{\partial I}{\partial t}(z,t). \quad (7.2)$$

Similarly, applying Kirchhoff's current law to the main node of the circuit in Fig. 7.3 gives

$$I(z,t) = I(z+\Delta\ell,t) + \Delta I$$
$$= I(z+\Delta\ell,t) + G\Delta\ell V(z+\Delta\ell/2,t) + C\Delta\ell\frac{\partial V}{\partial t}(z+\Delta\ell/2,t)$$

or

$$-\frac{I(z+\Delta\ell,t) - I(z,t)}{\Delta\ell} = GV(z+\Delta\ell/2,t) + C\frac{\partial V}{\partial t}(z+\Delta\ell/2,t). \quad (7.3)$$

As $\Delta\ell \to 0$, Eq. (7.3) becomes

$$-\frac{\partial I}{\partial z}(z,t) = GV(z,t) + C\frac{\partial V}{\partial t}(z,t). \tag{7.4}$$

Differentiating Eq. (7.2) with respect to z and Eq. (7.4) with respect to t, the two equations become

$$-\frac{\partial^2 V}{\partial z^2} = R\frac{\partial I}{\partial z} + L\frac{\partial^2 I}{\partial z \partial t} \tag{7.2a}$$

and

$$-\frac{\partial^2 I}{\partial t \partial z} = G\frac{\partial V}{\partial t} + C\frac{\partial^2 V}{\partial t^2}. \tag{7.4a}$$

Substituting Eqs. (7.4) and (7.4a) into Eq. (7.2a) gives

$$\frac{\partial^2 V}{\partial z^2} = LC\frac{\partial^2 V}{\partial t^2} + (RC+GL)\frac{\partial V}{\partial t} + RGV. \tag{7.5}$$

Similarly, we obtain the equation for current I as

$$\frac{\partial^2 I}{\partial z^2} = LC\frac{\partial^2 I}{\partial t^2} + (RC+GL)\frac{\partial I}{\partial t} + RGI. \tag{7.6}$$

Equations (7.5) and (7.6) have the same mathematical form, which in general may be written as

$$\frac{\partial^2 \Phi}{\partial z^2} = LC\frac{\partial^2 \Phi}{\partial t^2} + (RC+GL)\frac{\partial \Phi}{\partial t} + RG\Phi \tag{7.7}$$

where $\Phi(z,t)$ has replaced either $V(z,t)$ or $I(z,t)$.

Ignoring certain transmission-line parameters in Eq. (7.7) leads to the following special cases [12]:

(a) $L = C = 0$ yields

$$\frac{\partial^2 \Phi}{\partial z^2} = k_1 \Phi \tag{7.8}$$

where $k_1 = RG$. Equation (7.8) is the one-dimensional elliptic partial differential equation called Poisson's equation.

(b) $R = C = 0$ or $G = L = 0$ yields

$$\frac{\partial^2 \Phi}{\partial z^2} = k_2 \frac{\partial \Phi}{\partial t} \tag{7.9}$$

where $k_2 = GL$ or RC. Equation (7.9) is the one-dimensional parabolic partial differential equation called the diffusion equation.

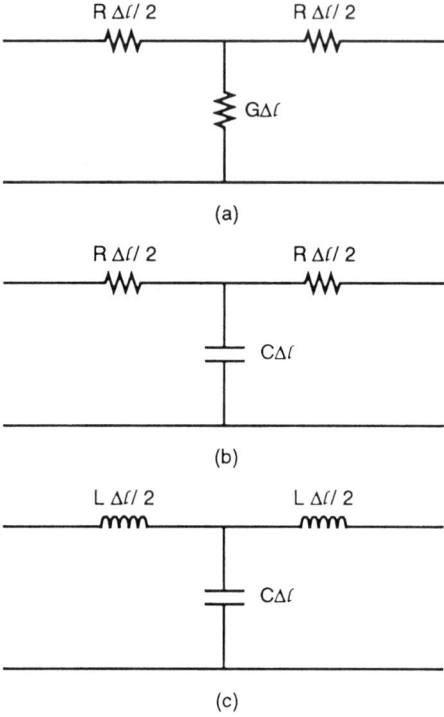

Figure 7.4 Transmission-line equivalent models for: (a) elliptic PDE, Poisson's equation, (b) parabolic PDE, diffusion equation, (c) hyperbolic PDE, wave equation.

(c) $R = G = 0$ (lossless line) yields

$$\frac{\partial^2 \Phi}{\partial z^2} = k_3 \frac{\partial^2 \Phi}{\partial t^2} \qquad (7.10)$$

where $k_3 = LC$. This is the one-dimensional hyperbolic partial differential equation called the Helmholtz equation, or simply the wave equation. Thus, under certain conditions, the one-dimensional transmission line can be used to model problems involving an elliptic, parabolic, or hyperbolic partial differential equation (PDE). The transmission line of Fig. 7.3 reduces to those in Fig. 7.4 for these three special cases.

Apart from the equivalent models, other transmission-line parameters are of interest. A detailed explanation of these parameters can be found in standard field theory texts, e.g., [13,14]. We briefly present these important parameters. For the lossless line in Fig. 7.4(c), the characteristic resistance

$$R_o = \sqrt{\frac{L}{C}}, \qquad (7.11a)$$

the wave velocity

$$u = \frac{1}{\sqrt{LC}}, \qquad (7.11b)$$

and the reflection coefficient at the load

$$\Gamma = \frac{R_L - R_o}{R_L + R_o}, \qquad (7.11c)$$

where R_L is the load resistance.

The generality of the TLM method has been demonstrated in this section. In the following sections, the method is applied specifically to diffusion [15,16] and wave propagation problems [10–12,17,18].

7.3 Solution of Diffusion Equation

We now apply the TLM method to the diffusion problem arising from current density distribution within the conducting region [15]. If the wire has a circular cross section with radius a and is infinitely long, then the problem becomes one-dimensional. We will assume sinusoidal source or harmonic fields (with time factor $e^{j\omega t}$).

The analytical solution of the problem has been treated in Example 2.3. For the TLM solution, consider the equivalent network of the cylindrical problem in Fig. 7.5, where $\Delta \ell$ is the distance between nodes or the mesh size. Applying Kirchhoff's laws to the network in Fig. 7.5 gives

$$\frac{\partial I_\rho}{\partial \rho} = -j\omega C V_\phi \qquad (7.12a)$$

$$\frac{\partial V_\phi}{\partial \rho} = -R I_\rho \qquad (7.12b)$$

where R and C are the resistance and capacitance per unit length.

Within the conductor, Maxwell's curl equations ($\sigma >> \omega \epsilon$) are

$$\nabla \times \mathbf{E} = -j\omega \mu \mathbf{H} \qquad (7.13a)$$

$$\nabla \times \mathbf{H} = \sigma \mathbf{E} \qquad (7.13b)$$

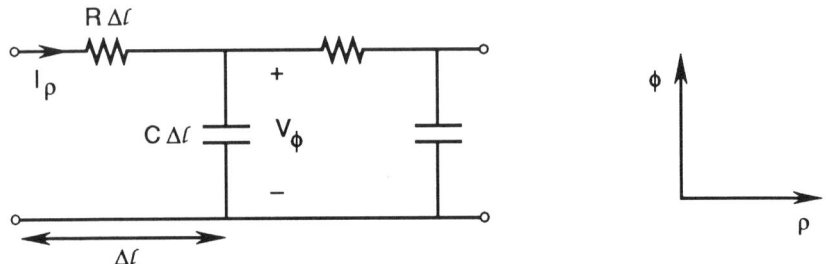

Figure 7.5 RC equivalent network.

where **E** and **H** are assumed to be in phasor forms. In cylindrical coordinates, Eq. (7.13) becomes

$$-\frac{\partial E_z}{\partial \rho} = -j\omega\mu H_\phi$$

$$\frac{1}{\rho}\frac{\partial}{\partial \rho}(\rho H_\phi) = \sigma E_z.$$

These equations can be written as

$$\frac{\partial E_z}{\partial \rho} = j\omega(\mu/\rho)(\rho H_\phi) \tag{7.14a}$$

$$\frac{\partial}{\partial \rho}(\rho H_\phi) = (\sigma\rho)E_z. \tag{7.14b}$$

Comparing Eq. (7.12) with Eq. (7.14) leads to the following analogy between the network and field quantities:

$$\boxed{\begin{aligned} I_\rho &\equiv -E_z \\ V_\phi &\equiv \rho H_\phi \\ C &\equiv \mu/\rho \\ R &\equiv \sigma\rho \end{aligned}}$$

(7.15a)
(7.15b)
(7.15c)
(7.15d)

Therefore, solving the impedance network is equivalent to solving Maxwell's equations.

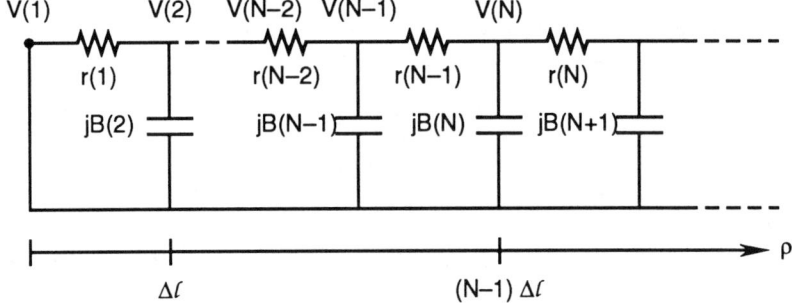

Figure 7.6 The overall equivalent network.

We can solve the overall impedance network in Fig. 7.6 by an iterative method. Since the network in Fig. 7.6 is in the form of a ladder, we apply the *ladder* method. By applying Kirchhoff's current law, the Nth nodal voltage $(N > 2)$ is related to $(N-1)$th and $(N-2)$th voltages according to

$$V(N) = \frac{r(N-1)}{r(N-2)}[V(N-1) - V(N-2)]$$
$$+ jB(N-1)r(N-1)V(N-1) + V(N-1) \qquad (7.16)$$

where the resistance r and susceptance B are given by

$$r(N) = R\Delta\ell = \sigma(N - 0.5)(\Delta\ell)^2, \qquad (7.17a)$$

$$B(N) = \omega C\Delta\ell = \frac{\omega\mu\Delta\ell}{(N-1)\Delta\ell} = \frac{\omega\mu}{N-1}. \qquad (7.17b)$$

We note that $V(1) = 0$ because the magnetic field at the center of the conductor $(\rho = 0)$ is zero. Also $V(2) = I(1) \cdot r(1)$, where $I(1)$ can be arbitrarily chosen, say $I(1) = 1$. Once $V(1)$ and $V(2)$ are known, we can use Eq. (7.16) to scan all nodes in Fig. 7.6 once from left to right to determine all nodal voltages $(\equiv \rho H_\phi)$ and currents $(\equiv E_z = J_z/\sigma)$.

Example 7.1

Develop a computer program to determine the relative (or normalized) current density $J_z(\rho)/J_z(a)$ in a round copper wire operated at 1 GHz. Plot the relative current density against the radial position ρ/a for cases $a/\delta = 1, 2,$ and 4. Take $\Delta\ell/\delta = 0.1$, $\mu = \mu_0$, $\sigma = 5.8 \times 10^7$ mhos/m.

510 Numerical Techniques in Electromagnetics

Solution

The computer program is presented in Fig. 7.7. It calculates the voltage at each node using Eqs. (7.16) and (7.17). The current on each $r(N)$ is found from Fig. 7.6 as

$$I(N-1) = \frac{V(N) - V(N-1)}{r(N-1)}.$$

Since $J = \sigma E$, we obtain $J_z(\rho)/J_z(a)$ as the ratio of $I(N)$ and $I(NMAX)$, where $I(NMAX)$ is the current at $\rho = a$.

To verify the accuracy of the TLM solution, we also calculate the exact $J_z(\rho)/J_z(a)$ using Eqs. (2.3.10) to (2.3.12). (For further details, see Example 2.3.) Table 7.1 shows a comparison between TLM results and exact results for the case $a/\delta = 4.0$. It is noticed that the percentage error is maximum (about 8%) at the center of the wire and diminishes to zero as we approach the surface of the wire. Figure 7.8 portrays the plot of the relative current density versus the radial position for cases $a/\delta = 1, 2,$ and 4.

```
0001      C===================================================================
0002      C USING THE TLM METHOD, THIS PROGRAM CALCULATES THE RELATIVE
0003      C CURRENT DENSITY JR IN A ROUND COPPER WIRE
0004      C THE EXACT SOLUTION JRE IS ALSO INCLUDED
0005      C===================================================================
0006
0007              REAL MIU, JR(50), JRE(50)
0008              COMPLEX J, V(50), I(50)
0009              DIMENSION BERO(50), BEIO(50)
0010      C
0011      C FUNCTIONS FOR RESISTANCE AND SUSCEPTANCE
0012      C
0013              R(N) = SIGMA*( FLOAT(N) - 0.5 )*(H**2)
0014              B(N) = OMEGA*MIU/FLOAT(N-1)
0015      C
0016      C SPECIFY INPUT DATA
0017      C
0018              F = 1.0E+9
0019              SIGMA = 5.8E+7
0020              PIE = 4.0*ATAN(1.0)
0021              MIU = 4.0*PIE*1.0E-7
0022              OMEGA = 2.0*PIE*F
0023              DELTA = SQRT( 2.0/( SIGMA*MIU*OMEGA ) )
0024              H = 0.1*DELTA
0025              A = 4.0*DELTA
0026              NMAX = A/H
0027              J = (0.0,1)
0028      C
0029      C INITIALIZE AND CALCULATE RELATIVE CURRENT DENSITY JR
0030      C USING TLM
0031      C
0032              I(1) = (0.1, 0.0)
0033              V(1) = (0.0,0.0)
0034              V(2) = I(1)*R(1)
0035              DO 10  N = 3,NMAX + 1
```

Table 7.1 Comparison of Relative Current Density Obtained from TLM and Exact Solutions ($a/\delta = 4.0$)

Radial position (ρ/a)	TLM result	Exact result
0.1	0.11581	0.10768
0.2	0.11765	0.11023
0.3	0.12644	0.12077
0.4	0.14953	0.14612
0.5	0.19301	0.19138
0.6	0.26150	0.26082
0.7	0.36147	0.36115
0.8	0.50423	0.50403
0.9	0.70796	0.70786
1.0	1.0	1.0

```
0036              V(N) = ( R(N-1)/R(N-2) )*( V(N-1) - V(N-2) )
0037             1          + J*B(N-1)*R(N-1)*V(N-1)  +  V(N-1)
0038        10   CONTINUE
0039              I(NMAX) = ( V(NMAX+1) - V(NMAX) )/R(NMAX)
0040              DO 20 N = 2,NMAX + 1
0041              I(N-1) = ( V(N) - V(N-1) )/R(N-1)
0042              JR(N-1) = CABS( I(N-1)/I(NMAX) )
0043        20   CONTINUE
0044        C
0045        C CALCULATE THE CURRENT DENSITY JRE
0046        C FROM THE EXACT SOLUTION
0047        C
0048              DO 40 NRO = 1,NMAX
0049              BERO(NRO) = 0.0
0050              BEIO(NRO) = 0.0
0051              X = ( FLOAT(NRO)/FLOAT(NMAX) )*A*SQRT(2.0)/DELTA
0052              DO 30 K=0,20
0053              CALL FACTORIAL(K,FA)
0054              FB = ( (X/2.0)**(2*K) )/(FA**2)
0055              BERO(NRO) = BERO(NRO) + FB*COS(K*PIE/2.0)
0056              BEIO(NRO) = BEIO(NRO) + FB*SIN(K*PIE/2.0)
0057        30   CONTINUE
0058              JRE(NRO) = SQRT( BERO(NRO)**2 + BEIO(NRO)**2 )
0059        40   CONTINUE
0060              WRITE(6,50)
0061        50   FORMAT(3X,'RADIAL POSITION',4X,'TLM J',12X,'EXACT J',/)
0062              DO 70 N=2,NMAX
0063              RHO = FLOAT(N)/FLOAT(NMAX)
0064              JRE(N) = JRE(N)/JRE(NMAX)      ! RELATIVE CURRENT DENSITY
```

```
0065            WRITE(6,60) RHO, JR(I), JRE(I)
0066      60    FORMAT(6X,F5.2,8X,F9.5,8X,F9.5,/)
0067      70    CONTINUE
0068            STOP
0069            END

0001      C ********************************************
0002      C SUBROUTINE FOR CALCULATE K!
0003      C
0004            SUBROUTINE FACTORIAL(K,F)
0005
0006            F = 1.0
0007            IF(K.GE.2) THEN
0008            DO 10 J = 2,K
0009            F = F*FLOAT(J)
0010      10    CONTINUE
0011            ENDIF
0012            RETURN
0013            END
```

Figure 7.7 Computer program for Example 7.1.

7.4 Solution of Wave Equations

In order to show how Maxwell's equations may be represented by the transmission-line equations, the differential length of the lossless transmission line between two nodes of the mesh is represented by lumped inductors and capacitors as shown in Fig. 7.9 for two-dimensional wave propagation problems [17,18]. At the nodes, pairs of transmission lines form impedance discontinuity. The complete network of transmission-line-matrix is made up of a large number of such building blocks as depicted in Fig. 7.10. Notice that in Fig. 7.10 single lines are used to represent a transmission-line pair. Also, a uniform internodal distance of $\Delta\ell$ is assumed throughout the matrix (i.e., $\Delta\ell = \Delta x = \Delta z$). We shall first derive equivalences between network and field quantities.

7.4.1 Equivalence Between Network and Field Parameters

We refer to Fig. 7.9 and apply Kirchhoff's current law at node O to obtain

$$I_x(x - \Delta\ell/2) - I_x(x + \Delta\ell/2) + I_z(z - \Delta\ell/2) - I_z(z + \Delta\ell/2) = 2C\Delta\ell\frac{\partial V_y}{\partial t}.$$

Dividing through by $\Delta\ell$ gives

$$\frac{I_x(x - \Delta\ell/2) - I_x(x + \Delta\ell/2)}{\Delta\ell} + \frac{I_z(z - \Delta\ell/2) - I_z(z + \Delta\ell/2)}{\Delta\ell} = 2C\frac{\partial V_y}{\partial t}.$$

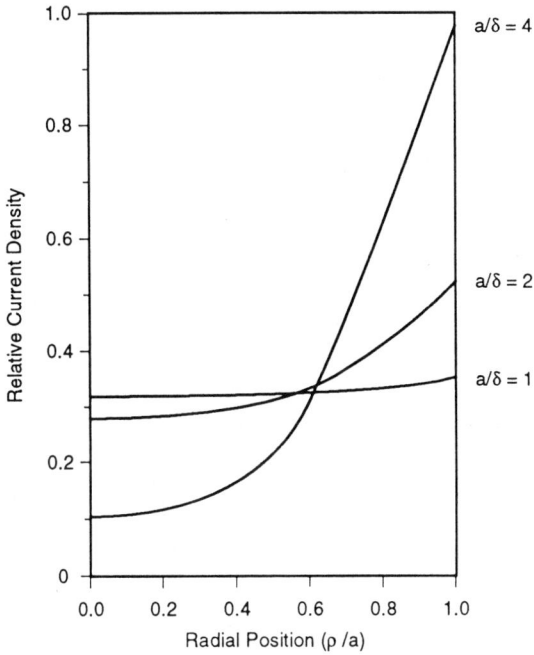

Figure 7.8 Relative current density versus radial position.

Taking the limit as $\Delta\ell \to 0$ results in

$$-\frac{\partial I_z}{\partial z} - \frac{\partial I_x}{\partial x} = 2C\frac{\partial V_y}{\partial t}. \qquad (7.18a)$$

Applying Kirchhoff's voltage law around the loop in the $x - y$ plane gives

$$V_y(x - \Delta\ell/2) - L\Delta\ell/2 \frac{\partial I_x(x - \Delta\ell/2)}{\partial t}$$
$$- L\Delta\ell/2 \frac{\partial I_x(x + \Delta\ell/2)}{\partial t} - V_y(x + \Delta\ell/2) = 0.$$

Upon rearranging and dividing by $\Delta\ell$, we have

$$\frac{V_y(x - \Delta\ell/2) - V_y(x + \Delta\ell/2)}{\Delta\ell} = \frac{L}{2}\frac{\partial I_x(x - \Delta\ell/2)}{\partial t} + \frac{L}{2}\frac{\partial I_x(x + \Delta\ell/2)}{\partial t}.$$

Again, taking the limit as $\Delta\ell \to 0$ gives

$$\frac{\partial V_y}{\partial x} = -L\frac{\partial I_x}{\partial t}. \qquad (7.18b)$$

514 Numerical Techniques in Electromagnetics

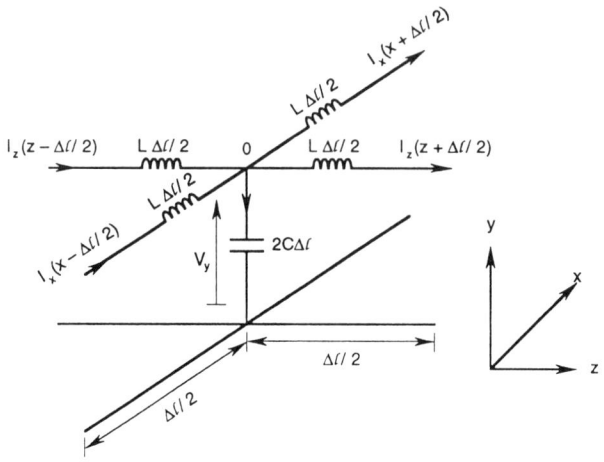

Figure 7.9 Equivalent network of a two-dimensional TLM shunt node.

Taking similar steps on the loop in the $y - z$ plane yields

$$\frac{\partial V_y}{\partial z} = -L\frac{\partial I_z}{\partial t}. \tag{7.18c}$$

These equations will now be combined to give a wave equation. Differentiating Eq. (7.18a) with respect to t, Eq. (7.18b) with respect to x, and Eq. (7.18c) with respect to z, we have

$$-\frac{\partial^2 I_z}{\partial z \partial t} - \frac{\partial^2 I_x}{\partial x \partial t} = 2C\frac{\partial^2 V_y}{\partial t^2} \tag{7.19a}$$

$$\frac{\partial^2 V_y}{\partial x^2} = -L\frac{\partial^2 I_x}{\partial t \partial x} \tag{7.19b}$$

$$\frac{\partial^2 V_y}{\partial z^2} = -L\frac{\partial^2 I_z}{\partial t \partial z}. \tag{7.19c}$$

Substituting Eqs. (7.19b) and (7.19c) into Eq. (7.19a) leads to

$$\frac{\partial^2 V_y}{\partial x^2} + \frac{\partial^2 V_y}{\partial z^2} = 2LC\frac{\partial^2 V_y}{\partial t^2}. \tag{7.20}$$

Equation (7.20) is the Helmholtz wave equation in two-dimensional space.

In order to show the field theory equivalence of Eqs. (7.19) and (7.20), consider Maxwell's equations

$$\nabla \times \mathbf{E} = -\mu\frac{\partial \mathbf{H}}{\partial t} \tag{7.21a}$$

Transmission-line-matrix Method

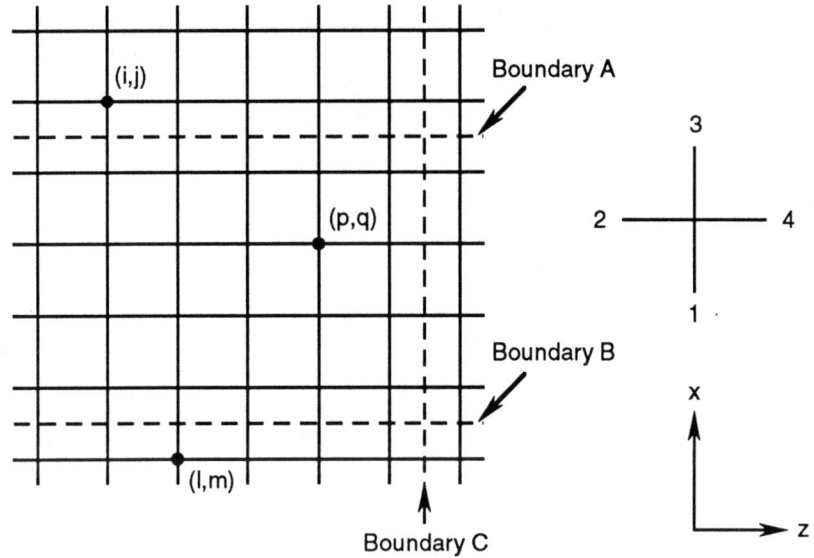

Figure 7.10 Transmission-line matrix and boundaries.

and

$$\nabla \times \mathbf{H} = \epsilon \frac{\partial \mathbf{E}}{\partial t}. \tag{7.21b}$$

Expansion of Eq. (7.21) in the rectangular coordinate system yields

$$\frac{\partial E_z}{\partial y} - \frac{\partial E_y}{\partial z} = -\mu \frac{\partial H_x}{\partial t}, \tag{7.22a}$$

$$\frac{\partial E_x}{\partial z} - \frac{\partial E_z}{\partial x} = -\mu \frac{\partial H_y}{\partial t}, \tag{7.22b}$$

$$\frac{\partial E_y}{\partial x} - \frac{\partial E_x}{\partial y} = -\mu \frac{\partial H_z}{\partial t}, \tag{7.22c}$$

$$\frac{\partial H_z}{\partial y} - \frac{\partial H_y}{\partial z} = \epsilon \frac{\partial E_x}{\partial t}, \tag{7.22d}$$

$$\frac{\partial H_x}{\partial z} - \frac{\partial H_z}{\partial x} = \epsilon \frac{\partial E_y}{\partial t}, \tag{7.22e}$$

$$\frac{\partial H_y}{\partial x} - \frac{\partial H_x}{\partial y} = \epsilon \frac{\partial E_z}{\partial t}. \tag{7.22f}$$

Consider the situation for which $E_x = E_z = H_y = 0, \frac{\partial}{\partial y} = 0$. It is noticed at once that this mode is a transverse electric (TE) mode with respect to the z-axis but a transverse magnetic (TM) mode with respect to the y-axis. Thus by the principle of duality, the network in Fig. 7.9 can be used for E_y, H_x, H_z fields as well as E_x, E_z, H_y fields. A network capable of reproducing TE waves is also capable of reproducing TM waves. For TE waves, Eq. (7.22) reduces to

$$\frac{\partial H_x}{\partial z} - \frac{\partial H_z}{\partial x} = \epsilon \frac{\partial E_y}{\partial t}, \tag{7.23a}$$

$$\frac{\partial E_y}{\partial x} = -\mu \frac{\partial H_z}{\partial t}, \tag{7.23b}$$

$$\frac{\partial E_y}{\partial z} = \mu \frac{\partial H_x}{\partial t}. \tag{7.23c}$$

Taking similar steps on Eqs. (7.23a–c) as were taken for Eqs. (7.18a–c) results in another Helmholtz equation

$$\frac{\partial^2 E_y}{\partial x^2} + \frac{\partial^2 E_y}{\partial z^2} = \mu\epsilon \frac{\partial^2 E_y}{\partial t^2}. \tag{7.24}$$

Comparing Eqs. (7.23) and (7.24) with Eqs. (7.18) and (7.20) yields the following equivalence between the parameters

$$\boxed{\begin{array}{c} E_y \equiv V_y \\ H_x \equiv -I_z \\ H_z \equiv I_x \\ \mu \equiv L \\ \epsilon \equiv 2C \end{array}} \tag{7.25}$$

Thus in the equivalent circuit:

- the voltage at shunt node is E_y,
- the current in the z direction is $-H_x$,
- the current in the x direction is H_z,
- the inductance per unit length represents the permeability of the medium,
- twice the capacitance per unit length represents the permittivity of the medium.

7.4.2 Dispersion Relation of Propagation Velocity

For the basic transmission line in the TLM which has $\mu_r = \epsilon_r = 1$, the inductance and capacitance per unit length are related by [8]

$$\frac{1}{\sqrt{(LC)}} = \frac{1}{\sqrt{(\mu_0 \epsilon_0)}} = c, \qquad (7.26)$$

where $c = 3 \times 10^8$ m/s is the speed of light in vacuum. Notice from Eq. (7.26) that for the equivalence made in Eq. (7.25), if voltage and current waves on each transmission line component propagate at the speed of light c, the complete network of intersecting transmission lines represents a medium of relative permittivity twice that of free space. This implies that as long as the equivalent circuit in Fig. 7.9 is valid, the propagation velocity in the TLM mesh is $1/\sqrt{2}$ of the velocity of light. The manner in which wave propagates on the mesh is now derived as follows.

If the ratio of the length of the transmission-line element to the free-space wavelength of the wave is $\theta/2\pi = \Delta \ell / \lambda$ (θ is called the *electrical length* of the line), the voltage and current at node i is related to those at node $i+1$ by the transfer-matrix equation (see Prob. 7.2) given as [19]

$$\begin{bmatrix} V_i \\ I_i \end{bmatrix} = \begin{bmatrix} (\cos\theta/2) & (j\sin\theta/2) \\ (j\sin\theta/2) & (\cos\theta/2) \end{bmatrix} \begin{bmatrix} 1 & 0 \\ (2j\tan\theta/2) & 1 \end{bmatrix} \begin{bmatrix} (\cos\theta/2) & (j\sin\theta/2) \\ (j\sin\theta/2) & (\cos\theta/2) \end{bmatrix} \begin{bmatrix} V_{i+1} \\ I_{i+1} \end{bmatrix}. \qquad (7.27)$$

If the waves on the periodic structure have a propagation constant $\gamma_n = \alpha_n + j\beta_n$, then

$$\begin{bmatrix} V_i \\ I_i \end{bmatrix} = \begin{bmatrix} e^{\gamma_n \Delta \ell} & 0 \\ 0 & e^{\gamma_n \Delta \ell} \end{bmatrix} \begin{bmatrix} V_{i+1} \\ I_{i+1} \end{bmatrix}. \qquad (7.28)$$

Solution of Eqs. (7.27) and (7.28) gives

$$\cosh(\gamma_n \Delta \ell) = \cos(\theta) - \tan(\theta/2)\sin(\theta). \qquad (7.29)$$

This equation describes the range of frequencies over which propagation can take place (passbands), i.e.,

$$|\cos(\theta) - \tan(\theta/2)\sin(\theta)| < 1, \qquad (7.30a)$$

and the range of frequencies over which propagation cannot occur (stopbands), i.e.,

$$|\cos(\theta) - \tan(\theta/2)\sin(\theta)| > 1. \qquad (7.30b)$$

For the lowest frequency propagation region,
$$\gamma_n = j\beta_n \qquad (7.31a)$$
and
$$\theta = \frac{2\pi\Delta\ell}{\lambda} = \frac{\omega}{c}\Delta\ell. \qquad (7.31b)$$
Introducing Eq. (7.31) in Eq. (7.29), we obtain
$$\cos(\beta_n \Delta\ell) = \cos\left(\frac{\omega\Delta\ell}{c}\right) - \tan\left(\frac{\omega\Delta\ell}{2c}\right)\sin\left(\frac{\omega\Delta\ell}{c}\right). \qquad (7.32)$$
Applying trigonometric identities
$$\sin(2A) = 2\sin(A)\cos(A)$$
and
$$\cos(2A) = 1 - 2\sin^2(A)$$
to Eq. (7.32) results in
$$\sin\left(\frac{\beta_n \Delta\ell}{2}\right) = \sqrt{2}\sin\left(\frac{\omega\Delta\ell}{2c}\right) \qquad (7.33)$$
which is a transcendental equation. If we let r be the ratio of the velocity u_n of the waves on the network to the free-space wave velocity c, then
$$r = u_n/c = \frac{\omega}{\beta_n c} = \frac{2\pi}{\lambda\beta_n} \qquad (7.34a)$$
or
$$\beta_n = \frac{2\pi}{\lambda r}. \qquad (7.34b)$$
Substituting Eqs. (7.34) into Eq. (7.33), the transcendental equation becomes
$$\sin\left(\frac{\pi}{r} \cdot \frac{\Delta\ell}{\lambda}\right) = \sqrt{2}\sin\left(\frac{\pi\Delta\ell}{\lambda}\right). \qquad (7.35)$$
By selecting different values of $\Delta\ell/\lambda$, the corresponding values of $r = u_n/c$ can be obtained numerically as in Fig. 7.11 for two-dimensional problems. From Fig. 7.11, we conclude that the TLM can only represent Maxwell's equations over the range of frequencies from zero to the first network cutoff frequency, which occurs at $\omega\Delta\ell/c = \pi/2$ or $\Delta\ell/\lambda = 1/4$. Over this range, the velocity of the waves behaves according to the characteristic of Fig. 7.11. For frequencies much smaller than the network cutoff frequency, the propagation velocity approximates to $1/\sqrt{2}$ of the free-space velocity.

Following the same procedure, the dispersion relation for three-dimensional problems can be derived as
$$\sin\left(\frac{\pi}{r} \cdot \frac{\Delta\ell}{\lambda}\right) = 2\sin\left(\pi\frac{\Delta\ell}{\lambda}\right). \qquad (7.36)$$
Thus for low frequencies ($\Delta\ell/\lambda < 0.1$), the network propagation velocity in three-dimensional space may be considered constant and equal to $c/2$.

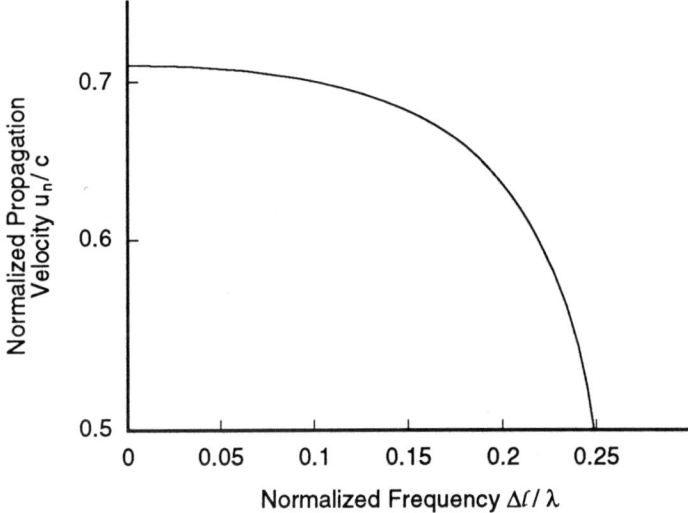

Figure 7.11 Dispersion of the velocity of waves in a two-dimensional TLM network.

7.4.3 Scattering Matrix

If $_kV_n^i$ and $_kV_n^r$ are the voltage impulses incident upon and reflected from terminal n of a node at time $t = k\Delta\ell/c$, we derive the relationship between the two quantities as follows. Let us assume that a voltage impulse function of unit magnitude is launched into terminal 1 of a node, as shown in Fig. 7.12(a), and that the characteristic resistance of the line is normalized. A unit-magnitude delta function of voltage and current will then travel towards the junction with unit energy ($S_i = 1$). Since line 1 has three other lines joined to it, its effective terminal resistance is 1/3. With the knowledge of its reflection coefficient, both the reflected and transmitted voltage impulses can be calculated. The reflection coefficient is

$$\Gamma = \frac{R_L - R_o}{R_L + R_o} = \frac{1/3 - 1}{1/3 + 1} = -\frac{1}{2} \qquad (7.37)$$

so that the reflected and transmitted energies are

$$S_r = S_i\Gamma^2 = \frac{1}{4} \qquad (7.38a)$$

$$S_t = S_i(1 - \Gamma^2) = \frac{3}{4} \qquad (7.38b)$$

520   Numerical Techniques in Electromagnetics

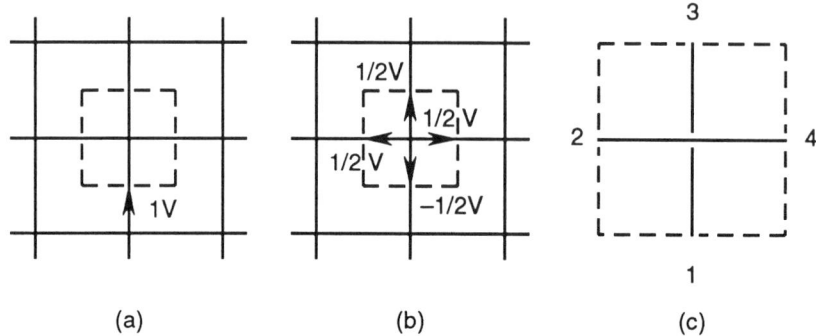

Figure 7.12 The impulse response of a node in a matrix.

where subscripts i, r, and t indicate incident, reflected, and transmitted quantities, respectively. Thus a voltage impulse of $-1/2$ is reflected back in terminal 1, while a voltage impulse of $1/2 = [\frac{3}{4} \div 3]^{1/2}$ will be launched into each of the other three terminals as shown in Fig. 7.12(b).

The more general case of four impulses being incident on four branches of a node can be obtained by applying the superposition principle to the previous case of a single pulse. Hence, if a time $t = k\Delta\ell/c$, voltage impulses $_kV_1^i, {}_kV_2^i, {}_kV_3^i$, and $_kV_4^i$ are incident on lines 1 to 4, respectively, at any junction node as in Fig. 7.12(c), the combined voltage reflected along line 1 at time $t = (k+1)\Delta\ell/c$ will be [9,10]

$$_{k+1}V_1^r = \frac{1}{2}\left\langle {}_kV_2^i + {}_kV_3^i + {}_kV_4^i - {}_kV_1^i \right\rangle. \qquad (7.39)$$

In general, the total voltage impulse reflected along line n at time $t = (k+1)\Delta\ell/c$ will be

$$_{k+1}V_n^r = \frac{1}{2}\left[\sum_{m=1}^{4} {}_kV_m^i\right] - {}_kV_n^i, \quad n = 1, 2, 3, 4. \qquad (7.40)$$

This idea is conveniently described by a *scattering matrix* equation relating the reflected voltages at time $(k+1)\Delta\ell/c$ to the incident voltages at the previous time step $k\Delta\ell/c$:

$$\begin{bmatrix} V_1 \\ V_2 \\ V_3 \\ V_4 \end{bmatrix}_{k+1}^r = \frac{1}{2} \begin{bmatrix} -1 & 1 & 1 & 1 \\ 1 & -1 & 1 & 1 \\ 1 & 1 & -1 & 1 \\ 1 & 1 & 1 & -1 \end{bmatrix} \begin{bmatrix} V_1 \\ V_2 \\ V_3 \\ V_4 \end{bmatrix}_k^i \qquad (7.41)$$

Also an impulse emerging from a node at position (z,x) in the mesh (reflected impulse) becomes automatically an incident impulse at the neighboring node. Hence

$$\boxed{\begin{aligned} {}_{k+1}V_1^i(z,x+\Delta\ell) &= {}_{k+1}V_3^r(z,x) \\ {}_{k+1}V_2^i(z+\Delta\ell,x) &= {}_{k+1}V_4^r(z,x) \\ {}_{k+1}V_3^i(z,x-\Delta\ell) &= {}_{k+1}V_1^r(z,x) \\ {}_{k+1}V_4^i(z-\Delta\ell,x) &= {}_{k+1}V_2^r(z,x) \end{aligned}} \qquad (7.42)$$

Thus by applying Eqs. (7.41) and (7.42), the magnitudes, positions, and directions of all impulses at time $(k+1)\Delta\ell/c$ can be obtained at each node in the network provided that their corresponding values at time $k\Delta\ell/c$ are known. The impulse response may, therefore, be found by initially fixing the magnitude, position, and direction of travel of impulse voltages at time $t = 0$, and then calculating the state of the network at successive time intervals.

The propagation of pulses in the TLM model is illustrated in Fig. 7.13, where the first two iterations following an initial excitation pulse in a two-dimensional shunt-connected TLM are shown. We have assumed free-space propagation for the sake of simplicity. The scattering process forms the basic algorithm of the TLM method [10,17].

7.4.4 Boundary Representation

Boundaries are usually placed halfway between two nodes, in order to ensure synchronism. In practice, this is achieved by making the mesh size $\Delta\ell$ an integer fraction of the structure's dimensions.

Any resistive load at boundary C (see Fig. 7.10) may be simulated by introducing a reflection coefficient Γ

$$_{k+1}V_4^i(p,q) = {}_kV_2^r(p+1,q) = \Gamma\left[{}_kV_4^r(p,q)\right] \qquad (7.43)$$

where

$$\Gamma = \frac{R_s - 1}{R_s + 1} \qquad (7.44)$$

and R_s is the surface resistance of the boundary normalized by the line characteristic impedance. If, for example, a perfectly conducting wall ($R_s = 0$) is to be simulated along boundary C, Eq. (7.44) gives $\Gamma = -1$, which represents a short circuit, and

$$_{k+1}V_4^i(p,q) = -{}_kV_4^r(p,q) \qquad (7.45)$$

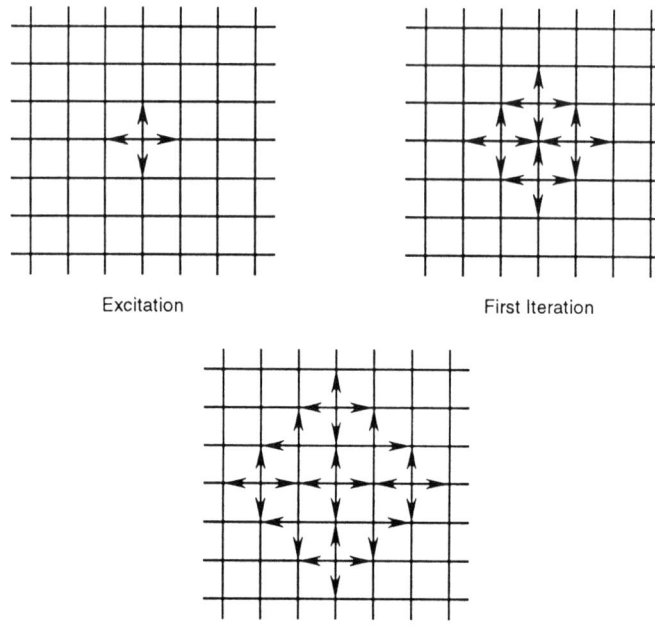

Figure 7.13 Scattering in a two-dimensional TLM network excited by a Dirac impulse.

is used in the simulation. For waves striking the boundary at arbitrary angles of incidence, a method for modeling free-space boundaries is discussed in [20].

7.4.5 Computation of Fields and Frequency Response

We continue with the TE mode of Eq. (7.23) as our example and calculate E_y, H_x, and H_z. E_y at any point corresponds to the node voltage at the point, H_z corresponds to the net current entering the node in the x direction (see Eq. (7.25)), while H_x is the net current in the negative z direction. For any point $(z = m, x = n)$ on the grid of Fig. 7.10, we have for each kth transient time

$$_k E_y(m,n) = \frac{1}{2}[_k V_1^i(m,n) + _k V_2^i(m,n) + _k V_3^i(m,n) + _k V_4^i(m,n)] \quad (7.46)$$

$$- _k H_x(m,n) = _k V_2^i(m,n) - _k V_4^i(m,n), \quad (7.47)$$

and

$$_k H_z(m,n) = _k V_3^i(m,n) - _k V_1^i(m,n). \quad (7.48)$$

Transmission-line-matrix Method

Thus, a series of discrete delta-function of magnitudes E_y, H_x, and H_z corresponding to time intervals of $\Delta\ell/c$ are obtained by the iteration of Eqs. (7.41) and (7.42). (Notice that reflections at the boundaries A and B in Fig. 7.10 will cancel out, thus $H_z = 0$.) Any point in the mesh can serve as an output or observation point. Equations (7.46) to (7.48) provide the output-impulse functions for the point representing the response of the system to an impulsive excitation. These output functions may be used to obtain the output waveform. For example, the output waveform corresponding to a pulse input may be obtained by convolving the output-impulse function with the shape of the input pulse.

Sometimes we are interested in the response to a sinusoidal excitation. This is obtained by taking the Fourier transform of the impulse response. Since the response is a series of delta functions, the Fourier transform integral becomes a summation, and the real and imaginary parts of the output spectrum are given by [9,10]

$$\boxed{\text{Re}\,[F(\Delta\ell/\lambda)] = \sum_{k=1}^{N} {}_kI \cos\left(\frac{2\pi k \Delta\ell}{\lambda}\right)} \quad (7.49\text{a})$$

$$\boxed{\text{Im}\,[F(\Delta\ell/\lambda)] = \sum_{k=1}^{N} {}_kI \sin\left(\frac{2\pi k \Delta\ell}{\lambda}\right)} \quad (7.49\text{b})$$

where $F(\Delta\ell/\lambda)$ is the frequency response, ${}_kI$ is the value of the output impulse response at time $t = k\Delta\ell/c$, and N is the total number of time intervals for which the calculation is made. Henceforth, N will be referred to as the number of iterations.

7.4.6 Output Response and Accuracy of Results

The output impulse function, in terms of voltage or current, may be taken from any point in the TLM mesh. It consists of a train of impulses of varying magnitude in the time domain separated by a time interval $\Delta\ell/c$. Thus, the frequency response obtained by taking the Fourier transform of the output response consists of series of delta functions in the frequency domain corresponding to the discrete modal frequencies for which a solution exists. For practical reasons, the output response has to be truncated, and this results in a spreading of the solution delta function $\sin x/x$ type of curves.

To investigate the accuracy of the result, let the output response be truncated after N iterations. Let $V_{out}(t)$ be the output impulse function taken within $0 < t < N\Delta\ell/c$. We may regard $V_{out}(t)$ as an impulse function

524 Numerical Techniques in Electromagnetics

$V_\infty(t)$, taken within $0 < t < \infty$, multiplied by a unit pulse function $V_p(t)$ of width $N\Delta\ell/c$, i.e.,

$$V_{out}(t) = V_\infty(t) \times V_p(t) \qquad (7.50)$$

where

$$V_p = \begin{cases} 1, & 0 \le t \le N\Delta\ell/c \\ 0, & \text{elsewhere.} \end{cases} \qquad (7.51)$$

Let $S_{out}(f)$, $S_\infty(f)$, and $S_p(f)$ be the Fourier transform of $V_{out}(t)$, $V_\infty(t)$, and $V_p(t)$, respectively. The Fourier transform of Eq. (7.50) is the convolution of $S_\infty(f)$ and $S_p(f)$. Hence

$$S_{out}(f) = \int_{-\infty}^{\infty} S_\infty(\alpha) S_p(f-\alpha)\, d\alpha \qquad (7.52)$$

where

$$S_p(f) = \frac{N\Delta\ell}{c} \frac{\sin\dfrac{\pi N\Delta\ell f}{c}}{\dfrac{\pi N\Delta\ell f}{c}} e^{-(\pi N\Delta\ell f)/c} \qquad (7.53)$$

which is of the form $\sin x/x$. Equations (7.52) and (7.53) show that $S_p(f)$ is placed in each of the positions of the exact response $S_\infty(f)$. Since the greater the number of iterations N the sharper the maximum peak of the curve, the accuracy of the result depends on N. The following examples are taken from Johns's work [9,18].

Example 7.2

The FORTRAN program in Fig. 7.14 is for the numerical calculations of one-dimensional TEM wave problems. It should be mentioned that the computer program in this example and the following ones are modified versions of those in Agba [21]. The calculations were carried out on a 25 by 11 rectangular matrix. TEM field-continuation boundaries were fixed along $x = 2$ and $x = 10$, producing boundaries, in effect, along the lines $x = 1.5$ and $x = 10.5$. The initial impulse excitation was on all points along the line $z = 4$, and the field along this line was set to zero at all subsequent time intervals. In this way, interference from boundaries to the left of the excitation line was avoided. Calculations in the z direction were terminated at $z = 24$, so that no reflections were received from points at $z = 25$ in the matrix, and the boundary C in Fig. 7.10, situated at $z = 24.5$, was therefore matched to free space. The output-impulse response for E_y and H_x was taken at the point $z = 14$, $x = 6$, which is 10.5 mesh points away from the boundary C, for 100, 150, and 200 iterations.

Since the velocity of waves on the matrix is less than that in free space by a factor u_n/c (see Fig. 7.11), the effective intrinsic impedance presented by the network matrix is less by the same factor. The magnitude of the wave impedance on the matrix, normalized to the intrinsic impedance of free space, is given by $Z = |E_y|/|H_x|$ and is tabulated in Table 7.2, together with Arg(Z), for the various numbers of iterations made. A comparison is made with the exact impedance values [13].

```
0001      C ************************************************************
0002      C THIS PROGRAM APPLIES THE TLM METHOD TO SOLVE
0003      C ONE-DIMENSIONAL WAVE PROBLEMS. THE SPECIFIC EXAMPLE
0004      C SOLVED HERE IS DESCRIBED AS FOLLOWS:-
0005      C
0006      C THE TEM WAVES ON A 25 X 11 MATRIX
0007      C THE BOUNDARIES ARE AT X = 2 AND X = 10.
0008      C INITIAL IMPULSE EXCITATION IS ALONG Z = 4 AT t = 0
0009      C AND SUBSEQUENTLY THIS LINE IS SET TO ZERO. THE GRID
0010      C IS TERMINATED AT Z = 25. OUTPUT IS TAKEN AT Z = 14,
0011      C X = 6 FOR Ey AND Hx FOR 100,150,200 ITERATIONS
0012      C VI(IT,I,J,K) -- ARRAY FOR INCIDENT VOLTAGE
0013      C VR(IT,I,J,K) -- ARRAY FOR REFLECTED VOLTAGE
0014      C IT = 1       -- FOR PREVIOUS PULSE VALUE
0015      C    = 2       -- FOR CURRENT PULSE VALUE
0016      C I,J      -- CORRESPOND TO NODE LOCATION (Z,X)
0017      C K = 1 ..4  -- FOR TERMINALS, SEE FIG. 6C
0018      C NX       -- INDEX OF NODES IN X-DIRECTION
0019      C NZ       -- INDEX OF NODES IN Z-DIRECTION
0020      C NX/NZ B,E  -- INDEX OF BEGINNING,END NODE
0021      C NX/NZ O    -- INDEX OF OUTPUT NODE
0022      C      GAMMA       -- REFLECTION COEFFICIENT AT
0023      C                     THE BOUNDARY C
0024      C      DELTA       -- MESH SIZE DIVIDED BY LAMDA
0025      C      ITRATE      -- NO. OF ITERATIONS
0026      C************************************************************
0027
0028            IMPLICIT INTEGER*2(I-N),REAL*8(A,D-H,O-Y),
0029           1          COMPLEX*8(C,Z)
0030            DIMENSION VI(2,25,11,4),VR(2,25,11,4),
0031           1          OUT(20,10),EFI(20),EFR(20),
0032           2 HFI(20),HFR(20)
0033
0034            DATA NXB,NXE,NZB,NZE,NT,ITRATE/2,10,4,24,4,200/
0035            DATA NXO,NZO,PIE,GAMMA,DELTA/6,14,3.1415927,0,.002/
0036
0037      C STEP #1 ***********************************************
0038      C INSERT INITIAL PULSE EXCITATION ALONG LINE Z = 4
0039      C
0040            DO 10 J = NXB, NXE
0041       10   VI(1,NZB+1,J,2) = 1.0
0042      C
0043      C STEP #2 ***********************************************
0044      C USING EQS. (7.40) TO (7.42), CALCULATE THE
0045      C REFLECTED VOLTAGE AND SUBMIT IT DIRECTLY
0046      C TO THE NEIGHBORING NODE.
0047      C
0048            DO 90 ITIME = 1, ITRATE
0049            IT = 2
```

$\Delta\ell/\lambda$	$|Z|$	Arg(Z)	TLM	results			Exact	results
			$|Z|$	Arg(Z)	$|Z|$	Arg(Z)	$|Z|$	Arg(Z)
Number of iterations	100		150		200			
0.002	0.9789	−0.1368	0.9730	−0.1396	0.9781	−0.1253	0.9747	−0.1282
0.004	0.9028	−0.2432	0.8980	−0.2322	0.9072	−0.2400	0.9077	−0.2356
0.006	0.8114	−0.3068	0.8229	−0.2979	0.8170	−0.3046	0.8185	−0.3081
0.008	0.7238	−0.3307	0.7328	−0.3457	0.7287	−0.3404	0.7256	−0.3390
0.010	0.6455	−0.3201	0.6367	−0.3350	0.6396	−0.3281	0.6414	−0.3263
0.012	0.5783	−0.2730	0.5694	−0.2619	0.5742	−0.2680	0.5731	−0.2707
0.014	0.5272	−0.1850	0.5313	−0.1712	0.5266	−0.1797	0.5255	−0.1765
0.016	0.4993	−0.0609	0.5043	−0.0657	0.5009	−0.0538	0.5018	−0.0545
0.018	0.5002	−0.0790	0.4987	−0.0748	0.5057	−0.0785	0.5057	0.0768

Table 7.2 Normalized Impedance of a TEM Wave with Free-space Discontinuity.

```
0050              DO 50 I = NZB+1, NZE
0051              DO 50 J = NXB, NXE
0052              SUM = 0.0
0053              DO 20 K = 1, NT
0054    20        SUM = SUM + VI(IT-1,I,J,K)
0055              DO 30 K = 1, NT
0056              VR(IT,I,J,K) = 0.5*SUM - VI(IT-1,I,J,K)
0057    30    CONTINUE
0058    40    CONTINUE
0059              VI(IT,I,J-1,3) = VR(IT,I,J,1)
0060              VI(IT,I-1,J,4) = VR(IT,I,J,2)
0061              VI(IT,I,J+1,1) = VR(IT,I,J,3)
0062              VI(IT,I+1,J,2) = VR(IT,I,J,4)
0063      C
0064      C STEP #3 *******************************************
0065      C USING EQS. (7.43) AND (7.44), INSERT BOUNDARY CONDITIONS
0066      C
0067              IF(J .EQ. NXE)VI(IT,I,NXE,3)=VR(IT,I,NXE,1)
0068              IF(J .EQ. NXB)VI(IT,I,NXB,1)=VR(IT,I,NXB,3)
0069              IF(I .EQ. NZE)VI(IT,NZE,J,4)=GAMMA*VR(IT,NZE,J,4)
0070
0071    50    CONTINUE
0072      C
0073      C STEP #4 *******************************************
0074      C USING EQS. (7.46) - (7.49), CALCULATE IMPULSE RESPONSE
0075      C OF Ey and Hx AT Z=NZO,X=NXO
0076      C
0077              EI = 0.0
0078              DO 60 K = 1, NT
0079    60        EI = EI + VI(IT,NZO,NXO,K) * 0.5
0080              HI = VI(IT,NZO,NXO,2) - VI(IT,NZO,NXO,4)
0081      C
0082      C SUM THE FREQUENCY RESPONSE (imaginary and real
0083      C parts) FOR DIFFERENT VALUES OF MESH-SIZE DIVIDED
0084      C BY WAVELENGTH
0085      C
0086              DEL = DELTA
0087              II = 1
0088              DO 70 K = 1, 20
0089              T = DFLOAT(ITIME)
0090              EFI(K) = EFI(K) + EI * SIN(2 * PIE * T * DEL)
0091              EFR(K) = EFR(K) + EI * COS(2 * PIE * T * DEL)
0092              HFI(K) = HFI(K) + HI * SIN(2 * PIE * T * DEL)
0093              HFR(K) = HFR(K) + HI * COS(2 * PIE * T * DEL)
0094              OUT(K,II) = DEL
0095    70        DEL = DEL + 0.002
0096      C
0097      C SAVE THE CURRENT PULSE MAGNITUDE FOR NEXT ITERATION
0098      C
0099              DO 80 I = NZB, NZE
0100              DO 80 J = NXB, NXE
0101              DO 80 K = 1, NT
0102              VI(IT-1,I,J,K) = VI(IT,I,J,K)
0103              VR(IT-1,I,J,K) = VR(IT,I,J,K)
0104    80    CONTINUE
0105              IT = ITIME
0106              IF((IT .EQ. 100) .OR.( IT .EQ. 150)
0107     1                         .OR.(IT .EQ. 200))THEN
0108              PRINT*, IT
0109              IF(IT .EQ. 100) II = 2
0110              IF(IT .EQ. 150) II = 3
0111              IF(IT .EQ. 200) II = 4
```

```
0112        C
0113        C STEP #5 ********************************************
0114        C CALCULATE MAGNITUDE & ARGUMENT OF IMPEDANCE
0115        C
0116                    DO K = 1, 20
0117                    CEF = CMPLX(EFR(K),EFI(K))
0118                    CHF = CMPLX(HFR(K),HFI(K))
0119                    OUT(K,II) = CABS(CEF)/CABS(CHF) ! MAGNITUDE
0120
0121                    ZARG = CEF/CHF
0122                    XZ   = REAL( ZARG )
0123                    YZ   = AIMAG( ZARG )
0124                    XYZ  = YZ / XZ
0125                    OUT(K,II+4) = -ATAN( XYZ ) ! ARGUMENT
0126
0127                    END DO
0128                    END IF
0129        90          CONTINUE
0130        C
0131        C STEP #6 ********************************************
0132        C CALCULATE EXACT VALUE OF IMPEDANCE [ REF. 13]
0133        C
0134                    DEL = DELTA
0135                    DO 100 K = 1, 20
0136                    R2 = 1.0/TRANSC(DEL)
0137                    PRINT*,R2
0138                    R3 = TAN(21.0*R2*PIE*DEL)
0139                    CNUM = CMPLX(R2, R3)
0140                    RIG = R2 * R3
0141                    RR = 1.0
0142                    CDEM = CMPLX(RR, RIG)
0143                    OUT(K,5) = CABS(CNUM)/(CABS(CDEM)*R2)
0144                    ZARG = CNUM/CDEM
0145                    XZ   = REAL( ZARG )
0146                    YZ   = AIMAG( ZARG )
0147                    XYZ  = YZ / XZ
0148                    OUT(K,9) = ATAN( XYZ )
0149                    DEL = DEL + .002
0150        100         CONTINUE
0151                    DO 110 K = 1, 20
0152        110         WRITE(6,120) (OUT(K,J),J = 1, 10)
0153        120         FORMAT(2X,10F10.4)
0154                    STOP
0155                    END

0001        C ********************************************
0002        C THIS SUBPROGRAM USE EQ. (7.35) TO CALCULATE
0003        C RATIO Vn/c, GIVEN (DELTA/LAMDA).
0004        C
0005                    FUNCTION TRANSC(DELTA)
0006                    REAL LAMDA,LAMDAC
0007                    PIE = 3.14159265
0008                    TETA = PIE * DELTA
0009                    TET  = SQRT(2.0) * SIN(TETA)
0010                    TRANSC = TETA/ASIN(TET)
0011                    RETURN
0012                    END
```

Figure 7.14 Computer program for Example 7.2.

Table 7.3 Normalized Impedance of a Rectangular Waveguide with Simple Load

$\Delta\ell/\lambda$	TLM results		Exact results	
	$\|Z\|$	$Arg(Z)$	$\|Z\|$	$Arg(Z)$
0.020	1.9391	0.8936	1.9325	0.9131
0.021	2.0594	0.6175	2.0964	0.6415
0.022	1.9697	0.3553	2.0250	0.3603
0.023	1.7556	0.1530	1.7800	0.1438
0.024	1.5173	0.0189	1.5132	0.0163
0.025	1.3036	−0.0518	1.2989	−0.0388
0.026	1.1370	−0.0648	1.1471	−0.0457
0.027	1.0297	−0.0350	1.0482	−0.0249
0.028	0.9776	0.0088	0.9900	0.0075
0.029	0.9620	0.0416	0.9622	0.0396
0.030	0.9623	0.0554	0.9556	0.0632

Example 7.3

The second example is on a rectangular waveguide with a simple load. The FORTRAN program used for the numerical analysis is basically similar to that of one-dimensional simulation. A 25 x 11 matrix was used for the numerical analysis of the waveguide. Short-circuit boundaries were placed at $x = 2$ and $x = 10$, the width between the waveguide walls thus being 9 mesh points. The system was excited at all points along the line $z = 2$, and the impulse function of the output was taken from the point $(x = 6, z = 12)$. The C boundary at $z = 24$ represented an abrupt change to the intrinsic impedance of free space. The minor changes in the program of Fig. 7.14 are shown in Fig. 7.15. The cutoff frequency for the waveguide occurs [19] at $\Delta\ell/\lambda_n = 1/18$, λ_n is the network-matrix wavelength, which corresponds to $\Delta\ell/\lambda = \sqrt{2}/18$ since

$$\frac{\lambda_n}{\lambda} = \frac{u_n}{c} = \frac{\sqrt{\mu_o \epsilon_o}}{\sqrt{\mu_n \epsilon_n}} = \frac{\sqrt{LC}}{\sqrt{2LC}} = \frac{1}{\sqrt{2}}.$$

A comparison between the results for the normalized guide impedance using this method is made with exact results in Table 7.3.

530 Numerical Techniques in Electromagnetics

```
   :
   :
   :
0034         DATA IXB,IXE,IZB,IZE,IT,ITRATE/2,10,2,24,4,200/
0035         DATA IX0,IZ0,PIE,GAMMA,DELTA/6,12,3.1415927,0,.02/
   :
   :
   :
0064      C STEP #3 *********************************************
0065      C USING EQS. (7.43) AND (7.44), INSERT BOUNDARY CONDITIONS
0066      C
0067              IF(J .EQ. IXE)VI(IT,I,IXE,3)= - VR(IT,I,IXE,1)
0068              IF(J .EQ. IXB)VI(IT,I,IXB,1)=VR(IT,I,IXB,1)
0069              IF(I .EQ. IZE)VI(IT,IZE,J,4)=GAMMA*VR(IT,IZE,J,4)
   :
   :
0095   70        DEL = DEL + 0.001
   :
   :
```

Figure 7.15 Modification in the program in Fig. 7.14 for simulating waveguide problem in Example 7.3.

7.5 Inhomogeneous and Lossy Media in TLM

In our discussion on the transmission-line-matrix (TLM) method of numerical analysis in the last section, it was assumed that the medium in which wave propagates was homogeneous and lossless. In this section, we consider media that are inhomogeneous or lossy or both. This necessitates that we modify the equivalent network of Fig. 7.9 and the corresponding transmission line matrix of Fig. 7.10. Also, we need to draw the corresponding equivalence between the network and Maxwell's equations and derive the scattering matrix. We will finally consider how lossy boundaries are represented.

7.5.1 General Two-dimensional Shunt Node

To account for the inhomogeneity of a medium (where ϵ is not constant), we introduce additional capacitance at nodes to represent an increase in permittivity [17,22-24]. We achieve this by introducing an additional length of line or stub to the node as shown in Fig. 7.16(a). The stub of length $\Delta \ell/2$

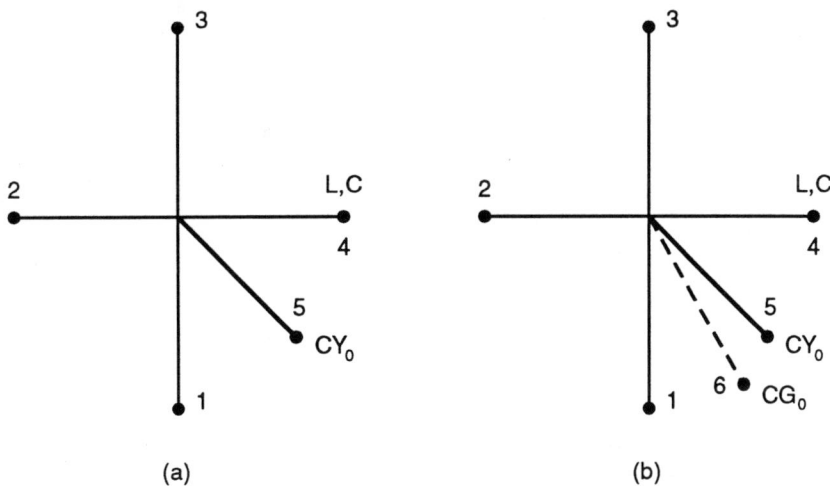

Figure 7.16 A two-dimensional node with: (a) Permittivity stub, (b) permittivity and loss stub.

is open circuited at the end and is of variable characteristic admittance Y_o relative to the unity characteristic admittance assumed for the main transmission line. At low frequencies, the effect of the stub is to add to each node an additional lumped shunt capacitance $CY_o\Delta\ell/2$, where C is the shunt capacitance per unit length of the main lines that are of unity characteristic admittance. Thus at each node, the total shunt capacitance becomes

$$C' = 2C\Delta\ell + CY_o\Delta\ell/2$$

or

$$C' = 2C\Delta\ell(1+Y_o/4). \quad (7.54)$$

To account for the loss in the medium, we introduce a power-absorbing line at each node, lumped into a single resistor, and this is simulated by an infinite or matched line of characteristic admittance G_o normalized to the characteristic impedance of the main lines as illustrated in Fig. 7.16(b).

Due to these additional lines, the equivalent network now becomes that shown in Fig. 7.17. (Compare Fig. 7.17 with Fig. 7.9.) Applying Kirchhoff's current law to shunt node O in the x-z plane in Fig. 7.17 and taking limits as $\Delta\ell \to 0$ results in

$$-\frac{\partial I_z}{\partial z} - \frac{\partial I_x}{\partial x} = \frac{G_o V_y}{Z_o \Delta\ell} + 2C(1+Y_o/4)\frac{\partial V_y}{\partial t}. \quad (7.55)$$

Expanding Maxwell's equations $\nabla \times \mathbf{E} = -\mu\dfrac{\partial \mathbf{H}}{\partial t}$ and $\nabla \times \mathbf{H} = \sigma\mathbf{E} + \epsilon\dfrac{\partial \mathbf{E}}{\partial t}$

532   Numerical Techniques in Electromagnetics

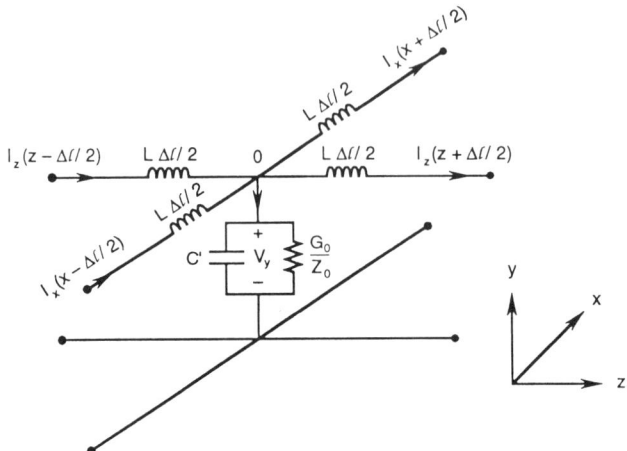

Figure 7.17   General two-dimensional shunt node.

for $\partial/\partial y \equiv 0$ leads to

$$\frac{\partial H_x}{\partial z} - \frac{\partial H_z}{\partial x} = \sigma E_y + \epsilon_0 \epsilon_r \frac{\partial E_y}{\partial t}. \tag{7.56}$$

This may be considered as denoting TE_{m0} modes with field components H_z, H_x, and E_y. From Eqs. (7.55) and (7.56), the following equivalence between the TLM equations and Maxwell's equations can be drawn:

$$\boxed{\begin{aligned} E_y &\equiv V_y \\ H_x &\equiv -I_z \\ H_z &\equiv I_x \\ \epsilon_0 &\equiv 2C \\ \\ \epsilon_r &\equiv \frac{4+Y_o}{4} \\ \\ \sigma &\equiv \frac{G_o}{Z_o \Delta \ell} \end{aligned}} \tag{7.57}$$

where $Z_o = \sqrt{L/C}$. From Eq. (7.57), the normalized characteristic admittance G_o of the loss stub is related to the conductivity of the medium by

$$G_o = \sigma \Delta \ell Z_o. \tag{7.58}$$

Thus losses on the matrix can be varied by altering the value of G_o. Also from Eq. (7.57), the variable characteristic admittance Y_o of the permittivity stub is related to the relative permittivity of the medium as

$$Y_o = 4(\epsilon_r - 1). \tag{7.59}$$

7.5.2 Scattering Matrix

We now derive the impulse response of the network comprising of the interconnection of many generalized nodes such as that in Fig. 7.17. As in the previous section, if $_kV_n(z,x)$ is unit voltage impulse reflected from the node at (z,x) into the nth coordinate direction ($n = 1, 2, \cdots, 5$) at time $k\Delta\ell/c$, then at node (z,x),

$$\begin{bmatrix} V_1(z,x) \\ V_2(z,x) \\ V_3(z,x) \\ V_4(z,x) \\ V_5(z,x) \end{bmatrix}_{k+1}^r = [S] \begin{bmatrix} V_3(z, x - \Delta\ell) \\ V_4(z - \Delta\ell, x) \\ V_1(z, x + \Delta\ell) \\ V_2(z + \Delta\ell, x) \\ V_5(z, x + \Delta\ell) \end{bmatrix}_k^i \tag{7.60}$$

where $[S]$ is the scattering matrix given by

$$[S] = \frac{2}{Y} \begin{bmatrix} 1 & 1 & 1 & 1 & Y_o \\ 1 & 1 & 1 & 1 & Y_o \\ 1 & 1 & 1 & 1 & Y_o \\ 1 & 1 & 1 & 1 & Y_o \\ 1 & 1 & 1 & 1 & Y_o \end{bmatrix} - [I] \tag{7.61}$$

$[I]$ is a unit matrix and $Y = 4 + Y_o + G_o$. The coordinate directions 1, 2, 3, and 4 correspond to $-x, -z, +x$, and $+z$, respectively (as in the last section), and 5 refers to the permittivity stub. Notice that the voltage V_6 (see Fig. 7.16) scattered into the loss stub is dropped across G_o and not returned to the matrix. We apply Eq. (7.60) just as Eq. (7.41).

As in the last section, the output impulse function at a particular node in the mesh can be obtained by recording the amplitude and the time of the stream of pulses as they pass through the node. By taking the Fourier transform of the output impulse function using Eq. (7.49), the required information can be extracted.

The dispersion relation can be derived in the same manner as in the last section. If $\gamma_n = \alpha_n + j\beta_n$ is the network propagation constant and $\gamma = \alpha + j\beta$ is the propagation constant of the medium, the two propagation constants are related as

$$\frac{\beta}{\beta_n} = \frac{\theta/2}{\sin^{-1}[\sqrt{2(1+Y_o/4)}\sin\theta/2]} \quad (7.62a)$$

$$\frac{\alpha}{\alpha_n} = \frac{\sqrt{1-2(1+Y_o/4)\sin^2\theta/2}}{\sqrt{2(1+Y_o/4)}\cos\theta/2} \quad (7.62b)$$

where $\theta = 2\pi\Delta\ell/\lambda$ and

$$\alpha = \frac{G_o}{8\Delta\ell(1+Y_o/4)}. \quad (7.63)$$

In arriving at Eq. (7.62), we have assumed that $\alpha_n \Delta\ell \ll 1$. For low frequencies, the attenuation constant α_n and phase constant β_n of the network are fairly constant so that Eq. (7.62) reduces to

$$\gamma_n = \sqrt{2(1+Y_o/4)}\gamma. \quad (7.64)$$

From this, the network velocity $u_n (= \omega/\beta_n = \beta c/\beta_n)$ of waves on the matrix is readily obtained as

$$u_n^2 = \frac{c^2}{2(1+Y_o/4)} \quad (7.65)$$

where c is the free-space velocity of waves.

7.5.3 Representation of Lossy Boundaries

The above analysis has incorporated conductivity σ of the medium in the TLM formulation. To account for a lossy boundary [24,26], we define the reflection coefficient

$$\Gamma = \frac{Z_s - Z_o}{Z_s + Z_o} \quad (7.66)$$

where $Z_o = \sqrt{\mu_o/\epsilon_o}$ is the characteristic impedance of the main lines and Z_s is the surface impedance of the lossy boundary given by

$$Z_s = \sqrt{\frac{\mu\omega}{2\sigma_c}}(1+j) \quad (7.67)$$

where μ and σ_c are the permeability and conductivity of the boundary. It is evident from Eqs. (7.66) and (7.67) that the reflection coefficient Γ is in general complex. However, complex Γ implies that the shape of the pulse functions is altered on reflection at the conducting boundary, and this cannot be accounted for in the TLM method [22]. Therefore, assuming that Z_s is small compared with Z_o and that the imaginary part of Γ is negligible,

$$\Gamma \simeq -1 + \sqrt{\frac{2\epsilon_o \omega}{\sigma_c}} \qquad (7.68)$$

where $\mu = \mu_o$ is assumed. We notice that Γ is slightly less than -1. Also, we notice that Γ depends on the frequency ω and hence calculations involving lossy boundaries are only accurate for the specific frequency; calculations must be repeated for a different value of $\Delta \ell / \lambda$. The following example is taken from Akhtarzad and Johns [23].

Example 7.4

Consider the lossy homogeneously filled waveguide shown in Fig. 7.18. The guide is 6 cm wide and 13 cm long. It is filled with a dielectric of relative permittivity $\epsilon_r = 4.9$ and conductivity $\sigma = 0.05$ mhos/m and terminated in an open circuit discontinuity. Calculate the normalized wave impedance.

Solution

The computer program for this problem is in Fig. 7.19. It is an extension of the program in Fig. 7.14 with the incorporation of new concepts developed in this section. Enough comments are added to make it self-explanatory. The program is suitable for a two-dimensional TE_{m0} mode.

The waveguide geometry shown in Fig. 7.18 is simulated on a matrix of 12 x 26 nodes. The matrix is excited at all points along line $z = 1$ with impulses corresponding to E_y. The impulse function of the output at point $(z, x) = (6, 6)$ is taken after 700 iterations. Table 7.4 presents both the TLM and theoretical values of the normalized wave impedance and shows a good agreement between the two.

```
0001      C=================================================
0002      C THIS PROGRAM SOLVES A TYPICAL TWO-DIMENSIONAL
0003      C WAVE PROPAGATION PROBLEM AS STATED BELOW:
0004      C A WAVE GUIDE 6cm X 13cm LONG IS FILLED WITH A
0005      C DIELECTRIC OF RELATIVE PERMITTIVITY EQUAL TO 4.9
0006      C AND CONDUCTIVITY 0.05 MHO/M. THIS GEOMETRY IS
0007      C BEING SIMULATED ON A MATRIX OF 12 X 26 NODES.
0008      C THE MATRIX WAS EXCITED AT ALL POINTS ALONG THE
0009      C LINE Z = 1 WITH IMPULSES CORRESPONDING TO Ey.
0010      C THE IMPULSE FUNCTION OF THE OUTPUT WAS TAKEN FROM
0011      C THE POINT (Z = 6, X = 6) AFTER 750 ITERATIONS.
0012      C THE BOUNDARIES WERE SHORT CIRCUITED AT X = 0.5
0013      C AND X = 12.5 AND TERMINATED AT Z = 26.5 IN AN
0014      C OPEN CIRCUIT DISCONTINUITY.
0015      C VI -- ARRAY FOR INCIDENT VOLTAGE
```

536 Numerical Techniques in Electromagnetics

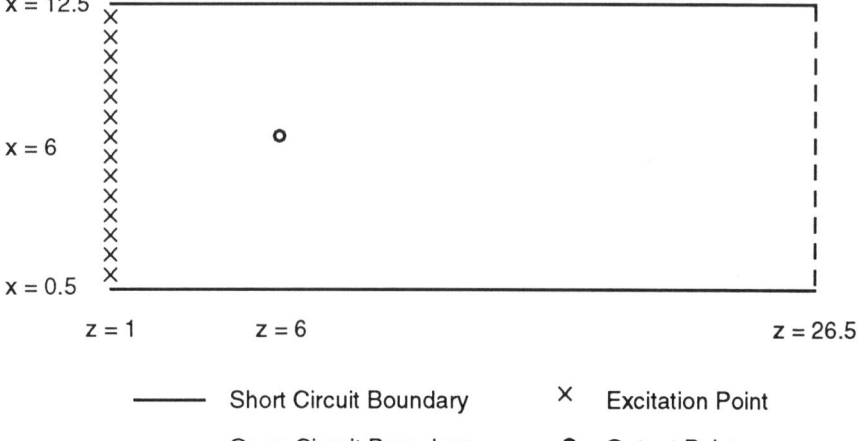

Figure 7.18 A lossy homogeneously filled waveguide.

```
0016      C VR -- ARRAY FOR REFLECTED VOLTAGE
0017      C NX -- INDEX OF NODES IN X-DIRECTION
0018      C NZ -- INDEX OF NODES IN Z-DIRECTION
0019      C RO -- REFLECTION COEFFIENT
0020      C DELTA  -- MESH SIZE (SPACING)
0021      C DELTA  -- MESH SIZE DIVIDED BY LAMBDA
0022
0023             IMPLICIT REAL*8(A,D-H,O-Y),COMPLEX*8(C,Z)
0024             DIMENSION VI(0:1,27,13,5),VR(0:1,27,13,5)
0025           1 ,EFI(50),EFR(50),HFI(50),HFR(50),OUT(10,5)
0026
0027             DATA NXB,NXE,NZB,NZE,NT,ITRATE/1,12,1,26,5,750/
0028             DATA NXI,NZI,NXO,NZO,RoS,RoC/0,1,6,6,-1.0,1.0/
0029             DATA Eo,Er,SIGMA,DELTA,PI/8.854E-12,4.9,0.05,
0030           1             0.005,3.1415927/
0031
0032             Yo = 4.0 * (Er/2.0 - 1)
0033             Ur = 1.0
0034             Uo = 4.0 * PI * 1.0E-07
0035             Go = SIGMA * DELTA * SQRT( Uo/Eo )
0036             Y = 4.0 + Yo + Go
0037      C
0038      C SINCE REFLECTION COEFFICIENT, RO, DEPENDS ON
0039      C FREQUENCY, THE ITERATIONS MUST BE REPEATED FOR
0040      C EACH VALUE OF THE MESH-SIZE DIVIDED BY WAVELENGTH.
0041      C
0042             DELTA = 0.0
0043             DO 90 L = 1, 10
0044             DELTA = DELTA + 0.003
0045      C
0046      C INITIALIZE ALL NODES (BOTH FREE AND FIXED)
0047      C
0048             DO 10 I = NZB, NZE
0049             DO 10 J = NXB, NXE
0050             DO 10 K = 1, NT
0051             VI(0,I,J,K) = 0.0
```

Table 7.4 Impedance of a Homogeneously Filled Waveguide with Losses

$\Delta\ell/\lambda$	TLM results		Exact results	
	$|Z|$	$\text{Arg}(Z)$	$|Z|$	$\text{Arg}(Z)$
0.003	0.0725	1.5591	0.0729	1.5575
0.006	0.1511	1.5446	0.1518	1.5420
0.009	0.2446	1.5243	0.2453	1.5205
0.012	0.3706	1.4890	0.3712	1.4840
0.015	0.5803	1.4032	0.5792	1.3977
0.018	1.0000	1.0056	0.9979	1.0065
0.021	1.1735	0.5156	1.1676	0.5121
0.024	0.5032	−0.1901	0.5093	−0.2141
0.027	0.6766	0.6917	0.6609	0.6853
0.030	0.8921	−0.3869	0.8921	−0.4185

```
0052      10        VR(0,I,J,K) = 0.0
0053      C
0054      C START THE ITERATION
0055      C
0056                DO 90 ITime = 0, ITRATE
0057                IT = 1
0058                DO 60 I = NZB, NZE
0059                DO 50 J = NXB, NXE
0060                SUM = 0.0
0061
0062                DO 30 K = 1, 4
0063      30        SUM = SUM + VI(IT-1,I,J,K)
0064                SUM = (SUM + Yo*VI(IT-1,I,J,5)) * 2.0/Y
0065      C
0066      C INSERT THE INITIAL CONDITION AT I = 1
0067      C
0068                IF(ITIME .EQ. 0 .AND. I .EQ. NZI) SUM = 1.0
0069
0070                DO 40 K = 1, NT
0071      40        VR(IT,I,J,K) = SUM - VI(IT-1,I,J,K)
0072
0073                IF(J .NE. NXB) VI(IT,I,J-1,3) = VR(IT,I,J,1)
0074                IF(I .NE. NZB+1)VI(IT,I-1,J,4)= VR(IT,I,J,2)
0075                IF(J .NE. NXE) VI(IT,I,J+1,1) = VR(IT,I,J,3)
0076                IF(I .NE. NZE) VI(IT,I+1,J,2) = VR(IT,I,J,4)
0077                VI(IT,I,J,5)  = VR(IT,I,J,5)
0078      C
0079      C INSERT THE BOUNDARY CONDITIONS
0080      C FOR THE SHORT CIRCUIT AT X = 0.5
0081      C
0082                IF(J .EQ. NXB) VI(IT,I,1,1) =RoS*VR(IT,I,1,1)
0083      C
0084      C FOR THE SHORT CIRCUIT AT X = 12.5
0085      C
```

```
0086              IF(J .EQ. NXE) VI(IT,I,J,3) =RoS*VR(IT,I,J,3)
0087        C
0088        C FOR THE OPEN CIRCUIT DISCONTINUITY AT Z = 26.5
0089        C
0090              IF(I .EQ. NZE) VI(IT,I,J,4) =RoC*VR(IT,I,J,4)
0091
0092        50    CONTINUE
0093        60    CONTINUE
0094        C
0095        C IN ORDER TO CONSERVE SPACE, THE ARRAYS
0096        C HAVE TO BE UPDATED
0097        C
0098              DO 70 I = 1, NZE
0099              DO 70 J = 1, NXE
0100              DO 70 K = 1, NT
0101              VI(IT-1,I,J,K) = VI(IT,I,J,K)
0102        70    VR(IT-1,I,J,K) = VR(IT,I,J,K)
0103        C
0104        C CALCULATE IMPULSE RESPONSE AT Z=NZO, X=NXO
0105        C
0106              EI = 0.0
0107              DO 80 K = 1, 4
0108        80    EI = EI + VI(IT,NZO,NXO,K)
0109              EI = (EI + Yo * VI(IT,NZO,NXO,5)) * 2.0/Y
0110              HI = -(VI(IT,NZO,NXO,2)-VI(IT,NZO,NXO,4))
0111        C
0112        C SUM THE FREQUENCY RESPONSE (imaginary and
0113        C real parts) FOR DIFFERENT VALUES OF
0114        C MESH-SIZE DIVIDED BY WAVELENGTH
0115        C
0116              T = DFLOAT(ITIME)
0117              EFI(L)=EFI(L)+EI*DSIN(2.0 * PI * T * DELTA)
0118              EFR(L)=EFR(L)+EI*DCOS(2.0 * PI * T * DELTA)
0119              HFI(L)=HFI(L)+HI*DSIN(2.0 * PI * T * DELTA)
0120              HFR(L)=HFR(L)+HI*DCOS(2.0 * PI * T * DELTA)
0121              OUT(L,1) = DELTA
0122              IF( ITIME .EQ. ITRATE ) THEN
0123              CEF = CMPLX(EFR(L),EFI(L))
0124              CHF = CMPLX( HFR(L), HFI(L))
0125              OUT(L,2) = CABS(CEF)/CABS(CHF)
0126        C
0127        C CALCULATE ARGUMENT Z
0128        C
0129              ZARG = CEF/CHF
0130              XZ   = REAL( ZARG )
0131              YZ   = AIMAG( ZARG )
0132              XYZ  = YZ / XZ
0133              OUT(L,3) = -ATAN( XYZ )
0134              print*,out(1,1),out(1,2),out(1,3)
0135              END IF
0136        90    CONTINUE
0137              DO 100 K = 1, L-1
0138        100   WRITE(6,110) (OUT(K,J),J = 1, 3)
0139        110   FORMAT(2X,10F10.4)
0140              STOP
0141              END
```

Figure 7.19 Computer program for Example 7.4.

7.6 Three-dimensional TLM Mesh

The TLM mesh considered in Sections 7.4 and 7.5 is two-dimensional. The choice of shunt-connected nodes to represent the two-dimensional wave propagation was quite arbitrary; the TLM mesh could have equally been made up of series-connected nodes. To represent a three-dimensional space, however, we must apply a hybrid TLM mesh consisting of three shunt and three series nodes to simultaneously describe all the six field components. First of all, we need to understand what a series-connected node is.

7.6.1 Series Nodes

Figure 7.20 portrays a lossless series-connected node that is equipped with a short-circuited stub called the permeability stub. The corresponding network representation is illustrated in Fig. 7.21. The input impedance of the short-circuited stub is

$$Z_{in} = jZ_o\sqrt{\frac{L}{C}}\tan\left(\frac{\omega\Delta\ell}{2c}\right) \simeq j\omega L Z_o \Delta\ell/2 \qquad (7.69)$$

where Eq. (7.26) has been applied. This represents an impedance with inductance

$$L' = L\frac{\Delta\ell}{2}Z_o. \qquad (7.70)$$

Hence the total inductance on the side in which the stub is inserted is $L\Delta\ell(1+Z_o)/2$ as in Fig. 7.21. We now apply Kirchhoff's voltage law around the series node of Fig. 7.21 and obtain

$$V_z + L\frac{\Delta\ell}{2}(1+Z_o)\frac{\partial I_x}{\partial t} + V_y + \frac{\partial V_y}{\partial z}\Delta\ell - \left(V_z + \frac{\partial V_z}{\partial y}\Delta\ell\right) + 3L\frac{\Delta\ell}{2}\frac{\partial I_x}{\partial t} - V_y = 0.$$

Dividing through by $\Delta\ell$ and rearranging terms leads to

$$\frac{\partial V_z}{\partial y} - \frac{\partial V_y}{\partial z} = 2L(1+Z_o/4)\frac{\partial I_x}{\partial t}. \qquad (7.71)$$

Note that the series node is oriented in the y-z plane. Equations for series nodes in the x-y and x-z planes can be obtained in a similar manner as

$$\frac{\partial V_y}{\partial x} - \frac{\partial V_x}{\partial y} = 2L(1+Z_o/4)\frac{\partial I_z}{\partial t} \qquad (7.72)$$

and

$$\frac{\partial V_x}{\partial z} - \frac{\partial V_z}{\partial x} = 2L(1+Z_o/4)\frac{\partial I_y}{\partial t}, \qquad (7.73)$$

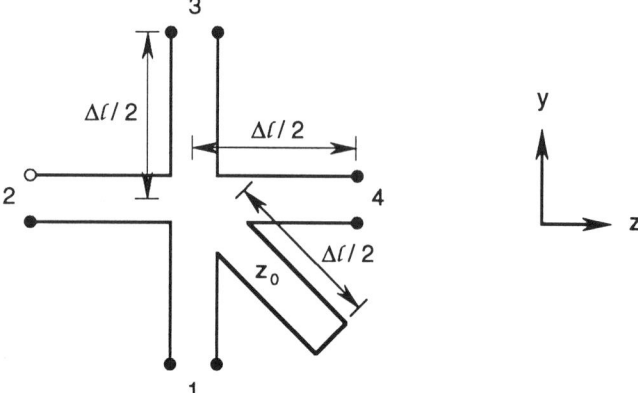

Figure 7.20 A lossless series connected node with permeability stub.

respectively.

Comparing Eqs. (7.71) to (7.73) with Maxwell's equations, the following equivalences can be identified:

$$\boxed{\begin{aligned} E_x &\equiv V_x \\ E_z &\equiv V_z \\ \mu &\equiv 2L \\ \mu_r &\equiv \frac{4+Z_o}{4} \end{aligned}} \qquad (7.74)$$

A series-connected two-dimensional TLM mesh is shown in Fig. 7.22(a) while the equivalent one-dimensional mesh is in Fig. 7.22(b). A voltage impulse incident on a series node is scattered in accordance with Eq. (7.60), where the scattering matrix is now

$$[S] = \frac{2}{Z} \begin{bmatrix} -1 & 1 & 1 & -1 & -1 \\ 1 & -1 & -1 & 1 & 1 \\ 1 & -1 & -1 & 1 & 1 \\ -1 & 1 & 1 & -1 & -1 \\ -Z_o & Z_o & Z_o & -Z_o & -Z_o \end{bmatrix} + [I] \qquad (7.75)$$

$Z = 4 + Z_o$, and $[I]$ is the unit matrix. The velocity characteristic for the two-dimensional series matrix is the same as for the shunt node [23]. For low frequencies ($\Delta \ell/\lambda < 0.1$) the velocity of the waves on the matrix is approximately $1/\sqrt{2}$ of the free-space velocity. This is due to the fact that the stubs have twice the inductance per unit length, while the capacitance per unit length remains unchanged. This is the dual of the two-dimensional

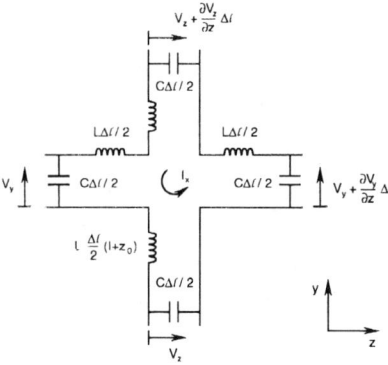

Figure 7.21 Network representation of a series node.

shunt case in which the capacitance was doubled and the inductance was unchanged.

7.6.2 Three-dimensional Node

A three-dimensional TLM node [26] consists of three shunt nodes in conjunction with three series nodes. The voltages at the three shunt nodes represent the three components of the **E** field, while the currents of the series nodes represent the three components of the **H** field. In the x-z plane, for example, the voltage-current equations for the shunt node are

$$\frac{\partial I_x}{\partial z} - \frac{\partial I_z}{\partial x} = 2C\frac{\partial V_y}{\partial t} \tag{7.76a}$$

$$\frac{\partial V_y}{\partial x} = -L\frac{\partial I_x}{\partial t} \tag{7.76b}$$

$$\frac{\partial V_y}{\partial z} = -L\frac{\partial I_z}{\partial t} \tag{7.76c}$$

and for the series node in the x-z plane, the equations are

$$\frac{\partial V_x}{\partial z} - \frac{\partial V_z}{\partial x} = 2L\frac{\partial I_y}{\partial t} \tag{7.77a}$$

$$\frac{\partial I_y}{\partial x} = -C\frac{\partial V_z}{\partial t} \tag{7.77b}$$

$$\frac{\partial I_y}{\partial z} = -C\frac{\partial V_x}{\partial t}. \tag{7.77c}$$

Maxwell's equations $\nabla \times \mathbf{E} = \dfrac{\partial \mathbf{B}}{\partial t}$ and $\nabla \times \mathbf{H} = \epsilon\dfrac{\partial \mathbf{E}}{\partial t}$ for $\dfrac{\partial}{\partial y} \equiv 0$ give

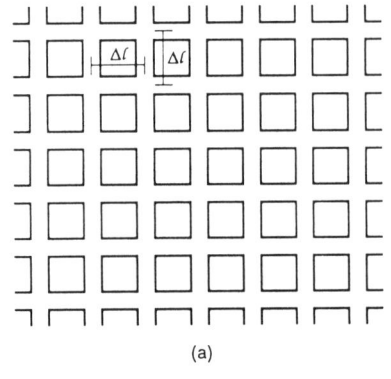

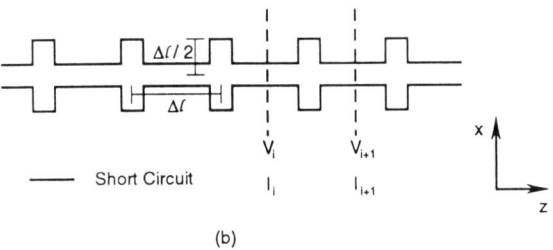

Figure 7.22 (a) A two-dimensional series-connected TLM mesh. (b) A one-dimensional series-connected TLM mesh.

$$\frac{\partial H_x}{\partial z} - \frac{\partial H_z}{\partial x} = \epsilon \frac{\partial E_y}{\partial t} \qquad (7.78a)$$

$$\frac{\partial E_y}{\partial x} = \mu \frac{\partial H_x}{\partial t} \qquad (7.78b)$$

$$\frac{\partial E_y}{\partial z} = -\mu \frac{\partial H_z}{\partial t} \qquad (7.78c)$$

and

$$\frac{\partial E_x}{\partial z} - \frac{\partial E_z}{\partial x} = -\mu \frac{\partial H_y}{\partial t} \qquad (7.79a)$$

$$\frac{\partial H_y}{\partial x} = -\epsilon \frac{\partial E_x}{\partial t} \qquad (7.79b)$$

$$\frac{\partial H_y}{\partial z} = -\epsilon \frac{\partial E_z}{\partial t}. \qquad (7.79c)$$

A similar analysis for shunt and series nodes in the x-y and y-z planes will yield the voltage-current equations and the corresponding Maxwell's equations. The three sets of two-dimensional shunt and series nodes oriented

Transmission-line-matrix Method

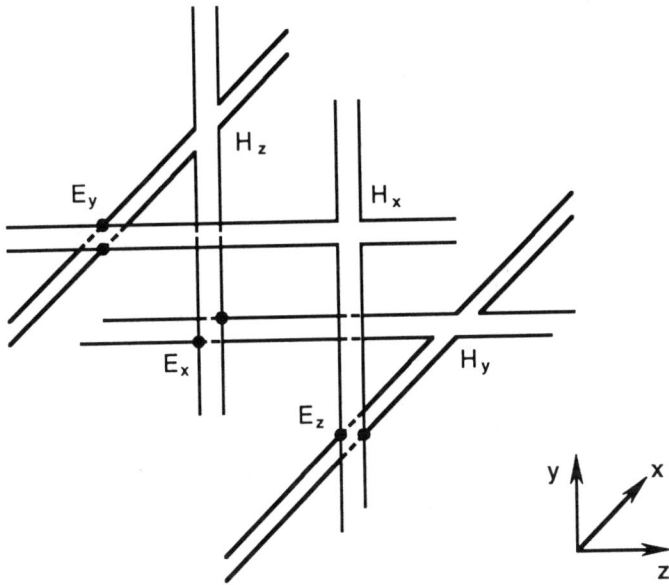

Figure 7.23 A three-dimensional node consisting of three shunt nodes and three series nodes.

in the x-y, y-z, and z-x planes form a three-dimensional model. The two-dimensional nodes must be connected in such a way as to correctly describe Maxwell's equations at each three-dimensional node. Each of the shunt and series nodes has a spacing of $\Delta\ell/2$ so that like nodes are spaced $\Delta\ell$ from each other.

Figure 7.23 illustrates a three-dimensional node representing a cubical volume of space $\Delta\ell/2$ long in each direction. A close examination shows that if the voltage between lines represents the **E** field and the current in the lines represents the **H** field, then the following Maxwell's equations are satisfied:

$$\frac{\partial H_x}{\partial z} - \frac{\partial H_z}{\partial x} = \epsilon \frac{\partial E_y}{\partial t} \tag{7.80a}$$

$$\frac{\partial E_z}{\partial y} - \frac{\partial E_y}{\partial z} = -\mu \frac{\partial H_x}{\partial t} \tag{7.80b}$$

$$\frac{\partial E_y}{\partial x} - \frac{\partial E_x}{\partial y} = -\mu \frac{\partial H_z}{\partial t} \tag{7.80c}$$

$$\frac{\partial E_x}{\partial z} - \frac{\partial E_z}{\partial x} = -\mu \frac{\partial H_y}{\partial t} \tag{7.80d}$$

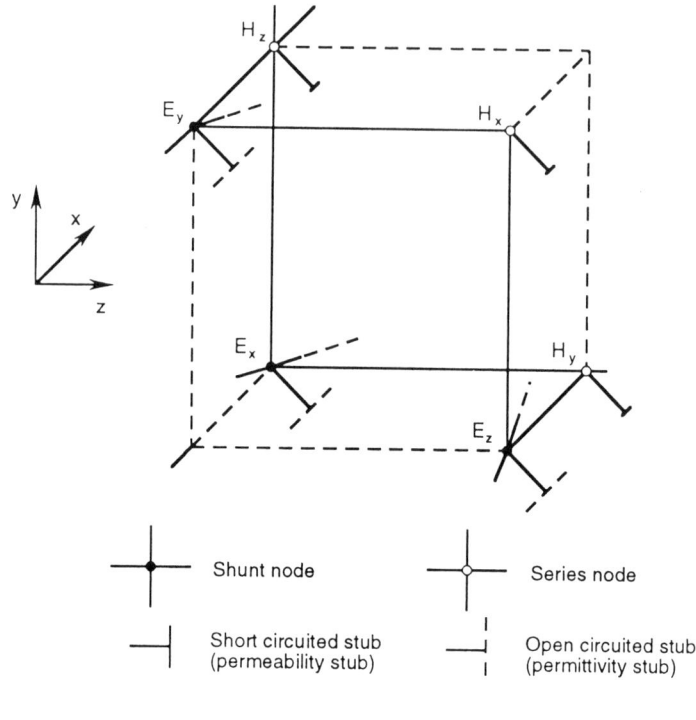

Figure 7.24 A general three-dimensional node.

$$\frac{\partial H_z}{\partial y} - \frac{\partial H_y}{\partial z} = \epsilon \frac{\partial E_x}{\partial t} \qquad (7.80\text{e})$$

$$\frac{\partial H_y}{\partial x} - \frac{\partial H_x}{\partial y} = \epsilon \frac{\partial E_z}{\partial t}. \qquad (7.80\text{f})$$

In the upper half of the node in Fig. 7.23, we have a shunt node in the x-z plane (representing Eq. (7.80a)) connected to a series node in the y-z plane (representing Eq. (7.80b)) and a series node in the x-y plane (representing Eq. (7.80c)). In the lower half of the node, a series node in the x-z plane (representing Eq. (7.80d)) is connected to a shunt node in the y-z plane (representing Eq. (7.80e)) and a shunt node in the x-y plane (representing Eq. (7.80f)). Thus Maxwell's equations are completely satisfied at the three-dimensional node. A three-dimensional TLM mesh is obtained by stacking similar nodes in x, y, and z directions (see Fig. 7.25, for example).

The wave characteristics of the three-dimensional mesh are similar to those of the two-dimensional mesh with the difference that low-frequency velocity is now $c/2$ instead of $c/\sqrt{2}$.

Figure 7.24 shows a schematic diagram of a three-dimensional node using single lines to represent pairs of transmission lines. It is more general than the representation in Fig. 7.23 in that it includes the permittivity, permeability, and loss stubs. Note that the dotted lines making up the corners of the cube are guidelines and do not represent transmission lines or stubs. It can be shown that for the general node the following equivalences apply [27]:

$$\begin{aligned}
E_x &\equiv \text{the common voltage at shunt node } E_x \\
E_y &\equiv \text{the common voltage at shunt node } E_y \\
E_z &\equiv \text{the common voltage at shunt node } E_z \\
H_x &\equiv \text{the common current at series node } H_x \\
H_y &\equiv \text{the common current at series node } H_y \\
H_z &\equiv \text{the common current at series node } H_z \\
\epsilon_o &\equiv C \text{ (the capacitance per unit length of lines)} \\
\epsilon_r &\equiv 2(1+Y_o/4) \\
\mu_o &\equiv L \text{ (the inductance per unit length of lines)} \\
\mu_r &\equiv 2(1+Z_o/4) \\
\sigma &\equiv \frac{G_o}{\Delta \ell \frac{L}{C}}
\end{aligned} \quad (7.81)$$

where Y_o, Z_o, and G_o remain as defined in Sections 7.4 and 7.5. Interconnection of many of such three-dimensional nodes forms a TLM mesh representing any inhomogeneous media. The TLM method for three-dimensional problems is therefore concerned with applying Eq. (7.60) in conjunction with Eqs. (7.61) and (7.75) and obtaining the impulse response. Any of the field components may be excited initially by specifying initial impulses at the appropriate nodes. Also, the response at any node may be monitored by recording the pulses that pass through the node.

7.6.3 Boundary Conditions

Boundary conditions are simulated by short-circuiting shunt nodes (electric wall) or open-circuiting series nodes (magnetic wall) situated on a boundary. The tangential components of **E** must vanish in the plane of an electric wall, while the tangential components of **H** must be zero in the plane of a

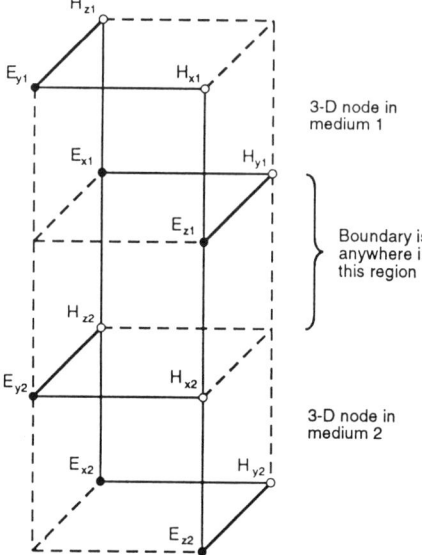

Figure 7.25 A dielectric/dielectric boundary in TLM mesh.

magnetic wall. For example, to set E_x and E_y equal to zero in a particular plane, all shunt nodes E_x and E_y lying in that plane are shorted. Similarly, to set H_y and H_z equal to zero in some plane, the series nodes H_y and H_z in that plane are simply open-circuited.

The continuity of the tangential components of **E** and **H** fields across a dielectric/dielectric boundary is automatically satisfied in the TLM mesh. For example, for a dielectric/dielectric boundary in the x-z plane such as shown in Fig. 7.25, the following equations valid for a transmission-line element joining the nodes on either side of the boundaries are applicable:

$$E_{z1} = E_{z2} + \frac{\partial E_{z2}}{\partial y}\Delta\ell$$

$$E_{x1} = E_{x2} + \frac{\partial E_{x2}}{\partial y}\Delta\ell$$

$$H_{x1} = H_{x2} + \frac{\partial H_{x2}}{\partial y}\Delta\ell$$

$$H_{z1} = H_{z2} + \frac{\partial H_{z2}}{\partial y}\Delta\ell. \qquad (7.82)$$

Finally, wall losses are included by introducing imperfect reflection coefficients as discussed in Section 7.5. The three-dimensional TLM mesh will

Table 7.5 Resonant Wavenumber ($k_c a$) of an Empty Rectangular Cavity, where $k_c a = 4\pi a/c$ and λ is the Free-space Wavelength

Modes	Exact results	TLM results	Error %
TM_{110}	5.6636	5.6400	0.42
TE_{101}	7.0249	6.9819	0.61
TM_{210}, TE_{011}	7.8540	7.8112	0.54

be applied in solving the three-dimensional problems of resonant cavities in the following examples, taken from Akhtarzad and Johns [26].

Example 7.5

Determine the resonant frequency of an $a \times b \times d$ empty rectangular cavity using the TLM method. Take $a = 12\Delta\ell$, $b = 8\Delta\ell$, and $d = 6\Delta\ell$.

Solution

The exact solution [13,14] for TE_{mnp} or TM_{mnp} mode is

$$f_r = \frac{c}{2}\sqrt{(m/a)^2 + (n/b)^2 + (p/d)^2}$$

from which we readily obtain

$$k_c = \frac{w_r}{c} = \frac{2\pi f_r}{c} = \pi\sqrt{(m/a)^2 + (n/b)^2 + (p/d)^2}.$$

The TLM program, the modified version of the program in [21], is shown in Fig. 7.26. The program initializes all field components by setting them equal to zero at all nodes in the $12\Delta\ell \times 8\Delta\ell \times 8\Delta\ell$ TLM mesh and exciting one field component. With subroutine COMPUTE, it applies Eq. (7.60) in conjunction with Eq. (7.61) and (7.75) to calculate the reflected **E** and **H** field components at all nodes. It applies the boundary conditions and calculates the impulse response at a particular node in the mesh.

The results of the computation along with the exact analytical values for the first few modes in the cavity are shown in Table 7.5.

```
0001        C************************************************************
0002        C  THIS PROGRAM ANLYZES THREE-DIMENSIONAL WAVE PROBLEMS
0003        C  USING THE TLM METHOD
0004        C  THE SPECIFIC EXAMPLE SOLVED HERE IS THE DETERMINATION
0005        C  OF THE TM DOMINANT MODE OF RECTANGULAR LOSSLESS CAVITY
0006        C  OF DIMENSION 12 X 8 X 6
0007        C          E      ...  E - field
0008        C          H      ...  H - field
0009        C          X      ...  X - component
0010        C          Y      ...  Y - component
0011        C          Z      ...  Z - component
0012        C          I      ...  incident impulse
0013        C          R      ...  reflected impulse
0014        C          NX     ...  number of nodes in X-direction
0015        C          NY     ...  number of nodes in Y-direction
0016        C          NZ     ...  number of nodes in Z-direction
0017        C  SHUNT and SERIES  ...  scattering matrix
0018        C    DELTA = DELTA/LAMDA   (FREE SPACE)
0019
0020               IMPLICIT   REAL*8(A,B,D-H,O-Z),COMPLEX*8(C),
0021        1      INTEGER*2(I-N)
0022               DIMENSION EXR(2,7,9,13,5), EXI(2,7,9,13,5),
0023        1               EYR(2,7,9,13,5), EYI(2,7,9,13,5),
0024        2               EZR(2,7,9,13,5), EZI(2,7,9,13,5),
0025        3               HXR(2,7,9,13,5), HXI(2,7,9,13,5),
0026        4               HYR(2,7,9,13,5), HYI(2,7,9,13,5),
0027        5               HZR(2,7,9,13,5), HZI(2,7,9,13,5),
0028        6               EX(2,1000),EY(2,1000),EZ(2,1000),
0029        7               HX(2,1000),HY(2,1000),HZ(2,1000),
0030        8               OUT(12,1000),SHUNT(5,5),SERIES(5,5)
0031
0032               COMMON / PART1 /NT,NXB,NXE,NYB,NYE,NZB,NZE,IT
0033               SQMAG(CM) = ( CABS(CM) ) ** 2
0034               DATA ITRATE,NXB,NXE,NYB,NYE,NZB,NZE,NT,DELTA
0035        1         /1000, 1,  12, 1,  8,  1,  6,  4, 1.0/
0036               DATA NXI,NYI,NZI,NXO,NYO,NZO,RoH,NUMPT,DEL
0037        1         /11, 7,  0,  2,  2, 2,-1.0,1000,0.0001/
0038               DATA PI,          Eo,         Er, Ur,SIGMA,DELTA
0039        1         /3.1415927,8.854E-12,1.0,1.0,0.0,0.01 /
0040               DATA SERIES /-1.0,1.0,1.0,-1.0,-1.0,
0041        1                    1.0,-1.0,-1.0,1.0,1.0,
0042        2                    1.0,-1.0,-1.0,1.0,1.0,
0043        3                   -1.0,1.0,1.0,-1.0,-1.0,
0044        4                   -1.0,1.0,1.0,-1.0,-1.0/
0045               DATA SHUNT / 1.0,1.0,1.0,1.0,1.0,
0046        1                   1.0,1.0,1.0,1.0,1.0,
0047        2                   1.0,1.0,1.0,1.0,1.0,
0048        3                   1.0,1.0,1.0,1.0,1.0,
0049        4                   1.0,1.0,1.0,1.0,1.0/
0050
0051               Uo = 4.0 * PI * 1.0E-07
0052               Yo = 4.0 * (Er - 1.0)
0053               Zo = 4.0 * (Ur - 1.0)
0054               Go = SIGMA * DELTA * SQRT(Uo/Eo)
0055               Y  = 4.0 + Yo + Go
0056               Z  = 4.0 + Zo
0057               DO 5 I = 1,5
0058               SHUNT(I,5)  = SHUNT(I,5) * Yo
0059               SERIES(5,I) = SERIES(5,I) * Zo
0060        5      CONTINUE
0061        C
```

```
0062        C  INITIALIZE ALL THE NODES
0063        C
0064                  DO 10 IT = 1,2
0065                  DO 10 K = NZB, NZE
0066                  DO 10 J = NYB, NYE
0067                  DO 10 I = NXB, NXE
0068                  DO 10 L = 1, NT
0069                     EXI(IT,K,J,I,L) = 0.0
0070                     EYI(IT,K,J,I,L) = 0.0
0071                     EZI(IT,K,J,I,L) = 0.0
0072                     HXI(IT,K,J,I,L) = 0.0
0073                     HYI(IT,K,J,I,L) = 0.0
0074                     HZI(IT,K,J,I,L) = 0.0
0075                     EXR(IT,K,J,I,L) = 0.0
0076                     EYR(IT,K,J,I,L) = 0.0
0077                     EZR(IT,K,J,I,L) = 0.0
0078                     HXR(IT,K,J,I,L) = 0.0
0079                     HYR(IT,K,J,I,L) = 0.0
0080        10           HZR(IT,K,J,I,L) = 0.0
0081        C
0082        C  INSERT THE INITIAL EXCITATION
0083        C
0084                  DO 15 K = NZB, NZE
0085                  DO 15 L = 1, 4
0086                     EZR(2,K,NYI,NXI,L) = 1.0
0087        15        CONTINUE
0088                  IT = 2
0089        C
0090        C  START ITERATION
0091        C
0092                  DO 60 ITime = 1, ITRATE
0093                     IF(ITIME .EQ. 1) GOTO 25
0094        C
0095        C  COMPUTE THE REFLECTED E-FIELD AT ALL NODES
0096        C
0097                     ISIGN = -1
0098                     CALL COMPUTE(EXR,EYR,EZR,SHUNT
0099       1                          ,Y,EXI,EYI,EZI,ISIGN)
0100        C
0101        C  RE-INITIALIZE MATRIX AND FORCE TANGENTIAL
0102        C  E-FIELD TO ZERO AT BOUNDARY
0103        C
0104                     DO 20 K = NZB, NZE
0105                     DO 20 J = NYB, NYE
0106                     DO 20 I = NXB, NXE
0107                     DO 20 L = 1, NT
0108                        HXI(IT-1,K,J,I,L) = 0.0
0109                        HYI(IT-1,K,J,I,L) = 0.0
0110                        HZI(IT-1,K,J,I,L) = 0.0
0111                        IF(I .EQ. NXB) THEN
0112                           EYR(IT,K,J,I,L) = 0.0
0113                           EZR(IT,K,J,I,L) = 0.0
0114                           HXR(IT,K,J,I,L) = 0.0
0115                        END IF
0116                        IF(J .EQ. NYB)THEN
0117                           EXR(IT,K,J,I,L) = 0.0
0118                           EZR(IT,K,J,I,L) = 0.0
0119                           HYR(IT,K,J,I,L) = 0.0
0120                        END IF
0121                        IF(K .EQ. NZB)THEN
0122                           EXR(IT,K,J,I,L) = 0.0
0123                           EYR(IT,K,J,I,L) = 0.0
0124                           HZR(IT,K,J,I,L) = 0.0
```

```
0125              END IF
0126       20     CONTINUE
0127       25     CONTINUE
0128              DO 30 K = NZB, NZE
0129              DO 30 J = NYB, NYE
0130              DO 30 I = NXB, NXE
0131              IF(J .NE. NYB)HZI(IT-1,K,J-1,I,4)=EXR(IT,K,J,I,1)
0132              IF(K .NE. NZB)HYI(IT-1,K-1,J,I,4)=EXR(IT,K,J,I,2)
0133                           HZI(IT-1,K,J,I,2)   =EXR(IT,K,J,I,3)
0134                           HYI(IT-1,K,J,I,2)   =EXR(IT,K,J,I,4)
0135              IF(I .NE. NXB)HZI(IT-1,K,J,I-1,3)=EYR(IT,K,J,I,1)
0136              IF(K .NE. NZB)HXI(IT-1,K-1,J,I,4)=EYR(IT,K,J,I,2)
0137                           HZI(IT-1,K,J,I,1)   =EYR(IT,K,J,I,3)
0138                           HXI(IT-1,K,J,I,2)   =EYR(IT,K,J,I,4)
0139              IF(I .NE. NXB)HYI(IT-1,K,J,I-1,3)=EZR(IT,K,J,I,1)
0140              IF(J .NE. NYB)HXI(IT-1,K,J-1,I,3)=EZR(IT,K,J,I,2)
0141                           HYI(IT-1,K,J,I,1)   =EZR(IT,K,J,I,3)
0142                           HXI(IT-1,K,J,I,1)   =EZR(IT,K,J,I,4)
0143       30     CONTINUE
0144       C
0145       C INSERT THE BOUNDARY CONDITIONS AT ALL BOUNDARIES
0146       C
0147              DO 35 K = NZB, NZE
0148              DO 35 J = NYB, NYE
0149              DO 35 I = NXB, NXE
0150                IF(I .EQ. NXB)THEN
0151                  HYI(IT-1,K,J,I,1) = RoH * HYR(IT,K,J,I,1)
0152                  HZI(IT-1,K,J,I,1) = RoH * HZR(IT,K,J,I,1)
0153                END IF
0154                IF(I .EQ. NXE)THEN
0155                  HZI(IT-1,K,J,I,3) = RoH * HZR(IT,K,J,I,3)
0156                  HYI(IT-1,K,J,I,3) = RoH * HYR(IT,K,J,I,3)
0157                END IF
0158                IF(J .EQ. NYB)THEN
0159                  HXI(IT-1,K,J,I,1) = RoH * HXR(IT,K,J,I,1)
0160                  HZI(IT-1,K,J,I,2) = RoH * HZR(IT,K,J,I,2)
0161                END IF
0162                IF(J .EQ. NYE)THEN
0163                  HXI(IT-1,K,J,I,3) = RoH * HXR(IT,K,J,I,3)
0164                  HZI(IT-1,K,J,I,4) = RoH * HZR(IT,K,J,I,4)
0165                END IF
0166                IF(K .EQ. NZB)THEN
0167                  HXI(IT-1,K,J,I,2) = RoH * HXR(IT,K,J,I,2)
0168                  HYI(IT-1,K,J,I,2) = RoH * HYR(IT,K,J,I,2)
0169                END IF
0170                IF(K .EQ. NZE)THEN
0171                  HXI(IT-1,K,J,I,4) = RoH * HXR(IT,K,J,I,4)
0172                  HYI(IT-1,K,J,I,4) = RoH * HYR(IT,K,J,I,4)
0173                END IF
0174       35     CONTINUE
0175       C
0176       C  COMPUTE THE H-FIELDS AT ALL THE NODES
0177       C
0178              ISIGN = 1
0179              CALL COMPUTE(HXR,HYR,HZR,SERIES,
0180             1             Z,HXI,HYI,HZI,ISIGN)
0181       C
0182       C  RE-INITIALIZE ALL THE NODES
0183       C
0184              DO 40 K = NZB, NZE
0185              DO 40 J = NYB, NYE
0186              DO 40 I = NXB, NXE
0187              DO 40 L = 1, NT
```

Transmission-line-matrix Method   551

```
0188                EXI(IT-1,K,J,I,L) = 0.0
0189                EYI(IT-1,K,J,I,L) = 0.0
0190                EZI(IT-1,K,J,I,L) = 0.0
0191       40    CONTINUE
0192             DO 45 K = NZB, NZE
0193             DO 45 J = NYB, NYE
0194             DO 45 I = NXB, NXE
0195                EZI(IT-1,K,J,I,4)  =HXR(IT,K,J,I,1)
0196                EYI(IT-1,K,J,I,4)  =HXR(IT,K,J,I,2)
0197             IF(J .NE. NYE)EZI(IT-1,K,J+1,I,2)=HXR(IT,K,J,I,3)
0198             IF(K .NE. NZE)EYI(IT-1,K+1,J,I,2)=HXR(IT,K,J,I,4)
0199                EZI(IT-1,K,J,I,3)  =HYR(IT,K,J,I,1)
0200                EXI(IT-1,K,J,I,4)  =HYR(IT,K,J,I,2)
0201             IF(I .NE. NXE)EZI(IT-1,K,J,I+1,1)=HYR(IT,K,J,I,3)
0202             IF(K .NE. NZE)EXI(IT-1,K+1,J,I,2)=HYR(IT,K,J,I,4)
0203                EYI(IT-1,K,J,I,3)  =HZR(IT,K,J,I,1)
0204                EXI(IT-1,K,J,I,3)  =HZR(IT,K,J,I,2)
0205             IF(I .NE. NXE)EYI(IT-1,K,J,I+1,1)=HZR(IT,K,J,I,3)
0206             IF(J .NE. NYE)EXI(IT-1,K,J+1,I,1)=HZR(IT,K,J,I,4)
0207       45    CONTINUE
0208       C
0209       C   CALCULATE THE IMPULSE RESPONSE AT NXO,NYO,NZO
0210       C
0211             EXT = 0.0
0212             EYT = 0.0
0213             EZT = 0.0
0214             HXT = 0.0
0215             HYT = 0.0
0216             HZT = 0.0
0217             DO 50 L = 1, NT
0218                EXT = EXT + EXI(IT-1,NZO,NYO,NXO,L) * (2.0/Y)
0219                EYT = EYT + EYI(IT-1,NZO,NYO,NXO,L) * (2.0/Y)
0220                EZT = EZT + EZI(IT-1,NZO,NYO,NXO,L) * (2.0/Y)
0221                HXT = HXT + HXI(IT-1,NZO,NYO,NXO,L) * (2.0/Y)
0222                HYT = HYT + HYI(IT-1,NZO,NYO,NXO,L) * (2.0/Y)
0223       50       HZT = HZT + HZI(IT-1,NZO,NYO,NXO,L) * (2.0/Y)
0224       C
0225       C   SUM THE FREQUENCY RESPONSE(imaginary and real
0226       C   parts) FOR DIFFERENT VALUES OF MESH-SIZE DIVIDED
0227       C   BY WAVELENGTH BUT FIRST CONVOLVE IMPULSE RESPONSE
0228       C   WITH HANNING PROFILE
0229       C
0230             DINCR = DELTA
0231             DO 55 L = 1, NUMPT
0232                T = DFLOAT(ITIME)
0233                ANT = DFLOAT(ITRATE)
0234                EXTH = EXT * (1.0 + DCOS(PI * T/ANT)) * 0.5
0235                EYTH = EYT * (1.0 + DCOS(PI * T/ANT)) * 0.5
0236                EZTH = EZT * (1.0 + DCOS(PI * T/ANT)) * 0.5
0237                HXTH = HXT * (1.0 + DCOS(PI * T/ANT)) * 0.5
0238                HYTH = HYT * (1.0 + DCOS(PI * T/ANT)) * 0.5
0239                HZTH = HZT * (1.0 + DCOS(PI * T/ANT)) * 0.5
0240                EX(1,L) = EX(1,L) + EXTH * DCOS(2.0*PI*T*DINCR)
0241                EX(2,L) = EX(2,L) + EXTH * DSIN(2.0*PI*T*DINCR)
0242                EY(1,L) = EY(1,L) + EYTH * DCOS(2.0*PI*T*DINCR)
0243                EY(2,L) = EY(2,L) + EYTH * DSIN(2.0*PI*T*DINCR)
0244                EZ(1,L) = EZ(1,L) + EZTH * DCOS(2.0*PI*T*DINCR)
0245                EZ(2,L) = EZ(2,L) + EZTH * DSIN(2.0*PI*T*DINCR)
0246                HX(1,L) = HX(1,L) + HXTH * DCOS(2.0*PI*T*DINCR)
0247                HX(2,L) = HX(2,L) + HXTH * DSIN(2.0*PI*T*DINCR)
0248                HY(1,L) = HY(1,L) + HYTH * DCOS(2.0*PI*T*DINCR)
0249                HY(2,L) = HY(2,L) + HYTH * DSIN(2.0*PI*T*DINCR)
```

```
0250              HZ(1,L) = HZ(1,L) + HZTH * DCOS(2.0*PI*T*DINCR)
0251              HZ(2,L) = HZ(2,L) + HZTH * DSIN(2.0*PI*T*DINCR)
0252              OUT(1,L) = DINCR
0253     55       DINCR = DINCR + DEL
0254     60       CONTINUE
0255              DO L = 1, NUMPT
0256                 CEX = CMPLX(EX(1,L),EX(2,L))
0257                 CEY = CMPLX(EY(1,L),EY(2,L))
0258                 CEZ = CMPLX(EZ(1,L),EZ(2,L))
0259                 CHX = CMPLX(HX(1,L),HX(2,L))
0260                 CHY = CMPLX(HY(1,L),HY(2,L))
0261                 CHZ = CMPLX(HZ(1,L),HZ(2,L))
0262                 OUT(2,L) = CABS(CEX)
0263                 OUT(3,L) = CABS(CEY)
0264                 OUT(4,L) = CABS(CEZ)
0265                 OUT(5,L) = CABS(CHX)
0266                 OUT(6,L) = CABS(CHY)
0267                 OUT(7,L) = CABS(CHZ)
0268              OUT(8,L)=SQRT(SQMAG(CEX)+SQMAG(CEY)+SQMAG(CEZ))
0269              OUT(9,L)=SQRT(SQMAG(CHX)+SQMAG(CHY)+SQMAG(CHZ))
0270              END DO
0271     C
0272     C  PICK THE MODES
0273     C
0274              DO 65 L = 2, 9
0275              DO 65 K = 1, NUMPT
0276                 IF(K .NE. 1 .AND. K .NE. NUMPT) THEN
0277                    IF(OUT(L,K) .GT. OUT(L,K-1) .AND.
0278     1                 OUT(L,K) .GT. OUT(L,K+1)) THEN
0279                       WRITE(6,80) L,K,(OUT(J,K),J=1,9)
0280                    END IF
0281                 END IF
0282     65       CONTINUE
0283     C
0284     C  WRITE OUT DATA FOR PLOTTING
0285     C
0286              DO 70 L = 2, 9
0287              DO 70 K = 1, NUMPT
0288                 WRITE(6,75) K, OUT(1,K), OUT(L,K)
0289     70       CONTINUE
0290     75       FORMAT(I5,4F15.8)
0291     80       FORMAT(2I5,10F8.4)
0292              STOP
0293              END

0001     C*************************************************
0002     C  THIS SUBROUTINE COMPUTES THE REFLECTED PULSES
0003     C
0004              SUBROUTINE COMPUTE(AXR,AYR,AZR,SCATTER,
0005     1                           W,AXI,AYI,AZI,ISIGN)
0006              IMPLICIT REAL*8(A-H,O-Z), INTEGER*2(I-N)
0007              DIMENSION AXR(2,NZE,NYE,NXE,NT),
0008     1        AYR(2,NZE,NYE,NXE,NT),SCATTER(5,5),
0009     2        AZR(2,NZE,NYE,NXE,NT),AXI(2,NZE,NYE,NXE,NT),
0010     3        AYI(2,NZE,NYE,NXE,NT),AZI(2,NZE,NYE,NXE,NT)
0011
0012              COMMON / PART1 /NT,NXB,NXE,NYB,NYE,NZB,NZE,IT
0013
0014              DO 20 K = NZB, NZE
0015              DO 20 J = NYB, NYE
0016              DO 20 I = NXB, NXE
0017     C  INSERT FEW LINES HERE FOR INHOMOGENEOUS CAVITY
0018              DO 15 L = 1, NT
```

Transmission-line-matrix Method 553

```
0019                   AXRS = 0.0
0020                   AYRS = 0.0
0021                   AZRS = 0.0
0022                   DO 10 M = 1, NT
0023              AXRS = AXRS+2./W*SCATTER(L,M)*AXI(IT-1,K,J,I,M)
0024              AYRS = AYRS+2./W*SCATTER(L,M)*AYI(IT-1,K,J,I,M)
0025              AZRS = AZRS+2./W*SCATTER(L,M)*AZI(IT-1,K,J,I,M)
0026        10         CONTINUE
0027              AXR(IT,K,J,I,L)=AXRS+ISIGN*AXI(IT-1,K,J,I,L)
0028              AYR(IT,K,J,I,L)=AYRS+ISIGN*AYI(IT-1,K,J,I,L)
0029              AZR(IT,K,J,I,L)=AZRS+ISIGN*AZI(IT-1,K,J,I,L)
0030        15         CONTINUE
0031        20    CONTINUE
0032              RETURN
0033              END
```

Figure 7.26 Computer program for Example 7.5.

Example 7.6

Modify the TLM program in Fig. 7.26 to calculate the resonant wavenumber $k_c a$ of the inhomogeneous cavities in Fig. 7.27. Take $\epsilon_r = 16$, $a = \Delta \ell$, $b = 3a/10$, $d = 4a/10$, $s = 7a/12$.

Solution

The main program in Fig. 7.26 is applicable to this example. Only the subroutine COMPUTE requires slight modification to take care of the inhomogeneity of the cavity. The modifications in the subprogram for the cavities in Fig. 7.27(a) and (b) are shown in Fig. 7.28(a) and (b), respectively. For each modification, the few lines in Fig. 7.28 are inserted in between lines 15 and 17 in subroutine COMPUTE of Fig. 7.26. The results are shown in Table 7.6 for TE_{101} mode.

```
          :
          :
          :
0001       C****************************************************
0002       C   THIS SUBROUTINE COMPUTES THE REFLECTED PULSES
0003       C
0004              SUBROUTINE COMPUTE(AXR,AYR,AZR,SCATTER,
0005           1                      WW,AXI,AYI,AZI,ISIGN)
          :
          :
          :
0017       C   INSERT FEW LINES HERE FOR INHOMOGENEOUS CAVITY
0018              IF ((I.GE.9).AND.(I.LE.13)
0019           1              .AND.(ISIGN.EQ.-1)) THEN
0020              NT = 5
0021              W = WW
```

554 Numerical Techniques in Electromagnetics

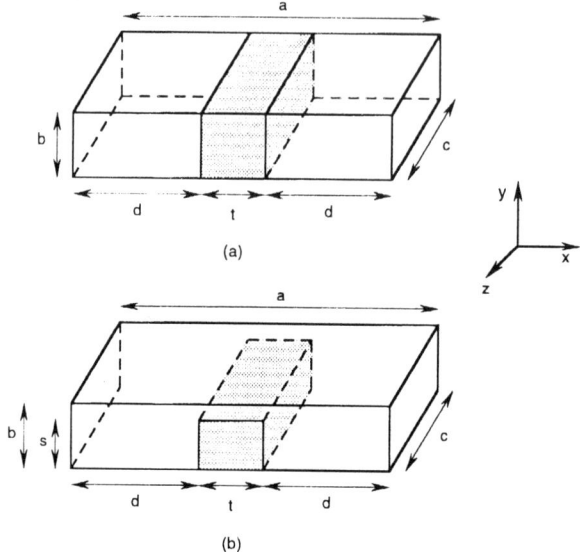

Figure 7.27 Rectangular cavity loaded with dielectric slab.

```
0022                    ELSE
0023            NT = 4
0024            W = 4.0
0025            END IF
```
⋮

Figure 7.28(a) Modification in subroutine COMPUTE for the inhomogeneous cavity of Fig.7.27(a).

```
0001    C***************************************************
0002    C  THIS SUBROUTINE COMPUTES THE REFLECTED PULSES
0003    C
0004            SUBROUTINE COMPUTE(AXR,AYR,AZR,SCATTER,
0005        1                     WW,AXI,AYI,AZI,ISIGN)
```
⋮
```
0017    C  INSERT FEW LINES HERE FOR INHOMOGENEOUS CAVITY
0018            IF (J.GT.4) GO TO 5
0019            IF ((I.GE.9).AND.(I.LE.13)
0020        1                     .AND.(ISIGN.EQ.-1)) THEN
0021              NT = 5
0022              W = WW
0023                    GO TO 6
```

Table 7.6 Resonant Wavenumber ($k_c a$) for TE_{101} Mode of Inhomogeneous Rectangular Cavities, where $k_c a = 4\pi a/c$, and λ is the Free-space Wavelength

Modes	Exact results	TLM results	Error %
Fig. 7.27(a)	2.589	2.5761	0.26
Fig. 7.27(b)	(none)	3.5387	

```
0024            END IF
0025       5    CONTINUE
0026            NT = 4
0027            W = 4.0
0028       6    CONTINUE
0029            DO 15 L = 1, NT
  :
  :
  :
```

Figure 7.28(b) Modification in subroutine COMPUTE for the inhomogeneous cavity of Fig. 7.27 (b).

7.7 Error Sources and Correction

As in all approximate solutions such as the TLM technique, it is important that the error in the final result be minimal. In the TLM method, four principal sources of error can be identified [10,27]:

- truncation error,
- coarseness error,
- velocity error,
- misalignment error.

Each of these sources of error and ways of minimizing it will be discussed.

7.7.1 Truncation Error

The truncation error is due to the need to truncate the impulse response in time. As a result of the finite duration of the impulse response, its Fourier transform is not a line spectrum but rather a superposition of $\sin x/x$ functions, which may interfere with each other and cause a slight shift in their maxima. The maximum truncation error is given by

$$e_T = \frac{\Delta S}{\Delta \ell / \lambda_c} = \frac{3\lambda_c}{SN^2\pi^2\Delta\ell}, \qquad (7.83)$$

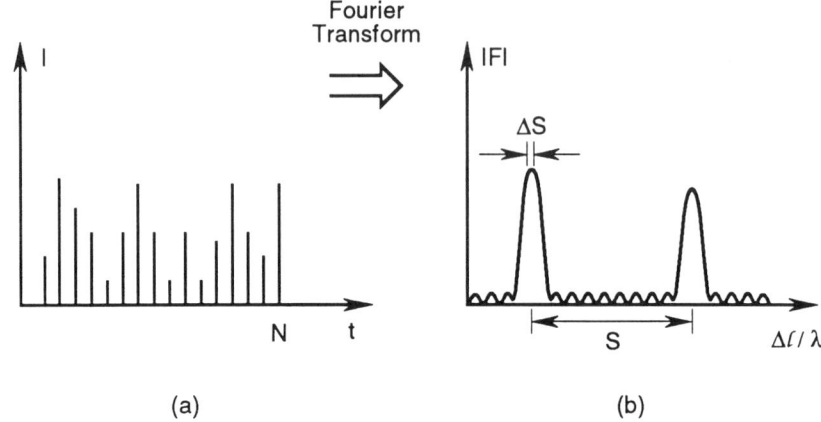

Figure 7.29 Source of truncation error: (a) Truncated output impulse, (b) resulting truncation error in the frequency domain.

where λ_c is the cutoff wavelength to be calculated, ΔS is the absolute error in $\Delta \ell/\lambda_c$, S is the frequency separation (expressed in terms of $\Delta\ell/\lambda_c$, λ_c being the free-space wavelength) between two neighboring peaks as shown in Fig. 7.29, and N is the number of iterations. Equation (7.83) indicates that e_T decreases with increasing N and increasing S. It is therefore desirable to make N large and suppress all unwanted modes close to the desired mode by carefully selecting the input and output points in the TLM mesh. An alternative way of reducing the truncation error is to use a Hanning window in the Fourier transform. For further details on this, one should consult [10,30].

7.7.2 Coarseness Error

This occurs when the TLM mesh is too coarse to resolve highly nonuniform fields as can be found at corners and edges. An obvious solution is to use a finer mesh ($\Delta \ell \to 0$), but this would lead to large memory requirements and there are limits to this refinement. A better approach is to use variable mesh size so that a higher resolution can be obtained in the nonuniform field region [45]. This approach requires more complicated programming.

7.7.3 Velocity Error

This stems from the assumption that propagation velocity in the TLM mesh is the same in all directions and equal to $u_n = u/\sqrt{2}$, where u is the propagation velocity in the medium filling the structure. The assumption is only valid if the wavelength λ_n in the TLM mesh is large compared with the mesh size $\Delta\ell$ ($\Delta\ell/\lambda_n < 0.1$). Thus the cutoff frequency f_{cn} in the TLM mesh is related to the cutoff frequency f_c of the real structure according to $f_c = f_{cn}\sqrt{2}$. If $\Delta\ell$ is comparable with λ_n, the velocity of propagation depends on the direction and the assumption of constant velocity results in a velocity error in f_c. Fortunately, a measure to reduce the coarseness error takes care of the velocity error as well.

7.7.4 Misalignment Error

This error occurs in dielectric interfaces in three-dimensional inhomogeneous structures such as microstrip or fin line. It is due to the manner in which boundaries are simulated in a three-dimensional TLM mesh; dielectric interfaces appear halfway between nodes, while electric and magnetic boundaries appear across such nodes. If the resulting error is not acceptable, one must make two computations, one with recessed and one with protruding dielectric, and take the average of the results.

7.8 Concluding Remarks

This chapter has described the transmission-line-matrix (TLM) method, which is a modeling process rather than a numerical method for solving differential or integral equations. The flexibility, versatility, and generality of the time-domain method have been demonstrated. Our discussion in this chapter has been introductory, and one is advised to consult [31,32] for a more in-depth treatment. A generalized treatment of TLM in the curvilinear coordinate system is presented in [33].

Although the application of the TLM method in this chapter has been limited to diffusion and wave propagation problems, the method has a wide range of applications. The technique has been applied to other problems such as:

- cutoff frequencies in fin lines [28,29],
- transient analysis of strip lines [34],
- linear and nonlinear lumped networks [35-40],
- microstrip lines on anisotropic substrates [17,41],
- antenna problems [33,35,42,43],
- induced currents in biological bodies exposed to EM fields [44], and
- cylindrical and spherical waves [35,45,46].

Many more applications can be found, not only in EM, but also in other fields such as electrical machines [47-50], filters [51], structures [52], optics, and acoustics.

A comparison of the TLM method with the finite difference method can be interesting. While TLM provides a physical model, finite difference provides a mathematical model. According to Johns, the two methods complement each other rather than compete with each other [53].

A major advantage of the TLM method, as compared with other numerical techniques, is the ease with which complicated structures can be analyzed. The great flexibility and versatility of the method reside in the fact that the TLM mesh incorporates the properties of EM fields and their interaction with the boundaries and material media. Hence, the EM problem need not be formulated for every new structure. Thus a general-purpose program such as in [54] can be developed such that only the parameters of the structure need be entered for computation. Another advantage of using the TLM method is that certain stability properties can be deduced by inspection of the circuit. There are no problems with convergence, stability, or spurious solutions. The method is limited only by the amount of memory storage required, which depends on the complexity of the TLM mesh.

References

[1] G. Kron, "Numerical solution of ordinary and partial differential equations by means of equivalent circuits," *J. Appl. Phys.*, Vol. 16, Mar. 1945, pp. 172–186.

[2] C. H. Durney and C. C. Johnson, *Introduction to Modern Electromagnetics.* New York: McGraw-Hill, 1969, pp. 286–287.

[3] P. P. Silvester and F. L. Ferrari, *Finite Elements for Electrical Engineers.* Cambridge: Cambridge University Press, 1983, p. 24.

[4] R. H. Park, "Definition of an ideal synchronous machine and formulators for armature flux linkages," *Gen. Elect. Rev.*, vol. 31, 1928, pp. 332–334.

[5] G. Kron, "Equivalent circuit of the field equations of Maxwell," *Proc. IRE*, May 1944, pp. 289–299.

[6] G. Kron, *Equivalent Circuits of Electrical Machinery.* New York: John Wiley, 1951.

[7] N. Marcovitz and J. Schwinger, "On the reproduction of the electric and magnetic fields produced by currents and discontinuity in wave guides, I," *J. Appl. Phys.*, vol. 22, no. 6, June 1951, pp. 806–819.

[8] J. Schwinger and D. S. Saxon, *Discontinuities in Waveguides.* New York: Gordon and Breach, 1968.

[9] P. B. Johns and R. L. Beurle, "Numerical solution of 2-dimensional

scattering problems using a transmission line matrix," *Proc. IEEE*, vol. 118, no. 9, Sept. 1971, pp. 1203–1208.

[10] W. J. R. Hoefer, "The transmission-line matrix method—theory and applications," *IEEE Trans. Microwave Theory Tech.*, vol. MTT-33, no. 10, Oct. 1985, pp. 882–893.

[11] M. N. O. Sadiku and L. C. Agba, "A simple introduction to the transmission-line modeling," *IEEE Trans. Circ. Sys.*, vol. CAS- 37, no. 8, Aug. 1990, pp. 991–999.

[12] B. J. Ley, *Computer Aided Analysis and Design for Electrical Engineers.* New York: Holt, Rinehart and Winston, 1970, pp. 815–817.

[13] M. N. O. Sadiku, *Elements of Electromagnetics.* New York: Holt, Rinehart and Winston, 1989, pp. 494–611.

[14] C. R. Paul and S. A. Nasar, *Introductory Electromagnetic Fields*, 2nd ed. New York: McGraw-Hill, 1987, pp. 368–384.

[15] C. C. Wong, "Solution of the network analog of one-dimensional field equations using the ladder method," *IEEE Trans. Educ.*, vol. E-28, no. 3, Aug. 1985, pp. 176–179.

[16] C. C. Wong and W. S. Wong, "Multigrid TLM for diffusion problems," *Int. J. Num. Model.*, vol. 2, no. 2, 1989, pp. 103–111.

[17] G. E. Marike and G. Yek, "Dynamic three-dimensional T.L.M. analysis of microstrip lines on anisotropic substrate," *IEEE Trans. Micro. Theo. Tech.*, vol. MTT-33, no. 9, Sept. 1985, pp. 789–799.

[18] P. B. Johns, "Applications of the transmission-line matrix method to homogeneous waveguides of arbitrary cross-section," *Proc. IEEE.* vol. 119, no. 8, Aug. 1972, pp. 1086–1091.

[19] R. A. Waldron, *Theory of Guided Electromagnetic Waves.* London: Van Nostrand Reinhold Co., 1969, pp. 157–172.

[20] N. R. S. Simons and E. Bridges, "Method for modelling free space boundaries in TLM situations," *Elect. Lett.*, vol. 26, no. 7, March 1990, pp. 453–455.

[21] L. C. Agba, "Transmission-line-matrix modeling of inhomogeneous rectangular waveguides and cavities," M.S. thesis, Department of Electrical and Computer Engr., Florida Atlantic University, Boca Raton, Aug. 1987.

[22] P. B. Johns, "The solution of inhomogeneous waveguide problems using a transmission line matrix," *IEEE Trans. Micro. Theo. Tech.*, vol. MTT-22, no. 3, Mar. 1974, pp. 209–215.

[23] S. Akhtarzad and P. B. Johns, "Generalized elements for t.l.m. method of numerical analysis," *Proc. IEE*, vol. 122, no. 12, Dec. 1975,

pp. 1349–1352.

[24] S. Akhtarzad and P. B. Johns, "Numerical solution of lossy wave guides: T.L.M. computer programs," *Elect. Lett.*, vol. 10, no. 15, July 25, 1974, pp. 309–311.

[25] S. Akhtarzad and P. B. Johns, "Transmission line matrix solution of waveguides with wall losses," *Elect. Lett.*, vol 9. no. 15, July 1973, pp. 335–336.

[26] S. Akhtarzad and P. B. Johns, "Solution of Maxwell's equations in three space dimensions and time by the T.L.M. method of numerical analysis," *Proc. IEE*, vol. 122, no. 12, Dec. 1975, pp. 1344–1348.

[27] S. Akhtarzad and P. B. Johns, "Three-dimensional transmission-line Matrix computer analysis of microstrip resonators," *IEEE Trans. Micro. Theo. Tech.*, vol. MTT-23, no. 12, Dec. 1975, pp. 990–997.

[28] Y. C. Shih and W. J. R. Hoefer, "The accuracy of TLM analysis of finned rectangular waveguides," *IEEE Trans. Micro. Theo. Tech.*, vol. MTT-28, no. 7, July 1980, pp. 743–746.

[29] Y. C. Shih and W. J. R. Hoefer, "Dominant and second-order mode cutoff frequencies in fin lines calculated with a two-dimensional TLM program," *IEEE Trans. Micro. Theo. Tech.*, vol. MTT-28, no. 12, Dec. 1980, pp. 1443–1448.

[30] N. Yoshida, et al., "Transient analysis of two-dimensional Maxwell's equations by Bergeron's method," *Trans. IECE* (Japan), vol. J62B, June 1979, pp. 511–518.

[31] T. Itoh (ed.), *Numerical Techniques for Microwave and Millimeter-wave Passive Structures.* New York: John Wiley, 1989, pp. 496–591.

[32] P. B. Johns, "Simulation of electromagnetic wave interactions by transmission-line modeling (TLM)," *Wave Motion*, vol. 10, no. 6, 1988, pp. 597–610.

[33] A. K. Bhattacharyya and R. Garg, "Generalised transmission line model for microstrip patches," *IEE Proc.*, vol. 132, Pt. H, no. 2, April 1985, pp. 93–98.

[34] N. Yoshida and I. Fukai, "Transient analysis of a stripline having corner in three-dimensional space," *IEEE Trans. Micro. Theo. Tech.*, vol. MTT-32, no. 5, May 1984, pp. 491–498.

[35] W. R. Zimmerman, "Network analog of Maxwell's field equations in one and two dimensions," *IEEE Trans. Educ.*, vol. E-25, no. 1, Feb. 1982, pp. 4–9.

[36] P. B. Johns and M. O'Brien, "Use of the transmission-line modelling

(t.l.m.) method to solve non-linear lumped networks," *Radio Electron. Engr.*, vol. 50, no. 1/2, Jan./Feb., 1980, pp. 59–70.

[37] J. W. Bandler, et al., "Transmission-line modeling and sensitivity evaluation for lumped network simulation and design in the time domain," *J. Franklin Inst.*, vol. 304, no. 1, 1971, pp. 15–23.

[38] C. R. Brewitt-Taylor and P. B. Johns, "On the construction and numerical solution of transmission line and lumped network models of Maxwell's equations," *Inter. J. Numer. Meth. Engr.*, vol. 15, 1980, pp. 13–30.

[39] P. Saguet and W. J. R. Hoefer, "The modelling of multiaxial discontinuities in quasi-planar structures with the modified TLM method," *Int. J. Num. Model.*, vol. 1, 1988, pp. 7–17.

[40] E. M. El-Sayed and M. N. Morsy, "Analysis of microwave ovens loaded with lossy process materials using the transmission-line matrix method," *Int. J. Num. Meth. Engr.*, vol. 20, 1984, pp. 2213–2220.

[41] N. G. Alexopoulus, "Integrated-circuit structures on anisotropic substrates," *IEEE Trans. Micro. Theo. Tech.*, vol. MTT-33, no. 10, Oct. 1985, pp. 847–881.

[42] I. Palocz and N. Marcovitz, "A network-oriented approach in the teaching of electromagnetics," *IEEE Trans. Educ.*, vol. E-28, no. 3, Aug. 1985, pp. 150–154.

[43] H. Pues and A. Van de Capelle, "Accurate transmission-line model for the rectangular microstrip antenna," *IEE Proc.*, vol. 131, Pt. H, no. 6, Dec. 1984, pp. 334–340.

[44] J. F. Deford and O. P. Gandhi, "An impedance method to calculate currents induced in biological bodies exposed to quasi-static electromagnetic fields," *IEEE Trans. Elect. Comp.*, vol. EMC-27, no. 3, Aug. 1985, pp. 168–173.

[45] D. A. Al-Mukhtar and T. E. Sitch, "Transmission-line matrix method with irregularly graded space," *IEE Proc.*, vol. 128, Pt. H, no. 6, Dec. 1981, pp. 299–305.

[46] H. L. Thal, "Exact circuit analysis of spherical waves," *IEEE Trans. Ant. Prog.*, vol. AP-26, no. 2, Mar. 1978, pp. 282–287.

[47] C. V. Jones and D. L. Prior, "Unification of fields and circuit theories of electrical machines," *Proc. IEE*, vol. 119, no. 7, July 1972, pp. 871–876.

[48] P. Hammond and G. J. Rogers, "Use of equivalent circuits in electrical-machine studies," *Proc. IEE*, vol. 121, no. 6, June 1974, pp. 500–507.

[49] E. M. Freeman, "Equivalent circuits from electromagnetic theory: low-frequency induction devices," *Proc. IEE*, vol. 121, no. 10, Oct. 1974, pp. 1117–1121.

[50] W. J. Karplus and W. W. Soroka, *Analog Methods: Computation and Simulation*. New York: McGraw-Hill, 1959.

[51] G. L. Ragan (ed.), *Microwave Transmission Circuits*. New York: McGraw-Hill, 1948, pp. 544–547.

[52] R. H. MacNeal, *Electric Circuit Analogies for Elastic Structures*, vol. 2. New York: John Wiley & Sons, 1962.

[53] P. B. Johns, "On the relationship between TLM and finite-difference methods for Maxwell's equations," *IEEE Trans. Micro. Theo. Tech.*, vol. MTT-35, no. 1, Jan. 1987, pp. 60, 61.

[54] S. Akhtarzad, "Analysis of lossy microwave structures and microstrip resonators by the TLM method," Ph.D. thesis, University of Nottingham, England, July 1975.

Problems

7.1 Verify Eq. (7.16).

7.2 For the two-port network in Fig. 7.30(a), the relation between the input and output variables can be written in matrix form as

$$\begin{bmatrix} V_1 \\ I_1 \end{bmatrix} = \begin{bmatrix} A & B \\ C & D \end{bmatrix} \begin{bmatrix} V_2 \\ -I_2 \end{bmatrix}.$$

For the lossy line in Fig. 7.30(b), show that the ABCD matrix (also called the cascaded matrix) is

$$\begin{bmatrix} \cosh \gamma \ell & Z_o \sinh \gamma \ell \\ \dfrac{1}{Z_0} \sinh \gamma \ell & \cosh \gamma \ell \end{bmatrix}.$$

7.3 Consider an EM wave propagation in a lossless medium in TEM mode ($E_y = 0 = E_z = H_z = H_x$) along the z direction. Using one-dimensional TLM mesh, derive the equivalencies between network and field quantities.

Transmission-line-matrix Method 563

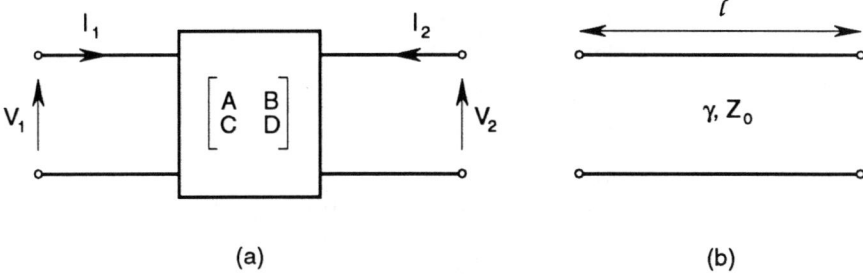

Figure 7.30 For Problem 7.2.

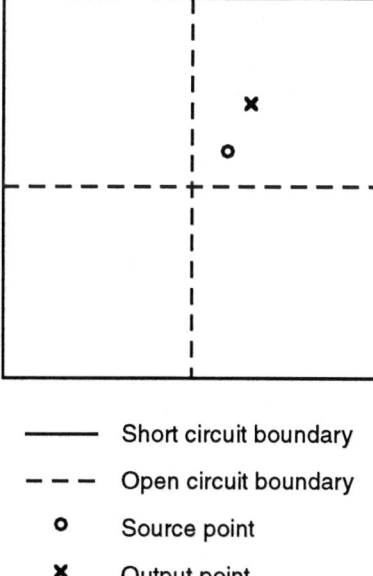

——— Short circuit boundary
- - - Open circuit boundary
○ Source point
× Output point

Figure 7.31 Square cross section waveguide of Problem 7.4.

564 Numerical Techniques in Electromagnetics

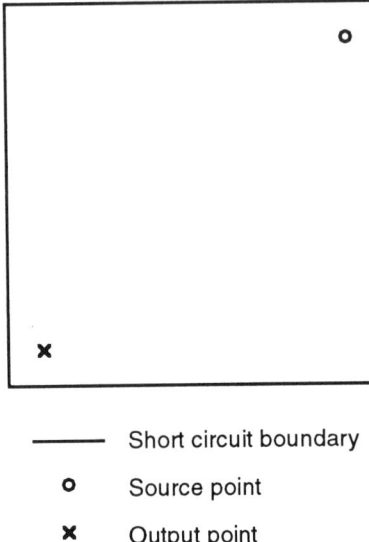

Figure 7.32 Square cross section waveguide of Problem 7.5.

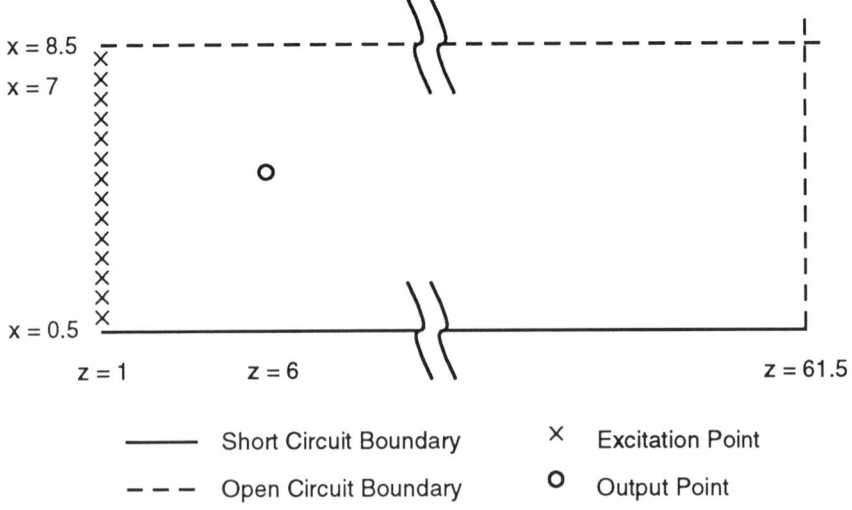

Figure 7.33 The 61 x 8 TLM mesh of Problem 7.11.

7.4 Modify the program in Fig. 7.14 to calculate the cutoff frequency (expressed in terms of $\Delta\ell/\lambda$) in a square section waveguide of size $10\Delta\ell$. Perform the calculation for the TM_{11} mode by using open-circuit symmetry boundaries to suppress even-order modes and by taking the excitation and output points as in Fig. 7.31 to suppress the TM_{13}, TM_{33}, and TM_{15} modes. Use $N = 500$.

7.5 Repeat Prob. 7.4 of higher-order modes but take excitation and output points as in Fig. 7.32.

7.6 For the waveguide with a free space discontinuity considered in Example 7.2, plot the variation of the magnitude of the normalized impedance of the guide with $\Delta\ell/\lambda$. The plot should be for frequencies above and below the cutoff frequency, i.e., including both evanescent and propagating modes.

7.7 Rework Example 7.4, but take the output point at $(x = 6, z = 13)$.

7.8 Verify Eq. (7.62).

7.9 For transverse waves on a stub-loaded transmission-line matrix, the dispersion relation is given by

$$\sin^2(\frac{\beta_n \Delta\ell}{2}) = 2(1 + Y_o/4)\sin^2(\frac{\omega \Delta\ell}{2c}).$$

Plot the velocity characteristic similar to that in Fig. 7.11 for $Y_o = 0$, 1, 2, 10, 20, 100.

7.10 Verify Eq. (7.68).

7.11 Consider the 61 x 8 rectangular matrix with boundaries at $x = 0.5$ and $x = 8.5$ as in Fig. 7.33. By making one of the boundaries, say $x = 8.5$, an open circuit, a waveguide of twice the width can be simulated. For the TE_{m0} family of modes, excite the system at all points on line $z = 1$ with impulses corresponding to E_y and take the impulse function of the output at point $x = 7$, $z = 6$. Calculate the normalized wave impedance $Z = E_y/H_x$ for frequencies above cutoff, i.e., $\Delta\ell/\lambda = 0.023, 0.025, 0.027, \cdots, 0.041$. Take $\sigma = 0$, $\epsilon_r = 2$, $\mu_r = 1$.

7.12 Repeat Prob. 7.11 for a lossy waveguide with $\sigma = 278$ mhos/m, $\epsilon_r = 1$, $\mu_r = 1$.

566 Numerical Techniques in Electromagnetics

7.13 Using the TLM method, determine the cutoff frequency (expressed in terms of $\Delta\ell/\lambda$) of the lowest order TE and TM modes for the square waveguide with cross section shown in Fig. 7.34. Take $\epsilon_r = 2.45$.

7.14 For the dielectric ridge waveguide of Fig. 7.35, use the TLM method to calculate the cutoff wavenumber k_c of the dominant mode. Express the result in terms of $k_c a$ ($= \omega a/c$) and try $\epsilon_r = 2$ and $\epsilon_r = 8$. Take $a = 10\Delta\ell$.

7.15 Rework Example 7.6 for the inhomogeneous cavity of Fig. 7.36. Take $\epsilon_r = 16$, $a = 12\Delta\ell$, $b = 3a/10$, $d = 4a/10$, $s = 7a/12$, $u = 3d/8$.

7.16 Consider a single microstrip line whose cross section is shown in Fig. 7.37. Dispersion analysis of the line by the TLM method involves resonating a section of the transmission line by placing shorting planes along the axis of propagation (the z-axis in this case). Write a TLM computer program and specify the input data as:

$$E_x = 0 = E_z \quad \text{along} \quad y = 0, \; y = b,$$
$$E_x = 0 = E_z \quad \text{along} \quad x = 2a,$$
$$E_x = 0 = E_z \quad \text{for} \quad y = H, \; -W \leq x \leq W,$$
$$H_y = 0 = H_z \quad \text{along} \quad x = 0.$$

Plot the dispersion curves depicting the phase constant β as a function of frequency f for cases when the line is air-filled and dielectric filled. The distance L ($= \pi/\beta$) between the shorting planes is the variable. Assume the dielectric substrate and the walls of the enclosure are lossless. Take $\epsilon_r = 4.0$, $a = 2$ mm, $H = 1.0$ mm, $W = 1.0$ mm, $b = 2$ mm, $\Delta\ell = a/8$.

7.17 For the cubical cavity of Fig. 7.38, use the TLM technique to calculate the time taken for the total power in the lossy dielectric cavity to decay to $1/e$ of its original value. Consider cases when the cavity is completely filled with dielectric material and half-filled. Take $\epsilon_r = 2.45$, $\sigma = 0.884$ mhos/m, $\mu_r = 1$, $\Delta\ell = 0.3$ cm, $2a = 7\Delta\ell$.

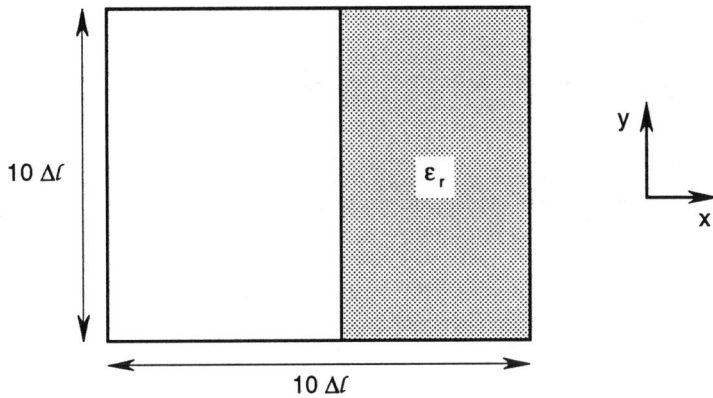

Figure 7.34 For Problem 7.13.

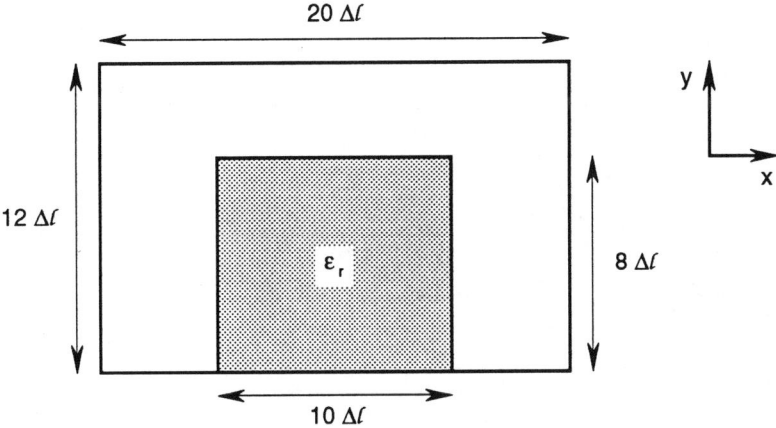

Figure 7.35 For Problem 7.14.

568 Numerical Techniques in Electromagnetics

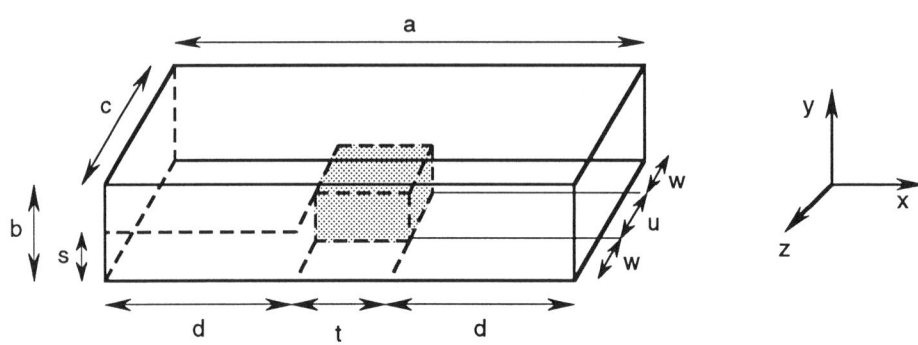

Figure 7.36 The inhomogeneous cavity of Problem 7.15

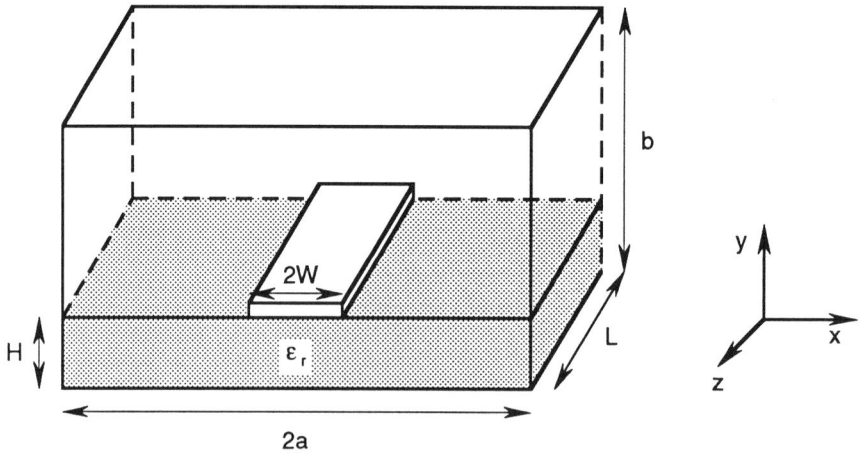

Figure 7.37 The microstrip line of Problem 7.16.

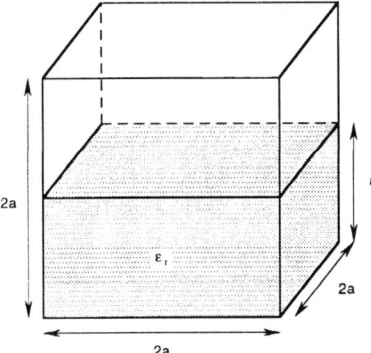

Figure 7.38 The lossy cavity of Problem 7.17.

Chapter 8
Monte Carlo Methods

> "One thing I have learned in a long life: that all our science, measured against reality, is primitive and childlike—and yet is the most precious thing we have." —Albert Einstein

8.1 Introduction

Unlike the deterministic numerical methods covered in the foregoing chapters, Monte Carlo methods are nondeterministic (probabilistic or stochastic) numerical methods employed in solving mathematical and physical problems. The Monte Carlo method (MCM), also known as the *method of statistical trials*, is the marriage of two major branches of theoretical physics: the probabilistic theory of random process dealing with Brownian motion or random-walk experiments and potential theory, which studies the equilibrium states of a homogeneous medium [1]. It is a method of approximately solving problems using sequences of random numbers. It is a means of treating mathematical problems by finding a probabilistic analog and then obtaining approximate answers to this analog by some experimental sampling procedure. The solution of a problem by this method is closer in spirit to physical experiments than to classical numerical techniques.

It is generally accepted that the development of Monte Carlo techniques as we presently use them dates from about 1944, although there are a number of undeveloped instances on much earlier occasions. Credit for the development of MCM goes to a group of scientists, particularly von Neumann and Ulam, at Los Alamos during the early work on nuclear weapons. The groundwork of the Los Alamos group stimulated a vast outpouring of literature on the subject and encouraged the use of MCM for a variety of problems [2–4]. The name "Monte Carlo" comes from the city in Monaco, famous for its gambling casinos.

Monte Carlo methods are applied in two ways: simulation and sampling. Simulation refers to methods of providing mathematical imitation of real random phenomena. A typical example is the simulation of a neutron's motion into a reactor wall, its zigzag path being imitated by a random walk. Sampling refers to methods of deducing properties of a large set of elements by studying only a small, random subset. For example, the average value of $f(x)$ over $a < x < b$ can be estimated from its average over a finite number of points selected randomly in the interval. This amounts to a Monte Carlo method of numerical integration. MCMs have been applied successfully for solving differential and integral equations, for finding eigenvalues, for inverting matrices, and particularly for evaluating multiple integrals.

The simulation of any process or system in which there are inherently random components requires a method of generating or obtaining numbers that are random. Examples of such simulation occur in random collisions of neutrons, in statistics, in queueing models, in games of strategy, and in other competitive enterprises. Monte Carlo calculations require having available sequences of numbers which appear to be drawn at random from particular probability distributions.

8.2 Generation of Random Numbers and Variables

Various techniques for generating random numbers are discussed fully in [5-10]. The almost universally used method of generating random numbers is to select a function $g(x)$ that maps integers into random numbers. Select x_0 somehow, and generate the next random number as $x_{k+1} = g(x_k)$. The commonest function $g(x)$ takes the form

$$g(x) = (ax + c) \mod m \tag{8.1}$$

where

$$\begin{aligned}
x_0 &= \text{starting value or a seed } (x_0 > 0), \\
a &= \text{multiplier } (a \geq 0), \\
c &= \text{increment } (c \geq 0), \\
m &= \text{the modulus.}
\end{aligned}$$

The modulus m is usually 2^t for t-digit binary integers. For a 31-bit computer machine, for example, m may be 2^{31-1}. Here x_0, a, and c are integers in the same range as $m > a$, $m > c$, $m > x_0$. The desired sequence of random numbers $\{x_n\}$ is obtained from

$$x_{n+1} = (ax_n + c) \bmod m \tag{8.2}$$

This is called a *linear congruential sequence*. For example, if $x_0 = a = c = 7$ and $m = 10$, the sequence is

$$7, 6, 9, 0, 7, 6, 9, 0, ... \tag{8.3}$$

It is evident that congruential sequences always get into a loop; i.e., there is ultimately a cycle of numbers that is repeated endlessly. The sequence in Eq. (8.3) has a period of length 4. A useful sequence will of course have a relatively long period. The terms *multiplicative congruential method* and *mixed congruential method* are used by many authors to denote linear congruential methods with $c = 0$ and $c \neq 0$, respectively. Rules for selecting $x_0, a, c,$ and m can be found in [6,10].

Here we are interested in generating random numbers from the uniform distribution in the interval (0,1). These numbers will be designated by the letter U and are obtained from Eq. (8.2) as

$$U = \frac{x_{n+1}}{m}. \tag{8.4}$$

Thus U can only assume values from the set $\{0, 1/m, 2/m, ..., (m-1)/m\}$. For generating random numbers X uniformly distributed in the interval (a, b), we use

$$X = a + (b - a)\, U \tag{8.5}$$

For random numbers in the interval (0,1), a quick test of the randomness is that the mean is 0.5. Other tests can be found in [3,6].

Random numbers produced by a computer code (using Eqs. (8.2) and (8.4)) are not truly random; in fact, given the seed of the sequence, all numbers U of the sequence are completely predictable. Some authors emphasize this point by calling such computer-generated sequences *pseudorandom numbers*. However, with a good choice of $a, c,$ and m, the sequences of U appear to be sufficiently random in that they pass a series of statistical tests of randomness. They have the advantage over truly random numbers of being generated in a fast way and of being reproducible, when desired, especially for program debugging.

It is usually necessary in a Monte Carlo procedure to generate random variable X from a given probability distribution $F(x)$. This can be accomplished using several techniques [6,13–15] including the *direct method* and *rejection method*.

The direct method, otherwise known as inversion or transform method, entails inverting the cumulative probability function $F(x) = \text{Prob}(X \leq x)$ associated with the random variable X. The fact that $0 \leq F(x) \leq 1$ intuitively suggests that by generating random number U uniformly distributed over $(0,1)$, we can produce a random sample X from the distribution of $F(x)$ by inversion. Thus to generate random X with probability distribution $F(x)$, we set $U = F(x)$ and obtain

$$X = F^{-1}(U) \tag{8.6}$$

where X has the distribution function $F(x)$. For example, if X is a random variable that is exponentially distributed with mean μ, then

$$F(x) = 1 - e^{-x/\mu}, \quad 0 < x < \infty. \tag{8.7}$$

Solving for X in $U = F(X)$ gives

$$X = -\mu \ln(1 - U). \tag{8.8}$$

Since $(1 - U)$ is itself a random number in the interval $(0,1)$, we simply write

$$X = -\mu \ln U. \tag{8.9}$$

Sometimes the inverse $F^{-1}(x)$ required in Eq. (8.6) does not exist or is difficult to obtain. This situation can be handled using the rejection method. Let $f(x) = \dfrac{dF(x)}{dx}$ be the probability density function of the random variable X. Let $f(x) = 0$ for $a > x > b$, and $f(x)$ is bounded by M (i.e., $f(x) \leq M$) as shown in Fig. 8.1. We generate two random numbers (U_1, U_2) in the interval $(0,1)$. Then

$$X_1 = a + (b-a)U_1 \text{ and } f_1 = U_2 M \tag{8.10}$$

are two random numbers with uniform distributions in (a,b) and $(0, M)$, respectively. If

$$f_1 \leq f(X_1) \tag{8.11}$$

then X_1 is accepted as choice of X, otherwise X_1 is rejected and a new pair (U_1, U_2) is tried again. Thus in the rejection technique all points falling above $f(x)$ are rejected, while those points falling on or below $f(x)$ are utilized to generate X_1 through $X_1 = a + (b-a)U_1$.

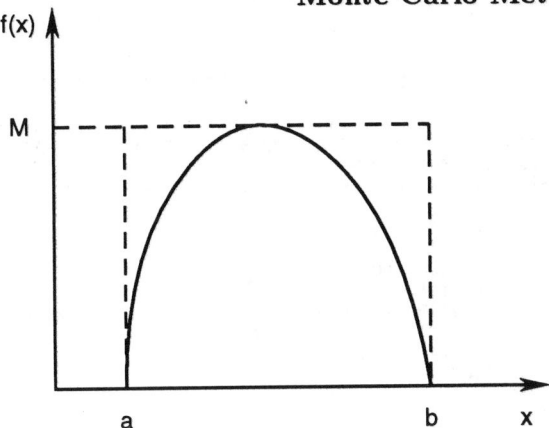

Figure 8.1 The rejection method of generating a random variable with probability density function $f(x)$.

Example 8.1

Develop a subroutine for generating random number U uniformly distributed between 0 and 1. Using this subroutine, generate random variable Θ with probability distribution given by

$$T(\theta) = \frac{1}{2}(1 - \cos \theta), \quad 0 < \theta < \pi.$$

Solution

The subroutine for generating U is shown in Fig. 8.2. In this subroutine, $m = 2^{21} - 1 = 2147483647, c = 0,$ and $a = 7^5 = 16807$. By supplying a seed (e.g., 1234), the subroutine provides one random number U per call in the main program. The seed is selected as any integer between 1 and m.

The subroutine in Fig. 8.2 is meant to illustrate the concepts explained in this section. Most computers have routines for generating random numbers.

To generate the random variable Θ, set

$$U = T(\Theta) = \frac{1}{2}(1 - \cos \Theta),$$

then

$$\Theta = T^{-1}(U) = \cos^{-1}(1 - 2U).$$

Using this, a sequence of random numbers Θ with the given distribution is generated in the main program of Fig. 8.2.

```
0001     ***************************************************
0002     C  PROGRAM FOR GENERATING RANDOM VARIABLES WITH A GIVEN
0003     C  PROBABILITY DISTRIBUTION
0004     C***************************************************
0005
0006           DOUBLE PRECISION ISEED
0007
0008           ISEED = 1234.D0
0009           DO 10 I=1,100
0010           CALL RANDOM (ISEED,R)
0011           THETA = ACOSD(1.0 - 2.0*R)
0012           WRITE(6,*) I,THETA
0013     10    CONTINUE
0014           STOP
0015           END

0001     C++++++++++++++++++++++++++++++++++++++++++++++++++
0002     C  SUBROUTINE FOR GENERATING RANDOM NUMBERS IN
0003     C  THE INTERVAL (0,1)
0004     C++++++++++++++++++++++++++++++++++++++++++++++++++
0005
0006           SUBROUTINE RANDOM (ISEED,R)
0007           DOUBLE PRECISION  ISEED, DEL, A
0008           DATA DEL,A/2147483647.D0, 16807.D0/
0009
0010           ISEED = DMOD( A*ISEED, DEL )
0011           R = ISEED/DEL
0012           RETURN
0013           END
```

Figure 8.2 Random number generator; for Example 8.1.

8.3 Evaluation of Error

Monte Carlo procedures give solutions which are averages over a number of tests. For this reason, the solutions contain fluctuations about a mean value, and it is impossible to ascribe a 100% confidence in the results. To evaluate the statistical uncertainty in Monte Carlo calculations, we must resort to various statistical techniques associated with random variables. We briefly introduce the concepts of expected value and variance, and utilize the central limit theorem to arrive at an error estimate [13,16].

Suppose that X is a random variable. The *expected value* or *mean value* $\bar{x}$ of X is defined as

$$\bar{x} = \int_{-\infty}^{\infty} x f(x) dx \qquad (8.12)$$

where $f(x)$ is the probability density distribution of X. If we draw random and independent samples, $x_1, x_2, ..., x_N$ from $f(x)$, our estimate of x would take the form of the mean of N samples, namely,

$$\hat{x} = \frac{1}{N} \sum_{n=1}^{N} x_n. \tag{8.13}$$

While $\bar{x}$ is the true mean value of X, $\hat{x}$ is the unbiased estimator of $\bar{x}$, an unbiased estimator being one with the correct expectation value. Although the expected value of $\hat{x}$ is equal to $\bar{x}$, $\hat{x} \neq \bar{x}$. Therefore, we need a measure of the spread in the values of $\hat{x}$ about $\bar{x}$.

To estimate the spread of values of X, and eventually of $\hat{x}$ about $\bar{x}$, we introduce the *variance* of X defined as the expected value of the square of the deviation of X from $\bar{x}$, i.e.,

$$\text{Var}(x) = \sigma^2 = \overline{(x-\bar{x})^2} = \int_{-\infty}^{\infty} (x-\bar{x})^2 f(x) dx. \tag{8.14}$$

But $(x - \bar{x})^2 = x^2 - 2x\bar{x} + \bar{x}^2$. Hence

$$\sigma^2(x) = \int_{-\infty}^{\infty} x^2 f(x) dx - 2\bar{x} \int_{-\infty}^{\infty} x f(x) dx + \bar{x}^2 \int_{-\infty}^{\infty} f(x) dx \tag{8.15}$$

or

$$\sigma^2(x) = \overline{x^2} - \bar{x}^2. \tag{8.16}$$

The square root of the variance is called the *standard deviation*, i.e.,

$$\sigma(x) = (\overline{x^2} - \bar{x}^2)^{1/2}. \tag{8.17}$$

The standard deviation provides a measure of the spread of x about the mean value $\bar{x}$; it yields the order of magnitude of the error. The relationship between the variance of $\hat{x}$ and the variance of x is

$$\sigma(\hat{x}) = \frac{\sigma(x)}{\sqrt{N}}. \tag{8.18}$$

This shows that if we use $\hat{x}$ constructed from N values of x_n according to Eq. (8.13) to estimate $\bar{x}$, then the spread in our results of $\hat{x}$ about $\bar{x}$ is proportional to $\sigma(x)$ and falls off as the number of N of samples increases.

In order to estimate the spread in $\hat{x}$, we define the *sample variance*

$$S^2 = \frac{1}{N-1} \sum_{n=1}^{N} (x_n - \hat{x})^2. \tag{8.19}$$

Again, it can be shown that the expected value of S^2 is equal to $\sigma^2(x)$. Therefore the sample variance is an unbiased estimator of $\sigma^2(x)$. Multiplying out the square term in Eq. (8.19), it is readily shown that the *sample standard deviation* is

$$S = \left(\frac{N}{N-1}\right)^{1/2} \left[\frac{1}{N}\sum_{n=1}^{N} x_n^2 - \hat{x}^2\right]^{1/2}. \qquad (8.20)$$

For large N, the factor $N/(N-1)$ is set equal to one.

As a way of arriving at the *central limit theorem*, a fundamental result in probability theory, consider the binomial function

$$B(M) = \frac{N!}{M!(N-M)!} p^M q^{N-M} \qquad (8.21)$$

which is the probability of M successes in N independent trials. In Eq. (8.21), p is the probability of success in a trial, and $q = 1 - p$. If M and $N - M$ are large, we may use *Stirling's formula*

$$n! \sim n^n e^{-n} \sqrt{2\pi n} \qquad (8.22)$$

so that Eq. (8.21) is approximated [17] as the normal distribution:

$$B(M) \simeq f(\hat{x}) = \frac{1}{\sigma(\hat{x})\sqrt{2\pi}} \exp\left[-\frac{(\hat{x} - \bar{x})^2}{2\sigma^2(\hat{x})}\right] \qquad (8.23)$$

where $\bar{x} = Np$ and $\sigma(\hat{x}) = \sqrt{Npq}$. Thus as $N \to \infty$, the central limit theorem states that the probability density function which describes the distribution of $\hat{x}$ that results from N Monte Carlo calculations is the normal distribution $f(\hat{x})$ in Eq. (8.23). In other words, the sum of a large number of random variables tends to be normally distributed. Inserting Eq. (8.18) into Eq. (8.23) gives

$$f(\hat{x}) = \sqrt{\frac{N}{2\pi}} \frac{1}{\sigma(x)} \exp\left[-\frac{N(\hat{x} - \bar{x})^2}{2\sigma^2(x)}\right]. \qquad (8.24)$$

The normal (or Gaussian) distribution is very useful in various problems in engineering, physics, and statistics. The remarkable versatility of the Gaussian model stems from the central limit theorem. For this reason, the Gaussian model often applies to situations in which the quantity of interest results from the summation of many irregular and fluctuating components.

Monte Carlo Methods 579

In Example 8.2, we present an algorithm based on central limit theorem for generating Gaussian random variables.

Since the number of samples N is finite, absolute certainty in Monte Carlo calculations is unattainable. We try to estimate some limit or interval around $\bar{x}$ such that we can predict with some confidence that $\hat{x}$ falls within that limit. Suppose we want the probability that $\hat{x}$ lies between $\bar{x} - \delta$ and $\bar{x} + \delta$. By definition,

$$\text{Prob } \{\bar{x} - \delta < \hat{x} < \bar{x} + \delta\} = \int_{\bar{x}-\delta}^{\bar{x}+\delta} f(\hat{x}) d\hat{x}. \tag{8.25}$$

By letting $\lambda = \dfrac{(\hat{x} - \bar{x})}{\sqrt{2/N}\,\sigma(x)}$,

$$\text{Prob}\{\bar{x} - \delta < \hat{x} < \bar{x} + \delta\} = \frac{2}{\sqrt{\pi}} \int_0^{(\sqrt{N/2})(\delta/\sigma)} e^{-\lambda^2} d\lambda$$

$$= \text{erf}\left(\sqrt{N/2}\,\frac{\delta}{\sigma(x)}\right) \tag{8.26a}$$

or

$$\text{Prob}\{\bar{x} - z_{\alpha/2}\frac{\sigma}{\sqrt{N}} \leq \hat{x} \leq \bar{x} + z_{\alpha/2}\frac{\sigma}{\sqrt{N}}\} = 1 - \alpha \tag{8.26b}$$

where erf(x) is the error function and $z_{\alpha/2}$ is the upper $\alpha/2 \times 100$ percentile of the standard normal deviation. Equation (8.26) may be interpreted as follows: if the Monte Carlo procedure of taking random and independent observations and constructing the associated random interval $\bar{x} \pm \delta$ is repeated for large N, approximately $\text{erf}\left(\sqrt{\frac{N}{2}}\frac{\delta}{\sigma(x)}\right) \times 100$ percent of these random intervals will contain $\hat{x}$. The random interval $\hat{x} \pm \delta$ is called a *confidence interval* and $\text{erf}\left(\sqrt{\frac{N}{2}}\frac{\delta}{\sigma(x)}\right)$ is the *confidence level*. Most Monte Carlo calculations use error $\delta = \sigma(x)/\sqrt{N}$, which implies that $\hat{x}$ is within one standard deviation of $\bar{x}$, the true mean. From Eq. (8.26), it means that the probability that the sample mean $\hat{x}$ lies within the interval $\hat{x} \pm \sigma(x)/\sqrt{N}$ is 0.6826 or 68.3%. If higher confidence levels are desired, two or three standard deviations may be used. For example,

$$\text{Prob}\left(\bar{x} - M\frac{\sigma(x)}{\sqrt{N}} < \hat{x} < \bar{x} + M\frac{\sigma(x)}{\sqrt{N}}\right) = \begin{cases} 0.6826, & M = 1 \\ 0.954, & M = 2 \\ 0.997, & M = 3 \end{cases} \tag{8.27}$$

where M is the number of standard deviations.

580   Numerical Techniques in Electromagnetics

In Eqs. (8.26) and (8.27), it is assumed that the population standard deviation σ is known. Since this is rarely the case, σ must be estimated by the sample standard deviation S calculated from Eq. (8.20) so that the normal distribution is replaced by the student's t-distribution. It is well known that the t-distribution approaches the normal distribution as N becomes large, say $N > 30$. Equation (8.26) is equivalent to

$$\text{Prob}\{\bar{x} - \frac{S\, t_{\alpha/2;N-1}}{\sqrt{N}} \leq \hat{x} \leq \bar{x} + \frac{S\, t_{\alpha/2;N-1}}{\sqrt{N}}\} = 1 - \alpha \tag{8.28}$$

where $t_{\alpha/2;N-1}$ is the upper $100 \times \alpha/2$ percentage point of the student's t-distribution with $(N-1)$ degrees of freedom; and its values are listed in any standard statistics text. Thus the upper and lower limits of a confidence interval are given by

$$\boxed{\begin{aligned} \text{upper limit} &= \bar{x} + \frac{S\, t_{\alpha/2;N-1}}{\sqrt{N}} \\ \text{lower limit} &= \bar{x} - \frac{S\, t_{\alpha/2;N-1}}{\sqrt{N}} \end{aligned}} \quad \begin{aligned} (8.29) \\ (8.30) \end{aligned}$$

For further discussion on error estimates in Monte Carlo computations, consult [18,19].

Example 8.2

A random variable X with Gaussian (or normal) distribution is generated using the central limit theorem. According to the central limit theorem, the sum of a large number of independent random variables about a mean value approaches a Gaussian distribution regardless of the distribution of the individual variables. In other words, for any random numbers, $Y_i, i = 1, 2, ..., N$ with mean $\overline{Y}$ and variance $\text{Var}(Y)$,

$$Z = \frac{\sum_{i=1}^{N} Y_i - N\overline{Y}}{\sqrt{N\, \text{Var}(Y)}} \tag{8.2.1}$$

converges asymptotically with N to a normal distribution with zero mean and a standard deviation of unity. If Y_i are uniformly distributed variables (i.e., $Y_i = U_i$), then $\overline{Y} = 1/2$, $\text{Var}(Y) = 1/\sqrt{12}$, and

$$Z = \frac{\sum_{i=1}^{N} U_i - N/2}{\sqrt{N/12}} \tag{8.2.2}$$

and the variable
$$X = \sigma Z + \mu \tag{8.2.3}$$
approximates the normal variable with mean μ and variance σ^2. A value of N as low as 3 provides a close approximation to the familiar bell-shaped Gaussian distribution. To ease computation, it is a common practice to set $N = 12$ since this choice eliminates the square root term in Eq. (8.2.2). However, this value of N truncates the distribution at $\pm 6\sigma$ limits and is unable to generate values beyond 3σ. For simulations in which the tail of the distribution is important, other schemes for generating Gaussian distribution must be used [20–22].

Thus, to generate a Gaussian variable X with mean μ and standard deviation σ, we follow these steps:

(1) Generate 12 uniformly distributed random numbers $U_1, U_2, \ldots, U_{12}$.

(2) Obtain $Z = \sum_{i=1}^{12} U_i - 6$.

(3) Set $X = \sigma Z + \mu$.

8.4 Numerical Integration

For one-dimensional integration, several quadrature formulas such as presented in Section 3.10 exist. The number of such formulas is relatively few for multidimensional integration. It is for such multidimensional integrals that a Monte Carlo technique becomes valuable for at least two reasons. The quadrature formulas become very complex for multiple integrals, while the MCM remains almost unchanged. The convergence of Monte Carlo integration is independent of dimensionality, which is not true for quadrature formulas. The statistical method of integration has been found to be an efficient way to evaluate two- or three-dimensional integrals in antenna problems, particularly those involving very large structures [23]. Two types of Monte Carlo integration procedures, the crude MCM and the MCM with antithetic variates, will be discussed. For other types, such as hit-or-miss and control variates, see [24–26]. Application of MCM to improper integrals will be covered briefly.

8.4.1 Crude Monte Carlo Integration

Suppose we wish to evaluate the integral

$$I = \int_R f \tag{8.31}$$

where R is an n-dimensional space. Let $\mathbf{X} = (X^1, X^2, ..., X^n)$ be a random variable that is uniformly distributed in R. Then $f(\mathbf{X})$ is a random variable whose mean value is given by [27,28]

$$\overline{f(\mathbf{X})} = \frac{1}{|R|} \int_R f = \frac{I}{|R|} \tag{8.32}$$

and the variance by

$$\text{Var}(f(\mathbf{X})) = \frac{1}{|R|} \int_R f^2 - \left(\frac{1}{|R|} \int_R f\right)^2 \tag{8.33}$$

where

$$|R| = \int_R d\mathbf{X}. \tag{8.34}$$

If we take N independent samples of $\mathbf{X}$, i.e., $\mathbf{X}_1, \mathbf{X}_2, ... \mathbf{X}_N$, all having the same distribution as $\mathbf{X}$ and form the average

$$\frac{f(\mathbf{X}_1) + f(\mathbf{X}_2) + ... + f(\mathbf{X}_N)}{N} = \frac{1}{N} \sum_{i=1}^{N} f(\mathbf{X}_i) \tag{8.35}$$

we might expect this average to be close to the mean of $f(\mathbf{X})$. Thus, from Eqs. (8.32) and (8.35),

$$\boxed{I = \frac{|R|}{N} \sum_{i=1}^{N} f(\mathbf{X}_i)} \tag{8.36}$$

This Monte Carlo formula applies to any integration over a finite region R. For the purpose of illustration, we now apply Eq. (8.36) to one- and two-dimensional integrals.

For a one-dimensional integral, suppose

$$I = \int_a^b f(x) dx. \tag{8.37}$$

Applying Eq. (8.36) yields

$$\boxed{I = \frac{b-a}{N} \sum_{i=1}^{N} f(X_i)} \qquad (8.38)$$

where x_i is a random number in the interval (a,b), i.e.,

$$X_i = a + (b-a)\, U, \quad 0 < U < 1. \qquad (8.39)$$

For a two-dimensional integral

$$I = \int_a^b \int_c^d f(X^1, X^2)\, dX^1\, dX^2, \qquad (8.40)$$

the corresponding Monte Carlo formula is

$$\boxed{I = \frac{(b-a)(d-c)}{N} \sum_{i=1}^{N} f(X_i^1, X_i^2)} \qquad (8.41)$$

where
$$\begin{aligned} X_i^1 &= a + (b-a)\, U^1, \quad 0 < U^1 < 1 \\ X_i^2 &= c + (d-c)\, U^2, \quad 0 < U^2 < 1. \end{aligned} \qquad (8.42)$$

The convergence behavior of the unbiased estimator I in Eq. (8.36) is slow since the variance of the estimator is of the order $1/N$. Accuracy and convergence is increased by reducing the variance of the estimator using an improved method, the method of antithetic variates.

8.4.2 Monte Carlo Integration with Antithetic Variates

The term *antithetic variates* [29,30] is used to describe any set of estimators which mutually compensate each other's variations. For convenience, we assume that the integral is over the interval $(0,1)$. Suppose we want an estimator for the single integral

$$I = \int_0^1 g(U)\, dU. \qquad (8.43)$$

We expect the quantity $\frac{1}{2}[g(U) + g(1-U)]$ to have smaller variance than $g(U)$. If $g(U)$ is too small, then $g(1-U)$ will have a good chance of being too large and conversely. Therefore, we define the estimator

$$I = \frac{1}{N} \sum_{i=1}^{N} \frac{1}{2}[g(U_i) + g(1-U_i)] \qquad (8.44)$$

where U_i are random numbers between 0 and 1. The variance of the estimator is of the order $\frac{1}{N^4}$, a tremendous improvement over Eq. (8.36). For two-dimensional integral,

$$I = \int_0^1 \int_0^1 g(U^1, U^2) \, dU^1 \, dU^2, \tag{8.45}$$

and the corresponding estimator is

$$I = \frac{1}{N} \sum_{i=1}^{N} \frac{1}{4}[g(U_i^1, U_i^2) + g(U_i^1, 1 - U_i^2) + g(1 - U_i^1, U_i^2) + g(1 - U_i^1, 1 - U_i^2)]. \tag{8.46}$$

Following similar lines, the idea can be extended to higher order integrals. For intervals other than (0,1), transformations such as in Eqs. (8.38) to (8.42) should be applied. For example,

$$\int_a^b f(x) \, dx = (b-a) \int_0^1 g(U) \, dU$$

$$\simeq \frac{b-a}{N} \sum_{i=1}^{N} \frac{1}{2}[g(U_i) + g(1 - U_i)] \tag{8.47}$$

where $g(U) = f(X)$ and $X = a + (b-a)U$. It is observed from Eqs. (8.44) and (8.46) that as the number of dimensions increases, the minimum number of antithetic variates per dimension required to obtain an increase in efficiency over crude Monte Carlo also increases. Thus the crude Monte Carlo method becomes preferable in many dimensions.

8.4.3 Improper Integrals

The integral

$$I = \int_0^\infty g(x) \, dx \tag{8.48}$$

may be evaluated using Monte Carlo simulations [31]. For a random variable X having probability density function $f(x)$, where $f(x)$ integrates to 1 on interval $(0, \infty)$,

$$\int_0^\infty \frac{g(x)}{f(x)} \, dx = \int_0^\infty g(x) \, dx. \tag{8.49}$$

Hence, to compute I in Eq. (8.48), we generate N independent random variables distributed according to a probability density function $f(x)$ integrating to 1 on the interval $(0, \infty)$. The sample mean

$$\overline{g(x)} = \frac{1}{N} \sum_{i=1}^{N} \frac{g(x_i)}{f(x_i)} \tag{8.50}$$

gives an estimate for I.

Example 8.3

Evaluate the integral

$$I = \int_0^1 \int_0^{2\pi} e^{j\alpha\rho\cos\phi}\, \rho\, d\rho\, d\phi$$

using the Monte Carlo method.

Solution

This integral represents radiation from a circular aperture-antenna with a constant amplitude and phase distribution. It is selected because it forms at least part of every radiation integral. The solution is available in the closed form, which can be used to assess the accuracy of the Monte Carlo results. In closed form,

$$I(\alpha) = \frac{2\pi J_1(\alpha)}{\alpha}$$

where $J_1(\alpha)$ is Bessel function of the first order.

A simple program for evaluating the integral employing Eqs. (8.41) and (8.42), where $a = 0, b = 1, c = 0,$ and $d = 2\pi$, is shown in Fig. 8.3. The program calls the routine RANDU in Vax 11/780 to generate random numbers U^1 and U^2. For different values of N, both the crude and antithetic variate Monte Carlo methods are used in evaluating the radiation integral, and the results are compared with the exact value in Table 8.1 for $\alpha = 5$. In applying Eq. (8.46), the following correspondences are used:

$$U^1 \equiv X^1,\ U^2 \equiv X^2,\ 1 - U^1 \equiv b - X^1 = (b-a)(1-U^1),$$
$$1 - U^2 \equiv d - X^2 = (d-c)(1-U^2).$$

```
0001      C***********************************************
0002      C   INTEGRATION USING CRUDE MONTE CARLO
0003      C   AND ANTITHETIC METHODS
0004      C
0005      C ONLY FEW LINES NEED BE CHANGED TO USE THIS
0006      C PROGRAM FOR ANY MULTI-DIMENSIONAL INTEGRATION
0007      C***********************************************
0008
0009            DATA  IS1,IS2,IS3,IS4/1234,5678,9012,3456/
0010            DATA  A,B,C/0.0,1.0,0.0/
0011                              ! LIMITS OF INTEGRATION
0012            COMPLEX F,SUM1, SUM2, J, AREA1, AREA2
0013
```

Table 8.1 Results of Example 8.3 on Monte Carlo Integration of Radiation Integral

N	Crude MCM	Antithetic variates MCM
500	$-0.2892 - j0.0742$	$-0.2887 - j0.0585$
1000	$-0.5737 + j0.0808$	$-0.4982 - j0.0080$
2000	$-0.4922 - j0.0040$	$-0.4682 - j0.0082$
4000	$-0.3999 - j0.0345$	$-0.4216 - j0.0323$
6000	$-0.3608 - j0.0270$	$-0.3787 - j0.0440$
8000	$-0.4327 - j0.0378$	$-0.4139 - j0.0241$
10,000	$-0.4229 - j0.0237$	$-0.4121 - j0.0240$
Exact: $-0.4116 + j0$		

```
0014       C
0015       C   SPECIFY THE INTEGRAND
0016       C
0017           F(RHO,PHI) = RHO*CEXP(J*ALPHA*RHO*COS(PHI))
0018
0019           J = (0.0,1.0)
0020           ALPHA = 5.0
0021           PIE = 3.1415927
0022           D = 2.0*PIE
0023           DO 30 NRUN = 500,10000,500   ! NO. OF RUNS
0024           SUM1 = (0.0,0.0)
0025           SUM2 = (0.0,0.0)
0026           DO 10 I=1,NRUN
0027           CALL RANDU(IS1,IS2,U1)
0028           CALL RANDU(IS3,IS4,U2)
0029           X1 = A + (B - A)*U1
0030           X2 = C + (D - C)*U2
0031           X3 = (B - A)*(1.0 - U1)
0032           X4 = (D - C)*(1.0 - U2)
0033           SUM1 = SUM1 + F(X1,X2)
0034           SUM2 = SUM2 + F(X1,X2) + F(X1,X4) + F(X3,X2)
0035          1             + F(X3,X4)
0036       10  CONTINUE
0037           AREA1 = (B-A)*(D-C)*SUM1/FLOAT(NRUN)
0038           AREA2 = (B-A)*(D-C)*SUM2/(4.0*FLOAT(NRUN))
0039           PRINT *,NRUN, AREA1, AREA2
0040           WRITE(6,*) NRUN, AREA1,AREA2
0041           WRITE(6,20) NRUN,AREA1,AREA2
0042       20  FORMAT(2X,'NRUN =',I5,3X,'AREA1 = ',F12.6,3X,F12.6,'AREA2 = ',
0043          1             F12.6,3X,F12.6,/)
0044       30  CONTINUE
0045           STOP
0046           END
```

Figure 8.3 Program for Monte Carlo evaluation of two-dimensional integral; for Example 8.3.

8.5 Solution of Potential Problems

The connection between potential theory and Brownian motion (or random walk) was first shown in 1944 by Kakutani [32]. Since then the resulting so-called probabilistic potential theory has been applied to problems in many disciplines such as heat conduction [33–38], electrostatics [39–46], and electrical power engineering [47,48]. An underlying concept of the probabilistic or Monte Carlo solution of differential equations is the random walk. Different types of random walk lead to different Monte Carlo methods. The most popular types are the *fixed-random walk* and *floating random walk*. Other types that are less popular include the *Exodus method, shrinking boundary method, inscribed figure method*, and the *surface density method*.

8.5.1 Fixed Random Walk

Suppose, for concreteness, that the MCM with fixed random walk is to be applied to solve Laplace's equation

$$\nabla^2 V = 0 \quad \text{in region } R \tag{8.51a}$$

subject to Dirichlet boundary condition

$$V = V_p \text{ on boundary } B. \tag{8.51b}$$

We begin by dividing R into mesh and replacing ∇^2 by its finite difference equivalent. The finite difference representation of Eq. (8.51a) in two-dimensional R is given by Eq. (3.26), namely,

$$\boxed{\begin{aligned} V(x,y) = &p_{x+}V(x+\Delta,y) + p_{x-}V(x-\Delta,y) \\ &+ p_{y+}V(x,y+\Delta) + p_{y-}V(x,y-\Delta) \end{aligned}} \tag{8.52a}$$

where

$$\boxed{p_{x+} = p_{x-} = p_{y+} = p_{y-} = \frac{1}{4}} \tag{8.52b}$$

In Eq. (8.52), a square grid of mesh size Δ, such as in Fig. 8.4, is assumed. The equation may be given a probabilistic interpretation. If a random-walking particle is instantaneously at the point (x,y), it has probabilities p_{x+}, p_{x-}, p_{y+}, and p_{y-} of moving from (x,y) to $(x+\Delta,y), (x-\Delta,y), (x,y+\Delta)$, and $(x,y-\Delta)$, respectively. A means of determining which way the particle should move is to generate a random number $U, 0 < U < 1$ and instruct the particle to walk as follows:

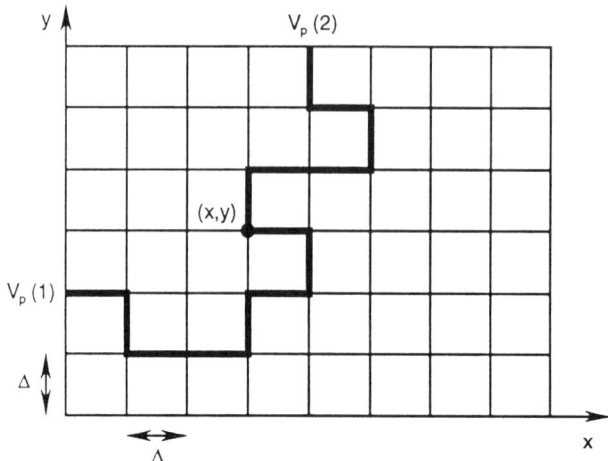

Figure 8.4 Configuration for fixed random walks.

$$\begin{array}{ll} (x,y) \longrightarrow (x+\Delta, y) & \text{if } 0 < U < 0.25 \\ (x,y) \longrightarrow (x-\Delta, y) & \text{if } 0.25 < U < 0.5 \\ (x,y) \longrightarrow (x, y+\Delta) & \text{if } 0.5 < U < 0.75 \\ (x,y) \longrightarrow (x, y-\Delta) & \text{if } 0.75 < U < 1 \end{array} \quad (8.53)$$

If a rectangular grid rather than a square grid is employed, then $p_{x+} = p_{x-}$ and $p_{y+} = p_{y-}$, but $p_x \neq p_y$. Also for a three-dimensional problem in which cubical cells are used, $p_{x+} = p_{x-} = p_{y+} = p_{y-} = p_{z+} = p_{z-} = \frac{1}{6}$. In both cases, the interval $0 < U < 1$ is subdivided according to the probabilities.

To calculate the potential at (x, y), a random-walking particle is instructed to start at that point. The particle proceeds to wander from node to node in the grid until it reaches the boundary. When it does, the walk is terminated and the prescribed potential V_p at that boundary point is recorded. Let the value of V_p at the end of the first walk be denoted by $V_p(1)$, as illustrated in Fig. 8.2. Then a second particle is released from (x, y) and allowed to wander until it reaches a boundary point, where the walk is terminated and the corresponding value of V_p is recorded as $V_p(2)$. This procedure is repeated for the third, fourth,...., and Nth particle released from (x, y), and the corresponding prescribed potential $V_p(3), V_p(4), ..., V_p(N)$ are noted. According to Kakutani [32], the expected value of $V_p(1), V_p(2), ..., V_p(N)$ is the solution of the Dirichlet problem at (x, y), i.e.,

$$\boxed{V(x,y) = \frac{1}{N}\sum_{i=1}^{N} V_p(i)} \tag{8.54}$$

where N, the total number of walks, is large. The rate of convergence varies as $\sqrt{N}$ so that many random walks are required to ensure accurate results.

If it is desired to solve Poisson's equation

$$\nabla^2 V = -g(x,y) \quad \text{in } R \tag{8.55a}$$

subject to

$$V = V_p \quad \text{on } B, \tag{8.55b}$$

then the finite difference representation is in Eq. (3.25), namely,

$$\boxed{\begin{aligned} V(x,y) = & p_{x+}V(x+\Delta, y) + p_{x-}V(x-\Delta, y) \\ & + p_{y+}V(x, y+\Delta) + p_{y-}V(x, y-\Delta) + \frac{\Delta^2 g}{4} \end{aligned}} \tag{8.56}$$

where the probabilities remain as stated in Eq. (8.52b). The probabilistic interpretation of Eq. (8.56) is similar to that for Eq. (8.52). However, the term $\Delta^2 g/4$ in Eq. (8.56) must be recorded at each step of the random walk. If m_i steps are required for the ith random walk originating at (x,y) to reach the boundary, then one records

$$V_p(i) + \frac{\Delta^2}{4}\sum_{j=1}^{m_i-1} g(x_j, y_j). \tag{8.57}$$

Thus the Monte Carlo result for $V(x,y)$ is

$$\boxed{V(x,y) = \frac{1}{N}\sum_{i=1}^{N} V_p(i) + \frac{\Delta^2}{4N}\sum_{i=1}^{N}[\sum_{j=1}^{m_i-1} g(x_j, y_j)]} \tag{8.58}$$

An interesting analogy to the MCM just described is the walking drunk problem [15,35]. We regard the random-walking particle as a "drunk," the squares of the mesh as the "blocks in a city," the nodes as "crossroads," the boundary B as the "city limits," and the terminus on B as the "policeman." Though the drunk is trying to walk home, he is so intoxicated that he wanders randomly throughout the city. The job of the policeman is to seize the drunk in his first appearance at the city limits and ask him to pay a

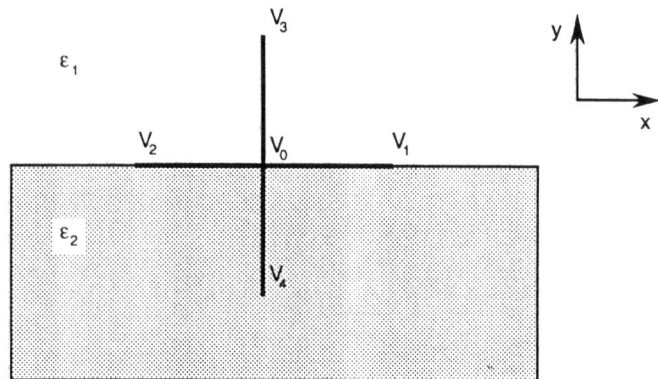

Figure 8.5 Interface between media of dielectric permittivities ϵ_1 and ϵ_2.

fine V_p. What is the expected fine the drunk will receive? The answer to this problem is in Eq. (8.54).

On the dielectric boundary, the boundary condition $D_{1n} = D_{2n}$ is imposed. Consider the interface along y = constant plane as shown in Fig. 8.5. According to Eq. (3.46), the finite difference equivalent of the boundary condition at the interface is

$$V_o = p_{x+}V_1 + p_{x-}V_2 + p_{y+}V_3 + p_{y-}V_4 \qquad (8.59a)$$

where

$$p_{x+} = p_{x-} = \frac{1}{4}, \quad p_{y+} = \frac{\epsilon_1}{2(\epsilon_1 + \epsilon_2)}, \quad p_{y-} = \frac{\epsilon_2}{2(\epsilon_1 + \epsilon_2)}. \qquad (8.59b)$$

An interface along x = constant plane can be treated in a similar manner.

On a line of symmetry, the condition $\frac{\partial V}{\partial n} = 0$ must be imposed. If the line of symmetry is along the y-axis as in Fig. 8.6(a), according to Eq. (3.48)

$$V_o = p_{x+}V_1 + p_{y+}V_3 + p_{y-}V_4 \qquad (8.60a)$$

where

$$p_{x+} = \frac{1}{2} \quad p_{y+} = p_{y-} = \frac{1}{4}. \qquad (8.60b)$$

The line of symmetry along the x-axis, shown in Fig. 8.6(b), is treated similarly following Eq. (3.49).

This MCM is called *fixed random walk* type since the step size Δ is fixed and the steps of the walks are constrained to lie parallel to the coordinate axes. Unlike in the finite difference method (FDM), where the potential at all mesh points are determined simultaneously, MCM is able to solve for the potential at any isolated point in the solution region. One disadvantage of this MCM is that it is slow if potential at many points is required and is therefore recommended for solving problems for which only a few potentials are required. It shares a common difficulty with FDM in connection with irregularly shaped bodies having Neumann boundary conditions. This drawback is fully removed by employing MCM with floating random walk.

8.5.2 Floating Random Walk

The mathematical basis of the floating random walk method is the mean value theorem of potential theory. If S is a sphere of radius r, centered at (x, y, z), which lies wholly within region R, then

$$V(x, y, z) = \frac{1}{4\pi a^2} \int_S V(r') dS'. \qquad (8.61)$$

That is, the potential at the center of any sphere within R is equal to the average value of the potential taken over its surface. When the potential varies in two dimensions, $V(x, y)$ is given by

$$V(x, y) = \frac{1}{2\pi \rho} \oint_L V(\rho') dl' \qquad (8.62)$$

where the integration is around a circle of radius ρ centered at (x, y). It can be shown that Eqs. (8.61) and (8.62) follow from Laplace's equation. Also, Eqs. (8.61) and (8.62) can be written as

$$V(x, y, z) = \int_0^1 \int_0^1 V(a, \theta, \phi) dF dT \qquad (8.63)$$

$$V(x, y) = \int_0^1 V(a, \phi) dF \qquad (8.64)$$

where

$$F = \frac{\phi}{2\pi}, \quad T = \frac{1}{2}(1 - \cos\theta) \qquad (8.65)$$

and θ and ϕ are regular spherical coordinate variables. The functions F and T may be interpreted as the probability distributions corresponding to ϕ and θ. While $dF/d\phi = $ constant, $dT/d\theta = \frac{1}{2}\sin\theta$; i.e., all angles ϕ are equally probable, but the same is not true for θ.

The floating random walk MCM depends on the application of Eqs. (8.61) and (8.62) in a statistical sense. For a two-dimensional problem, suppose that a random-walking particle is at some point (x_j, y_j) after j

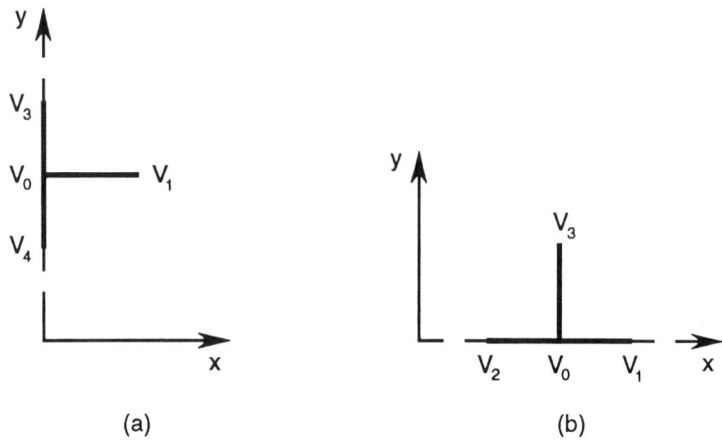

Figure 8.6 Satisfying symmetry conditions: (a) $\partial V/\partial x = 0$, (b) $\partial V/\partial y = 0$.

steps in the ith walk. The next $(j+1)$th step is taken as follows. First, a circle is constructed with center at (x_j, y_j) and radius ρ_j, which is equal to the short distance between (x_j, y_j) and the boundary. The ϕ coordinate is generated as a random variable uniformly distributed over $(0, 2\pi)$, i.e., $\phi = 2\pi U$, where $0 < U < 1$. Thus the location of the random-walking particle after the $(j+1)$th step is illustrated in Fig. 8.7 and given as

$$x_{j+1} = x_j + \rho_j \cos \phi_j \qquad (8.66a)$$
$$y_{j+1} = y_j + \rho_j \sin \phi_j \qquad (8.66b)$$

The next random walk is executed by constructing a circle centered at (x_{j+1}, y_{j+1}) and of radius ρ_{j+1}, which is the shortest distance between (x_{j+1}, y_{j+1}) and the boundary. This procedure is repeated several times, and the walk is terminated when the walk approaches some prescribed small distance τ of the boundary. The potential $V_p(i)$ at the end of this ith walk is recorded as in fixed random walk MCM and the potential at (x, y) is eventually determined after N walks using Eq. (8.54).

The floating random walk MCM can be applied to a three-dimensional Laplace problem by proceeding along lines similar to those outlined above. A random-walking particle at (x_j, y_j, z_j) will step to a new location on the surface of a sphere whose radius r_j is equal to the shortest distance between point (x_j, y_j, z_j) and the boundary. The ϕ coordinate is selected as a random number U between 0 and 1, multiplied by 2π. The coordinate θ is determined by selecting another random number U between 0 and 1,

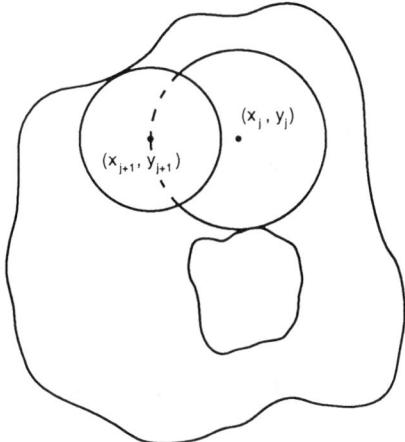

Figure 8.7 Configuration of floating random walks.

and solving for $\Theta = \cos^{-1}(1 - 2U)$ as in Example 8.1. Thus the location of the particle after its $(j+1)$th step in the ith walk is

$$x_{j+1} = x_j + r_j \cos\phi_j \sin\theta_j \qquad (8.67a)$$
$$y_{j+1} = y_j + r_j \sin\phi_j \sin\theta_j \qquad (8.67b)$$
$$z_{j+1} = z_j + r_j \cos\theta_j \qquad (8.67c)$$

Finally, we apply Eq. (8.54).

Solving Poisson's equation (8.55) for a two-dimensional problem requires only a slight modification. For a three-dimensional problem, $V(a, \theta, \phi)$ in Eq. (8.63) is replaced by $[V(a,\theta,\phi) + r^2 g/6]$. This requires that the term $gr_j^2/6$ at every jth step of the ith random walk be recorded.

An approach for handling a discretely inhomogeneous medium is presented in [39,43,44].

It is evident that in the floating random walk MCM, neither the step sizes nor the directions of the walk are fixed in advance. The quantities may be regarded as "floating" and hence the designation *floating random walk*. A floating random walk bypasses many intermediate steps of a fixed random walk in favor of a long jump. Fewer steps are needed to reach the boundary, and so computation is much more rapid than in fixed random walk.

8.5.3 Exodus Method

The *Exodus method*, first suggested in [49] and developed in [56], does not employ random numbers and is generally faster and more accurate than the fixed random walk. It basically consists of dispatching numerous walkers (say 10^6) simultaneously in directions controlled by the probabilities of going from one node to its neighbors. As these walkers arrive at new nodes, they are dispatched according to the probabilities until a set number (say 99.999%) have reached the boundaries. The advantage of the Exodus method is its independence of the random number generator.

To implement the Exodus method, we first divide the solution region R into mesh, such as in Fig. 8.4. Suppose p_k is the probability that a random walk starting from point (x,y) ends at node k on the boundary with prescribed potential $V_p(k)$. For M boundary nodes (excluding the corner points since a random walk never terminates at those points), the potential at the starting point (x,y) of the random walks is

$$\boxed{V(x,y) = \sum_{k=1}^{M} p_k V_p(k)} \tag{8.68}$$

If m is the number of different boundary potentials ($m = 4$ in Fig. 8.4), Eq. (8.68) can be simplified to

$$V(x,y) = \sum_{k=1}^{m} p_k V_p(k), \tag{8.69}$$

where p_k in this case is the probability that a random walk terminates on boundary k. Since $V_p(k)$ is specified, our problem is reduced to finding p_k. We find p_k using the Exodus method in a manner similar to the iterative process applied in Section 3.5.

Let $P(i,j)$ be the number of particles at point (i,j) in R. We begin by setting $P(i,j) = 0$ at all points (both fixed and free) except at point (x,y), where $P(i,j)$ assumes a large number N (say, $N = 10^6$ or more). By a scanning process, we dispatch the particles at each free node to its neighboring nodes according to the probabilities p_{x+}, p_{x-}, p_{y+}, and p_{y-} as illustrated in Fig. 8.8. Note that in Fig. 8.8(b), new $P(i,j) = 0$ at that node, while old $P(i,j)$ is shared among the neighboring nodes. When all the free nodes in R are scanned as illustrated in Fig. 8.8, we record the number of particles that have reached the boundary (i.e., the fixed nodes). We keep scanning the mesh until a set number of particles (say 99.99% of N) have reached the boundary, where the particles are absorbed. If N_k is the number of particles that reached side k, we calculate

$$p_k = \frac{N_k}{N}. \tag{8.70}$$

Monte Carlo Methods 595

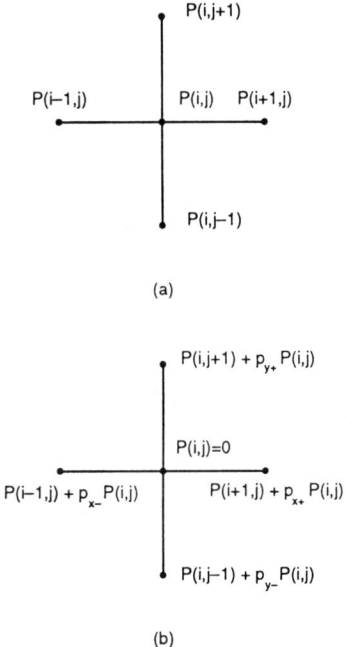

Figure 8.8 (a) Before the particles at (i,j) are dispatched, (b) after the particles at (i,j) are dispatched.

Hence Eq. (8.69) can be written as

$$V(x,y) = \frac{\sum_{k=1}^{m} N_k V_p(k)}{N} \tag{8.71}$$

Thus the problem is reduced to just finding N_k using the Exodus method, given N and $V_p(k)$. We notice that if $N \to \infty$, $\Delta \to 0$, and all the particles were allowed to reach the boundary points, the values of p_k and consequently $V(x,y)$ would be exact. It is easier to approach this exact solution using the Exodus method than any other MCMs or any other numerical techniques such as difference and finite element methods.

8.5.4 Regional Monte Carlo Method

The Monte Carlo methods discussed so far can only be recommended for single point calculations; they are generally inefficient for whole field calculations. There are several Monte Carlo procedures for whole field calculations [37,38,50–53].

596 Numerical Techniques in Electromagnetics

One regional Monte Carlo method involves solving a PDE using the separation of variables and then estimating the expansion coefficients using sampling techniques [52,53]. For example, consider the elliptic PDE

$$U_{xx} + U_{yy} = \alpha^2 U \quad \text{in region } R, \tag{8.72a}$$

where α^2 is constant, subject to the Dirichlet condition

$$V = V_p \quad \text{on boundary } B. \tag{8.72b}$$

A separation of variables in cylindrical coordinates gives [52]

$$U(\rho,\phi) = \frac{1}{2}a_0 I_0(\alpha\rho) + \sum_{n=1}^{\infty} I_n(\alpha\rho)[a_n \cos n\phi + b_n \sin n\phi], \tag{8.73}$$

where I_n is the modified Bessel function of order n. The expansion coefficients a_n and b_n are obtained in the usual manner as

$$a_n = \frac{2}{I_n(\alpha\rho)} \frac{1}{2\pi} \int_0^{2\pi} U(\rho,\phi) \cos n\phi \, d\phi, \tag{8.74a}$$

$$b_n = \frac{2}{I_n(\alpha\rho)} \frac{1}{2\pi} \int_0^{2\pi} U(\rho,\phi) \sin n\phi \, d\phi. \tag{8.74b}$$

These coefficients may be estimated by sampling θ uniformly on $[0, 2\pi]$ and using

$$\hat{a}_n = \frac{2}{I_n(\alpha r)} \frac{1}{M} \sum_{i=1}^{M} U(r,\theta_i) \cos n\theta_i, \tag{8.75a}$$

$$\hat{b}_n = \frac{2}{I_n(\alpha r)} \frac{1}{M} \sum_{i=1}^{M} U(r,\theta_i) \sin n\theta_i, \tag{8.75b}$$

where (ρ, ϕ) are the global coordinates, while (r, θ) are local as in Fig. 8.9. Select an arbitrary point P, as in Fig. 8.9, draw the largest circle (with center at P) that lies entirely in R, and proceed to sample Eq. (8.75). However, $U(r, \theta_i)$ is unknown. We use the single point theory to get a one-particle estimate $\hat{U}(r,\theta_i)$ of $U(r,\theta_i)$.

To get a one-particle estimate, we proceed as in floating random walk. To estimate $U(P_0)$, a point P_1 is sampled uniformly on largest circle (centered at P_0, radius r_0) lying entirely within R. If P_1 lies within some small distance τ of B, then $U(P_1)$ is known and $U(P_1)/I_0(\alpha r_0)$ is taken as a

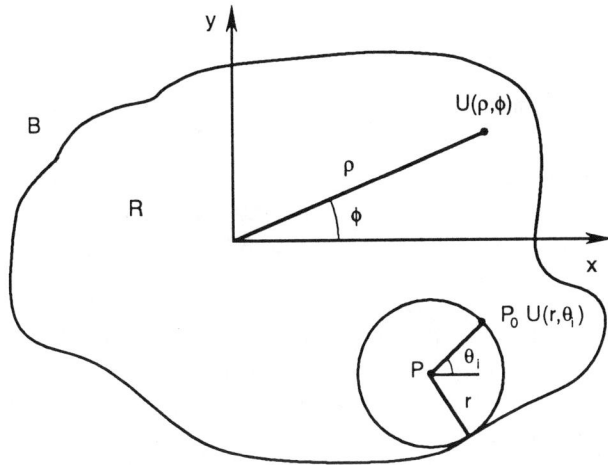

Figure 8.9 Global and local coordinates for a regional Monte Carlo procedure.

one-particle estimate of $U(P_0)$. If P_1 is not close to B, $U(P_1)$ is estimated in the same fashion as $U(P_0)$. That is, a point P_2 is sampled uniformly on the largest circle (centered at P_1, radius r_1) lying within R. If P_2 lies within distance τ of B, then $U(P_2)$ is known and $U(P_2)/I_0(\alpha r_1)$ of $U(P_1)$ or $U(P_2)/[I_0(\alpha r_0)I_0(\alpha r_1)]$ is a one-particle estimate of $U(P_0)$. If P_2 does not lie close to B, the procedure is repeated (as in floating random walk) until the walk terminates very close to the boundary B at the nth jump, whereupon

$$\hat{U}(r,\theta_i) = \hat{U}(P_0) = \frac{U(P_n)}{I_0(\alpha r_0)I_0(\alpha r_1)\cdots I_0(\alpha r_n)} \quad (8.76)$$

is taken as a one-particle estimate of $U(P_0)$. Notice that r in Eqs. (8.75) and (8.76) is fixed throughout the calculations, whereas the location of P_0 (and consequently $P_1, P_2, \cdots, P_n$) varies with random angle θ_i, $0 < \theta_i < 2\pi$.

Once $\hat{U}(r,\theta_i)$ is calculated, we find

$$\hat{a}_n = \frac{2}{I_n(\alpha r)} \frac{1}{M} \sum_{i=1}^{M} \hat{U}(r,\theta_i) \, \cos n\theta_i, \quad (8.77a)$$

$$\hat{b}_n = \frac{2}{I_n(\alpha r)} \frac{1}{M} \sum_{i=1}^{M} \hat{U}(r,\theta_i) \, \sin n\theta_i. \quad (8.77b)$$

Substitution of Eq. (8.77) into Eq. (8.73) gives the estimated solution at

598   Numerical Techniques in Electromagnetics

any point in R as

$$\hat{U}(\rho,\phi) = \frac{1}{2}\hat{a}_0 I_0(\alpha\rho) + \sum_{n=1}^{N} I_n(\alpha\rho)\left[\hat{a}_n \cos n\phi + \hat{b}_n \sin n\phi\right]. \qquad (8.78)$$

Note that the infinite summation has been truncated at $n = N$. Also note that the expansion coefficients are calculated once and for all by selecting point P and Eq. (8.78) is used to estimate $U(\rho,\phi)$ at any point in R.

In order to increase the efficiency and reduce the computation time of the Monte Carlo methods discussed above, various modifications have been proposed. A technique that involves using Green's function in the floating random walk is suggested in [42]. Other regional MCMs include the *shrinking boundary method* [37], the *inscribed figure method* [38], and *surface density method* [50,51].

The random walk MCMs applied to elliptic partial differential equations in this section can be applied to parabolic partial differential equations as well [54–55]. They may also be used to solve problems with nonrectangular solution regions [56–57] (see Probs. 8.19 to 8.21).

Example 8.4

Give a probabilistic interpretation using the finite difference form of the energy equation

$$u\frac{\partial T}{\partial x} + v\frac{\partial T}{\partial y} = \alpha\left(\frac{\partial^2 T}{\partial x^2} + \frac{\partial^2 T}{\partial y^2}\right).$$

Assume a square grid of size Δ.

Solution

Applying a backward difference to the left-hand side and a central difference to the right-hand side, we obtain

$$u\frac{T(x,y) - T(x-\Delta,y)}{\Delta} + v\frac{T(x,y) - T(x,y-\Delta)}{\Delta} =$$
$$\alpha\frac{T(x+\Delta,y) - 2T(x,y) + T(x-\Delta,y)}{\Delta^2}$$
$$+ \alpha\frac{T(x,y+\Delta) - 2T(x,y) + T(x,y-\Delta)}{\Delta^2}. \qquad (8.4.1)$$

Rearranging terms leads to

$$T(x,y) = p_{x+}T(x+\Delta,y) + p_{x-}T(x-\Delta,y) + p_{y+}T(x,y+\Delta) + p_{y-}T(x,y-\Delta) \qquad (8.4.2)$$

where

$$p_{x+} = p_{y+} = \frac{1}{\frac{u\Delta}{\alpha} + \frac{v\Delta}{\alpha} + 4} \quad (8.4.3a)$$

$$p_{x-} = \frac{(1+\frac{\Delta u}{\alpha})}{\frac{u\Delta}{\alpha} + \frac{v\Delta}{\alpha} + 4} \quad (8.4.3b)$$

$$p_{y-} = \frac{(1+\frac{\Delta v}{\alpha})}{\frac{u\Delta}{\alpha} + \frac{v\Delta}{\alpha} + 4}. \quad (8.4.3c)$$

Equation (8.4.2) is given probabilistic interpretation as follows: a walker at point (x,y) has probabilities p_{x+}, p_{x-}, p_{y+}, and p_{y-} of moving to point $(x+\Delta, y), (x-\Delta, y), (x, y+\Delta)$, and $(x, y-\Delta)$, respectively. With this interpretation, Eq. (8.4.2) can be used to solve the differential equation with fixed random MCM.

Example 8.5

Consider a conducting trough of infinite length with square cross section shown in Fig. 8.10. The trough wall at $y = 1$ is connected to $100V$, while the other walls are grounded as shown. We intend to find the potential within the trough using the fixed random walk MCM.

Solution

The problem is solving Laplace's equation subject to

$$V(0,y) = V(1,y) = V(x,0) = 0, V(x,1) = 100. \quad (8.5.1)$$

The exact solution obtained by the method of separation of variables is given in Eq. (2.31), namely,

$$V(x,y) = \frac{400}{\pi} \sum_{n=0}^{\infty} \frac{\sin k\pi x \sinh k\pi y}{k \sinh k\pi}, \quad k = 2n+1. \quad (8.5.2)$$

Applying the fixed random MCM, the flowchart in Fig. 8.11 was developed. Based on the flowchart, the program of Fig. 8.12 was developed. A built-in standard subroutine RANDU in VAX 750 (also in VAX 780) was used to generate random numbers U uniformly distributed between 0 and 1. The step size Δ was selected as 0.05. The results of the potential computation are listed in Table 8.2 for three different locations. The average number of random steps $\overline{m}$ taken to reach the boundary is also shown. It

600 Numerical Techniques in Electromagnetics

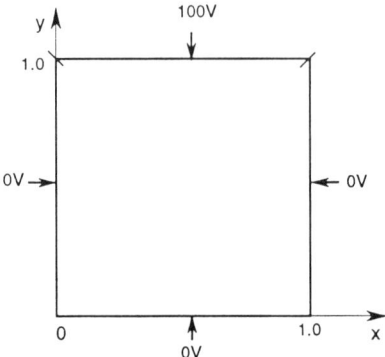

Figure 8.10 For Example 8.5.

is observed from Table 8.2 that it takes a large number of random steps for small step size and that the error in MCM results can be less than 1%.

Rather than using Eq. (8.54), an alternative approach of determining $V(x,y)$ is to calculate the probability of a random walk terminating at a grid point located on the boundary. The information is easily extracted from the program used for obtaining the results in Table 8.2. To illustrate the validity of this approach, the potential at $(0.25, 0.75)$ was calculated. For $N = 1000$ random walks, the number of walks terminating at $x = 0, x = 1, y = 0$ and $y = 1$ are 461, 62, 66, and 411, respectively. Hence, according to Eq. (8.71)

$$V(x,y) = \frac{461}{1000}(0) + \frac{62}{1000}(0) + \frac{66}{1000}(0) + \frac{411}{1000}(100) = 41.1. \quad (8.5.3)$$

The statistical error in the simulation can be found. In this case, the potential on the boundary takes values 0 or $V_o = 100$ so that $V(x,y)$ has a binomial distribution with mean $V(x,y)$ and variance

$$\sigma^2 = \frac{V(x,y)[V_o - V(x,y)]}{N}. \quad (8.5.4)$$

At point $(0.5, 0.5)$, for example, $N = 1000$ gives $\sigma = 1.384$ so that at 68% confidence interval, the error is $\delta = \sigma/\sqrt{N} = 0.04375$.

```
0001    C********************************************************
0002    C   MONTE CARLO SOLUTION OF POTENTIAL PROBLEM
0003    C   INVOLVING LAPLACE'S EQUATION
0004    C********************************************************
0005    C
0006    C   SPECIFY INPUT PARAMETERS
0007    C
```

Monte Carlo Methods 601

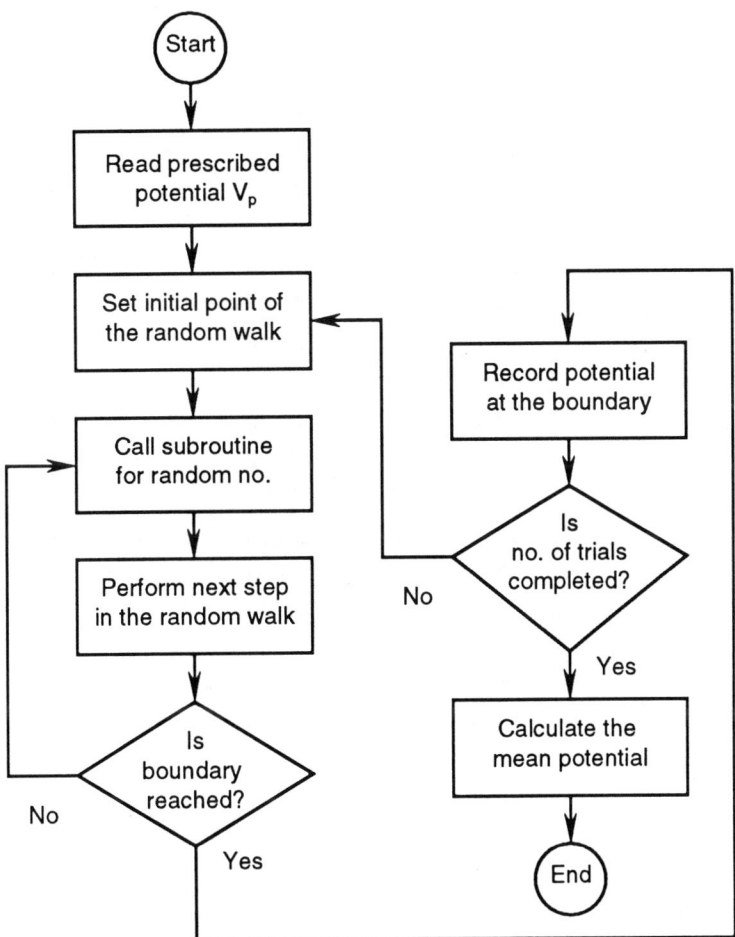

Figure 8.11 Flowchart for random walk of Example 8.5.

Numerical Techniques in Electromagnetics

Table 8.2 Results of Example 8.5

x	y	N	$\overline{m}$	Monte Carlo solution	Exact solution
0.25	0.75	250	66.20	42.80	43.20
		500	69.65	41.80	
		750	73.19	41.60	
		1000	73.95	41.10	
		1250	73.67	42.48	
		1500	73.39	42.48	
		1750	74.08	42.67	
		2000	74.54	43.35	
0.5	0.5	250	118.62	21.60	25.0
		500	120.0	23.60	
		750	120.27	25.89	
		1000	120.92	25.80	
		1250	120.92	25.92	
		1500	120.78	25.27	
		1750	121.5	25.26	
		2000	121.74	25.10	
0.75	0.25	250	64.82	7.60	6.797
		500	68.52	6.60	
		750	68.56	6.93	
		1000	70.17	7.50	
		1250	72.12	8.00	
		1500	71.78	7.60	
		1750	72.40	7.43	
		2000	72.40	7.30	

```
0008            DATA  V1,V2,V3,V4/0.0,0.0,100.0,0.0/
0009            DATA  IS1,IS2/1234,5678/
0010            DATA  P1,P2,P3/0.25,0.5,0.75/
0011            DATA  A,B/1.0,1.0/
0012
0013            NRUN = 1000   ! NO. OF RUNS
0014            DELTA = 0.05  ! STEP SIZE
0015      C
0016      C  INITIALIZE VARIABLES
0017      C
```

```
0018            X0 = 0.25
0019            Y0 = 0.75
0020            I0 = X0/DELTA
0021            J0 = Y0/DELTA
0022            IMAX = A/DELTA
0023            JMAX = B/DELTA
0024            SUM = 0.0
0025            NS = 0   ! NO. OF WALKS BEFORE REACHING BOUNDARY
0026            M1 = 0   ! NO. OF WALKS TERMINATING AT V1
0027            M2 = 0
0028            M3 = 0
0029            M4 = 0   ! NO. OF WALKS TERMINATING AT V4
0030    C
0031    C   START RUNNING MONTE CARLO SIMULATION
0032    C
0033            DO 70 K=1,NRUN
0034            I=I0
0035            J=J0
0036    10      CALL RANDU(IS1,IS2,R)
0037            NS = NS + 1
0038            IF( R.GE.0.0.AND. R.LT.P1)   I = I + 1
0039            IF( R.GE.P1. AND. R.LT.P2)   J = J + 1
0040            IF( R.GE.P2. AND. R.LT.P3)   I = I - 1
0041            IF( R.GE.P3)                 J = J - 1
0042    C
0043    C   CHECK IF (I,J) IS ON THE BOUNDARY
0044    C
0045            IF( I.EQ.0) THEN
0046            SUM = SUM + V4
0047            M4 = M4 +1
0048            GO TO 60
0049            ELSE
0050            ENDIF
0051            IF( IMAX - I) 20,20,30
0052    20        SUM = SUM +  V2
0053              M2 = M2 +1
0054               GO TO 60
0055    30      IF( J.EQ.0) THEN
0056            SUM = SUM + V1
0057            M1 = M1 + 1
0058            GO TO 60
0059            ELSE
0060            ENDIF
0061            IF( JMAX - J) 40,40,50
0062    40      SUM = SUM + V3
0063            M3 = M3 + 1
0064            GO TO 60
0065    50      GO TO 10
0066    60      IF( MOD(K,250).NE.0.0 ) GO TO 70
0067            V = SUM/FLOAT(K)
0068            STEPS = FLOAT(NS)/FLOAT(K)    ! AVERAGE NO. OF WALKS
0069            PRINT *,X0,Y0,K,V,STEPS
0070    70      CONTINUE
0071            PRINT *, V
0072            WRITE(6,80) NRUN,V,M1,M2,M3,M4
0073    80      FORMAT(2X,'NRUN=',I6,3X,'V=',F12.6,3X,'Ms =',4I6,/)
0074            STOP
0075            END
```

Figure 8.12 Program for Example 8.5.

Example 8.6

Use the floating random walk MCM to determine the potential at points (1.5, 0.5), (1.0, 1.5), and (1.5, 2.0) in the two-dimensional potential system in Fig. 8.13.

Solution

To apply the floating random walk, we use the flowchart in Fig. 8.11 except that we apply Eq. (8.66) instead of Eq. (8.53) at every step in the random walk. A program based on the modified flowchart was developed. The shortest distance ρ from (x,y) to the boundary was found by dividing the solution region in Fig. 8.13 into three rectangles and checking

if $\{(x,y) : 1 < x < 2, 0 < y < 1\}, \rho = \text{minimum}\{x - 1, 2 - x, y\}$

if $\{(x,y) : 0 < x < 1, 1 < y < 2.5\}, \rho = \text{minimum}\{x, y - 1, 2.5 - y\}$

if $\{(x,y) : 1 < x < 2, 1 < y < 2.5\}$,

$\rho = \text{minimum}\{2 - x, 2.5 - y, \sqrt{(x-1)^2 + (y-1)^2}\}.$

A prescribed tolerance $\tau = 0.05$ was selected so that if the distance between a new point in the random walk and the boundary is less than τ, it is assumed that the boundary is reached and the potential at the closest boundary point is recorded.

Table 8.3 presents the Monte Carlo result with the average number of random steps $\overline{m}$. It should be observed that it takes fewer walks to reach the boundary in floating random walk than in fixed random walk. Since no analytic solution exists, we compare Monte Carlo results with those obtained using finite difference with $\Delta = 0.05$ and 500 iterations. As evident in Table 8.3, the Monte Carlo results agree well with the finite difference results even with 1000 walks. Also, by dividing the solution region into 32 elements, the finite element results [58] at points (1.5,0.5), (1.0,1.5), and (1.5,2.0) are 11.265, 9.788, and 21.05 V, respectively.

Unlike the program in Fig. 8.12, where the error estimates are not provided for the sake of simplicity, the program in Fig. 8.14 incorporates evaluation of error estimates in the Monte Carlo calculations. Using Eq. (8.29), the error is calculated as

$$\delta = \frac{S \, t_{\alpha/2;n-1}}{\sqrt{n}}.$$

In the program in Fig. 8.14, the number of trials n (the same of N in Section 8.3), with different seed values, is taken as 5 so that $t_{\alpha/2;n-1} = 2.776$. The

Monte Carlo Methods

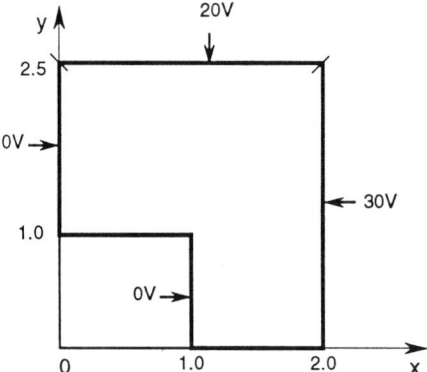

Figure 8.13 For Example 8.6.

sample variance S is calculated using Eq. (8.19). The values of δ are also listed in Table 8.3. Notice that unlike in Table 8.2, where $\overline{m}$ and V are the mean values after N walks, $\overline{m}$ and V in Table 8.3 are the mean values of n trials, each of which involves N walks, i.e., the "mean of the mean" values. Hence the results in Table 8.3 should be regarded as more accurate than those in Table 8.2.

```
0001      ********************************************************
0002      C   MONTE CARLO (FLOATING RANDOM WALK) SOLUTION OF
0003      C   POTENTIAL PROBLEM INVOLVING LAPLACE'S EQUATION
0004      C   ERROR ESTIMATES ARE ALSO EVALUATED
0005      C********************************************************
0006
0007            DATA   PIE/3.1415/
0008            DIMENSION  T(10),V(10),STEPS(10)
0009            DATA ( T(I), I=2,10 )/12.706,  4.303, 3.182, 2.776,
0010         1                  2.571,   2.447, 2.365, 2.306, 2.262/
0011            ! T(I) ARE THE   T-DISTRIBUTION PARAMETERS
0012
0013      C
0014      C  INITIALIZE VARIABLES
0015      C
0016            NTRIALS = 5 ! NO. OF TRIALS
0017            X0 = 1.5  !  STARTING POINT, WHERE POTENTIAL IS
0018            Y0 = 0.5  !                         REQUIRED
0019            TOL = 0.005   ! TOLERANCE
0020      C
0021      C  START RUNNING MONTE CARLO SIMULATION
0022      C
0023            DO 120 NRUN=250,2000,250   ! NO. OF RUNS
0024            DO 80 J=1,NTRIALS
0025            IS1 = 1000*FLOAT(J)
0026            IS2 = 2000*FLOAT(J)
0027            SUM = 0.0
0028            M = 0 ! NO. OF STEPS TAKEN TO REACH BOUNDARY
0029            DO 60 I=1,NRUN
0030            X = X0
```

Table 8.3 Results of Example 8.6

x	y	N	$\overline{m}$	Monte Carlo solution ($V \pm \delta$)	Finite Difference solution (V)
1.5	0.5	250	6.738	11.52 ± 0.8973	11.44
		500	6.668	11.80 ± 0.9378	
		750	6.535	11.83 ± 0.4092	
		1000	6.476	11.82 ± 0.6205	
		1250	6.483	11.85 ± 0.6683	
		1500	6.465	11.72 ± 0.7973	
		1750	6.468	11.70 ± 0.6894	
		2000	6.460	11.55 ± 0.5956	
1.0	1.5	250	8.902	10.74 ± 0.8365	10.44
		500	8.984	10.82 ± 0.3709	
		750	8.937	10.75 ± 0.5032	
		1000	8.928	10.90 ± 0.7231	
		1250	8.836	10.84 ± 0.7255	
		1500	8.791	10.93 ± 0.5983	
		1750	8.788	10.87 ± 0.4803	
		2000	8.811	10.84 ± 0.3646	
1.5	2.0	250	7.242	21.66 ± 0.7509	21.07
		500	7.293	21.57 ± 0.5162	
		750	7.278	21.53 ± 0.3505	
		1000	7.316	21.53 ± 0.2601	
		1250	7.322	21.53 ± 0.3298	
		1500	7.348	21.51 ± 0.3083	
		1750	7.372	21.55 ± 0.2592	
		2000	7.371	21.45 ± 0.2521	

```
0031              Y = Y0
0032       10     CALL RANDU(IS1,IS2,RN)
0033              PHI = 2.0*PIE*RN
0034       C      FIND THE SHORTEST DISTANCE
0035              RC = SQRT( (X-1.0)**2 + (Y-1.0)**2 )
0036              IF(Y.GT.1.0)  GO TO 20
0037              R = X - 1.0
0038              IF( R.GT.(2.0-X) ) R = 2.0 - X
0039              IF(R.GT.Y) R = Y
```

```
0040            GO TO 40
0041     20     IF(X.GT.1.0) GO TO 30
0042            R = X
0043            IF( R.GT.(Y-1.0) ) R = Y - 1.0
0044            IF( R.GT.(2.5-Y) ) R = 2.5 - Y
0045            GO TO 40
0046     30     R = 2.0 - X
0047            IF(R.GT.RC) R = RC
0048            IF( R.GT.(2.5-Y) ) R = 2.5 - Y
0049     40     X = X + R*COS(PHI)
0050            Y = Y + R*SIN(PHI)
0051            M = M + 1
0052     C
0053     C   CHECK IF (X,Y) IS ON THE BOUNDARY
0054     C
0055            IF( (X.LT.(1.0+TOL)).AND.(Y.LT.(1.0+TOL)) ) GO TO 50
0056                                  ! FOR THE CORNER POINT
0057            IF( X.GE.(2.0-TOL) ) THEN
0058              SUM = SUM + 30.0
0059              GO TO 50
0060            ELSE
0061            ENDIF
0062            IF( Y.GE.(2.5-TOL) ) THEN
0063              SUM = SUM + 20.0
0064              GO TO 50
0065            ELSE
0066            ENDIF
0067            IF(Y.GT.1.0. AND. X.LT.TOL) GO TO 50
0068            IF(Y.LT.1.0.AND.X.LT.(1.0-TOL)) GO TO 50
0069            IF(Y.LE.TOL.AND.X.GE.1.0) GO TO 50
0070            IF(Y.LE.(1.0+TOL).AND.X.LT.1.0) GO TO 50
0071            GO TO 10
0072     50     CONTINUE
0073     60     CONTINUE
0074            V(J) = SUM/FLOAT(NRUN)
0075            STEPS(J) = FLOAT(M)/FLOAT(NRUN)   ! AVERAGE NO. OF WALKS
0076            PRINT *,X0,Y0,V(J),STEPS(J)
0077            WRITE(6,*) X0,Y0,V(J),STEPS(J)
0078            WRITE(6,70) X0,Y0,NRUN,V(J),STEPS(J)
0079     70     FORMAT(2X,'X = ',F5.2,3X,'Y=',F5.2,3X,
0080          1        'NRUN=',I6,3X,'V=',F12.6,3X,'STEPS=',F10.3,/)
0081     80     CONTINUE
0082     C
0083     C   FIND THE MEAN VALUE OF V AND MEAN NO. OF STEP
0084     C
0085            SUM = 0.0
0086            SUM1 = 0.0
0087            DO 90 I=1,NTRIALS
0088            SUM = SUM + V(I)
0089            SUM1 = SUM1 + STEPS(I)
0090     90     CONTINUE
0091            VMEAN = SUM/FLOAT(NTRIALS)
0092            STEPM = SUM1/FLOAT(NTRIALS)
0093     C
0094     C   CALCULATE ERROR
0095     C
0096            SUM = 0.0
0097            DO 100 I=1,NTRIALS
0098            SUM = SUM + ( V(I) - VMEAN )**2
0099     100    CONTINUE
0100            STD = SQRT( SUM/FLOAT(NTRIALS-1) )
0101            ERROR = STD*T(NTRIALS)/SQRT( FLOAT(NTRIALS) )
0102            PRINT *,NTRIALS,VMEAN,STEPM,ERROR
```

```
0103            WRITE(6,110) NTRIALS,VMEAN,STEPM,ERROR
0104     110    FORMAT(2X,'NO. OF TRIALS',I6,3X,'MEAN V =',F12.6,3X,
0105       1          'MEAN M = ',F12.6,3X,'ERROR=',F12.6,//)
0106     120    CONTINUE
0107            STOP
0108            END
```

Figure 8.14 Applying floating random walk to solve the problem in Fig. 8.13; for Example 8.6.

8.6 Inverse Monte Carlo Analysis

As discussed in Section 5.2, a general inverse problem can be posed in the form of a Fredholm equation of the first kind, namely

$$f(x) = \int_{-\infty}^{\infty} K(x,t)\Phi(t)dt. \tag{8.79}$$

In Eq. (8.79), $K(x,t)$ is the kernel, Φ can be measured at n discrete values of y, yielding a vector $\mathbf{f}(x)$, and $\Phi(x)$ is the unknown function to be determined. Our goal is to obtain $m \leq n$ parameters of Φ by inversion. We let

$$\Phi(x) = \begin{cases} \alpha_i, & x \in (x_i, x_{i+1}), \quad i=1,2,\cdots,m \\ 0, & \text{otherwise} \end{cases} \tag{8.80}$$

where $x_1 \leq x_2 \leq \cdots \leq x_{m+1}$. If a suitable quadrature can be applied, we transform the problem to a system of n equations in m unknowns, leading to a quadrature approximation of the form [59]

$$\mathbf{f} = \frac{1}{N} \sum_{j=1}^{N} \mathbf{K}(\xi_j) \, \alpha \tag{8.81}$$

where $\mathbf{K}(\xi_j) = K(\xi_j, y)$ and ξ_j is a random sample of $x \in (x_i, x_{i+1})$. In Eq. (8.81), all quantities are known except the α; hence Eq. (8.81) can be cast into the matrix form

$$\mathbf{f} = A\alpha \tag{8.82}$$

from which α can be obtained by matrix inversion.

Example 8.7

Given
$$I(y) = \int_0^1 H(x-y)\Phi(x)\,dx, \quad y \in [0,1)$$

where
$$H(y-x) = \begin{cases} 0, & x < y \\ 1, & x \geq y \end{cases}$$

and the "measurements" $\bar{I}(0) = 1$, $\bar{I}(0.2) = 0.92$, $\bar{I}(0.4) = 0.76$, and $\bar{I}(0.8) = 0.08$, find $\Phi(x)$ using the inverse Monte Carlo technique.

Solution

Let $\Phi(x)$ assume the step-wise form

$$\Phi(x) = \begin{cases} \alpha_1, & x \in \Delta_1 = [0, 0.2) \\ \alpha_2, & x \in \Delta_2 = [0.2, 0.4) \\ \alpha_3, & x \in \Delta_3 = [0.4, 0.8) \\ \alpha_4, & x \in \Delta_4 = [0.8, 1.0) \\ 0, & \text{otherwise.} \end{cases}$$

We construct Monte Carlo estimate $\hat{I}$ of $\bar{I}$, each of the form

$$\hat{I}_j = \frac{1}{N} \sum_{i=1}^{N} H(\xi_i - y_j)\Phi(\xi_i), \quad j = 1, 2, 3, 4$$

in which y_j assumes the values $y_1 = 0$, $y_2 = 0.2$, $y_3 = 0.4$, and $y_4 = 0.8$. Substituting $\bar{I}$ for $\hat{I}$, we obtain

$$\bar{I} = A\alpha$$

where
$$A_{ij} = \frac{1}{N} \sum_{k=1}^{N_i} H(\xi_k - y_j)$$

and N_i is the number of sampled values, ξ_i, within Δ_i. Solving $\bar{I} = A\alpha$ by matrix inversion leads to the results in Table 8.4.

Table 8.4 Results of Example 8.7

Parameter	Sample 2,500	Population 10,000	N 40,000	Exact
α_1	0.408	0.392	0.418	0.4
α_2	0.666	0.803	0.765	0.8
α_3	1.74	1.7	1.7	1.7
α_4	0.414	0.389	0.405	0.4

8.7 Concluding Remarks

The Monte Carlo technique is essentially a means of estimating expected values and hence is a form of numerical quadrature. Although the technique can be applied to simple processes and estimating multidimensional integrals, the power of the technique rests in the fact that [59]:

- it is often more efficient than other quadrature formulas for estimating multidimensional integrals,
- it is adaptable in the sense that variance reduction techniques can be tailored to the specific problem, and
- it can be applied to highly complex problems for which the definite integral formulation is not obvious and standard analytic techniques are ineffective.

For rigorous mathematical justification for the methods employed in Monte Carlo simulations, one is urged to read [32,60]. As is typical with current MCMs, other numerical methods of solutions appear to be preferable when they may be used. Monte Carlo techniques often yield numerical answers of limited accuracy and are therefore employed as a last resort. However, there are problems for which the solution is not feasible using other methods. Problems that are probabilistic and continuous in nature (e.g., neutron absorption, charge transport in semiconductors, and scattering of waves by random media) are ideally suited to these methods and represent the most logical and efficient use of the stochastic methods. Since the recent appearance of vector machines, the importance of the Monte Carlo methods is growing.

It should be emphasized that in any Monte Carlo simulation, it is important to indicate the degree of confidence of the estimates or insert error bars in graphs illustrating Monte Carlo estimates. Without such information, Monte Carlo results are of questionable significance.

Applications of MCMs to other branches of science and engineering are summarized in [14,15,25,61]. EM-related problems, besides those covered in

Monte Carlo Methods 611

this chapter, to which Monte Carlo procedures have been applied include:
- diffusion problems [52,54,63]
- strip transmission lines [40]
- random periodic arrays [64]
- waveguide structures [65-69]
- scattering of waves by random media [70-75]
- noise in magnetic recording [76,77], and
- induced currents in biological bodies [78].

References

[1] R. Hersch and R. J. Griego, "Brownian motion and potential theory," *Sci. Amer.*, Mar. 1969, pp. 67–74.

[2] T. F. Irvine and J. P. Hartnett (eds.), *Advances in Heat Transfer*. New York: Academic Press, 1968.

[3] H. A. Meyer (ed.), *Symposium on Monte Carlo Methods*. New York: John Wiley, 1956.

[4] D. D. McCracken, "The Monte Carlo method," *Sci. Amer.*, vol. 192, May 1955, pp. 90-96.

[5] T. E. Hull and A. R. Dobell, "Random number generators," *SIAM Review*, vol. 4, no. 3, July 1982, pp. 230–254.

[6] D. E. Knuth, *The Art of Computer Programming*, vol. 2. Reading, MA: Addison-Wesley, 1969, pp. 9, 10, 78, 155.

[7] J. Banks and J. Carson, *Discrete Event System Simulation*. Englewood Cliffs, NJ: Prentice-Hall, 1984, pp. 257–288.

[8] P. A. W. Lewis, et al., "A pseudo-random number generator for the system/360," *IBM System Jour.*, vol. 8, no. 2, 1969, pp. 136–146.

[9] S. S. Kuo, *Computer Applications of Numerical Methods*. Reading, MA: Addison-Wesley, 1972, pp. 327–345.

[10] A. M. Law and W. D. Kelton, *Simulation Modeling and Analysis*. New York: McGraw-Hill, 1982, pp. 219–228.

[11] D. E. Raeside, "An introduction to Monte Carlo methods," *Amer. Jour. Phys.*, vol. 42, Jan. 1974, pp. 20–26.

[12] C. Jacoboni and L. Reggiani, "The Monte Carlo method for the solution of charge transport in semiconductors with applications to covalent materials," *Rev. Mod. Phys.*, vol. 55, no. 3, July 1983, pp. 645–705.

[13] H. Kobayashi, *Modeling and Analysis: An Introduction to System Performance Evaluation Methodology*. Reading, MA: Addison-Wesley, 1978, pp. 221–247.

[14] I. M. Sobol, *The Monte Carlo Method*. Chicago: University of Chicago Press, 1974, pp. 24–30.

[15] Y. A. Shreider, *Method of Statistical Testing (Monte Carlo Method)*, Amsterdam: Elsevier, 1964, pp. 39–83. Another translation of the same Russian text: Y. A. Shreider, *The Monte Carlo Method (The Method of Statistical Trials)*. Oxford: Pergamon, 1966.

[16] E. E. Lewis and W. F. Miller, *Computational Methods of Neutron Transport*. New York: John Wiley, 1984, pp. 296–360.

[17] I. S. Sokolinkoff and R. M. Redheffer, *Mathematics of Physics and Modern Engineering*. New York: McGraw-Hill, 1958, pp. 644–649.

[18] M. H. Merel and F. J. Mullin, "Analytic Monte Carlo error analysis," *J. Spacecraft*, vol. 5, no. 11, Nov. 1968, pp. 1304–1308.

[19] A. J. Chorin, "Hermite expansions in Monte-Carlo computation," *J. Comp. Phys.*, vol. 8, 1971, pp. 472–482.

[20] R. Y. Rubinstein, *Simulation and the Monte Carlo Method*. New York: John Wiley, 1981, pp. 20–90.

[21] W. J. Graybeal and U. W. Pooch, *Simulation: Principles and Methods*. Cambridge, MA: Winthrop Pub., 1980, pp. 77–97.

[22] B. J. T. Morgan, *Elements of Simulation*. London: Chapman & Hall, 1984, pp. 77–81.

[23] C. W. Alexion, et al., "Evaluation of radiation fields using statistical methods of integration,"*IEEE Trans. Ant. Prog.*. vol. AP-26, no. 2, Mar. 1979, pp. 288–293.

[24] R. C. Millikan, "The magic of the Monte Carlo method," *BYTE*, vol. 8, Feb. 1988, pp. 371–373.

[25] M. H. Kalos and P. A. Whitlook, *Monte Carlo Methods*. New York: John Wiley, 1986, pp. 89–116.

[26] J. M. Hammersley and D. C. Handscomb, *Monte Carlo Methods*. London: Methuen, 1964.

[27] S. Haber, "A modified Monte-Carlo quadrature II," *Math. Comp.*, vol. 21, July 1967, pp. 388–397.

[28] S. Haber, "Numerical evaluation of multiple integrals," *SIAM Rev.*, vol. 12, no. 4, Oct. 1970, pp. 481–527.

[29] J. M. Hammersley and K. W. Morton, "A new Monte Carlo technique: antithetic variates," *Proc. Camb. Phil. Soc.*, vol. 52, 1955, pp. 449–475.

[30] J. H. Halton and D. C. Handscomb, "A method for increasing the efficiency of Monte Carlo integration," *J. ACM*, vol. 4, 1957, pp. 329–340.

[31] S. J. Yakowitz, *Computational Probability and Simulation*. Reading, MA: Addison-Wesley, 1977, pp. 192, 193.

[32] S. Kakutani, "Two-dimensional Brownian motion harmonic functions," *Proc. Imp. Acad.* (Tokyo), vol. 20, 1944, pp. 706–714.

[33] A. Haji-Sheikh and E. M. Sparrow, "The floating random walk and its application to Monte Carlo solutions of heat equations," *J. SIAM Appl. Math.*, vol. 14, no. 2, Mar. 1966, pp. 370–389.

[34] A. Haji-Sheikh and E. M. Sparrow, "The solution of heat conduction problems by probability methods," *J. Heat Transfer*, Trans. ASME, Series C, vol. 89, no. 2, May 1967, pp. 121–131.

[35] G. E. Zinsmeiter, "Monte Carlo methods as an aid in teaching heat conduction," *Bull. Mech. Engr. Educ.*, vol. 7, 1968, pp. 77–86.

[36] R. Chandler, et al., "The solution of steady state convection problems by the fixed random walk method," *J. Heat Transfer*, Trans. ASME, Series C, vol. 90, Aug. 1968, pp. 361–363.

[37] G. E. Zinsmeiter and S. S. Pan, "A method for improving the efficiency of Monte Carlo calculation of heat conduction problems," *J. Heat Transfer*, Trans. ASME, Series C, vol. 96, 1974, pp. 246–248.

[38] G. E. Zinsmeiter and S. S. Pan, "A modification of the Monte Carlo method," *Inter. J. Num. Meth. Engr.*, vol. 10, 1976, pp. 1057–1064.

[39] G. M. Royer, "A Monte Carlo procedure for potential theory of problems," *IEEE Trans. Micro. Theo. Tech.*, vol. MTT-19, no. 10, Oct. 1971, pp. 813–818.

[40] R. M. Bevensee, "Probabilistic potential theory applied to electrical engineering problems," *Proc. IEEE*, vol. 61, no. 4, April 1973, pp. 423–437.

[41] R. L. Gibbs and J. D. Beason, "Solutions to boundary value problems of the potential type by random walk method," *Am. J. Phys.*, vol. 43, no. 9, Sept. 1975, pp. 782–785.

[42] J. H. Pickles, "Monte Carlo field calculations," *Proc. IEE*, vol. 124, no. 12, Dec. 1977, pp. 1271–1276.

[43] F. Sanchez-Quesada, et al., "Monte-Carlo method for discrete inhomogeneous problems," *Proc. IEE*, no. 125, no. 12, Dec. 1978, pp. 1400–1402.

[44] R. Schlott, "A Monte Carlo method for the Dirichlet problem of dielectric wedges," *IEEE Trans. Micro. Theo. Tech.*, vol. 36, no. 4, April 1988, pp. 724–730.

[45] M. N. O. Sadiku, "Monte Carlo methods in an introductory electromagnetic class," *IEEE Trans. Educ.*, vol. 35, no. 1, Feb. 1990, pp.

73–80.

[46] M. D. R. Beasley, et al., "Comparative study of three methods for computing electric fields," *Proc. IEE*, vol. 126, no. 1, Jan. 1979, pp. 126–134.

[47] J. R. Currie, et al., "Monte Carlo determination of the frequency of lightning strokes and shielding failures on transmission lines," *IEEE Trans. Power Appl. Syst.*, vol. PAS-90, 1971, pp. 2305–2310.

[48] R. S. Velazquez, et al., "Probabilistic calculations of lightning protection for tall buildings," *IEEE Trans. Indust. Appl.*, vol. IA-18, no. 3, May/June 1982, pp. 252–259.

[49] A. F. Emery and W. W. Carson, "A modification to the Monte Carlo method—the Exodus method," *J. Heat Transfer*, Trans. ASME, Series C, vol. 90, 1968, pp. 328–332.

[50] T. J. Hoffman and N. E. Banks, "Monte Carlo surface density solution to the Dirichlet heat transfer problem," *Nucl. Sci. Engr.*, vol. 59, 1976, pp. 205–214.

[51] T. J. Hoffman and N. E. Banks, "Monte Carlo solution to the Dirichlet problem with the double-layer potential density," *Trans. Amer. Nucl. Sci.*, vol. 18, 1974, pp. 136, 137. See also vol. 19, 1974, p. 164; vol. 24, 1976, p. 181.

[52] T. E. Booth, "Exact Monte Carlo solution of elliptic partial differential equations," *J. Comp. Phys.*, vol. 39, 1981, pp. 396–404.

[53] T. E. Booth, "Regional Monte Carlo solution of elliptic partial differential equations," *J. Comp. Phys.*, vol. 47, 1982, pp. 281–290.

[54] A. F. Ghoniem, "Grid-free simulation of diffusion using random walk methods," *J. Comp. Phys.*, vol. 61, 1985, pp. 1–37.

[55] E. S. Troubetzkoy and N. E. Banks, "Solution of the heat diffusion equation by Monte Carlo," *Trans. Amer. Nucl. Soc.*, vol. 19, 1974, pp. 163, 164.

[56] M. N. O. Sadiku and D. Hunt, "Solution of Dirichlet problems by the Exodus method," *IEEE Trans. Micro. Theo. Tech.*, vol. 40, no. 1, Jan. 1992, pp. 89–95.

[57] M. N. O. Sadiku, "Monte Carlo solution of axisymmetric potential problems," *Proc. of IEEE Industr. Appl. Soc. Annual Meeting*, Oct. 1990, pp. 1894–1990.

[58] M. N. O. Sadiku, *Elements of Electromagnetics*. New York: Holt, Rinehart & Winston, 1989.

[59] W. L Dunn, "Inverse Monte Carlo analysis," *J. Comp. Phys.*, vol. 41, 1981, pp. 154–166.

[60] A. W. Knapp, "Connection between Brownian motion and potential theory," *J. Math. Analy. Appl.*, vol. 12, 1965, pp. 328–349.

[61] J. H. Halton, "A retrospective and prospective survey of the Monte Carlo method," *SIAM Rev.*, vol. 12, no. 1, Jan. 1970, pp. 1–61.

[62] A. T. Bharucha-Reid (ed.), *Probabilistic Methods in Applied Mathematics*. New York: Academic Press, vol. 1, 1968; vol. 2, 1970; vol. 3, 1973.

[63] G. W. King, "Monte Carlo method for solving diffusion problems," *Ind. Engr. Chem.*, vol. 43, no. 11, Nov. 1951, pp. 2475–2478.

[64] Y. T. Lo, "Random periodic arrays," *Rad. Sci.*, vol. 3, no. 5, May 1968, pp. 425–436.

[65] T. Troudet and R. J. Hawkins, "Monte Carlo simulation of the propagation of single-mode dielectric waveguide structures," *Appl. Opt.*, vol. 27, no. 24, Feb. 1988, pp. 765–773.

[66] T. R. Rowbotham and P. B. Johns, "Waveguide analysis by random walks," *Elect. Lett.*, vol. 8, no. 10, May 1972, pp. 251–253.

[67] P. B. Johns and T. R. Rowbotham, "Solution of resistive meshes by deterministic and Monte Carlo transmission-line modelling," *IEE Proc.*, vol. 128, Part A, no. 6, Sept. 1981, pp. 453–462.

[68] R. G. Olsen, "The application of Monte Carlo techniques to the study of impairments in the waveguide transmission system," *B. S. T. J.*, vol. 50, no. 4, April 1971, pp. 1293–1310.

[69] H. E. Rowe and D. T. Young, "Transmission distortion in multimode random waveguides," *IEEE Trans. Micro. Theo. Tech.*, vol. MMT-20, no. 6, June 1972, pp. 349–365.

[70] M. Nieto-Vesperinas and J. M. Soto-Crespo, "Monte Carlo simulations for scattering of electromagnetic waves from perfectly conductive random rough surfaces," *Opt. Lett.*, vol. 12, no. 12, Dec. 1987, pp. 979–981.

[71] G. P. Bein, "Monte Carlo computer technique for one-dimensional random media," *IEEE Trans. Ant. Prop.*, vol. AP-21, no. 1, Jan. 1973, pp. 83–88.

[72] N. Garcia and E. Stoll, "Monte Carlo calculation for electromagnetic-wave scattering from random rough surfaces," *Phy. Rev. Lett.*, vol. 52, no. 20, May 1984, pp. 1798–1801.

[73] J. Nakayama, "Anomalous scattering from a slightly random surface," *Rad. Sci.*, vol. 17, no. 3, May-June 1982, pp. 558–564.

[74] A. Ishimaru, *Wave Propagation and Scattering in Random Media*. New York: Academic Press, vol. 2, 1978.

[75] A. K. Fung and M. F. Chen, "Numerical simulation of scattering from simple and composite random surfaces," *J. Opt. Soc. Am.*, vol. 2, no. 12, Dec. 1985, pp. 2274–2284.

[76] R. A. Arratia and H. N. Bertram, "Monte Carlo simulation of particulate noise in magnetic recording," *IEEE Trans. Mag.*, vol. MAG-20, no. 2, Mar. 1984, pp. 412–420.

[77] P. K. Davis, "Monte Carlo analysis of recording codes," *IEEE Trans. Mag.*, vol. MAG-20, no. 5, Sept. 1984, p. 887.

[78] J. H. Pickles, "Monte-Carlo calculation of electrically induced human-body currents," *IEE Proc.*, vol. 134, Pt. A, no. 9, Nov. 1987, pp. 705–711.

Problems

8.1 Write a program to generate 1000 pseudorandom numbers U uniformly distributed between 0 and 1. Calculate their mean and compare the calculated mean with the expected mean (0.5) as a test of randomness.

8.2 Generate 10,000 random numbers uniformly distributed between 0 and 1. Find the percentage of numbers between 0 and 0.1, between 0.1 and 0.2, etc., and compare your results with the expected distribution of 10% in each interval.

8.3 (a) Using the linear congruential scheme, generate 10 pseudorandom numbers with $a = 1573, c = 19, m = 10^3$, and seed value $X_0 = 89$.
(b) Repeat the generation with $c = 0$.

8.4 For $a = 13, m = 2^6 = 64$, and $X_0 = 1, 2, 3$, and 4, find the period of the random number generator using the multiplicative congruential method.

8.5 Develop a subroutine that uses the inverse transformation method to generate a random number from a distribution with the probability density function

$$f(x) = \begin{cases} 0.25, & 0 \leq x \leq 1 \\ 0.75, & 1 \leq x \leq 1. \end{cases}$$

8.6 It is not easy to apply the inverse transform method to generate normal distribution. However, by making use of the approximation

$$e^{-x^2/2} \simeq \frac{2\,e^{-kx}}{\left(1 + e^{-kx}\right)^2}, \quad x > 0$$

where $k = \sqrt{\frac{8}{\pi}}$, the inverse transform method can be applied. Develop a subroutine to generate normal deviates using inverse transform method.

8.7 Using the rejection method, generate a random variable from $f(x) = 5x^2$, $0 \leq x \leq 1$.

8.8 Use the rejection method to generate Gaussian (or normal) deviates in the truncated region $-a \leq X \leq a$.

8.9 Use sample mean Monte Carlo integration to evaluate:

(a) $\int_0^1 4\sqrt{1-x^2}\,dx$,

(b) $\int_0^1 \sin x\,dx$,

(c) $\int_0^1 e^x\,dx$,

(d) $\int_0^1 \frac{1}{\sqrt{x}}\,dx$.

8.10 Evaluate the following four-dimensional integrals:

(a) $\int_0^1 \int_0^1 \int_0^1 \int_0^1 exp(x^1 x^2 x^3 x^4 - 1)\,dx^1\,dx^2\,dx^3\,dx^4$,

(b) $\int_0^1 \int_0^1 \int_0^1 \int_0^1 \sin(x^1 + x^2 + x^3 + x^4)\,dx^1\,dx^2\,dx^3\,dx^4$.

8.11 The radiation from a rectangular aperture with constant amplitude and phase distribution may be represented by the integral

$$I(\alpha,\beta) = \int_{-1/2}^{1/2} \int_{-1/2}^{1/2} e^{j(\alpha x + \beta y)}\,dx\,dy.$$

Evaluate this integral using a Monte Carlo procedure and compare your result for $\alpha = \beta = \pi$ with the exact solution

$$I(\alpha,\beta) = \frac{\sin(\alpha/2)\sin(\beta/2)}{\alpha\beta/4}.$$

618 Numerical Techniques in Electromagnetics

8.12 Consider the differential equation

$$\frac{\partial^2 W}{\partial x^2} + \frac{\partial^2 W}{\partial y^2} + \frac{k}{y}\frac{\partial W}{\partial y} = 0$$

where k = constant. By finding its finite difference form, give a probabilistic interpretation to the equation.

8.13 Given the one-dimensional differential equation

$$y'' = 0, \qquad 0 \leq x \leq 1$$

subject to $y(0) = 0, y(1) = 10$, use an MCM to find $y(0.25)$ assuming $\Delta x = 0.25$ and the following 20 random numbers:

0.1306, 0.0422, 0.6597, 0.7905, 0.7695, 0.5106, 0.2961, 0.1428, 0.3666, 0.6543, 0.9975, 0.4866, 0.8239, 0.8722, 0.1330, 0.2296, 0.3582, 0.5872, 0.1134, 0.1403.

8.14 Use a Monte Carlo method to solve Laplace's equation in the triangular region $x \geq 0, y \geq 0, x + y \leq 1$ with the boundary condition $V(x,y) = x + y + 0.5$. Determine V at (0.4, 0.2), (0.35, 0.2), (0.4, 0.15), (0.45, 0.2), and (0.4, 0.25).

8.15 Use a Monte Carlo procedure to determine the potential at points (2,2), (3,3), and (4,4) in the problem shown in Fig. 8.15(a). By virtue of double symmetry, it is sufficient to consider a quarter of the solution region as shown in Fig. 8.15(b).

8.16 In the solution region of Fig. 8.16, $\rho_s = x(y-1)$ nC/m^2. Find the potential at the center of the region using a Monte Carlo method.

8.17 Consider the potential system shown in Fig. 8.17. Determine the potential at the center of the solution region. Take $\epsilon_r = 2.25$.

8.18 Apply an MCM to solve Laplace's equation in the three-dimensional region

$$|x| \leq 1, \quad |y| \leq 0.5, \quad |z| \leq 0.5$$

subject to the boundary condition

$$V(x,y,z) = x + y + z + 0.5.$$

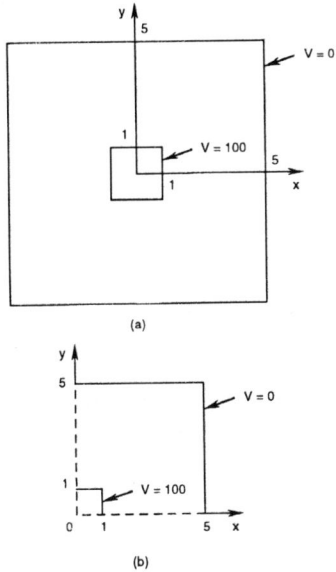

Figure 8.15 For Problem 8.15.

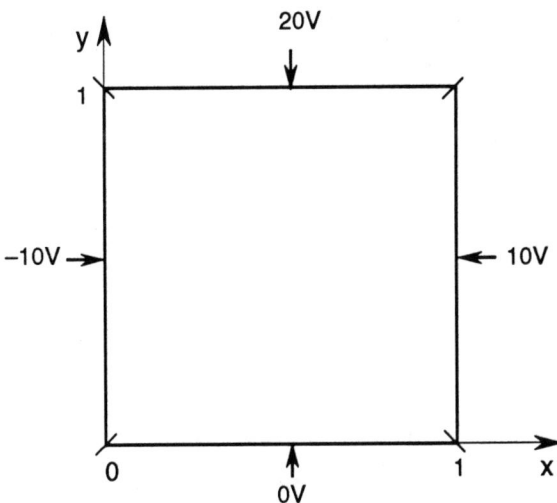

Figure 8.16 For Problem 8.16.

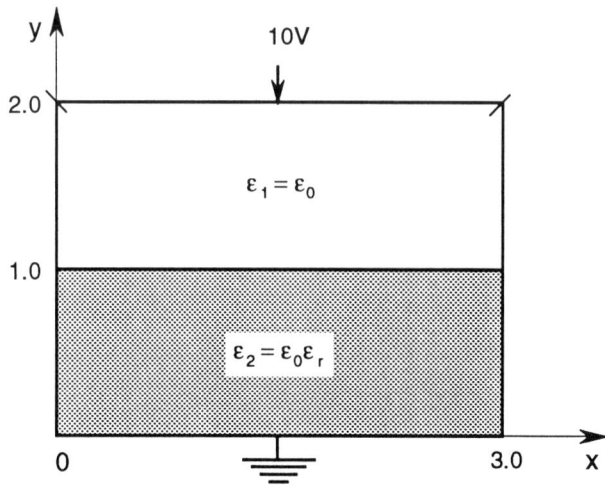

Figure 8.17 For Problem 8.17.

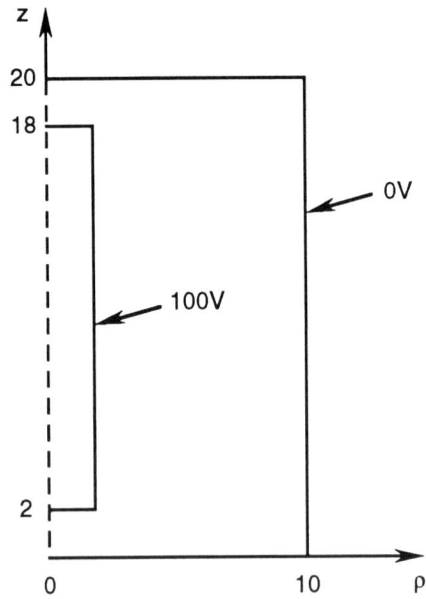

Figure 8.18 For Problem 8.20.

Find the solution at (0.5, 0.1, 0.1).

8.19 Show that the probabilistic interpretation to the finite difference equivalent of Laplace's equation for an axisymmetric potential problem, $V = V(\rho, z)$, is

$$V(\rho,z) = p_{\rho+}V(\rho+\Delta,z) + p_{\rho-}V(\rho-\Delta,z) + p_{z+}V(\rho,z+\Delta) + p_{z-}V(\rho,z-\Delta)$$

where $\Delta\rho = \Delta z = \Delta$. For $\rho \neq 0$,

$$p_{z+} = p_{z-} = \frac{1}{4}, \quad p_{\rho+} = \frac{1}{4} + \frac{\Delta}{8\rho}, \quad p_{\rho-} = \frac{1}{4} - \frac{\Delta}{8\rho}$$

and for $\rho = 0$ ($\partial V/\partial \rho = 0$),

$$p_{z+} = p_{z-} = \frac{1}{6}, \quad p_{\rho+} = \frac{2}{3}, \quad p_{\rho+} = 0.$$

8.20 Consider the finite cylindrical conductor held at $V = 100$ enclosed in a larger grounded cylinder as shown in Fig. 3.53. The axial symmetric problem is portrayed in Fig. 8.18 for your convenience. Using a Monte Carlo technique, write a program to determine the potential at points $(\rho, z) = (2, 10), (5, 10), (8, 10), (5, 2)$, and $(5,18)$.

8.21 Figure 3.54 shows a prototype of an electrostatic particle focusing system employed in a recoil-mass time-of-flight spectometer. It is essentially a finite cylindrical conductor that abruptly expands radius by a factor of 2. The axial symmetric problem is portrayed in Fig. 8.19 for your convenience. Write a program based on an MCM to calculate the potential at points $(\rho, z) = (5, 18), (5, 10), (5, 2), (10, 2)$, and $(15,2)$.

8.22 Consider the square region shown in Fig. 8.20. The transition probability $p(Q, S_i)$ is defined as the probability that a randomly walking particle leaving point Q will arrive at side S_i of the square boundary. Using the Exodus method, write a program to determine:
(a) $p(Q_1, S_i)$, $i = 1, 2, 3, 4$,
(b) $p(Q_2, S_i)$, $i = 1, 2, 3, 4$.

622   Numerical Techniques in Electromagnetics

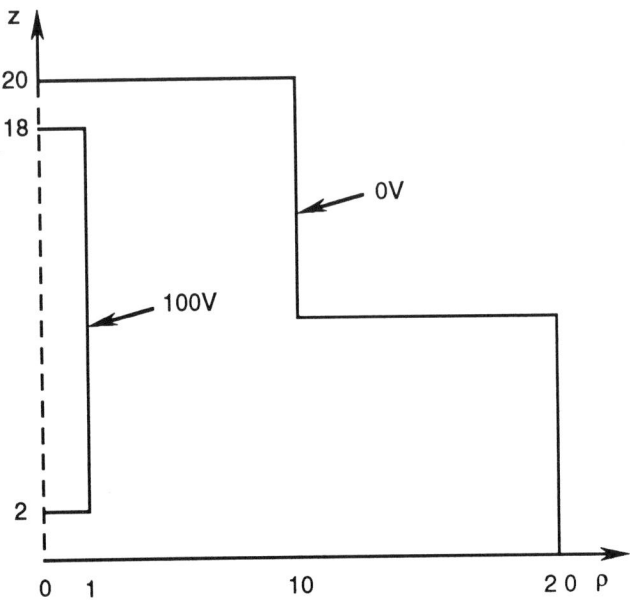

Figure 8.19   For Problem 8.21.

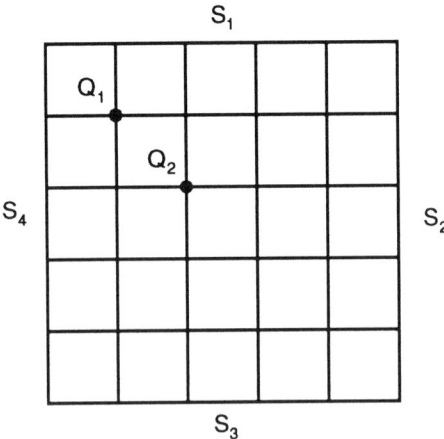

Figure 8.20   For Problem 8.22.

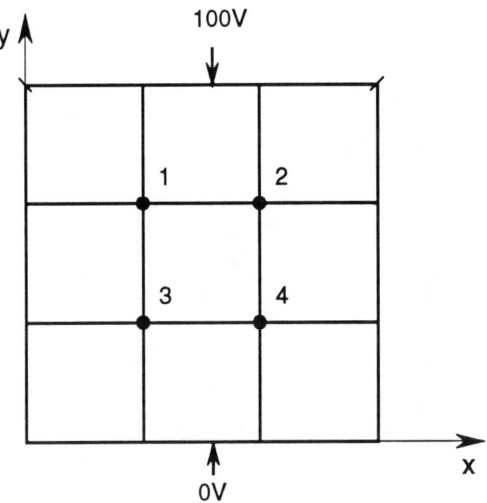

Figure 8.21 For Problem 8.24.

8.23 Given the one-dimensional differential equation

$$\frac{d^2\Phi}{dx^2} = 0, \quad 0 \le x \le 1$$

subject to $\Phi(0) = 0$, $\Phi(1) = 10$, use the Exodus method to find $\Phi(0.25)$ by injecting 256 particles at $x = 0.25$. You can solve this problem by hand calculation.

8.24 Use the Exodus method to find the potential at node 4 in Fig. 8.21. Inject 256 particles at node 4 and scan nodes in the order 1, 2, 3, 4. You can solve this problem by hand calculation.

8.25 Using the Exodus method, write a program to calculate $V(0.25, 0.75)$ in Example 8.5.

8.26 Write a program to calculate $V(1.0, 1.5)$ in Example 8.6 using the Exodus method.

8.27 Write a program that will apply the Exodus method to determine the potential at point $(0.2, 0.4)$ in the system shown in Fig. 8.22.

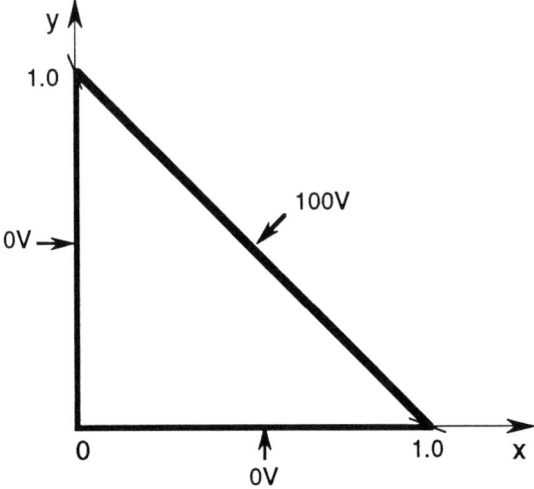

Figure 8.22 For Problem 8.27.

Appendix A
Vector Relations

A.1 Vector Identities

If **A** and **B** are vector fields while U and V are scalar fields, then

$$\nabla V(U+V) = \nabla U + \nabla V$$

$$\nabla(UV) = U\nabla V + V\nabla U$$

$$\nabla\left(\frac{U}{V}\right) = \frac{V(\nabla U) - U(\nabla V)}{V^2}$$

$$\nabla V^n = n\, V^{n-1}\, \nabla V \qquad (n = \text{integer})$$

$$\nabla(\mathbf{A}\cdot\mathbf{B}) = (A\cdot\nabla)\mathbf{B} + (\mathbf{B}\cdot\nabla)\mathbf{A} + \mathbf{A}\times(\nabla\times\mathbf{B}) + \mathbf{B}\times(\nabla\times\mathbf{A})$$

$$\nabla\cdot(\mathbf{A}+\mathbf{B}) = \nabla\cdot\mathbf{A} + \nabla\cdot\mathbf{B}$$

$$\nabla\cdot(\mathbf{A}\times\mathbf{B}) = \mathbf{B}\cdot(\nabla\times\mathbf{A}) - \mathbf{A}\cdot(\nabla\times\mathbf{B})$$

$$\nabla\cdot(V\mathbf{A}) = V\nabla\cdot\mathbf{A} + \mathbf{A}\cdot\nabla V$$

$$\nabla\cdot(\nabla V) = \nabla^2 V$$

$$\nabla\cdot(\nabla\times\mathbf{A}) = 0$$

$$\nabla\times(\mathbf{A}+\mathbf{B}) = \nabla\times\mathbf{A} + \nabla\times\mathbf{B}$$

$$\nabla\times(\mathbf{A}\times\mathbf{B}) = \mathbf{A}(\nabla\cdot\mathbf{B}) - \mathbf{B}(\nabla\cdot\mathbf{A}) + (\mathbf{B}\cdot\nabla)\mathbf{A} - (\mathbf{A}\cdot\nabla)\mathbf{B}$$

$$\nabla\times(V\mathbf{A}) = \nabla V\times\mathbf{A} + V(\nabla\times\mathbf{A})$$

$$\nabla\times(\nabla V) = 0$$

$$\nabla\times(\nabla\times\mathbf{A}) = \nabla(\nabla\cdot\mathbf{A}) - \nabla^2\mathbf{A}$$

A.2 Vector Theorems

If v is the volume bounded by the closed surface S, and $\mathbf{a}_n$ is a unit normal to S, then

$$\oint_S \mathbf{A} \cdot d\mathbf{S} = \int_v \nabla \cdot \mathbf{A}\, dv \quad \text{(Divergence theorem)}$$

$$\oint_S V\, d\mathbf{S} = \int_v \nabla V\, dv \quad \text{(Gradient theorem)}$$

$$\oint_S \mathbf{A} \times d\mathbf{S} = -\int_v \nabla \times \mathbf{A}\, dv$$

$$\oint_S \left[(\mathbf{A} \cdot d\mathbf{S})\mathbf{A} - \frac{1}{2} A^2\, d\mathbf{S} \right] = \int_v \left[(\nabla \times \mathbf{A}) \times \mathbf{A} + \mathbf{A} \nabla \cdot \mathbf{A} \right] dv$$

$$\oint_S U \nabla V \cdot d\mathbf{S} = \int_v \left[U \nabla^2 V + \nabla U \cdot \nabla V \right] dv \quad \text{(Green 1st identity)}$$

$$\oint_S \left[U \nabla V - V \nabla U \right] \cdot d\mathbf{S} = \int_v \left[U \nabla^2 V - V \nabla^2 U \right] dv \quad \text{(Green 2nd identity)}$$

where $d\mathbf{S} = dS\, \mathbf{a}_n$.

If S is the area bounded by the closed path L and the positive directions of elements $d\mathbf{S}$ and $d\mathbf{l}$ are related by the right-hand rule, then

$$\oint_L \mathbf{A} \cdot d\mathbf{l} = \int_S \nabla \times \mathbf{A} \cdot d\mathbf{S} \quad \text{(Stokes' theorem)}$$

$$\oint_L V\, d\mathbf{l} = -\int_S \nabla V \times d\mathbf{S}.$$

A.3 Orthogonal Coordinates

Rectangular Coordinates (x, y, z)

$$\nabla V = \frac{\partial V}{\partial x} \mathbf{a}_x + \frac{\partial V}{\partial y} \mathbf{a}_y + \frac{\partial V}{\partial z} \mathbf{a}_z$$

$$\nabla \cdot \mathbf{A} = \frac{\partial A_x}{\partial x} + \frac{\partial A_y}{\partial y} + \frac{\partial A_z}{\partial z}$$

$$\nabla \times \mathbf{A} = \left[\frac{\partial A_z}{\partial y} - \frac{\partial A_y}{\partial z} \right] \mathbf{a}_x + \left[\frac{\partial A_x}{\partial z} - \frac{\partial A_z}{\partial x} \right] \mathbf{a}_y + \left[\frac{\partial A_y}{\partial x} - \frac{\partial A_x}{\partial y} \right] \mathbf{a}_z$$

$$\nabla^2 V = \frac{\partial^2 V}{\partial x^2} + \frac{\partial^2 V}{\partial y^2} + \frac{\partial^2 V}{\partial z^2}$$

$$\nabla^2 \mathbf{A} = \nabla^2 A_x \mathbf{a}_x + \nabla^2 A_y \mathbf{a}_y + \nabla^2 A_z \mathbf{a}_z$$

Appendix A: Vector Relations

Cylindrical Coordinates (ρ, ϕ, z)

$$\nabla V = \frac{\partial V}{\partial \rho}\mathbf{a}_\rho + \frac{1}{\rho}\frac{\partial V}{\partial \phi}\mathbf{a}_\phi + \frac{\partial V}{\partial z}\mathbf{a}_z$$

$$\nabla \cdot \mathbf{A} = \frac{1}{\rho}\frac{\partial}{\partial \rho}(\rho A_\rho) + \frac{1}{\rho}\frac{\partial}{\partial \phi}A_\phi + \frac{\partial}{\partial z}A_z$$

$$\nabla \times \mathbf{A} = \left[\frac{1}{\rho}\frac{\partial}{\partial \phi}A_z - \frac{\partial}{\partial z}A_\phi\right]\mathbf{a}_\rho + \left[\frac{\partial}{\partial z}A_\rho - \frac{\partial}{\partial \rho}A_z\right]\mathbf{a}_\phi$$

$$+ \frac{1}{\rho}\left[\frac{\partial}{\partial \rho}(\rho A_\phi) - \frac{\partial}{\partial \phi}A_\rho\right]\mathbf{a}_z$$

$$\nabla^2 V = \frac{1}{\rho}\frac{\partial}{\partial \rho}\left(\rho\frac{\partial V}{\partial \rho}\right) + \frac{1}{\rho^2}\frac{\partial^2 V}{\partial \phi^2} + \frac{\partial^2 V}{\partial z^2}$$

$$\nabla^2 \mathbf{A} = \left[\nabla^2 A_\rho - \frac{2}{\rho^2}\frac{\partial}{\partial \phi}A_\phi - \frac{A_\rho}{\rho^2}\right]\mathbf{a}_\rho$$

$$+ \left[\nabla^2 A_\phi + \frac{2}{\rho^2}\frac{\partial}{\partial \phi}A_\rho - \frac{A_\phi}{\rho^2}\right]\mathbf{a}_\phi + \nabla^2 A_z \mathbf{a}_z$$

Spherical Coordinates (r, θ, ϕ)

$$\nabla V = \frac{\partial V}{\partial r}\mathbf{a}_r + \frac{1}{r}\frac{\partial V}{\partial \theta}\mathbf{a}_\theta + \frac{1}{r\sin\theta}\frac{\partial V}{\partial \phi}\mathbf{a}_\phi$$

$$\nabla \cdot \mathbf{A} = \frac{1}{r^2}\frac{\partial}{\partial r}(r^2 A_r) + \frac{1}{r\sin\theta}\frac{\partial}{\partial \theta}(\sin\theta A_\theta) + \frac{1}{r\sin\theta}\frac{\partial A_\phi}{\partial \phi}$$

$$\nabla \times \mathbf{A} = \frac{1}{r\sin\theta}\left[\frac{\partial}{\partial \theta}(\sin\theta A_\phi) - \frac{\partial}{\partial \phi}A_\theta\right]\mathbf{a}_r + \frac{1}{r}\left[\frac{1}{\sin\theta}\frac{\partial}{\partial \phi}A_r - \frac{\partial}{\partial r}(rA_\phi)\right]\mathbf{a}_\theta$$

$$+ \frac{1}{r}\left[\frac{\partial}{\partial r}(rA_\theta) - \frac{\partial}{\partial \theta}A_r\right]\mathbf{a}_\phi$$

$$\nabla^2 V = \frac{1}{r^2}\frac{\partial}{\partial r}\left(r^2\frac{\partial V}{\partial r}\right) + \frac{1}{r^2\sin\theta}\frac{\partial}{\partial \theta}\left(\sin\theta\frac{\partial V}{\partial \theta}\right) + \frac{1}{r^2\sin^2\theta}\frac{\partial^2 V}{\partial \phi^2}$$

$$\nabla^2 \mathbf{A} = \left[\nabla^2 A_r - \frac{2}{r^2\sin\theta}\frac{\partial}{\partial \theta}(\sin\theta A_\theta) - \frac{2}{r^2\sin\theta}\frac{\partial}{\partial \phi}A_\phi - \frac{2}{r^2}A_r\right]\mathbf{a}_r$$

$$+ \left[\nabla^2 A_\theta + \frac{2}{r^2}\frac{\partial}{\partial \theta}A_r - \frac{2\cos\theta}{r^2\sin^2\theta}\frac{\partial}{\partial \phi}A_\phi - \frac{1}{r^2\sin^2\theta}A_\theta\right]\mathbf{a}_\theta$$

$$+ \left[\nabla^2 A_\phi + \frac{2\cos\theta}{r^2\sin^2\theta}\frac{\partial}{\partial \phi}A_\theta + \frac{2}{r^2\sin\theta}\frac{\partial}{\partial \phi}A_r - \frac{1}{r^2\sin^2\theta}A_\phi\right]\mathbf{a}_\phi$$

Appendix B

Review of FORTRAN

This appendix is meant to review or perhaps add more to what the reader has already learned from an introductory course on FORTRAN (FORmula TRANslation) programming. For further treatment of rules that are not covered here, one may consult the Schaum's outline [1] or a FORTRAN reference manual.

B.1 Characters

The FORTRAN language has 49 characters comprising of 26 alphabets from A to Z, the 10 digits from 0 to 9, and 12 symbols:

$$+ \quad - \quad * \quad / \quad = \quad (\quad) \quad \$ \quad . \quad , \quad ' \quad :$$

and the blank space, often denoted by [ ]. The alphabets and digits are referred to as alphanumeric characters.

B.2 Variables

A variable is a symbolic name assigned to a memory location in the computer. It consists of 1 to 6 characters, an alphabet followed by alphanumeric characters. There are basically two types of variables. Variables starting with the 6 alphabets

$$I, J, K, L, M, N$$

are considered integers unless declared otherwise. Examples of integer variables are I2, LAD, and N27. Variables beginning with alphabets

$$A, B, \cdots, H, O, P, \cdots, Z$$

are presumed real unless stated otherwise. Examples of real variables are ABU, H20, P4Q, and S2.

B.3 Constants

There are two types of numerical constants or numbers: the integer (or fixed-point) number such as 9, −34, 456, etc., and the real (floating-point) numbers such as 9.8, −34.2, 456.8, etc. A real constant may be represented in the exponential form, e.g., 1.42×10^{-5} may be written as 1.42E−5. In double precision, the same constant may be written as 1.42D−5. A complex number $0.3 + j5.4$ is written as (0.3, 5.4). There are two logical constants: .TRUE. and .FALSE. A character constant is a string of characters enclosed by apostrophes, e.g., 'THE NO. is 409'.

B.4 Statement Declaration

If a variable is to be used other than in its presumed form, it must be declared, e.g.,

INTEGER A, H, X2
REAL I, JOHN, MM
COMPLEX J, U
DOUBLE PRECISION ABU, F2
LOGICAL N, F2, A

If a linear or multidimensional array is to be used, its dimension must be declared, e.g.,

DIMENSION I(5), A(0:10), X2(−5:5), Y(4,5), UGO(−5:0,2:10)

where I, A, and X2 are one-dimensional arrays with I(1), ⋯, I(15), A(0), ⋯, A(10), and X2(−5), ⋯, X(5), respectively, while Y and UGO are two-dimensional.

If constants are to be stored, they must be declared in the DATA statement having the general form

DATA list of variables/ list of constants /

e.g.,

DATA A/0.25/, PIE/3.14159/
DATA NAME/'TOM'/, K/50/
DATA (J(I), I=0,5) /2, −3, 4*8/

where 4*8 instructs the compiler to use 8 four times.

If variables or arrays in the main program and subprogram are to be related, they must be declared using COMMON statement. There are two types of COMMON statements: named and unnamed (or blank). The named type has the general form

COMMON / name / list

e.g.,

>COMMON /P/ X, Y, J(10)
>COMMON /BLK2/ T, PIE

The unnamed type has the form

>COMMON list

e.g.

>COMMON C(10),D

B.5 Arithmetic Statements and Mathematical Functions

An arithmetic statement instructs the compiler to perform a certain function which may involve any of the five arithmetic operators:

- +   addition
- −   subtraction
- *   multiplication
- /   division
- ** exponentiation

If parentheses are not used to specify the order of operation, the following order is presumed:

1. exponentiation,
2. multiplication and division,
3. addition and subtraction.

Thus A**B/C − D is assumed to mean $A^B/C - D$. Examples of arithmetic statements are:

```
N = N + 1
A(I) = ( B(I) + C(I)/2.0 )**4
F = Q*SIN(X) + F*H - 10.5
```

where SIN(X) is a built-in function. The FORTRAN language has a library of mathematical functions. Examples are:

Mathematics	FORTRAN
$\sqrt{x}$	SQRT(X)
e^x	EXP(X)
$\log_e(x)$	ALOG(X)
$\log_{10}(x)$	ALOG10(X)
$\tan^{-1}(x)$	ATAN(X)

632 Numerical Techniques in Electromagnetics

Some of these functions have generic and intrinsic types [2]. For example,

Y = SIN(X), where x is in radians,
Y = SIND(X), where x is in degrees,
Y = DSIN(X), where x and y are in double precision,
Y = CSIN(X), where x and y are complex.

For a more extensive list of FORTRAN mathematical functions, see Section B.12.

B.6 Transfer of Control

A transfer statement is used to alter the flow of logic. There are two types of transfer statements: unconditional and conditional. The conditional transfer of control has the form

$$\text{GO TO n}$$

where n is the number of the statement whom control should be transferred. For example,

> GO TO 5
> $\vdots$
> 5   N = N + 1

The conditional transfer statements are of two main types. One type, the computed GO TO, has the form

$$\text{GO TO } (n_1, n_2, \cdots, n_k), \text{ I}$$

where n_i are statement numbers and I is a variable which can only take values $1, 2, \cdots, k$. For example,

$$\text{GO TO } (10, 20, 50, 10), \text{ LIN}$$

means that if

> LIN = 1, 4,   GO TO 10
> LIN = 2,   GO TO 20
> LIN = 3,   GO TO 50

The other type of conditional transfer is the IF statement. The IF statement can assume two forms: arithmetic and logical. The arithmetic IF statement has the form

$$\text{IF (expr) } n_1, n_2, n_3$$

meaning that if

> expr < 0,   GO TO n_1
> expr = 0,   GO TO n_2

expr > 0,   GO TO n_3

The logical IF statement has the form

IF (logical expr)   statement X

meaning that if the logical expression is true, execute statement X, e.g.

IF (A.LT.B) GO TO 50
IF(J.NE.K) J = J + 1
IF (N.EQ.5.AND.M.LT.10) WRITE (6,10) X,Y

Common relational operators used in the logical IF statement are

.LT.	less than
.LE.	less than or equal to
.GT.	greater than
.GE.	greater than or equal to
.EQ.	equal to
.NE.	not equal to

while the logical operators are

.AND.	and
.OR.	or
.NOT.	not

Another type of IF statement is the IF THEN ELSE with the general form

IF (expr) THEN
⋮   (statements)
ELSE
⋮   (statements)
ENDIF

e.g.,

IF (N.EQ.5) THEN
   M = 2*N
   SHAD = .TRUE.
ELSE
   M = N
   SHAD = .FALSE.
ENDIF

The ENDIF statement terminates the block IF statement. The END statement terminates the main program or a subprogram. It must also be the last statement of a FORTRAN program.

B.7 DO Loops

The DO loop allows us to repeat a process in a FORTRAN program. It has the general form

DO   n   I = IN, IE, INC

⋮

⋮   (statements)

⋮

n   CONTINUE

where I denotes any integer variable, IN = initial value of I, IE = end value of I, and INC = increment (nonnegative). For example,

SUM = 0.0
DO 100 K = 1, 30, 2
SUM = SUM + FLOAT(K)/5.0
100   CONTINUE

B.8 Input/output Statement

The FORMAT statement instructs the compiler the format in which data are to be read, written, or printed. The unformatted (or format-free) input/output statements have the forms

READ *, list
PRINT *, list

where list is the list of items to be read or printed. The formatted input/output statements require field specifications. Common field specifications include X, I, F, and E. Their general forms are:

nX   leaves n blank spaces
Iw   denotes integer type of data with width w
Fw.d   indicates real type of data with width w, to d decimal places
Ew.d   specifies real type of data in exponential form with width w, to d decimal places.

Examples of input/output statements are

>   READ *, N, A
>   PRINT *, A
>   READ (5,10) T, K, B
>   10 FORMAT (F12.3, I4, E10.7)
>   WRITE (6,20) M, U, V
>   20 FORMAT (2X, 'M=', I3, 'U=', E9.3, 'V=', F8.3)

B.9 Subprograms

These are useful when a "package" of calculations is required within a program. There are two basic types of subprograms: FUNCTION and SUBROUTINE.

The FUNCTION subprogram has the following form:

FUNCTION NAME (arguments)

⋮

⋮   (statements)

⋮

NAME = (expression)
RETURN
END

where arguments are input variables and NAME is the output. As an example, if the calculation of $[x^2 + y^2]^{1/2} - 10$ is required several times in the main program, the following subprogram can be used:

FUNCTION FUN(X,Y)
R = X**2 + Y**2
FUN = SQRT(R) −10.0
RETURN
END

The SUBROUTINE subprogram has the general form

SUBROUTINE NAME (arguments)

⋮

⋮   (statements)

⋮

RETURN

END

where arguments are input and output variables and NAME is the name given to the subroutine. For example, $[x^2 + y^2]^{1/2} - 10$ can be calculated repeatedly by calling the following subprogram:

SUBROUTINE ABU(X,Y,FUN)
R = X**2 + Y**2
FUN = SQRT(R) −10.0
RETURN
END

where FUN is returned to the calling program. The major difference between SUBROUTINE and FUNCTION subprograms is the way they are used in the main program. While a value is assigned to NAME in FUNCTION subprogram, values are transmitted to the main program via the arguments in the SUBROUTINE subprogram. A CALL statement is required in the main program to use a SUBROUTINE. In the example given above, the main program will contain, for example,

A = 2.0
B = −3.0
U = FUN(A, B)

if FUNCTION subprogram is used or

A = 2.0
B = −3.0
CALL ABU(A, B, C)
U = C

if SUBROUTINE subprogram is used. Note that in terms of variable type A, B, and C in the main program must match X, Y, and FUN in the subprograms.

B.10 Other Statements

Other statements useful in writing a FORTRAN program include STOP, PARAMETERS, EQUIVALENCE, and IMPLICIT. The STOP statement is an executable statement which stops the program. As many STOP statements may be used as needed, and they may be placed anywhere in the main program or subprogram. A STOP statement must appear before the END statement in the main program to serve as the last logical statement.

The PARAMETER statement, usually at the beginning of a program, assigns numerical values to specified variables. It has the general form

PARAMETER (variable$_1$ = constant$_1$, variable$_2$ = constant$_2$, ⋯)

For example,

PARAMETER (DIMEN = 10, PIE = 3.14593, FLAG = .TRUE.)
DIMENSION A(DIMEN), B(DIMEN,DIMEN)

To alter the size of arrays A and B, only the numerical value assigned DIMEN needs to be changed.

The EQUIVALENCE statement is inserted at the beginning of a program to allow the same memory location be referred to by different names. It has the general form

EQUIVALENCE (list), (list), ···

For example,

DIMENSION T(10)
EQUIVALENCE (A, U, B), (T(1), D)

tells the compiler that A, U, and B will share the same memory location while T(1) and D will share the same storage location.

The IMPLICIT declaration, usually at the beginning of a program, overrides the implied type of all variables beginning with a declared letter or range of letter. It has the general form

IMPLICIT type$_1$ (list), type$_2$ (list),···

where the types are INTEGER, REAL, COMPLEX, LOGICAL, CHARACTER, and DOUBLE PRECISION. For example,

IMPLICIT COMPLEX (A,B), REAL (I-L)
REAL   ABET
INTEGER K2

declares all variables beginning with A or B to be complex except ABET, which is subsequently declared as real, and all variables beginning with L, J, K, or L as real except K2, which is declared as integer.

B.11 Program Organization

The 80 columns of each line of a FORTRAN program may be used as follows:

Columns	Contents
1	C or * is placed here for comments
1 – 5	statement numbering
6	a character is placed here to show continuation of the previous line
7 – 72	FORTRAN statement
73 – 80	used for identification, sequencing, or any other purpose (optional).

B.12 FORTRAN Library Functions

The following in-built subprograms are common to most FORTRAN systems [2]. The letters R, I, C, and D denote real, integer, complex, and double precision argument, respectively.

B.12.1 Integer Functions

Function name	Example	Explanation
IABS	J = IABS(I)	Absolute value of argument
INT	I = INT (R)	Convert floating-point to fixed-point
IFIX	I = IFIX (R)	Convert floating-point to fixed-point
IDINT	I = IDINT (D)	Convert double-precision to fixed-point
MOD	I = MOD (J,K)	Remaindering: arg 1 − [arg 1/ arg 2] arg 2, where [X] = integral part of X
MAX0	I = MAX0 (I,J,K,...)	Largest of set of arguments
MAX1	I = MAX1 (X,Y,Z,...)	Largest of set of arguments
MIN0	I = MIN0 (I,J,K,...)	Smallest of set of arguments
M1N1	I = MIN1 (X,Y,Z,...)	Smallest of set of arguments
ISIGN	I = ISIGN (I,J)	Sign of arg 2 × arg 1
IDIM	I = IDIM (I,J)	Positive difference: arg 1 − min (arg l,arg 2)

B.12.2 Real Functions

Function name	Example	Explanation
ABS	X = ABS (R)	Absolute value of argument
AINT	X = AINT (R)	Truncation: sign of argument times absolute value of the largest integer in argument

AMOD	X = AMOD (Y,Z)	Remaindering: arg 1 −[arg 1/arg 2] arg 2, where $[X]$ = integral part of X
AMAX0	X = AMAX0 (I,J,K,...)	Largest of set of arguments
AMAX1	X = AMAX1 (X,Y,Z,...)	Largest of set of arguments
AMIN0	X = AMIN0 (I,J,K,...)	Smallest of set of arguments
AMIN1	X = AMIN1 (X,Y,Z,...)	Smallest of set of arguments
FLOAT	X = FLOAT (I)	Convert fixed point to floating point
SIGN	X = SIGN (Y,Z)	Transfer of sign: sign of arg 2 times arg 1
DIM	X = DIM (Y,Z)	Positive difference: arg 1 − min (arg 1, arg 2)
SNGL	X = SNGL (D)	Convert double precision to single precision
REAL	X = REAL (C)	Obtaining the real part of a complex number
AIMAG	X = AIMAG (C)	Obtaining the imaginary part of a complex number
SQRT	X = SQRT (Y)	$x = \sqrt{y}$: the square root of y
EXP	X = EXP (Y)	$x = e^y$: the natural anti-logarithm of y
ALOG	X = ALOG (Y)	$x = \ln y$: the natural logarithm of y
ALOG10	X = ALOG10 (Y)	$x = \log_{10} y$: the common logarithm of y
SIN	X = SIN (Y)	$x = \sin y$: the trigonometric sine
COS	X = COS (Y)	$x = \cos y$: the trigonometric cosine
TAN	X = TAN (Y)	$x = \tan y$: the trigonometric tangent
ATAN	X = ATAN (Y)	$x = \tan^{-1} y$: the arctangent of y, result is placed in first two quadrants depending on sign of y
ATAN2	X = ATAN2 (Y,Z)	$x = \tan^{-1}(y/z)$: the same as ATAN except the result is placed in the proper quadrant
ASIN	X = ASIN (Y)	$x = \sin^{-1} y$: the arcsine of y

ACOS	X = ACOS (Y)	$x = \cos^{-1} y$: the arccosine of y
SINH	X = SINH (Y)	$x = \sinh y$: the hyperbolic sine
COSH	X = COSH (Y)	$x = \cosh y$: the hyperbolic cosine
TANH	X = TANH (Y)	$x = \tanh y$: the hyperbolic tangent
CABS	X = CABS (C)	Magnitude of a complex number: $x = \sqrt{a^2 + b^2}$, where $c = a + jb$

B.12.3 Double Precision Functions

Function name	Example	Explanation
DABS	D = DABS (D1)	Absolute value of argument
DMAX1	D = DMAX1 (D1,D2)	Largest of set of arguments
DMIN1	D = DMIN1 (D1,D2,···)	Smallest of set of arguments
DSIGN	D = SIGN (D1,D2)	Transfer of sign: sign of arg 2 times arg 1
DBLE	D = DBLE (X)	Convert single-precision to double precision
DMOD	D = DMOD (D1,D2)	Remaindering: arg 1 - [arg 1/arg 2] arg 2, where [X] = integral part of X
DSQRT	D = DSQRT (D1)	Double precision equivalent of SQRT
DEXP	D = DEXP (D1)	Double precision equivalent of EXP
DLOG	D = DLOG (D1)	Double precision equivalent of ALOG
DLOG10	D = DLOG10 (D1)	Double precision equivalent of ALOG10
DSIN	D = DSIN (D1)	Double precision equivalent of SIN
DCOS	D = DCOS (D1)	Double precision equivalent of COS
DTAN	D = DTAN (D1)	Double precision equivalent of TAN
DTAN2	D = DTAN2 (D1,D2)	Double precision equivalent of TAN2
DASIN	D = DASIN (D1)	Double precision equivalent of ASIN
DACOS	D = DACOS (D1)	Double precision equivalent of ACOS
DSINH	D = DSINH (D1)	Double precision equivalent of SINH
DCOSH	D = DCOSH (D1)	Double precision equivalent of COSH
DTANH	D = DTANH (D1)	Double precision equivalent of TANH

B.12.4 Complex Functions

Function name	Example	Explanation
CABS	C = CABS (C1)	Absolute value of argument, magnitude of a complex number
CANG	C = CANG(C1)	Argument of a complex number
CMPLX	C = CMPLX (X,Y)	Express two arguments in complex form, $c = x + jy$
CONJG	C = CONJG (C1)	Obtain complex conjugate of argument
CSQRT	C = CSQRT (C1)	Complex square-root function
CEXP	C = CEXP (C1)	Complex exponential
CLOG	C = CLOG (C1)	Complex natural logarithm
CSIN	C = CSIN (C1)	Complex sine
CCOS	C = CCOS (C1)	Complex cosine

References

[1] S. Lipschutz and A. Poe, *Programming with Fortran* (Schaum's outline series). New York: McGraw-Hill, 1978.

[2] W. Cheney and D. Kincaid, *Numerical Mathematics and Computing*, 2nd ed. Monterey, CA: Brooks/Cole, 1985, pp. 505–528.

Appendix C
How to Write Good Programs

A good computer program should be well documented, of reasonable size, and capable of performing a set of calculations with reasonable accuracy and within a reasonable amount of time. The following programming techniques should be kept in mind when preparing a program to improve documentation, reduce size, and enhance speed and accuracy.

C.1 Documentation

A computer program should be well documented in order to enable users other than the writer to follow the logic of the program. It should be easily understood from the point of view of the mathematical model. To achieve this, keep the following points in mind:

(1) Use comment statements to state purpose of the program, input and output variables, and their units, the solution method used, and any references to literature if, necessary.

(2) For the output pages, include a header so that the output makes sense to the reader without the need to refer to the program.

(3) Choose variable names to indicate relationship with their associated symbols, e.g.,
LAMBDA is better than WAVELEN to denote λ,
PIE is better than PI to denote π,
EPSILON is better than DIECON to denote ϵ.

(4) To facilitate future testing or modification, sample input and output data should be provided.

C.2 Size Reduction

For some time-sharing systems, the cost of running a program is based on the size of the program. The size of the program can be reduced by following these hints [1]:

(1) Avoid using arrays when they are not needed.
(2) When the same variables are used in the main program and the subprograms, use a COMMON statement to have the same memory locations reserved for the variables.
(3) When variables are used at different times in a program, use the EQUIVALENCE statement to save storage space.

C.3 Accuracy Enhancement

A specified level of accuracy is important in most engineering problems. The following techniques will ensure reasonably accurate results [2]:

(1) In programming a sum, rearrange the numbers so that smaller contributions are added first, e.g., if $A > B > C$, then $(C + B) + A$ is better than $(A + B) + C$.
(2) Rearrange formulas so that small differences of large numbers are avoided whenever possible. For example, if $X \ll R$, SQRT (2.0 * R * X + X ** 2) is better than SQRT ((R + X) ** 2 - R ** 2).
(3) If possible, rearrange your formulas to cut down on the number of arithmetic operations. For example, X - Y is better than (X**2 - Y**2)/(X + Y).
(4) For critical floating-point calculations, use double precision.
(5) If possible, work with integers. Although this is relatively slow, roundoff errors are avoided.
(6) Avoid mixing types of variables and constants. It is risky and is the source of many frustrating, hard-to-find programming bugs.

C.4 Speed Enhancement

The following guidelines can help to improve the speed of a computer program.

(1) Avoid repeating unnecessary calculations. For example,

```
        DO 1 I = 1,5
        DO 2 J = 1,10
        CALL FACTORIAL (I,FI)
```

Appendix C: How to Write Good Programs 645

```
           CALL FACTORIAL (J,FJ)
           A = FI*SIN(ALPHA)/FJ
    1      CONTINUE
    2      CONTINUE
```

can be improved as follows:

```
           SINAL = SIN(ALPHA)
           DO 1 I = 1,5
           CALL FACTORIAL (I,FI)
           DO 2 J = 1,10
           CALL FACTORIAL (J,FJ)
           A = FI*SINAL/FJ
    1      CONTINUE
    2      CONTINUE
```

(2) Arrange the program so that advantage may be taken of quantities already computed. Whenever possible, quantities should be reused rather than recomputed.

(3) Pay attention to language details regarding calculational speed. For example, in FORTRAN programming it is faster if mixed-mode arithmetic is avoided, e.g., $A = A + 1$ is slower than $A = A + 1.0$ if A is real. Also, it is faster to use multiplication rather than exponentiation if possible, e.g., X*X*X is faster than X**3. Finally, the loop

```
           DO 10 I=1,50
           DO 10 J=1,50
    10     A(I,J) = 0.0
```

is substantially faster than

```
           DO 10 J=1,50
           DO 10 I=1,50
    10     A(I,J) = 0.0
```

because when FORTRAN is used, the elements of A appear in memory in the following order

A(1,1), A(2,1), ⋯, A(50,1), A(1,2), A(2,2), ⋯

For most other programming languages, the order is the contrary one:

A(1,1), A(1,2), ⋯, A(1,50), A(2,1), A(2,2), ⋯

For this reason, a translation of a program in FORTRAN into ALGOL, PASCAL, or PL/1 requires an interchange of all indices.

References

[1] J. R. Wolberg, *Application of Computers to Engineering Analysis.* New York: McGraw-Hill, 1971, pp. 140–156.

[2] I. A. Dodes, *Numerical Analysis for Computer Science.* New York: North-Holland, 1978, pp. 49, 50.

Appendix D

Solution of Simultaneous Equations

Application of some numerical methods to EM problems often results in a set of simultaneous equations

$$\begin{bmatrix} a_{11} & a_{12} & \cdots & a_{1n} \\ a_{21} & a_{22} & \cdots & a_{2n} \\ \vdots & & & \vdots \\ a_{n1} & a_{n2} & \cdots & a_{nn} \end{bmatrix} \begin{bmatrix} x_1 \\ x_2 \\ \vdots \\ x_n \end{bmatrix} = \begin{bmatrix} b_1 \\ b_2 \\ \vdots \\ b_n \end{bmatrix} \quad \text{(D.1a)}$$

or

$$[A][X] = [B] \quad \text{(D.1b)}$$

where $[A]$ is the coefficient matrix, $[X]$ is the column matrix of the unknowns to be determined, and $[B]$ is the column matrix of constants. Familiarity with the various techniques for solving Eq. (D.1); is therefore vital. In this appendix, we provide a brief coverage of direct and iterative procedures for solving Eq. (D.1); direct methods are more versatile for linear problems, while iterative methods are suitable for non-linear problems. We also consider various techniques for solving eigenvalue systems $[A][X] = \lambda[X]$.

D.1 Elimination Methods

Elimination methods constitute the simplest direct approach to the solution of a set of simultaneous equations. They usually involve successive elimination of the unknowns by combining equations. Such methods include Gauss's method, Gauss-Jordan, Cholesky's or Crout's method, and the square-root method. Only Gauss's and Cholesky's methods will be discussed; the reader should consult [1-4] for the treatment of other methods.

D.1.1 Gauss's Method

This simple method involves eliminating one unknown at a time and proceeding with the remaining equations. This leads to a set of simultaneous equations in triangular form from which each unknown is determined by back-substitution. To describe this method, consider Eq. (D.1b), i.e.,

$$a_{11}x_1 + a_{12}x_2 + \cdots + a_{1n}x_n = b_1 \qquad \text{(D.2a)}$$
$$a_{21}x_1 + a_{22}x_2 + \cdots + a_{2n}x_n = b_2 \qquad \text{(D.2b)}$$
$$\vdots$$
$$a_{n1}x_1 + a_{n2}x_2 + \cdots + a_{nn}x_n = b_n. \qquad \text{(D.2c)}$$

We divide Eq. (D.2a) by a_{11} to give

$$x_1 + a'_{12}x_2 + \cdots + a'_{1n}x_n = b'_1 \qquad \text{(D.3)}$$

where the primes denote that the coefficients are new. We multiply Eq. (D.3) by $-a_{i1}$ for $i = 2, 3, \cdots, n$ and add Eq. (D.3) to the ith equation in (D.2) to eliminate x_1 from other equations so that Eq. (D.2) becomes

$$x_1 + a'_{12}x_2 + \cdots + a'_{1n}x_n = b'_1 \qquad \text{(D.4a)}$$
$$a'_{22}x_2 + \cdots + a'_{2n}x_n = b'_2 \qquad \text{(D.4b)}$$
$$\vdots$$
$$a'_{n2}x_2 + \cdots + a'_{nn}x_n = b'_n. \qquad \text{(D.4c)}$$

Equation (D.2a) used to eliminate x_1 from other equations is called the *pivot equation* and a_{11} is called the *pivot coefficient*. We now use Eq. (D.4b) as the pivot equation and we take similar steps to eliminate x_2 from all equations following the pivot equation. Continuing this reduction procedure eventually leads to an equivalent triangular set of equations:

$$\begin{aligned} x_1 + u_{12}x_2 + u_{13}x_3 + \cdots + u_{1n}x_n &= c_1 \\ x_2 + u_{23}x_2 + \cdots + u_{2n}x_n &= c_2 \\ x_3 + \cdots + u_{3n}x_n &= c_3 \\ &\vdots \\ x_n &= c_n. \end{aligned} \qquad \text{(D.5)}$$

This completes the first phase known as *forward elimination* in the Gauss algorithm, and the system in Eq. (D.5) is said to be in *upper triangular*

Appendix D: Solution of Simultaneous Equations 649

form. The second phase known as *back substitution* involves solving for the unknowns in Eq. (D.5) by starting at the bottom. That is,

$$x_n = c_n$$
$$x_{n-1} = c_{n-1} - u_{n-1,n} x_n$$
$$\vdots$$
$$x_1 = c_1 - u_{12} x_2 - \cdots - u_{1n} x_n.$$
(D.6)

In summary, this algorithm can be stated as:

Forward elimination

$$a'_{kj} = a_{kj}/a_{kk}, \quad b'_k = b_k/a_{kk}, \quad j = k, k+1, \cdots, n$$
$$a'_{ij} = a_{ij} - a_{ik} \, a'_{kj}, \quad i = k+1, \cdots, n$$
$$b'_i = b_i - a_{ik} \, b'_k, \quad i = k+1, \cdots, n$$
(D.7a)

Backward substitution

$$x_n = b_n, \quad \text{for the last row}$$
$$x_i = b_i - \sum_{j=i+1}^{n} a_{ij} \, x_j, \quad i = n-1, \cdots, 1.$$
(D.7b)

Based on the idea outlined above, a general FORTRAN subroutine for solving a set of simultaneous equations by Gaussian elimination is shown in Fig. D.1.

```
0001      C***********************************************************
0002      C   THIS SUBROUTINE EMPLOYS GAUSSIAN ELIMINATION METHOD
0003      C   TO SOLVE A SET OF SIMULTANEOUS EQUATIONS [A][X] = [B]
0004      C   IDM = DIMENSION OF [A] IN THE CALLING PROGRAM
0005      C    N  = ACTUAL SIZE OF [A] IN THE CALLING PROGRAM
0006      C***********************************************************
0007
0008                SUBROUTINE GAUSS(A,B,X,N,IDM)
0009                DIMENSION A(IDM,IDM), B(IDM), X(IDM)
0010      C
0011      C   FORWARD ELIMINATION
0012      C
0013                DO 40 K=1,N-1
0014                DO 30 I=K+1,N
0015                FACTOR = A(I,K)/A(K,K)
0016      C
0017      C   CALCULATE ELEMENTS OF THE NEW MATRIX
0018      C
0019                DO 20 J=K+1,N
0020                A(I,J) =A(I,J) - FACTOR*A(K,J)
0021           20   CONTINUE
```

```
0022            A(I,K) = FACTOR
0023            B(I) = B(I) - FACTOR*B(K)
0024      30 CONTINUE
0025      40 CONTINUE
0026      C
0027      C BACK SUBSTITUTION PROCESS BEGINS
0028      C
0029            X(N) = B(N)/A(N,N)  !1ST STEP
0030            DO 60 I=N-1,1,-1
0031            SUM = 0.0
0032            DO 50 J=I+1,N
0033            SUM = SUM + A(I,J)*X(J)
0034      50 CONTINUE
0035            X(I) = ( B(I) - SUM )/A(I,I)
0036      60 CONTINUE
0037            RETURN
0038            END
```

Figure D.1 Gauss elimination method of solving [A][X] = [B].

D.1.2 Cholesky's Method

This method, also known as Crout's method or the method of matrix decomposition, involves determining a lower triangular matrix that will reduce the original system in Eq. (D.1) to a unit upper triangular matrix. If the original system

$$[A][X] = [B] \tag{D.1a}$$

or

$$\begin{bmatrix} a_{11} & a_{12} & \cdots & a_{1n} \\ a_{21} & a_{22} & \cdots & a_{2n} \\ \vdots & & & \vdots \\ a_{n1} & a_{n2} & \cdots & a_{nn} \end{bmatrix} \begin{bmatrix} x_1 \\ x_2 \\ \vdots \\ x_n \end{bmatrix} = \begin{bmatrix} b_1 \\ b_2 \\ \vdots \\ b_n \end{bmatrix} \tag{D.1b}$$

can be redefined in the upper unit triangular matrix $[T]$ such that

$$[T][X] = [C] \tag{D.8a}$$

or

$$\begin{bmatrix} 1 & T_{12} & \cdots & T_{1n} \\ 0 & 1 & \cdots & T_{2n} \\ \vdots & & & \vdots \\ 0 & 0 & \cdots & 1 \end{bmatrix} \begin{bmatrix} x_1 \\ x_2 \\ \vdots \\ x_n \end{bmatrix} = \begin{bmatrix} c_1 \\ c_2 \\ \vdots \\ c_n \end{bmatrix} \tag{D.8b}$$

the unknown x_i can be obtained by back substitution. Let $[A]$ be a product of an upper unit triangular matrix $[T]$ and a lower triangular matrix $[L]$, i.e.,

$$[L][T] = [A]. \tag{D.9}$$

Appendix D: Solution of Simultaneous Equations

Since
$$[L][TX - C] = 0 = [AX - B], \quad (D.10)$$

it follows that
$$[L][C] = [B]. \quad (D.11)$$

For computational reasons, it is convenient to work with the augmented form of the matrices. The augmented matrix is obtained by adding the column vector of constants to the square coefficient matrix. Equations (D.8) and (D.11) may be combined to give

$$\begin{bmatrix} a_{11} & a_{12} & \cdots & a_{1n} & \vdots & b_1 \\ a_{21} & a_{22} & \cdots & a_{2n} & \vdots & b_2 \\ \vdots & & & \vdots & & \\ a_{n1} & a_{n2} & \cdots & a_{nn} & \vdots & b_n \end{bmatrix} =$$

$$\begin{bmatrix} L_{11} & 0 & \cdots & 0 \\ L_{21} & L_{22} & \cdots & 0 \\ \vdots & & & \\ L_{n1} & L_{n2} & \cdots & L_{nn} \end{bmatrix} \begin{bmatrix} 1 & T_{12} & \cdots & T_{1n} & \vdots & c_1 \\ 0 & 1 & \cdots & T_{2n} & \vdots & c_2 \\ \vdots & & & & & \\ 0 & 0 & \cdots & 1 & \vdots & c_n \end{bmatrix} \quad (D.12a)$$

or
$$[A \vdots B] = [L][T \vdots C]. \quad (D.12b)$$

The elements of $[L]$, $[T]$, and $[C]$ can be defined in terms of $[A]$ and $[B]$ as follows [1,2,5]:

$$L_{ij} = a_{ij} - \sum_{k=1}^{j-1} L_{ik} T_{kj}, \quad i \geq j, \; i = 1, 2, \cdots, n$$

$$L_{ij} = a_{i1}, \quad j = 1$$

$$T_{ij} = \frac{1}{L_{ii}} \left(a_{ij} - \sum_{k=1}^{i-1} L_{ik} T_{kj} \right), \quad i < j, \; j = 2, 3, \cdots, n \quad (D.13)$$

$$T_{ij} = a_{ij}/a_{11}, \quad i = 1$$

$$c_i = \frac{1}{L_{ii}} \left(b_i - \sum_{k=1}^{i-1} L_{ik} c_k \right), \quad i = 2, 3, \cdots, n$$

$$c_1 = b_1/L_{11}.$$

652 Numerical Techniques in Electromagnetics

The unknown x_i are obtained by back substitution as follows:

$$x_n = c_n$$
$$x_i = c_i - \sum_{j=i+1}^{n} T_{ij} x_j, \qquad i = 1, 2, \cdots, n-1. \tag{D.14}$$

Cholesky's method can easily be applied in calculating the determinant of $[A]$. Since

$$\det[A] = \det[L]\det[T] \tag{D.15}$$

and $\det[T] = 1$ due to the fact that $T_{ii} = 1$, it follows that

$$\det[A] = \det[L] = L_{11} L_{22} \cdots L_{nn}$$

or

$$\det[A] = \prod_{i=1}^{n} L_{ii}. \tag{D.16}$$

Figure D.2 shows a subroutine based on Cholesky's method of solving a set of simultaneous equations.

```
0001      C*************************************************************
0002      C     THIS SUBROUTINE APPLIES CHOLESKY ELIMINATION METHOD
0003      C     TO SOLVE THE SYSTEM [AA][X] = [B]
0004      C        [A]  = AUGMENTED MATRIX = [AA : B]
0005      C        MAX  = MAXIMUM ROW DIMENSION OF [AA] AND [X] MATRIX
0006      C        MAX1 = MAXIMUM COLUMN DIMENSION OF [AA] = MAX + 1
0007      C        N    = ACTUAL ROW SIZE OF [AA]
0008      C        M    = ACTUAL COLUMN SIZE OF [AA] = N+1
0009      C        [X]  = STORES THE RESULTS
0010      C
0011      C REF: [2, PP. 184, 185]
0012      C*************************************************************
0013            SUBROUTINE CHOLESKY(A,MAX,MAX1,N,M,X)
0014      C
0015      C  NOTE: THIS PROGRAM CAN BE USED TO SOLVE COMPLEX EQUATIONS
0016      C        BY REPLACING EVERY SUM = 0.0 WITH SUM = (0.0,0.0)
0017      C        USING THE NEXT LINE INSTEAD OF THE ONE FOLLOWING IT
0018      C        COMPLEX A,X,SUM
0019            REAL A,X,SUM
0020            DIMENSION A(MAX,MAX1), X(MAX)
0021
0022      C
0023      C  CALCULATE THE FIRST ROW OF THE [T] MATRIX
0024      C
0025            DO 10 J = 2, M
0026               A(1,J)=A(1,J)/A(1,1)     !CALCULATES T(1,J)
0027      10   CONTINUE
0028      C
0029      C CALCULATE OTHER ELEMENTS OF [T] AND [L] MATRICES
0030      C
```

Appendix D: Solution of Simultaneous Equations 653

```
0031              DO 60 I=2,N
0032                DO 30 II=I,N
0033                  SUM = 0.0
0034                  DO 20 K=1,I-1
0035                    SUM = SUM + A(II,K)*A(K,I)
0036      20        CONTINUE
0037                  A(II,I) = A(II,I) - SUM   !CALCULATES [L]
0038      30      CONTINUE
0039                DO 50 J = I+1,M
0040                  SUM = 0.0
0041                  DO 40 K=1,I-1
0042                    SUM= SUM + A(I,K)*A(K,J)
0043      40        CONTINUE
0044                  A(I,J) = ( A(I,J)-SUM )/A(I,I) !CALCULATES [T]
0045      50      CONTINUE
0046      60    CONTINUE
0047      C
0048      C SOLVE FOR [X] BY BACK SUBSTITUTION
0049      C
0050              X(N) = A(N,M)
0051              DO 80 NN = 1,N-1
0052                SUM = 0.0
0053                I= N - NN
0054                DO 70 J = I+1,N
0055                  SUM = SUM + A(I,J)*X(J)
0056      70      CONTINUE
0057                X(I) = A(I,M) - SUM
0058      80    CONTINUE
0059            RETURN
0060            END
```

Figure D.2 Cholesky's elimination method of solving [A][X] = [B].

D.2 Iterative Methods

The direct, or elimination, method for solving a system of simultaneous equations can be used for $n = 25$ to 60. This number can be greater if the system is well conditioned or the matrix is sparse. For very large systems, say $n = 100$ or even 1000, elimination methods become time-consuming and prove inadequate due to roundoff error. For these types of problems, indirect or iterative methods provide an alternative.

D.2.1 Jacobi's Method

This is the simplest iterative method. If the system in Eq. (D.1) is rearranged so that the ith equation is explicit in x_i, we obtain

$$x_1 = \frac{1}{a_{11}}\left[b_1 - a_{12}x_2 - a_{13}x_3 - \cdots - a_{1n}x_n\right] \quad \text{(D.16a)}$$

$$x_2 = \frac{1}{a_{22}}\left[b_2 - a_{21}x_1 - a_{23}x_3 - \cdots - a_{2n}x_n\right] \quad \text{(D.16b)}$$

$$\vdots$$

$$x_n = \frac{1}{a_{nn}}\left[b_n - a_{n1}x_1 - a_{n2}x_2 - \cdots - a_{n,n-1}x_{n-1}\right] \quad \text{(D.16c)}$$

assuming that the diagonal elements are all nonzero. We start the solution process by using guesses for the x's, say $x_1 = x_2 = \cdots = x_n = 0$. The first equation can be solved for x_1, the second for x_2, and so on. If we denote the estimates after the kth iteration as $x_1^k, x_2^k, \cdots, x_n^k$, the estimates after $(k+1)$th iteration can be obtained from Eq. (D.16) as

$$x_i^{k+1} = \frac{1}{a_{ii}}\left[b_i - \sum_{j=1, j\neq i}^{n} a_{ij} x_j^k\right], \quad i = 1, 2, \cdots, n. \quad \text{(D.17)}$$

The iteration process is continued until values of x_i at two successive iterations are within an allowable prescribed deviation.

Convergence is measured in terms of the change in x_i from the kth iteration to the next. If we compute

$$d_i = \left|\frac{x_i^{k+1} - x_i^k}{x_i^{k+1}}\right| \cdot 100\% \quad \text{(D.18)}$$

for each x_i, convergence can be checked using the criterion

$$d_i < \epsilon_s \quad \text{(D.19)}$$

where ϵ_s is a specified small quantity. A better test would be to compute

$$d = \frac{\sum_{i=1}^{n} |x_i^{k+1} - x_i^k|}{\sum_{i=1}^{n} |x_i^{k+1}|} \cdot 100\% \quad \text{(D.20)}$$

and require that $d < \epsilon_s$.

Appendix D: Solution of Simultaneous Equations   655

D.2.2 Gauss-Seidel Method

This is the most commonly used iterative method. In Jacobi's method the entire set of x_i from the kth iteration is used in calculating the new set during the $(k+1)$th iteration, whereas the most recently calculated value of each variable is used at each step in the Gauss-Seidel method. This makes the Gauss-Seidel method converge more rapidly than (about twice as) Jacobi's method and is always used in preference to it. Instead of Eq. (D.17), we use

$$x_i^{k+1} = \frac{1}{a_{ii}}\left[b_i - \sum_{j=1}^{i-1} a_{ij} x_j^{k+1} - \sum_{j=i+1}^{n} a_{ij}\, x_j^k\right], \qquad i=1,2,\cdots,n. \quad (D.21)$$

A computer program based on this method is displayed in Fig. D.3.

```
0001      C*************************************************************
0002      C  THIS SUBROUTINE EMPLOYS GAUSS-SEIDEL ITERATIVE METHOD
0003      C  TO SOLVE A SET OF SIMULTANEOUS EQUATIONS [A][X] = [B]
0004      C  IDM = DIMENSION OF [A] IN THE CALLING PROGRAM
0005      C    N = ACTUAL SIZE OF [A] IN THE CALLING PROGRAM
0006      C*************************************************************
0007
0008               SUBROUTINE GSEIDEL(A,B,X,N,IDM)
0009               DIMENSION A(IDM,IDM), B(IDM), X(IDM)
0010      C
0011      C  INITIALIZATION
0012      C
0013               DO 10 I=1,N
0014               X(I) = 0.0
0015        10     CONTINUE
0016               K = 0  !NO. OF ITERATIONS
0017               TOL = 1.0E-30  !TOLERANCE FOR ZERO
0018        20     RES = 0.0
0019               DO 40 I=1,N
0020               SUM = 0.0
0021               DO 30 J=1,N
0022               IF(J.EQ.I) GO TO 30
0023               SUM = SUM + A(I,J)*X(J)
0024        30     CONTINUE
0025               XNEW = ( B(I) - SUM )/A(I,I)
0026      C
0027      C  FIND THE LARGEST RESIDUE
0028      C
0029               DIFF = ABS( X(I) - XNEW )
0030               IF( DIFF.GT.RES ) RES = DIFF
0031      C
0032      C  REPLACE OLD VALUE WITH NEWLY COMPUTED VALUE
0033      C
0034               X(I) = XNEW
0035        40     CONTINUE
0036      C
0037      C  TEST FOR CONVERGENCE
0038      C
```

```
0039          K = K + 1
0040          PRINT *,K
0041          IF( RES.GE.TOL ) GO TO 20
0042          RETURN
0043          END
```

Figure D.3 Gauss-Seidel iterative method of solving $[A][X] = [B]$.

D.2.3 Relaxation Method

This is a slight modification of the Gauss-Seidel method and is designed to enhance convergence. If x_i^k is added to the right-hand side of Eq. (D.21) and $(a_{ii}\, x_i^k)/a_{ii}$ is subtracted from it, we obtain

$$x_i^{k+1} = x_i^k + \frac{1}{a_{ii}}\left[b_i - \sum_{j=1}^{i-1} a_{ij} x_j^{k+1} - \sum_{j=i}^{n} a_{ij}\, x_j^k\right], \qquad i = 1, 2, \cdots, n. \quad \text{(D.22)}$$

The second term on the right-hand side can be regarded as a correction term. The correction term tends to zero as convergence is approached. If this term is multiplied by ω, Eq. (D.22) becomes

$$x_i^{k+1} = x_i^k + \frac{\omega}{a_{ii}}\left[b_i - \sum_{j=1}^{i-1} a_{ij} x_j^{k+1} - \sum_{j=i}^{n} a_{ij}\, x_j^k\right], \qquad i = 1, 2, \cdots, n. \quad \text{(D.23)}$$

The *relaxation factor* ω is selected such that $1 < \omega < 2$. The choice of a proper value of ω is problem dependent and is often determined via trial and error. The added weight of ω is intended to improve the estimate by pushing it closer to the exact value.

D.2.4 Gradient Methods

The iterative methods considered above may be broadly classified as *stationary* while the ones to be presented now are *gradient* (or nonstationary) methods. The two common gradient methods are the *steepest method* and *conjugate gradients method* [6-8]. A major advantage gradient methods have over stationary methods is that convergence is faster; hence gradient methods are particularly useful when the number of simultaneous equations is very large.

A set of n simultaneous equations may be solved by finding the position of the minimum of an error function defined over an n-dimensional space. In each step of a gradient method, a trial set of values for the variables

Appendix D: Solution of Simultaneous Equations

is used to generate a new set corresponding to a lower value of the error function. If $\overline{X}$ is the trial vector, the vector residual is

$$R = B - A\overline{X} \tag{D.24}$$

where A is real, symmetric, and positive definite. If we define the error function as

$$e = R^t A^{-1} R, \tag{D.25}$$

then

$$e = \overline{X}^t A \overline{X} - 2B^t \overline{X} + B^t A^{-1} B \tag{D.26}$$

showing that e is quadratic in $\overline{X}$.

Starting from an arbitrary point X_o, we locate a sequence of points

$$X_{k+1} = X_k + \alpha_k D_k \tag{D.27}$$

which are successively closer to X, where X minimizes e in Eq. (D.26). The parameter α_k is proportional to the distance between X_i and X_{i+1} along the direction vector D_k. Substituting Eq. (D.27) into Eq. (D.26) and setting $\partial e / \partial \alpha_k$ equal to zero gives

$$\alpha_k = \frac{D_k^t R_k}{D_k^t A D_k}. \tag{D.28}$$

Both the methods of steepest descent and conjugate gradients use Eq. (D.28) but differ in the choice of D_k.

In the method of descent, D_k is taken as the direction of maximum gradient of e at X_k. This direction is proportional to X_k so that the iterative algorithm has the form:

(i) Select an arbitrary X_0.
(ii) Compute $R_0 = B - AX_0$.
(iii) Determine successively

$$\begin{aligned} U_k &= AR_k \\ \alpha_k &= \frac{R_k^t R_k}{R_k^t U_k} \\ X_{k+1} &= X_k + \alpha_k R_k \\ R_{k+1} &= R_k - \alpha_k U_k. \end{aligned} \tag{D.29}$$

(iv) Repeat step (iii) until residual vector ($R^T R$) becomes sufficiently small.

In the method of conjugate gradients, D_k are selected as n vectors P_k which are mutually conjugate. The vectors P_k are conjugate or orthogonal to A, i.e.,

$$P_k^t \, A \, P_k = 0, \quad i \neq j$$
$$\neq 0, \quad i = j. \quad \text{(D.30)}$$

Thus the conjugate gradients algorithm is as follows:

(i) Select an arbitrary X_0.
(ii) Set $P_0 = R_0 = B - AX_0$.
(iii) Determine successively

$$U_k = AR_k$$
$$\alpha_k = \frac{P_k^t \, R_k}{P_k^t \, U_k}$$
$$X_{k+1} = X_k + \alpha_k R_k$$
$$R_{k+1} = R_k - \alpha_k U_k \quad \text{(D.31)}$$
$$\beta_k = -\frac{R_{k+1}^t \, U_k}{P_k^t \, U_k}$$
$$P_{k+1} = R_k + \beta_k P_k.$$

(iv) Repeat step (iii) until $k = n - 1$ or the residual vector $(R^T R)$ becomes sufficiently small.

This algorithm is guaranteed to yield the true solution in no more than n iterations—a condition known as *quadratic convergence*. Because of this, the conjugate gradients method has the advantage of an iterative scheme in that the roundoff error is limited only to the final step of the solution and also the advantage of a direct method in that it converges to the exact solution in a finite number of steps.

The subroutine in Fig. D.4 applies the conjugate gradients method to solve a given set of simultaneous equations. Typical areas where the conjugate gradient methods have been applied in EM can be found in [9–12].

```
0001      C*************************************************************
0002      C  THIS SUBROUTINE APPLIES THE CONJUGATE GRADIENTS METHOD
0003      C  TO SOLVE A SET OF SIMULTANEOUS EQUATIONS [A][X] = [B]
0004      C  IDM = DIMENSION OF [A] IN THE CALLING PROGRAM
0005      C   N = ACTUAL SIZE OF [A] IN THE CALLING PROGRAM
0006      C*************************************************************
0007
0008               SUBROUTINE CONJUGATE(A,B,N,IDM,X)
0009
0010               DIMENSION A(IDM,IDM), B(IDM), X(IDM)
```

Appendix D: Solution of Simultaneous Equations   659

```
0011                DIMENSION U(100), R(100), P(100) !change 100
0012                                                 !if N.GT.100
0013        C
0014        C   INITIALIZATION
0015        C
0016                K=0 !K=NO. OF ITERATIONS
0017                TOL = 1.0E-30 !TOLERANCE FOR ZERO
0018                DO 10 I=1,N
0019                X(I) = 0.0
0020                P(I) = B(I)
0021                R(I) = B(I)
0022        10      CONTINUE
0023        C
0024        C   ITERATION BEGINS HERE
0025        C
0026        20      K = K + 1
0027                DO 30 I=1,N
0028                U(I) = 0.0
0029                DO 30 J=1,N
0030                U(I) = U(I) + A(I,J)*P(J)
0031        30      CONTINUE
0032                SUM1 = 0.0
0033                SUM2 = 0.0
0034                DO 40 I=1,N
0035                SUM1 = SUM1 + P(I)*R(I)
0036                SUM2 = SUM2 + P(I)*U(I)
0037        40      CONTINUE
0038                ALPHA = SUM1/SUM2
0039                DO 50 I=1,N
0040                X(I) = X(I) + ALPHA*P(I)
0041        C       PRINT *,K,I,X(I)
0042                R(I) = R(I) - ALPHA*U(I)
0043        50      CONTINUE
0044        C
0045        C   CALCULATE RESIDUALS AND DIRECTION VECTORS
0046        C
0047                SUM3 = 0.0
0048                SUM4 = 0.0
0049                DO 60 I=1,N
0050                SUM3 = SUM3 + R(I)*R(I)
0051                SUM4 = SUM4 + R(I)*U(I)
0052        60      CONTINUE
0053        C
0054        C   CHECK IF RESIDUALS ARE ALREADY SMALL
0055        C   IF SO ALGORITHM HAS CONVERGED
0056        C
0057                IF( (K.EQ.N).OR.(SUM3.LT.TOL) ) GO TO 80
0058                BETA = - SUM4/SUM2
0059                DO 70 I=1,N
0060                P(I) = R(I) + BETA*P(I)
0061        70      CONTINUE
0062                GO TO 20
0063        80      RETURN
0064                END
```

Figure D.4 This subroutine applies the conjugate gradients method to solve $[A][X] = [B]$.

D.3 Matrix Inversion

If $[A]$ is a square matrix, there is another matrix $[A]^{-1}$, called the *inverse* of $[A]$, such that

$$[A][A]^{-1} = [A]^{-1}[A] = [I] \tag{D.32}$$

where I is the *identity* or *unit matrix*. Matrix inversion can be used to solve a set of simultaneous equations in Eq. (D.1) as

$$[X] = [A]^{-1}[B]. \tag{D.33}$$

The solution of a system of simultaneous equations by matrix inversion and multiplication is most valuable when several systems are to be solved, all of which have the same coefficient matrix but different column matrices of constants. This situation requires calculating the inverse matrix only once and using it as a premultiplier of each of the column matrices of constants [2,13].

The inversion of matrices is closely related to the solution of sets of simultaneous equations. The inverse of $[A]$ can be determined from Eq. (D.32). If we let $[C] = [A]^{-1}$, then

$$[A][C] = [I] \tag{D.34a}$$

or

$$\begin{bmatrix} a_{11} & a_{12} & \cdots & a_{1n} \\ a_{21} & a_{22} & \cdots & a_{2n} \\ \vdots & & & \\ a_{n1} & a_{n2} & \cdots & a_{nn} \end{bmatrix} \begin{bmatrix} c_{11} & c_{12} & \cdots & c_{1n} \\ c_{21} & c_{22} & \cdots & c_{2n} \\ \vdots & & & \\ c_{n1} & c_{n2} & \cdots & c_{nn} \end{bmatrix} = \begin{bmatrix} 1 & 0 & \cdots & 0 \\ 0 & 1 & \cdots & 0 \\ \vdots & & & \\ 0 & 0 & \cdots & 1 \end{bmatrix}. \tag{D.34b}$$

This may be regarded as n sets of n simultaneous equations with identical coefficient matrix. The ith set of n simultaneous equations, for example, is

$$\begin{bmatrix} a_{11} & a_{12} & \cdots & a_{1n} \\ a_{21} & a_{22} & \cdots & a_{2n} \\ \vdots & & & \\ a_{ni} & a_{ni} & \cdots & a_{ni} \\ \vdots & & & \\ a_{n1} & a_{n2} & \cdots & a_{nn} \end{bmatrix} \begin{bmatrix} c_{1i} \\ c_{2i} \\ \vdots \\ c_{ii} \\ \vdots \\ c_{ni} \end{bmatrix} = \begin{bmatrix} 0 \\ 0 \\ \vdots \\ 1 \\ \vdots \\ 0 \end{bmatrix}. \tag{D.35}$$

Thus, the inversion of $[A]$ may be accomplished by solving n sets of equations like Eq. (D.35). A common approach for matrix inversion is applying

Appendix D: Solution of Simultaneous Equations

elimination method, with or without pivotal compensation. This implies that any elimination technique (Gauss, Gauss-Jordan, or Cholesky's method) can be modified to calculate an inverse matrix. Here we apply the Gauss-Jordan elimination method.

To apply the Gauss-Jordan method, we first augment the coefficient matrix by the identity matrix to obtain

$$[A \vdots I] = \begin{bmatrix} a_{11} & a_{12} & \cdots & a_{1n} & \vdots & 1 & 0 & \cdots & 0 \\ a_{21} & a_{22} & \cdots & a_{2n} & \vdots & 0 & 1 & \cdots & 0 \\ \vdots & & & & \vdots & \vdots & & & \\ a_{n1} & a_{n2} & \cdots & a_{nn} & \vdots & 0 & 0 & \cdots & 1 \end{bmatrix}. \quad (D.36)$$

The goal is to transform this augmented matrix to another augmented matrix of the form

$$[I \vdots C] = \begin{bmatrix} 1 & 0 & \cdots & 0 & \vdots & c_{11} & c_{12} & \cdots & c_{1n} \\ 0 & 1 & \cdots & 0 & \vdots & c_{21} & c_{22} & \cdots & c_{2n} \\ \vdots & & & & \vdots & \vdots & & & \\ 0 & 0 & \cdots & 1 & \vdots & c_{n1} & c_{n2} & \cdots & c_{nn} \end{bmatrix} \quad (D.37)$$

where $[C]$ is the inverse of $[A]$. The transformation is achieved using the Gauss-Jordan method, which involves applying the following equations in the order listed [2]:

$$\begin{aligned} a'_{kj} &= a_{kj}/a_{kk}, \quad j = 1, 2, \cdots, n, \ j \neq k \\ a'_{kk} &= 1/a_{kk}, \\ a'_{ij} &= a_{ij} - a_{ik} \, a'_{kj}, \quad i = 1, 2, \cdots, n, \ i \neq k \\ & \qquad\qquad\qquad\qquad j = 1, 2, \cdots, n, \ j \neq k \\ a'_{ik} &= -a_{ik} \, a'_{kk}, \quad i = 1, 2, \cdots, n, \ i \neq k. \end{aligned} \quad (D.38)$$

We apply Eq. (D.38) for $k = 1, 2, \cdots, n$. A computer program applying Eq. (D.38) is presented in Fig. D.5. An iterative method of correcting the elements of the inverse matrix is available in [14].

```
0001    C***********************************************************
0002    C       SUBROUTINE FOR MATRIX INVERSION USING GAUSS-JORDAN
0003    C       ELIMINATION METHOD
0004    C       A IS THE MATRIX TO BE INVERTED; IT IS DESTROYED
0005    C       IN THE COMPUTATION AND REPLACED BY THE INVERSE
```

```
0006      C     N  = THE ORDER OF A
0007      C     IDM = THE DIMENSION OF A
0008      C
0009      C     REF: [2, p. 197 ]
0010      C*************************************************************
0011            SUBROUTINE INVERSE(A,N,IDM)
0012            DIMENSION A(IDM,IDM)
0013      C
0014      C CALCULATE   ELEMENTS OF REDUCED MATRIX
0015      C
0016            DO 60 K=1,N
0017      C
0018      C CALCULATE NEW ELEMENTS OF PIVOT ROW
0019      C
0020            DO 40 J=1,N
0021            IF(J.EQ.K) GO TO 40
0022            A(K,J) = A(K,J)/A(K,K)
0023      40    CONTINUE
0024      C
0025      C   CALCULATE ELEMENT REPLACING PIVOT ELEMENT
0026      C
0027            A(K,K) = 1.0/A(K,K)
0028      C
0029      C   CALCULATE NEW ELEMENTS NOT IN PIVOT ROW OR PIVOT COLUMN
0030      C
0031            DO 50 I=1,N
0032            IF(I.EQ.K) GO TO 50
0033            DO 50 J=1,N
0034            IF(J.EQ.K) GO TO 50
0035            A(I,J) = A(I,J) - A(I,K)*A(K,J)
0036      50    CONTINUE
0037      C
0038      C   CALCULATE REPLACEMENT ELEMENTS FOR PIVOT
0039      C   COLUMN-EXCEPT PIVOT ELEMENT
0040      C
0041            DO 60 I=1,N
0042            IF(I.EQ.K) GO TO 60
0043            A(I,K) = - A(I,K)*A(K,K)
0044      60    CONTINUE
0045            RETURN
0046            END
```

Figure D.5 Matrix inversion using Gauss-Jordan elimination method.

D.4 Eigenvalue Problems

The nature of these problems is discussed in Section 1.3. Here we are concerned with the so-called standard eigenproblems

$$[A - \lambda I][X] = 0 \tag{D.39}$$

or the generalized eigenproblem

$$[A - \lambda B][X] = 0. \tag{D.40}$$

Appendix D: Solution of Simultaneous Equations

To show that Eqs.(D.39) and (D.40) are solved in the same way, we premultiply Eq. (D.40) by B^{-1} to obtain

$$[B^{-1}A - \lambda I][X] = 0. \tag{D.41}$$

Assuming $C = B^{-1}A$ gives

$$[C - \lambda I][X] = 0 \tag{D.42}$$

showing that Eq. (D.39) is a special case of Eq. (D.40) in which $B = I$. Thus, the procedure for solving Eq. (D.39) applies to Eq. (D.40) or (D.42).

The eigenvalue problems of Eqs. (D.39) and (D.40) are solved by either direct or indirect methods. In direct methods, such as Jacobi's method, the relevant matrix elements are stored in the computer, and an explicit procedure is used to obtain some or all of the eigenvalues $\lambda_1, \lambda_2, \cdots, \lambda_n$ and eigenvalues $X_1, X_2, \cdots, X_n$. Indirect methods are basically iterative, and the matrix elements are usually generated rather than stored.

D.4.1 Iteration (or Power) Method

The most commonly used iterative method is the *power method*. The method is suitable in situations where either the greatest or the least eigenvalue is required. Suppose that one of the eigenvalues of A, say λ_1, satisfies the condition

$$|\lambda_1| > |\lambda_i|, \quad i \neq 1, \tag{D.43}$$

then $|\lambda_1|$ is said to be the *dominant* eigenvalue of A. In many applications, the dominant eigenvalue is the most important and is probably the only eigenvalue we are interested in. The iteration method is specifically used for finding the dominant eigenvalues.

The iterative procedure is essentially based on the condition that should a trial vector $[X]_i$ be assumed, an approximate eigenvalue and a new trial eigenvector $[X]_{i+1}$ can be determined from Eq. (D.39) or Eq. (D.40). To find the largest eigenvalue $|\lambda_1|$, we rewrite Eq. (D.39) as

$$[A][X] = \lambda[X] \tag{D.44}$$

and follow these steps [2]:

(1) Assume a trial vector $[X]_0 = (x_1, x_2, \cdots, x_n)$, e.g., $[X]_0 = (1, 1, \cdots, 1)$. Substituting $[X]_0$ to the left-hand side of Eq. (D.44) gives the first approximation to the corresponding eigenvector, i.e.,

$$\lambda[X]_1 = (\lambda x_1, \lambda x_2, \cdots, \lambda x_n).$$

(2) Normalize the new vector $\lambda[X]$ by dividing it by the magnitude of its first component or by dividing the vector $[X]$ by the magnitude of the first component.
(3) Substitute the normalized vector into the left-hand side of Eq. (D.44) and obtain a new approximate eigenvector.
(4) Repeat steps (2) and (3) until the components of the new and previous eigenvectors differ by some prescribed tolerance. The normalizing factor will be the *largest eigenvalue* λ_1 while $[X]$ is the associated eigenvector.

To find the smallest eigenvalue, we first premultiply Eq. (D.44) by the inverse of $[A]$ to obtain

$$[X] = \lambda[A]^{-1}[X]$$

or

$$[A]^{-1}[X] = \frac{1}{\lambda}[X]. \tag{D.45}$$

Thus the iteration formula becomes

$$[A]^{-1}[X]_i = \frac{1}{\lambda}[X]_{i+1}. \tag{D.46}$$

In this form, the iteration converges to the largest value of $1/\lambda$, which corresponds to the smallest eigenvalue λ of $[A]$.

Once the largest eigenvalue is found, the method can be used to obtain the next largest eigenvalue by transforming $[A]$ to another matrix possessing only the remaining eigenvalues [2]. This so-called *matrix deflation* procedure assumes that $[A]$ is symmetric. The matrix deflation is continued until all the eigenvalues have been extracted. Error propagation from one stage of the deflation to the next leads to inaccurate results, especially for large eigenproblems. Jacobi's method, to be discussed in the next section, is recommended for large eigenproblems.

The subroutine in Fig. D.6 is useful for finding the largest eigenvalues of a matrix.

```
0001      C***********************************************************
0002      C     THIS SUBROUTINE FINDS THE LARGEST EIGENVALUE AND
0003      C     THE ASSOCIATED EIGENVECTOR OF
0004      C     A MATRIX EQUATION [A][X] = LAMBDA*[X]
0005      C     IDM = THE DIMENSION OF MATRIX [A]
0006      C     N   = THE ACTUAL SIZE OF MATRIX [A]
0007      C     IT  = THE NO. OF ITERATIONS USED
0008      C     REF: [2, p. 238,239]
0009      C***********************************************************
0010            SUBROUTINE POWER(A,LAMBDA,X,IDM,N,IT)
0011            REAL LAMBDA
0012            PARAMETER (ISIZE=100) !INCREASE ISIZE IF IDM > 100
```

Appendix D: Solution of Simultaneous Equations 665

```
0013                DIMENSION  A(IDM,IDM),X(IDM)
0014                DIMENSION  D(ISIZE), Z(ISIZE)
0015
0016                EPSI = 1.0E-30   !TOLERANCE FOR ZERO
0017       C
0018       C  INITIALIZATION
0019       C
0020                DO 20 I=1,N
0021                X(I) = 1.0
0022       20       CONTINUE
0023       C
0024       C  CALCULATE [A][X] AND CALL IT [D]
0025       C
0026                IT = 0
0027       30       DO 40 I=1,N
0028                D(I) = 0.0
0029                DO 40 J=1,N
0030                D(I) = D(I) + A(I,J)*X(J)
0031       40       CONTINUE
0032                IT = IT + 1
0033       C
0034       C  NORMALIZE VECTOR [D] AND CALL IT [Z]
0035       C
0036                DO 50 I=1,N
0037                Z(I) = D(I)/D(1)
0038       50       CONTINUE
0039       C
0040       C  CHECK IF RESULT IS SUFFICIENTLY ACCURATE
0041       C
0042                DO 60 I=1,N
0043                DIFF = X(I) - Z(I)
0044                IF( ABS(DIFF) - ESPI*ABS(Z(I)) ) 60, 60, 70
0045       60       CONTINUE
0046                GO TO 90
0047       C
0048       C  SUFFICIENT ACCURACY HAS NOT BEEN ATTAINED
0049       C  LET  [X] = [Z] AND REPEAT THE PROCESS
0050       C
0051       70       DO 80 I=1,N
0052                X(I) = Z(I)
0053       80       CONTINUE
0054                IF(IT.GE.100) GO TO 110
0055                GO TO 30
0056       C
0057       C  SUFFICIENT ACCURACY HAS BEEN ATTAINED
0058       C
0059       90       DO 100 I=1,N
0060                X(I) = Z(I)
0061       100      CONTINUE
0062       C
0063       C  OR MAXIMUM NUMBER OF ITERATIONS HAS BEEN REACHED
0064       C
0065       110      LAMBDA = D(1)
0066                RETURN
0067                END
```

Figure D.6 Subroutine for finding the largest eigenvalue of equation [A][X] = LAMBDA [X].

D.4.2 Jacobi's Method

Jacobi's method is perhaps the most reliable method for solving eigenvalue problems. Its major advantage is that it finds all eigenvalues and eigenvectors simultaneously with excellent accuracy.

The method transforms a symmetric matrix $[A]$ into a diagonal matrix having the same eigenvalues as $[A]$. This is achieved by eliminating one pair of off-diagonal elements of $[A]$ at a time. Given

$$[A][X] = \lambda[X], \tag{D.47}$$

let $\lambda_1, \lambda_2, \cdots, \lambda_n$ be the eigenvalues and $[V_1], [V_2], \cdots, [V_n]$ the corresponding eigenvectors. Then,

$$\begin{aligned} [A][V_1] &= \lambda_1[V_1] \\ [A][V_2] &= \lambda_2[V_2] \\ &\vdots \\ [A][V_n] &= \lambda_1[V_n] \end{aligned} \tag{D.48}$$

or simply

$$[A][V] = [V][\lambda] \tag{D.49}$$

where

$$[V] = \Big[[V_1], [V_2], \cdots, [V_n]\Big] \tag{D.50a}$$

$$[\lambda] = \begin{bmatrix} \lambda_1 & 0 & \cdots & 0 \\ 0 & \lambda_2 & \cdots & 0 \\ \vdots & \vdots & & \vdots \\ 0 & 0 & \cdots & \lambda_n \end{bmatrix}. \tag{D.50b}$$

From the theory of matrices, if $[A]$ is symmetric, $[V]$ is orthogonal, i.e.,

$$[V]^t = [V]^{-1}. \tag{D.51}$$

Hence, premultiplying Eq. (D.49) by $[V]^t$ leads to

$$[V]^t[A][V] = [\lambda] \tag{D.52}$$

signifying that the eigenvalues of $[V]^t[A][V]$, which is known as the orthogonal transformation of $[A]$, are the same as those of $[A]$. Thus the problem of finding the eigenvalues is reduced to finding the $[V]$ matrix.

Appendix D: Solution of Simultaneous Equations

The $[V]$ matrix is constructed iteratively by using a unitary matrix (or plane rotation matrix) $[R]$. If we let

$$[A_1] = [A]$$
$$[A_2] = [R_1]^t[A_1][R_1]$$
$$[A_3] = [R_2]^t[A_2][R_2] = [R_2]^t[R_1]^t[A][R_1][R_2] \qquad (D.53)$$
$$\vdots$$
$$[A_k] = [R_{k-1}]^t \cdots [R_1]^t[A][R_1] \cdots [R_{k-1}],$$

then as $k \to \infty$

$$[A_k] \to [\lambda]$$
$$[R_1][R_2] \cdots [R_{k-1}] \to [V]. \qquad (D.54)$$

The unitary transformation matrix $[R]$ eliminates the pair of equal elements a_{pq} and a_{qp}. It is given by [1,2,7]

$$[R_k] = \begin{bmatrix} 1 & & & & & \\ & 1 & & & & \\ & & \cos\theta & & -\sin\theta & \\ & & & 1 & & \\ & & \sin\theta & & \cos\theta & \\ & & & & & 1 \end{bmatrix} \begin{matrix} \\ \\ p \\ \\ q \\ \\ \end{matrix} \qquad (D.55a)$$

i.e.,

$$\begin{aligned} R_{qq} = R_{pp} &= \cos\theta \\ -R_{pq} = R_{qp} &= \sin\theta \\ R_{ii} &= 1, \quad i \neq p,q \\ R_{ij} &= 0, \quad \text{elsewhere.} \end{aligned} \qquad (D.55b)$$

The choice of θ in the transformation matrix must be such that new elements $a'_{pq} = a'_{qp} = 0$, i.e.,

$$a'_{pq} = (-a_{pp} + a_{qq})\cos\theta\sin\theta + a_{pq}(\cos^2\theta - \sin^2\theta) = 0. \qquad (D.56)$$

Hence

$$\tan 2\theta = \frac{2a_{pq}}{a_{pp} - a_{qq}}, \quad -45° < \theta < 45°. \qquad (D.57)$$

An alternative manipulation of Eq. (D.56) gives

$$\cos\theta = \left[\frac{\sqrt{(a_{pp}-a_{qq})^2+4a_{pq}^2}+(a_{pp}-a_{qq})}{2\sqrt{(a_{pp}-a_{qq})^2+4a_{pq}^2}}\right]^{1/2} \quad \text{(D.58a)}$$

$$\sin\theta = \frac{a_{pq}}{\sqrt{(a_{pp}-a_{qq})^2+4a_{pq}^2}\cos\theta}. \quad \text{(D.58b)}$$

Notice that Eq. (D.53) requires an infinite number of transformations because the elimination of elements a_{pq} and a_{qp} in one step will in general undo the elimination of previously treated elements in the same row or column. However, the transformation converges rapidly and ceases when all the off-diagonal elements become negligible in magnitude.

The program in Fig. D.7 determines all the eigenvalues and eigenvectors of symmetric matrices employing Jacobi's method.

```
0001      C***********************************************
0002      C   USING JACOBI METHOD,
0003      C   THIS SUBROUTINE FINDS ALL THE EIGENVALUES AND
0004      C   THE ASSOCIATED EIGENVECTORS OF
0005      C   A MATRIX EQUATION [A][X] = LAMBDA*[X]
0006      C   A        = SYMMETRIC MATRIX
0007      C   LAMBDA(J) = EIGENVALUES, ORDERED FROM ALGEBRAICALLY
0008      C              LARGEST TO SMALLEST
0009      C   RT(I,J)  = MATRIX THAT WILL EVENTUALLY CONTAIN THE
0010      C              EIGENVECTORS (J) ASSOCIATED WITH LAMBDA (J)
0011      C   IDM  = THE DIMENSION OF MATRIX [A]
0012      C   N    = THE ACTUAL SIZE OF MATRIX [A]
0013      C   IT   = THE NO. OF ITERATIONS USED
0014      C   REF: [2, p. 268-271]
0015      C***********************************************
0016            SUBROUTINE EIGEN(A,RT,IDM,N,LAMBDA)
0017            DIMENSION A(IDM,IDM),RT(IDM,IDM)
0018            REAL LAMBDA(IDM)
0019      C
0020      C   GENERATE AN N X N IDENTITY MATRIX RT WHICH WILL
0021      C   EVENTUALLY CONTAIN THE EIGENVECTORS
0022      C
0023            DO 20 I=1,N
0024            DO 10 J=1,N
0025            RT(I,J) = 0.0
0026      10    CONTINUE
0027            RT(I,I) = 1.0
0028      20    CONTINUE
0029            NSWEEP = 0
0030      30    NRSKIP = 0
0031      C
0032      C   BEGIN A SWEEP WHICH WILL TRANSFORM EACH OFF-DIAGONAL
0033      C   ELEMENT IN TURN TO ZERO
0034      C
0035            NMIN1 = N-1
0036            DO 130 I=1,NMIN1
0037            IP1 = I + 1
0038            DO 120 J=IP1,N
```

Appendix D: Solution of Simultaneous Equations 669

```
0039                AV = 0.5*( A(I,J) + A(J,I) )
0040                DIFF = A(I,I) - A(J,J)
0041                RAD = SQRT( DIFF*DIFF + 4.0*AV*AV )
0042       C
0043       C  CHECK IF RAD IS ZERO.  IF SO, NO NEED OF
0044       C  ROTATION
0045       C
0046                IF(RAD.EQ.0.0) GO TO 80
0047       C
0048       C  CHECK IF DIFF IS NEGATIVE.  IF SO, INTERCHANGE
0049       C  A(I,I) AND A(J,J) AND PERFORM ROTATION
0050       C
0051                IF(DIFF.LT.0.0) GO TO 60
0052                IF( ABS(A(I,I)).EQ.ABS(A(I,I))+100.*ABS(AV) ) GO TO 40
0053                GO TO 40
0054       40       IF( ABS(A(J,J)).EQ.ABS(A(J,J))+100.*ABS(AV) ) GO TO 80
0055       50       COSINE = SQRT( (RAD + DIFF)/(2.0*RAD) )
0056                SINE = AV/(RAD*COSINE)
0057                GO TO 70
0058       C
0059       C  FOR THE DIAGONAL ELEMENTS, PERFORM ROTATION
0060       C
0061       60       SINE = SQRT( (RAD-DIFF)/(2.0*RAD) )
0062                IF(AV.LT.0.0) SINE = -SINE
0063                COSINE = AV/(RAD*SINE)
0064       C
0065       C  CHECK IF SIN(THETA) IS NEGLIGIBLE
0066       C  IF SO, SKIP ROTATION
0067       C
0068       70       IF(1.0.LT.1.0 + ABS(SINE)) GO TO 90
0069       80       NRSKIP = NRSKIP + 1
0070                GO TO 120
0071       C
0072       C  PERFORM ROTATION
0073       C  PREMULTIPLY BY THE ROTATION MATRIX
0074       C
0075       90       DO 100 K=1,N
0076                Q = A(I,K)
0077                A(I,K) = COSINE*Q + SINE*A(J,K)
0078                A(J,K) = - SINE*Q + COSINE*A(J,K)
0079       100      CONTINUE
0080       C
0081       C  POSTMULTIPLY BY THE TRANSFORM OF THE ROTATION MATRIX
0082       C
0083                DO 110 K=1,N
0084                Q = A(K,I)
0085                A(K,I) = COSINE*Q + SINE*A(K,J)
0086                A(K,J) = - SINE*Q + COSINE*A(K,J)
0087       C
0088       C   POSTMULTIPLY THE CURRENT PRODUCT OF ALL THE RT MATRICES
0089       C   UP TO THIS POINT BY THE CURRENT RT MATRIX
0090       C
0091                Q = RT(K,I)
0092                RT(K,I) = COSINE*Q + SINE*RT(K,J)
0093                RT(K,J) = - SINE*Q + COSINE*RT(K,J)
0094       110      CONTINUE
0095       120      CONTINUE
0096       130      CONTINUE
0097       C
0098       C  KEEP A TALLY OF THE NUMBER OF SWEEPS
0099       C
0100                NSWEEP = NSWEEP + 1
0101                IF(NSWEEP.GT.50) GO TO 140
```

```
0102     C       WRITE(6,*)NRSKIP,NSWEEP
0103     C
0104     C    CHECKIF THE NUMBER OF ROTATIONS SKIPPED IS LESS/
0105     C    EQUAL TO THE NO. OF ELEMENTS ABOVE THE MAIN DIAGONAL
0106     C    IF EQUAL, CONVERGENCE HAS OCCURRED
0107     C
0108              IF(NRSKIP.LT.N*(N-1)/2) GO TO 30
0109     140      CONTINUE
0110     C        WRITE(6,*) NSWEEP
0111              DO 150 J=1,N
0112              LAMBDA(J) = A(J,J)
0113     150      CONTINUE
0114              RETURN
0115              END
```

Figure D.7 Subroutine for finding all the eigenvalues and eigenvectors of equation [A][X] = LAMBDA [X].

References

[1] A. W. Al-Khafaji and J. R. Tooley, *Numerical Methods in Engineering Practice*. New York: Holt, Rinehart and Winston, 1986, pp. 84–159, 203–270.

[2] M. L. James et al., *Applied Numerical Methods for Digital Computation*. 3rd ed., New York: Harper & Row, 1985, pp. 146–298.

[3] S. A. Hovanessian and L. A. Pipes, *Digital Computer Methods in Engineering*. New York: McGraw-Hill, 1969, pp. 1–48.

[4] W. Cheney and D. Kincaid, *Numerical Mathematics and Computing*, 2nd ed. Monterey, CA: Brooks/Cole, 1985, pp. 201–257.

[5] R. L. Ketter and S. P. Prawel, *Modern Methods of Engineering Computation*. New York: McGraw-Hill, 1969, pp. 66–117.

[6] A. Ralston and H. S. Wilf (eds.), *Mathematical Methods for Digital Computers*. New York: John Wiley, 1960, pp. 62–72.

[7] A. Jennings, *Matrix Computation for Engineers and Scientists*. New York: John Wiley, 1977, pp. 182–222, 250–254.

[8] J. C. Nash, *Compact Numerical Methods for Computers: Linear Algebra and Function Minimization*. New York: John Wiley, 1979, pp. 195–199.

[9] T. K. Sarkar, et al., "A limited survey of various conjugate gradient methods for solving complex matrix equations arising in electromagnetic wave interaction," *Wave Motion*, vol. 10, no. 6, 1988, pp. 527–546.

Appendix D: Solution of Simultaneous Equations   671

[10] A. F. Peterson and R. Mittra, "Method of conjugate gradients for the numerical solution of large-body electromagnetic scattering problems," *J. Opt. Soc. Am.*, Pt. A, vol. 2, no. 6, June 1985, pp. 971–977.

[11] T. K. Sarkar, "Application of the Fast Fourier transform and the conjugate gradient method for efficient solution of electromagnetic scattering from both electrically large and small conducting bodies," *Electromagnetics*, vol. 5, 1985, pp. 99–122.

[12] D. T. Borup and O. P. Gandhi, "Calculation of high-resolution SAR distributions in biological bodies using the FFT algorithm and conjugate gradient method," *IEEE Trans. Micro. Theo. Tech.*, vol. MTT-33, no. 5, May 1985, pp. 417–419.

[13] R. W. Southworth and S. L. Deleeuw, *Digital Computation and Numerical Methods*. New York: McGraw-Hill, 1965, pp. 247–251.

[14] S. Hovanessian, *Computational Mathematics in Engineering*. Lexington, MA: Lexington Books, 1976, p. 25.

Appendix E
Answers to Odd-numbered Problems

Chapter 1

1.1 Proof.

1.3 Proof.

1.5 They satisfy Maxwell's equations.

1.7 $\mathbf{H}_s = -j\dfrac{20}{\omega\mu_o}\left[k_y \sin(k_x x)\sin(k_y y)\mathbf{a}_x + k_x \cos(k_x x)\cos(k_y y)\mathbf{a}_y\right]$.

1.9 (a) $\mathbf{A} = x^2 \cos(\omega t - kz)\mathbf{a}_y$.
 (b) $\mathbf{B} = \sin x \cos(\omega + x - y)\mathbf{a}_z$,
 (c) $\mathbf{C} = 10\cos(\omega t - kz)\mathbf{a}_x - 5e^{\pi/4}\sin(\omega t + kz)\mathbf{a}_y$.

1.11 Proof.

1.13 (a) Parabolic, (b) elliptic, (c) elliptic, (d) elliptic.

Chapter 2

2.1 If a and d are functions of x only; c and e are functions of y only; $b = 0$; and f is the sum of a function of x and a function of y only.

2.3 (a)
$$V(x,y) = \frac{4V_o}{\pi} \sum_{n=1,3,5}^{\infty} \frac{\sinh(n\pi x/b)\sin(n\pi y/b)}{n\sinh(n\pi a/b)}$$
$$-\frac{4V_o}{\pi} \sum_{n=1,3,5}^{\infty} \frac{\sinh[n\pi(a-x)/b]\sin(n\pi y/b)}{n\sinh(n\pi a/b)}.$$

(b) $V(x,y) = \dfrac{4V_o}{\pi} \sum_{n=\text{odd}}^{\infty} \dfrac{1}{n} \sin(n\pi y/b) e^{-n\pi x/b}$

2.5 (a) $\pi/2 - \dfrac{4}{\pi} \sum_{n=\text{odd}}^{\infty} \dfrac{\cos nx \, \sinh ny}{n^2 \sinh n\pi}$,

(b) $x + \dfrac{2}{\pi} \sum_{n=1}^{\infty} \dfrac{(-1)^n}{n} e^{-n^2\pi^2 kt} \sin n\pi x$,

(c) $-\dfrac{2}{\pi^2 a^2} \sum_{n=1}^{\infty} \dfrac{\cos n\pi a}{n^2} \sin n\pi x \, \sin n\pi at$.

2.7 (a) $\Phi(\rho,\phi) = \dfrac{\sin\phi}{\rho}$,

(b) $\Phi(\rho,\phi,z) = \dfrac{4}{\pi} \sum_{n=1,3,5}^{\infty} \dfrac{I_0(n\pi\rho/L)}{I_0(n\pi a/L)} \dfrac{\sin(n\pi z/L)}{n}$,

(c) $\Phi(\rho,\phi,z) = 2\sum_{n=1}^{\infty} \dfrac{J_0(\rho x_n/a)}{x_n J_3(x_n)} \cos 2\phi \exp[-x_n^2 \, kt/a^2]$,

where $J_2(x_n) = 0$.

2.9 $\Phi(\rho,t) = \sum_{n=0}^{\infty} \dfrac{4J_2(\lambda_{0n})}{\lambda_{0n}^2 J_1(\lambda_{0n})} J_0(\lambda_{0n}\rho/a)\cos(\lambda_{0n}t/a)$.

2.11 Proof.

2.13 (a) 0, (b) $\dfrac{2}{9}$, (c) $\dfrac{1}{24}$.

2.15 (a) $\dfrac{1}{2} + \sum_{n=1}^{\infty} \dfrac{(2n+1)}{2n(n+1)} P_n^1(0)(r/a)^n P_n(\cos\theta)$,

(b) $-\dfrac{7}{5}\dfrac{a^3}{r^2} P_1(\cos\theta) - \dfrac{3}{10}\dfrac{a^5}{r^4} P_3(\cos\theta)$,

(c) $\frac{2}{3}[1 - \frac{r^2}{a^2} P_2(\cos\theta)]$.

2.17 For $r < a$, $0 \leq \theta \leq \pi/2$,

$$V = \frac{a\rho}{2\epsilon}\left[-z + \sum_{n=0}^{\infty} \frac{(-1)^n(2n)!}{(n!2^n)^2}(r/a)^{2n} P_{2n}(\cos\theta)\right],$$

for $r > a$,

$$V = \frac{a\rho}{2\epsilon}\sum_{n=0}^{\infty} \frac{(-1)^{n+1}(2n+2)!}{[(n+1)!2^{n+1}]^2}(r/a)^{2n+1} P_{2n}(\cos\theta).$$

2.19 Proof.

2.21 $V = V_o + E_o r\cos\theta[1 - (a/r)^3]$,

$\mathbf{E} = -E_o[1 + 2(a/r)^3]\cos\theta \mathbf{a}_r + E_o[1 - (a/r)^3]\sin\theta \mathbf{a}_\theta$, $E_{max} = 3E_o$.

2.23 Proof.

2.25 $\frac{1}{3}\cos 2\phi\, P_2^2(\cos\theta)$.

2.27 (a) $\frac{32}{\pi^3}\sum_{m=1}^{\infty}\sum_{n=1}^{\infty}\sum_{p=1}^{\infty} \frac{m[1-(-1)^m e^\pi]}{(m^2+1)(2n-1)(2p-1)}$

$\cdot \frac{\sin mx \sin(2n-1)y \sin(2p-1)z}{[m^2 + (2n-1)^2 + (2p-1)^2]}$,

(b) $-\frac{128}{\pi^3}\sum_{m=1}^{\infty}\sum_{n=1}^{\infty}\sum_{p=1}^{\infty} \frac{1}{(2m-3)(4m^2-1)(2n-1)(2p-1)}$

$\cdot \frac{\sin(2m-1)x \sin(2n-1)y \sin(2p-1)z}{[(2m-1)^2 + (2n-1)^2 + (2p-1)^2]}$.

2.29

$$2\Phi_o a^2 \sum_{n=1}^{\infty} \frac{J_0(\lambda_{0n}r/a)}{\lambda_{0n}^2 J_1(\lambda_{0n})}\left[\frac{(\lambda_{0n}/a)\sin\omega t - \omega\sin\lambda_{0n}t}{\lambda_{0n}^2 - \omega^2 a^2}\right].$$

2.31

$$V = \begin{cases} \sum_{k=1}^{\infty} \sin\beta x\, [a_n \sinh\beta y + b_n \cosh\beta y], & 0 \leq y \leq c \\ \sum_{k=1}^{\infty} c_n \sin\beta x \sinh\beta y, & c \leq y \leq b \end{cases}$$

where

$$\beta = \frac{n\pi}{a}, \quad n = 2k - 1$$

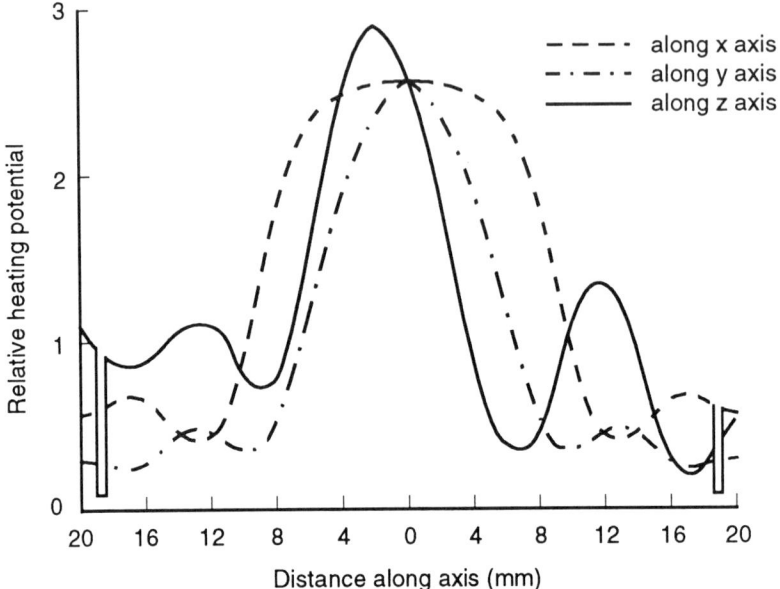

Figure E.1 For Problem 2.43.

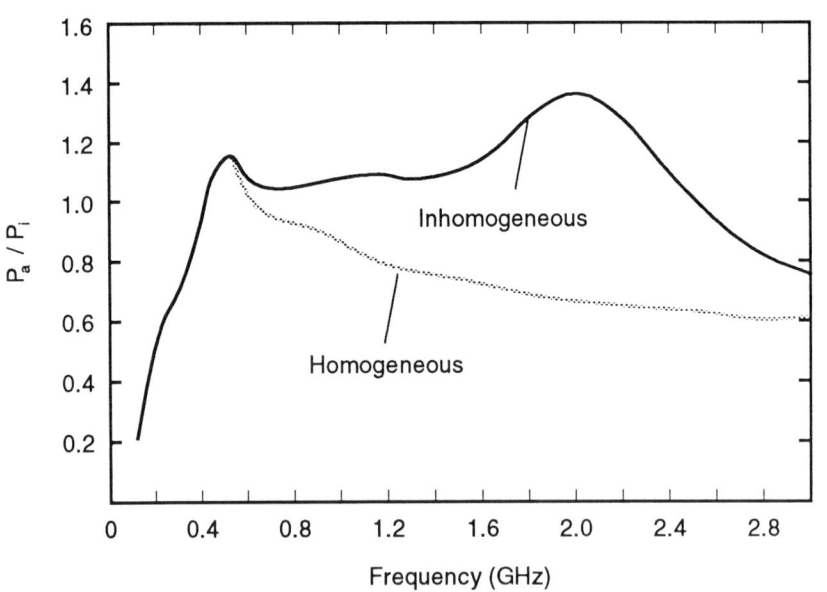

Figure E.2 For Problem 2.45.

$$a_n = 4V_0[\epsilon_1 \tanh \beta c - \epsilon_2 \coth \beta c]/d_n,$$
$$b_n = 4V_0(\epsilon_2 - \epsilon_1)/d_n,$$
$$c_n = 4V_0[\epsilon_1 \tanh \beta c - \epsilon_2 \coth \beta c + (\epsilon_2 - \epsilon_1) \coth \beta c]/d_n,$$
$$d_n = n\pi \sinh \beta b [\epsilon_1 \tanh \beta c - \epsilon_2 \coth \beta c + (\epsilon_2 - \epsilon_1) \coth \beta b].$$

2.33 $P_1^0 = 0.5, P_2^0 = 1.25, P_2^1 = 1.29904, P_2^2 = 2.25$.
$P_3^0 = -0.4375, P_3^1 = 0.32476, P_3^2 = 5.625, P_3^3 = 9.74279$, etc.

2.35 Proof.

2.37 Proof.

2.39 Proof.

2.43 See Fig. E.1.

2.45 See Fig. E.2.

Chapter 3

3.1 Proof.

3.3 $(1-a^2)[\Phi(i+1,j+1) + \Phi(i-1,j-1)]$
$-2\Big[\Phi(i,j+1) + \Phi(i,j-1) - a^2\Phi(i+1,j) - a^2\Phi(i-1,j)\Big]$
$+\Phi(i-1,j+1) + \Phi(i-1,j-1) - \Phi(i+1,j-1) - \Phi(i-1,j-1) = 0.$

3.5 (a) $\Phi(0.25, 0.25) = 8.897$
(b) $\Phi(0.25, 0.25) = 8.649, 8.623, 8.669.$

3.7 $V_A = 61.462 = V_E, \quad V_B = 21.954 = V_D, \quad V_C = 45.99$ V.

3.9 (a) $e^{kx/2} \dfrac{\sinh \sigma (1-x) \sin \pi y}{\sinh \sigma}, \quad \sigma = \dfrac{1}{2}[k^2 + 4\pi^2]^{1/2},$
(b) Proof, (c) $V(1/2, 1/2) = 0.8846,$
(d) (i) 0.658, (ii) 0.887, (iii) 0.992.

3.11 $r \leq 1/2$.

3.13 (a) Proof, (b) Proof, (c) Euler: $r \leq 1/4$ for stability, Leapfrog: unstable, Dufort-Frankel: unconditionally stable.

3.15 (a) After 5 iterations, $V_1 = 73.79, V_2 = 79.54, V_3 = 40.63, V_4 = 45.31, V_5 = 61.33$

(b) After 5 iterations, $V_1 = 19.93$, $V_2 = 19.96$, $V_3 = 6.634$, $V_4 = 6.649$.

3.17 42.78Ω.

3.19 $k_c = 4.4429$ (exact), $k_c = 3.5124$ (exact).

3.21 Proof.

3.23 See Yee's paper [Ref. 42 in this chapter].

3.25 65.85, 23.32, 6.4, 10.23, 10.34.

3.27 Proof.

3.29 3.667 (exact).

3.31 0.6321 (exact).

3.33 $a_0 = 2/9$, $a_1 = 16/9 = a_2$, $a_3 = 2/9$.

3.35 (a) 1.724, (b) 3.963, (c) 15.02.

Chapter 4

4.1 (a) 1.3333, (b) -4.667, (c) 157.08.

4.3 (a) $y'' = 0$,   (b) $1 + y'^2 - yy'' = 0$,
(c) $xy' \cos(xy') + \sin(xy') = 0$,   (d) $y'' + y = 0$,
(e) $2y^{iv} - 10y$,   (f) $3u + 2v'' = 0$.

4.5 Proof.

4.7 $\nabla^2 \Phi - \Phi_{tt} = \lambda \Phi$.

4.9 $\frac{1}{2} \int \left[(y')^2 + y^2 - 2y \sin \pi x \right] dx$.

4.11 For exact solution, $\Phi = 2.1667x - 0.1667x^3$,
for $N = 1$, $\tilde{\Phi} = 2.25x - 0.25x^2$,
for $N = 2, 3$, $\tilde{\Phi}$ is the same as the exact solution.

4.13 (a) $a_1 = 10.33$, $a_2 = -1.46$, $a_3 = 0.48$,
(b) $a_1 = 10.44$, $a_2 = -1.61$, $a_3 = 0.67$,
(c) $a_1 = 10.21$, $a_2 = -1.32$, $a_3 = 0.35$,
(d) $a_1 = 10.21$, $a_2 = -1.32$, $a_3 = 0.35$.

Appendix E: Answers to Odd-numbered Problems

4.15

Method	a_1	a_2
Collocation	0.9292	−0.05115
Subdomain	0.9237	−0.05991
Galerkin	0.9334	−0.05433
Least squares	0.9327	−0.06813
Rayleigh-Ritz	0.9334	−0.05433

4.17 $\tilde{\Phi} = (1-x)(1 - 0.2090x - 0.789x^2 + 0.2090x^3)$.

4.19 $\tilde{\lambda}_0 = 0.2$, $\lambda_0 = 0.1969$ (exact).

4.21 $a/\lambda_c = 0.2948$.

Chapter 5

5.1 (a) Nonsingular, Fredholm IE of the 2nd kind,

  (b) Nonsingular, Volterra IE of the 2nd kind,

  (c) Nonsingular, Fredholm IE of the 2nd kind.

5.3 (a) $y = x - \int_0^x (x-t)y\,dt$,

  (b) $y = 1 + x - \cos x - \int_0^x (x-t)y(t)\,dt$.

5.5 Proof.

5.7 $G(x,y;x',y') =$
$$-\frac{4}{ab}e^{x'-x}\sum_{m=1}^{\infty}\sum_{n=1}^{\infty}\left[\frac{\sin\frac{m\pi x}{a}\sin\frac{n\pi y}{b}\sin\frac{m\pi x'}{a}\sin\frac{n\pi y'}{b}}{1+(m\pi/a)^2+(n\pi/b)^2}\right].$$

5.11 Proof.

5.13 (a) 62.712Ω, (b) 26.755Ω .

5.19 $B_1 = 0.0094035 - j0.0035738$, $B_2 = 0.0004534 - j0.0020312$.

5.21 Proof.

5.23 See Fig. E.3.

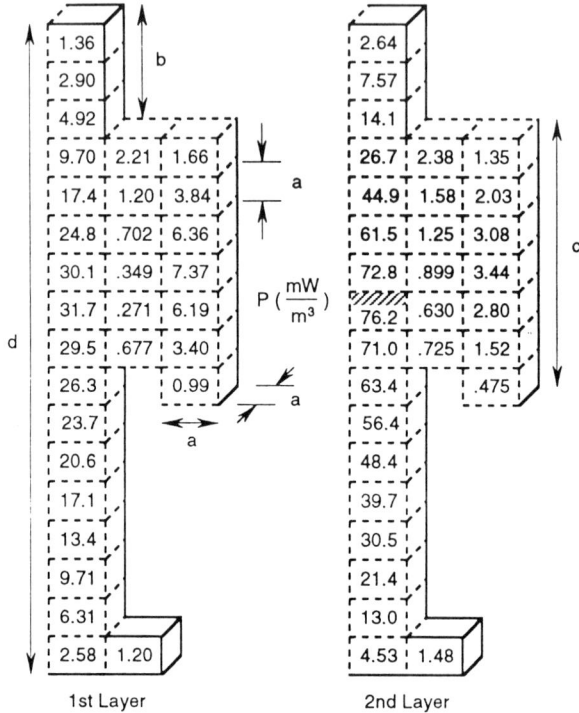

Figure E.3 For Problem 5.23.

Chapter 6

6.1 (a)
$$\begin{bmatrix} 0.5909 & -0.1364 & -0.4545 \\ -0.1364 & 0.4545 & -0.3182 \\ -0.4545 & -0.3182 & 0.7727 \end{bmatrix},$$

(b)
$$\begin{bmatrix} 0.6667 & -0.6667 & 0 \\ -0.6667 & 1.042 & -0.375 \\ 0 & -0.375 & 0.375 \end{bmatrix}.$$

6.3 $\alpha_1 = \dfrac{1}{23}[4x+3y-24]$, $\alpha_2 = \dfrac{1}{23}[-5x+2y+30]$, $\alpha_3 = \dfrac{1}{23}[x-5y+17]$.

6.5 See Example 3.3.

6.7 See Example 6.2.

6.9 See Example 3.4.

6.11 Exact: 4.0, 2.0, 1.789, 1.414, 1.333, 1.109, 0.9701, 0.8, 0.6667, 0.5714 cm.

6.13 (a) 4, (b) 3, see Fig. E.4.

6.17 (a) $A\bar{x}$, (b) $\frac{A}{12}(x_1^2 + x_2^2 + x_3^2 + 9\bar{x}^2)$,

(c) $\frac{A}{12}(x_1 y_1 + x_2 y_2 + x_3 y_3 + 9\bar{x}\ \bar{y})$,

where $\bar{x} = \frac{1}{3}(x_1 + x_2 + x_3)$, $\bar{y} = \frac{1}{3}(y_1 + y_2 + y_3)$.

6.19 (a)

$$\frac{1}{2}\begin{bmatrix} 1 & 0 & -1 \\ 0 & 0 & 0 \\ -1 & 0 & 1 \end{bmatrix},\qquad \frac{1}{2}\begin{bmatrix} 0 & 1 & -1 \\ 0 & -1 & 1 \\ 0 & 0 & 0 \end{bmatrix},$$

(b)

$$\frac{1}{6}\begin{bmatrix} 3 & 0 & -4 & 0 & 0 & 1 \\ 0 & 8 & 0 & 0 & -8 & 0 \\ -4 & 0 & 8 & 0 & 0 & -4 \\ 0 & 0 & 0 & 0 & 0 & 0 \\ 0 & -8 & 0 & 0 & 8 & 0 \\ 1 & 0 & -4 & 0 & 0 & 3 \end{bmatrix},\qquad \frac{1}{6}\begin{bmatrix} 3 & 1 & 0 & 1 & 0 & 0 \\ -4 & 8 & 0 & -4 & 0 & 0 \\ 0 & 0 & 8 & 0 & -8 & 0 \\ 1 & -4 & 0 & 3 & 0 & 0 \\ 0 & 0 & -8 & 0 & 8 & 0 \\ 0 & 0 & 0 & 0 & 0 & 0 \end{bmatrix}.$$

6.21 Proof.

6.23 $\frac{v}{20}(z_1^2 + z_2^2 + z_3^2 + z_4^3)$.

Chapter 7

7.1 Proof.

7.3 $V \equiv E_x$, $I \equiv H_y$, $C \equiv \epsilon$, $L \equiv \mu$.

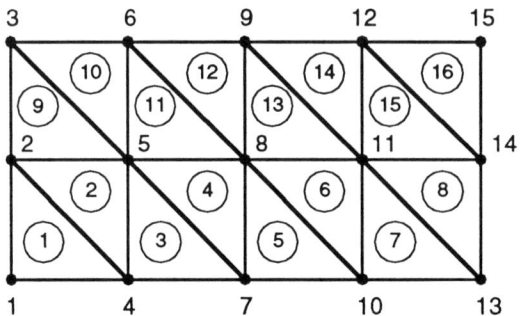

Figure E.4 For Problem 6.13(b).

7.5

Mode	TLM($\Delta\ell/\lambda$)	Analytic($\Delta\ell/\lambda$)
TM_{11}	0.0501	0.05
TM_{12}	0.0788	0.0791
TM_{22}	0.0999	0.1
TM_{13}	0.1106	0.1118
TM_{23}	0.1269	0.1275
TM_{14}	0.1422	0.1458
TM_{33}	0.1497	0.150
TM_{24}	0.1558	0.1581
TM_{34}	0.1759	0.1768
TM_{15}	0.1726	0.1803
TM_{25}	0.1841	0.1904
TM_{44}	0.1995	0.20

7.9 See Fig. E.5.

Appendix E: Answers to Odd-numbered Problems

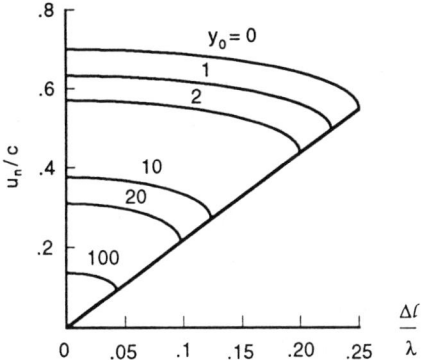

Figure E.5 For Problem 7.9.

7.11

	$\|Z\|$		$\arg(Z)$	
$(\Delta\ell/\lambda)$	TLM	Analytic	TLM	Analytic
0.023	6.2296	2.0396	−0.9263	−1.5708
0.025	2.2531	2.6599	1.8694	1.5708
0.027	0.2531	0.2527	−1.5309	−1.5708
0.029	6.8085	6.6448	1.5923	1.5708
0.031	0.2706	0.2826	−1.5582	−1.5708
0.033	1.8526	1.8320	1.5787	1.5708
0.035	0.8573	0.8521	−1.5569	−1.5708
0.037	0.4875	0.4796	1.5548	1.5708
0.039	5.9766	6.0828	−1.5633	−1.5708
0.041	0.2076	0.2118	−1.5488	−1.5708

7.13   0.03721 and 0.03838 (TLM),

0.03730 and 0.03850 (exact).

7.15 $k_c a = 5.5290$.

7.17   0.493 ns,   0.724 ns.

Chapter 8

8.3   (a) 16, 187, 170, 429, 836, 47, 950, 369, 456, 307,

       (b) 997, 281, 13, 449, 277, 721, 133, 209, 757, 761 .

8.7 $M = 5$, $a = 0$, $b = 1$. Generate the random variable as follows:

       (1) Generate two uniformly distributed random variables U_1 and U_2 from (0, 1).

       (2) Check if $U_2 \leq f_X(U_2)/M = U_2$.

       (3) If the inequality holds, accept U_2.

       (4) Otherwise, reject U_1 and U_2 and repeat (1) to (3).

8.9   (a) 3.14159 (exact),

       (b) 0.4597 (exact),

       (c) 1.71828 (exact),

       (d) 2.0 .

8.11   0.4053 (exact).

8.13   2.5 (exact).

8.15   42.59, 17.72, 4.33.

8.17   23.43 (exact).

8.19   Proof.

8.21   10.35, 23.56, 10.40, 1.30, 0.1124 V (for $\Delta = 0.25$).

8.23   2.461.

8.25   43.20 (exact).

8.27   35.55 V.

Index

A

ABCD Matrix, 562
Absorbing boundary conditions, 184
Acceleration factor, 175
Accuracy
    of computer program, 644
    of Monte Carlo integration, 583
    of finite difference, 161, 184
    of TLM solution, 523
Ampere's law, 4
Amplification factor, 162
Analytical methods, 1, 29-133
Area coordinates, 460
Attenuation, 104-107

B

Band matrix method, 154, 418-419, 432

Bandwidth reduction, 454-457
Basis function, 266, 354-355
Bei function, 54
Ber function, 54
Bessel's equation, 48, 77
Bessel's functions, 48-52, 76-79
    computation of, 227-230
    spherical type of, 65-67
Biot-Savart law, 4
Boundary conditions, 7
    between two media, 7
    classification of, 19-21
    for TLM, 521, 522, 534-535, 545-547
Boundary element method, 481-482

C

Cauchy-Euler equation, 45, 61
Central limit theorem, 578, 580

Chebyshev functions, 76-82, 219
Cholesky's method, 650-653
Coarseness error, 556
Coefficient matrix, see
    Element/Global Coefficient
    matrix
Collocation method, 274-275, 279
Computational molecule, 141, 153, 167-170, 218-219, 238-239
Conductivity, 5
Confidence interval, 579
Conjugate gradients method, 656-658
Continuity equation, 6, 13-14
Constitutive parameters, 7
Constitutive relations, 6
Convergence of solution, 267, 277, 583, 589, 654-656, 658
Coulomb's law, 3
Crank-Nicholson algorithm, 142, 238

D

Determinant, 652
Diffusion equation, 20, 505
    analytical solution of, 42
    finite difference solution of, 140-146
    functional for, 265
    TLM solution of, 507-512
Dirichlet condition, 19-21, 284
Discretization errors, 161
Divergence theorem, 2, 626, see also Gauss's theorem
Dufort-Frankel scheme, 242-243
Dyadic Green's function, 369

E

Eigenfunction expansion, 323-324
Eigenproblem, 19, 283, 435, 662-670

Eigenvalue, 19, 174, 283, 663-667
Eigenvector, 19, 174, 663-667
Electrical length, 517
Element coefficient matrix, 413
Elimination methods, 647-653
Elliptic PDE, 17-20, 505
    finite difference solution of, 152-158
Error function, 253, 579
Errors, 161-162
    evaluation of, 576-580, 604
    in TLM, 555-557
Euler scheme, 243
Euler's constant, 50
Euler's equation, 261-263
Euler's rule, 214-216
Exodus method, 593-595
Expansion function, 278, see also Basis function
Expected value, see Mean value

F

Finite difference approximation, 136-140, 180
    for boundary points, 210
Finite difference method (FDM), 135-254, 275, 407
Finite difference solution
    accuracy of, 161-162
    in cylindrical coordinates, 205-209
    of elliptic PDEs, 152-158
    of hyperbolic PDEs, 147-151
    of parabolic PDEs, 140-146
    in spherical coordinates, 209-210
    stability of, 161-164, 184
Finite-difference time domain (FD-TD) method, 179-204
Finite element method (FEM), 407-499

Finite elements, 408
  higher order, 457-473
  for three dimensions, 474-479
Fixed nodes, 154, 418
Fixed random walk, 587-591
Floating random walk, 591-593
FORTRAN, 629-641
Fredholm integral, 310, 358, 608
Free nodes, 154, 418
Functional, 259
Fundamental matrices, 463-473

G

Gamma function, 49
Galerkin method, 276, 281
Gaussian distribution, see
  Normal distribution
Gaussian rules, 221-224
Gauss's law, 3, 4, 168, 171
Gauss's method, 648-650
Gauss's theorem, 2-3, see also
  Divergence theorem
Gauss-Seidel method, 655
Global coefficient matrix, 414, 454
Gradient methods, 656-659
Green's functions, 315-334, 369
Green's theorem, 13, 626

H

Hallen integral equation, 353-354, 358
Hankel functions, 51, 341
Helmholtz equation, 8, 12, 341, 506
  analytical solution of, 37-38, 47-52
  finite difference solution of, 173-176
  finite element solution of, 432-439, 474-479
  Green's function for, 320
  TLM solution of, 512-530
  variational solution of, 284
Hermite function, 76-79, 84-86, 221-222, 225
Higher order elements, 457-473
Human body, 366-385
Hybrid methods, 479-482
Hyperbolic PDE, 17-20
  finite difference solution of, 147-151
Hyperthermia, 366

I

Infinite element method, 479-480
Inner product, 256
Integral equations, 310-312
Integration molecule, 224, 226, 227
Interpolation functions, see
  Shape functions
Inverse Monte Carlo analysis, 608
Inversion, see Matrix inversion
Iterative methods, 154, 174, 418, 431, 509, 653-656, 663

J

Jacobi's method, 653-654, 666-668

K

Kelvin functions, 54
Kronecker delta, 70

L

Laguerre functions, 76-81, 221, 223
Laplace's equation, 4-5, 12, 20

analytical solution of, 32-37, 40-42, 43-46, 59-64
finite difference solution of, 153-156, 166, 204-207
finite element solution of, 394-429
functional for, 265, 412
Green's function for, 319-320
variational solution of, 290-293
Leapfrog scheme, 242-243
Least squares method, 277, 281, 290-292
Legendre's equation, 61, 67, 77
Legendre functions, 61-63, 76-79, 221-222
associated type of, 67-69
Leibnitz rule, 331
Local coordinates, 459-461, 475
Lorentz condition, 10
Lorentz force equation, 6

M

Material medium, 7
Matrix inversion, 660-662
Maxwell's equations, 2, 6, 340-341, 367-368, 501
finite difference solution of, 179
TLM solution of, 507-508, 514-515, 531-532, 541-544
Mean value, 576
Mesh generation, 439-454
Mesh size, 152, 162, 507, 521, 556, 587
Method of descent, 657
Method of images, 321-323, 327
Mie scattering, 96-101
Misalignment error, 557
Modeling errors, 161, 555
Moment methods, 309-404
Monte Carlo integration, 581-586

Monte Carlo methods, 571-624
Multiple integration, 224-227

N

Neumann condition, 20-21, 284
Neumann function, 48
Newton-Cotes rules, 218-220, 224, 227-229, 252
Node numbering, 414-415
Norm, 257
Normal distribution, 578, 579-581
Numerical integration, 214-228, 581-586

O

Ohm's law, 5
Operator equation, 257, 263, 273
Orthogonal functions, 76-81

P

Parabolic PDE, 17-20, see also Diffusion equation
Partial differential equation (PDE), 15-19, 135
Pascal triangle, 458
Permeability, 5
Permittivity, 4
Phase shift, 105-107
Pocklington's integral equation, 354
Point-matching method, see Collocation method
Poisson's equation, 4-5, 12, 505
analytical solution of, 87-90
finite difference solution of, 152-155, 207
finite element solution of, 429-432
functional for, 263, 265

Green's function for, 320
Rayleigh-Ritz solution of, 271-273
Poisson's integral formula, 329
Poynting theorem, 24
Program documentation, 643
Pseudorandom numbers, 573, see also Random numbers

R

Radar cross section, 103, see also scattering cross section
Radiation, 350
Random numbers, 572-576
Random walk, 587-593
Rayleigh-Ritz method, 255, 266-268
Relaxation factor, 155, 656
Relaxation method, 656
Residual, 154, 175, 274, 292, 656
Retarded potential, 11
Roundoff errors, 161-162, 653, 658

S

Scattering, 340-350
   by a cylinder, 57, 340-344
   by a sphere, 96-101
   by parallel wires, 344-348
Scattering cross sections, 101-104, 343
Scattering matrix, 519-521, 533-534
Separation of variables, 29-32, 326
   in cylindrical coordinates, 43-53, 203
   in rectangular coordinates, 32-38
   in spherical coordinates, 59-68
Series expansion, 86-94

Series node, 539-541
Shape functions, 412, 444, 461-462, 480
Shunt node, 530-533
Simpson's rule, 217-218, 224-227, 253
Simultaneous equations, 647-671
Solution region, 15, see also Material medium
Specific absorption rate, 366
Stability of finite difference, 161-164, 184
Standard deviation, 577-578
Step size, 207, 590, 593, 599, see also Mesh size
Stokes's theorem, 3, 626
Subdomain method, 275, 280
Successive over-relaxation (SOR), 155
Superposition principle, 22, 40, 520
Symmetry condition, 169-170, 194

T

Taylor's series, 138
Tensor integral equation, 367, 370
Tetrahedral element, 474-477
Testing function, see Weighting function
Three-dimensional elements, 474-479
Transmission-line modeling (TLM), 501-569
Transmission line
   analytical solution of, 90-94
   finite difference solution of, 165-172, 176-179
   moment method solution of, 334-340

Trapezoidal rule, 216-217, 253
Truncation conditions, 185
Truncation errors, 161-162, 555-556

U

Uniqueness theorem, 23-24

V

Variance, 577
Variational methods, 255-307
Variational principle, see Functional
Vector identities, 625-627
Velocity error, 556
Volterra integral, 310-312
Von Neumann's method, 163

W

Wave equation, 8-9, 20, see also Helmholtz equation
    analytical solution of, 37-38, 47-52, 64-68
    finite difference solution of, 147-151, 173-176
    finite element solution of, 432-439
    functional for, 265
    Green's function for, 320
    TLM solution of, 512-530
Waveguides
    finite difference solution of, 173-176
    variational solution of, 287-290
Weighted residual method, 255, 273-278, 309
Weighting function, 274-276, 354, 357-358

Y

Yee's algorithm, 179-184, 247-249, 251